RESIDUATED LATTICES: AN ALGEBRAIC GLIMPSE AT SUBSTRUCTURAL LOGICS

STUDIES IN LOGIC

AND

THE FOUNDATIONS OF MATHEMATICS

VOLUME 151

ELSEVIER

AMSTERDAM • BOSTON • HEIDELBERG • LONDON • NEW YORK • OXFORD
PARIS • SAN DIEGO • SAN FRANCISCO • SINGAPORE • SYDNEY • TOKYO

RESIDUATED LATTICES: AN ALGEBRAIC GLIMPSE AT SUBSTRUCTURAL LOGICS

Nikolaos GALATOS
School of Information Science
Japan Advanced Institute of Science and Technology
Nomi, Ishikawa 923–1292, Japan

Peter JIPSEN
Department of Mathematics and Computer Science
Chapman University
Orange, CA 92866, USA

Tomasz KOWALSKI
Research School of Information Sciences and Engineering
Australian National University
Canberra 0200, ACT, Australia

Hiroakira ONO
School of Information Science
Japan Advanced Institute of Science and Technology
Nomi, Ishikawa 923–1292, Japan

ELSEVIER

AMSTERDAM • BOSTON • HEIDELBERG • LONDON • NEW YORK • OXFORD
PARIS • SAN DIEGO • SAN FRANCISCO • SINGAPORE • SYDNEY • TOKYO

Elsevier
Radarweg 29, PO Box 211, 1000 AE Amsterdam, The Netherlands
The Boulevard, Langford Lane, Kidlington, Oxford OX5 1GB, UK

First edition 2007

Library of Congress Cataloging-in-Publication Data
A catalog record for this book is available from the Library of Congress

British Library Cataloguing in Publication Data
A catalogue record for this book is available from the British Library

ISBN-13: 978-0-444-52141-5

Transferred to Digital Printing 2007

To Wim Blok, Bjarni Jónsson and Helena Rasiowa

Hiroakira: *To my wife Kazue and to the memory of my father Katuzi*

Nick: *To my mentor Costas and to my wife Smarula*

Peter: *To my mentor Bjarni and to my wife Julie*

Tomasz: *To my mentor Andrzej and to the memory of my father Henryk*

Contents

Detailed Contents

List of Figures

List of Tables

Introduction

This book is about residuated lattices and substructural logics. Although the corresponding fields of study originated independently, they are now considered complementary and fundamentally interlinked through algebraic logic. Substructural logics encompass among many others, classical logic, intuitionistic logic, relevance logics, many-valued logics, fuzzy logics, linear logic and their non-commutative versions. Examples of residuated lattices include Boolean algebras, Heyting algebras, MV-algebras, basic logic algebras and lattice-ordered groups; a variety of other algebraic structures can be rendered as residuated lattices. Applications of substructural logics and residuated lattices span across proof theory, algebra and computer science.

Substructural logics and residuated lattices. Originally, substructural logics were introduced as logics which, when formulated as Gentzen-style systems, lack some (including 'none' as a special case) of the three basic *structural rules* of *contraction, weakening* and *exchange*. For example, relevance logics and linear logic lack the weakening rule, many-valued logics, fuzzy logics and linear logic lack the contraction rule, and hence all of them can be regarded as substructural logics. These logics have been studied extensively and various theories have been developed for their investigation. However their study has been carried out independently, mainly due to the different motivations behind them, avoiding comparisons between different substructural logics.

On the other hand, the general study of substructural logics has a comparative character, focusing on the absence or presence of structural rules. As such, at least in the initial stages of research, it was a study on how structural rules affect logical properties. This naturally led to a syntactic or proof-theoretic approach yielding deep results about properties of particular logics, provided that they can be formalized in systems, like cut-free Gentzen calculi. An obvious limitation of this study comes from the fact that not all logics have such a formulation.

Semantical methods, in contrast, provide a powerful tool for analyzing substructural logics from a more uniform perspective. Both Kripke-style semantics and algebraic semantics for certain subclasses of substructural logics, e.g. relevant logics, were already introduced in the 70s and 80s and were studied to various extents. However, it took more than a decade until researchers began to understand, via an algebraic analysis, why substructural logics, especially when formulated in Gentzen-style sequent systems, encompass most of the interesting classes of nonclassical logics. The key observation is that these logics all share the *residuation property.* Though this is usually not noticed, it is revealed explicitly in a sequent formulation by the use of extralogical symbols, denoted by commas. To be precise, consider the following equivalence, concerning implication $\to$ in sequent systems for classical, intuitionistic logic and various other logics: A formula γ follows from formulas α and β if and only if the implication $\alpha \to \gamma$ follows from β alone. Here 'follows from' is given by the particular substructural logic, and is usually captured by a sequent separator such as $\Rightarrow$. Even more formally,

$$\alpha, \beta \Rightarrow \gamma \text{ is provable} \quad \text{iff} \quad \beta \Rightarrow \alpha \to \gamma \text{ is provable},$$

where provability is taken with respect to a sequent calculus for the particular substructural logic. Put differently, the sequent calculus proves (in the form of sequents) the valid derivations of the particular logic.
If we replace the auxiliary symbol 'comma' by a new logical connective $\cdot$, called *fusion*, we have:

$$\alpha \cdot \beta \Rightarrow \gamma \text{ is provable} \quad \text{iff} \quad \beta \Rightarrow \alpha \to \gamma \text{ is provable}.$$

In algebraic models, this can be expressed as

$$a \cdot b \leq c \text{ iff } b \leq a \to c,$$

which is known as the *law of residuation* in residuated (ordered) structures.

The realization that *substructural logics are logics whose algebraic models are residuated structures* provides a new perspective on propositional logic. Implication, admittedly the most important logical connective, can be understood as the *residual* of fusion, a connective that behaves like a semigroup or groupoid operation. From a mathematical point of view, it is easier to discuss the latter than the former, just like developing a theory of multiplication of numbers is easier than developing a theory of division.

The addition of the connectives of conjunction and disjunction corresponds, algebraically, to considering lattice-ordered structures, thus producing residuated lattices, our main objects of interest. These were introduced by Ward and Dilworth in 1930s, as the main tool in the abstract study of ideal theory in rings. Under a more general definition, residuated lattices have been studied extensively in recent years, firmly establishing the area as a fully-fledged research field.

About the book and the intended audience. We present recent research on residuated structures and substructural logics, emphasizing their connections. The research field is developing rapidly, so it was impossible to cover all important topics. The book itself originated from a short monograph, written by the last two coauthors and published in 2001 under the title *Residuated Lattices: An algebraic glimpse at logics without contraction.* The focus was on logics with exchange and weakening, algebraic models of which are integral commutative residuated lattices. Due to significant advances in the study of residuated lattices and to a deeper understanding of their connections with substructural logics, our book develops the general theory and includes recent results in the setting without structural rules. Furthermore, it refers to a broader perspective, proposes new directions of research, and has attained a stronger algebraic flavor, partly due to the background of the first two coauthors.

The main aim of the book is to consider both the algebraic and logical perspective within a common framework, to show that the interplay between algebra and logic is very natural and that it provides a deeper understanding of the subject. We hope that researchers from algebra and proof theory will find helpful even the introductory chapters which aim to illustrate the connections between the two fields in a familiar setting. Moreover, we hope that specialists in particular logics or algebraic structures will find a framework for a uniform study of substructural logics and residuated lattices.

The first part of the book, consisting of the first three chapters, does not assume knowledge of logic or universal algebra and is intended to be accessible to graduate students. In particular, it can be used as a textbook for a graduate course in the subject complemented by some of the next four chapters. Therefore, we have included many examples, exercises, explanatory remarks about the connections of algebra and logic, and suggested research directions. The later chapters contain many recent results of interest to the specialists in the area, and are to a large extent mutually independent as it is implied by the dependency chart.

As a short overview of the history of this book, we would like to mention that the last coauthor started working in the research area of the book more than 20 years ago. Initially, his approach to substructural logics was mainly proof-theoretical, and part of his work was done in collaboration with Y. Komori and with his own students at Hiroshima University. In the middle of the 1990s, he incorporated algebraic methods to his study of substructural logics and shortly after the third coauthor joined the project at JAIST. At the same time a group of algebraists at Vanderbilt University were working on residuated structures. Through the contact of the second and third coauthor, it was realized that these algebraic structures were related to substructural logics and that they are of interests to both logicians and algebraists. This stimulated the initiation of the workshop series "Algebra and Substructural Logics," of which three have been organized so far, and the writing of the 70-page monograph [KO01]. The first two coauthors, originating from Vanderbilt University, participated in the workshop series and the first coauthor moved to JAIST. The participation of the first two authors in the project has strengthened its algebraic side and has lead to the current book.

Contents. In Chapter 1 we cover the basic definitions and results of first-order logic and universal algebra to assist readers who may not be familiar with some of these aspects. This is followed by presentations of (propositional) classical and intuitionistic logic, both via a Hilbert system and a Gentzen system. A section on the connections between these logics and their algebraic counterparts, namely Boolean algebras and Heyting algebras, is followed by a discussion of the cut elimination theorem for the Gentzen systems. We conclude with a section on consequence relations and logical matrices.

Chapter 2 begins with defining the substructural logic **FL** via a Gentzen-style sequent calculus and several extensions that are obtained by adding certain basic structural rules. A discussion about the general concept of substructural logics over **FL** and their algebraic models leads to the central definition of FL-algebras and residuated lattices. Some of their basic algebraic properties are given and we examine the most prominent examples of substructural logics and their corresponding algebras, summarized in a schematic diagram of the lattice of logics. The parametrized local deduction theorem and an equivalent Hilbert system for **FL** are derived and, after a brief discussion of algebraization, we show that the variety of FL-algebras is the equivalent algebraic semantics of the Hilbert system for **FL**.

Chapter 3 considers the many aspects of residuation, starting from pairs of residuated maps and Galois connections, to residuated pogroupoids and involutive residuated structures. A long section of examples and constructions of residuated algebraic structures is followed by a list of varieties of FL-algebras and diagrams of the subvariety lattice. The last section covers the recently developed structure theory of residuated lattices, which is one of the reasons for many of the successes of algebraic methods in the analysis of substructural logics.

Chapter 4 starts with detailed proofs of cut elimination for various sequent systems, organized according to the presence or absence of the basic structural rules. Decidability of the systems is obtained as an easy consequence of cut elimination for systems that lack the contraction rule and the main idea in proof of decidability for systems with contraction and exchange is given. After a brief survey of various decidability results in other logics and in varieties of residuated lattices, we prove the undecidability of the quasiequational theory of various distributive varieties of FL-algebras and residuated lattices, as a consequence of the unsolvability of the word problem; this corresponds to the undecidability of the deducibility relation for certain relevance logics.

In Chapter 5 we discuss some important logical properties of substructural logics from both a proof-theoretic and an algebraic perspective. These include the disjunction property, Halldén completeness, Maksimova's variable separation property and Craig interpolation property. For the basic substructural logics, these properties are usually obtained as consequences of cut elimination. Their algebraic counterparts, however, are applicable to *all* substructural logics, and provide a description of logical properties in a general setting. Among algebraic characterizations we obtain the correspondence between amalgamation property (algebraic) and Robinson property (logical), as well as between generalized versions of the amalgamation property (algebraic) on the one hand and Craig interpolation property and deductive interpolation property (logical) on the other.

In Chapter 6 we discuss completions of residuated lattices and focus on canonical extensions and Dedekind-MacNeille completions. A third way of completing a residuated lattice, by means of a nucleus, is shown to be isomorphic to its Dedekind-MacNeille completion. After mentioning some negative results about completions of particular subvarieties, we describe an embedding of a partial residuated lattice into a full one. This embedding preserves finiteness in certain subvarieties, yielding their the finite embeddability property.

An alternative, algebraic, proof of cut elimination is presented in Chapter 7, by means of a quasiembedding of a Gentzen matrix, which is very much related to the embeddings of Chapter 6. We proceed to prove the cut elimination property and the decidability for a two-sided sequent calculus for involutive FL-algebras that do not assume the cyclicity axiom, and we conclude the chapter with a discussion of tableau systems and the finite model property.

Chapter 8 deals with a generalization of the Glivenko translation of intuitionistic propositional logic in classical logic, by means of a double negation interpretation. We show that this translation holds between many other pairs of substructural logics and we characterize all such pairs for which one of the logics is involutive. We show that the symmetric key notion of Glivenko equivalence between two logics divides the whole subvariety in equivalence classes that are intervals. The largest and smallest logics in the equivalence class of a given logic are characterized and help provide equivalent conditions to when a substructural logic is involved in a Glivenko-type theorem.

In Chapter 9 we study various aspects of the subvariety lattices of residuated lattices and FL-algebras, or dually to the lattice of substructural logics. More than half of the chapter is devoted to the investigation of minimal (atoms) and almost minimal non-trivial varieties, or equivalently of maximal and almost maximal consistent substructural logics. Particular properties which specify areas of the lattice where we focus include commutativity, representability, idempotency (or n-potency), cancellativity, integrality and their combinations. We proceed by providing axiomatizations of the join of two varieties of residuated lattices, or equivalently of the intersection of

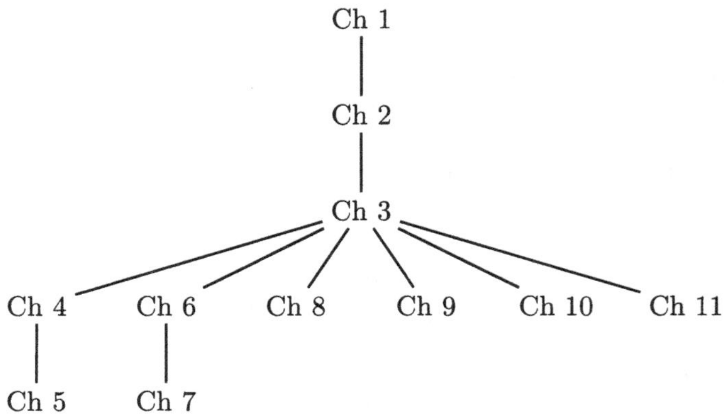

FIGURE 0.1. Dependencies between chapters.

two substructural logics. In the presence of properties like commutativity, these axiomatizations simplify considerably. After a brief reference to cases where the join of two varieties is their direct product, we concentrate on the subvariety lattice of lattice-ordered groups, which we show to be equivalent to the the one of their negative cones.

In Chapter 10 we discuss the structure of subvariety lattices a little more, focusing on splittings. As the name suggests, a splitting is a way of dividing a lattice in half. If possible, this allows for studying these two parts separately. Splitting methods proved to be quite successful in investigating lattices of modal logics, but our results here are negative, even in the somewhat narrow setting of integral, commutative residuated lattices that we adopt. Namely, we prove that there is only one splitting and of a trivial nature, since the lattice splits into its bottom element and everything else. The argument however develops an extension of Jankov's characteristic formulas, which may be applicable in other contexts.

Finally, Chapter 11 deals with the universal algebraic properties of being semisimple and possessing a discriminator term. As in the previous chapter the setting is narrowed to the commutative and integral case, but the results are positive. In the first part we prove that free algebras are semisimple. Somewhat surprisingly, the result is obtained by proof-theoretical methods. In the second part, we show that all semisimple varieties are discriminator varieties, which implies that semisimplicity is a Malcev property in the commutative integral setting.

Topics not covered in the book. The rapid and ongoing development of the field, as well as the fact that it reaches out to many diverse areas of mathematics, philosophy, computer science and linguistics, prohibits us from including large parts of the theory and many interesting applications. Furthermore, our selection has been limited to the areas where we have contributed most. Therefore, the book is not intended to be comprehensive and complete in all respects.

In particular, the book does not cover details of the theories of specific substructural logics, like relevance logics, fuzzy logics, paraconsistent logics, many-valued logics and linear logics, as we have focused on comparative study. The interested reader can to refer to more specialized books for further reading. Likewise, we did not include details on specific classes of algebras, including lattice-ordered groups, Boolean, Heyting and relation algebras, ordered monoids, (generalized) MV and BL-algebras and various special residuated structures (like cone algebras, bricks, complementary semigroups and sectionally residuated lattices).

Furthermore, the book does not cover many results on reducts of residuated lattices or fragments of substructural logics. Despite the strong interest of the authors in these topics and the rich relevant literature, time constraints have forced us to leave many such results out of the book.

We do not discuss categorical approaches to substructural logic, nor investigations into first-order substructural logics or of substructural logics with modalities. Although associativity is one of the structural rules and the study of non-associative structures has motivation from linguistics and algebra, we do not include such structures in the book, since their study is still at a very early stage. The investigation of relational semantics for substructural logics is also under development and is omitted from the book.

A biased survey of algebraic logic. The prehistory of algebraic logic began with Boole, who in [Boo48, Boo54] created what we now know as Boolean algebras. Boolean algebras are algebras of sets, or equivalently of unary relations. Algebras of binary relations were introduced in one form or another by de Morgan [DM56, DM64], Schröder [Sch95a, Sch95b] and Peirce [Pei70, Pei80]. Frege [Fre79] and Peano developed aspects of first-order logic. This was followed by the axiomatic approach of Whitehead and Russel's monumental *Principia Mathematica*, and Hilbert and Ackermann's *Grundzüge der Logik*.

With Birkhoff [Bir35] universal algebra emerged and algebraic logic gained a solid foundation. The breakthrough that began the historical era was marked by Lindenbaum and Tarski's construction of what is now known as a *Lindenbaum-Tarski algebra*, explaining in precise terms why the connection between propositions and algebraic terms observed by Boole is so tight. In the first half of 20th Century, abstract algebra became to a large extent an *alter ego* of logic, especially in the form of *logical matrices* (algebras with a distinguished subset). The work of Stone [Sto36] revealed the connection to topology, which was explored extensively later on.

Algebraic investigations into logic were further boosted by the emergence of nonclassical logics, such as intuitionistic (Brouwer, Heyting), many-valued (Łukasiewicz, Post), and modal logics (Lewis, Łukasiewicz). Various kinds of logical matrices were investigated in order to give semantics to these logics, and two classes of these: Heyting algebras (matrices for intuitionistic logic) and MV-algebras (matrices for Łukasiewicz's many-valued logics) became well-established and survive to this day.

Proof theoretic approaches to classical and intuitionistic logic, in the form of sequent calculi, are due to Gentzen [Gen35], and these methods were refined and widely disseminated in Kleene's well-known "Introduction to Metamathematics" [Kle52]. Curry [Cur63], and independently Ohnishi and Matsumoto [OM57], developed sequent systems for basic modal logics. Although originally not algebraic in nature, we mention these developments, because, as it turns out, they can now be incorporated into algebraic logic.

In the 1940s and 1950s algebraic logic was advocated by Tarski and acquired wide following among his students and collaborators. They created and advanced theories of modal algebras (e.g., McKinsey [McKi40, MT44, MT46, MT48]), Boolean algebras with operators (e.g., Jónsson [JT51, JT52]), relation algebras (e.g., Chin [CT51] and Lyndon [Lyn50, Lyn56]) and MV-algebras (e.g., Chang [Cha58, Cha59]).

One outstanding achievement of this period was Jónsson's Lemma: a surprising link between congruence distributivity (a universal algebraic property) and ultraproducts (a model theoretical construction). Although more general in scope than algebraic logic, it is of primary importance there, as a vast majority of logically motivated algebras are congruence distributive. Another was abstract algebraic rendering of topological Stone duality, and extending it beyond Boolean algebras. Carried out by Jónsson and Tarski in [JT51, JT52], it anticipated Kripke relational semantics and motivated many generalizations. It still remains inspiring.

But algebraic logic was also developed in other quarters. For example, logics between intuitionistic and classical, known then as intermediate and nowadays as superintuitionistic, were extensively studied by algebraic methods in the 1960s, following Umezawa [Ume55]. Results obtained in this area until the beginning of the 1970s are collected in a survey paper [HO73]. The seminal [RS63], by Rasiowa and Sikorski, was one of the first attempts at a systematic presentation of algebraic logic and exerted strong influence on this direction of research.

After this period of prosperity, in the 1960s and 1970s algebraic logic lost some ground to Kripke semantics, which then became almost synonymous with nonclassical logics. This was especially evident in modal logics, but also in relevant logics. Algebraic semantic was regarded (at best) as useful but unintuitive. The development of algebraic logic slowed down somewhat, but did not come to a halt. Henkin, Monk and Tarski [HMT85a, HMT85b] (originally published in 1971) algebraized first-order predicate logic (with arbitrarily many variables). Prucnal and Wroński in a short but influential note [PW74], investigated a precursor to the notion of *algebraizable* logic. Maksimova [Mak77] analyzed algebraically the logical property of interpolation, establishing a standard for finding algebraic equivalents of logical notions. Blok [Blo78] used algebraic methods extensively in investigating the lattice of modal logics. In fact, although modal logicians were the first to abandon algebra in favor of relational semantics, they were also the first to discover the phenomenon of its incompleteness. The tables started turning again.

In 1989 Blok and Pigozzi published [BP89] summarizing their work on algebraizability of logics. It marks the beginning of the area now known as *abstract algebraic logic*, and has become a classic in the field. Also in the 1980s systematic algebraic analyses of substructural logics (not yet known by this name) began, with the work of Ono and Komori [OK85] as a landmark. Although algebraic research of substructural logics already existed, for example in relevant logics, algebraic structures were used in a limited way. Typically, they were used to prove completeness by matrices or a Lindenbaum-Tarski construction. The approach of [OK85] was different in that it defined a class of structures and showed that these can serve as semantics for different kinds of nonclassical logics that had so far been studied independently. A general research project in this direction is announced in [Ono90].

A remarkable step toward such systematic study was made at the conference on "logics with restricted structural rules" in Tübingen in October 1990. This was the very first conference devoted to substructural logics. Twelve speakers gave talks divided into four categories: Lambek calculus, relevant logic, linear logic, and BCK logic. During the conference K. Došen proposed to use "substructural logics" as a generic term for all these. The term immediately caught on, witness the title of the proceedings: "Substructural Logics" ([SHD93]). One of the papers it contains is a survey by Došen entitled "A historical introduction to substructural logics" ([Doš93]).

In the 1990s algebraic logic matured and began to diversify, with several main trends emerging.

- *Abstract algebraic logic.* It deals with several notions of algebraizability and their connections to various logics, or various *presentations* of logics, as it turns out that one and the same logic can be algebraizable in one presentation and not algebraizable in another. See Blok and Pigozzi [BP89], Czelakowski [Cze01], Andréka, Németi and Sain [ANS01], and Font, Jansana and Pigozzi [FJP03].

- *Algebraic equivalents of logical notions.* Initiated by [Mak77] where equivalence of Craig Interpolation Property for superintuitionistic logics and Amalgamation Property for varieties of Heyting algebras was proved. A long list of similar equivalences could now be compiled, here are a few of these:
  - logics $\longleftrightarrow$ varieties
  - lattices of extensions $\longleftrightarrow$ dual lattices of subvarieties
  - finite model property $\longleftrightarrow$ generation by finite algebras
  - tabularity $\longleftrightarrow$ finite generation
  - interpolation $\longleftrightarrow$ amalgamation
  - Beth definability $\longleftrightarrow$ surjectivity of epimorphisms
  - deduction theorem $\longleftrightarrow$ EDPC
- *Logically motivated classes of algebras.* Often referred to as *varieties of logic*, these are classes of algebras that arose as semantics for some logic but have acquired life of their own. A typical example is that of MV-algebras as presented in the monograph [CDM00]. Another is the treatment of polymodal algebras in [Gol00]. For varieties connected to substructural logics, both [KO01] and [JT02] are examples. This trend emerged out of the oldest part of algebraic logic.

The line of investigation that we pursue in this book involves aspects of each of these trends as they relate to substructural logics and residuated lattices.

Acknowledgements.

We are indebted to researchers working in the area for their substantial contributions in the rapid development of the field in recent years. We are grateful to many of them for valuable discussions, numerous suggestions and encouragement, and in particular to all the participants of the workshop "Algebra and Substructural Logics."

We would like to thank Dov Gabbay for supporting our project, Rob Goldblatt, Franco Montagna and Constantine Tsinakis for their helpful suggestions on early versions of the manuscript and for providing useful directions for improvement. We thank Josep-Maria Font, Tadeusz Litak, Alessandra Palmigiano, James Raftery, Yde Venema and Jacob Vosmaer for useful critical comments. The students of the last coauthor helped correct many oversights during the weekly seminars organized at JAIST and we are grateful to them. The writing of our book was greatly facilitated by the resources available to us from our home institutions: Australian National University, Chapman University and Japan Advanced Institute of Science and Technology; in particular Chapman University has hosted our working repository for files of the book. We also would like to acknowledge the importance of on-line communication tools in making our collaboration, spanning three continents, efficient and enjoyable.

Last, but not least, we would each like to express our gratitude to our families for their understanding and constant support.

Nick Galatos, Peter Jipsen,
Tomasz Kowalski, Hiroakira Ono
January, 2007

CHAPTER 1

Getting started

As mentioned in the introduction, this book is about substructural logics and their algebraic semantics, residuated lattices. Consequently, our exposition naturally uses the terminology, basic facts, methods and ideas of both logic and algebra. In this chapter we give all the necessary relevant background in a self-contained way, without assuming any previous knowledge. Our metatheory for discussing the main objects of our study will be standard first-order logic with the usual model-theoretical semantics given by classes of structures. After a quick review of first-order logic syntax and semantics and of the basics of universal algebra, we present a gentle introduction to classical and intuitionistic propositional logics, by both Hilbert systems and sequent calculi, and to their algebraic semantics, the varieties of Boolean and Heyting algebras. Of course, readers familiar with both algebra and logic are advised to skim through this chapter and move to the next one.

1.1. First-order languages and semantics

A *language* (or *signature*) $\mathcal{L}$ is the disjoint union of a set $\mathcal{L}^{\mathrm{o}}$ of *operation symbols* and $\mathcal{L}^{\mathrm{r}}$ of *relation symbols*, each with a fixed non-negative *arity*. Operation symbols of arity 0 are called *constant symbols*.

For a set X, the set of *$\mathcal{L}$-terms* over X is denoted by $Tm_{\mathcal{L}}(X)$ and is defined as the smallest set T such that $X \subseteq T$, and if $f \in \mathcal{L}^{\mathrm{o}}$ has arity n and $t_1, \ldots, t_n \in T$ then $f(t_1, \ldots, t_n) \in T$. Note that terms are simply strings of symbols. We fix a countable set of symbols (disjoint from $\mathcal{L}$) called *variables*, and we denote the set of all $\mathcal{L}$-terms over this set of symbols simply by $Tm_{\mathcal{L}}$. The actual names of the symbols in this set are not important, and we use the letters u, v, w, x, y, z, with possible subscripts, to range over variables (i.e., as "metavariables"). Later we may also omit reference to the language $\mathcal{L}$ when confusion is unlikely.

The set of *atomic $\mathcal{L}$-formulas* consists of all expressions $t_1 = t_2$ and $R(t_1, \ldots, t_n)$, where $t_1, \ldots, t_n \in Tm_{\mathcal{L}}$ and $R \in \mathcal{L}^{\mathrm{r}}$ has arity n. In particular, the expressions of the form $t_1 = t_2$ are called *identities* or *equations*. The set of all *first-order $\mathcal{L}$-formulas* is the smallest set F that contains all atomic formulas and for all $\varphi, \psi \in F$ and any variable x we have

$$(\varphi \text{ and } \psi),\ (\varphi \text{ or } \psi),\ (\text{not } \varphi),\ (\varphi \Rightarrow \psi),\ (\varphi \Leftrightarrow \psi),\ (\forall x \varphi),\ (\exists x \varphi) \in F.$$

Parentheses are usually omitted with the convention that not , $\forall x$, $\exists x$ bind more tightly than and, followed by or, $\Rightarrow$ and $\Leftrightarrow$. A variable x in a first-order formula φ is *bound* if it occurs within the scope of a quantifier $\forall x$ or $\exists x$, else it is *free* in φ. A *first-order sentence* is a first-order formula containing no free variables. When $x_1, \ldots, x_n$ are all free variables in φ, $\forall x_1 \ldots \forall x_n \varphi$ is a first-order sentence, which is called a *universal closure* of φ. A *quantifier-free* formula (or a *open* formula) is a first-order formula that does not contain any universal ($\forall$) or existential ($\exists$) quantifier. A *universal* formula is of the form $\forall x_1 \ldots \forall x_n \varphi$ where φ is quantifier-free. A *strict universal Horn formula* is a first-order $\mathcal{L}$-formula of the form

$$\varphi_1 \text{ and } \varphi_2 \text{ and } \cdots \text{ and } \varphi_n \ \Rightarrow \ \psi$$

where $\varphi_1, \varphi_2, \ldots, \varphi_n, \psi$ are atomic formulas. If $\varphi_1, \varphi_2, \ldots, \varphi_n, \psi$ are all equations then the displayed first order formula is called a *quasiequation* or a *quasi-identity*.

An *$\mathcal{L}$-structure* $\mathbf{A} = (A, (l^{\mathbf{A}})_{l \in \mathcal{L}})$ is a *nonempty* set A, called the *universe*, together with an $\mathcal{L}$-tuple of operations and relations defined on A, where $l^{\mathbf{A}}$ has the same arity as l. We also refer to $l^{\mathbf{A}}$ as a *basic operation* or a *basic relation* to distinguish it from definable operations and relations. Recall that an *operation* of arity n on a set A is simply a map from A^n to A and that a *relation* of arity n on a set A is subset of A^n. If $\mathcal{L}$ is finite, the operations and relations are usually listed explicitly in some fixed order, say $(A, f_1^{\mathbf{A}}, \ldots, f_m^{\mathbf{A}}, R_1^{\mathbf{A}}, \ldots, R_n^{\mathbf{A}})$, and when confusion is unlikely the superscript $^{\mathbf{A}}$ is often omitted. An *algebra* is a structure without any relations. Two algebras $\mathbf{A}$ and $\mathbf{B}$ are of the *same type* when both of them are $\mathcal{L}$-structures for some language $\mathcal{L}$. An algebra is *finite* if the universe is a finite set, and is *trivial* (or *degenerate*) if the universe is a singleton set.

A *sublanguage* $\mathcal{K}$ of a language $\mathcal{L}$ is simply a subset of $\mathcal{L}$, where every symbol retains its arity. The *$\mathcal{K}$-reduct* of an $\mathcal{L}$-structure $\mathbf{A} = (A, (l^{\mathbf{A}})_{l \in \mathcal{L}})$ is the $\mathcal{K}$-structure $(A, (l^{\mathbf{A}})_{l \in \mathcal{K}})$ on the same universe, where $\mathcal{K}$ is a sublanguage of $\mathcal{L}$. In this case $\mathbf{A}$ is called an *expansion* of $(A, (l^{\mathbf{A}})_{l \in \mathcal{K}})$. The $\mathcal{K}$-reduct of $\mathbf{A}$ is also called the $\mathcal{M}$-free reduct of $\mathbf{A}$, where $\mathcal{M}$ is the complement of $\mathcal{K}$ in $\mathcal{L}$. If $\mathcal{K}$ is clear from the context, we simply refer to the $\mathcal{K}$-reduct as the reduct of $\mathbf{A}$.

An *assignment* or *valuation* into $\mathbf{A}$ is a function h from the set of variables to A. Any such function h extends uniquely to a function (also denoted by h) from $Tm_{\mathcal{L}}$ to A by defining $h(f(t_1, \ldots, t_n)) = f^{\mathbf{A}}(h(t_1), \ldots, h(t_n))$ for each operation symbol $f \in \mathcal{L}^{\mathrm{o}}$ with arity n. For a first-order $\mathcal{L}$-formula φ one defines $\mathbf{A}, h \models \varphi$ (read as $\mathbf{A}$ *satisfies* φ with respect to h) inductively in the standard way by: [1]

$\mathbf{A}, h \models s = t$	if $h(s) = h(t)$
$\mathbf{A}, h \models R(t_1, \ldots, t_n)$	if $(h(t_1), \ldots, h(t_n)) \in R^{\mathbf{A}}$ for each $R \in \mathcal{L}^{\mathrm{r}}$
$\mathbf{A}, h \models \varphi \textsf{ and } \psi$	if $\mathbf{A}, h \models \varphi$ and $\mathbf{A}, h \models \psi$
$\mathbf{A}, h \models \varphi \textsf{ or } \psi$	if $\mathbf{A}, h \models \varphi$ or $\mathbf{A}, h \models \psi$
$\mathbf{A}, h \models \textsf{not } \varphi$	if it is not the case that $\mathbf{A}, h \models \varphi$
$\mathbf{A}, h \models \varphi \Rightarrow \psi$	if either not $\mathbf{A}, h \models \varphi$ or $\mathbf{A}, h \models \psi$
$\mathbf{A}, h \models \varphi \Leftrightarrow \psi$	if $\mathbf{A}, h \models \varphi$ iff $\mathbf{A}, h \models \psi$
$\mathbf{A}, h \models \forall x \varphi$	if $\mathbf{A}, h' \models \varphi$ for all assignments h' such that $h'(u) = h(u)$ for all variables u distinct from x
$\mathbf{A}, h \models \exists x \varphi$	if $\mathbf{A}, h' \models \varphi$ for some assignment h' such that $h'(u) = h(u)$ for all variables u distinct from x

A term s is sometimes written $s(x_1, \ldots, x_n)$ or $s(\bar{x})$ where the list of variables is assumed to include all variables in s. In this case the value $h(s)$ of a term under an assignment h into $\mathbf{A}$ is also denoted $s^{\mathbf{A}}(\bar{a})$ where $\bar{a} = (a_1, \ldots, a_n)$ and $h(x_i) = a_i$ for $i = 1, \ldots, n$. We call $s^{\mathbf{A}}$ the term function induced by s on $\mathbf{A}$. Moreover this notation extends to first-order formulas, where $\varphi(\bar{x})$ indicates that all free variables in φ appear in the list $\bar{x}$, and $\varphi^{\mathbf{A}}(\bar{a})$ is true iff $\mathbf{A}, h \models \varphi$.

We say that a first-order formula φ is *valid* in $\mathbf{A}$ or that $\mathbf{A}$ is a *model* of φ, in symbols $\mathbf{A} \models \varphi$, if $\mathbf{A}$ satisfies φ with respect to h (i.e., $\mathbf{A}, h \models \varphi$) for all assignments h into $\mathbf{A}$. Clearly, $\mathbf{A} \models \varphi$ iff $\mathbf{A} \models \tilde{\varphi}$, where $\tilde{\varphi}$ is any universal closure of φ. Note that for any first-order *sentence* φ, $\mathbf{A} \models \varphi$ iff $\mathbf{A}, h \models \varphi$ for *some* assignment h. The assumption that φ is closed is obviously crucial. For a class $\mathcal{K}$ of $\mathcal{L}$-structures and a set F of first-order $\mathcal{L}$-formulas, we write $\mathcal{K} \models F$ if $\mathbf{A} \models \varphi$ for all $\mathbf{A} \in \mathcal{K}$ and all $\varphi \in F$. Analogously, one defines $\mathcal{K} \models \varphi$ and $\mathbf{A} \models F$.

The *equational theory of* $\mathcal{K}$, denoted by $\mathrm{Th}_e(\mathcal{K})$, is the set of equations that are valid in all members of $\mathcal{K}$. In general, a collection of equations is called an *equational theory* if it is the equational theory of some class of $\mathcal{L}$-structures.

Likewise the *first-order theory of* $\mathcal{K}$, denoted by $\mathrm{Th}(\mathcal{K})$, the *universal theory of* $\mathcal{K}$, denoted by $\mathrm{Th}_u(\mathcal{K})$, and the *quasiequational theory* of $\mathcal{K}$, denoted by $\mathrm{Th}_q(\mathcal{K})$, is respectively the set of first-order formulas, universal formulas and quasi-identities that are valid in all members of $\mathcal{K}$. Also a *first-order*, (*universal*, *quasiequational*) *theory* is a set of $\mathcal{L}$-formulas that is the first-order (universal, quasiequational) theory of some class of $\mathcal{L}$-structures.

[1] We do not distinguish between the formal symbol for equality and the actual equality relation, relying on context to tell them apart, and use the same symbol $=$.

For a set F of first-order $\mathcal{L}$-formulas, $\mathrm{Mod}(F)$ denotes the class of all $\mathcal{L}$-structures in which all the first-order formulas in F are valid, i.e., the class of all structures that are models of universal closures of the first-order formulas in F. Any class of the form $\mathrm{Mod}(F)$ is called a *first-order class* (or an *elementary class*). A *universal class*, a *quasiequational class* or an *equational class* results if F contains only universal formulas, quasi-identities or identities respectively. In this case we refer to F as a *universal, quasiequational* or *equational basis*, respectively. The next two results follow immediately from these definitions. In fact, using terminology from later sections, the various forms of Th Mod and Mod Th are closure operators (on classes), see Section 1.6.1, associated with the Galois connection, see Section 3.1.2, determined by $\models$ on sets of formulas and classes of structures[2].

THEOREM 1.1. *For any set of equations E, $\mathrm{Th}_e\,\mathrm{Mod}(E)$ is the smallest equational theory containing E. Likewise $\mathrm{Th}\,\mathrm{Mod}(F)$, $\mathrm{Th}_u\,\mathrm{Mod}(F)$ and $\mathrm{Th}_q\,\mathrm{Mod}(F)$ are the smallest first-order (universal, quasiequational) theories that contain a set F of first-order formulas or universal formulas or quasi-identities respectively.*

THEOREM 1.2. *For any class $\mathcal{K}$ of $\mathcal{L}$-structures, $\mathrm{Mod}\,\mathrm{Th}(\mathcal{K})$ is the smallest first-order class, $\mathrm{Mod}\,\mathrm{Th}_u(\mathcal{K})$ is the smallest universal class, $\mathrm{Mod}\,\mathrm{Th}_q(\mathcal{K})$ is the smallest quasiequational class and $\mathrm{Mod}\,\mathrm{Th}_e(\mathcal{K})$ is the smallest equational class that contains $\mathcal{K}$.*

As mentioned before, a structure is a set with some finitary operations and relations on it: any collection of operations and relations will do, including none, so structures are many and varied. We will mention here some classes of structures that will be of use later, hence our list will be rather biased toward ordered algebraic structures.

1.1.1. Preorders. A structure $\mathbf{Q} = (Q, \leq)$ is a *preordered set* (also called a *quasiordered set*) if $\leq$ is a binary relation on Q such that, for all $x, y, z \in Q$ the following hold:

$x \leq x$ (reflexivity),
$x \leq y$ and $y \leq z$ imply $x \leq z$ (transitivity).

1.1.2. Posets. A structure $\mathbf{P} = (P, \leq)$ is a partially ordered set, or a *poset*, if it is a preordered set and for all $x, y \in P$

$x \leq y$ and $y \leq x$ imply $x = y$ (antisymmetry).

A relation with these properties is called a *partial order*. If $\mathbf{P} = (P, \leq)$ is a poset, then $\mathbf{P}^\partial = (P, \geq)$ is also a poset and is called the *dual* of $\mathbf{P}$. Let a be an element of P and X be a subset of P. We say that a is

[2]Sometimes, theories are defined to be sets of the first-order sentences. But, as mentioned before, the validity of a given first-order formula is the same as that of its universal closure which is a first-order sentence. Therefore, there are no essential differences between these two definitions.

- the *smallest (or least) element in* X (or the *minimum of* X) if $a \in X$ and $a \leq x$ holds for every $x \in X$,
- *minimal* in X if $a \in X$ and $x \leq a$ implies $x = a$, for every $x \in X$,
- a *lower bound* of X if $a \leq x$, for every $x \in X$.

These notions have their natural counterparts, called respectively *greatest* (or *maximum*), *maximal* and an *upper bound*, defined by replacing $\leq$ with $\geq$ everywhere. Note that the maximum element of a poset is the minimum element of the dual poset. The same holds for the other pairs of *dual* properties. Another pair of dual notions is that of *infimum* or *meet*, and *supremum* or *join* of X, defined as the *greatest lower bound of* X and the *smallest (least) upper bound of* X respectively and denoted by $\inf X$ and $\sup X$ (if they exist). The smallest element of X and the greatest element of X are denoted by $\min X$ and $\max X$ (if they exist). Note that when $a = \inf X$, $a \in X$ if and only if $a = \min X$. When the set P itself has a smallest element, it is sometimes denoted by $\perp$ (read *bottom*), and similarly the greatest element or *top* of P, whenever it exists, is denoted by $\top$. We will meet more duals soon.

A nonempty subset F of a poset $\mathbf{P}$ is called an *order filter* or an *upset* if $x \in F$ and $x \leq y$ imply $y \in F$. The order filter *generated* by a set X is defined to be the smallest order filter containing X and is denoted by $\uparrow X$. We abbreviate $\uparrow\{x\}$ by $\uparrow x$ and call it the *principal filter generated by* x. It is easy to see that $\uparrow X = \{y \in P : x \leq y \text{ for some } x \in X\}$, i.e., that $\uparrow X = \bigcup_{x \in X} \uparrow x$. The dual notions of *order ideal* or *downset* and of the order ideal $\downarrow X$ *generated* by a set X are defined in a natural way. A subset S of P is called *convex*, if for all $x, z \in S$ and $y \in P$, $x \leq y \leq z$ implies $y \in S$.

1.1.3. Lattices. An algebra $\mathbf{A} = (A, \wedge, \vee)$ is a *lattice* if the binary operations, called respectively *meet* and *join*, are *commutative*, *associative* and mutually *absorptive*, that is, if for all $x, y, z \in A$ the following hold:

$x \wedge y = y \wedge x$ (commutativity of meet)
$x \vee y = y \vee x$ (commutativity of join)
$x \wedge (y \wedge z) = (x \wedge y) \wedge z$ (associativity of meet)
$x \vee (y \vee z) = (x \vee y) \vee z$ (associativity of join)
$x \vee (x \wedge y) = x$ (absorption of meet by join)
$x \wedge (x \vee y) = x$ (absorption of join by meet)

Although we have not yet defined Birkhoff's deductive system for equational logic (Section 1.2.7), the following simple equational proofs use the absorption laws to show that in any lattice, meet (and dually join) are *idempotent*:

$$x \wedge x = x \wedge (x \vee (x \wedge y)) = x$$
$$x \vee x = x \vee (x \wedge (x \vee y)) = x$$

It is clear that if $\mathbf{A} = (A, \wedge, \vee)$ is a lattice, then the algebra $\mathbf{A}^\partial = (A, \vee, \wedge)$ is also a lattice that is called the *dual* of $\mathbf{A}$. Again the equations above come in *dual* pairs, and it is true in general that if an equation $s = t$ holds in all lattices then so does its dual, that is the equation $s^\partial = t^\partial$, where the terms s^∂ and t^∂ come from s and t by uniformly replacing meet by join and *vice versa*. Indeed, if $s^\partial = t^\partial$ fails in a lattice, then $s = t$ fails in the dual lattice. Thus, from now on we will not bother stating the duals explicitly.

That the names "meet" and "join" are also used for suprema and infima in posets is not a confusing coincidence, but rather an indication that these concepts are indeed the same. Namely, for any lattice $\mathbf{A}$ we can define a binary relation $\leq^{\mathbf{A}}$ on A by $x \leq^{\mathbf{A}} y$ iff $x \wedge y = x$ (or by duality, $x \geq^{\mathbf{A}} y$ iff $x \vee y = x$, but we promised not to bother). On the other hand, if a poset $\mathbf{P}$ has the property that for all $x, y \in P$ the infimum and supremum of $\{x, y\}$ exist (equivalently, infima and suprema of finite nonempty subsets exist), we can define two binary operations $\wedge^{\mathbf{P}}$ and $\vee^{\mathbf{P}}$ by $x \wedge^{\mathbf{P}} y = \inf\{x, y\}$ and $x \vee^{\mathbf{P}} y = \sup\{x, y\}$.

LEMMA 1.3. *Given a lattice* $\mathbf{A}$ *and a poset* $\mathbf{P}$*, the structure* $(A, \leq^{\mathbf{A}})$ *is a poset and the algebra* $(P, \wedge^{\mathbf{P}}, \vee^{\mathbf{P}})$ *is a lattice.*

Thus, a poset in which infima and suprema of finite nonempty subsets exist is a lattice. (Even though the languages are technically not the same, we identify the two types of structures and freely choose whichever perspective is most appropriate. Note that if we view a lattice as a poset, then the dual lattice and the dual poset match. Moreover, we identify the expression $x \leq y$ with the equation $x = x \wedge y$ and refer to inequalities as equations or identities.) This includes the case of a poset $\mathbf{C}$ for which $\inf X$ and $\sup X$ exist for *any* subset X of C. Such a poset is known as a *complete lattice*. Since $\inf X$ and $\sup X$ for a finite X, say, $X = \{x_1, \ldots, x_n\}$, are just $x_1 \wedge \cdots \wedge x_n$ and $x_1 \vee \cdots \vee x_n$ (parenthetically, by associativity we can forget the parentheses), we often write $\bigwedge X$ and $\bigvee X$ for $\inf X$ and $\sup X$. When X is expressed as $\{x_i : i \in I\}$ by an index set I, they are written also as $\bigwedge_{i \in I} x_i$ and $\bigvee_{i \in I} x_i$.

According to the above definition, in a complete lattice $\mathbf{C}$ both $\inf X$ and $\sup X$ must exist for any subset X. However note that $\sup X = \inf\{u \in C \colon u$ is an upper bound of $X\}$, hence the existence of $\inf X$ for every subset X implies the existence of $\sup X$ for every subset X, and vice versa by duality. Thus, the existence of either all infima or all suprema is enough for a lattice to be complete.

A lattice that has both a smallest element and a greatest element is said to be *bounded*. Note that any complete lattice $\mathbf{A}$ is bounded since $\inf \emptyset$ and $\inf A$ are equal to $\top$ and $\bot$ respectively, where $\emptyset$ denotes the empty set.

Many posets can be represented graphically by *Hasse diagrams* where the elements are denoted by dots (or circles) and an element a is connected by an upward sloping solid line to b iff $a \leq b$ and there is no element strictly between a and b. In this situation b is called a *cover* (or *upper cover*) of a, and dually a is a *subcover* (or *lower cover*) of b. In symbols we write $a \prec b$ and refer to $\prec$ as the *covering relation*. An *atom* is a cover of the

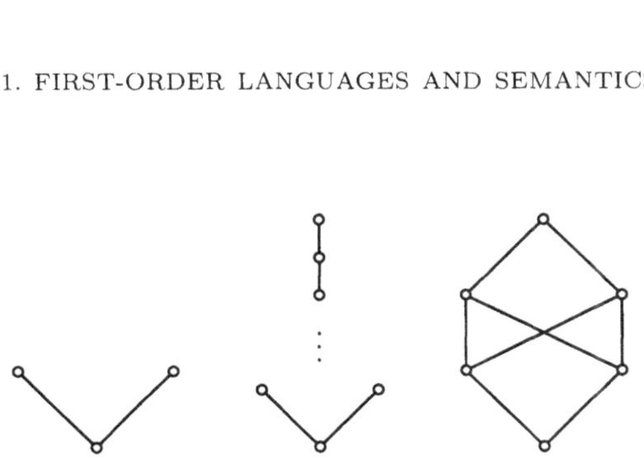

FIGURE 1.1. Examples of posets that are not lattices.

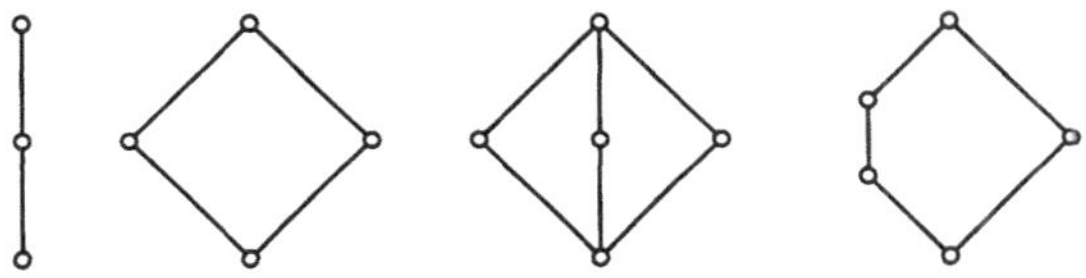

FIGURE 1.2. Examples of lattices.

bottom element, and a *coatom* is a subcover of the top element of a lattice (if the lattice has a bottom and/or top). Figure 1.1 shows some examples of Hasse diagrams that are not lattices since they contain two-element subsets without a supremum.

A lattice $\mathbf{A}$ is *distributive* if the following distributive laws hold:

$$x \wedge (y \vee z) = (x \wedge y) \vee (x \wedge z)$$
$$x \vee (y \wedge z) = (x \vee y) \wedge (x \vee z)$$

Either of the above equations can be derived from the other. E.g. assuming the first one we have $(x \vee y) \wedge (x \vee z) = ((x \vee y) \wedge x) \vee ((x \vee y) \wedge z) = x \vee (y \wedge x) \vee (x \wedge z) \vee (y \wedge z) = x \vee (y \wedge z)$. Note that the first two lattices in Figure 1.2 are distributive while the others are not.

A notion weaker than distributivity is modularity. A lattice is *modular* if $x \leq y$ implies $(x \vee z) \wedge y = x \vee (z \wedge y)$. One way to remember the modularity condition is to note that it stipulates that the two natural ways of projecting the element z in the interval $[x, y]$ coincide. Clearly, modularity can be stated as an equation (simply replace y by $x \vee y$ in the second expression). Note that in Figure 1.2 only the first two lattices are distributive, but the last is the only non-modular.

An important lattice concept is that of a *filter*. For any lattice $\mathbf{A}$, a nonempty subset F of A is a filter if

- $a, b \in F$ implies $a \wedge b \in F$,

- $a \in F, b \in A$ and $a \leq b$ imply $b \in F$.

Clearly, A itself is a filter of $\mathbf{A}$. Other filters are called *proper filters* of $\mathbf{A}$. When a lattice $\mathbf{A}$ has a smallest element 0, we can show the existence of maximal ones among the proper filters of $\mathbf{A}$ by using Zorn's lemma. These filters are called *maximal filters* or *ultrafilters*.

For any set X, a construction of fundamental importance is the *powerset* $\mathcal{P}(X) = \{S : S \subseteq X\}$. It is the universe of a complete distributive lattice $(\mathcal{P}(X), \cap, \cup)$ as well as the basis for several other algebras discussed subsequently.

A partial order on a set P is a *total order* or *linear order*, if for all $x, y \in P$ the following holds:

$$x \leq y \text{ or } y \leq x.$$

Totally ordered sets are also called *chains*. Every chain forms a lattice. In fact, $x \wedge y = \min\{x, y\}$ and $x \vee y = \max\{x, y\}$ hold. It is easy to see that it is in fact a distributive lattice. A poset $(P, \leq)$ is called an *antichain* if $\leq$ is the equality relation on P. Clearly, antichains with two or more elements are not lattices.

Infinite analogues of the distributive laws in a complete lattice are the following:

$$x \wedge \bigvee_{i \in I} y_i = \bigvee_{i \in I} (x \wedge y_i) \quad \text{(join-infinite distributive law)},$$

$$x \vee \bigwedge_{i \in I} y_i = \bigwedge_{i \in I} (x \vee y_i) \quad \text{(meet-infinite distributive law)}.$$

The distributivity of a complete lattice does not always imply either of these laws, and neither of them imply the other. For example if $A \subseteq \mathcal{P}(\mathbb{N})$ contains all finite sets of $\mathbb{N} = \{0, 1, 2, \ldots\}$, as well as the set $\mathbb{N}$ itself, then $(A, \cap, \cup)$ is a complete lattice. Note that the supremum of infinitely many elements of A is always $\mathbb{N}$ (since this is the only member of A that has infinitely many subsets), hence the first law fails with $x = \{0\}$ and $y_i = \{i\}$ for $i = 1, 2, 3, \ldots$, while it is fairly easy to check that the second law holds.

An element in a lattice is *join-irreducible* if it is not the finite join of strictly smaller elements. Note that $\bot$ is the empty join of strictly smaller elements, hence it is not join-irreducible. If an element is not the supremum of all elements strictly below it, or equivalently if it has a unique subcover, then it is said to be *completely join-irreducible*. Even stronger is the notion of a *join-prime* element a, defined by the condition that for any set of elements X, if $a \leq \bigvee X$ then $a \leq x$ for some $x \in X$. The concepts of *meet-irreducible*, *completely meet-irreducible*, and *meet-prime* element are defined dually.

1.1.4. Heyting algebras and Boolean algebras. We say that an algebra $\mathbf{A} = (A, \wedge, \vee, \to, 0)$ is a *Heyting algebra* if $\mathbf{A} = (A, \wedge, \vee, 0)$ is a lattice with a smallest element 0 and for all $x, y, z \in A$ the following holds:

$$x \wedge y \leq z \text{ iff } y \leq (x \to z) \quad \text{(the law of } \wedge\text{-residuation)}.$$

The law of $\wedge$-residuation says that

$$x \to z = \max\{y : x \wedge y \leq z\}.$$

This is to be contrasted with the law of residuation that will be discussed in the next chapter. The element $x \to z$ is called the $\wedge$-*residual* of z by x, or the *pseudocomplement* of x *relative* to z. Every Heyting algebra is bounded. In fact, 0 is its smallest element by definition, and moreover $0 \to 0$, which is denoted by 1, is the greatest. To see this, take any element $x \in A$. Then it is clear that $0 \wedge x \leq 0$. Thus, by the law of residuation, $x \leq 0 \to 0$.

Although the law of $\wedge$-residuation is not an identity, it is equivalent to the following pair of (less illuminating) identities:

$$x \wedge y \wedge (x \to z) \leq z$$
$$y \leq x \to ((x \wedge y) \vee z).$$

Therefore, the class HA of Heyting algebras is an equational class. To see this, note that we always have $y \wedge (x \to z) \leq x \to z$, hence the first identity follows from the $\wedge$-residuation law. Conversely, if this identity holds and we assume that $y \leq x \to z$ then $y = y \wedge (x \to z)$ so $x \wedge y \leq z$ follows from this identity. Similarly the second identity and the forward implication are mutually deducible. In Chapters 3 and 4, we will see this phenomenon and the next lemma again in the more general setting of residuated lattices.

LEMMA 1.4.

(1) *The distributive law holds in every Heyting algebra. In fact, the join-infinite distributive law holds for all existing infinite joins. More precisely, if* $\bigvee_{i \in I} y_i$ *exists, then* $\bigvee_{i \in I}(x \wedge y_i)$ *exists also and* $x \wedge \bigvee_{i \in I} y_i$ *is equal to* $\bigvee_{i \in I}(x \wedge y_i)$.
(2) *Conversely, for any complete lattice, if the join-infinite distributive law holds in it, then residuals always exist and hence it is also a Heyting algebra. In particular, every finite distributive lattice is a reduct of a Heyting algebra.*

PROOF. We show the second part of (1). It is obvious that $x \wedge \bigvee_{i \in I} y_i$ is an upper bound of the set $\{x \wedge y_i : i \in I\}$. Suppose that z is an upper bound of this set. Then, for each $j \in I$, $x \wedge y_j \leq z$ and hence $y_j \leq x \to z$. Since $\bigvee_{i \in I} y_i$ exists, $\bigvee_{i \in I} y_i \leq x \to z$ and therefore $x \wedge \bigvee_{i \in I} y_i \leq z$. This means that $x \wedge \bigvee_{i \in I} y_i$ is the least upper bound of $\{x \wedge y_i : i \in I\}$. Thus, $x \wedge \bigvee_{i \in I} y_i = \bigvee_{i \in I}(x \wedge y_i)$. □

The following hold in every Heyting algebra $\mathbf{A}$. For all $x, y, z \in A$:

$$\begin{aligned}
&\text{if } x \leq y \text{ then } z \to x \leq z \to y \text{ and } y \to z \leq x \to z,\\
&x \to (y \wedge z) = (x \to y) \wedge (x \to z),\\
&(x \vee y) \to z = (x \to z) \wedge (y \to z).\\
&x \wedge (x \to y) = x \wedge y,
\end{aligned} \tag{1.1}$$

Define a unary operation $\neg$ by $\neg x = x \to 0$. From the above equations it follows that

$$\begin{aligned}
x \wedge \neg x &= 0,\\
\neg(x \vee y) &= \neg x \wedge \neg y.
\end{aligned}$$

A Heyting algebra $\mathbf{A}$ is a *Boolean algebra* if for all $x \in A$

$$x \vee \neg x = 1.$$

Thus $\neg x$ is a *complement* of x, i.e., the elements $x, \neg x$ meet to 0 and join to 1. In addition the following hold in every Boolean algebra $\mathbf{A}$. For all $x, y \in A$

$$\begin{aligned}
\neg(\neg x) &= x,\\
\neg(x \wedge y) &= \neg x \vee \neg y,\\
x \to y &= \neg x \vee y
\end{aligned}$$

Alternatively, Boolean algebras can be defined as algebras of the form $\mathbf{A} = (A, \wedge, \vee, \neg, 0, 1)$ such that $(A, \wedge, \vee)$ is a distributive lattice satisfying the following conditions:

$$\begin{aligned}
x \wedge \neg x = 0 \text{ and } x \vee \neg x = 1,\\
x \wedge 0 = 0 \text{ and } x \vee 1 = 1.
\end{aligned} \tag{1.2}$$

Although the operation $\to$ from the previous definition is not part of the language here, it can be defined by the term $\neg x \vee y$, just as $\neg x$ was defined by the term $x \to 0$. Since the defining identities in each language can be derived in both cases, the two definitions are said to be *term-equivalent*. From an algebraic point of view there is no reason to distinguish between term-equivalent algebras. We denote the equational class of Boolean algebras by BA. A typical example of Boolean algebras is the Boolean algebra $(\mathcal{P}(X), \cap, \cup, ^c, \emptyset, X)$ defined by the powerset of a set X with set-theoretical operations, where S^c means the complement of a subset S with respect to X.

The smallest nontrivial Boolean algebra is based on the two-element chain and is denoted by $\mathbf{2}$. Its underlying set is $\{0, 1\}$, and the operations are defined by:

$$x \wedge y = \min\{x, y\},\ x \vee y = \max\{x, y\} \text{ and } \neg x = 1 - x.$$

1.1.5. Semigroups, monoids and other groupoids. An algebra of the form $\mathbf{B} = (B, \cdot)$ is a *groupoid* (or *binar*) if $\cdot$ is a binary operation on B. A *semigroup* is an *associative* groupoid, i.e., if for all $x, y, z \in B$ the following holds:

$$(x \cdot y) \cdot z = x \cdot (y \cdot z) \quad \text{(associativity)}.$$

A groupoid $\mathbf{B}$ is *commutative*, if for all $x, y \in B$

$$x \cdot y = y \cdot x \quad \text{(commutativity)}.$$

A *unital groupoid* is an algebra of the form $\mathbf{B} = (B, \cdot, 1)$ such that $(B, \cdot)$ is a groupoid and 1 is a *unit element* for $\cdot$, i.e., for all $x \in B$

$$x \cdot 1 = 1 \cdot x = x.$$

In this terminology, a *monoid* is a unital semigroup.

A *group* is an algebra $\mathbf{G} = (G, \cdot, {}^{-1}, 1)$, where $(G, \cdot, 1)$ is a monoid and ${}^{-1}$ is an inverse operation, i.e., for all $x \in G$

$$x \cdot x^{-1} = x^{-1} \cdot x = 1.$$

Usually $x \cdot y$ is simply written as xy, and x^n is defined by $x^0 = 1$, $x^{n+1} = x^n x$ and $x^{-n} = (x^n)^{-1}$ for $n > 0$.

An algebra $\mathbf{B} = (B, \cdot)$ is a *semilattice* if it is a commutative semigroup in which $\cdot$ is *idempotent*, i.e., satisfies $x^2 = x$. A binary relation $\leq$ may be defined on a groupoid B by

$$x \leq y \text{ if and only if } xy = y.$$

It is a good exercise to show that B is a semilattice if and only if $\leq$ is a partial order on B and xy is equal to the meet of x and y with respect to this order. Conversely a poset in which the infimum of any finite nonempty subset exists is obviously a semilattice if one defines $xy = \inf\{x, y\}$. Such a semilattice is called a *meet semilattice*, and the notion of *join semilattice* is defined dually.

A structure $\mathbf{B} = (B, \cdot, \leq)$ is a *partially ordered groupoid* or *pogroupoid* if $(B, \cdot)$ is a groupoid, $(B, \leq)$ a poset, and moreover, the following *monotonicity* holds for all $x, y, z \in B$ between $\cdot$ and $\leq$:

$$x \leq y \text{ implies both } xz \leq yz \text{ and } zx \leq zy.$$

In case $(B, \leq)$ is a lattice, the structure $\mathbf{B}$ is called a *lattice-ordered groupoid* or *ℓ-groupoid*, and if the order is a chain, $\mathbf{B}$ is called a *totally ordered groupoid*. These notions are naturally extended to semigroups, monoids and groups.

For example the structure $([0, 1], \times, 1, \leq)$ is a totally ordered monoid, where $[0, 1]$ is the unit interval and $\times$ is the multiplication of real numbers.

1.2. Concepts from universal algebra

We now give a concise introduction (or reminder) of basic notions and facts from universal algebra. We omit proofs of theorems that are involved and do not fall within the scope of the book. The interested reader is referred to the original sources, or to the excellent books on the subject [BS81] and [MMT87].

1.2.1. Homomorphisms, subalgebras, substructures, direct products. Universal algebra considers a few fundamental notions that occur in many different situations. One of the most basic is that of a structure preserving map. A function $h : A \to B$ between universes of algebras of the same type is a *homomorphism*, in symbols $h\colon \mathbf{A} \to \mathbf{B}$, if it *preserves operations*, that is,

$$h(f^{\mathbf{A}}(a_1, \ldots, a_n)) = f^{\mathbf{B}}(h(a_1), \ldots, h(a_n)),$$

for each basic operation symbol f and for each n-tuple $(a_1, \ldots, a_n)$ of elements of A, where $f^{\mathbf{A}}$ and $f^{\mathbf{B}}$ denote the interpretation of f respectively in $\mathbf{A}$ and $\mathbf{B}$. This notation is sometimes used for clarity, but in practice the superscripts are usually omitted, since interpretations are typically clear from the context. An injective homomorphism is called an *embedding* and a bijective one an *isomorphism*. A homomorphism $h : \mathbf{A} \to \mathbf{A}$ is an *endomorphism* and if it is bijective it is an *automorphism*.

There are three basic ways of obtaining new structures from old ones. We say that $\mathbf{A}$ is a *subalgebra* of $\mathbf{B}$, and often denote it by $\mathbf{A} \subseteq \mathbf{B}$, if $A \subseteq B$ and the inclusion map from A to B is a homomorphism. In other words, $\mathbf{A}$ is a subalgebra of $\mathbf{B}$ if and only if $A \subseteq B$ and for each basic operation f and for each n-tuple $(a_1, \ldots, a_n)$ of elements of A, $f^{\mathbf{B}}(a_1, \ldots, a_n) \in A$, i.e., A is *closed under* all basic operations of $\mathbf{B}$. In this case, the universe A of $\mathbf{A}$ is called a *subuniverse* of B. The intersection of a nonempty set of subuniverses of $\mathbf{B}$ is again a subuniverse of $\mathbf{B}$. Thus, for a given nonempty subset S of B, there is a smallest subalgebra $\mathbf{C}$ of $\mathbf{B}$ containing S. Then $\mathbf{C}$ (the set C) is called the subalgebra (the subuniverse, respectively) of $\mathbf{B}$ *generated* by the set S. A *partial algebra* $\mathbf{A}$ is a set A with partial operations, i.e., functional relations. If for an n-ary such functional relation f and for elements $a_1, a_2, \ldots, a_{n-1}$ of A, there is no element $b \in A$ such that $(a_1, a_2, \ldots, a_{n-1}, b) \in f$, then we say that $f(a_1, a_2, \ldots, a_{n-1})$ is *undefined.* If there is a (unique) such element b, we write $f(a_1, a_2, \ldots, a_{n-1}) = b$, as usual. If $\mathbf{A}$ is a (partial) algebra, then every subset B of A gives rise to a *partial subalgebra* $\mathbf{B}$ of $\mathbf{A}$: if $f^{\mathbf{A}}(a_1, a_2, \ldots, a_{n-1}) = b$ and $a_1, a_2, \ldots, a_{n-1}, b \in B$, then $f^{\mathbf{B}}(a_1, a_2, \ldots, a_{n-1}) = b$.

We say that $\mathbf{A}$ is a *homomorphic image* of $\mathbf{B}$, if there is a homomorphism from $\mathbf{B}$ *onto* $\mathbf{A}$. An *indexed system* $(\mathbf{A}_i : i \in I)$ of algebras of the same type is a sequence of algebras indexed by a set I. Formally, it is a map from I to the set $\{\mathbf{A}_i : i \in I\}$. An algebra $\mathbf{A}$ is a *direct product* of the indexed system $(\mathbf{A}_i : i \in I)$, written $\prod_{i\in I} \mathbf{A}_i$, if $A = \prod_{i\in I} A_i$ and the operations on A are defined coordinatewise, i.e., for each basic operation f and each element $(a_i)_{i\in I}$ of $\prod_{i\in I} A_i$, $f^{\mathbf{A}}((a_i)_{i\in I}) = (f^{\mathbf{A}_i}(a_i))_{i\in I}$.

These notions are defined for structures as well. A homomorphism must also preserve relations, that is,

$$(a_1, \ldots, a_n) \in R^{\mathbf{A}} \qquad \text{implies} \qquad (h(a_1), \ldots, h(a_n)) \in R^{\mathbf{B}}$$

for each relation symbol R. A *substructure* $\mathbf{A}$ of a structure $\mathbf{B}$ is defined in the same way as the above. Thus, $R^{\mathbf{A}}$ is the restriction of $R^{\mathbf{B}}$ to A for each relation symbol R. A relation $R^{\mathbf{A}}$ on the direct product $\mathbf{A}$ of $(\mathbf{A}_i : i \in I)$ is given by the direct product of $(R^{\mathbf{A}_i} : i \in I)$.

The following examples illustrate the above definitions: The structure $((0,1), \times, \leq)$ is a substructure of the totally ordered semigroup $([0,1], \times, \leq)$, where $(0,1)$ is the open interval of reals. Also, for a given a such that $0 < a < 1$, the semigroup $([0,a], \min)$ is a subalgebra of the semigroup $([0,1], \min)$. However the monoid $([0,a], \min, a)$ is not a submonoid of the monoid $([0,1], \min, 1)$, since in this case the unit element in the former algebra differs from that in the latter. Define a map $h : [0,1] \to [0,a]$ by $h(x) = ax$. Then, h is a homomorphism from $([0,1], \min, 1)$ onto $([0,a], \min, a)$, and hence the latter is a homomorphic image of the former.

Therefore, sometimes it is essential to mention explicitly what language we use. For example, when we are talking about homomorphisms between lattices, we call them *lattice homomorphisms*, to stress the language.

Let $\mathbf{P}$ and $\mathbf{Q}$ be posets. A map $h : P \to Q$ is *order-preserving* or, *monotone (increasing)*, if it is a homomorphism $h\colon \mathbf{P} \to \mathbf{Q}$, i.e, if for all $x, y \in P$

$$x \leq^{\mathbf{P}} y \text{ implies } h(x) \leq^{\mathbf{Q}} h(y).$$

It is said to *reflect the order* if for all $x, y \in P$, $h(x) \leq h(y)$ implies $x \leq y$. It is an *order-embedding* if for all $x, y \in P$

$$x \leq^{\mathbf{P}} y \text{ if and only if } h(x) \leq^{\mathbf{Q}} h(y).$$

Also, a map h *preserves existing meets* if for all $X \subseteq P$, whenever $\inf X$ exists then $\inf h(X)$ $(= \inf\{h(p) : p \in X\})$ exists and $h(\inf X) = \inf h(X)$. Dually, maps *preserving existing joins* are defined. When both $\mathbf{P}$ and $\mathbf{Q}$ are lattices, a map $h : P \to Q$ is a *complete lattice isomorphism* if it is a lattice isomorphism which preserves existing infinite meets and joins.

THEOREM 1.5. *Suppose that both* $\mathbf{P}$ *and* $\mathbf{Q}$ *are lattices. Then a map* $h : P \to Q$ *is an order-isomorphism if and only if it is a complete lattice isomorphism.*

For a class $\mathcal{K}$ of algebras, we define $\mathsf{H}(\mathcal{K})$, $\mathsf{S}(\mathcal{K})$, $\mathsf{P}(\mathcal{K})$ and $\mathsf{I}(\mathcal{K})$ to be, respectively, the class of all homomorphic images of algebras from $\mathcal{K}$, the class of all subalgebras of algebras from $\mathcal{K}$, the class of all direct products of algebras from $\mathcal{K}$ and the class of all algebras isomorphic to some algebra from $\mathcal{K}$. Since H, S, P and I operate on classes of algebras (that may fail to be sets) we refer to them as *class operators*.

LEMMA 1.6. [Bir35] *The class operators* H, S, P *and* I *preserve identities, i.e., if an identity is valid in a class* $\mathcal{K}$ *of algebras, then it is valid in* $\mathsf{H}(\mathcal{K})$, $\mathsf{S}(\mathcal{K})$, $\mathsf{P}(\mathcal{K})$ *and* $\mathsf{I}(\mathcal{K})$.

Class operators can be applied successively to any class of algebras, potentially yielding larger and larger classes. It turns out however, that there are only finitely many such enlargements.

LEMMA 1.7. [Tar46] *For any composition* O *of* H, S, P *and* I, *and any class* $\mathcal{K}$ *of similar algebras,* $\mathsf{O}(\mathcal{K}) \subseteq \mathsf{HSP}(\mathcal{K})$.

We say that $\mathcal{K}$ is a *variety* if it is closed under H, S and P. Clearly, if $\mathcal{K}$ is a variety then $\mathsf{HSP}(\mathcal{K}) = \mathcal{K}$. Conversely, when $\mathsf{HSP}(\mathcal{K}) = \mathcal{K}$, it is a variety by the lemma above. Thus, we have the following.

THEOREM 1.8. [Tar46] *A class* $\mathcal{K}$ *of algebras is a variety if and only if* $\mathsf{HSP}(\mathcal{K}) = \mathcal{K}$.

From the same lemma it also follows that for any class $\mathcal{K}$ of algebras, $\mathsf{HSP}(\mathcal{K})$ is a variety, in fact the smallest variety containing $\mathcal{K}$. Thus, it is called the variety *generated by* $\mathcal{K}$ and often denoted by $\mathsf{V}(\mathcal{K})$. We write $\mathsf{V}(\mathbf{A})$ for $\mathsf{V}(\{\mathbf{A}\})$.

One class operator which is of particular importance is that of *subdirect product*, abbreviated $\mathsf{P_S}$. It combines S and P in a particular way that proves to be useful for representing big algebras by means of smaller ones. We say that an algebra $\mathbf{A}$ is a subdirect product of an indexed system $(\mathbf{A}_i : i \in I)$, if $\mathbf{A} \subseteq \prod_{i\in I} \mathbf{A}_i$ and all the coordinate projections restricted to $\mathbf{A}$ are onto (in other words, each $\mathbf{A}_i$ is a homomorphic image of $\mathbf{A}$). If this is the case, we call $(\mathbf{A}_i : i \in I)$ a *subdirect representation* of $\mathbf{A}$. This is not unique, but of course always exists, since $\mathbf{A}$ itself is its own subdirect representation. For certain algebras, this trivial representation is the best we can do. Namely, we say that $\mathbf{A}$ is *subdirectly irreducible* if every subdirect representation $(\mathbf{A}_i : i \in I)$ of $\mathbf{A}$ contains (an isomorphic copy of) $\mathbf{A}$ as a factor. If $\mathcal{K}$ is a class of algebras we denote by $(\mathsf{V}(\mathcal{K}))_{\mathsf{SI}}$ the class of all subdirectly irreducible members of $\mathsf{V}(\mathcal{K})$. Subdirectly irreducible algebras are the prime numbers of universal algebra. The theorem below due to Birkhoff [Bir44] is one of the cornerstones of universal algebra. A proof of this result can also be found e.g., in [BS81].

THEOREM 1.9. [Bir44] *Every algebra has a subdirect representation with subdirectly irreducible factors. In particular, every finite algebra has a subdirect representation with finitely many subdirectly irreducible finite factors.*

Notice that if $\mathbf{A} \in \mathcal{V}$ and $\mathcal{V}$ is a variety (actually, closure under H suffices) then all subdirectly irreducible factors of $\mathbf{A}$ also belong to $\mathcal{V}$.

COROLLARY 1.10. *Every variety is generated by its subdirectly irreducible members.*

1.2.2. Congruences. An *equivalence relation* on a set A is a reflexive, symmetric and transitive binary relation on A. If θ is such a relation, the set A is partitioned into *equivalence classes* $[a]_\theta = \{x\colon a\ \theta\ x\}$; we also use the notation a/θ. Each homomorphism $h\colon \mathbf{A} \to \mathbf{B}$ induces a natural equivalence relation $\ker_h$ on A defined by $a \ker_h b$ if $h(a) = h(b)$. This relation is called the *kernel* of h and it partitions A into equivalence classes having the additional property that

$$a_1 \ker_h b_1, \dots, a_n \ker_h b_n \text{ implies } f(a_1, \dots, a_n) \ker_h f(b_1, \dots, b_n)$$

for any n-ary basic operation f. We say that the operations are *compatible* with the equivalence classes. An equivalence relation θ on $\mathbf{A}$ with this property is called a *congruence*, and the equivalence classes of such a relation are called *congruence classes*.

To express the fact that a pair (x, y) belongs to a congruence θ, in addition to set-theoretical $(x, y) \in \theta$ and the usual shorthand $x\ \theta\ y$, we will often write $x \equiv_\theta y$.

Each congruence θ on $\mathbf{A}$ defines a *quotient algebra* $\mathbf{A}/\theta$. Its elements are equivalence classes of θ and operations are defined naturally by putting $f([a_1]_\theta, \dots, [a_n]_\theta) = [f(a_1, \dots, a_n)]_\theta$. Since θ preserves the operations, this does not depend on the choice of representatives. Each congruence θ on $\mathbf{A}$ induces a *natural homomorphism* $h_\theta : \mathbf{A} \to \mathbf{A}/\theta$, taking each element $a \in A$ to its equivalence class $[a]_\theta$. We can show that $\ker_{(h_\theta)} = \theta$ for each congruence θ. Moreover, we have the following homomorphism theorem.

THEOREM 1.11. [BS81] *For each surjective homomorphism $k : \mathbf{A} \to \mathbf{B}$, the quotient algebra $\mathbf{A}/\ker_k$ is isomorphic to $\mathbf{B}$. The isomorphism $i : \mathbf{A}/\ker_k \to \mathbf{B}$ satisfies $k = i \circ h_{(\ker_k)}$.*

Thus, homomorphisms and congruences are two sides of the same coin. The set of all congruences on $\mathbf{A}$ ordered by inclusion forms a complete lattice $\mathbf{Con\,A}$, called the *congruence lattice* of $\mathbf{A}$. In line with the distinction between algebras and their universes, $\mathrm{Con}\,\mathbf{A}$ stands for the bare set of all congruences of an algebra $\mathbf{A}$. Further, $\mathrm{Cg}^{\mathbf{A}}(X)$ and $\mathrm{Sg}^{\mathbf{A}}(Y)$ stand for the *congruence generated* by the set $X \subseteq A^2$ (the smallest congruence containing X) and the subuniverse generated by $Y \subseteq A$, respectively. We will write $\mathrm{Cg}^{\mathbf{A}}(a, b)$ instead of $\mathrm{Cg}^{\mathbf{A}}(\{a, b\})$ in the case of *principal* congruences, i.e., congruences generated by a single pair of elements. The bottom element of $\mathbf{Con\,A}$ is the identity relation Δ on A, the top element is the full relation ∇ on A, and meets are set-theoretical intersections. It remains to construct joins in an explicit way. Recall that for any two binary relations θ and η, the *relational product* $\theta \circ \eta$ is a binary relation defined by $(a, b) \in \theta \circ \eta$ if and only if there exists an element c such that $(a,c) \in \theta$ and $(c,b) \in \eta$. Clearly, the relational product is associative. Suppose that a set of congruences $\{\theta_i : i \in I\}$ on **A** is given. Then the join $\bigvee_{i\in I} \theta_i$ is the set-theoretical union of all relational products of the form $\eta_1 \circ \eta_2 \circ \cdots \circ \eta_m$ such that each $\eta_j \in \{\theta_i : i \in I\}$.

Surprisingly perhaps, many facts about **A** can be deduced from this lattice alone. For instance, each subdirect representation of **A** is given by a system $(\theta_i : i \in I)$ of congruences on **A** such that $\bigcap_{i\in I} \theta_i = \Delta$. Furthermore, we have the following.

THEOREM 1.12. [BS81] *An algebra* **A** *is subdirectly irreducible if and only if* Δ *is* completely meet irreducible *in* **Con A**, *i.e., for all systems* $(\theta_i : i \in I)$ *of congruences on* **A**, $\bigcap_{i\in I} \theta_i = \Delta$ *implies* $\theta_j = \Delta$ *for some* $j \in I$.

Hence, for every non-trivial subdirectly irreducible algebra **A**, the congruence lattice **Con A** has a unique atom μ, called the *monolith.* In fact, μ is given by the intersection of all congruences in Con **A** except Δ. An algebra **A** is *simple* if Con **A** is equal to $\{\Delta, \nabla\}$. In other words, **A** is simple iff any nontrivial algebra which is a homomorphic image of **A** is isomorphic to **A**. An algebra **A** is *semisimple* if it is isomorphic to a subdirect product of simple algebras. A variety is *semisimple* if all of its algebras are semisimple.

LEMMA 1.13.

(1) *If an algebra* **A** *is simple then it is subdirectly irreducible.*
(2) *A variety* $\mathcal{V}$ *is semisimple if and only if all of the subdirectly irreducible members of* $\mathcal{V}$ *are simple.*

If **Con A** is a distributive lattice, the algebra **A** is called *congruence distributive* . If for $\theta, \eta \in$ **Con A** we have $\theta\circ\eta = \eta\circ\theta$, **A** is called *congruence permutable.* When this is the case, we also have $\theta \vee \eta = \theta \circ \eta$. An algebra is *arithmetical* iff it is both congruence distributive and permutable. If there is an element $e \in A$ such that for any $\theta, \eta \in$ Con **A** we have $[e]_\theta = [e]_\eta$ implies $\theta = \eta$, then **A** is called *e-regular*, and if **A** is e-regular for any $e \in A$, then **A** is called *regular*. On the other hand, if e is definable by a term (in particular, if e is a constant in the type), **A** is called *point-regular*. If any of the above properties is true of all algebras in a class $\mathcal{K}$, the corresponding notion is applied to the whole class.

THEOREM 1.14. *The variety of lattices is congruence distributive.*

PROOF. Although this result follows from general facts on Mal'cev conditions (see Section 1.2.5), we will prove it here directly because it illustrates some properties of congruences on lattices (and lattices themselves) that will later on be assumed without mention. Firstly, observe that $(x \wedge y) \vee (x \wedge z) \leq x \wedge (y \vee z)$ holds in any lattice, so we really are after the reverse inclusion. Secondly, if α is a congruence on a lattice **A** and

$a, b \in A$, then $a \equiv_\alpha b$ iff $a \wedge b \equiv_\alpha a \vee b$. It follows that congruence classes on lattices are *convex*, i.e., $a \in [b]_\alpha$ iff $\{x \in A\colon a \wedge b \leq x \leq a \vee b\} \subseteq [b]_\alpha$. Thirdly, as we have seen, for congruences θ and η (on any algebra, not necessarily a lattice) we have

$$\theta \vee \eta = \theta \cup (\theta \circ \eta) \cup (\theta \circ \eta \circ \theta) \cup (\theta \circ \eta \circ \theta \circ \eta) \ldots$$

and thus $(x, y) \in \theta \vee \eta$ iff there are elements $c_0, \ldots, c_{k+1}$ with $c_0 = x$, $c_{k+1} = y$ and $(c_i, c_{i+1}) \in \theta$ or $(c_i, c_{i+1}) \in \eta$.

After this prelude, let α, β, γ be congruences on a lattice $\mathbf{A}$. We want to show $\alpha \wedge (\beta \vee \gamma) \subseteq (\alpha \wedge \beta) \vee (\alpha \wedge \gamma)$. Since meets of congruences are set-theoretical intersections, it amounts to $\alpha \cap (\beta \vee \gamma) \subseteq (\alpha \cap \beta) \vee (\alpha \cap \gamma)$. Let $(x, y) \in \alpha \cap (\beta \vee \gamma)$. By remarks above, we can assume $x \leq y$. As $(x, y) \in \beta \vee \gamma$, we have $x \;\beta\; c_1 \;\gamma\; c_2 \ldots c_k \;\beta\; y$, for some $c_1, \ldots, c_k \in A$. Consider the elements $x \vee c_i$ ($i \in \{1, \ldots, k\}$). Since $x \vee x = x$ and $x \vee y = y$, we still have $x \;\beta\; (x \vee c_1) \;\gamma\; (x \vee c_2) \ldots (x \vee c_k) \;\beta\; y$. By similar argument, we get $x \;\beta\; y \wedge (x \vee c_1) \;\gamma\; y \wedge (x \vee c_2) \ldots y \wedge (x \vee c_k) \;\beta\; y$. But now we have $x \leq y \wedge (x \vee c_i) \leq y$, for every $i \in \{1, \ldots, k\}$. Since $(x, y) \in \alpha$ and congruence classes are convex, we get that $y \wedge (x \vee c_i) \in [x]_\alpha = [y]_\alpha$ for every $i \in \{1, \ldots, k\}$. Therefore, each pair $(y \wedge (x \vee c_i), y \wedge (x \vee c_{i+1}))$ belongs to α, and thus $x \equiv_{\alpha\cap\beta} (y \wedge (x \vee c_1)) \equiv_{\alpha\cap\gamma} (y \wedge (x \vee c_2)) \ldots (y \wedge (x \vee c_k)) \equiv_{\alpha\cap\beta} y$, concluding the argument. □

The proof above depends only on the presence of lattice operations, so we immediately obtain the next corollary.

COROLLARY 1.15. *Every algebra that has a lattice reduct is congruence distributive.*

A class $\mathcal{K}$ of algebras has the *congruence extension property* (CEP) iff for any $\mathbf{A} \subseteq \mathbf{B} \in \mathcal{K}$ and any $\theta \in \mathbf{Con}\,\mathbf{A}$, there is a $\eta \in \mathbf{Con}\,\mathbf{B}$ such that θ is the restriction of η to $A \times A$.

A quite well-behaved subclass of congruence distributive varieties are varieties with *(equationally) definable principal congruences* (cf. [BP82], [BKP84], [BP94a], [BP94b]). Among these, an even smaller subclass of *discriminator* varieties (cf. [Wer78]) is very well-behaved. We define both concepts below.

A variety $\mathcal{V}$ has *definable principal congruences* (DPC) iff there is a first-order formula $\varphi(x, y, z, w)$ in the language of $\mathcal{V}$ (with no other free variables than the ones displayed) such that, for any $a, b, c, d \in \mathbf{A} \in \mathcal{V}$, we have: $(c, d) \in \theta(a, b)$ iff $\mathbf{A} \models \varphi(a, b, c, d)$. If $\varphi(x, y, z, w)$ can be taken to be a conjunction (equivalently, a finite set) of equations, then $\mathcal{V}$ has *equationally definable principal congruences* (EDPC).

A *ternary discriminator* on a set A is an operation t on A defined by:

$$t(x, y, z) = \begin{cases} x, & \text{for } x \neq y, \\ z, & \text{for } x = y. \end{cases}$$

A *discriminator variety* is a variety $\mathcal{V}$, such that the ternary discriminator is a term-operation on every subdirectly irreducible algebra in $\mathcal{V}$.

LEMMA 1.16. *Any algebra* $\mathbf{A}$ *with a discriminator term is simple. Thus, every discriminator variety is semisimple.*

PROOF. Consider $\mathrm{Cg}^{\mathbf{A}}(a,b)$ for arbitrary distinct $a, b \in A$. Take any $(c,d) \in A^2$. We have $t(a,b,c) = a$ and $t(a,b,d) = a$, so $t(t(a,b,c), t(a,b,d), d) = t(a,a,d) = d$. On the other hand, $t(a,a,c) = c$ and $t(a,a,d) = d$, so $t(t(a,a,c), t(a,a,d), d) = t(c,d,d) = c$. Letting $p(x)$ be the polynomial $t(t(a,x,c), t(a,x,d), d)$ we obtain $c = p(a)$ and $d = p(b)$. Therefore $(c,d) \in \mathrm{Cg}^{\mathbf{A}}(a,b)$. □

An example of a discriminator variety is BA, with discriminator defined by $t(x,y,z) = ((x \leftrightarrow y) \wedge z) \vee (\neg(x \leftrightarrow y) \wedge x)$. It is easily seen that this term is the discriminator on the two element Boolean algebra, which is the only subdirectly irreducible member of BA. Since every variety of Heyting algebras that is strictly bigger than BA contains the three element algebra $\mathbf{H}_3$, which is subdirectly irreducible but not simple, we conclude that BA is the only discriminator subvariety of HA.

LEMMA 1.17. *Every discriminator variety has EDPC. Every variety with EDPC has CEP.*

PROOF. For the first part, let $\mathcal{V}$ be a discriminator variety with a ternary discriminator term $t(x,y,z)$. We claim that for every $\mathbf{A} \in \mathcal{V}$ and any $a,b,c,d \in A$ we have $(c,d) \in \mathrm{Cg}^{\mathbf{A}}(a,b)$ iff $\mathbf{A} \models t(a,b,c) = t(a,b,d)$. Let $\theta = \mathrm{Cg}^{\mathbf{A}}(a,b)$.

For the 'if' direction suppose $\mathbf{A} \models t(a,b,c) = t(a,b,d)$. We want to show $[c]_\theta = [d]_\theta$. Consider a subdirect representation $\prod_{i\in I} \mathbf{B}_i$ of $\mathbf{A}/\theta$. By assumption we have, for any $i \in I$

$$\mathbf{B}_i \models t(\pi_i([a]_\theta), \pi_i([b]_\theta), \pi_i([c]_\theta)) = t(\pi_i([a]_\theta), \pi_i([b]_\theta), \pi_i([d]_\theta))$$

where π_i is the i-th projection. But $[a]_\theta = [b]_\theta$, by definition of θ, and so

$$\begin{aligned}\mathbf{B}_i \models \pi_i([c]_\theta) &= t(\pi_i([a]_\theta), \pi_i([b]_\theta), \pi_i([c]_\theta)) \\ &= t(\pi_i([a]_\theta), \pi_i([b]_\theta), \pi_i([d]_\theta)) \\ &= \pi_i([d]_\theta)\end{aligned}$$

because t is the discriminator on $\mathbf{B}_i$. Since i was arbitrary, $[c]_\theta = [d]_\theta$ as needed. As the proof for the 'only if' direction simply reverses the above reasoning, we ask the reader to carry it out as an exercise.

For the second statement, let $\mathcal{V}$ have EDPC, with $\varphi(x,y,z,t)$ being the principal congruence formula (a conjunction of equations). Observe first that since every congruence is a join of principal congruences it suffices to prove that every principal congruence extends. So let $\mathbf{A} \subseteq \mathbf{B}$ and $\mathrm{Cg}^{\mathbf{A}}(a,b)$

for some $a, b \in A$. Consider $\mathrm{Cg}^{\mathbf{B}}(a, b)$ and any pair $(c, d) \in A^2 \cap \mathrm{Cg}^{\mathbf{B}}(a, b)$. By EDPC we obtain $\mathbf{B} \models \varphi(a, b, c, d)$ and since φ is a conjunction of equations, it follows that $\mathbf{A} \models \varphi(a, b, c, d)$ as well. Thus, $(c, d) \in \mathrm{Cg}^{\mathbf{A}}(a, b)$ as required. □

1.2.3. Free algebras. For any language $\mathcal{L}$, the set $Tm_{\mathcal{L}}(X)$ of all terms over the set X is an algebra, with the operation symbols interpreted in an obvious way. This algebra, which we denote by $\mathbf{Tm}_{\mathcal{L}}(X)$[3] , is called the *absolutely free algebra over* X. If X is the countable set of variables, then we denote it simply by $\mathbf{Tm}_{\mathcal{L}}$ and call it the *absolutely free algebra.* It is easy to see that $\mathbf{Tm}_{\mathcal{L}}(X)$ is *generated* by the set X, i.e., it can be obtained by applying basic operations repeatedly to X and that it has the following *universal mapping property over* X.

> For any $\mathcal{L}$-algebra $\mathbf{A}$ and any map $h : X \to A$, extends uniquely to a homomorphism (also denoted by h) from $\mathbf{Tm}_{\mathcal{L}}(X)$ to $\mathbf{A}$.

Many semantical and syntactical notions of logic can be given algebraic meaning using this algebra. One such notion is that of *assignment* or *valuation*, which can be thought of as being just one of the functions above, for the case when X happens to be the set of variables. Another is *substitution*, which is a function from the set of variables into $Tm_{\mathcal{L}}$ (taking each variable to the term substituted for it) and expanded to a unique endomorphism of $\mathbf{Tm}_{\mathcal{L}}$. A *substitution instance* of a term is simply the image of this term under some substitution. An arbitrary congruence θ on $\mathbf{Tm}_{\mathcal{L}}$ need not preserve substitutions, for it may happen that $s\ \theta\ t$ but not $\sigma(s)\ \theta\ \sigma(t)$ for some endomorphism σ. The congruences that *do* preserve them are called *fully invariant.* Since any congruence on $\mathbf{Tm}_{\mathcal{L}}$ identifies certain terms, a fully invariant one identifies terms in such a way that all their substitutions are also identified.

Recall that a class of algebras is an *equational class* if it is of the form $\mathrm{Mod}(E)$, for some set of equations E. Let θ_E be the smallest fully invariant congruence on $\mathbf{Tm}_{\mathcal{L}}$ such that for each equation $(s = t) \in E$ we have $s\,\theta_E\,t$ (it can be shown that such a congruence exists for any E).

LEMMA 1.18. [BS81] *The algebra* $\mathbf{Tm}_{\mathcal{L}}/\theta_E$ *belongs to* $\mathrm{Mod}(E)$. *Moreover,* $\mathbf{Tm}_{\mathcal{L}}/\theta_E$ *has the universal mapping property (over the set* X *of variables) with respect to* $\mathrm{Mod}(E)$. *That is, for any* $\mathbf{A} \in \mathrm{Mod}(E)$, *every map* $h : X \to A$ *can be extended to a homomorphism* h *from* $\mathbf{Tm}_{\mathcal{L}}/\theta_E$ *to* $\mathbf{A}$.

An algebra $\mathbf{F}$ that has the universal mapping property with respect to a class $\mathcal{K}$ of algebras of the same type is said to be *free for* $\mathcal{K}$. If $\mathbf{F}$ itself belongs to $\mathcal{K}$ it is said to be *free in* $\mathcal{K}$. Thus, the lemma above states that $\mathbf{F} = \mathbf{Tm}_{\mathcal{L}}/\theta_E$ is free in $\mathrm{Mod}(E)$.

[3]Later we will also use $\mathbf{Fm}_{\mathcal{L}}(X)$ for that algebra, if the language $\mathcal{L}$ is a language of some (propositional) logic. We ask the reader to wait until Section 1.3.1 for a justification.

Notice that by Lemma 1.6 we have $\mathsf{V}(\mathrm{Mod}(E)) = \mathrm{Mod}(E)$, so all equational classes are varieties. This leads to another cornerstone of universal algebra, due to Birkhoff [Bir35] (see e.g., [BS81] for a proof). Let $\mathcal{K}$ be any class of algebras of the same type.

THEOREM 1.19. [Bir35] *A class $\mathcal{K}$ of algebras is a variety if and only if $\mathcal{K}$ is an equational class.*

When a variety $\mathcal{K}$ is equal to $\mathrm{Mod}(E)$ for a set of equations E, it is called the variety *axiomatized* by E. For example the free semigroup generated by a set X consists of elements of the form $x_1x_2\cdots x_m$ with $m > 0$ such that $x_i \in X$ for each i. The semigroup operation is just concatenation; for example we get $x_1x_2\cdots x_my_1y_2\cdots y_n$ from $x_1x_2\cdots x_m$ and $y_1y_2\cdots y_n$. Therefore, free semigroups are always infinite. On the other hand, the free semilattice generated by X is finite when X is finite. In fact if X has n elements then the free semilattice has $2^n - 1$ elements.

COROLLARY 1.20. *The following classes of algebras form equational classes, and hence varieties: lattices, distributive lattices, semigroups, semilattices, monoids, groups, Heyting algebras and Boolean algebras.*

1.2.4. More on Heyting and Boolean algebras. Every Heyting algebra $\mathbf{A}$ is 1-regular. This follows from the fact that for each $\theta \in \mathrm{Con}\,\mathbf{A}$, $(a, b) \in \theta$ iff $(1, (a \to b) \wedge (b \to a)) \in \theta$. The notion of (lattice) filter can be applied to Heyting algebras, since they have lattice reducts. Each Heyting algebra $\mathbf{A}$ has a smallest filter $\{1\}$, where 1 is the greatest element of $\mathbf{A}$. (Thus the non-emptiness of a filter F can be replaced by the condition that $1 \in F$.) The set of all filters of $\mathbf{A}$ forms a complete lattice with meets given by set-theoretical intersections. For each nonempty subset S of A, there exists a smallest filter $\mathrm{Fg}^{\mathbf{A}}(S)$ containing S, which can be described explicitly as follows:

$$\{x \in A : a_1 \wedge \ldots \wedge a_m \leq x \text{ for some } a_i \in S\}.$$

The filter $\mathrm{Fg}^{\mathbf{A}}(S)$ is called the filter *generated* by S. In particular, for each element $c \in A$, the filter $\mathrm{Fg}^{\mathbf{A}}(c)$ generated by $\{c\}$ is equal to the order filter $\uparrow c = \{x \in A \mid c \leq x\}$. Recall that filters of the form $\uparrow c$ are called *principal* filters. A proper filter F is said to be *prime* when the following condition holds:

$$a \vee b \in F \text{ implies } a \in F \text{ or } b \in F.$$

Maximal filters are prime in every distributive lattice, and the converse holds also in every Boolean algebra. The dual notions of a (lattice) *ideal*, the ideal *generated* by S, a *prime* ideal and the *principal* ideal $\downarrow c$ are defined in a natural way. We can show the following *prime filter theorem*, by using Zorn's lemma.

THEOREM 1.21. *Let G be a proper filter of a distributive lattice $\mathbf{A}$ such that $G \cap H = \emptyset$ for a subset H of A. Then there exists a prime filter F of $\mathbf{A}$ such that $G \subseteq F$ and $F \cap H = \emptyset$.*

There is a close correspondence between congruences and filters of Heyting algebras. For each congruence θ on $\mathbf{A}$, let $\mathrm{F_c}(\theta) = \{x \in A\colon (1, x) \in \theta\}$. Then, $\mathrm{F_c}(\theta)$ is a filter of $\mathbf{A}$, called the filter *determined* by a congruence θ. Conversely, for each filter F, $\Theta_\mathrm{f}(F) = \{(x, y) \in A \times A : (x \rightarrow y) \wedge (y \rightarrow x) \in F\}$ is a congruence, called the congruence *determined* by a filter F. Moreover the following result holds.

THEOREM 1.22. *For all congruences θ, η and all filters F, G of a Heyting algebra $\mathbf{A}$, the following hold:*

- *$\theta \subseteq \eta$ implies $\mathrm{F_c}(\theta) \subseteq \mathrm{F_c}(\eta)$,*
- *$F \subseteq G$ implies $\Theta_\mathrm{f}(F) \subseteq \Theta_\mathrm{f}(G)$,*
- *$\Theta_\mathrm{f}(\mathrm{F_c}(\theta)) = \theta$ and $\mathrm{F_c}(\Theta_\mathrm{f}(F)) = F$.*

Thus, the map $\mathrm{F_c}$ is an order-isomorphism, and therefore is a complete lattice isomorphism between the congruence lattice and the lattice of all filters of $\mathbf{A}$.

Because of this correspondence, the quotient algebra $\mathbf{A}/\Theta_\mathrm{f}(F)$ is sometimes represented as $\mathbf{A}/F$, for a given filter F of $\mathbf{A}$. Using the above result, we can show that the class HA of Heyting algebras has EDPC. It suffices to give an equation $\varepsilon(a, b)$ such that for any $a, b \in \mathbf{A} \in \mathsf{HA}$, we have $a \in \uparrow b$ iff $\mathbf{A} \models \varepsilon(a, b)$. For this $\varepsilon(a, b)$, we can take $b \leq a$, or more precisely $b \wedge a = b$. By Lemma 1.17 we conclude that HA has CEP as well. The same argument shows that the variety BA of Boolean algebras has EDPC and CEP.

Recall that any non-trivial Heyting algebra $\mathbf{A}$ is subdirectly irreducible iff it has a unique minimal nontrivial congruence. By Theorem 1.22 the latter is equivalent to saying that it has a unique minimal filter F. Obviously any such filter F must be a principal filter $\uparrow d$ for some $d < 1$ and moreover $d \in \uparrow e$ for any $e < 1$. This means that d is the greatest element in the set $A \backslash \{1\}$. Thus we have the following.

THEOREM 1.23. *A non-trivial Heyting algebra is subdirectly irreducible iff it has a unique subcover of 1. Hence, it is simple iff it is isomorphic to the two-element Boolean algebra **2**.*

1.2.5. Mal'cev conditions. Many properties of (classes of) algebras depend on the behavior of congruences rather then the algebras themselves. In fact, this simple observation is yet another cornerstone of universal algebra. In turn, many of these *congruence conditions* are associated with the presence of certain terms. Such properties are known as *Mal'cev properties* or *Mal'cev conditions*. Possessing a ternary discriminator or equationally definable principal congruences are in fact examples of Mal'cev properties. Other, classical, examples of Mal'cev conditions include congruence distributivity and congruence permutability.

LEMMA 1.24. *For any variety $\mathcal{V}$ the following hold:*

(1) *$\mathcal{V}$ is congruence permutable if and only if there exists a ternary term $p(x,y,z)$ such that $\mathcal{V} \models p(x,y,y) = x = p(y,y,x)$.*
(2) *If there exists a ternary term $M(x,y,z)$ such that $\mathcal{V} \models M(x,x,y) = M(x,y,x) = M(y,x,x) = x$, then $\mathcal{V}$ is congruence distributive.*

PROOF. For the left-to-right direction of (1), suppose $\mathcal{V}$ is congruence permutable. Let $\mathbf{F}$ be the free algebra in $\mathcal{V}$ on generators x, y, z. Let $\alpha = \mathrm{Cg}^{\mathbf{F}}(\{x,y\})$ and $\beta = \mathrm{Cg}^{\mathbf{F}}(\{y,z\})$. We then have $(x,z) \in \alpha \circ \beta$ and thus $(x,z) \in \beta \circ \alpha$. Therefore, there is a term $p(x,y,z)$ such that we have $x \equiv_\beta p(x,y,z) \equiv_\alpha z$. Consider now the equation $p(x,y,y) = x$. It holds throughout $\mathcal{V}$ if and only if, for any algebra $\mathbf{A} \in \mathcal{V}$ and any homomorphism $h\colon \{x,y,z\} \mapsto A$, we have the following:

$$\text{if } h(y) = h(z), \text{ then } h(p(x,y,y)) = h(x).$$

By correspondence between congruences and kernels of homomorphisms, this means that for any congruence θ on $\mathbf{F}$ we have:

$$\text{if } y \equiv_\theta z, \text{ then } p(x,y,y) \equiv_\theta x.$$

By definition of β, the antecedent of the above can be expressed as $\beta \subseteq \theta$. Therefore, the condition reduces to

$$p(x,y,y) \equiv_\beta x.$$

But, $p(x,y,z) \equiv_\beta x$ and $z \equiv_\beta y$, so $p(x,y,y) \equiv_\beta x$ and thus $\mathcal{V} \models p(x,y,y) = x$ indeed holds. Similar reasoning shows that $\mathcal{V} \models p(y,y,z) = z$.

For the converse, let $p(x,y,z)$ satisfy the conditions of our Lemma, and α and β be congruences on some algebra $\mathbf{A} \in \mathcal{V}$. Then, $(a,b) \in \alpha \circ \beta$ implies $a \equiv_\alpha c \equiv_\beta b$ for some $c \in A$. It follows that $b = p(c,c,b) \equiv_\alpha p(a,c,b) \equiv_\beta p(a,b,b) = a$ (it is an instructive exercise to verify it by drawing a picture). Therefore $(a,b) \in \beta \circ \alpha$ as needed.

For (2), take congruences $\alpha, \beta, \gamma \in \mathrm{Con}\,\mathbf{A}$. It suffices to show $\alpha \wedge (\beta \vee \gamma) \leq (\alpha \wedge \beta) \vee (\alpha \wedge \gamma)$ since the converse holds in every lattice as shown in Section 2.1. Take $(a,b) \in \alpha \wedge (\beta \vee \gamma)$. Then $(a,b) \in \alpha$ and there are $c_1, \dots, c_n \in A$ such that $a \equiv_\beta c_1 \equiv_\gamma c_2 \equiv \cdots \equiv_\beta c_n \equiv_\gamma b$. Now observe that for each i we have $M(a,c_i,b) \equiv_\alpha M(a,c_i,a) = a$. Thus, we obtain $a = M(a,a,b) \equiv_{\alpha\wedge\beta} M(a,c_1,b) \equiv_{\alpha\wedge\gamma} M(a,c_2,b) \equiv \cdots \equiv_{\alpha\wedge\beta} M(a,c_n,b) \equiv_{\alpha\wedge\gamma} M(a,b,b) = b$. Therefore, $(a,b) \in (\alpha \wedge \beta) \vee (\alpha \wedge \gamma)$ as required. □

The term p above in known as *Mal'cev term* and the term M as *majority term*. For example, $M(x,y,z) = (x \vee y) \wedge (y \vee z) \wedge (z \vee x)$ is a majority term for lattices. Thus, the variety of lattices is congruence distributive (as we have already shown directly, cf. Theorem 1.14).

Notice that the condition for congruence distributivity, unlike the one for congruence permutability, is sufficient but not necessary. There is a necessary and sufficient Mal'cev condition for congruence distributivity, connecting that property with *Jónsson terms*. We will not present it here, because the varieties we deal with will all be arithmetical, and for these the following lemma holds. The term m in the lemma is called $\frac{2}{3}$*-minority term*.

LEMMA 1.25. *For a variety $\mathcal{V}$ the following are equivalent:*

(1) *$\mathcal{V}$ is arithmetical.*
(2) *$\mathcal{V}$ has a Mal'cev term and a majority term.*
(3) *There exists a ternary term $m(x, y, z)$ such that $\mathcal{V} \models m(x, y, x) = m(x, y, y) = m(y, y, x) = x$.*

PROOF. Here is an outline of the proof. Clearly (3) implies (1). We will show that (1) implies (2) implies (3), and that (3) implies (2)

Suppose (1) holds. Then $\mathcal{V}$ is arithmetical and so it has a Mal'cev term p. To show that a suitable majority term M exists, consider congruences $\alpha = \mathrm{Cg}^{\mathbf{F}}(\{x, y\})$, $\beta = \mathrm{Cg}^{\mathbf{F}}(\{y, z\})$, and $\gamma = \mathrm{Cg}^{\mathbf{F}}(\{x, z\})$ on the three-generated free algebra $\mathbf{F}$. By congruence distributivity and permutability, we have $(x, z) \in \gamma \wedge (\alpha \vee \beta)$ iff $(x, z) \in (\gamma \wedge \alpha) \vee (\gamma \wedge \beta)$ iff $(x, z) \in (\gamma \wedge \alpha) \circ (\gamma \wedge \beta)$. Take $M(x, y, z)$ with $x \equiv_{\gamma\wedge\alpha} M(x, y, z) \equiv_{\gamma\wedge\beta} z$. Conclude that $\mathcal{V} \models M(x, x, y) = M(x, y, x) = M(y, x, x) = x$. Thus (2) holds.

Define $m(x, y, z) = p(x, M(x, y, z), z)$ and conclude that (2) implies (3).

If (3) holds, we define the terms $p(x, y, z) = m(x, y, z)$ and $M(x, y, z) = m(x, m(x, y, z), z)$ and verify that they are respectively the Mal'cev term and majority term. □

THEOREM 1.26. *All discriminator varieties are arithmetical, have EDPC and CEP, and are semisimple.*

PROOF. The ternary discriminator for a variety $\mathcal{V}$ is a $\frac{2}{3}$-minority term for $\mathcal{V}$. This is easily seen from the fact that the relevant equations hold on all subdirectly irreducible members of $\mathcal{V}$. The rest follows by Lemmas 1.17 and 1.16. □

1.2.6. Ultraproducts and Jónsson's Lemma. We consider mostly varieties in which all members have lattice reducts, hence they are congruence distributive. This allows us to make frequent use of a result of Jónsson (cf. [Jón67]), sometimes referred to as *Jónsson's Lemma*, which characterizes subdirectly irreducible algebras in congruence distributive varieties. Before we state it, a technical notion of *ultraproduct* is needed. Although in full generality it applies to any set of relational structures of a given type, we will only introduce it in the context of algebras. Thus, let $(\mathbf{A}_i : i \in I)$ be an indexed family of algebras of the same type. By an *ultrafilter on I* we mean an ultrafilter of $\mathcal{P}(I)$, viewed as a Boolean algebra. Let U be any ultrafilter on I. For elements a and b of the direct product $\prod_{i\in I}\mathbf{A}_i$, we define their *equalizer*, $\|a=b\|$ to be $\{i\in I : a(i)=b(i)\}$. Now, let $\equiv_U$ be the binary relation on $\prod_{i\in I}A_i$ defined by $a\equiv_U b$ if $\|a=b\|\in U$.

LEMMA 1.27. *The relation $\equiv_U$ is a congruence on $\prod_{i\in I}\mathbf{A}_i$.*

The quotient algebra $\mathbf{A}=(\prod_{i\in I}\mathbf{A}_i)/{\equiv_U}$ is called the *ultraproduct* of the family $\{\mathbf{A}_i : i\in I\}$ with respect to the ultrafilter U. As usual, we will write $(\prod_{i\in I}\mathbf{A}_i)/U$ for $(\prod_{i\in I}\mathbf{A}_i)/{\equiv_U}$. Similarly, for an element $a\in\prod_{i\in I}A_i$, we write $[a]_U$ for its equivalence class. If all the factors $\mathbf{A}_i$ are identical, say $\mathbf{A}_i=\mathbf{A}$, then $(\prod_{i\in I}\mathbf{A}_i)/U$ is called an *ultrapower* of $\mathbf{A}$ and is denoted by $\mathbf{A}^I/U$. To state succinctly the most important property of ultraproducts, we will expand the equalizer notation as follows. Let φ be any first-order formula in the appropriate language, and $\overline{a}$ be a tuple of elements of $\prod_{i\in I}A_i$ of suitable length. By $\|\varphi(\overline{a})\|$ we denote the set $\{i\in I : \varphi^{\mathbf{A}_i}(\overline{a}(i)) \text{ is true}\}$.

THEOREM 1.28. *For any first-order formula φ with k free variables and any k-tuple $\overline{a}$ of elements of $\prod_{i\in I}A_i$*

$$(\prod_{i\in I}\mathbf{A}_i)/U\models\varphi([\overline{a}]_U) \quad \textit{iff} \quad \|\varphi(\overline{a})\|\in U.$$

COROLLARY 1.29. *Every first-order sentence is preserved under the ultraproduct construction. More precisely, for any first-order sentence φ if $\mathbf{A}_i\models\varphi$ for each $i\in I$ then $(\prod_{i\in I}\mathbf{A}_i)/U\models\varphi$, where $(\prod_{i\in I}\mathbf{A}_i)/U$ is an ultraproduct of $\{\mathbf{A}_i : i\in I\}$.*

The ultraproduct construction proved to be so important that it deserves a class operator of its own. The class of all ultraproducts of algebras from $\mathcal{K}$ is denoted $\mathsf{P_U}(\mathcal{K})$. By the preceding corollary any first-order class of structures is closed under this operator. In fact the following fundamental result characterizes first-order classes semantically.

THEOREM 1.30. [CK90] *A class of structures is axiomatized by a first-order sentence if and only if the class is closed under ultraproducts and the complement of the class is closed under ultrapowers.*

THEOREM 1.31. (Jónsson's Lemma) [Jón67] *Let $\mathcal{V}$ be a congruence distributive variety generated by a class $\mathcal{K}$. Then, we have $(\mathsf{V}(\mathcal{K}))_{\mathsf{SI}}\subseteq\mathsf{HSP_U}(\mathcal{K})$.*

Now assume furthermore that $\mathcal{K}$ is a finite set of finite algebras. In this case, one can show that every ultraproduct of members of $\mathcal{K}$ is isomorphic to a member of $\mathcal{K}$. Thus, we have $\mathsf{HSP_U}(\mathcal{K})=\mathsf{HS}(\mathcal{K})$. By the above theorem, $\mathcal{V}$ has only finitely many (up to isomorphism) subdirectly irreducible members, each of which is finite. Hence we have the following.

COROLLARY 1.32. *Let $\mathcal{V}$ be a congruence distributive variety generated by a finite set $\mathcal{K}$ of finite algebras. Then $\mathcal{V}$ has finitely many subvarieties.*

1.2.7. Equational logic. At the beginning of this chapter an equational theory was defined as a set of $\mathcal{L}$-equations that are true in some class of $\mathcal{L}$-algebras. In fact, given a set E of $\mathcal{L}$-equations, the smallest equational theory that contains E is given by $\mathrm{Th}_e \mathrm{Mod}(E)$ (Theorem 1.1). Birkhoff [Bir35] showed how the equations in this theory can be obtained syntactically. Here we give a presentation of Birkhoff's *equational deductive system* known as *equational logic*.

Let E be a set of equations. An *equational proof of* $s = t$ *from* E is defined inductively by the following two clauses.

- If $s = t$ is a substitution instance of a member of $E \cup \{x = x\}$ then $s = t$ is a (one-step) equational proof of $s = t$ from E.
- If P_i is an equational proof of $s_i = t_i$ from E for $i = 1, 2, \ldots, n$ then

$$\frac{\mathsf{P}_1 \quad \mathsf{P}_2 \quad \cdots \quad \mathsf{P}_n}{s = t}(name)$$

is an equational proof of $s = t$ from E provided

$$\frac{s_1 = t_1 \quad s_2 = t_2 \quad \cdots \quad s_n = t_n}{s = t}(name)$$

matches one of the following three rules:

$$\frac{t = s}{s = t}\ (\mathrm{sym}) \qquad \frac{s = r \quad r = t}{s = t}\ (\mathrm{tran})$$

$$\frac{s_1 = t_1 \quad s_2 = t_2 \quad \cdots \quad s_n = t_n}{f(s_1, s_2, \ldots, s_n) = f(t_1, t_2, \ldots, t_n)}\ (\mathrm{cong}_f)$$

where $f \in \mathcal{L}^{\mathrm{o}}$ is of arity n.

Here each rule is represented in such a way that the premises are listed above the line (separated by spaces), the conclusion below, and a name of the rule in parentheses. So proofs can be considered as trees, with formulas as labels of the nodes, and each non-terminal node has children labeled by the premises of the rule needed to deduce the formula at this node. Of course, the tree order can always be extended to a linear order, making these formal proofs resemble more closely the informal way of presenting proofs.

Our presentation of equational logic is chosen to be similar to other logical systems considered later. We note that traditionally equational proof systems also include a substitution rule. This rule could be added here for convenience, but does not produce a stronger proof system since we already allow any substitution instances of equations in E. In other words, the equations in E are taken as *axioms* of the system and not as *assumptions* in a given proof.

The symbols $\vdash_E s = t$ are used to indicate that there exists an equational proof of $s = t$ from E. Birkhoff [Bir35] proved the following completeness theorem for equational logic.

THEOREM 1.33. *For a set of equations E and an equation $s = t$,*

$$\mathrm{Mod}(E) \models s = t \text{ if and only if } \vdash_E s = t.$$

1.2.8. Quasivarieties. As we have seen, algebras are naturally classified into varieties, which correspond to equational theories. However, some classes of algebras are defined only by *quasiequations*, that is first-order formulas of the form ε_0 and ... and $\varepsilon_{n-1} \Rightarrow \varepsilon_n$, where $\varepsilon_0, \ldots, \varepsilon_n$ are equations. Recall that a class $\mathcal{K}$ of algebras is a quasiequational class if $\mathcal{K}$ is precisely the class of models of some set Q of quasi-equations. A class $\mathcal{K}$ is called a *quasivariety* if $\mathcal{K}$ is closed under isomorphic images, subalgebras, direct products, and ultraproducts, or equivalently, if $\mathcal{K} = \mathsf{ISPP_U}(\mathcal{K})$. Mal'cev proved the following analogue of Birkhoff's Theorem (Theorem 1.19).

THEOREM 1.34. [Mal71] *A class $\mathcal{K}$ of algebras is a quasivariety iff $\mathcal{K}$ is a quasi-equational class.*

For a class of algebras $\mathcal{K}$ of the same type, the smallest quasivariety that contains $\mathcal{K}$ is given by $\mathsf{ISPP_U}(\mathcal{K})$. It is called the quasivariety generated by $\mathcal{K}$ and is denoted $\mathsf{Q}(\mathcal{K})$ for short. *Subvarieties* and *subquasivarieties* of a variety $\mathcal{V}$ form complete lattices under the operations of set-theoretical intersection and *(quasi) varietal join*, where the latter stands for the smallest (quasi)variety containing the union of them. We will write $\mathbf{\Lambda}_q(\mathcal{V})$ for the lattice of subquasivarieties of $\mathcal{V}$, and $\mathbf{\Lambda}(\mathcal{V})$ for the lattice of subvarieties of $\mathcal{V}$.

1.3. Logic

1.3.1. Hilbert calculus for classical logic. We introduce first a Hilbert-style calculus **HK** for classical logic. Anticipating its algebraization, we choose the language $\mathcal{L}$ of classical propositional logic to be that of Boolean algebras, with $\mathcal{L}^\circ = \{\wedge, \vee, \to, \neg\}$.[4] However in logic it is traditional to call variables *propositional variables*, the operation symbols *logical connectives*, and terms are called *$\mathcal{L}$-formulas*. The set of all $\mathcal{L}$-formulas is denoted by $Fm_{\mathcal{L}}$. In order to avoid confusion with first-order $\mathcal{L}$-formulas, we will always use the adjective "first-order" on the few occasions when we consider such formulas. Most of this book is concerned with propositional substructural logics, and therefore $\mathcal{L}$-formulas and $\mathcal{L}$-terms are regarded as two names for the same syntactic objects, so $Fm_{\mathcal{L}} = Tm_{\mathcal{L}}$. Roman letters p, q, r etc. are used for propositional variables.

There are several standard ways of introducing syntactic calculi for classical logic, e.g. Hilbert-style calculi, Gentzen's sequent calculus, tableaux systems and Gentzen's natural deduction system. One of the special features common to Hilbert-style calculi is that they have only a few rules of

[4]Note that the constants 0 and 1 are definable in Boolean algebras by $1 = x \to x$ and $0 = \neg 0$, so they can be omitted from the language.

inference and many axioms schemes, though there are various ways of choosing the axiom schemes. The following calculus is a standard one with *modus ponens* as its single rule. We state the axioms (axiom schemes) and rule (scheme) using metavariables φ, ψ, and χ; instances of the axioms and the rule are obtained by replacing the metavariables with arbitrary formulas.

Rule of inference: modus ponens

$$\frac{\varphi \quad \varphi \to \psi}{\psi}\ (\mathrm{mp})$$

Axiom schemes of HK

(A1) $\varphi \to (\chi \to \varphi)$
(A2) $(\varphi \to (\chi \to \psi)) \to ((\varphi \to \chi) \to (\varphi \to \psi))$
(A3) $(\varphi \wedge \psi) \to \varphi$
(A4) $(\varphi \wedge \psi) \to \psi$
(A5) $(\chi \to \varphi) \to ((\chi \to \psi) \to (\chi \to (\varphi \wedge \psi)))$
(A6) $\varphi \to (\varphi \vee \psi)$
(A7) $\psi \to (\varphi \vee \psi)$
(A8) $(\varphi \to \chi) \to ((\psi \to \chi) \to ((\varphi \vee \psi) \to \chi))$
(A9) $(\varphi \to \chi) \to ((\varphi \to \neg\chi) \to \neg\varphi)$
(A10) $\varphi \to (\neg\varphi \to \chi)$
(A11) $\neg\neg\varphi \to \varphi$

Modus ponens can be read as "from $\varphi \to \psi$ and φ, infer ψ." *Proofs* in **HK** and *end formulas*, also known as *conclusions*, of proofs are defined simultaneously as follows:

(P1) Every *instance* φ of one of the axiom schemes of **HK** is itself a proof, whose end formula is φ.

(P2) Suppose that P and Q are proofs whose end formulas are of the form $\varphi \to \psi$ and φ, respectively. Then, the following figure is a proof whose end formula is ψ:

$$\frac{\mathsf{P} \quad \mathsf{Q}}{\psi}$$

Thus, each proof is of the form of a finite labeled tree whose single root is labeled by the end formula. A proof with the end formula φ is called a *proof of* φ (in **HK**). If there is a proof of a formula φ in **HK**, φ is *provable* in **HK**. Using the notion of "provability", we can rephrase the above condition in the following way:

> Every substitution instance of an axiom scheme is provable. If both $\varphi \to \psi$ and φ are provable then ψ is also provable.

In syntactic arguments, induction on the *length* of a proof is often used to get a result. Intuitively, the length of a proof R is the length of the longest branch in the tree R. More precisely, when R consists of a single formula, its length is 1, and when R is of the form in the above (2), the length of R is the maximum of the lengths of P and Q plus 1. From this definition, it follows that if P is a proper *subproof* of a proof R then the length of P is smaller than that of R.

Here is an example of a proof of $p \to p$ in **HK**. Note that uppermost formulas in the proof are substitution instances of axiom schemes (2), (1) and (1) (from left to right), respectively.

$$\dfrac{p \to (p \to p) \qquad \dfrac{p \to ((p \to p) \to p) \qquad (p \to ((p \to p) \to p)) \to ((p \to (p \to p)) \to (p \to p))}{(p \to (p \to p)) \to (p \to p)}}{p \to p}$$

A more traditional way of defining proofs in Hilbert-style calculi is obtained by *linearizing* the proofs defined here. That is, a *proof* in this second definition of a formula α in **HK** is a sequence of formulas $\beta_1, \beta_2, \ldots, \beta_n$ ($n > 0$) with $\beta_n = \alpha$, which satisfies the following condition:

> for each $i \leq n$, either β_i is a substitution instance of one of axiom schemes of **HK**, or there are $j, k < i$ such that β_k is of the form $\beta_j \to \beta_i$.

The *length* of a proof under the second definition is defined to be n. The "linearizing" process can be understood as a transformation of a given proof of the first definition into a proof of the second definition, which is determined by an order-preserving map from a tree into a chain. Obviously, if P is a proper *subproof* of a proof R in the first definition, then the length of the transformed P is smaller than that of the transformed R by the second definition of lengths. The above proof of $p \to p$ can be transformed for example into the following "linear" proof. Comparison between these two proofs shows some of the merits and demerits of the two approaches to defining proofs.

(1) $p \to ((p \to p) \to p)$,
(2) $(p \to ((p \to p) \to p)) \to ((p \to (p \to p)) \to (p \to p))$,
(3) $(p \to (p \to p)) \to (p \to p)$,
(4) $p \to (p \to p)$,
(5) $p \to p$.

We can easily observe that for any formula α, the formula $\alpha \to \alpha$ is provable in **HK**, by replacing every p by α in the above proof of $p \to p$. This argument can be easily generalized and thus we have the following.

LEMMA 1.35. *If a formula α is provable in* **HK**, *then every substitution instance of α is also provable in it.*

We have introduced the Hilbert calculus **HK** by using axiom schemes, and modus ponens as a single rule. The above lemma suggests an alternative way of introducing **HK**. This alternative calculus has (finitely many) axioms which are obtained from the original ones by replacing φ, ψ, χ in each axiom scheme by distinct fixed propositional variables p, q, r, respectively. The notion of proof is also modified to allow substitution instances of axioms in (P1) and of rules in (P2), rather than instances of metavariables φ, ψ and χ. We chose our presentation of the Hilbert calculus in order to conform with the Gentzen calculus to be introduced in the next section.

Finding proofs in **HK** of a given formula is sometimes unintuitive and cumbersome. It will be convenient to have more flexibility here. Let Γ be a set of formulas. *Deductions (proofs) from assumptions* Γ in **HK** and their end formulas (conclusions), are defined similarly to proofs in **HK**, but we also add:

(P3) Every formula β in Γ is itself a deduction from assumptions Γ, whose end formula is β.

Note that (P3) does not allow arbitrary substitution instances of formulas in Γ. A deduction from Γ in **HK** with end formula α is called simply a deduction, or a proof, of α from Γ, and if there exists such a deduction, we say that α is a *consequence* of Γ in **HK**, or that α is *deducible* (or provable) from Γ in **HK** (in symbols $\Gamma \vdash_{\mathbf{HK}} \alpha$). The notion of the *length of a deduction* with assumptions is defined exactly as for proofs without assumptions. We may omit the subscript **HK** of the consequence relation $\vdash_{\mathbf{HK}}$ when confusion is unlikely. Clearly the concept of deduction reduces to that of proof when the set of assumptions is empty. Thus $\vdash_{\mathbf{HK}} \alpha$ means exactly that α is provable in **HK**. Traditionally, set notation on the left of $\vdash$ is abbreviated as follows:

$\Gamma \cup \Delta \vdash \psi$ is written $\Gamma, \Delta \vdash \psi$,
$\Gamma \cup \{\alpha\} \cup \Delta \vdash \psi$ is written $\Gamma, \alpha, \Delta \vdash \psi$ and
$\{\varphi_1, ..., \varphi_n\} \vdash \psi$ is written $\varphi_1, ..., \varphi_n \vdash \psi$.

Since any deduction consists of a finite number of formulas, we have the following *compactness* property.

LEMMA 1.36. *Suppose that* Γ *is a (possibly infinite) set of formulas and* α *is a formula. Then* $\Gamma \vdash_{\mathbf{HK}} \alpha$ *iff* $\Delta \vdash_{\mathbf{HK}} \alpha$ *for some finite subset* Δ *of* Γ.

The argument used in Lemma 1.35 extends to deductions.

LEMMA 1.37. *For every set* $\Gamma \cup \{\alpha\}$ *of formulas, if* $\Gamma \vdash_{\mathbf{HK}} \alpha$, *then for every substitution* σ, $\sigma[\Gamma] \vdash_{\mathbf{HK}} \sigma(\alpha)$.

We refer to Lemma 1.37 by saying that $\vdash_{\mathbf{HK}}$ is *substitution invariant*. We note that if $\alpha \in \Gamma$ and σ is a substitution, we may not have $\Gamma \vdash_{\mathbf{HK}} \sigma(\alpha)$; for example $p \nvdash_{\mathbf{HK}} q$, for distinct propositional variables p, q. This emphasizes the fact that in propositional deductions (as opposed to equational deductions in Birkhoff's equational logic) one cannot freely use substitution instances of the assumptions. In other words, we allow substitution for axioms of the system, but not for the assumptions of a proof.

We give here two examples of deductions in **HK** (in an informal way). By applying modus ponens to $\alpha \to \beta$ and α, we have $\alpha, \alpha \to \beta \vdash \beta$. Another deductions is given by

$$\alpha \to (\beta \to \gamma), \alpha \to \beta \vdash \alpha \to \gamma$$

In fact, from $(\alpha \to (\beta \to \gamma)) \to ((\alpha \to \beta) \to (\alpha \to \gamma))$ which is a substitution instance of (A2) and the assumption $\alpha \to (\beta \to \gamma)$, we infer $(\alpha \to \beta) \to (\alpha \to \gamma)$ by modus ponens. Using the second assumption, we then deduce $\alpha \to \gamma$.

LEMMA 1.38.

(1) *If both* $\Gamma \vdash \alpha$ *and* $\alpha, \Delta \vdash \beta$ *hold, then* $\Gamma, \Delta \vdash \beta$ *holds.*
(2) *If* $\Gamma \vdash \alpha \to \beta$ *holds then* $\alpha, \Gamma \vdash \beta$ *holds.*

PROOF. For the first statement, let P and Q be deductions of α from Γ and of β from α, Δ, respectively. If α does not occur in Q, then Q itself is a deduction of β from Γ, Δ. Otherwise, we replace each occurrence of the assumption α in Q by P, thus obtaining a deduction of β from Γ, Δ.

The second statement can be obtained from the first by applying it to $\Gamma \vdash \alpha \to \beta$ and $\alpha, \alpha \to \beta \vdash \beta$. □

The converse of Lemma 1.38(2) is called the *deduction theorem* for **HK**.

THEOREM 1.39. *If* $\alpha, \Gamma \vdash \beta$ *holds then* $\Gamma \vdash \alpha \to \beta$ *holds.*

PROOF. Let P be a deduction of β from α, Γ. Roughly speaking, a required deduction Q of $\alpha \to \beta$ from Γ can be obtained from P by replacing each formula δ in it by $\alpha \to \delta$ and inserting some suitable formulas to make it a correct deduction. More precisely, the argument goes as follows. Take a formula δ in P. Using induction on the length of the (sub)deduction of P to δ, we show that there exists a deduction of $\alpha \to \delta$ from Γ.

First let δ be one of the uppermost formulas in P. If δ is α, then we replace this δ by a proof of $\alpha \to \alpha$. Otherwise, δ is either a member of Γ or a substitution instance of one of axiom schemes. In either case we replace δ by a deduction of $\alpha \to \delta$ from δ, which is obtained from a substitution instance of the axiom scheme (A1) by applying modus ponens. Next, suppose that δ is inferred from $\gamma \to \delta$ and γ by modus ponens. By the induction hypothesis we can assume that there are deductions of $\alpha \to (\gamma \to \delta)$ and $\alpha \to \gamma$ from Γ. Since $\alpha \to (\gamma \to \delta), \alpha \to \gamma \vdash \alpha \to \delta$ as shown earlier, by combining the deductions together we get a deduction of $\alpha \to \delta$ from Γ. Finally, taking β for δ, we have a deduction of $\alpha \to \beta$ from Γ. □

We mention some examples of consequences in **HK**.

- $\alpha \to (\beta \to \gamma) \vdash \beta \to (\alpha \to \gamma)$,
- $\alpha \vdash \beta \to \alpha$,
- $\alpha \to (\alpha \to \gamma) \vdash \alpha \to \gamma$.

The first relation is an easy consequence of the deduction theorem, and the second and the third come from the axiom schemes (A1) and (A2), respectively.

Sometimes it is convenient to introduce a logical constant 1 for the *true* proposition in our language. In this case we add 1 as an axiom of **HK**.

(A12) 1.

Also, it is possible to introduce a logical constant 0 for the *false* proposition and define a negation $\neg\alpha$ by $\alpha \to 0$. In this case, we replace the axiom scheme (A10) by $0 \to \chi$, and delete the axiom scheme (A9), since it becomes an instance of (A2).

Axiom scheme (A11) is called the *law of double negation*. Here we show that (A11) can be replaced by the following *law of excluded middle* under the assumption of other axiom schemes of **HK**:

(A13) $\varphi \vee \neg\varphi$.

It is easy to show that $(\varphi \vee \neg\varphi) \to (\neg\neg\varphi \to \varphi)$ is provable, by applying (A8) to the instances $\varphi \to (\neg\neg\varphi \to \varphi)$, $\neg\varphi \to (\neg\neg\varphi \to \varphi)$ of (A1) and (A10). To show the converse, we note that the following consequence holds.

$\alpha \to \beta \vdash \neg\beta \to \neg\alpha.$

In fact, this is obtained from (A1) and (A9), by modus ponens and the deduction theorem. Applying this consequence to (A6), we get that $\neg(\varphi \vee \neg\varphi) \to \neg\varphi$ is provable. Moreover, since $\neg\varphi \to (\varphi \vee \neg\varphi)$ is an instance of (A7), we have that $\neg(\varphi \vee \neg\varphi) \to (\varphi \vee \neg\varphi)$ is provable, using (MP) twice on an instance of (A2). Applying (A9) to this formula with $\neg(\varphi \vee \neg\varphi) \to \neg(\varphi \vee \neg\varphi)$, we show that $\neg\neg(\varphi \vee \neg\varphi)$ is provable, and hence by (A11) $\varphi \vee \neg\varphi$ is provable.

1.3.2. Gentzen's sequent calculus for classical logic. The Hilbert calculus for classical logic, which we saw in the previous section, provides a satisfactory formal notion of proof in a system with a single rule of inference. Nevertheless, this system is not well suited for proof-theoretical analysis, since there is no normal form for proofs, nor a natural proof search algorithm. To this end, proof theory introduces the notion of a *standard* proof, which allows the derivation of logical properties from the analysis of the structure of proofs. In most cases, such analysis is based on combinatorial and constructive arguments and logical properties are reflected in the structure of the proofs.

However, the amount of information that we get from such syntactic analyses depends highly on the way a logic is formalized. In some formulations the structure of proofs reflects logical properties in a more sensitive way than in others. A formulation of logics particularly amenable to proof analysis was devised by G. Gentzen in the 1930s in the form of *sequent calculi* for classical and intuitionistic logic. The main advantage of sequent calculi comes from the fact that there is a *standard* or *cut-free* proof, of every provable formula. This is an important consequence of the celebrated *cut elimination theorem.* Moreover, it turns out that standard proofs reflect certain logical properties well: for example, we obtain a decision procedure.

The language $\mathcal{L}$ of **LK** and the formulas of $\mathcal{L}$ are defined in the same way as in the previous section. A *sequent* of **LK** is an expression of the form $\alpha_1, \ldots, \alpha_m \Rightarrow \beta_1, \ldots, \beta_n$, with $m, n \geq 0$, where α_is and β_js are formulas.[5] While formulas are the basic syntactic objects in Hilbert calculi, in Gentzen's sequent calculi this role is played by sequents. In the following, Greek capital letters like $\Sigma, \Lambda, \Gamma, \Delta$ denote finite (possibly empty) sequences of formulas *separated by commas.* Ina sequent $\Gamma \Rightarrow \Theta$ its *antecedent* (*succedent*) is Γ (Θ). Though we use the same commas in the antecedent and the succedent of a sequent in **LK**, their meaning is different: commas in the former correspond to *conjunction* while in the latter to *disjunction.* Moreover, the meaning of comma varies in different substructural logic, as will be explained at the beginning of the next chapter. Keeping in mind that here Greek capital letters denote sequences of formulas, we use similar notation as in the context of $\vdash_{\mathbf{HK}}$. Precisely speaking, Γ, Δ means the sequence of formulas obtained by concatenating Γ and Δ (in that order), as in the sequent $\Gamma, \Delta \Rightarrow \Lambda, \Theta$. Similarly, Λ, Θ is such a concatenation. Also, Γ, α means the sequence of formulas obtained by concatenating Γ and a single formula α.

In the following, we will introduce a formulation of the sequent calculus **LK**, which is slightly different, but not in any essential way, from the original one due to Gentzen. The system **LK** consists of the *initial sequents* and three types of rules of inference, i.e., *structural rules*, the *cut rule*, and *rules for logical connectives.* Below they are expressed as schemes, so lower case letters are metavariables for formulas and capital letters are metavariables for (possibly empty) sequences of formulas. An *instance of a rule* is obtained by replacing lowercase letters by formulas and uppercase letters by sequences of formulas.

Initial sequents:

$$\alpha \Rightarrow \alpha.$$

Structural rules: Weakening rule:

$$\frac{\Gamma, \Delta \Rightarrow \Pi}{\Gamma, \alpha, \Delta \Rightarrow \Pi}\ (w \Rightarrow) \qquad\qquad \frac{\Gamma \Rightarrow \Lambda, \Theta}{\Gamma \Rightarrow \Lambda, \alpha, \Theta}\ (\Rightarrow w)$$

Contraction rule:

[5]Note that we use the same symbol $\Rightarrow$ for the separator in sequents and for implication in first order logic. This will cause no confusion, as the usage of the symbol will be clear from the context.

$$\frac{\Gamma, \alpha, \alpha, \Delta \Rightarrow \Pi}{\Gamma, \alpha, \Delta \Rightarrow \Pi}\ (c \Rightarrow) \qquad \frac{\Gamma \Rightarrow \Lambda, \alpha, \alpha, \Theta}{\Gamma \Rightarrow \Lambda, \alpha, \Theta}\ (\Rightarrow c)$$

Exchange rule:

$$\frac{\Gamma, \alpha, \beta, \Delta \Rightarrow \Pi}{\Gamma, \beta, \alpha, \Delta \Rightarrow \Pi}\ (e \Rightarrow) \qquad \frac{\Gamma \Rightarrow \Lambda, \alpha, \beta, \Theta}{\Gamma \Rightarrow \Lambda, \beta, \alpha, \Theta}\ (\Rightarrow e).$$

Cut rule:

$$\frac{\Gamma \Rightarrow \alpha, \Theta \quad \Sigma, \alpha, \Delta \Rightarrow \Pi}{\Sigma, \Gamma, \Delta \Rightarrow \Pi, \Theta}$$

Rules for logical connectives:

$$\frac{\Gamma, \alpha, \Delta \Rightarrow \Pi \quad \Gamma, \beta, \Delta \Rightarrow \Pi}{\Gamma, \alpha \vee \beta, \Delta \Rightarrow \Pi}\ (\vee \Rightarrow)$$

$$\frac{\Gamma \Rightarrow \Lambda, \alpha, \Theta}{\Gamma \Rightarrow \Lambda, \alpha \vee \beta, \Theta}\ (\Rightarrow \vee 1) \qquad \frac{\Gamma \Rightarrow \Lambda, \beta, \Theta}{\Gamma \Rightarrow \Lambda, \alpha \vee \beta, \Theta}\ (\Rightarrow \vee 2)$$

$$\frac{\Gamma, \alpha, \Delta \Rightarrow \Pi}{\Gamma, \alpha \wedge \beta, \Delta \Rightarrow \Pi}\ (\wedge 1 \Rightarrow) \qquad \frac{\Gamma, \beta, \Delta \Rightarrow \Pi}{\Gamma, \alpha \wedge \beta, \Delta \Rightarrow \Pi}\ (\wedge 2 \Rightarrow)$$

$$\frac{\Gamma \Rightarrow \Lambda, \alpha, \Theta \quad \Gamma \Rightarrow \Lambda, \beta, \Theta}{\Gamma \Rightarrow \Lambda, \alpha \wedge \beta, \Theta}\ (\Rightarrow \wedge)$$

$$\frac{\Gamma \Rightarrow \alpha, \Theta \quad \Sigma, \beta, \Delta \Rightarrow \Pi}{\Sigma, \Gamma, \alpha \rightarrow \beta, \Delta \Rightarrow \Pi, \Theta}\ (\rightarrow \Rightarrow) \qquad \frac{\alpha, \Gamma \Rightarrow \beta, \Theta}{\Gamma \Rightarrow \alpha \rightarrow \beta, \Theta}\ (\Rightarrow \rightarrow)$$

$$\frac{\Gamma \Rightarrow \alpha, \Theta}{\neg\alpha, \Gamma \Rightarrow \Theta}\ (\neg \Rightarrow) \qquad \frac{\Gamma, \alpha \Rightarrow \Theta}{\Gamma \Rightarrow \Theta, \neg\alpha}\ (\Rightarrow \neg)$$

When the logical constant 1 is in our language, we add the sequent $\Rightarrow 1$ as an initial sequent. Likewise, when 0 is in our language then we add $0 \Rightarrow$ as initial sequent. Actually, we the systems obtained by adding 0, or $\neg$, or both (together with their rules or axioms) lead to equivalent systems (see Exercise 27). Therefore, we will use **LK** for any of these systems (allowing also the insertion or inclusion of 1).

In each of the above rules of inference, the sequents over the horizontal line are *upper sequents* of the rule and the sequent under the horizontal line is its *lower sequent*. Formulas appearing within the sequences denoted by capital Greek letters are called *side formulas* and the formula of the lower sequent involving small Greek letters is the *main formula* of the rules. The formula to which the cut rule is applied (the formula α in the above) is the *cut formula* of a given application of the cut rule. We define *proofs* in **LK** and *end sequents* of proofs almost in the same way as those in **HK**. Here is the definition.

(Q1) Every initial sequent is a proof itself, whose end sequent is the initial sequent.

(Q2) Suppose that P and Q are proofs whose end sequents are s_1 and s_2, respectively. Moreover suppose that there exists an instance of a rule of **LK** with upper sequents s_1 and s_2 such that the lower sequent is s. Then, the following figure is a proof whose end sequent is s:

$$\dfrac{\mathsf{P} \quad \mathsf{Q}}{s}$$

(When a rule has a single upper sequent, we define a proof in the same way but by neglecting s_2 and Q in the above.) Thus, each proof of **LK** is a finite labeled tree, like proofs of **HK**, but here each node is labeled by a sequent, not by a formula. The *length* of a proof is defined as in **HK**. A proof with end sequent s is called a *proof of* s (in **LK**). If there exists a proof of a sequent s, we say that s is *provable* in **LK** ($\vdash^{seq}_{\mathbf{LK}} s$, in symbols). In particular, when the sequent $\Rightarrow \alpha$ is provable, we say simply that the formula α is provable. When our language contains logical constants 1 and 0, we add $\Rightarrow 1$ and $0 \Rightarrow$ as initial sequents.

We give here two examples of proofs in **LK**, in which we indicate which rule is used.

1. A proof of the formula $((\alpha \to \beta) \to \alpha) \to \alpha$:

$$\dfrac{\dfrac{\dfrac{\dfrac{\dfrac{\alpha \Rightarrow \alpha}{\alpha \Rightarrow \beta, \alpha}\,(\Rightarrow w)}{\Rightarrow \alpha \to \beta, \alpha} \qquad \alpha \Rightarrow \alpha}{(\alpha \to \beta) \to \alpha \Rightarrow \alpha, \alpha}}{(\alpha \to \beta) \to \alpha \Rightarrow \alpha}\,(\Rightarrow c)}{\Rightarrow ((\alpha \to \beta) \to \alpha) \to \alpha}$$

2. A proof of the *distributive law*:

$$\dfrac{\dfrac{\dfrac{\dfrac{\dfrac{\dfrac{\dfrac{\alpha\Rightarrow\alpha}{\alpha,\beta\Rightarrow\alpha}\,(\Rightarrow w)\quad\dfrac{\beta\Rightarrow\beta}{\alpha,\beta\Rightarrow\beta}\,(\Rightarrow w)}{\alpha,\beta\Rightarrow\alpha\wedge\beta}}{\alpha,\beta\Rightarrow(\alpha\wedge\beta)\vee(\alpha\wedge\gamma)}\qquad\dfrac{\dfrac{\dfrac{\alpha\Rightarrow\alpha}{\alpha,\gamma\Rightarrow\alpha}\,(\Rightarrow w)\quad\dfrac{\gamma\Rightarrow\gamma}{\alpha,\gamma\Rightarrow\gamma}\,(\Rightarrow w)}{\alpha,\gamma\Rightarrow\alpha\wedge\gamma}}{\alpha,\gamma\Rightarrow(\alpha\wedge\beta)\vee(\alpha\wedge\gamma)}}{\alpha,\beta\vee\gamma\Rightarrow(\alpha\wedge\beta)\vee(\alpha\wedge\gamma)}}{\alpha\wedge(\beta\vee\gamma),\beta\vee\gamma\Rightarrow(\alpha\wedge\beta)\vee(\alpha\wedge\gamma)}}{\alpha\wedge(\beta\vee\gamma),\alpha\wedge(\beta\vee\gamma)\Rightarrow(\alpha\wedge\beta)\vee(\alpha\wedge\gamma)}}{\alpha\wedge(\beta\vee\gamma)\Rightarrow(\alpha\wedge\beta)\vee(\alpha\wedge\gamma)}\,(\Rightarrow c)$$

The following lemma will justify our intuitive understanding of the meaning of sequents and the different role of commas in antecedents and succedents, mentioned before. Let Σ be a finite sequence $\delta_1, \ldots, \delta_k$ of formulas. We define the formulas $\Sigma^\wedge$ and $\Sigma^\vee$ as follows:

- $\Sigma^\wedge$ is $\delta_1 \wedge \ldots \wedge \delta_k$ if $k > 0$, and is 1, otherwise,
- $\Sigma^\vee$ is $\delta_1 \vee \ldots \vee \delta_k$ if $k > 0$, and is 0, otherwise.

Here $\delta_1 \wedge \ldots \wedge \delta_k$ stands for, say, the left associated corresponding formula. Since both conjunction and disjunction will be interpreted as associative (and commutative) operations, we will not worry much about the parenthesization (or the order) of the formulas.

LEMMA 1.40. *For each sequent $\Gamma \Rightarrow \Theta$, the following are mutually equivalent:*

(1) *the sequent $\Gamma \Rightarrow \Theta$ is provable in* **LK**,
(2) *the sequent $\Gamma^\wedge \Rightarrow \Theta^\vee$ is provable in* **LK**,
(3) *the formula $\Gamma^\wedge \to \Theta^\vee$ is provable in* **LK**.

1.3.3. Calculi for intuitionistic logic. Mathematical arguments often contain infinitary and non-constructive reasoning. Existence theorems are sometimes proved by showing that by assuming the non-existence we can derive a contradiction, not by giving a concrete way of finding the existing objects. In the beginning of the 20th Century, a group of mathematicians now called *intuitionists* criticized non-constructive arguments in traditional mathematics. L.E.J. Brouwer in particular proposed a program of rebuilding mathematics on a constructive basis, that is, on the principle that a proposition about the existence of a mathematical object cannot be considered true unless the object in question has been constructed.

This requirement, taken seriously, must lead to substantial modifications in logic. For example, according to intuitionistic principles, the formula $\varphi \vee \neg\varphi$ is true if and only if, for any proposition φ, there either is a construction of the object described by φ, or the assumption of there being such a construction results in a contradiction. Since there are formulas φ for which neither is the case (e.g., 'every even integer greater than two is a sum of two primes'), the law of excluded middle is not a theorem of intuitionistic logic. Also implication is understood in a constructivist spirit, as being true if and only if there is a construction such that given a construction of (the object described by) the antecedent produces (the object described by) the succedent. Thus, trivially, $\varphi \rightarrow \varphi$ is an intuitionistic theorem. Less trivially, it immediately follows that implication is not definable by $\neg\varphi \vee \psi$ as in classical logic. Also, such a constructivist implication does not make $\neg\neg\varphi \rightarrow \varphi$ a theorem. For the intuitionist reading of the formula $\neg\neg\varphi$ is something like 'it is contradictory to assume that any construction of φ leads to a contradiction'. So a construction of φ is not (yet) ruled out, it is possible that φ can be constructed, we do not (yet) know of a counterexample. But not knowing whether a counterexample exists does not imply that φ can be constructed. By the same token, *reductio ad absurdum*, a way of proving φ by showing that $\neg\varphi$ derives a contradiction, cannot be accepted either, since for a constructivist it only implies the provability of $\neg\neg\varphi$.

In the late 1920s, A. Heyting introduced a logical system underlying intuitionistic mathematics, which is called *intuitionistic logic*. In the following, we introduce both a Hilbert calculus **HJ** and a Gentzen calculus **LJ** for intuitionistic logic. In fact, both are modified versions of the corresponding calculi for classical logic.

The Hilbert calculus **HJ** is obtained from **HK** by deleting the law of double negation (A11), (or, deleting the law of excluded middle (A13)). Then, every result on **HK** mentioned in the section 3.1, including the deduction theorem (Theorem 1.39), holds also for **HJ**.

In Gentzen calculus **LJ**, every sequent must be of the form $\alpha_1, \ldots, \alpha_m \Rightarrow \beta$, where $m \geq 0$ and β is a formula or the empty sequence. That is, succedents of sequents must be either a single formula or empty. Initial sequents of **LJ** are the same as those of **LK**, but each rule is restricted so that both the upper sequent(s) and the lower sequent satisfy the above requirement on their succedents. More precisely, we need to make the following changes to the rules of **LK**, given in the previous subsection.

- Delete both ($\Rightarrow c$) and ($\Rightarrow e$),
- Λ and Θ must be empty,
- Π is either a single formula or empty.

For example, ($\rightarrow\Rightarrow$) becomes

$$\frac{\Gamma \Rightarrow \alpha \quad \Sigma, \beta, \Delta \Rightarrow \delta}{\Sigma, \alpha \rightarrow \beta, \Gamma, \Delta \Rightarrow \delta} \ (\rightarrow\Rightarrow)$$

Proofs and the provability of sequents are defined in the same way. We can see that the proof of $\Rightarrow ((\alpha \rightarrow \beta) \rightarrow \alpha) \rightarrow \alpha$ in **LK** given before is not a proof in **LJ**. Actually, we will develop tools (cut elimination and algebraization) that will allow us to show that there exists no proof of the formula $((p \rightarrow q) \rightarrow p) \rightarrow p$ in **LJ**, if p, q are distinct propositional variables.

A rule of the sequent calculus **LK** (**LJ**) is *invertible*, if whenever the *lower* sequent of the rule is provable in **LK** (**LJ**) then so are the *upper* sequents. Among rules for logical connectives of **LK**, $(\vee \Rightarrow), (\Rightarrow \wedge), (\Rightarrow \rightarrow), (\neg \Rightarrow)$ and $(\Rightarrow \neg)$ are invertible. Except $(\neg \Rightarrow)$, they are also invertible as rules of **LJ**. Let us see the invertibility of $(\Rightarrow \rightarrow)$ in **LJ**. Suppose $\Gamma \Rightarrow \alpha \rightarrow \beta$ is provable. Since $\alpha \rightarrow \beta, \alpha \Rightarrow \beta$ is provable, we have $\Gamma, \alpha \Rightarrow \beta$ in **LJ** by an application of cut. In particular, $\gamma \Rightarrow \alpha \rightarrow \beta$ is provable iff $\gamma, \alpha \Rightarrow \beta$ is provable, iff $\gamma \wedge \alpha \Rightarrow \beta$ is provable by using Lemma 1.40 (for **LJ**). This equivalence resembles the law of $\wedge$-residuation of Heyting algebras. In fact, there is a strong connection between these facts. We will show later that $\alpha \Rightarrow \beta$ is provable in **LJ** iff $\alpha \leq \beta$ is valid in HA.

1.3.4. Provability in Hilbert and Gentzen calculi. In this and the next subsections, we show that provability of formulas in **HK** and in **LK** is the same. This holds between **HJ** and **LJ**, as well. Additionally, we introduce an algebraic interpretation of formulas and sequents in Boolean and Heyting algebras.

LEMMA 1.41. *If a formula is provable in* **HK**, *it is provable in* **LK**.

PROOF. It suffices to show that

- the effect of modus ponens can be simulated in **LK**, i.e., if both α and $\alpha \rightarrow \beta$ are provable in **LK** then β is also provable in it,
- every substitution instance of each axiom scheme of **HK** is provable in **LK**.

The lemma follows then by induction. □

Similarly, we can show the following.

LEMMA 1.42. *If a formula is provable in* **HJ**, *it is provable in* **LJ**.

In the Hilbert calculus **HK**, we introduced the notion of deducibility, and of a deduction with assumptions, by extending the notion of provability and of a proof, respectively. Similarly, we can define deducibility (proof with assumptions) in any given sequent calculus. If $S \cup \{s\}$ is a set of sequents, we say that s is *deducible* or *provable* from assumptions S in **LK**, in symbols $S \vdash^{seq}_{\mathbf{LK}} s$, if there is a *deduction* or *proof* of s in **LK** from S. Here, deductions or proofs from assumptions are defined in the same way as proofs in **LK** but also by adding the following.

(Q3) Every sequent in S is a deduction, whose end sequent is itself.

We stress that substitution instances of the assumptions, i.e., formulas in S, can not be used freely in the deduction. Though this notion has its own interest, we will not discuss it much further in the present book. (For more information on this topic, see [GO].) We consider here a particular case. We say that a formula ψ is *deducible* from the (possibly infinite) set of formulas Γ in **LK** (in symbols, $\Gamma \vdash_{\mathbf{LK}} \psi$), when the sequent $\Rightarrow \psi$ is deducible from sequents $\Rightarrow \varphi_1, \ldots, \Rightarrow \varphi_m$ in **LK**, for some $\varphi_1, \ldots, \varphi_m \in \Gamma$, i.e., $\{\Rightarrow \varphi : \varphi \in \Gamma\} \vdash^{seq}_{\mathbf{LK}} \Rightarrow \psi$. We use the same abbreviations as for the deducibility relation determined by **HK**. For instance, $\Gamma \vdash_{\mathbf{LK}} \psi$ is written also as $\varphi_1, \ldots, \varphi_m \vdash_{\mathbf{LK}} \psi$, when Γ is a finite set $\{\varphi_1, \ldots, \varphi_m\}$. We give here a deduction of $\Rightarrow \alpha \to ((\beta \to \gamma) \to \gamma)$ from $\Rightarrow \alpha \to \beta$.

$$\dfrac{\dfrac{\dfrac{\Rightarrow \alpha \to \beta \quad \dfrac{\alpha \Rightarrow \alpha \quad \dfrac{\beta \Rightarrow \beta \quad \gamma \Rightarrow \gamma}{\beta \to \gamma, \beta \Rightarrow \gamma}}{\alpha \to \beta, \alpha, \beta \to \gamma \Rightarrow \gamma}}{\alpha, \beta \to \gamma \Rightarrow \gamma}}{\alpha \Rightarrow (\beta \to \gamma) \to \gamma}}{\Rightarrow \alpha \to ((\beta \to \gamma) \to \gamma)}$$

The deducibility relation $\vdash_{\mathbf{LK}}$ of **LK** is called the *external consequence relation* of $\vdash^{seq}_{\mathbf{LK}}$. The *internal relation* $\vdash^{int}_{\mathbf{LK}}$ of $\vdash^{seq}_{\mathbf{LK}}$ is defined by

$$\varphi_1, \ldots, \varphi_m \vdash^{int}_{\mathbf{LK}} \psi \text{ iff } \vdash^{seq}_{\mathbf{LK}} \varphi_1, \ldots, \varphi_m \Rightarrow \psi.$$

Unlike the external relation the internal one is not a consequence relation for a very simple reason: its first argument is a finite sequence of formulas not a (possibly infinite) set of formulas. This mismatch of data types seems trivial for the case of **LK**, but it is essential for the systems we will consider in the next chapter. If we view the expression $\varphi_1, \ldots, \varphi_m$ as either a set or a sequence, we can see that the internal and the external relations are related for **LK** (for finite sets/sequences), but in the context of substructural logics this is just a coincidence. This is due to the fact that our metalogic is classical.

LEMMA 1.43. *$\varphi_1, \ldots, \varphi_m \vdash_{\mathbf{LK}} \psi$ iff $\varphi_1, \ldots, \varphi_m \vdash^{int}_{\mathbf{LK}} \psi$.*

PROOF. The implication from the right-hand side to the left-hand is obvious by the cut rule. To show the converse, take each occurrence of a sequent $\Delta \Rightarrow \Pi$ in a given deduction P of $\Rightarrow \psi$ from $\Rightarrow \varphi_1, \ldots, \Rightarrow \varphi_m$. Then we show by induction on the length of the subdeduction of P to $\Delta \Rightarrow \Pi$ that the sequent $\varphi_1, \ldots, \varphi_m, \Delta \Rightarrow \Pi$ is provable in **LK**. In particular, we have that $\varphi_1, \ldots, \varphi_m \Rightarrow \psi$ is provable in **LK**. □

Clearly, the above lemma implies the deduction theorem for **LK**. Careful inspection of its proof reveals that the presence of structural rules is indispensable in the argument. An alternative way of showing the lemma is to use the deduction theorem for **HK** and the fact that the deducibility relation of **LK** is the same as that of **HK**, which is shown later in this section. We will see that the deduction theorem fails for most *substructural logics*; this is the main reason for the failure of the above lemma for those logics. Obviously, we can introduce the notion of deducibility for **LJ** in the same way. Like in Lemmas 1.41 and 1.42, we can show the following.

LEMMA 1.44. *For each Γ and ψ, $\Gamma \vdash_{\mathbf{HK}} \psi$ implies $\Gamma \vdash_{\mathbf{LK}} \psi$. The same implication holds for* **HJ** *and* **LJ**.

1.4. Logic and algebra

1.4.1. Validity of formulas in algebras. We now give algebraic interpretations for formulas and sequents. We will refer to the notion of free algebra, introduced in the previous sections.

Recall that $Fm_{\mathcal{L}}$ denotes the set of formulas in the language $\mathcal{L}$ (that is, $\mathcal{L}$-formulas, or equivalently, $\mathcal{L}$-terms, cf. the discussion in Section 1.3.1) and BA, HA denote the varieties of Boolean algebras and of Heyting algebras, respectively.

Let $\mathbf{A}$ be a Heyting algebra and f an assignment in A, i.e., a map from the set of variables to A. Recall that f is expanded uniquely to a homomorphism $f : \mathbf{Fm}_{\mathcal{L}} \to \mathbf{A}$. As mentioned in the first section of this chapter, for a pair of terms s, t we say that $\mathbf{A}, f \models s = t$ iff $f(s) = f(t)$ and that $\mathbf{A} \models s = t$ iff $\mathbf{A}, f \models s = t$ for each assignment f in $\mathbf{A}$. For a class $\mathcal{K}$ of Heyting algebras, we write $\mathcal{K} \models s = t$ iff $\mathbf{A} \models s = t$ for all $\mathbf{A} \in \mathcal{K}$. In particular, when $\mathbf{A}, f \models \varphi = 1$ holds for a formula φ, we say that φ is *true* in $\mathbf{A}$ by the assignment f. Also, when $\mathbf{A} \models \varphi = 1$ ($\mathcal{K} \models \varphi = 1$), we say that φ is *valid* in $\mathbf{A}$ (in $\mathcal{K}$, respectively).[6] (Note that in the next chapter, we introduce algebras where 1 is not the greatest element. In these cases, we enlarge the set of true elements of A to include all elements above 1 and say that φ is valid in $\mathbf{A}$ if $\mathbf{A} \models \varphi \geq 1$, i.e., $\mathbf{A} \models \varphi \wedge 1 = 1$.) In view of Lemma 1.40, we extend the notion of validity to sequents. A sequent $\Gamma \Rightarrow \Theta$ of **LK** is *valid* in a Boolean algebra $\mathbf{A}$, if the formula $\Gamma^{\wedge} \to \Theta^{\vee}$ is valid in it, i.e., if $\mathbf{A} \models \Gamma^{\wedge} \leq \Theta^{\vee}$. Similarly, a sequent $\Gamma \Rightarrow \alpha$ of **LJ** is *valid* in a Heyting algebra $\mathbf{A}$, if the formula $\Gamma^{\wedge} \to \alpha$ is valid in it.

Now, we have the following soundness results.

THEOREM 1.45.
(1) *If a sequent of* **LK** *is provable in* **LK**, *then it is valid in* BA.
(2) *If a sequent of* **LJ** *is provable in* **LJ**, *then it is valid in* HA.

PROOF. The theorem can be shown by using induction on the length of a proof of a given sequent. This can be done by showing first that every initial sequent is valid and next that in each rule if the upper sequents are valid then the lower sequent is also valid. In fact, we can show a stronger result, i.e., for a given algebra $\mathbf{A}$ and a given assignment f on $\mathbf{A}$, if the images under f of the upper sequents are true in $\mathbf{A}$ then so is the image of the lower sequent. In these calculations, the following facts are useful.

- In Heyting algebras, $x \wedge y \leq z$ iff $x \leq y \to z$, and hence $x \leq y$ iff $x \to y = 1$.

[6]Notice that sometimes they are written as $\mathbf{A} \models \varphi$ and $\mathcal{K} \models \varphi$, respectively, in the literature.

- In Boolean algebras, $y \to z = \neg y \vee z$ and $\neg\neg y = y$ hold. Therefore, $x \wedge y \leq z$ iff $x \leq \neg y \vee z$. By replacing y by $\neg y$, we have also $x \wedge \neg y \leq z$ iff $x \leq y \vee z$.

We show the above for the rule $(\to \Rightarrow)$ of **LK**, for example. It suffices to show essentially that in each Boolean algebra, $(a \to b) \wedge x \wedge y \leq w \vee v$ follows from $x \leq a \vee w$ and $b \wedge y \leq v$. We have $(a \to b) \wedge x \wedge y = (\neg a \vee b) \wedge x \wedge y \leq (\neg a \wedge x) \vee (b \wedge y)$, by distributivity. As our first assumption implies $\neg a \wedge x \leq w$, $(\neg a \wedge x) \vee (b \wedge y) \leq w \vee v$. □

1.4.2. Lindenbaum-Tarski algebras. We now show the completeness of the Hilbert calculi **HK** and **HJ**, from which the completeness of the Gentzen calculi **LK** and **LJ** follows. The proofs for **HK** and **HJ** are similar, so we consider only the case of **HK**.

We define a binary relation $\equiv$ on $\mathbf{Fm}_{\mathcal{L}}$, putting $\alpha \equiv \beta$ if the formulas $\alpha \to \beta$ and $\beta \to \alpha$ are both provable in **HK**. We can show the following.

LEMMA 1.46. *The relation $\equiv$ is a fully invariant congruence on $\mathbf{Fm}_{\mathcal{L}}$.*

Consider the set $E = \{\alpha = \beta \colon \alpha \equiv \beta\}$ of equations between formulas. By Lemma 1.18, the algebra $\mathbf{D} = \mathbf{Fm}_{\mathcal{L}}/_{\equiv}$ is free in the variety $\mathrm{Mod}(E)$ axiomatized by E. This algebra is known as the *Lindenbaum-Tarski algebra* of $\mathrm{Mod}(E)$. We can show that $\mathbf{D}$ is a Boolean algebra. Let $\mathsf{V}(\mathbf{D})$ be the variety generated by $\mathbf{D}$. Using the soundness of **HK** (Theorem 1.45(1)), we have that for all terms s, t, $s \equiv t$ implies $\mathsf{BA} \models s = t$, which in turn implies $\mathbf{D} \models s = t$. In other words, $\mathsf{V}(\mathbf{D}) \subseteq \mathsf{BA} \subseteq \mathrm{Mod}(E)$.

LEMMA 1.47. *Let X be a set and θ be a fully invariant congruence on $\mathbf{Fm}_{\mathcal{L}}(X)$. Then, for all terms s, t, $\mathbf{Fm}_{\mathcal{L}}(X)/\theta \models s = t$ iff $s\theta t$. In particular, $\mathbf{D} \models s = t$ iff $s \equiv t$. Thus, $\mathrm{Mod}(E) \subseteq \mathsf{V}(\mathbf{D})$.*

COROLLARY 1.48. *The varieties BA, $\mathrm{Mod}(E)$ and $\mathsf{V}(\mathbf{D})$ are equal. In particular, $\mathsf{BA} \models s = t$ if and only if both $s \to t$ and $t \to s$ are provable in* **HK**.

One of the consequences of the above result is that the variety BA is generated by a single algebra $\mathbf{D}$. This can be generalized as follows.

> If a variety $\mathcal{V}$ has a nontrivial member and X is an infinite set, then $\mathcal{V}$ is generated by the free algebra (in $\mathcal{V}$) over X.

COROLLARY 1.49. *The following three conditions are mutually equivalent. For any formula α,*

(1) *α is provable in* **HK**,
(2) *α is provable in* **LK**,
(3) *α is valid in* BA.

Similar equivalences hold among **HJ**, **LJ** *and* HA.

The above corollary together with the definition of validity of formulas and sequents tell us how to understand provability in Hilbert and Gentzen calculi algebraically. Provability of a formula α in a Hilbert calculus is the same as the validity of the equation $\alpha = 1$ (the inequality $1 \leq \alpha$, in general) in the corresponding class of algebras. On the other hand, the provability of a sequent $\alpha_1, \dots, \alpha_m \Rightarrow \beta_1, \dots, \beta_n$ in the Gentzen calculus **LK** (in **LJ** with $n \leq 1$) means the validity of the inequality $\alpha_1 \wedge \dots \wedge \alpha_m \leq \beta_1 \vee \dots \vee \beta_n$ in the class of Boolean algebras (Heyting algebras, respectively). Thus, the arrow $\Rightarrow$ in sequents is interpreted as the inequality $\leq$ in algebras, but as shown in the next chapter the interpretation of *commas* will be different.

THEOREM 1.50. *The variety* BA *is generated by the single finite algebra* **2**. *That is, for terms* s, t, $s = t$ *is valid in* BA *iff it is valid in the two-valued Boolean algebra* **2**.

PROOF. The only-if part is trivial. Suppose that for a Boolean algebra $\mathbf{A}$ and an assignment f in it, $f(s) = f(t)$ does not hold. Let $f(s) = a$ and $f(t) = b$ for $a, b \in A$. Without loss of generality, we can assume that $a \not\leq b$. Then, the principal filter $\uparrow a$ generated by a is proper since $b \notin \uparrow a$. By using Zorn's lemma, we have the *ultrafilter theorem* for Boolean algebras, which says that there exists an ultrafilter G of $\mathbf{A}$ which includes $\uparrow a$ but does not contain b (cf. Theorem 1.21). Then, it is obvious that the quotient algebra $\mathbf{A}/G$ is isomorphic to **2**. Let g be an assignment on $\mathbf{A}/G$ such that $g(x) = [f(x)]_G$ for each variable x. Then, $g(u) = [f(u)]_G$ for every term u, by induction. In particular, $g(s) = [a]_G = 1$ while $g(t) = [b]_G = 0$. Thus, $s = t$ is not valid in **2**. □

A formula is a *tautology* iff it is valid in the two-element Boolean algebra **2**. Theorem 1.50 shows that the theorems of **HK** coincide with tautologies. This provides an effective way of checking if a formula is a theorem of **HK**: we check if it is valid in **2**. Things are different for **HJ**, as observed by Gödel.

THEOREM 1.51. *The variety* HA *of Heyting algebras is not generated by any single finite Heyting algebra.*

PROOF. [Göd33] Let $x_0, \dots, x_n$ be mutually distinct variables. For each $n > 0$, define a term s_n by $\bigvee_{0 \leq i < j \leq n} (x_i \leftrightarrow x_j)$, where $y \leftrightarrow z$ is an abbreviation of $(y \to z) \wedge (z \to y)$. It is easy to see that if a Heyting algebra consists of at most n elements then s_n is valid in it. On the other hand, we consider a totally ordered set with $n+1$ elements, which naturally forms a Heyting algebra $\mathbf{H}_{n+1}$. For each $n > 0$, s_n is not valid in $\mathbf{H}_{n+1}$, and a fortiori not valid in HA. If HA is generated by a single finite Heyting algebra, then some s_m must be valid in it, which is a contradiction. □

By Corollary 1.15, the variety HA is congruence distributive. Therefore, every subvariety of HA generated by a finite set of finite Heyting algebras has only finitely many subvarieties, by Corollary 1.32, while HA itself has infinitely many. In Chapter 6, we show that HA is generated by the set of all finite Heyting algebras.

1.4.3. Algebraization. The method of using Lindenbaum-Tarski algebras is a basic tool in discussing relations between logic and algebra. Corollary 1.49 can be understood as a semantical characterization of provability. This can be extended to a characterization of deducibility via *equational consequence*, called *algebraization in the sense of Blok-Pigozzi*, while the weaker results shown in the previous subsection are called *algebraization in the sense of Lindenbaum-Tarski.* Algebraization in the sense of Blok-Pigozzi will be discussed in general in the next chapter. Here, we give a brief introduction to the topic, restricting our attention to the intuitionistic logic.

If $\mathcal{K}$ is a class of algebras, we define the *equational consequence* $\models_{\mathcal{K}}$ determined by $\mathcal{K}$ by

$E \models_{\mathcal{K}} u = w$ iff for each $\mathbf{A} \in \mathcal{K}$ and each assignment g in A, $g(u) = g(w)$ holds whenever $g(s) = g(t)$ holds for all $(s = t) \in E$.

where $E \cup \{u = w\}$ is a set of equations. When E is a finite set $\{s_1 = t_1, \ldots, s_m = t_m\}$, $E \models_{\mathcal{K}} u = w$ iff, for each $\mathbf{A} \in \mathcal{K}$, the quasi-equation $(s_1 = t_1 \text{ and } \ldots \text{ and } s_m = t_m) \Rightarrow u = w$ is valid in $\mathbf{A}$. By modifying slightly the proof of Theorem 1.45, we have the following.

LEMMA 1.52. *For any set of formulas* $\Gamma \cup \{\alpha\}$, *if* $\Gamma \vdash_{\mathbf{LJ}} \alpha$ *then* $\{\gamma = 1 : \gamma \in \Gamma\} \models_{\mathsf{HA}} \alpha = 1$.

Next, we show the following.

THEOREM 1.53. *For any set of formulas* Γ *and formula* α, *the following three conditions are mutually equivalent.*

(1) $\Gamma \vdash_{\mathbf{HJ}} \alpha$,
(2) $\Gamma \vdash_{\mathbf{LJ}} \alpha$,
(3) $\{\gamma = 1 : \gamma \in \Gamma\} \models_{\mathsf{HA}} \alpha = 1$.

By taking into account the fact that $\gamma \vdash_{\mathbf{HJ}} (\gamma \leftrightarrow 1)$ and $(\gamma \leftrightarrow 1) \vdash_{\mathbf{HJ}} \gamma$ for any formula γ and also that both $s = t \models_{\mathsf{HA}} (s \leftrightarrow t) = 1$ and $(s \leftrightarrow t) = 1 \models_{\mathsf{HA}} s = t$, this theorem is equivalent to the following theorem. These two results together with the above facts form the *algebraization theorem*.

THEOREM 1.54. *For any set of equations* $\{s_i = t_i : i \in I\} \cup \{u = w\}$, *the following three conditions are mutually equivalent.*

(1) $\{s_i \leftrightarrow t_i : i \in I\} \vdash_{\mathbf{HJ}} u \leftrightarrow w$,
(2) $\{s_i \leftrightarrow t_i : i \in I\} \vdash_{\mathbf{LJ}} u \leftrightarrow w$,
(3) $\{s_i = t_i : i \in I\} \models_{\mathsf{HA}} u = w$.

PROOF. By Lemma 1.44, the first condition implies the second. The second implies the third by Lemma 1.52 and the remark just above the present theorem. The proof of the first from the third goes essentially in the same way as the argument in the previous subsection. First, we fix a set of equations $\{s_i = t_i : i \in I\}$. We change the definition of the relation $\equiv$ by defining $\alpha \equiv \beta$ by $\{s_i \leftrightarrow t_i : i \in I\} \vdash_{\mathbf{HJ}} \alpha \leftrightarrow \beta$. Contrary to Lemma 1.46, this time we cannot assume in general that the relation $\equiv$ is fully invariant. But the other arguments work well. In particular, corresponding to Corollary 1.48, we can show that the third statement implies the first. □

The algebraization theorem says that the consequence relation $\vdash_{\mathbf{HJ}}$ can be translated into the equational consequence relation $\models_{\mathsf{HA}}$ determined by HA, and vice versa. The same result holds not only for **HK**, **LK** and BA, but also for every *axiomatic extension* of intuitionistic logic and the corresponding subvariety of HA in general, as shown in the next subsection. Using the terminology of algebraic logic, our result can be expressed as follows. The consequence relation $\vdash_{\mathbf{HJ}}$ (or equivalently $\vdash_{\mathbf{LJ}}$) is *algebraizable* with *defining equation* $x = 1$ and *equivalence formula* $u \leftrightarrow w$, and an *equivalent algebraic semantics* for $\vdash_{\mathbf{HJ}}$ is the variety HA of Heyting algebras.

The compactness of $\models_{\mathsf{BA}}$ and $\models_{\mathsf{HA}}$ follows directly from the algebraization theorem and from the compactness of $\vdash_{\mathbf{HK}}$, see Lemma 1.36, and $\vdash_{\mathbf{HJ}}$.

COROLLARY 1.55. *Suppose that E is a (possibly infinite) set of equations, ε is an equation and K is either one of BA and HA. Then $E \models_{\mathsf{K}} \varepsilon$ iff $E_0 \models_{\mathsf{K}} \varepsilon$ for some finite subset E_0 of E.*

Once again, we clarify the relation between provability of sequents and the deducibility relation from an algebraic point of view. As we mentioned before, the provability of a sequent $\alpha_1, \ldots, \alpha_m \Rightarrow \beta$ in **LJ** is equivalent to the validity of the inequation $\alpha_1 \wedge \ldots \wedge \alpha_m \leq \beta$ in every Heyting algebra. On the other hand, it can be shown that in view of Theorem 1.21, Theorem 1.53 essentially states that the deduction $\alpha_1, \ldots, \alpha_m \vdash_{\mathbf{HJ}} \beta$ holds iff in every Heyting algebra $\mathbf{A}$ and for every assignment h into $\mathbf{A}$ the filter generated by $\{h(\alpha_1), \ldots, h(\alpha_m)\}$ contains the element $h(\beta)$, i.e., $h(\alpha_1) \wedge \ldots \wedge h(\alpha_m) \leq h(\beta)$. Thus, the two notions coincide; in other words, Lemma 1.43 holds for **LJ**, as well. In view of this $\vdash^{seq}_{\mathbf{LJ}} \alpha_1, \ldots, \alpha_n \Rightarrow b$ iff $\vdash_{\mathbf{HJ}} \alpha_1 \wedge \cdots \wedge \alpha_n \to b$. Moreover, this illustrates an algebraic algebraic aspect of the deduction theorem. Connections between filter generation and various forms of deduction theorem will be clarified in the next chapter.

1.4.4. Superintuitionistic logics. Until now, we have discussed only classical and intuitionistic logics. In this subsection, we introduce *superintuitionistic logics*, namely logics stronger than or equal to intuitionistic logic. The first question should be "what are logics?".

We fix a set of formulas Γ, and define $T_\Gamma = \{\alpha : \Gamma \vdash_{\mathbf{HJ}} \alpha\}$. We say that Γ is *closed under deducibility* (of **HJ**), namely under $\vdash_{\mathbf{HJ}}$, if and only if $T_\Gamma = \Gamma$. Obviously, T_Γ is closed under the deducibility for any Γ. Now a set of formulas Γ is a *superintuitionistic logic* (or, a logic over intuitionistic logic **Int**) if and only if it is closed under both substitution and deducibility. (One may feel uneasy about this definition of logics as sets of formulas. Our intention is to identify a given logic with the set of *theorems* of the logic.) See also Note 2, in the end of the chapter. The set $Fm_{\mathcal{L}}$ of all formulas and the set **Int** of all formulas which are provable in **HJ** are two of extreme examples of superintuitionistic logics. The logic $Fm_{\mathcal{L}}$ is said to be *inconsistent*, and all other superintuitionistic logics are called *consistent*.

For a given set of formulas Γ, let $\Sigma[\Gamma]$ be the closure of Γ under substitution. Then, the set $T_{\Sigma[\Gamma]}$ is the smallest superintuitionistic logic containing Γ and is called the *axiomatic extension* of the intuitionistic logic **Int** by Γ, or the superintuitionistic logic *axiomatized by* Γ. If a superintuitionistic logic can be expressed as $T_{\Sigma[\Gamma]}$ for some Γ, it is called simply an axiomatic extension of **Int**. In particular, when a logic is of the form $T_{\Sigma[\Gamma]}$ with a finite Γ, it is said to be *finitely axiomatizable* over **Int**. We can show the following.

LEMMA 1.56. *Let Γ be a set of formulas. Then the following three conditions are mutually equivalent.*

(1) Γ *is a superintuitionistic logic,*
(2) Γ *is an axiomatic extension of* **Int**,
(3) Γ *contains* **Int** *and is closed under both substitution and modus ponens.*

From now on we use letters $\mathbf{L}, \mathbf{K}, \mathbf{H}$ etc. to denote logics. Note that $T_{\Sigma[\Gamma]}$ can be also represented as the set $\{\alpha\colon \{\delta = 1\colon \delta \in \Sigma[\Gamma]\} \models_{\mathsf{HA}} \alpha = 1\}$ by the algebraizability result, since it is closed under deducibility. Thus, for any Heyting algebra $\mathbf{A}$, the set $\mathbf{L}(\mathbf{A})$ defined by $\{\alpha : \alpha \text{ is valid in } \mathbf{A}\}$ is also a superintuitionistic logic. In general, for any class $\mathcal{K}$ of Heyting algebras, the set $\mathbf{L}(\mathcal{K})$ defined by $\{\alpha : \mathcal{K} \models \alpha = 1\}$ is a superintuitionistic logic.

Conversely, suppose that a superintuitionistic logic $\mathbf{L}$ is given. For any Heyting algebra $\mathbf{A}$, $\mathbf{L} \subseteq \mathbf{L}(\mathbf{A})$ iff $\mathbf{A} \models \delta = 1$ for every formula $\delta \in \mathbf{L}$, i.e., $\mathbf{A} \in \mathrm{Mod}(\{\delta = 1 : \delta \in \mathbf{L}\})$. We denote the variety $\mathrm{Mod}(\{\delta = 1 : \delta \in \mathbf{L}\})$ by $\mathsf{V}(\mathbf{L})$.

For a given superintuitionistic logic $\mathbf{L}$, we define its *deducibility* relation $\vdash_{\mathbf{L}}$ by

$$\Gamma \vdash_{\mathbf{L}} \alpha \text{ iff } \Gamma \cup \mathbf{L} \vdash_{\mathbf{HJ}} \alpha \quad \text{for every set of formulas } \Gamma \cup \{\alpha\}.$$

It is clear that the deducibility relation of a superintuitionistic logic is a substitution invariant consequence relation. Then, using the algebraization theorem, it can be shown that $\vdash_{\mathbf{L}}$ is algebraizable and an equivalent algebraic semantics for $\vdash_{\mathbf{L}}$ is the variety $\mathsf{V}(\mathbf{L})$ for every superintuitionistic logic $\mathbf{L}$, including classical logic.

For each superintuitionistic logic $\mathbf{L}$, we can introduce the Lindenbaum-Tarski algebra of $\mathbf{L}$ as before. More precisely, we define a binary relation $\equiv_{\mathbf{L}}$ on the set $Fm_{\mathcal{L}}$ of formulas, putting $\alpha \equiv_{\mathbf{L}} \beta$ if both formulas $\alpha \rightarrow \beta$ and $\beta \rightarrow \alpha$ belong to $\mathbf{L}$. As before, we can show that the relation $\equiv_{\mathbf{L}}$ is a fully invariant congruence on $\mathbf{Fm}_{\mathcal{L}}$, and the quotient algebra $\mathbf{D_L} = \mathbf{Fm}_{\mathcal{L}}/_{\equiv_{\mathbf{L}}}$ is free in $\mathsf{V}(\mathbf{L})$ and is called the *Lindenbaum-Tarski algebra* of $\mathbf{L}$. Obviously, the variety $\mathsf{V}(\mathbf{L})$ is generated by $\mathbf{D_L}$.

Let $\{\mathbf{L}_i : i \in I\}$ be a set of superintuitionistic logics. Then the intersection $\bigcap_{i \in I} \mathbf{L}_i$ is also a superintuitionistic logic. Thus, the set of all superintuitionistic logics forms a complete lattice with respect to set inclusion. Note that if a variety $\mathcal{V}$ is generated by a class $\mathcal{K}$ then $\mathbf{L}(\mathcal{V}) = \mathbf{L}(\mathcal{K})$. In particular, when $\mathcal{K}$ is the set $\{\mathbf{A}_i : i \in I\}$, $\mathbf{L}(\mathcal{V}) = \bigcap_{i \in I} \mathbf{L}(\mathbf{A}_i)$.

Let $Eq_{\mathcal{L}}$ be the set of all equations in our language $\mathcal{L}$. A subset E of $Eq_{\mathcal{L}}$ is *closed* (with respect to HA) iff for any equation $u = v$, $E \models_{\mathsf{HA}} u = v$ implies that $u = v \in E$. Then, there exists a bijection between the collection of all subvarieties of HA and the set of all closed subsets of $Eq_{\mathcal{L}}$. Thus, the collection $\mathbf{\Lambda}(\mathsf{HA})$ of all subvarieties of HA is in fact a set. Suppose that $\{\mathcal{V}_i : i \in I\}$ is a collection of subvarieties of HA. Then $\bigcap_{i \in I} \mathcal{V}_i$, defined by $\mathrm{Mod} \bigcup_{i \in I} \mathrm{Th}_e(\mathcal{V}_i)$, is a variety which is the infimum of that collection. Hence, $\mathbf{\Lambda}(\mathsf{HA})$ forms a complete lattice.

For subvarieties $\mathcal{V}, \mathcal{W}$ of HA, if $\mathcal{W}$ is a proper subclass of $\mathcal{V}$, then $\mathbf{L}(\mathcal{V}) \supset \mathbf{L}(\mathcal{W})$. Also, for any subvariety $\mathcal{V}$ of HA, $\mathsf{V}((\mathbf{L}(\mathcal{V}))) = \mathrm{Mod}(\{\delta = 1 : \delta \in \mathbf{L}(\mathcal{V})\}) = \mathrm{Mod}(Th_e(\mathcal{V})) = \mathcal{V}$. Thus, the map $\mathbf{L}\colon \mathcal{V} \mapsto \mathbf{L}(\mathcal{V})$ is a complete dual lattice isomorphism from the lattice $\mathbf{\Lambda}(\mathsf{HA})$ of subvarieties of HA to the lattice of all superintuitionistic logics.

Among all of its subvarieties, HA is the greatest and the variety generated by the *degenerate* Heyting algebra is the smallest with respect to class inclusion. A subvariety $\mathcal{V}$ of HA is (non-trivial) *minimal* if it is greater than the smallest (trivial) and there are no subvarieties between them, i.e., it is a cover of the smallest variety (an atom in the lattice of subvarieties).

LEMMA 1.57. *The variety* BA *of Boolean algebras is the single minimal element of* $\mathbf{\Lambda}(\mathsf{HA})$. *Equivalently, classical logic is the maximum element among consistent superintuitionistic logics.*

PROOF. Suppose that a formula α is not in classical logic but belongs to a superintuitionistic logic $\mathbf{L}$. Then there exists an assignment f on $\mathbf{2}$ such that $f(\alpha) = 0$. Let $p_1, \ldots, p_m$ be an enumeration of the variables in α. Define a substitution σ in such a way that σ replaces p_i by $q \rightarrow q$ if $f(p_i) = 1$ and by $\neg(q \rightarrow q)$ otherwise. Then the substitution instance $\sigma(\alpha)$ of α by the substitution σ must belong to $\mathbf{L}$. On the other hand, by using induction we can show that $\sigma(\alpha) \leftrightarrow \neg(q \rightarrow q)$ is in intuitionistic logic and hence in $\mathbf{L}$. Thus, $\neg(q \rightarrow q)$ also belongs to $\mathbf{L}$. Therefore, $\mathbf{L}$ must be inconsistent. $\square$

LEMMA 1.58. *The variety $V(\mathbf{H}_3)$ generated by the 3-valued Heyting algebra $\mathbf{H}_3$ is the smallest among subvarieties of* HA *which properly contain* BA. *Equivalently, the logic $\mathbf{L}(\mathbf{H}_3)$ is the greatest among superintuitionistic logics which is properly contained in classical logic.*

PROOF. Let $\mathbf{L}$ be a superintuitionistic logic which is properly contained in classical logic. Then, by the subdirect representation theorem and the fact that our language is countable, we can find at most countable subdirectly irreducible Heyting algebras $\{\mathbf{A}_i : i \in I\} \subseteq \mathsf{V}(\mathbf{L})$ such that $\mathbf{L} = \bigcap_{i \in I} \mathbf{L}(\mathbf{A}_i)$. Since $\mathbf{L}$ is consistent, at least one $\mathbf{A}_i$ is non-trivial, and therefore we can assume without loss of generality that they all are. If all of them are two element Boolean algebras, $\mathbf{L}$ is equal to classical logic. But this contradicts our assumption. Hence, there exists $j \in I$ such that $\mathbf{A}_j$ contains more than two elements. By Theorem 1.23, every non-trivial subdirectly irreducible Heyting algebra has a second greatest element. Let a be the second greatest element of $\mathbf{A}_j$. Then the subset $\{0, a, 1\}$ of $\mathbf{A}_j$ forms a subalgebra, which is isomorphic to $\mathbf{H}_3$. Hence, $\mathbf{L} \subseteq \mathbf{L}(\mathbf{A}_j) \subseteq \mathbf{L}(\mathbf{H}_3)$. □

THEOREM 1.59. [Jan68] *There exist continuum many superintuitionistic logics.*

Lattices of logics and subvarieties will be studied in Chapter 9 in a general setting.

1.5. Cut elimination in sequent calculi

1.5.1. Cut elimination. Gentzen proved the following *cut elimination theorem* for both **LK** and **LJ**, which is one of the most important results on the proof theory based on sequent calculi.

THEOREM 1.60. *For any sequent s, if s is provable in* **LK**, *it is provable in* **LK** *without using the cut rule. This holds also for* **LJ**.

A proof without any applications of the cut rule is called a *cut-free proof*. When the cut elimination holds for a sequent system **L**, we say that **L** is a *cut-free system*. In a cut-free system, we can restrict our attention only to cut-free proofs, and analyses of cut-free proofs often bring us important logical properties hidden in proofs. This is the reason why cut elimination is considered to be fundamental in the proof-theoretical approach.

In this subsection, we give a simple semantical proof of the cut elimination theorem for **LK**. A syntactic proof of the cut elimination theorem, based on Gentzen's original one, including that for **LJ**, is briefly outlined in a general setting in Chapter 4. An algebraic proof of cut elimination is given in Chapter 7.

For simplicity, we assume here that our language does not contain any logical constant, and also modify the notion of sequents so that every sequent is a formal expression of the form $\Gamma \Rightarrow \Pi$ with Γ, Π (finite) *multisets of formulas*, instead of sequences of formulas.[7] By taking multisets, we disregard the order of occurrences of formulas in sequences and pay attention only to the *multiplicity* of each formula. Thus, $\alpha, \beta, \gamma, \alpha$ is identified with $\alpha, \alpha, \gamma, \beta$, but distinguished from α, γ, β, for example. By introducing the latter convention, the exchange rules become unnecessary.

Next, we introduce an auxiliary sequent calculus called **LK***. Sequents of **LK*** are the same as those of **LK** with the above modification. The system **LK*** has neither a cut rule nor structural rules in an explicit form. It has essentially the same rules for logical connectives, except $(\rightarrow\Rightarrow), (\wedge\Rightarrow)$ and $(\Rightarrow\vee)$ as shown below:

Initial sequents:

$$\alpha, \Gamma \Rightarrow \Delta, \alpha.$$

Rules for logical connectives:

$$\frac{\alpha, \Gamma \Rightarrow \Delta \quad \beta, \Gamma \Rightarrow \Delta}{\alpha \vee \beta, \Gamma \Rightarrow \Delta} (\vee \Rightarrow) \qquad \frac{\Gamma \Rightarrow \Delta, \alpha, \beta}{\Gamma \Rightarrow \Delta, \alpha \vee \beta} (\Rightarrow \vee)$$

$$\frac{\alpha, \beta, \Gamma \Rightarrow \Delta}{\alpha \wedge \beta, \Gamma \Rightarrow \Delta} (\wedge \Rightarrow) \qquad \frac{\Gamma \Rightarrow \Delta, \alpha \quad \Gamma \Rightarrow \Delta, \beta}{\Gamma \Rightarrow \Delta, \alpha \wedge \beta} (\Rightarrow \wedge)$$

$$\frac{\Gamma \Rightarrow \Delta, \alpha \quad \beta, \Gamma \Rightarrow \Delta}{\alpha \rightarrow \beta, \Gamma \Rightarrow \Delta} (\rightarrow \Rightarrow) \qquad \frac{\alpha, \Gamma \Rightarrow \Delta, \beta}{\Gamma \Rightarrow \Delta, \alpha \rightarrow \beta} (\Rightarrow \rightarrow)$$

$$\frac{\Gamma \Rightarrow \Delta, \alpha}{\neg\alpha, \Gamma \Rightarrow \Delta} (\neg \Rightarrow) \qquad \frac{\alpha, \Gamma \Rightarrow \Delta}{\Gamma \Rightarrow \Delta, \neg\alpha} (\Rightarrow \neg)$$

Proofs and the provability of a sequent in **LK*** are defined in the same way as those in **LK**. The calculus **LK*** was discussed by Kanger [Kan57].

LEMMA 1.61. *Every proof of a sequent* $\Gamma \Rightarrow \Delta$ *in* **LK*** *can be transformed into a cut-free proof of the same sequent in* **LK**.[8]

PROOF. This can be proved by induction on the length of the proof of $\Gamma \Rightarrow \Delta$ in **LK***. If it is an initial sequent of **LK***, it can be proved by applying weakening rules several times to an initial sequent of **LK**. In the remaining case, it is enough to show how to get a cut-free proof (in **LK**) of the lower sequent of each rule of **LK***, when cut-free proofs of

[7]We use the same Greek capital letters for sets of formulas, sequences of formulas and multisets of formulas to avoid unnecessary complications and burden to readers, and will indicate the usage of this notation in each context.

[8]More precisely, for a given sequent $\Gamma \Rightarrow \Delta$ with multisets Γ and Δ, we consider a sequent of the form $\Gamma^* \Rightarrow \Delta^*$ of **LK** where Γ^* and Δ^* are arbitrary sequences of formulas consisting of the same formulas with the same multiplicity of Γ and Δ, respectively. Because of the exchange rules in **LK**, the choice of these sequences is inessential as long as provability is concerned.

upper sequents are given. This is trivial for rules other than $(\rightarrow\Rightarrow), (\wedge\Rightarrow)$ and $(\Rightarrow\vee)$, as these are exactly the same in **LK**. The three rules above, which are different, are shown to be derivable in **LK** only by inserting some applications of contraction rules. For instance, the following shows how $(\Rightarrow\vee)$ of **LK*** is transformed into a part of a proof in **LK** without using cut.

$$\dfrac{\dfrac{\dfrac{\Gamma \Rightarrow \Delta, \alpha, \beta}{\Gamma \Rightarrow \Delta, \alpha \vee \beta, \beta}\,(\Rightarrow \vee 1)}{\Gamma \Rightarrow \Delta, \alpha \vee \beta, \alpha \vee \beta}\,(\Rightarrow \vee 2)}{\Gamma \Rightarrow \Delta, \alpha \vee \beta}\,(\Rightarrow c)$$

□

A characteristic feature of the calculus **LK*** is given in the following lemma. Recall that every rule for logical connectives of **LK** except $(\rightarrow\Rightarrow), (\wedge\Rightarrow)$ and $(\Rightarrow\vee)$ is invertible. The second item of the following lemma can be shown in the same way as Theorem 1.45, by using the fact that validity in BA equals being a tautology, as shown in Theorem 1.50.

LEMMA 1.62.

(1) *Every rule for logical connectives of* **LK*** *is invertible.*
(2) *For every rule for logical connectives of* **LK***, *each of the upper sequent is a tautology iff the lower sequent is a tautology.*

Recall also that a proof in a sequent calculus is a tree, labeled by sequents, and obtained by applying rules of the calculus *downward*. We introduce here a dual notion, called a *decomposition* of a given sequent s (in **LK***), which is an (incomplete) proof of s in our proof search, starting from s and applying rules *upward*. More precisely, a decomposition of a sequent s and its *leaves*, which form a set of (occurrences of) sequents in it, are defined simultaneously as follows:

(1) The sequent s itself is a decomposition with a single leaf s.
(2) Suppose that R is a decomposition of s and that s_0 is one of its leaves. Moreover suppose that there exists an instance of a rule I of **LK*** with upper sequents s_1, s_2 and the lower sequent s_0. Then, the figure R′ obtained from R by putting s_1 and s_2 over this occurrence of s_0 is also a decomposition of s, which is called a decomposition obtained from R by applying I to s_0. The leaves of R′ is obtained from those of R by deleting s_0 and then adding s_1 and s_2. A decomposition of s obtained from R by applying any rule I of **LK*** with a single upper sequent is defined similarly.

A decomposition of a sequent s is *complete* when no rule can be applied to it any more. This happens when and only when any of its leaves is a sequent containing no logical connectives, i.e., it is of the form $p_1, \ldots, p_m \Rightarrow q_1, \ldots, q_n$ for variables $p_1, \ldots, p_m, q_1, \ldots, q_n$. It is easy to see that such a sequent is a tautology iff p_i is equal to q_j for some i, j. Here is an example of a complete decomposition of a sequent $\Rightarrow ((p \to q) \to p) \to p$.

$$\dfrac{\dfrac{\dfrac{p \Rightarrow p, q}{\Rightarrow p, p \to q} \quad p \Rightarrow p}{(p \to q) \to p \Rightarrow p}}{\Rightarrow ((p \to q) \to p) \to p}$$

Note that a complete decomposition R of a sequent s is in fact a proof of s in **LK*** if all of leaves of R are initial sequents. Also, it is easy to see that in any rule of logical connectives of **LK***, the total number of occurrences of logical connectives in (each of) the upper sequent(s) is smaller than that of the lower sequent. Therefore, we can get eventually a complete decomposition, by repeating decompositions.

Now, suppose that a sequent s is given. Let us take an arbitrary complete decomposition R of s. Then, take any leave $p_1, \ldots, p_m \Rightarrow q_1, \ldots, q_n$ of R and check whether p_i is equal to q_j for some i, j. If it is so for every leave, then all of the leaves of R are initial sequents. Thus, R is a proof of s in **LK***, which can be transformed into a cut-free proof in **LK** in finitely many steps by Lemma 1.61. If otherwise, such a leave $p_1, \ldots, p_m \Rightarrow q_1, \ldots, q_n$ is not a tautology. Thus, s is not a tautology by Lemma 1.62, and hence is not provable in **LK** by Theorem 1.45.

Such an invertible sequent calculus is not known for intuitionistic logic.

1.5.2. Decidability and subformula property. There are many important consequences of cut elimination, which are discussed comprehensively in Chapter 4. Here, we discuss two of its consequences.

Decidability. A given logic **L** is said to be *decidable* if there exists an algorithm that decides whether a formula α is provable in **L** or not, for any α; **L** is *undecidable* if it is not decidable. The *decision problem* of a given logic **L** is the question as whether **L** is decidable. As we have already shown, a formula is provable in classical logic if and only if it is a tautology. Since we can check in finitely many steps whether a given formula is a tautology, we can conclude that classical logic is decidable. Decidability of intuitionistic logic can also be obtained as a consequence of cut elimination for **LJ**. This can be shown by using Gentzen's original method, which is explained in Chapter 4 (and applies to **LK**, too). The algorithm described at the end of the previous subsection gives us a stronger result on classical logic than just decidability. It is sometimes called *Wang's algorithm* (see e.g., [Wan63]), and is a standard technique in automated theorem proving. For more information on automated theorem proving, see e.g., [RV01].

COROLLARY 1.63. *There exists an algorithm that for a given sequent s, decide whether it is provable in* **LK** *or not, and produces moreover a cut-free proof of s when it is provable.*

Subformula property. Another important consequence of cut elimination is the *subformula property.*

THEOREM 1.64. *If a sequent s is provable in* **LK** *then there exists a proof of s such that any formula appearing in it is a subformula of some formula in s. In fact, any cut-free proof of s has this property.*

PROOF. To show this, it is enough to check that in every rule of inference of **LK** except for cut, every formula appearing in the upper sequent(s) is a subformula of some formulas in the lower sequent. (This does not hold always for the cut rule, as the cut formula may not be a subformula of some formulas in the lower sequent.) □

We say that a proof in a sequent calculus **L** has the *subformula property* if it contains only formulas which are subformulas of some formulas in its endsequent. Also, we say that the *subformula property* holds for **L** when any provable sequent s in **L** has a proof of s with the subformula property.

As a consequence of the subformula property, we can show the following result on the conservativity of **LK**. Suppose that $\mathcal{K}$ is a sublanguage of $\mathcal{L}$. By the *$\mathcal{K}$-fragment* of **LK**, we mean the sequent calculus whose initial sequents and rules of inference are the same as those of **LK** except that only rules of inference for logical connectives in $\mathcal{K}$ are allowed. A formula α is said to be a *$\mathcal{K}$-formula* when it contains only logical connectives from $\mathcal{K}$. We can show the following.

COROLLARY 1.65. *For any nonempty sublanguage $\mathcal{K}$ of $\mathcal{L}$,* **LK** *is a conservative extension of the $\mathcal{K}$-fragment of* **LK**. *More precisely, for any sequent s consisting only of $\mathcal{K}$-formulas, s is provable in* **LK** *if and only if it is provable in the $\mathcal{K}$-fragment of* **LK**.

PROOF. Obviously, the only-if part is essential. Suppose that a sequent s which consists only of $\mathcal{K}$-formulas is provable in **LK**. Consider a cut-free proof P of s in **LK**. By the subformula property of **LK**, P contains only $\mathcal{K}$-formulas. Hence, there is no chance of applying a rule of inference for a logical connective other than those from $\mathcal{K}$. Thus, P is regarded as a proof in the $\mathcal{K}$-fragment of **LK**. □

1.6. Consequence relations and matrices

The issues we discuss here are based on *abstract algebraic logic* and are considerably more abstract than the rest of the chapter. Therefore the reader may skip this section until a reference to it is made in the following chapters.

1.6.1. Consequence relations. Let Q be a set. A subset $\vdash$ of $\mathcal{P}(Q) \times Q$ is called *consequence relation* over Q, if for every subset $X \cup Y \cup \{x, z\}$ of Q

- if $x \in X$, then $X \vdash x$ and
- if $X \vdash Y$ and $Y \vdash z$, then $X \vdash z$,

where $X \vdash x$ stands for $(X, x) \in \vdash$ and $X \vdash Y$ stands for: $X \vdash y$, for all $y \in Y$. Also, we use $X \dashv\vdash Y$ for the conjunction of $X \vdash Y$ and $Y \vdash X$.

A *closure operator on* the powerset $\mathcal{P}(Q)$ is a map $C : \mathcal{P}(Q) \to \mathcal{P}(Q)$ that is *increasing*, *monotone* and *idempotent*, i.e., $X \subseteq C(X)$, $X \subseteq Y$ implies $C(X) \subseteq C(Y)$, and $CC(X) = C(X)$, for $X, Y \subseteq Q$. We also refer to C as a closure operator *over* Q.

For a consequence relation $\vdash$ over Q, we define the map $C_\vdash$ on $\mathcal{P}(Q)$, by $C_\vdash(X) = \{x \in Q : X \vdash x\}$. Also, given a closure operator C on $\mathcal{P}(Q)$, we define the relation $\vdash_C$ over Q by $X \vdash_C x$ iff $x \in C(X)$. Exercise 33 asks the reader to verify that $C_\vdash$ is a closure operator and that $\vdash_C$ is consequence relation.

A consequence relation $\vdash$ is called *finitary*, if $X \vdash x$ implies that $X_0 \vdash x$, for some finite subset X_0 of X. Exercise 34 identifies the closure operators associated with finitary consequence relations.

Recall that $\mathbf{Fm}_\mathcal{L}$ denotes the absolutely free algebra (of formulas) over the language $\mathcal{L}$. A consequence relation on $(Fm_\mathcal{L})^k$ will be called a *k-dimensional consequence relation on $Fm_\mathcal{L}$*. We will be interested in the cases where k is equal to 1 or 2. Examples of 1-dimensional consequence relations are the deducibility relations $\vdash_{\mathbf{HK}}$, $\vdash_{\mathbf{HJ}}$ and $\vdash_{\mathbf{L}}$, for every superintuitionistic logic $\mathbf{L}$. Examples of 2-dimensional consequence relations include $\models_{\mathsf{BA}}$, $\models_{\mathsf{HA}}$ and Birkhoff's calculus for equational logic. In fact, they are relations over the set $(Fm_\mathcal{L})^2$, where each element (s, t) of the set is usually represented as $s = t$. By considering k-dimensional consequence relations, these two types of consequence relations can be treated in a uniform manner.

A k-dimensional consequence relation $\vdash$ on $\mathbf{Fm}_\mathcal{L}$ (we use boldface to indicate the additional structure) is called *substitution invariant*, if $T \vdash t$ implies that $\sigma[T] \vdash \sigma(t)$, for every substitution σ of $\mathbf{Fm}_\mathcal{L}$. Sometimes the term 'structural' is used for substitution invariance, but we avoid it as it clashes with the notion of 'substructurality', related to the presence of structural rules.

1.6.2. Inference rules. By a (k-dimensional) *rule* over $Fm_\mathcal{L}$ we understand a pair $r = (T, t)$, where $T \cup \{t\}$ is a subset of $(Fm_\mathcal{L})^k$. If T is empty, then r is called *axiomatic* or an *axiom*. We usually write rules in fractional notation $\frac{T}{t}(r)$. If $T = \{t_1, \ldots, t_n\}$ we write

$$\frac{t_1 \quad t_2 \quad \cdots \quad t_n}{t}\ (r)$$

The Hilbert systems **HK** and **HJ** specify sets of 1-dimensional rules. The Gentzen systems **LK** and **LJ** each give rise to both a consequence relation ($\vdash_{\mathbf{LK}}$ and $\vdash_{\mathbf{HJ}}$) over the set $Fm_{\mathcal{L}}$ of formulas and a consequence relation ($\vdash^{seq}_{\mathbf{LK}}$ and $\vdash^{seq}_{\mathbf{HJ}}$) over the set of sequents. The rules of these Gentzen systems do not fit in the definition of k-dimensional rules for any fixed k, since the length of a sequent can vary. Moreover, a Gentzen rule cannot be considered a union of k-dimensional rules for different lengths of k, as there are rules in which the sequents involved have different lengths.

Nevertheless, an expanded notion of rule can be defined. A *Gentzen rule* over $\mathbf{Fm}_{\mathcal{L}}$ is a pair $r = (T, t)$, where $T \cup \{t\}$ is a set of $\mathcal{L}$-sequents; the fractional notation convention applies also for Gentzen rules. Elements of $(Fm_{\mathcal{L}})^k$ can be viewed as sequents with empty left-hand side and k formulas on the right-hand side. From this point of view k-dimensional rules and consequence relations can be viewed as rules and consequence relations over sequents.

Note that we have used two different notions of sequent: one *classical* for **LK** and one *intuitionistic* for **LJ**. Moreover, we have also considered a different version of a classical sequent in **LK***, where the two sides are multisets rather than sequences. Risking loosing the reader until the end of this paragraph, we mention, without being precise, that a generalized notion of a sequent can be defined as an atomic formula in a first order language (for example for every pair of lengths of the two sides of a classical sequent we include a relational symbol in the first order language), and capture non-associative sequents or hyper-sequents, but this is beyond the scope of this book; see for example [GO]. Nevertheless, it is clear that we can consider consequence relations over such syntactic objects over the set $Fm_{\mathcal{L}}$ of formulas and naturally extend every substitution over $Fm_{\mathcal{L}}$ to such objects. Even more generally, we can consider a set Q, intuitively understood as a set of generalized sequents over a set of formulas, and a set Σ, intuitively understood as the set of substitutions, acting on Q. Then one can define the notion of a substitution invariant consequence relation over Q (and Σ). For details, see [BJ06] and [GT]. In the following we will assume that the rules and consequence relations are in such a general setting (and substitutions are elements of Σ), but the reader can safely assume that the presentation is specialized to Gentzen rules and consequence relations over classical or intuitionistic sequents, since we will make use of these notions only in those cases.

A *rule* over a set Q is a pair $r = (T, t)$, where $T \cup \{t\}$ is a subset of Q. We say that a rule $r = (T, t)$ is *derivable* in a consequence relation $\vdash$ over a set Q, if $T \vdash t$. All of the consequence relations we will consider are substitution invariant (and finitary) so for a derivable rule we will have $\sigma[T] \vdash \sigma(t)$, for every substitution σ (of $\mathbf{Fm}_{\mathcal{L}}$). The (substitution invariant) consequence relation $\vdash_R$ *presented* by a set R of rules is the least substitution invariant consequence relation in which all rules from R are derivable. In particular if R is the set of all instances of the axiom schemes and rule scheme in **HK**, then $\vdash_R$ is what we denoted by $\vdash_{\mathbf{HK}}$. Actually, just one instance of each axiom scheme or rule scheme (where every metavariable is replaced by a different variable) is enough, because substitutions are build into the notion of a proof (recall that axioms are just rules without assumptions).

Alternatively, given a set R of rules over a set Q we can define $\vdash_R$ by first defining the notion of a proof from assumptions. This is done in the same way as for **HK**, **LK** and for Birkhoff's system for equational logic, but we repeat it for completeness in the general setting that we consider.

1.6.3. Proofs and theorems. A *proof* of s (*conclusion*) from (the set of) *assumptions* S in a set of rules R is defined inductively as follows.

(1) Every element of S is a proof with that element as assumption and conclusion.
(2) If σ is a substitution,

$$\frac{s_1 \quad s_2 \quad \cdots \quad s_n}{s}\ (r)$$

is a rule in R and $\mathsf{P}_1, \mathsf{P}_2, \ldots, \mathsf{P}_n$ are proofs with conclusions $\sigma(s_1), \sigma(s_2), \ldots, \sigma(s_n)$ and (sets of) assumptions $S_1, S_2, \ldots, S_n$, respectively, then

$$\frac{\mathsf{P}_1 \quad \mathsf{P}_2 \quad \cdots \quad \mathsf{P}_n}{\sigma(s)}\ (r)$$

is a proof with set of assumptions $S_1 \cup S_2 \cup \cdots \cup S_n$ and conclusion $\sigma(s)$.

It is easy to see that $S \vdash_R s$ iff there is a proof of s from S in R.

Given a (substitution invariant) consequence relation $\vdash$ over a set Q, the set of *theorems* $Thm(\vdash)$ of $\vdash$ is defined to be the set $\{q \in Q : \emptyset \vdash q\}$.

A rule $r = (T, t)$ is called *admissible* in a consequence relation $\vdash$ over a set Q, if $T \subseteq Thm(\vdash)$ implies $t \in Thm(\vdash)$. In other words, r is admissible in $\vdash$ iff adding the rule r to $\vdash$ does not change the set of theorems, or, more precisely iff $Thm(\vdash + (r)) = Thm(\vdash)$; here $\vdash + (r)$ denotes the least (substitution invariant) consequence relation containing $\vdash$ and (r). The cut elimination theorem means that the cut rule is admissible in the cut-free system; nevertheless, it is not derivable in it.

Given a subset Q' of Q, the *Q'-fragment* of R is defined as the restriction of R to Q'. In particular, if $\mathcal{K}$ is a sublanguage of $\mathcal{L}$, the $\mathcal{K}$-fragment of a Hilbert-style system, like **HK** and **HJ**, or of a sequent calculus, like **LK** and **LJ**, is defined as its restriction to $\mathcal{K}$-formulas. If exactly the same $\mathcal{K}$-formulas are theorems of a system and its $\mathcal{K}$-fragment, then the system is said to be a *conservative extension* of its fragment. If the restriction of the consequence relation of a system to $\mathcal{K}$-formulas and the consequence relation of its $\mathcal{K}$-fragment coincide, then the system is said to be a *strong conservative extension* of its fragment.

1.6.4. Matrices. A k-dimensional *$\mathcal{L}$-matrix* is a pair $(\mathbf{A}, F)$, where $\mathbf{A}$ is an $\mathcal{L}$-algebra and $F \subseteq A^k$. For example, if $\mathbf{A}$ is a Boolean algebra, then $(\mathbf{A}, \{1\})$ is a 1-dimensional matrix. If $(\mathbf{A}, F)$ is a k-dimensional $\mathcal{L}$-matrix and $f : \mathbf{Fm}_{\mathcal{L}} \to \mathbf{A}$ is a homomorphism (or a valuation), we say that a rule $r = (T, t)$ is *valid* in $(\mathbf{A}, F, f)$, in symbols

$$T \models_{(\mathbf{A},F,f)} t,$$

if $f[T] \in F$ implies $f(t) \in F$. Here $f(t)$ is calculated coordinatewise. We say that a rule $r = (T, t)$ is *valid* in $(\mathbf{A}, F)$, in symbols

$$T \models_{(\mathbf{A},F)} t,$$

if $T \models_{(\mathbf{A},F,f)} t$, for all homomorphisms $f : \mathbf{Fm}_{\mathcal{L}} \to \mathbf{A}$.

We extend the definitions to sets; if M is a set of triples $(\mathbf{A}, F, f)$ or of matrices $(\mathbf{A}, F)$, we write $\models_M$ for the intersection of all the relations $\models_{(\mathbf{A},F,f)}$, where $(\mathbf{A}, F, f) \in M$. Obviously, $\models_{(\mathbf{A},F)} = \models_M$, where $M = \{(\mathbf{A}, F, f) \mid f : \mathbf{Fm}_{\mathcal{L}} \to \mathbf{A}$ is a homomorphism$\}$. Relations relative to matrices are usually called *global* in the literature, while relations relative to triples are called *local.*

We say that $(\mathbf{A}, F, f)$ is a *model* of a k-dimensional consequence relation $\vdash$, if $\vdash \subseteq \models_{(\mathbf{A},F,f)}$. Also, we say that $(\mathbf{A}, F)$ is a *(matrix) model* of $\vdash$, if $\vdash \subseteq \models_{(\mathbf{A},F)}$; in this case F is called a *deductive filter* of $\mathbf{A}$ with respect to $\vdash$. The deductive filters of the formula algebra $\mathbf{Fm}_{\mathcal{L}}$ are called *theories*.

The notion of Gentzen matrices can be defined by analogy and will be stated explicitly in Chapter 7.

1.6.5. Examples. For every operation symbol $f \in \mathcal{L}$ of arity n, we consider the (1-dimensional) rule

$$\frac{x_1 \quad x_2 \quad \cdots \quad x_n}{f(x_1, x_2, \ldots, x_n)} \ (f)$$

where $x_1, x_2, \ldots, x_n$ are distinct variables. Consider the consequence relation $\vdash_L$ on $\mathbf{Fm}_{\mathcal{L}}$ presented by the set L of all rules (f), for $f \in \mathcal{L}$. The deductive filters of an $\mathcal{L}$-algebra $\mathbf{A}$ with respect to $\vdash_L$ are exactly the subalgebras of $\mathbf{A}$.

It is not hard to verify that the deductive filters of a Boolean algebra $\mathbf{A}$ with respect to $\vdash_{\mathbf{HK}}$ are exactly the non-empty lattice filters of $\mathbf{A}$. Likewise, the deductive filters of Heyting algebras with respect to $\vdash_{\mathbf{HJ}}$ are exactly their non-empty lattice filters.

Recall that an equation is a pair of terms (s, t), which we usually denote by $s = t$. Consider the 2-dimensional consequence relation $\vdash_{Con}$ on $\mathbf{Fm}_{\mathcal{L}}$ presented by the set Con (for Congruence) of the following 2-dimensional rules.

$$\frac{}{x = x} \ (\text{refl}) \qquad \frac{x = y}{y = x} \ (\text{sym}) \qquad \frac{x = y \quad y = z}{x = z} \ (\text{tran})$$

$$\frac{x_1 = y_1 \quad x_2 = y_2 \quad \cdots \quad x_n = y_n}{f(x_1, x_2, \ldots, x_n) = f(y_1, y_2, \ldots, y_n)} \ (\mathrm{cong}_f)$$

where f ranges over all operation symbols in $\mathcal{L}$, and n is the arity of f. Note that Con is essentially Birkhoff's system for equational logic without substitution; so we will refer to the relation $\vdash_{Con}$ as *substitution-free equational logic*.

The deductive filters of an $\mathcal{L}$-algebra $\mathbf{A}$ with respect to $\vdash_{Con}$ are exactly the congruences of $\mathbf{A}$. If we do not include the rules (cong_f), then we obtain all equivalence relations on A. It is easy to see that the rules (cong_f) can be replaced by the rules

$$\frac{x = y}{t[x] = t[y]} \ (\mathrm{rep}_t)$$

where t ranges over all terms (or even just over all fundamental operations) and $t[x]$ and $t[y]$ denotes the replacement of a particular occurrence of a variable in t by x and y, respectively.

For every endomorphism σ of an $\mathcal{L}$-algebra $\mathbf{A}$, we introduce a new unary operation symbol f_σ. We extend $\mathcal{L}$ to a language $\mathcal{L}'$ that contains the symbol f_σ, for every endomorphism σ. For every $a \in A$, we define $f_\sigma^{\mathbf{A}}(a) = \sigma(a)$. So, A becomes the universe of an $\mathcal{L}'$-algebra $\mathbf{A}'$. We extend Con to the set FIC (for Fully Invariant Congruence) containing the rules

$$\frac{x = y}{f_\sigma(x) = f_\sigma(y)} \ (\mathrm{sub}_\sigma)$$

where σ ranges over all endomorphisms of $\mathbf{A}$. The deductive filters of the algebra $\mathbf{A}'$ with respect to $\vdash_{FIC}$ are the fully invariant congruences of $\mathbf{A}$. We will be interested in the fully invariant congruences of the algebra $\mathbf{Fm}_{\mathcal{L}}$. The rule (sub_σ) may look trivial, but recall that f_σ is a new symbol; so $f_\sigma(x) \neq \sigma(x)$, but $f_\sigma^{\mathbf{Fm}_{\mathcal{L}}}(x) = \sigma(x)$ (and $f_\sigma^{\mathbf{Fm}_{\mathcal{L}}}(t) = \sigma(t)$, for every term t). Since the system FIC is used mainly for the algebra $\mathbf{Fm}_{\mathcal{L}}$, we define $E \vdash_{Eq} \varepsilon$ iff $E \vdash_{FIC} \varepsilon'$ and ε is the result of evaluating all the symbols f_σ in ε' to their corresponding substitutions in $\mathbf{Fm}_{\mathcal{L}}$.

It is easy to see that Birkhoff's system for equational logic is essentially equivalent to the $\vdash_{Eq}$; the main difference is that in Birkhoff's system all substitutions are performed in the beginning of the proof; see Exercise 39.

1.6.6. First-order and (quasi)equational logic. Recall the definition of satisfaction in first-order logic (FOL) $\mathbf{A}, f \models \varphi$ from the beginning of the chapter. We write

$$\Phi \models \varphi \text{ (or } \Phi \models_{FOL} \varphi \text{ for clarity)}$$

iff for all $(\mathbf{A}, f)$, if $\mathbf{A}, f \models \Phi$, then $\mathbf{A}, f \models \varphi$.

Also recall that a syntactic relation relation $\vdash_{FOL}$ can be defined between sets of first-order formulas and first order formulas; some books define this relation only for the special case of first-order sentences. Axioms are all the axioms for **HK** applied to first-order formulas and

(P1) $(\forall x\varphi(x)) \Rightarrow \varphi(t)$, where t is a first-order term with no variables that are bound in $\varphi(t)$.
(P2) $(\forall x(\varphi \Rightarrow \psi)) \Rightarrow (\varphi \Rightarrow (\forall x\psi))$, where x is not a free variable of φ.
(Eq1) $x = x$
(Eq2) $x = y \Rightarrow y = x$
(Eq3) $x = y$ and $y = z \Rightarrow x = z$
(Eq4) $x_1 = y_1$ and $\cdots$ and $x_n = y_n \Rightarrow t(x_1, \ldots, x_n) = t(y_1, \ldots, y_n)$
(Eq5) $x_1 = y_1$ and $\cdots$ and $x_n = y_n \Rightarrow (\varphi(x_1, \ldots, x_n) \Rightarrow \varphi(y_1, \ldots, y_n))$

where x, y, z, x_i, y_i range over variables, t over first-order terms and φ over first-order formulas.[9] The rules of inference are Modus Ponens for first-order formulas and the Generalization Rule which states that from φ we obtain $\forall x\varphi$, *with a certain restriction to its application.*

In detail, a *proof in first-order logic* (FOL) of φ (conclusion) from (the set of) assumptions Φ is defined inductively as follows.

(1) Each axiom and each element of Φ is a proof with conclusion and assumption itself.
(2) If P_1 is a proof with conclusion ψ and assumptions Φ_1, and P_2 is a proof with conclusion $\psi \Rightarrow \chi$ and assumptions Φ_2, then (modus ponens)
$$\frac{\mathsf{P}_1 \quad \mathsf{P}_2}{\chi}\ (\text{mp})$$
is a proof of χ from assumptions $\Phi_1 \cup \Phi_2$.
(3) If P is a proof of ψ from assumptions Ψ *and x is not a free variable of* Ψ, then (generalization)
$$\frac{\mathsf{P}}{\forall x\psi}\ (\text{gen})$$
is a proof of $\forall x\psi$ from assumptions Ψ.

If there is a proof in FOL of φ from Φ, then we write $\Phi \vdash_{FOL} \varphi$. It is well known from classical first-order logic that the relations $\models_{FOL}$ and $\vdash_{FOL}$ coincide (for first-order formulas in general, not just for sentences).

Here we would like to clarify a point that frequently causes confusion. If $\Phi \cup \{\varphi\}$ is a set of open (in particular atomic) formulas, then $\Phi \models \varphi$ has been used in the literature of universal algebra for $\forall\bar{x}\Phi \models \forall\bar{x}\varphi$. We want to emphasize that these two uses are not the same; they coincide only for closed first-order formulas (sentences). For example, if x, y, z, w are distinct variables then $\forall x, y, z, w(x = y) \models \forall x, y, z, w(z = w)$, but $x = y \not\models z = w$; of course we also have $x = y \nvdash_{FOL} z = w$. Actually, it can be shown that $x = y \models z = w$ iff $\models \forall x, y, z, w(x = y \Rightarrow z = w)$; more generally, $\Phi \models \varphi$ iff $\models \forall\bar{x}(\mathsf{AND}\ \Phi \Rightarrow \varphi)$, if Φ is finite (where $\mathsf{AND}\ \Phi$ denotes the conjunction of

[9] In this context $\Rightarrow$ denotes the implication connective of first-order logic and not the separator of Gentzen sequents.

the formulas in Φ). The notion of $\Phi \models \varphi$ adopted in this book is exactly the one used in first order logic (this is why we write also $\Phi \models_{FOL} \varphi$). We will use $\Phi \models_{Eq} \varphi$ for $\forall \bar{x}\Phi \models \forall \bar{x}\varphi$.

Birkhoff [Bir35] proved that if $\Phi \cup \{\varphi\}$ is a set of equations, then $\Phi \models_{Eq} \varphi$ iff $\Phi \vdash_{Eq} \varphi$. Here FIC is the set of rules, which is introduced in the previous subsection. In other words $\models_{Eq}$ and $\vdash_{Eq}$ are the same consequence relation, which is called *equational logic*. It is often stated that equational logic is a fragment of first order logic, but this should not be taken to mean that if $\Phi \cup \{\varphi\}$ is a set of equations, then $\Phi \models_{Eq} \varphi$ iff $\Phi \models_{FOL} \varphi$, but rather that $\Phi \models_{Eq} \varphi$ iff $\forall \bar{x}\Phi \models_{FOL} \forall \bar{x}\varphi$. This confusion is caused by the dual life of an identity as an equation and as its universal closure. Another reason for the confusion is our writing $\mathbf{A} \models \varphi$ for $\mathbf{A} \models \forall \bar{x}\varphi$. Note that $\Phi \models_{Eq} \varphi$ iff for all $\mathcal{L}$-algebras $\mathbf{A}$, $\mathbf{A} \models \Phi$ implies $\mathbf{A} \models \varphi$.

For every quasiequation q of the form ε_1 and ε_2 and ... and $\varepsilon_n \Rightarrow \varepsilon_0$, we consider the inference rule

$$\frac{\varepsilon_1 \quad \varepsilon_2 \quad \ldots \quad \varepsilon_n}{\varepsilon_0} \ (q).$$

If Q is a set of quasiequations, we denote by $\vdash_Q$ the 2-dimensional consequence relation presented by Con together with the rules (q), for all $q \in Q$. The following theorem is an easy consequence of a theorem due to Selman (see Theorem 2.2.5 of [Gor98]).

THEOREM 1.66. *Let Q be a set of quasiequations. For any quasiequation q, the rule (q) is derivable in $\vdash_Q$ iff $Q \vdash_{FOL} q$.*

As a consequence, we obtain the completeness theorem for equational logic (Theorem 1.33).

COROLLARY 1.67. *For every set $E \cup \{\varepsilon\}$ of equations, $\vdash_E \varepsilon$ iff $E \vdash_{FOL} \varepsilon$.*

Exercises

(1) Prove Theorem 1.1 and Theorem 1.2.
(2) Give an example of a preordered set that is not ordered.
(3) Given a preordered set $\mathbf{Q} = (Q, \leq)$, we write $x \sim y$ if $x \leq y$ and $y \leq x$. Show that $\sim$ is an equivalence relation. We define the relation $\leq'$ on the set $Q/{\sim}$ of all equivalence classes by $X \leq' Y$ iff $x \leq y$, for some $x \in X$ and $y \in Y$. Show that $\mathbf{Q}/{\sim} = (Q/{\sim}, \leq')$ is a poset.
(4) Show that a set X in a poset $\mathbf{P}$ may have none or more than one minimal elements; consequently a minimal element need not be the minimum. Give an example where a lower bound of a set X is not a minimal element of X.
(5) Give an example of a poset such that none of its subsets, except for the singletons, have a minimum element. Also, give an example of an infinite totally-ordered poset such that none of its infinite subsets have

a minimum element. Finally, give an example of a poset such that every three-element set has a minimum but there is a two-element set that does not.

(6) Lemma 1.3 associates a poset to a lattice and a lattice to a poset in which infima and suprema of two-element subsets exist. Prove the lemma and also show that the lattice associated with the poset associated with a lattice is isomorphic to the original lattice. Conversely, show that the poset associated with the lattice associated with a poset in which infima and suprema of two-element subsets exist is isomorphic to the original poset.

(7) Draw the Hasse diagrams of all lattices with 6 or less elements.

(8) Observe that every finite lattice is complete. Give examples of lattices that are not distributive nor complete. Try to find all the 5-element non-distributive lattices.

(9) Prove that every chain is a distributive lattice.

(10) Show that every order preserving and reflecting onto map between two posets is an isomorphism. Conclude that if the map is between two lattices, it is a lattice isomorphism.

(11) Show that the free semilattice on an n-element set has $2^n - 1$ elements. Draw its Hasse diagram (assuming that it is an upper semilattice) and compare it with the diagram of Boolean algebra with n atoms.

(12) Show that the concepts of meet-irreducible and meet-prime coincide in distributive lattices.

(13) Prove the properties (1.1) for Heyting algebras. Give an example of a Heyting algebra where $\neg(x \wedge y) = \neg x \vee \neg y$ fails.

(14) Prove that the equations (1.2) provide an alternative definition for Boolean algebras.

(15) Show that a groupoid $(A, \cdot)$ is a (join) semilattice if and only if the relation $\leq$ defined by $x \leq y$ iff $xy = y$ is a partial order on A and xy is the greatest lower bound of x and y with respect to $\leq$.

(16) Prove Theorem 1.5. Then give an example of a lattice and a subposet of it that is not a sublattice.

(17) Prove Lemma 1.6, Lemma 1.7 and Theorem 1.8. [Hint: We denote the composition of two class operators O_1 and O_2 by $\mathsf{O}_1\mathsf{O}_2$ and we write $\mathsf{O}_1 \leq \mathsf{O}_2$ if $\mathsf{O}_1(\mathcal{K}) \subseteq \mathsf{O}_2(\mathcal{K})$, for every class $\mathcal{K}$ of similar algebras. Show that the class operators H, S and IP are idempotent and that $\mathsf{SH} \leq \mathsf{HS}$, $\mathsf{PS} \leq \mathsf{SP}$ and $\mathsf{PH} \leq \mathsf{HP}$.]

(18) Let A be the set of all finite and cofinite subsets of $\mathbb{N}$. Verify that A is the universe of a Boolean algebra $\mathbf{A}$ by checking the equations (1.2). Represent $\mathbf{A}$ as a subdirect product of copies of the two-element Boolean algebra. [Hint: Represent a subset of $\mathbb{N}$ by its characteristic function.]

(19) Let $f\colon \mathbf{A} \to \mathbf{B}$ be a homomorphism and let η be a congruence on $\mathbf{A}$. Verify that $\ker_f$ is a congruence and that h_η is a homomorphism. Also, show that $\ker_{h_\eta} = \eta$ and $h_{\theta_f} = f$.

(20) Congruence permutability can be generalized as follows. Call an algebra **A** congruence 3-permutable if for any $\alpha, \beta \in \mathbf{Con\,A}$ we have $\alpha \circ \beta \circ \alpha = \beta \circ \alpha \circ \beta$. Call **A** *congruence n-permutable* if the above holds for n alternating occurrences of α and β. A class of algebras is congruence n-permutable is every algebra in the class is. Prove that lattices are not congruence n-permutable for any n.
(21) Prove Theorem 1.11.
(22) Show that groups are regular, but monoids are not even 1-regular.
(23) Given an Heyting algebra **A**, a congruence θ on **A** and a filter F of **A**, verify that $\Theta_f(F)$ is a congruence and $\mathrm{F}_c(\theta)$ is a filter. Then prove Theorem 1.22.
(24) Prove the CEP and EDPC for Heyting algebras in detail using Theorem 1.22.
(25) Prove that the law of $\wedge$-residuation for Heyting algebras can be replaced by the following three equations.
- $x \wedge (y \to x) = x$,
- $x \wedge (x \to y) = x \wedge y$,
- $x \to (y \wedge z) = (x \to y) \wedge (x \to z)$.

(26) Verify the details in the last paragraph in the proof of Lemma 1.24.
(27) Show that the systems obtained by adding 0, or $\neg$, or both (together with their rules or axioms) in **LK** lead to equivalent systems. The term equivalence is given by $0 = \neg 1$ and $\neg\varphi = \varphi \to 0$.
(28) Prove Lemma 1.27.
(29) Let $\mathbf{LK}_1$ be a sequent calculus obtained from **LK** only by restricting the initial sequents to $p \Rightarrow p$ for a propositional variable p. Show that every sequent which is provable in **LK** is provable also in $\mathbf{LK}_1$. (Note that the converse holds trivially.)
(30) Let $\mathbf{LJ}_1$ be a sequent calculus obtained from **LJ** by first deleting weakening rules and then by taking arbitrary sequents of the form $\Gamma, \alpha, \Delta \Rightarrow \alpha$ for the initial sequents. Show that for any sequent $\Sigma \Rightarrow \beta$, it is provable in **LJ** iff it is provable in $\mathbf{LJ}_1$. (This fact implies that weakening rules are admissible but neither of them are derivable in $\mathbf{LJ}_1$.)
(31) For each formula α containing neither implication $\to$ nor the constant 1, define the *dual* α^∂ of α inductively as follows: $\alpha^\partial = p$ if α is a propositional variable p, $\alpha^\partial = \neg(\beta^\partial)$ if $\alpha = \neg\beta$, $\alpha^\partial = (\beta^\partial) \wedge (\gamma^\partial)$ if $\alpha = \beta \vee \gamma$, and $\alpha^\partial = (\beta^\partial) \vee (\gamma^\partial)$ if $\alpha = \beta \wedge \gamma$. Show that if a sequent $\alpha \Rightarrow \beta$ is provable in **LK** then $\beta^\partial \Rightarrow \alpha^\partial$ is provable in **LK**. (Hint: Use the cut elimination of **LK** and observe forms of rules for logical connectives of **LK**, except rules for implication.)
(32) Let $\mathcal{K}$ be the language consisting only of $\neg, \wedge$ and $\vee$ as logical connectives. Show that for $\mathcal{K}$-formulas α and β, a sequent $\alpha \Rightarrow \beta$ is provable in **LJ** if and only if the inequality $\alpha \leq \beta$ is valid in the variety of distributive lattices.

(33) Show that if $\vdash$ is a consequence relation on a set Q and C is a closure operator on $\mathcal{P}(Q)$, then $C_\vdash$ is a closure operator and $\vdash_C$ is a consequence relation. Moreover, verify that $C_{\vdash_C} = C$ and that $\vdash_{C_\vdash} = \vdash$.

(34) A closure operator C over a set Q is called *algebraic*, if we have $C(X) = \bigcup\{C(Y) \colon Y \text{ is a finite subset of } X\}$, for all $X \subseteq Q$. Show that C is algebraic iff $\vdash_C$ is finitary.

(35) Verify that the deductive filters of a Boolean algebra $\mathbf{A}$ with respect to $\vdash_{\mathbf{HK}}$ are exactly the non-empty lattice filters of $\mathbf{A}$. Show the same for Heyting algebras and $\vdash_{\mathbf{HJ}}$.

(36) Define a presentation for a consequence relation that defines the normal subgroups of a group.

(37) It is possible to define matrices $(\mathbf{A}, F)$, where $\mathbf{A}$ is an $\mathcal{L}$-structure. In this case we expand the definition of the rule to be a pair $(T \cup \Phi, t)$, where $T \cup \{t\}$ is as before a set of (sequences of) $\mathcal{L}$-terms and Φ is a set of atomic first order $\mathcal{L}$-formulas. Use this definition to verify that the order filters and the convex subsets of a poset $\mathbf{P}$ are exactly the deductive filters of $\mathbf{P}$ with respect to the (1-dimensional) consequence relation presented respectively by the rules

$$\frac{x \quad x \le y}{y} \text{ (filter)} \qquad \frac{x \quad y \quad x \le y \le z}{y} \text{ (convex)}$$

(38) Show that $S \vdash_R s$ iff there is a proof of s from S in R.

(39) Show that if E is a set of equations, then an equation ε is derivable in Birkhoff's system from E ($\vdash_E \varepsilon$) iff $E \vdash_{Eq} \varepsilon$.

Notes

(1) We have observed that the set of propositional formulas over a first order language $\mathcal{L}$ coincides with the terms of $\mathcal{L}$. Thus, the first order logic over $\mathcal{L}$ serves as the main underlying logic of the algebraic study (that allows for equations and quasiequations), but is not directly related to the propositional logic we discuss (since the connectives of the latter are the function symbols of the former). Traditionally, in propositional logic (classical and non-classical) the language of the first order metalogic is never explicitly stated. In this case we consider a propositional language; i.e., a set of connectives (with their associated arity). Although we could introduce two types of languages (first order and propositional), we do not do so in order to avoid confusion.

(2) In Abstract Algebraic Logic (AAL) a logic is simply a (substitution invariant) consequence relation over formulas. As a result a logic in the context of AAL is *not* determined by a set of formulas (its theorems). On the other hand, we have defined superintuitionistic logics $\mathbf{L}$ as sets of formulas closed under substitution and $\vdash_{\mathbf{HJ}}$. Nevertheless, $\mathbf{L}$ and its deducibility relation $\vdash_{\mathbf{L}}$ are interdefinable, since we have fixed the basic consequence relation $\vdash_{\mathbf{HJ}}$. Diverging superficially from the

terminology of AAL we are using the term logic for a set of formulas bearing in mind the close connection between **L** and $\vdash_{\mathbf{L}}$. We will do the same in the next chapter by fixing a different consequence relation and talking about logics *over* that consequence relation (or the system specifying it).

(3) We have identified propositional formulas with first order terms and propositional connectives with functional symbols of our first order language. This approach has the advantage that we consider only one type of language: a first-order language. An alternative approach would be to define a *propositional language* that consists of propositional variables and propositional connectives and distinguish it from a first-order language. We did not chose this option to avoid the confusion of having two different kinds of language.

(4) We end with a brief list of standard references for some of the material covered in this chapter. Extensive background information and proofs of results that we have stated without proof can generally be found in the appropriate book listed here. For lattices and order: [DP02], [Grä98], [Bly05], and also [Fuc63] and [Bir79]; for universal algebra: [BS81], [MMT87] and [Gor98]; for abstract algebraic logic: [FJP03]; for Boolean algebras: [Kop89]; for introduction to logic: [vD04]; for proof theory: [TS00], and more advanced proof theory [Sch77] and [Tak87]; for intuitionistic logic: [vD02] and [TvD88]; for model theory: [CK90], [Hod93] and [Hod97].

CHAPTER 2

Substructural logics and residuated lattices

Substructural logics are generalizations of classical and of intuitionistic logic obtained by removing some structural rules from the corresponding sequent calculi and adding further axioms. We will define a sequent calculus **FL**, a Hilbert-style system **HFL** and a class of algebraic models FL, and prove that they determine three equivalent consequence relations; actually, we will show that the first two relations $\vdash_{\mathbf{FL}}$ and $\vdash_{\mathbf{HFL}}$ are equal. In particular, the equivalence of the syntactic relation $\vdash_{\mathbf{FL}}$ or $\vdash_{\mathbf{HFL}}$ to the semantic relation $\models_{\mathsf{FL}}$ is what is known as *algebraization*. Substructural logics (over **FL**) will be then defined as axiomatic extensions of (the theorems of) $\vdash_{\mathbf{FL}}$ and, similar to the case of superintuitionistic logics, they will be shown to form a lattice dually isomorphic to the subvariety lattice of FL.

After defining the logic **FL**, its algebraic models, closely related to residuated lattices, and the notion of a substructural logic (over **FL**), we give an extensive list of examples, consisting of logics that have been studied independently before the realization that they are simply special cases of a general concept. These logics include relevant, linear, superintuitionistic, many-valued and fuzzy logics. At the same time, we discuss their algebraic semantics, paving the way for the algebraization of substructural logics.

In the previous chapter we proved that $\vdash_{\mathbf{HK}} \varphi$ implies $\vdash_{\mathbf{LK}} \varphi$ implies $\models_{\mathsf{BA}} \varphi$ implies $\vdash_{\mathbf{HK}} \varphi$, for every formula φ. In particular, the last implication was obtained using what is known as the *Lindenbaum construction*; we factor the absolutely free algebra (of formulas) by the fully invariant congruence given by mutual provability in $\vdash_{\mathbf{HK}} \varphi$ (equivalently, by the provability of the bi-implication of the two formulas). The deductive form of the above series of equivalences, i.e, $\Gamma \vdash_{\mathbf{HK}} \varphi$ iff $\Gamma \vdash_{\mathbf{LK}} \varphi$ iff $\{\psi = 1 : \psi \in \Gamma\} \models_{\mathsf{BA}} \varphi = 1$, for every formula φ and set of formulas Γ, as well as the compactness and deduction theorems (see Corollary 1.55 and Theorem 1.39) then follow from the above result.

We will see that only a restricted version of the deduction theorem (called parametrized local deduction theorem) holds for $\vdash_{\mathbf{FL}}$, so we will have to follow a different approach than in Chapter 1. We will show the equivalence of $\vdash_{\mathbf{FL}}$ and $\vdash_{\mathbf{HFL}}$ directly using a syntactic argument. For the algebraization, we will also need to handle the deductive case directly. One could in principle use the Lindenbaum algebra and factor it further by the congruence determined by a theory (generated by a deductively closed set of formulas) of $\vdash_{\mathbf{FL}}$. But the structure theory of the algebras in FL, needed for understanding congruence generation, will only come in the next chapter, so we cannot make use of it at this point. Instead, we will use the strong completeness of FL with respect to a system related to substitution-free equational logic and a syntactic argument to connect the latter to $\vdash_{\mathbf{HFL}}$, obtaining in this way the crucial implication for the algebraization.

2.1. Sequent calculi and substructural logics

In this section we arrive at the sequent calculus **FL** by removing the structural rules from in **LJ** and we define substructural logics as axiomatic extensions of **FL**. Among those we distinguish a few *basic* substructural logics, obtained by adding some of the structural rules back to **FL**.

The system **FL** will contain the identity axiom and the cut rule (written in an appropriate form). Anticipating the algebraic semantics we mention that the presence of this axiom and rule suggest that $\Rightarrow$ behaves like a preorder. We have that $\alpha \Rightarrow \alpha$ is provable, by the identity axiom, and that if both $\alpha \Rightarrow \beta$ and $\beta \Rightarrow \gamma$ are provable, then $\alpha \Rightarrow \gamma$ is provable, by the cut rule. Antisymmetry does not follow, but as in the Lindenbaum-Tarski construction, we identify logically equivalent formulas, namely we identify α and β if both $\alpha \Rightarrow \beta$ and $\beta \Rightarrow \alpha$ are provable. Consequently, we expect the semantics to be based on posets. The content of the cut rule is more than just transitivity, but we will discuss this later in conjunction with the algebraic interpretation of the other rules.

2.1.1. Structural rules. In Chapter 2, while discussing Gentzen systems, we mentioned the classification of inference rules into *logical* (introducing connectives), *structural* (all others, except cut) and the cut rule, which is a category of its own. While there is practically no controversy about admitting the logical rules or the cut rule, the structural rules can be easily challenged, if we wish to analyze a notion of inference which is looser than the deductive or mathematical one.

Probabilistic reasoning typically does not obey weakening. If Kowalski lives in Canberra, then Kowalski is probably Australian. However, if Kowalski lives in Canberra, grew up in a small town in Poland, and learned logic at the Jagiellonian University in Cracow, he is probably not Australian. Situations involving finitary resources typically do not obey contraction. One five thousand yen note and another can buy you a train ticket from Kanazawa to Osaka. But, sadly enough, you cannot use the *only* five thousand yen note in your wallet *twice* to achieve the same result. Certain everyday uses of language do not obey exchange. 'A candle was burning on the table and the room was brightly lit' is not the same as 'The room was brightly lit and a candle was burning on the table'. Likewise, 'I put on my socks and I put on my shoes' is different from 'I put on my shoes and I put on my socks'.

To analyze the function of the structural rules we consider how antecedents of sequents, i.e., the left-hand sides of sequents, are controlled by *left* structural rules in the case of **LJ**.

1) Exchange rule (e $\Rightarrow$): (also denoted by (e))

$$\frac{\Gamma, \alpha, \beta, \Delta \Rightarrow \varphi}{\Gamma, \beta, \alpha, \Delta \Rightarrow \varphi} \text{ (e)}$$

By the exchange rule (e), we can use antecedents in an arbitrary order.

2) Contraction rule (c $\Rightarrow$): (also denoted by (c))

$$\frac{\Gamma, \alpha, \alpha, \Delta \Rightarrow \varphi}{\Gamma, \alpha, \Delta \Rightarrow \varphi} \text{ (c)}$$

By the contraction rule, we can use each (occurrence of an) antecedent multiple times. Loosely speaking, we can manufacture spare copies of assumptions as we go up the proof tree. We do not have that possibility in a calculus without contraction. There, when a sequent $\Sigma \Rightarrow \Theta$ is proved, each (occurrence of an) antecedent in Σ is used *at most once* in the proof.

3) Left weakening rule (w $\Rightarrow$): (also denoted by (i))

$$\frac{\Gamma, \Delta \Rightarrow \varphi}{\Gamma, \alpha, \Delta \Rightarrow \varphi} \text{ (i)}$$

By the left weakening rule, we can add any redundant formula as an antecedent. Thus, when a sequent is proved in a calculus with no weakening rule, each (occurrence of) an antecedent is used *at least once* in its proof.

Thus, in sequent calculi with all three of the above left structural rules a sequent $\alpha_1, \ldots, \alpha_m \Rightarrow \beta$ is provable iff β *follows from* $\alpha_1, \ldots, \alpha_m$ by using antecedents $\alpha_1, \ldots, \alpha_m$ in arbitrary order and an arbitrary number of times (possibly none). To make our list of structural rules complete, we need to mention the following right structural rule.

4) Right weakening rule ($\Rightarrow$ w): (also denoted by (o))

$$\frac{\Gamma \Rightarrow}{\Gamma \Rightarrow \alpha} \text{ (o)}$$

These considerations have led to the notion of *substructural* logics, namely logics which, when formulated as Gentzen calculi, lack some of the structural rules. The above analysis of left structural rules tells us that substructural logics are *sensitive* to the number and order of occurrences of assumptions, or, as it is sometimes put, they are *resource-sensitive*. As a special case, let us consider a sequent calculus having only the exchange rule, and having neither the weakening rules nor the contraction rule. In such a case, every assumption must be used once and only once to derive a conclusion. This is a basic feature of *linear logic* introduced by Girard [Gir87a], which is obtained from **LK** by deleting both weakening rules and contraction rule. *Relevant logics* form another important class of substructural logics with a long history. Relevant logicians do not allow irrelevant assumptions and thus require that every assumption should be used in derivations. This means that weakening rules are rejected in relevant logics. We will have more on linear and relevant logics shortly.

We mentioned above that the provability of a sequent $\alpha_1, \ldots, \alpha_m \Rightarrow \beta$ in **LJ** encodes the fact that β *follows* (in the sense of **Int**) from $\alpha_1, \ldots, \alpha_m$, where the antecedents $\alpha_1, \ldots, \alpha_m$ can be used in arbitrary order and an arbitrary number of times (possibly none). This is precisely why we can identify the derivability of β from $\alpha_1, \ldots, \alpha_m$ in **Int** ($\alpha_1, \ldots, \alpha_m \vdash_{\mathbf{LJ}} \beta$) by the provability of the sequent $\alpha_1, \ldots, \alpha_m \Rightarrow \beta$ in **LJ** and is the reason behind the validity of the deduction theorem for $\vdash_{\mathbf{LJ}}$. Now, in other substructural logics the multiplicity or the order of $\alpha_1, \ldots, \alpha_m$ matters. For example, the derivation $\alpha, \beta \vdash \beta$ is not valid *internally* in relevant logic and $\alpha, \alpha \vdash \alpha$ is not valid *internally* in linear logic.

Here 'not valid internally in a logic' means that a being whose *metalogic* is not classical or intuitionistic, but linear or relevant would reject such a derivation. An important consequence of this is that the *internal* relation in a proper substructural logic is not a consequence relation, since the left-hand side for $\vdash$ is not a set and moreover the conditions of a consequence relation fail (see the example for relevant logic above). Therefore, the internal relation of such a logic cannot possibly be equal to the derivability relation (usually given by a Hilbert system) of such a logic, since the latter one is a consequence relation. A sequent calculus **C** for a particular substructural logic **L** (we will define substructural logics in the next sections) is designed in such a way that a sequent $\alpha_1, \ldots, \alpha_m \Rightarrow \beta$ is provable in the calculus iff β follows (in the specific sense dictated by the logic **L**) from $\alpha_1, \ldots, \alpha_m$, namely iff $\alpha_1, \ldots, \alpha_m \vdash^{int}_{\mathbf{L}} \beta$ is valid *internally* in the logic. Actually, $\vdash_{\mathbf{C}} \alpha_1, \ldots, \alpha_m \Rightarrow \beta$ is usually taken as the definition of the internal relation $\alpha_1, \ldots, \alpha_m \vdash^{int}_{\mathbf{C}} \beta$ of a sequent calculus; see page 50. In other words, the separator $\Rightarrow$ in sequents (in the sense of the provability of the sequent) represents *in a logical level* the *metalogical* internal relation $\vdash^{int}_{\mathbf{L}}$ of a substructural logic **L**. The fact that we will use the symbol $\vdash$ for (the consequence relation denoting) derivations between *sequents* (in order to reason *about* substructural logics) should not create the impression that substructural logics *internally* (namely their internal relations) satisfy the conditions for a consequence relation.

Finally, we mention that substructural logics satisfy *internally* the deduction theorem, namely their *internal* relations, although not consequence relations, satisfy the following. If **L** is a (commutative) substructural logic, then

$$\alpha, \Gamma \vdash_{\mathbf{L}}^{int} \beta \text{ iff } \Gamma \vdash_{\mathbf{L}}^{int} \alpha \to \beta$$

that is $\vdash_{\mathbf{C}} \alpha, \Gamma \Rightarrow \beta$ iff $\vdash_{\mathbf{C}} \Gamma \Rightarrow \alpha \to \beta$, where **C** is a sequent calculus for **L**; in the algebraic models this corresponds to the residuation property.

Before going into further mathematical details of substructural logics, we continue with some informal considerations on the effects of the lack of structural rules.

2.1.2. Comma, fusion and implication. A typical use of structural rules is demonstrated in the proof of the distributive law in Chapter 2.3, where both contraction and weakening are used in an essential way. In fact, the left weakening rule is used to derive a sequent of the form $\alpha, \beta \Rightarrow \alpha \wedge \beta$. On the other hand, the contraction rule is used to derive a sequent of the form $\alpha \wedge \beta \Rightarrow \gamma$ from a sequent $\alpha, \beta \Rightarrow \gamma$. Thus, we can expect that in sequent calculi having both left weakening and contraction rules, $\alpha, \beta \Rightarrow \gamma$ is provable if and only if $\alpha \wedge \beta \Rightarrow \gamma$ is provable and is shown for **LK** in Lemma 1.40. More precisely, in **LK** *commas in the left-hand side of a sequent correspond to conjunctions*, and *commas in the right-hand side correspond to disjunctions*. The former holds also in **LJ**, but neither holds in sequent calculi that lack some structural rules. To be precise, commas on the left can still be interpreted as conjunctions if the calculus lacks only (o), in all other cases commas are not interpretable this way (note that (i) + (c) imply (e); see Exercise 3). Then, it is natural to ask the following questions.

(1) What do commas mean in a sequent calculus lacking (left) weakening or contraction or both?
(2) In general, how and in which respects does the presence of structural rules affect logical properties?

To answer the first question, let us introduce a new logical connective called *fusion* $\cdot$, which will end up representing commas[1]. In sequent calculi for sequents with single succedents, we take the following rules for fusion.

$$\frac{\Gamma, \alpha, \beta, \Delta \Rightarrow \varphi}{\Gamma, \alpha \cdot \beta, \Delta \Rightarrow \varphi}\ (\cdot \Rightarrow) \qquad \frac{\Gamma \Rightarrow \alpha \quad \Sigma \Rightarrow \beta}{\Gamma, \Sigma \Rightarrow \alpha \cdot \beta}\ (\Rightarrow \cdot)$$

Then, $\alpha \cdot \beta \Rightarrow \varphi$ follows from $\alpha, \beta \Rightarrow \varphi$, by $(\cdot \Rightarrow)$. On the other hand, by $(\Rightarrow \cdot)$ we can show that $\alpha, \beta \Rightarrow \alpha \cdot \beta$ is provable. So, if $\alpha \cdot \beta \Rightarrow \varphi$ is provable, then by applying cut to $\alpha, \beta \Rightarrow \alpha \cdot \beta$ and $\alpha \cdot \beta \Rightarrow \varphi$ we get $\alpha, \beta \Rightarrow \varphi$. Thus, in general we have the following, which says that commas can be represented faithfully by fusion.

[1]The connective $\cdot$ is sometimes called the *multiplicative conjunction* and is expressed by the symbol $\otimes$. See [Gir87a] and also [Tro92]. In this case, to distinguish the usual conjunction $\wedge$ from the multiplicative one, the former is called the *additive* conjunction.

THEOREM 2.1. *In any sequent calculus having at least rules for $\cdot$ and the cut rule, the sequent $\alpha_1, \ldots, \alpha_m \Rightarrow \beta$ is provable if and only if $\alpha_1 \cdot \cdots \cdot \alpha_m \Rightarrow \beta$ is provable.*

The addition of fusion in the language and of the two rules in the system will not produce any new provable sequents among the ones that do not involve fusion. This is usually expressed by saying that fusion can be added *conservatively*.

We now turn to the algebraic interpretation of sequents in sequent calculi lacking some structural rules. Recall that a sequent $\alpha_1, \ldots, \alpha_m \Rightarrow \beta$ of **LJ** is interpreted as the equation $(\alpha_1 \wedge \ldots \wedge \alpha_m) \to \beta = 1$, or equivalently the inequation $\alpha_1 \wedge \ldots \wedge \alpha_m \leq \beta$ in Heyting algebras. Rules for fusion and Theorem 2.1 suggest the possibility of introducing algebraic structures with a semigroup operation (also denoted by $\cdot$) in order to give an algebraic interpretation to the logical connective *fusion*. By interpreting the arrow $\Rightarrow$ in sequents by inequality $\leq$ as we did before, a sequent $\alpha_1, \ldots, \alpha_m \Rightarrow \beta$ can be now understood as an inequation $\alpha_1 \cdot \ldots \cdot \alpha_m \leq \beta$.

To model fusion we add a binary operation to the intended poset semantics, that is assumed to be associative in order to capture the associativity of comma, implicit in sequences. The rules for fusion automatically force the assumption of monotonicity, which states that $\alpha \leq \gamma$ and $\beta \leq \delta$ imply $\alpha \cdot \beta \leq \gamma \cdot \delta$. Here $\leq$ denotes the order relation in the poset. Actually, the content of the cut rule that goes beyond transitivity is subsumed by monotonicity. In this way, we are naturally led to partially ordered semigroups (or even lattice-ordered semigroups if

ordinary conjunctions and disjunctions are taken into account),[2] in order to give algebraic interpretations of sequents in sequent calculi for substructural logics. We still need to give the algebraic interpretation of sequents with empty antecedent or succedent. We will do so later by introducing the constants 1 and 0.

Having defined the algebraic interpretation $\varepsilon(s)$ of a sequent s of the form $\alpha_1, \ldots, \alpha_m \Rightarrow \beta$ (for the moment s has non-empty antecedent and succedent) as the equation $\alpha_1 \cdot \ldots \cdot \alpha_m \leq \beta$, we define the interpretation $\varepsilon(r)$ of an inference rule of the form

$$\frac{s_1 \quad s_2 \qquad s_n}{s}\ (r)$$

as the quasi-equation

$$\varepsilon(s_1) \text{ and } \varepsilon(s_2) \text{ and } \ldots \text{ and } \varepsilon(s_n) \Rightarrow \varepsilon(s).$$

What is the algebraic meaning of each structural rule, then? Obviously, the algebraic interpretation of the exchange rule is equivalent (as a

[2]Partially ordered semigroups will be defined later as posets with an additional monotone and associative binary operation. In lattice-ordered semigroups the poset is a lattice.

quasiequation), relative to the theory of partially ordered semigroups, to the commutativity of the semigroup operation; see Exercise 1. Likewise, the contraction rule means that $\alpha \cdot \alpha \leq \varphi$ implies $\alpha \leq \varphi$, which can be restated as the property of being *square increasing*, i.e., $\alpha \leq \alpha \cdot \alpha$. Also, the left weakening rule ($w \Rightarrow$) means that $\gamma \cdot \alpha \leq \gamma$ and $\alpha \cdot \gamma \leq \gamma$.

Assuming the left weakening rule corresponds to adding $\alpha \cdot \beta \leq \alpha$ and $\alpha \cdot \beta \leq \beta$, for all α and β, to the theory of partially ordered semigroups, and hence $\alpha \cdot \beta$ is a lower bound of α and β. Moreover, when we assume the contraction rule, $\gamma \leq \alpha$ and $\gamma \leq \beta$ imply $\gamma \leq \gamma \cdot \gamma \leq \alpha \cdot \beta$. Therefore, $\alpha \cdot \beta$ is equal to the greatest lower bound of α and β and hence is equal to $\alpha \wedge \beta$, if we have both the left weakening and contraction rules. This explains why commas in the left-hand side of a sequent of **LJ** can be interpreted as conjunctions.

In sequent calculi without the exchange rule, we introduce two implication-like connectives $\backslash$ and $/$, which we call *left division* and *right division*[3]. We have the following rules for left and right division:

$$\frac{\alpha, \Gamma \Rightarrow \beta}{\Gamma \Rightarrow \alpha\backslash\beta}\ (\Rightarrow \backslash) \qquad\qquad \frac{\Gamma \Rightarrow \alpha \quad \Delta, \beta, \Sigma \Rightarrow \varphi}{\Delta, \Gamma, \alpha\backslash\beta, \Sigma \Rightarrow \varphi}\ (\backslash \Rightarrow)$$

$$\frac{\Gamma, \alpha \Rightarrow \beta}{\Gamma \Rightarrow \beta/\alpha}\ (\Rightarrow /) \qquad\qquad \frac{\Gamma \Rightarrow \alpha \quad \Delta, \beta, \Sigma \Rightarrow \varphi}{\Delta, \beta/\alpha, \Gamma, \Sigma \Rightarrow \varphi}\ (/ \Rightarrow)$$

The lack of symmetry in the statement of the rules ($\Rightarrow \backslash$) and ($\Rightarrow /$) justifies the need for both the connectives. Notice that asymmetry disappears, if exchange is present. Using these rules, it is easy to show that the sequents $\alpha, \alpha\backslash\beta \Rightarrow \beta$ and $\beta/\alpha, \alpha \Rightarrow \beta$ are provable. We say that a formula φ is *provably equivalent* to another formula ψ in a given sequent calculus, when the sequents $\varphi \Rightarrow \psi$ and $\psi \Rightarrow \varphi$ are provable in it.

Now, let us consider any sequent calculus having the cut rule, rules for fusion and rules for left and right division (as well as the identity axiom). Using Theorem 2.1, we can show that these three new connectives form a trio linked by the following *residuation equivalences*.

LEMMA 2.2. *The sequent $\alpha \cdot \beta \Rightarrow \gamma$ is provable iff $\alpha \Rightarrow \gamma/\beta$ is provable iff $\beta \Rightarrow \alpha\backslash\gamma$ is provable. Furthermore, these three sequents are even pairwise mutually derivable.*

We say that two sequents s_1 and s_2 are *mutually derivable* in a consequence relation $\vdash$ ($s_1 \dashv\vdash s_2$, in symbol) if $s_1 \vdash s_2$ and $s_2 \vdash s_1$.

[3]Left division is sometimes called *right residual* and right division *left residual*. This somewhat confusing terminology is due to the fact that, algebraically, we have $x \cdot y \leq z$ iff $y \leq x\backslash z$ (left division), but also $x\backslash z$ is the largest element y such that x multiplied *on the right* by y is smaller than z. So $x\backslash z$ is the right residual of x with respect to z. To make things worse, left (right) residual has also been used for left (right) division.

LEMMA 2.3. *The formula $\alpha\backslash\beta$ is provably equivalent to β/α for all α and β iff the exchange rule is derivable iff $\gamma\cdot\delta$ is provably equivalent to $\delta\cdot\gamma$ for all γ and δ.*

PROOF. Assume first that $\alpha\backslash\beta$ is provably equivalent to β/α for all α and β. Since $\delta,\gamma\Rightarrow\delta\cdot\gamma$ is provable, $\gamma\Rightarrow\delta\backslash(\delta\cdot\gamma)$ is provable by $(\Rightarrow\backslash)$ and so is $\gamma\Rightarrow(\delta\cdot\gamma)/\delta$ by the assumption and cut. Thus, $\gamma\cdot\delta\Rightarrow\delta\cdot\gamma$ is provable by the cut rule. Conversely, suppose that the exchange rule is derivable. Then, since $\alpha,\alpha\backslash\beta\Rightarrow\beta$ is provable, $\alpha\backslash\beta,\alpha\Rightarrow\beta$ is also provable. Therefore, $\alpha\backslash\beta\Rightarrow\beta/\alpha$ is provable.

The equivalence to the derivability of the exchange rule is left as an exercise. □

Symbols $\rightarrow$ and $\leftarrow$ are sometimes used for $\backslash$ and $/$, respectively. But as shown in Lemma 2.3 $\alpha\backslash\beta$ and β/α are provably equivalent, when the exchange rule is derivable. Thus, we reserve the symbol $\rightarrow$ for implication in that case. We intend it as a mnemonic, too: whenever $\rightarrow$ occurs, we assume (e) holds. Lemma 2.2 together with Theorem 2.1 show that deletion or addition of structural rules has a significant effect on logical properties of implication via fusion.

The algebraic meaning of the rules for left and right division should be clear. As we have already mentioned, lattice-ordered semigroups are employed for giving an algebraic interpretation of substructural logics. We need more when implication-like operations are considered. Our semigroups must then be *residuated*, which means that operations $\backslash$ and $/$ (called the left division and right division) are defined and the following conditions are stipulated for all x, y and z (cf. Lemma 2.2):

$$x\cdot y\leq z \text{ iff } y\leq x\backslash z \text{ iff } x\leq z/y.$$

In Chapter 1 (see page 43) we mentioned the logical constants 1 and 0 for the true and the false proposition, respectively. The constant 0 was used to define negation. In sequent calculi **LK** and **LJ**, the following initial sequents are assumed.

(C1) $\Rightarrow 1 \qquad 0\Rightarrow$

With the help of the weakening rules, the following sequents are provable:

$$\Gamma\Rightarrow 1 \qquad \Gamma,0,\Delta\Rightarrow\varphi$$

which means that every formula implies 1 and is implied by 0, respectively. But, *we do not assume these when a sequent calculus under consideration lacks the weakening rules.*

Then, what should we assume in general for the constants 1 and 0? The initial sequents in (C1) say that 1 is provable and 0 is contradictory. Our answer to this question is to assume that 1 is the *weakest* among provable formulas, and 0 is the *strongest* among contradictory formulas. That is, the constant 1 implies any provable formula, and the constant 0 is implied by any contradictory formula, i.e., a formula φ such that $\varphi \Rightarrow$ is provable. More precisely, we assume the initial sequents (C1) and the following rules of inference for 1 and 0:

$$\frac{\Gamma, \Delta \Rightarrow \varphi}{\Gamma, 1, \Delta \Rightarrow \varphi} \text{ (1w)} \qquad\qquad \frac{\Gamma \Rightarrow}{\Gamma \Rightarrow 0} \text{ (0w)}$$

Note that the constants 1 and 0 behave exactly as if they were 'an empty antecedent' and 'an empty succedent', respectively. Thus, the algebraic rendering of the sequent $\Rightarrow \beta$ is given by $1 \leq \beta$, and that of $\alpha_1, \ldots, \alpha_m \Rightarrow$ by $\alpha_1 \cdot \cdots \cdot \alpha_m \leq 0$, respectively. So, introducing constants 1 and 0 is indispensable if we want to capture algebraically the entire sequent calculus.

Since in general we have two implication-like connectives $\backslash$ and $/$, it is natural to introduce also two negation-like connectives $\sim$ and $-$ by putting ${\sim}\alpha = \alpha\backslash 0$ and $-\alpha = 0/\alpha$, respectively. When the exchange rule is assumed, they are provably equivalent and both of them are denoted by $\neg\alpha$.

Sometimes it is also convenient to add propositional constants for the *strongest true* and the *weakest false* propositions, i.e., propositions for which the pair of sequents following (C1) hold. To distinguish these propositions from 1 and 0, we use symbols $\top$ and $\bot$ and assume the following initial sequents:

(C2) $\qquad \Gamma \Rightarrow \top \qquad\qquad \Gamma, \bot, \Delta \Rightarrow \varphi$

Note that $\top$ is provably equivalent to $\sim \bot$ and also to $- \bot$, and that 0 is provably equivalent to both of ${\sim}1$ and -1[4]. Using left (right) weakening rule, we can show that $\top$ ($\bot$) is provably equivalent to 1 (0). Conversely, if $\top$ is equal to 1, then the left weakening rule (i) follows from the initial sequent for $\top$, the rule (1w) and the cut rule. Similarly, the right weakening rule follows from the assumption that $\bot$ is equal to 0.

We can show that for any formula α, both $\alpha \cdot 1$ and $1 \cdot \alpha$ are provably equivalent to α. In algebraic terms, this means that 1, viewed as a constant operation, is the unit element with respect to the semigroup operation. We will show in Section 2.6 that the algebraic models, a notion that we will make precise later, for substructural logics are *residuated lattice-ordered monoids* with an additional constant 0. These, called *FL-algebras* later on, will be the basic algebraic structures of this book. At times we will also consider richer signatures, with $\top$ and $\bot$ satisfying the above initial sequents. The corresponding algebraic structures are *bounded* FL-algebras and they have a least element $\bot$ and a greatest element $\top$. These algebraic structures are defined and discussed in a more systematic way in Section 2.2.

[4]A warning note: The reader should be aware of the usage of symbols for propositional constants in the literature, as the same symbols sometimes denote different constants. For example, linear logic has all four constants, but usually $\bot$ and 0 are swapped with respect to our usage. See e.g. Table 2.1, [Gir87a] and [Tro92].

2.1.3. Sequent calculus for the substructural logic FL. We will now formally define the sequent calculus **FL** (for Full Lambek Calculus) that will form the basis for all other logics considered in the book; we will use **FL** for both the system and the set of provable formulas in the system. Accordingly, we will call a logic *substructural* over **FL** if it is an *axiomatic* extension of **FL**. For more details, see the next section. The language of **FL** consists of constants 0, 1 and binary connectives $\wedge$, $\vee$, $\cdot$, $\backslash$ and $/$. Two unary connectives of left and right negation can be defined by putting: ${\sim}\alpha = \alpha\backslash 0$ and $-\alpha = 0/\alpha$.

By a *sequent* (of **FL**) we mean a list of the form $\alpha_1, \ldots, \alpha_m \Rightarrow \beta$, where $\alpha_1, \ldots, \alpha_m$ are formulas, $m \geq 0$, and β is a formula or the empty sequence. The system **FL** consists of two initial sequents on constants in addition to the initial 'identity' sequents, and rules for logical connectives and the cut rule. As before, rules of **FL** are specified by metarules, in which lower case letters denote formulas, and upper case letters, (possibly empty) sequences of formulas.

Initial sequents:

$$\Rightarrow 1 \qquad 0 \Rightarrow \qquad \alpha \Rightarrow \alpha.$$

Cut rule:

$$\frac{\Gamma \Rightarrow \alpha \quad \Sigma, \alpha, \Xi \Rightarrow \varphi}{\Sigma, \Gamma, \Xi \Rightarrow \varphi} \;(\text{cut})$$

Rules for logical connectives:

$$\frac{\Gamma, \Delta \Rightarrow \varphi}{\Gamma, 1, \Delta \Rightarrow \varphi} \;(1\text{w}) \qquad \frac{\Gamma \Rightarrow}{\Gamma \Rightarrow 0} \;(0\text{w})$$

$$\frac{\Gamma, \alpha, \Delta \Rightarrow \varphi \quad \Gamma, \beta, \Delta \Rightarrow \varphi}{\Gamma, \alpha \vee \beta, \Delta \Rightarrow \varphi} \;(\vee \Rightarrow)$$

$$\frac{\Gamma \Rightarrow \alpha}{\Gamma \Rightarrow \alpha \vee \beta} \;(\Rightarrow \vee) \qquad \frac{\Gamma \Rightarrow \beta}{\Gamma \Rightarrow \alpha \vee \beta} \;(\Rightarrow \vee)$$

$$\frac{\Gamma, \alpha, \Delta \Rightarrow \varphi}{\Gamma, \alpha \wedge \beta, \Delta \Rightarrow \varphi} \;(\wedge \Rightarrow) \qquad \frac{\Gamma, \beta, \Delta \Rightarrow \varphi}{\Gamma, \alpha \wedge \beta, \Delta \Rightarrow \varphi} \;(\wedge \Rightarrow)$$

$$\frac{\Gamma \Rightarrow \alpha \quad \Gamma \Rightarrow \beta}{\Gamma \Rightarrow \alpha \wedge \beta} \;(\Rightarrow \wedge)$$

$$\frac{\Gamma, \alpha, \beta, \Delta \Rightarrow \varphi}{\Gamma, \alpha \cdot \beta, \Delta \Rightarrow \varphi}\ (\cdot \Rightarrow) \qquad \frac{\Gamma \Rightarrow \alpha \quad \Delta \Rightarrow \beta}{\Gamma, \Delta \Rightarrow \alpha \cdot \beta}\ (\Rightarrow \cdot)$$

$$\frac{\Gamma \Rightarrow \alpha \quad \Xi, \beta, \Delta \Rightarrow \varphi}{\Xi, \Gamma, \alpha \backslash \beta, \Delta \Rightarrow \varphi}\ (\backslash \Rightarrow) \qquad \frac{\alpha, \Gamma \Rightarrow \beta}{\Gamma \Rightarrow \alpha \backslash \beta}\ (\Rightarrow \backslash)$$

$$\frac{\Gamma \Rightarrow \alpha \quad \Xi, \beta, \Delta \Rightarrow \varphi}{\Xi, \beta / \alpha, \Gamma, \Delta \Rightarrow \varphi}\ (/ \Rightarrow) \qquad \frac{\Gamma, \alpha \Rightarrow \beta}{\Gamma \Rightarrow \beta / \alpha}\ (\Rightarrow /)$$

The notion of a proof (from assumptions) in **FL** is defined as for the systems **LK** and **LJ** and is a special case of the general definition in Section 1.6.3, keeping in mind that every metarule of **FL** gives rise to infinitely many rules. If there is a proof in **FL** of a sequent s from assumptions S, then we write $S \vdash^{seq}_{\mathbf{FL}} s$.

Provability of a given sequent in **FL** is also defined in the same way as that in **LK** or **LJ**. Namely, we say that the sequent $\Gamma \Rightarrow \alpha$ is *provable* in **FL** if it can be obtained from initial sequents of **FL** by repeated applications of the rules of **FL**; i.e., if $\vdash^{seq}_{\mathbf{FL}} \Gamma \Rightarrow \alpha$. A formula α is *provable* in **FL**, if the sequent $\Rightarrow \alpha$ is provable in **FL**.

Logical connectives of **FL** are divided into two groups according to the form of the rules involving the connective. For connectives in the first group, the lower sequent of any of the corresponding rules has always the same *environment* or *context*, i.e., the same side formulas, as the upper sequent(s). The second group contains the remaining connectives. Connectives $\cdot, \backslash$ and $/$ are in the first group, because of the form of the rules $(\Rightarrow \cdot)$, $(\backslash \Rightarrow)$ and $(/ \Rightarrow)$. Sometimes, logical connectives of the second group are called *multiplicative* connectives, and those of the first group are called *additive*.

Some extensions of **FL** can be defined naturally by adding combinations of structural rules, i.e., exchange, contraction and the left and right weakening rules to the set of rules of **FL**. Typically, the left and right weakening rules are clustered together and referred to simply as weakening. Clearly, the rules (1w) and (0w) become redundant in presence of weakening.

We will refer to these calculi by a mnemonic devised in [Ono93] and [Ono90]. Let (e), (c), (i), (o) and (w) stand, respectively, for exchange, contraction, left-weakening, right-weakening and weakening.

$$\frac{\Gamma, \alpha, \beta, \Delta \Rightarrow \varphi}{\Gamma, \beta, \alpha, \Delta \Rightarrow \varphi}\ (\mathrm{e}) \quad \frac{\Gamma, \alpha, \alpha, \Delta \Rightarrow \varphi}{\Gamma, \alpha, \Delta \Rightarrow \varphi}\ (\mathrm{c}) \quad \frac{\Gamma, \Delta \Rightarrow \varphi}{\Gamma, \alpha, \Delta \Rightarrow \varphi}\ (\mathrm{i}) \quad \frac{\Gamma \Rightarrow}{\Gamma \Rightarrow \alpha}\ (\mathrm{o})$$

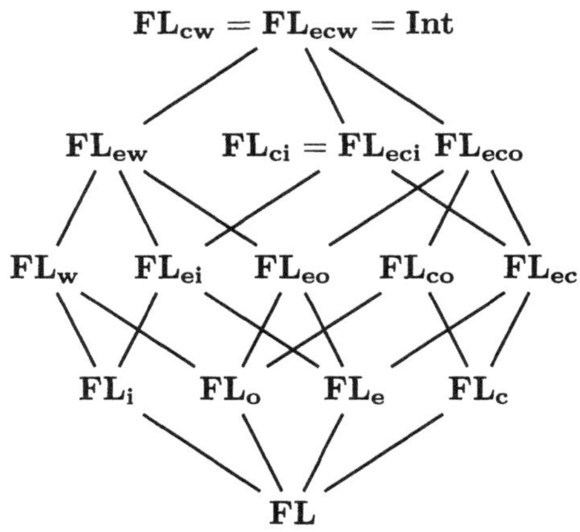

FIGURE 2.1. The join-semilattice of logics generated by $\mathbf{FL_i}$, $\mathbf{FL_o}$, $\mathbf{FL_e}$, $\mathbf{FL_c}$.

Let S be any subset of $\{e, c, i, o\}$. Then $\mathbf{FL}_S$ will stand for the extension of **FL** obtained by adding the structural rules from S with (i) + (o) abbreviated to (w). For instance, $\mathbf{FL_{ew}}$ will be the Gentzen system with exchange and weakening but without contraction. This system will be of some importance toward the end of the book, since its axiomatic extensions contain many important nonclassical logics and thus many interesting results have been obtained on them (see Section 2.3). We would like to add that several algebraic notions of general importance acquire a fairly tractable form in the variety of algebras corresponding to $\mathbf{FL_{ew}}$.

The systems $\mathbf{FL_S}$, for $S \subseteq \{e, c, i, o\}$, will often be our landmarks, because of their naturality as Gentzen calculi. We will call them *basic substructural systems* and call the associated logics, i.e., the set of theorems (provable formulas) in each system, *basic substructural logics*. In general we will use the same symbol for the Gentzen systems and their sets of theorems, with two traditional exceptions; namely, **LK** and **LJ** will always stand for Gentzen calculi whose sets of theorems are, respectively, **Cl** (classical logic) and **Int** (intuitionistic logic). At places (e.g., in Chapter 11), we will use superscripts and similar devices to distinguish between Gentzen calculi having the same set of theorems.

Basic substructural logics form a subsemilattice of the lattice of *logics over* **FL**, which is discussed in the next section. Figure 2.1 shows the join semilattice that they form. Recall that exchange is derivable from contraction and left-weakening; see Exercise 3 again. The corresponding algebraic statement is mentioned in Exercise 11. Thus, the join semilattice in question is generated by elements (e), (c), (i), (o) subject to the relation (e) + (c) + (i) = (c) + (i).

As we have mentioned already, there is some ambiguity as to whether we include constants $\top$ and $\bot$ in the language of substructural logics. To make the distinction clear, basic sequent calculi in the language with $\bot$, $\top$ and initial sequents for them, are indicated by adding $\bot$ as the subscript. For example, $\mathbf{FL}_\bot$ is the sequent calculus obtained from **FL** in this way, with both $\bot$ and $\top$ present.

For the corresponding algebraic models the difference between having and not having $\bot$ and $\top$ seems to be more substantial. For instance, we show in Chapter 4 that each lattice-ordered group is an unbounded FL-algebra, unless it is trivial. We will come back to the question of constants later on, see Section 3.6.7, where it will be demonstrated that the difference is in some respects not as big as it appears even from an algebraic point of view.

2.1.4. Deducibility and substructural logics over FL. Like in intuitionistic and classical logic, we can define logics over **FL** and their deducibility relations.

We say that a formula ψ is *deducible* or *provable* from a set of formulas Γ in **FL** (in symbols, $\Gamma \vdash_{\mathbf{FL}} \psi$), when the sequent $\Rightarrow \psi$ is provable in the calculus obtained from **FL** by adding all sequents $\Rightarrow \delta$ for $\delta \in \Gamma$ as initial sequents; i.e., if $\{\ \Rightarrow \delta : \delta \in \Gamma\} \vdash^{seq}_{\mathbf{FL}}\ \Rightarrow \psi$. Such a proof of the sequent $\Rightarrow \psi$ is also called a *deduction* of $\Rightarrow \psi$ in **FL** from Γ. The relation $\vdash_{\mathbf{FL}}$ is a finitary and substitution invariant consequence relation called the *external consequence relation* of **FL**; the terminology is due to [Avr88].

If the sequent $\varphi_1, \ldots, \varphi_k \Rightarrow \psi$ is provable in **FL**, then $\{\varphi_1, \ldots, \varphi_k\} \vdash_{\mathbf{FL}} \psi$ holds. Although the converse holds for **LJ** (and **LK**), it does not hold for **FL**. Still, it is immediate from the definition that $\vdash_{\mathbf{FL}} \psi$ holds if and only if $\Rightarrow \psi$ is provable in **FL**.

For a set of formulas Γ, we define $\Sigma_{\mathcal{L}}[\Gamma]$ to be the closure of Γ under substitution in the language $\mathcal{L}$. The *axiomatic extension* of the system **FL** by Γ, denoted by $\mathbf{FL} + \Gamma$, is the calculus obtained from **FL** by adding $\Rightarrow \varphi$ as initial sequents for all formulas $\varphi \in \Sigma_{\mathcal{L}}[\Gamma]$. We would like also to denote by $\mathbf{FL} + \Gamma$ the smallest set of formulas containing Γ that is closed under both substitution and the consequence relation $\vdash_{\mathbf{FL}}$ and call that the axiomatic extension of **FL**, viewed as a set of formulas. It turns out that these two definitions are reconciled by the following lemma, which is similar to Lemma 1.56. The lemma claims that substitutions can be performed first, therefore the set of formulas $\mathbf{FL} + \Gamma$ is the set of theorems of the system $\mathbf{FL} + \Gamma$. The dual meanings of both $\mathbf{FL} + \Gamma$ parallels the use of **FL** as a sequent systems and a sets of theorems. Although the lemma can be proved directly, it will also be an easy consequence of Theorem 2.14 and Corollary 2.17, proved later.

LEMMA 2.4. *If* $\Gamma \cup \{\psi\}$ *is a set of formulas, then* $\mathbf{FL}+\Gamma \vdash_{\mathbf{FL}} \psi$ *iff* $\Sigma_{\mathcal{L}}[\Gamma] \vdash_{\mathbf{FL}} \psi$.

With the above in mind, we define a *substructural logic* over **FL** as an axiomatic extension of **FL** (by some set of axiom schemes). As this will be our default notion of substructural logic in the rest of the book, a few cautionary notes are in order[5]. Substructural logics as defined above are sets of formulas closed under substitution and the consequence relation $\vdash_{\mathbf{FL}}$. This is at variance with the usage common in *abstract algebraic logic*, where a logic is defined as a consequence relation (or operation). An even tighter notion identifies a logic with a *calculus*, or a *presentation* of a logic (as a Gentzen-style sequent system, or as a Hilbert system, for example). So we are standing on a three-runged ladder. The lower rung is *logic as a calculus* (or system). In this sense the Hilbert system **HK** and Gentzen's **LK** are different logics. Indeed, **LK** and cut-free **LK** are different, too. This notion is perhaps most suitable for investigating proof systems or axiom systems. The middle rung is *logic as a consequence relation.* In this sense, all the three examples above are the same logic, but for instance relevant logic **R** (see Section 2.3) and its extension by the rule

$$\frac{\neg\varphi \vee \psi \quad \varphi}{\psi}$$

known as (γ), are different. This, as we already mentioned, is the usual notion in abstract algebraic logic because consequence relations are crucial to the notion of *algebraization.* We will see more of algebraization shortly. The upper rung is *logic as a set of theorems.* In this sense not only is **HK** the same as **LK**, but **R** and **R** augmented by (γ) are the same as well. This is because adding (γ) to the usual Hilbert system for **R** does not change the set of provable formulas. Recall from Section 1.6.3 that rules with this property (relative to a given logic) are called *admissible* (in that logic). Another example is the rule

$$\frac{\neg\varphi \to (\psi \vee \delta)}{(\neg\varphi \to \psi) \vee (\neg\varphi \to \delta)}$$

known as Harrop rule, which is admissible but not derivable in intuitionistic logic; see [Har60].

It will be clear from Lemma 2.5 below that our stance enables us to view basic substructural logics (see page 86) as logics over **FL** in our present technical sense. Sometimes we will call a substructural logic over **FL** simply a substructural logic, or even a logic. When **L** is the smallest substructural logic over **FL** containing a set of formulas Γ, or equivalently it is the axiomatic extension with the set of axiom schemes Γ, $\mathbf{L} = \mathbf{FL} + \Gamma$ is said to

[5]See also the endnotes in this and the previous chapter for a discussion on the use of the term 'logic' in literature.

be *axiomatized* by Γ. Moreover, it is said to be *finitely axiomatizable* over **FL**, if it is axiomatized by a finite set of axiom schemes.

LEMMA 2.5. *Let S be a subset of* $\{e, c, i, o\}$. *Then,* $\mathbf{FL}_S$ *is finitely axiomatizable over* **FL**.

PROOF. It is enough to show that for each S there exists a corresponding finite set Γ_S of formulas such that for any formula φ, φ is provable in the sequent calculus $\mathbf{FL}_S$ if and only if φ is deducible from the set of substitution instances of formulas in Γ_S. The required correspondences are given in the table below:

rule	axiom	alt. axiom
(e)	$(\alpha \cdot \beta)\backslash(\beta \cdot \alpha)$	$[\alpha\backslash(\beta\backslash\delta)]\backslash[\beta\backslash(\alpha\backslash\delta)]$
(c)	$\alpha\backslash(\alpha \cdot \alpha)$	$[\alpha\backslash(\alpha\backslash\beta)]\backslash(\alpha\backslash\beta)$
(i)	$\alpha\backslash 1$	
(o)	$0\backslash\alpha$	
(w)	$\alpha\backslash 1,\ 0\backslash\alpha$	

It is an informative exercise to work this out in some detail. □

With each substructural logic **L** we associate the *deducibility relation* $\vdash_{\mathbf{L}}$ of **L** defined by $\Gamma \vdash_{\mathbf{L}} \varphi$ if and only if $\mathbf{L} \cup \Gamma \vdash_{\mathbf{FL}} \varphi$, for any set of formulas $\Gamma \cup \{\varphi\}$.[6] Clearly, the relation $\vdash_{\mathbf{L}}$ is again a finitary substitution invariant consequence relation. Note that the relation $\vdash_{\mathbf{L}}$ where **L** is **FL** defined here is exactly the same as the relation $\vdash_{\mathbf{FL}}$ defined above, and hence no ambiguities occur. Clearly, **L** and $\vdash_{\mathbf{L}}$ are interdefinable, as $\mathbf{L} = Thm(\vdash_{\mathbf{L}})$. Obviously, $\vdash_{\mathbf{L}} \varphi$ is equivalent to $\varphi \in \mathbf{L}$ for any formula φ and any substructural logic **L**.

We can easily extend the notion of substructural logics over **FL** to those over **L**, for any given substructural logic **L** over **FL**. That is, **K** is a *substructural logic over* **L** iff it is a set of formulas closed under both substitution and the deducibility relation $\vdash_{\mathbf{L}}$. Since any set of formulas closed under $\vdash_{\mathbf{L}}$ is also closed under $\vdash_{\mathbf{FL}}$, substructural logics over a given logic **L** are exactly the same as substructural logics over **FL** containing **L**. We denote by $\mathbf{L} + \Gamma$ the axiomatic extension of **L** by the set of formulas Γ, namely the least logic that contains $\mathbf{L} \cup \Gamma$. Clearly, substructural logics over **L** are just the axiomatic extensions of **L**.

Substructural logics over $\mathbf{FL_w}$ will also be called substructural logics *with weakening*. Likewise we define *commutative*, *contractive*, *integral*, and *zero bounded* substructural logics, as logics over $\mathbf{FL_e}$, $\mathbf{FL_c}$, $\mathbf{FL_i}$ and $\mathbf{FL_o}$, respectively.

A substructural logic is called *involutive*, if the formulas $-{\sim}\varphi \leftrightarrow \varphi$ and ${\sim}{-}\varphi \leftrightarrow \varphi$ are provable; here $\varphi \leftrightarrow \psi$ can be taken to be $(\varphi\backslash\psi) \wedge (\psi\backslash\varphi) \wedge 1$, or

[6]Note that $\mathbf{L} + \Gamma \vdash_{\mathbf{FL}} \varphi$, define below, means a different thing, as formulas from Γ would be taken as axioms, not assumptions, and their substitutions would be allowed.

its variant without $\wedge 1$. We consider now sequent calculi for *involutive basic substructural logics.* A typical example of such a basic substructural logic is **Cl** which has **LK** as sequent calculus. Sequents of **LK** are expressions of the form $\Gamma \Rightarrow \Delta$ where both Γ and Δ are finite sequences of formulas (separated by commas). Analogously, we can introduce a sequent calculus $\mathbf{InFL_e}$ for the involutive commutative substructural logic, using these sequents, as follows. Note that here we can replace the connectives $\backslash$ and $/$ by the connective $\to$, due to the exchange rule.

- $\mathbf{InFL_e}$ has the same initial sequents as those of **FL**,
- it has the same exchange rules $(e \Rightarrow)$ and $(\Rightarrow e)$, and cut rule as **LK** (see Subsection 1.3.2),
- it has the same rules for logical connectives as those of **LK**, except $(\neg \Rightarrow)$ and $(\Rightarrow \neg)$,
- in addition, it has the following rules for constants and for fusion:

$$\frac{\Gamma, \Delta \Rightarrow \Pi}{\Gamma, 1, \Delta \Rightarrow \Pi}\ (1\mathrm{w}) \qquad \frac{\Gamma \Rightarrow \Pi, \Sigma}{\Gamma \Rightarrow \Pi, 0, \Sigma}\ (0\mathrm{w})$$

$$\frac{\Gamma, \alpha, \beta, \Delta \Rightarrow \Pi}{\Gamma, \alpha \cdot \beta, \Delta \Rightarrow \Pi}\ (\cdot \Rightarrow) \qquad \frac{\Gamma \Rightarrow \alpha, \Pi \quad \Delta \Rightarrow \beta, \Sigma}{\Gamma, \Delta \Rightarrow \alpha \cdot \beta, \Pi, \Sigma}\ (\Rightarrow \cdot)$$

Similarly, we can introduce $\mathbf{InFL_{ec}}$ (and $\mathbf{InFL_{ew}}$) by adding $(c \Rightarrow)$ and $(\Rightarrow c)$ ($(w \Rightarrow)$ and $(\Rightarrow w)$, respectively) to the calculus $\mathbf{InFL_e}$. In the non-commutative case, it is not obvious how to arrange side formulas properly on the right-hand side of sequents, in $(\Rightarrow \cdot)$ for example. In Chapter 7, we will introduce the sequent calculus **InFL** for an involutive extension of **FL**. The system **InFL** will be in some respects simpler and in other respects more complicated than $\mathbf{InFL_e}$, but it will address the the above problem.

By our definition, a substructural logic over **FL** is a set of formulas closed under both substitution and the consequence relation $\vdash_{\mathbf{FL}}$. In Corollary 2.17 we show that closure under $\vdash_{\mathbf{FL}}$ can be replaced by the conditions given below.

A set of formulas **L** is a substructural logic over **FL** iff it satisfies the following conditions;

- **L** includes all formulas in **FL**,
- if $\varphi, \varphi\backslash\psi \in \mathbf{L}$ then $\psi \in \mathbf{L}$,
- if $\varphi \in \mathbf{L}$ then $\varphi \wedge 1 \in \mathbf{L}$,
- if $\varphi \in \mathbf{L}$ and ψ is an arbitrary formula then $\psi\backslash\varphi\psi, \psi\varphi/\psi \in \mathbf{L}$,
- **L** is closed under substitution.

We can replace the third condition above by the following:

- if $\varphi, \psi \in \mathbf{L}$ then $\varphi \wedge \psi \in \mathbf{L}$.

Obviously, this implies the third condition above. On the other hand, since the formula $(\varphi \wedge 1)\backslash((\psi \wedge 1)\backslash(\varphi \wedge \psi))$ is in **L**, we have the converse by using of the second condition. Note that when **L** is a substructural logic over $\mathbf{FL_e}$ the fourth condition is redundant, and moreover when it is over $\mathbf{FL_{ei}}$ also the third condition is redundant.

Recall the definition of the $\mathcal{K}$-fragment of a sequent calculus or Hilbert system **S**. We will apply the same convention about a substructural logic **L** to mean the restriction of its theorems to $\mathcal{K}$-formulas, and to the consequence relation $\vdash_{\mathbf{S}}$ to mean the restriction obtained by considering only $\mathcal{K}$-formulas in **S**. If $\mathcal{K}$ contains all the connectives of **S** except for 0 and negation(s), then the $\mathcal{K}$-fragment of **S** will be denoted by $\mathbf{S}^+$ and will be called the *positive fragment* of **S**.

2.2. Residuated lattices and FL-algebras

We saw in the previous chapter that Boolean algebras are algebraic models for classical propositional logic and that Heyting algebras are models for intuitionistic logic. In particular, we have seen that these algebras are the *equivalent algebraic semantics* in the sense of Blok and Pigozzi for the corresponding logics; i.e., we have *strong* completeness and *inverse* strong completeness of the models: for every set of formulas $\Phi \cup \{\varphi, \psi\}$,

- $\Phi \vdash_{\mathbf{HK}} \psi$ iff $\{1 = \varphi \mid \varphi \in \Phi\} \models_{\mathsf{BA}} 1 = \psi$, and
- $\varphi = \psi \mathrel{=\!\!|\!\models}_{\mathsf{BA}} 1 = \varphi \leftrightarrow \psi$

Usual completeness (soundness and adequacy) is just the first statement with Φ equal to the empty set. *Strong* completeness refers to the presence of assumptions Φ. Note that the above two conditions imply also that for all sets $E \cup \{\varphi = \psi\}$ of equations

- $E \models_{\mathsf{BA}} \varphi = \psi$ iff $\{\varphi' \leftrightarrow \psi' \mid \varphi' = \psi' \in E\} \vdash_{\mathbf{HK}} \varphi \leftrightarrow \psi$

and, by the second condition, the composition of the transition from an equation $\varphi = \psi$ to the formula $\varphi \leftrightarrow \psi$ and from a formula φ to the equation $\varphi = 1$ produces an equation $\varphi \leftrightarrow \psi = 1$ equivalent to the original one $\varphi = \psi$ in the setting of Boolean algebras. *Inverse* refers to the third bullet and the fact that the translations are mutually inverse.

In other words the first condition states that $\models_{\mathsf{BA}}$ is at least as expressive as $\vdash_{\mathbf{HK}}$ (every deduction is expressible by, or understood in terms of, equational consequence), but it is conceivable that there is a consequence between equations given in terms of $\models_{\mathsf{BA}}$ that is not expressible by $\vdash_{\mathbf{HK}}$. The third condition states that this does not happen; the two relations are equally expressive. The same situation holds between **HJ** and Heyting algebras.

It is worth noting that the right-hand side of the first condition is simply a quasi-equation if Φ is finite. Because of the deduction theorem for **HK** and for **HJ** (and the fact that quasiequations are equivalent to certain equations in BA and in HA), the content of the direct and inverse completeness is the same as its strong version. We will see that the deduction theorem does not hold for most substructural logics, including **FL**. Therefore, in the setting of substructural logics the strong version of completeness is more informative than the weak one.

We will now define the class FL of algebras that will end up forming the equivalent algebraic semantics for substructural logics. The exact description of algebraizability will not be in the same terms as for Boolean and for Heyting algebras (the equation $\varphi = 1$ will be replaced by $\varphi \wedge 1 = 1$ and the biconditional $\leftrightarrow$ will be defined in a different way), but still the syntactic ($\vdash_{\mathbf{FL}}$) and the semantic ($\models_{\mathsf{FL}}$, the equational consequence relative to FL) relations will be equally expressive.

An algebra $\mathbf{A} = (A, \wedge, \vee, \cdot, \backslash, /, 1, 0)$ is called a *full Lambek algebra* or an *FL-algebra*, if

- $(A, \wedge, \vee)$ is a lattice (i.e., $\wedge$, $\vee$ are commutative, associative and mutually absorptive),
- $(A, \cdot, 1)$ is a monoid (i.e., $\cdot$ is associative, with unit element 1),
- $x \cdot y \leq z$ iff $y \leq x\backslash z$ iff $x \leq z/y$, for all $x, y, z \in A$,
- 0 is an arbitrary element of A.

Residuated lattices are exactly the 0-free reducts of FL-algebras. So, for an FL-algebra $\mathbf{A} = (A, \wedge, \vee, \cdot, \backslash, /, 1, 0)$, the algebra $\mathbf{A}_r = (A, \wedge, \vee, \cdot, \backslash, /, 1)$ is a residuated lattice and 0 is an arbitrary element of A. The maps $\backslash$ and $/$ are called the *left* and *right division*. We read $x\backslash y$ as 'x under y' and y/x as 'y over x'; in both expressions y is said to be the *numerator* and x the *denominator*. We will show that FL-algebras are the equivalent algebraic semantics for $\vdash_{\mathbf{FL}}$ (whence the name) in Section 2.6, give some examples of FL-algebras and residuated lattices in Section 2.3 and a comprehensive list of examples in Section 3.4, after we establish some more terminology and notation. We denote by RL and FL the classes of residuated lattices and FL-algebras, respectively. Residuated lattices were first introduced in [WD39] under a restricted definition; the concept arose naturally in the study of the lattice of ideals of a (commutative) ring (with unit); see Section 3.4.9 for more on this example.

We adopt the usual convention of writing xy for $x \cdot y$, and in the absence of parentheses, we assume that multiplication is performed first followed by the division operations and the lattice operations. So, for example, $x/yz \wedge u\backslash v$ simplifies $[x/(yz)] \wedge (u\backslash v)$. We also define $x^0 = 1$ and $x^{n+1} = x^n \cdot x$.

Recall that we denote by $Fm_{\mathcal{L}}$ the set of terms over the language $\mathcal{L}$ of FL-algebras. By $t \leq s$ we denote both the equality $t = t \wedge s$, if t, s are elements of a residuated lattice or an FL-algebra, and the equation $t = t \wedge s$, if t, s are terms. It is easy to see that in an FL-algebra the equality $s = t$ is equivalent to the inequality $1 \leq s\backslash t \wedge t\backslash s$.

The *opposite* t^{op} of a term t in the language of residuated lattices or FL-algebras, is defined inductively on the complexity of t. We define $1^{\mathrm{op}} = 1$, $0^{\mathrm{op}} = 0$, $x^{\mathrm{op}} = x$, if x is a variable, and $(s \cdot t)^{\mathrm{op}} = t \cdot s$, $(s\backslash t)^{\mathrm{op}} = t/s$, $(t/s)^{\mathrm{op}} = s\backslash t$, $(s \wedge t)^{\mathrm{op}} = t \wedge s$, and $(s \vee t)^{\mathrm{op}} = t \vee s$, for terms t, s. Intuitively, the opposite of a term t is the same term read backwards or its 'mirror image'. We extend the definition to equations by $(s = t)^{\mathrm{op}} = (t^{\mathrm{op}} = s^{\mathrm{op}})$, and to metalogical statements in the obvious way. Note that $(s \leq t)^{\mathrm{op}} = (t^{\mathrm{op}} \geq s^{\mathrm{op}})$. Examples of mutually opposite equations can be seen in each statement of the following lemma.

The mirror image symmetry in the axiomatization makes a simple *mirror image principle* hold. Namely, in the language of residuated lattices

the inequality $t \leq s$ holds if and only if $t^{\mathrm{op}} \leq s^{\mathrm{op}}$ holds.

This obviously carries over to equalities and other metalogical statements, hence it suffices to state results in only one form. As an example, consider the inequality $x(y/z) \leq xy/z$, which holds in any residuated monoid; see Lemma 2.6(5). Then by taking opposites, we have another inequality $(z\backslash y)x \leq z\backslash yx$, which holds also in any residuated monoid by our mirror image principle.

The mirror image principle is justified as follows. If $\mathbf{A}$ is a residuated lattice with multiplication $\cdot$ and division operations $\backslash$ and $/$, we define a new operation $*$ on A by $x * y = y \cdot x$ for all x, y. Then, the law of residuation states that the original right-division $/$ (left-division $\backslash$) is the left-division (the right-division, respectively) with respect to $*$; i.e., $x * y \leq z$ iff $y \leq z/x$ iff $x \leq y\backslash z$. The algebra $\mathbf{A}^{\mathrm{op}}$ that has multiplication $*$ and the division operations interchanged is, consequently, also a residuated lattice, which we call the *opposite* of $\mathbf{A}$. It is clear that if an equation ε fails in a residuated lattice $\mathbf{A}$, then its opposite equation $\varepsilon^{\mathrm{op}}$ fails in the opposite residuated lattice $\mathbf{A}^{\mathrm{op}}$.

The following lemma contains some basic properties of residuated lattices and FL-algebras. All of them are easy consequences of the definition and they will be shown in the next chapter in a more general setting. The reader is asked to verify them as an easy exercise.

LEMMA 2.6. [BT03] [GO06a] *The following identities hold in all residuated lattices and all FL-algebras.*

(1) $x(y \vee z) = xy \vee xz$ *and* $(y \vee z)x = yx \vee zx$;
(2) $x\backslash(y \wedge z) = (x\backslash y) \wedge (x\backslash z)$ *and* $(y \wedge z)/x = (y/x) \wedge (z/x)$;
(3) $x/(y \vee z) = (x/y) \wedge (x/z)$ *and* $(y \vee z)\backslash x = (y\backslash x) \wedge (z\backslash x)$;
(4) $(x/y)y \leq x$ *and* $y(y\backslash x) \leq x$;
(5) $x(y/z) \leq (xy)/z$ *and* $(z\backslash y)x \leq z\backslash(yx)$;
(6) $(x/y)/z = x/(zy)$ *and* $z\backslash(y\backslash x) = (yz)\backslash x$;
(7) $x\backslash(y/z) = (x\backslash y)/z$;
(8) $x/1 = x = 1\backslash x$;
(9) $1 \leq x/x$ *and* $1 \leq x\backslash x$;

(10) $x \leq y/(x\backslash y)$ *and* $x \leq (y/x)\backslash y$;
(11) $y/((y/x)\backslash y) = y/x$ *and* $(y/(x\backslash y))\backslash y = x\backslash y$;
(12) $x/(x\backslash x) = x$ *and* $(x/x)\backslash x = x$;
(13) $(z/y)(y/x) \leq z/x$ *and* $(x\backslash y)(y\backslash z) \leq x\backslash z$.

Moreover, multiplication is order preserving in both coordinates; each division operation is order preserving in the numerator and order reversing in the denominator.

Actually, multiplication distributes over arbitrary existing joins. We would like to point out that residuated lattices or FL-algebras need not be bounded. Nevertheless, if a residuated lattice has a lower bound $\bot$, then $\bot \leq \bot/x$, so $x \cdot \bot \leq \bot$. Consequently, $x \cdot \bot = \bot$ and likewise $\bot \cdot x = \bot$, for all x. Hence $x \leq \bot\backslash\bot$ and $x \leq \bot/\bot$, for all x, that is the algebra has a greatest element $\bot\backslash\bot = \bot/\bot$. Therefore, every lower-bounded residuated monoid or FL-algebra is actually bounded.

The following theorem says that in the presence of lattice operations the law of residuation can be captured by equations. The proof of the statement is left as an exercise.

THEOREM 2.7. [BT03] [JT02] *An algebra (of the appropriate type) is a residuated lattice or an FL-algebra if and only if it satisfies the equations defining lattices, the equations defining monoids, and the following four equations:*

(1) $x(x\backslash z \wedge y) \leq z$,
(2) $(y \wedge z/x)x \leq z$,
(3) $y \leq x\backslash(xy \vee z)$,
(4) $y \leq (z \vee yx)/x$.

Thus, RL *and* FL *are equational classes.*

Consequently, RL and FL are varieties. Moreover, they are congruence distributive since their algebras have lattice reducts, by Corollary 1.15, and congruence permutable with Mal'cev term (Lemma 1.24)

$$p(x, y, z) = x/(z\backslash y \wedge 1) \wedge z/(x\backslash y \wedge 1).$$

The constant 0 allows us to define two negation operations in FL-algebras. In detail, we define two unary *negation* operations, by putting $\sim x = x\backslash 0$ and $-x = 0/x$. We assume that the negation operations have priority over all other operations; for example, $-y/x$ means $(-y)/x$. Notice that $(\sim t)^{\mathrm{op}} = -t^{\mathrm{op}}$ and *vice versa*, so the mirror image of a term t is the term read backwards with the two negations interchanged.

LEMMA 2.8. [GO06b] *If* $\mathbf{A}$ *is an FL-algebra and* x, y *in* A, *then*

(1) $\sim(x \vee y) = \sim x \wedge \sim y$ *and* $-(x \vee y) = -x \wedge -y$.
(2) *If* $x \leq y$, *then* $\sim y \leq \sim x$ *and* $-y \leq -x$.
(3) $x \leq -\sim x$ *and* $x \leq \sim - x$.
(4) $\sim - \sim x = \sim x$ *and* $- \sim - x = -x$.

(5) $-{\sim}x = -{\sim}y$ *iff* ${\sim}x = {\sim}y$.
(6) ${\sim}{-}x = {\sim}{-}y$ *iff* $-x = -y$.
(7) $-{\sim}(x/y) \cdot y \leq -{\sim}x$ *and* $y \cdot {\sim}{-}(y\backslash x) \leq {\sim}{-}x$.
(8) $-y/x = -(xy)$ *and* $x\backslash{\sim}y = {\sim}(yx)$.
(9) $x\backslash -y = {\sim}x/y$.
(10) $x\backslash -y = -{\sim}x\backslash -y$ *and* ${\sim}y/x = {\sim}y/{\sim}{-}x$.
(11) $-{\sim}(x\backslash y) \leq -{\sim}(-{\sim}x\backslash -{\sim}y)$ *and*
${\sim}{-}(y/x) \leq {\sim}{-}({\sim}{-}y/{\sim}{-}x)$.
(12) $-{\sim}(-{\sim}y/x) = -{\sim}y/x$ *and* ${\sim}{-}(x\backslash{\sim}{-}y) = x\backslash{\sim}{-}y$.
(13) $xy \leq -z$ *iff* $-{\sim}x \cdot y \leq -z$. *Also,* $xy \leq {\sim}z$ *iff* $x \cdot {\sim}{-}y \leq {\sim}z$.
(14) ${\sim}1 = 0 = -1$ *and* $1 \leq {\sim}0 \wedge -0$.
(15) $xy \leq z$ *iff* $y({\sim}z) \leq {\sim}x$ *iff* $(-z)x \leq -y$.

PROOF. The first statement is a direct consequence of Lemma 2.6(3); the second statement follows from the first one. Statements (3) and (4) are consequences of statements (10) and (11) in Lemma 2.6, for $y = 0$. Moreover, (5) and (6) follow from (4). For (7), note that $(x/y)y(x\backslash 0) \leq x(x\backslash 0) \leq 0$, by (4) of Lemma 2.6. We have successively,

$$y(x\backslash 0) \leq (x/y)\backslash 0 = [0/((x/y)\backslash 0)]\backslash 0, \text{ by (4)},$$

$$[0/((x/y)\backslash 0)]y(x\backslash 0) \leq 0,$$

$$[0/((x/y)\backslash 0)]y \leq 0/(x\backslash 0), \text{ and}$$

$$-{\sim}(x/y) \cdot y \leq -{\sim}x.$$

Likewise, we prove the opposite identity. To obtain (8), note that, by (6) of Lemma 2.6, we have $-y/x = (0/y)/x = 0/xy = -(xy)$. For (9), we use (7) of Lemma 2.6 to obtain $x\backslash -y = x\backslash(0/y) = (x\backslash 0)/y = {\sim}x/y$. Using (9) and (4), we have $x\backslash -y = {\sim}x/y = {\sim}{-}{\sim}x/y = -{\sim}x\backslash -y$, so we obtain (10). By (3) and (10), $x\backslash y \leq x\backslash -{\sim}y = -{\sim}x\backslash -{\sim}y$; then, (11) follows by (2). For (12), we use (8) and (4) to obtain $-{\sim}(-{\sim}y/x) = -{\sim}{-}(x \cdot {\sim}y) = -(x \cdot {\sim}y) = -{\sim}y/x$. For (13) we have $xy \leq -z = 0/z$ iff $xyz \leq 0$ iff $yz \leq x\backslash 0 = -{\sim}x\backslash 0$ iff $-{\sim}x \cdot yz \leq 0$ iff $-{\sim}x \cdot y \leq -z$. Finally, (14) follows directly from the definition of the negation operations and (15) follows from the residuation law. □

To avoid repetition we establish the convention that any definition for residuated lattices applies to FL-algebras, as well. The same holds for definitions for FL-algebras that do not involve the constant 0.

A residuated lattice is called *commutative* if the monoid operation $\cdot$ is commutative; i.e., if it satisfies the identity $xy = yx$. If $\mathbf{A}$ is a commutative residuated lattice, then for all $x, y, z \in A$, we have $x \leq z/y$ iff $xy \leq z$ iff $yx \leq z$ iff $x \leq y\backslash z$. Thus, $z\backslash y = y/z$ for all $y, z \in A$; i.e., the two divisions collapse to one operation which we denote by $z \to y$. Thus, each commutative residuated lattice can be expressed as an algebra of the type

$(A, \wedge, \vee, \cdot, \rightarrow, 1)$. Various additional properties hold for commutative residuated lattices, like $x \rightarrow (y \rightarrow z) = y \rightarrow (x \rightarrow z)$. We denote the variety of all commutative residuated lattices by CRL. Anticipating their correspondence with $\mathbf{FL_e}$, commutative FL-algebras are also called *FL_e-algebras* and their variety is denoted by $\mathsf{FL_e}$.

A residuated lattice is called *integral* if it has a greatest element and that element is the multiplicative unit 1; i.e., $x \leq 1$, for all x. It is called *square-increasing* or *contractive* if it $x \leq x^2$, for all x. Both integrality and contraction can be captured by identities (actually, they are identities by treating $s \leq t$ as short for $s = s \wedge t$) and we denote the corresponding subvarieties of RL by IRL and KRL. Again, integral and contractive FL-algebras will also be called FL_i and FL_c-algebras, respectively; the corresponding varieties are denoted by $\mathsf{FL_i}$ and $\mathsf{FL_c}$. It is easy to see that in integral residuated lattices we have $xy \leq x \wedge y$, while in square-increasing ones we have $x \wedge y \leq xy$; actually these are equivalent formulations of the two properties in case meet is present in the type. FL-algebras that satisfy the property $x^{n+1} = x^n$, for some (fixed) natural number n, are called *n-potent*; the corresponding subvarieties are denoted by $\mathsf{P_nFL}$ and $\mathsf{P_nRL}$. In particular, we will write $\mathsf{E_n}$ for the variety $\mathsf{P_nFL_{ew}}$.

A *zero-bounded* FL-algebra or an *FL_o-algebra* is an FL-algebra that satisfies the identity $0 \leq x$; an *FL_w-algebra* is an integral and zero-bounded FL-algebra; the corresponding varieties are denoted by $\mathsf{FL_o}$ and $\mathsf{FL_w}$. We will apply combinations of the subscripts for algebras that enjoy multiple properties, so, for instance, an integral, zero-bounded and commutative FL-algebra will be called an FL_{ew}-algebra, and the variety of such algebras will be denoted by $\mathsf{FL_{ew}}$. Note that these naming conventions correspond to the ones adopted for basic substructural logics. As one may guess the above varieties will be shown to form equivalent algebraic semantics for the latter.

An FL-algebra is called *left involutive* (*right involutive*), if it satisfies the identity $-{\sim}x = x$ (${\sim}{-}x = x$, respectively). It is called *involutive*, if it is both left and right involutive; it is called *cyclic*, if it satisfies ${\sim}x = -x$. We denote the varieties of involutive FL-algebras and cyclic FL-algebras by InFL and CyFL, respectively. Note that every commutative pointed residuated lattice is cyclic.

LEMMA 2.9. *If $\mathbf{A}$ is a cyclic FL-algebra, then for all $x, y, z \in A$, we have*
(1) $xy \leq 0$ *iff* $yx \leq 0$,
(2) $xyz \leq 0$ *iff* $yzx \leq 0$ *iff* $zxy \leq 0$,
(3) $xyz \leq 0$ *iff* ${\sim}{\sim}x \cdot {\sim}{\sim}y \cdot {\sim}{\sim}z \leq 0$,
(4) $xy \leq {\sim}z$ *iff* ${\sim}{\sim}x \cdot {\sim}{\sim}y \leq {\sim}z$.

PROOF. For (1), if $xy \leq 0$, then $x \leq 0/y = y\backslash 0$, so $yx \leq 0$. Condition (2) is a direct consequence of (1).

For (3), note that by Lemma 2.8(3) ${\sim}{\sim}x \cdot {\sim}{\sim}y \cdot {\sim}{\sim}z \leq 0$ implies $xyz \leq 0$. Conversely, assume that $xyz \leq 0$. Then, we have $yz \leq x\backslash 0 = {\sim}x =$

$\sim\sim\sim x$, by Lemma 2.8(4). So, we obtain $\sim\sim x \cdot y \cdot z \leq 0$ and $y \cdot z \cdot \sim\sim x \leq 0$, by (2). By repeating the same argument twice, we obtain $\sim\sim x \cdot \sim\sim y \cdot \sim\sim z \leq 0$.

Finally, (4) follows easily from (3). □

Observe that RL is term equivalent to the subvariety of FL defined by the identity $1 = 0$. We will identify these two varieties and view RL as a subvariety of FL.

Residuated lattices or FL-algebras that can be represented as subdirect products of totally ordered algebras are called *representable* and the corresponding classes are denoted by RRL and RFL. We will see (Theorem 9.75) that these classes are actually varieties. In particular it will follow that the variety $\mathsf{RFL_{ei}}$ of representable FL_{ei}-algebras is axiomatized relative to $\mathsf{FL_{ei}}$ by

$$(x \to y) \vee (y \to x) = 1.$$

Finally, *distributive* residuated lattices or FL-algebras are simply the ones with a distributive lattice reduct; the corresponding varieties are denoted by DRL and DFL. Obviously, every representable algebra is distributive. Likewise, the varieties of *modular* residuated lattices and FL-algebras are denoted by MRL and MFL. We allow combinations of the preceding notations, so for example InCDRL denotes the variety of involutive commutative distributive residuated lattices.

The smallest non-trivial FL-algebra contains two elements: 0 and 1. The order and multiplication are the usual ones. The arrow (multiplication is commutative) is defined as in Boolean algebras. We denote by $\mathbf{2}$ this FL-algebra and by $\mathbf{2}_r$ its residuated lattice reduct. These algebras are both integral, commutative, contractive, representable; the first one is also zero-bounded, cyclic and involutive.

We also consider expansions of the signature of FL-algebras that include a constant $\bot$ that is evaluated as the least element in (lower) bounded algebras. In view of the remark after Lemma 2.6, if a residuated lattice or FL-algebra is lower bounded then it is bounded, so the inclusion of a constant for the bottom element suffices. In this expanded signature we define *bounded* FL-algebras as algebras whose lattice reduct is bounded and $\bot$ interprets the least element. The corresponding variety is denoted by $\mathsf{FL}_\bot$. The definition of bounded residuated lattices is covered by FL_o-algebras and $\mathsf{FL_o}$.

2.3. Important subclasses of substructural logics

It is fair to say that the most important classes of substructural logics ever considered fall, with very few exceptions, in the category of substructural logics over **FL**. To see that, we employ the notion of *algebraizability*[7]

[7]The reader may now have grown impatient with us mentioning algebraizability every so often and never giving a decent proof. We ask for patience. The proof from first principles will come in Section 2.6. Here we only try to bring on some intuition.

that we mentioned in the previous section, due to Blok and Pigozzi [BP89]. They define a consequence relation $\vdash$ (over the set of formulas $Fm_{\mathcal{L}}$ of some language $\mathcal{L}$) to be *algebraizable*, and a class $\mathcal{K}$ to be an *equivalent algebraic semantics* for $\vdash$ if there exists a (finite) set $\{\Delta_i : i \in I\}$ of binary definable connectives of $\mathcal{L}$ and a (finite) set $\{\delta_j = \varepsilon_j : j \in J\}$ of $\mathcal{L}$-equations in one variable such that for all sets of formulas $\Phi \cup \{\psi\}$ and equations $s = t$, we have

(1) $\Phi \vdash \psi$ iff $\{\delta_j(\varphi) = \varepsilon_j(\varphi) : j \in J, \varphi \in \Phi\} \models_{\mathcal{K}} \delta_j(\psi) = \varepsilon_j(\psi)$, for all $j \in J$, and
(2) $s = t \mathrel{=\!\!|\!\models_{\mathcal{K}}} \{\delta_j(s\Delta_i t) = \varepsilon_j(s\Delta_i t) : i \in I, j \in J\}$.

Let $\tau(\psi)$ abbreviate $\delta_j(\psi) = \varepsilon_j(\psi)$ for all $j \in J$, and $\tau[\Phi]$ abbreviate $\{\tau(\varphi) : \varphi \in \Phi\}$. Further, for an equation $s = t$ let $\rho(s = t)$ abbreviate $\{s\Delta_i t : i \in I\}$, and for a set S of equations, let $\rho[S]$ abbreviate $\cup\{\rho(u = v) : \{u = v\} \in S\}$. Notice that ρ can be viewed as a *translation from (sets of) equations to (sets of) formulas* and τ as a *translation from (sets of) formulas to (sets of) equations.* Indeed in Section 2.6 they will be defined that way. We can then restate the conditions above as follows

(1) $\Phi \vdash \psi$ iff $\tau[\Phi] \models_{\mathcal{K}} \tau(\psi)$, and
(2) $s = t \mathrel{=\!\!|\!\models_{\mathcal{K}}} \tau[\rho(s = t)]$.

Notice that every member of the set $\tau(\psi)$ is a semantic consequence of the set $\tau[\Phi]$ and likewise for $\tau[\rho(s = t)]$. Then, for every set of equations $S \cup \{s = t\}$ we have

$$
\begin{aligned}
S \models_{\mathcal{K}} s = t \text{ iff } & \tau[\rho[S]] \models_{\mathcal{K}} \tau[\rho(s = t)] && \text{by (2)} \\
\text{iff } & \rho[S] \vdash \rho(s = t) && \text{by (1)}
\end{aligned}
$$

and for every formula φ we calculate

$$
\begin{aligned}
\varphi \dashv\vdash \rho[\tau(\varphi)] \text{ iff } & \tau(\varphi) \mathrel{=\!\!|\!\models_{\mathcal{K}}} \tau[\rho[\tau(\varphi)]] && \text{by (1)} \\
\text{iff } & \tau(\varphi) \mathrel{=\!\!|\!\models_{\mathcal{K}}} \tau(\varphi) && \text{by (2)}
\end{aligned}
$$

Therefore, the following two conditions hold as well

(3) $S \models_{\mathcal{K}} s = t$ iff $\rho[S] \vdash \rho(s = t)$, and
(4) $\varphi \dashv\vdash \rho[\tau(\varphi)]$.

Thus, by (2) and (4), τ and ρ can be lifted to mutually inverse maps between sets of formulas and sets of equations. By (1) and (3), they establish a correspondence, often referred to as *strong completeness* between the syntactic consequence $\vdash$ and the semantic consequence $\models_{\mathcal{K}}$; see also the beginning of Section 2.2. More details will come in Section 2.6.

To apply the above to **FL** and FL we can use a single binary (definable) connective given by $\varphi\Delta\psi = \varphi\backslash\psi \wedge \psi\backslash\varphi$ and a single equation $1 = 1 \wedge \varphi$, so that $\rho(s = t)$ is $(s\backslash t) \wedge (t\backslash s)$ and $\tau(\varphi)$ is $1 = 1 \wedge \varphi$. Alternatively[8], $\rho(s = t)$ can be taken to be $\{s\backslash t, t\backslash s\}$. One can verify that both of these choices satisfy all the requirements for algebraizability. Thus, **FL** is algebraizable and its equivalent algebraic semantics is FL. This will be stated as Theorem 2.27 in Section 2.6, where we give a detailed account of how algebraization is effected using a more concise formulation. The existence of an equivalent algebraic semantics for a logic, essentially states that the consequence relation of the logic is equally expressible with the algebraic consequence of the semantics, and that the translations between the logic and the semantics are mutually inverse, modulo equivalence. An important property of algebraizability is that extensions of algebraizable logics are also algebraizable with the same equivalence formulas and defining equations. This will follow from Theorem 2.29. Furthermore, the lattice of axiomatic extensions of the logic is dually isomorphic to the subvariety lattice of the algebraic semantics, if the latter is a variety. From Theorem 2.29 it also follows that all fragments of logics over **FL** that contain at least conjunction and one of the divisions (implications) will be algebraizable. It is something of a curiosity that the non-algebraizability of all well-known non-algebraizable fragments of logics over **FL** can be traced down to some failure of algebraizability in the vicinity of **BCI**, to be defined shortly. Below we present eight classes of logics, each with rather different origin, most of which can be recognized as (fragments of) logics over **FL**.

2.3.1. Lambek calculus. It was some time around the 14th Century A.D. when logicians got down to analyzing (in quite extraordinary detail) the ways meanings are conveyed by expressions of natural language. One immediate observation they made was that some expressions, like *John* or *son*, or *a student*, mean something by themselves, whereas others, like *of* or *is*, mean something only together with other expressions: *son of John* or *John is a student* or a *son of John is a student*, for instance. The former they classified as *categorematic*, the latter as *syncategorematic*. Time passed. In 1930s, Ajdukiewicz observed that each word and phrase in natural language can be assigned what he called a *syntactic category* and what is now known as a *syntactic type*. Basic syntactic types include n for *noun phrase* and s for *sentence*, so 'John' has type n and 'John plays' has type s. Now 'plays' can be seen as a function that takes a noun on the left, and produces a sentence. Ajdukiewicz wrote it as $n\backslash s$. This device suffices to produce syntactic types for all words we normally care about: for example all intransitive verbs are assigned type $n\backslash s$. We can even assign types to *phrases*, or strings of words,

[8]If we wish to deal with implicational reducts, for instance, but the presence of meet in $\tau(\varphi)$ is a tougher nut. We will shortly see examples of logics (e.g. **R**) where the presence of $\wedge$ matters for algebraizability. The notion of *order-algebraizability*, i.e., algebraizability with respect to ordered algebras, overcomes this difficulty because we can replace $1 = 1 \wedge \varphi$ with $1 \leq \varphi$.

by juxtaposing the corresponding types (sometimes using commas to separate the types of the individual words): for example the phrase 'John plays' has type $n(n\backslash s)$ or $n, n\backslash s$. It *also* has type s as we have just seen, so we must allow for words and phrases to have multiple types.

It took Lambek, however, and his famous [Lam58], to realize that there is a logical calculus hidden behind syntactic types. Take again the phrase 'John plays.' As we said, it has type s and also type $n, n\backslash s$. But then *any* sentence has type s, so we would have to stipulate infinitely many assignments of types one for each sentence. It is thus desirable to have a way of deriving new types for words or phrases from existing ones. For example, we would like to have that every phrase of type $n(n\backslash s)$ has also type s: in symbols $n(n\backslash s) \Rightarrow s$.

John	plays	
n	$n\backslash s$	$n(n\backslash s) \Rightarrow s$

Also we would like 'John plays football' to have type s; here 'plays' functions as a transitive verb, namely it has type $(n\backslash s)/n$.

(John	plays)	football	
n	$(n\backslash s)/n$	n	$[n((n\backslash s)/n)]n \Rightarrow s$

One thing we may not like above is the need for parentheses around 'John plays'. It is not how things are done in natural language, although at times it might be desirable, like in 'Shakespeare plays rock' or in 'John ate the hamburger with relish'. In general, however, parentheses are undesirable. To eliminate them, we need first to assign to transitive verbs also type $n\backslash(s/n)$ and second to assume that if we parse the phrase in a different way, then we still obtain a sentence.

John	(plays	football)	$(n\backslash s)/n \Rightarrow n\backslash(s/n)$
n	$n\backslash(s/n)$	n	$n[(n\backslash(s/n))n] \Rightarrow s$

The same issue comes up when we consider adverbs of type $s\backslash s$

(John	plays)	here	
n	$n\backslash s$	$s\backslash s$	$[n(n\backslash s)](s\backslash s) \Rightarrow s(s\backslash s) \Rightarrow s$

which also needs to be be given type $(n\backslash s)\backslash(n\backslash s)$.

John	(plays	here)	
n	$n\backslash s$	$(n\backslash s)\backslash(n\backslash s)$	$s\backslash s \Rightarrow (n\backslash s)\backslash(n\backslash s)$

Clearly, we also need to assign the type s/s to adverbs for the positioning at the beginning of the sentence. Also, for placing in the middle of the sentence, like 'John *definitely* plays football', note that we need to have $s\backslash s \Rightarrow (n\backslash s)/(n\backslash s)$.

Such considerations undoubtedly clarify the syntax of natural language. What is more important to Lambek's idea though, is that all the desired 'deductions' of types can be considered as sequents and that all of them are derivable in Lambek calculus; we leave it as an exercise to the reader to verify that all of the above sequents are provable (in **FL**). So it is enough to assign type n to all nouns, type $(n\backslash s)$ to intransitive and $(n\backslash s)/n$ to transitive verbs (or both to verbs that function in both ways) and $s\backslash s$ to adverbs. Then, Lambek calculus will prove for us all the other 'derived' types that are assigned to these parts of speech and, therefore, we do not need to list them next to the entry in a dictionary.

Lambek's original calculus was non-associative in order to address the ambiguities of the non-associative syntax of natural language. The non-associative case is also important in linguistics, due partly to connections with context-free grammars (see, e.g., [Moo88] and the references in [BM02]). For more algebraically oriented studies of non-associative full Lambek calculus, see [Doš88, Doš89] and [GO]. We should also mention that in the original formulation, Lambek restricted the sequents to the form $\Gamma \Rightarrow \alpha$ with *nonempty* Γ. So, for instance $\Rightarrow \alpha\backslash\alpha$ is not a provable sequent (because it is not a sequent at all under his definition), and thus **FL** is not a conservative extension of Lambek's original calculus. However, if we allow empty left-hand sides, **FL** is a conservative extension of Lambek calculus *in that formulation.* Lambek calculus (in either formulation) is not algebraizable, but it is *order-algebraizable*[9] (i.e., algebraizable with respect to ordered algebras) and its order-algebraic semantics is the class of residuated (partially ordered) semigroups.

There are two ways of extending Lambek calculus. One is to expand the language by adding new connectives: this leads eventually to **FL**. The other is to add new axioms or rules. A mixture of the two is also possible, and in fact the most standard substructural logics considered in the literature can be obtained by extending Lambek calculus to **FL** first, and then adding some extra postulates, be they axioms or rules. An important exception to this are *implicational fragments* of various logics, an area that forms itself a chapter in nonclassical logic, not without connections to universal algebra. One important logic whose only connective is implication, is called **BCI** and can be defined as an extension of the implicational fragment of a commutative version of Lambek calculus by the exchange rule. Another one is **BCK**, extending **BCI** by adding the weakening rule.

2.3.2. BCK logic and algebras. **BCK** logic owes its cryptic name to *combinators*, which are higher order functions that apply to other functions. They were introduced in 1924 by M. Schönfinkel (who wrote only one paper on the subject) and then taken up by H. Curry in order to eliminate the need for variables in logic (cf. [CHS72]). Without going into details, the way combinators work can be explained in the following example. We want to express the fact that addition is commutative, i.e., $x + y = y + x$. Addition, viewed from a combinatory perspective, is a function, say f, that applied to another function, say x produces a function fx that itself applies to yet

[9]For a definition of order-algebraizability and its properties, see e.g., [Raf06].

another function, say y, producing $(fx)y$. Now, there is a combinator called C, whose action is $((Ca)b)c = (ac)b$. Using it we have $((Cf)x)y = (fy)x$, or, associating parentheses to the left, $Cfxy = fyx$. Commutativity of f is then expressible by $Cf = f$, with the intended meaning that an application of f to whatever it applies to results in the same thing as an application of Cf to that (in particular, $fyx = Cfxy = fxy$ in our example). Other combinators include the identity I, with action $Ia = a$, associativity B, with action $Babc = a(bc)$, first projection K, with action $Kab = a$, and contraction W, with action $Wab = abb$. Sloppy talk about 'functions applying to whatever they are applicable to' can be made precise by a notion of *type* and the familiar notation $f\colon \varphi \to \psi$, meaning that f takes an argument of type φ and returns an object (function) of type ψ. Since combinators are functions, we can calculate their types, too. Take for example K. From the definition $Kab = a$, we know that Kab has the type of a. Let it be φ. We can write that as $Kab\colon \varphi$. Now consider Ka. This is a function that applied to a function b of some type, say ψ, returns a function of type φ, so $Ka\colon \psi \to \varphi$. Finally, K itself is a function that takes an argument of type φ and returns a function of type $\psi \to \varphi$, in our notation $K\colon \varphi \to (\psi \to \varphi)$. So the type of K looks just like an implicational formula, and not just any formula, but actually a theorem of **Int**. This is an instance of the *Curry-Howard isomorphism* (for combinators) which states that there is a one to one correspondence between typed combinators and implicational formulas such that for a given type α, a combinator of that type, composed of B, C, K and W, exists if and only if α is a theorem of the implicational fragment of **Int**. So, the system with exactly these combinators is equivalent to the implicational fragment of **Int**. Notice that function application is an incarnation of *modus ponens* in that setting, for if $a\colon \varphi$ and $f\colon \varphi \to \psi$, then $fa\colon \psi$.

Weaker combinatory logics have also been considered, in particular **BCK**. As the name suggests **BCK** logic is (constructed via the Curry-Howard correspondence from) the system with combinators B, C and K, but without W. This is a form of logic without contraction. We leave it as an exercise for the reader to show that combinators correspond to the following implicational formulas:

(B) $(\varphi \to \psi) \to ((\chi \to \varphi) \to (\chi \to \psi))$,
(C) $(\varphi \to (\psi \to \chi)) \to (\psi \to (\varphi \to \psi))$,
(K) $\varphi \to (\psi \to \varphi)$,
(I) $\varphi \to \varphi$,
(W) $(\psi \to (\psi \to \varphi) \to (\psi \to \varphi)$.

The first three, together with *modus ponens* constitute the implicational logic **BCK**, which also turns out to be algebraizable; the axiom (I) follows from the others. Its equivalent quasivariety semantics is precisely the quasivariety BCK of *BCK-algebras*, i.e., algebras $(A, \to, 1)$ satisfying the following three equations and a quasiequation:

(1) $(x \to y) \to ((y \to z) \to (x \to z)) = 1$,
(2) $1 \to x = x$,
(3) $x \to 1 = 1$,
(4) $x \to y = 1$ and $y \to x = 1 \Rightarrow x = y$.

BCK was introduced by Imai and Iseki in [II66], and Wroński showed in [Wro83] that it is a proper quasivariety. The logic **BCK** is therefore a rather natural example of a deductive system whose equivalent algebraic semantics is not a variety. Idziak proved in [Idz84a] (see also [Idz84b]) that BCK-algebras are precisely implication subreducts of FL_{ew}-algebras, and that subreducts of FL_{ew}-algebras containing either meet or join form a variety. Blok and Raftery showed in [BR93] that BCK lacks the congruence extension property. The only minimal[10] variety of BCK-algebras is term equivalent to the $\{\to, 1\}$-reduct of the variety GBA of generalized Boolean algebras, that we will meet again later. It is shown in [Kow95] that it has precisely two covers in the lattice of subvarieties of BCK.

Implicational logics weaker than **BCK**, for example **BCI** that we mentioned at the end of Section 2.3.1, have also been studied. From an algebraic point of view, a disadvantage of **BCI** is that it is not algebraizable (as a sublogic of the implicational fragment, which we introduce in Section 2.3.3, of **R**), just order-algebraizable. However, both **BCI** and **BCK** have been considered as possible logical foundations for a formulation of naive set theory. An important observation in this direction was made by Grišin in [Gri82]. Considering Russell's paradox Grišin observed that although the definition $Z = \{x : x \notin x\}$ immediately yields a formula φ (namely $Z \in Z$) such that $\varphi \leftrightarrow \neg\varphi$, to derive a contradiction from that we need to reason somehow as follows. Assume φ. Then by *modus ponens* we get $\neg\varphi$ and that *together with* φ yields $\varphi \wedge \neg\varphi$. In this reasoning φ is used twice, which suggests that contraction may be involved. Another way of demonstrating that is by exhibiting a possible Gentzen-style proof of the sequent $\neg\varphi \to \varphi, \varphi \to \neg\varphi \Rightarrow$. It can go as follows.

$$
\dfrac{\dfrac{\dfrac{\dfrac{\varphi \Rightarrow \varphi \quad \dfrac{\varphi \Rightarrow \varphi}{\neg\varphi, \varphi \Rightarrow}}{\varphi \to \neg\varphi, \varphi, \varphi \Rightarrow}}{\varphi \to \neg\varphi, \varphi \Rightarrow}}{\varphi \to \neg\varphi \Rightarrow \neg\varphi} \quad \dfrac{\dfrac{\dfrac{\varphi \Rightarrow \varphi \quad \dfrac{\varphi \Rightarrow \varphi}{\neg\varphi, \varphi \Rightarrow}}{\varphi \to \neg\varphi, \varphi, \varphi \Rightarrow}}{\varphi \to \neg\varphi, \varphi \Rightarrow}}{\varphi, \varphi \to \neg\varphi \Rightarrow}}{\dfrac{\neg\varphi \to \varphi, \varphi \to \neg\varphi, \varphi \to \neg\varphi \Rightarrow}{\neg\varphi \to \varphi, \varphi \to \neg\varphi \Rightarrow}}
$$

This proof is valid for instance in **LJ** (with $\neg\varphi$ abbreviating $\varphi \to \bot$) and it contains three applications of contraction. Grišin showed that in fact *any* such proof must use contraction. Hence, the derivation of Russell's paradox in a logic without contraction is blocked.

[10]A variety is minimal if it is nontrivial, but it has no nontrivial proper subvarieties. We will investigate minimal varieties of FL-algebras in Chapter 9.

(id)	$\varphi \to \varphi$	
(wper)	$(\varphi \to ((\psi \to \chi) \to \delta)) \to ((\psi \to \chi) \to (\varphi \to \delta))$	(weak permutation)
(sf)	$(\varphi \to \psi) \to ((\psi \to \chi) \to (\varphi \to \chi))$	(suffixing)
(con)	$(\varphi \to (\varphi \to \psi)) \to (\varphi \to \psi)$	(contraction)
($\wedge\to$)	$(\varphi \wedge \psi) \to \varphi$	
($\wedge\to$)	$(\varphi \wedge \psi) \to \psi$	
($\to\wedge$)	$((\varphi \to \psi) \wedge (\varphi \to \chi)) \to (\varphi \to (\psi \wedge \chi))$	
($\to\vee$)	$\varphi \to (\varphi \vee \psi)$	
($\to\vee$)	$\psi \to (\varphi \vee \psi)$	
($\vee\to$)	$((\varphi \to \chi) \wedge (\psi \to \chi)) \to ((\varphi \vee \psi) \to \chi)$	
(d)	$(\varphi \wedge (\psi \vee \chi)) \to ((\varphi \wedge \psi) \vee (\varphi \wedge \chi))$	(distributivity)
(cp)	$(\varphi \to \neg\psi) \to (\psi \to \neg\varphi)$	(contraposition)
(dn)	$\neg\neg\varphi \to \varphi$	(double negation)

$$\frac{\varphi \quad \varphi \to \psi}{\psi}\ (\mathrm{mp}) \qquad\qquad \frac{\varphi \quad \psi}{\varphi \wedge \psi}\ (\mathrm{adj})$$

FIGURE 2.2. System **E** of entailment.

2.3.3. Relevant logics. Known also as *relevance* logics, this departure from classical logic was occasioned by the so-called paradoxes of implication. Consider the following reasoning: 'If $2 + 2 = 4$, then the fact the Moon is made of Camembert implies that $2 + 2 = 4$. But $2 + 2 = 4$. Therefore, by *modus ponens*, the fact that the Moon is made of Camembert implies that $2 + 2 = 4$'. This is a classically valid reasoning that leads to a conclusion that is counterintuitive to everyone but the trained (brainwashed, relevant logicians would say) mathematician. Short of giving up *modus ponens*, the only remaining culprit is the classical theorem $\varphi \to (\psi \to \varphi)$. Other similarly suspicious formulas include $\neg\varphi \to (\varphi \to \psi)$ and $(\varphi \to \psi) \vee (\psi \to \chi)$. The problem with all these is that antecedents are irrelevant to succedents. Relevant logicians counter this with various versions of the *variable sharing property* (or *relevance principle*) stating (in its weakest form) that an implication $\varphi \to \psi$ can only be a theorem if φ and ψ have a variable in common. We will come back to the variable sharing property in Chapter 5.

Typically, relevant logics include the distributivity axiom $(\varphi \wedge (\psi \vee \chi)) \to ((\varphi \wedge \psi) \vee (\varphi \wedge \chi))$, so, in our setting, they are extensions of $\mathbf{InDFL_{ec}}$. Of a number of relevant logics that have been extensively studied we mention three. The first is the system **E** of *entailment* of [AB75], whose Hilbert-style presentation is given in Figure 2.2. Note that certain axioms (and their labels) coincide with axioms in the Hilbert system $\mathbf{HFL_e}$ for $\mathbf{FL_e}$ to be given in Section 2.5.1.

PROPOSITION 2.10. [BP89] *The logic* **E** *is not algebraizable.*

$\to_{\mathbf{E}}$	0	$\frac{1}{2}$	1
0	1	1	1
$\frac{1}{2}$	0	1	1
1	0	0	1

	$\neg_{\mathbf{E}}$
0	1
$\frac{1}{2}$	$\frac{1}{2}$
1	0

$\to_{\mathbf{R}}$	$\bot$	f	t	$\top$
$\bot$	$\top$	$\top$	$\top$	$\top$
f	$\bot$	t	$\bot$	$\top$
t	$\bot$	f	t	$\top$
$\top$	$\bot$	$\bot$	$\bot$	$\top$

FIGURE 2.3. Tables for implication and negation.

PROOF. The proof is a variation on the one from [BP89]. Theorem 5.1 of [BP89] states (in a slightly more involved fashion) that for an algebraizable logic with equivalent quasivariety semantics $\mathcal{K}$, the lattice of deductive filters and $\mathcal{K}$-congruences are isomorphic on any algebra $\mathbf{A}$. So let $\mathbf{A}$ be the three element chain $\{0, \frac{1}{2}, 1\}$ with implication and negation defined as in Figure 2.3 (the two tables on the left). It is then not difficult to verify that A, $\{\frac{1}{2}, 1\}$ and $\{1\}$ are all the deductive filters of $\mathbf{A}$ (with respect to $\mathbf{E}$), so the lattice of deductive filters of $\mathbf{A}$ is the three element chain. But $\mathbf{A}$ is simple, so $\mathbf{Con}\,\mathbf{A}$ is the two element chain. □

Although $\mathbf{E}$ is not algebraizable, Maksimova showed that if we add the admissible rule

$$\frac{\varphi}{(\varphi \to \varphi) \to \varphi}$$

the resulting system $\mathbf{E}'$ is algebraizable. For the algebraic semantics, which are special non-associative groupoids, see [Mak71].

Notice also that $\mathbf{E}$ permits only a restricted version of permutation of the antecedents (wper), so strictly speaking we should have used the left division rather than the arrow, but we decided against it to keep the presentation closer to the original. The relevant logic $\mathbf{T}$ (of *ticket entailment*) is a logic weaker than $\mathbf{E}$ which is obtained from $\mathbf{E}$ by replacing (wper) by the following axiom, also known as (B):

$$\text{(pf)} \quad (\varphi \to \psi) \to ((\chi \to \varphi) \to (\chi \to \psi)) \quad \text{(prefixing)}$$

Another relevant logic system, often regarded as *the* basic relevant logic, is $\mathbf{R}$ of [AB75]. It extends $\mathbf{E}$ by the addition of the constant 1 in the language and the inclusion of the axioms

$$\begin{array}{rll} \text{(as)} & \varphi \to ((\varphi \to \psi) \to \psi) & \text{(assertion)} \\ (1) & 1 & \text{(unit)} \\ (1\to) & 1 \to (\varphi \to \varphi) & \text{(unit implication)} \end{array}$$

Although the original definition of $\mathbf{R}$ is as above, often in the literature of relevant logic, $\mathbf{R}$ is used for the 1-free fragment of $\mathbf{R}$, and $\mathbf{R}^t$ is used for $\mathbf{R}$ itself. One reason for that may be the fact that $(p \wedge 1) \to (1 \vee q)$ is a theorem of $\mathbf{R}$, so the relevance principle fails. Of course the latter could be weakened to the demand that the antecedents and succedents share a propositional variable or a propositional constant. The addition or removal of 1 does not cause any complications, since **R** is a (strong) conservative extension of its 1-free fragment, namely the two systems have the same deductions with assumptions over the 1-free formulas.

It can be shown that permutation, the axiom (C) of **BCK**, is a theorem of **R**; a proof of this fact, which is rather involved, can be found in the end notes of [GR04] (if the reader wishes to try their own hand at it, we propose Exercise 20). Therefore, **R** is an extension of **BCIW**. The logic **R** *is* algebraizable and its equivalent algebraic semantics is (term equivalent to) the variety of subreducts of $\mathsf{InDFL_{ec}}$, that is of involutive, distributive $\mathbf{FL_{ec}}$-algebras. The term equivalence simply replaces negation $\neg$ by the constant 0. The implicational fragment of **R**, which coincides with the implicational fragment of $\mathbf{FL_{ec}}$, is **BCIW**. By a famous result of Urquhart (cf. [Urq84]) **R** is undecidable, and therefore $\mathsf{InDFL_{ec}}$ has undecidable equational theory. We will discuss related decidability issues in Chapter 4. Bounded algebras from this variety, or, alternatively, members of $\mathsf{InDFL_{ec\bot}}$ are known as *De Morgan monoids*. The variety $\mathsf{DFL_{ec}}$ has also acquired a name, members of this variety are called *Dunn monoids*.

PROPOSITION 2.11. [BP89] *The implicational fragment of* **R** *is not algebraizable.*

PROOF. Our algebra is defined in Figure 2.3 (the table on the right). We leave it to the reader to verify that $\{\top, t\}$ and $\{\top, t, f\}$ are deductive filters but the algebra is simple. □

The algebra defined in the proof above can be expanded to a lattice-ordered algebra, by making $\top$ the greatest element, $\bot$ the smallest element, and t and f incomparable. Notice however that in this algebra $\{\top, t, f\}$ is *not* a deductive filter with respect to full **R**, because it is not closed under adjunction. This is a manifestation of the difference in algebraizability between **R** and its implicational reduct. As we have mentioned before, the implicational fragment of **R** is order-algebraizable, so the main obstacle in its algebraizability is the lack of lattice connectives that would make the ordering relation definable.

Among the extensions of **R**, one very often considered is axiomatized relatively to **R** by the formula

$$\text{(M)} \quad \varphi \to (\varphi \to \varphi) \quad \text{(mingle)}$$

This logic is called **RM** (often pronounced *R-mingle* after the name of the axiom). Algebraically, mingle is equivalent to the property of being *square decreasing*, i.e., the identity $x^2 \leq x$. As an extension of **R**, the logic **RM** is also algebraizable and its equivalent quasivariety semantics its precisely the variety generated by *Sugihara algebras* (defined shortly). This is easily derived from Meyer's proof in [Mey75] of completeness of **RM** with respect to the set of finite Sugihara algebras $\mathbf{S_n}$. A Sugihara algebra is an algebra with universe $S_n = \{a_{-n}, a_{-n+1}, \ldots, a_{-1}, a_0, a_1, \ldots, a_{n-1}, a_n\}$

for some natural number n, or $S_\infty = \{a_i : i \in \mathbb{Z}\}$, with lattice operations determined by the natural total ordering of the indices and multiplication defined by

$$a_i \cdot a_j = \begin{cases} a_i & \text{if } |i| > |j|, \\ a_j & \text{if } |i| < |j|, \\ a_i \wedge a_j & \text{if } |i| = |j|. \end{cases}$$

Intuitively, the result of multiplying a_i by a_j is that element of the pair (a_i, a_j) which is farther from 0, or the smallest of the two if they are equally far from 0. It is not difficult to see that multiplication defined this way is residuated. Each Sugihara algebra is then a residuated lattice $(S_\alpha, \wedge, \vee, \cdot, \to, 1)$ whose identity element 1 is a_0. Adding 0 to the signature and setting $1 = 0 = a_0$ makes Sugihara algebras into involutive $\mathbf{FL_{ec}}$-algebras satisfying $x^2 \leq x$. Observe that the 'natural' candidate for algebraic semantics for **RM** is the subvariety of $\mathsf{InDFL_{ec}}$ satisfying $x^2 \leq x$; it follows that this variety contains $\mathsf{InRFL_{ec}}$. Since representability follows from distributivity in this case, we get that the two varieties coincide. For more on **RM**, see [BR04] and its references.

Lattices of axiomatic extensions of **R** and **RM** have not received very much attention. One reason for this is that many of the extensions lack the relevance principle. We will see in Chapter 9 that there are continuum many maximal extensions of **R**, yet Świrydowicz shows in [Swi99] that there are only two such extensions with the relevance principle.

Sometimes *contraction-free* relevant logics are also considered. The most well-known of them is called **RW**, for **R**$-$(W), and turns out to be algebraizable, see [GR04], with equivalent algebraic semantics $\mathsf{InDFL_e}$; note that in **RW** the constant 0 is replaced (in a term equivalent way) by a negation connective. In the literature **RW** sometimes does not include the unit 1. Its Hilbert style formulation is obtained from **R** by dropping the contraction axiom. Another example, particularly suitable for this book, is Slaney and Meyer's *Abelian logic* (see, e.g., [MS02]). It can be defined as a Hilbert system by adding

$$\text{(A)} \quad ((\varphi \to \psi) \to \psi) \to \varphi \quad \text{(relativization)}$$

to **RW**. This formula is known as *relativization axiom*, because under the interpretation of $\varphi \to \psi$ as the negation of φ relative to ψ, it amounts to relative double negation. The corresponding equation is $(x \to y) \to y = x$ (as $x \leq (x \to y) \to y$ is always true, by Lemma 2.6(10)) and, by Lemma 3.25(3), it axiomatizes (commutative) lattice-ordered groups, to be defined in Chapter 3. Indeed, as its name suggests, abelian logic is algebraizable, and its equivalent semantics is the variety CLG of *abelian lattice-ordered groups*. Since CLG is generated by the lattice-ordered group $\mathbb{Z}$ of the integers (with

$x \to y = -x + y$, see Figure 2.5), Abelian logic is complete with respect to that particular model.

Nondistributive, and noncommutative versions of relevant logics are known; nonassociative ones have also been considered, **E** being one of them. See e.g., [AB75], [MR73c], [Dun86], [DR02] for more on entailment and relevant logics.

2.3.4. Linear logic. Conceived by Girard in [Gir87a], *linear logic* is the first (the first widely known and successful at the very least) logical calculus set up with machines rather than humans in mind. Linear implication, typically written $\varphi \multimap \psi$, is thought of as a causal process of bringing about the succedent ψ by means of the antecedent φ. So the formula $\varphi \multimap \psi$ acquires the intuitive sense of 'you can get ψ at the cost of φ'. Accordingly, an assumption, once used, is then 'used up' and thus no longer available. In particular, you cannot use φ twice, unless there are two copies of φ at your disposal. A persuasive analogy is that of *resources* in a (finite) storage space of a computer. Indeed, linear logic is often described as a logic of resource management, or to use a catch-phrase: a *resource sensitive* logic. Thinking of formulas as resources has an immediate consequence of allowing neither contraction nor weakening in proofs. For contraction amounts to using the same resource twice, which is only possible if two copies of the resource are stored. And weakening boils down to adding extra resources, which is only possible if some storage space is free.

Fully fledged linear logic is formulated in a language with four constants ($\bot$, $\top$, 0, 1) three unary connectives ($^\bot$, !, ?) and five binary ones (&, $\oplus$, $\otimes$, $⅋$, $\multimap$) with the following interpretations:

- & is conjunction.
- $\oplus$ is disjunction.
- $\otimes$ is fusion (corresponds to comma on the left of the turnstile).
- $\multimap$ is linear implication.
- $⅋$ is the dual of fusion (corresponds to comma on the right; in classical linear logic it is definable by $\varphi ⅋ \psi = (\varphi^\bot \otimes \psi^\bot)^\bot$).
- $\bot$ is the unit for $⅋$ and a dualizing element ($\varphi^\bot = \varphi \multimap \bot$).
- $\top$ is the largest element and the unit for &.
- 0 is the smallest element and the unit for $\oplus$.
- 1 is the unit for $\otimes$.
- $^\bot$ is linear negation.
- ! is a connective whose function is to allow contraction and weakening on the left of turnstile ($!\varphi$ is read *of course* φ).
- ? allows contraction and weakening on the right ($?\varphi$ is read *why not* φ and is definable by $?\varphi = (!(\varphi^\bot))^\bot$).

Since the symbols for the connectives are somewhat idiosyncratic in linear logic and, worse still, some of them clash with ours, in Table 2.3.4 we spell out the translation. The connectives $\top$, 0, & and $\oplus$ are called *additive*; the connectives 1, $\bot$, $\parr$ and $\otimes$ are called *multiplicative*; the connectives ! and ? are known as *exponentials*. This distinction leads to three natural reducts of linear logic, namely, *multiplicative-additive linear logic*, known by the acronym **MALL** (and quite thoroughly investigated), multiplicative linear logic (**MLL**) and additive linear logic (**ALL**), the latter two rather less studied.

Linear logic is typically formulated as a Gentzen system either "classically", (**CLL**, *classical linear logic*, with a list or formulas on either side of the sequent turnstile), or "intuitionistically" (**ILL**, *intuitionistic linear logic*, with a single formula on the right). **CLL** is an involutive version of **ILL**. Either way, the resulting Gentzen system lacks contraction and weakening and so is located somewhere in the vicinity of $\mathbf{FL_e}$. A Hilbert style formulation of classical linear logic (cf. [Avr88]) is given in Figure 2.4.

In this formulation $\parr$ is regarded as an abbreviation $\varphi \parr \psi = (\varphi^{\bot} \otimes \psi^{\bot})^{\bot}$. Classical linear logic is algebraizable and its equivalent quasivariety semantics is the variety of *girales* (cf. [Agl96]). The algebraic approach to linear logic is however much less advanced than other semantical approaches, mostly employing *phase spaces* and techniques from category theory. This is partly because interests of linear logicians focus more on *computation* (concrete way proofs are carried out) and less on abstract *provability*. Thus, as algebraically there seems to be no easy way of telling apart two different proofs of one and the same formula, the use of algebra is not appreciated among linear logicians.

Because of the presence of exponentials, full linear logic does not fit into our framework. But **MALL** turns out to be equivalent to $\mathbf{InFL_{e\bot}}$, that is $\mathbf{InFL_e}$ with bounds. **MALL** is also algebraizable, and its equivalent quasivariety semantics is the variety of *arabesques* (cf. [Agl96]), which is term-equivalent to $\mathsf{InFL_{e\bot}}$. Recalling that $\mathsf{InDFL_e}$ is the equivalent quasivariety semantics for the relevant logic **RW** we get that **MALL** can be seen as **RW** plus bounds minus distributivity. The intuitionistic version of **MALL**, often referred to as **IMALL**, has also been considered. In that logic, $\parr$ is no longer definable as de Morgan dual of fusion and thus only the $\parr$-free fragment of **IMALL** fits in out framework, but then it corresponds exactly to $\mathbf{FL_{e\bot}}$.

As the literature of linear logic is wide and diverse, we only mention [Tro92], which is a good survey and starting point for the interested reader.

2.3.5. Łukasiewicz logic and MV-algebras. Another contender for the title of the first nonclassical logic is the three valued system of Łukasiewicz, introduced in 1920 in [Łu20]. Although intuitionism predates it by a decade,

(id)	$\varphi \multimap \varphi$	
(sf)	$(\varphi \multimap \psi) \multimap ((\psi \multimap \chi) \multimap (\varphi \multimap \chi))$	
(per)	$(\varphi \multimap (\psi \multimap \chi)) \multimap (\psi \multimap (\varphi \multimap \chi))$	
$(\rightarrow\cdot)$	$\varphi \multimap (\psi \multimap (\varphi \otimes \psi))$	
$(\cdot\rightarrow)$	$\varphi \multimap (\psi \multimap \chi)) \multimap ((\varphi \otimes \psi) \multimap \chi)$	
$(\rightarrow\wedge)$	$((\varphi \multimap \psi) \,\&\, (\varphi \multimap \chi)) \multimap (\varphi \multimap (\psi \,\&\, \chi))$	
$(\wedge\rightarrow)$	$(\varphi \,\&\, \psi) \multimap \varphi$	
$(\wedge\rightarrow)$	$(\varphi \,\&\, \psi) \multimap \psi$	
$(\rightarrow\vee)$	$\varphi \multimap (\varphi \oplus \psi)$	
$(\rightarrow\vee)$	$\psi \multimap (\varphi \oplus \psi)$	
$(\vee\rightarrow)$	$((\varphi \multimap \chi) \,\&\, (\psi \multimap \chi)) \multimap ((\varphi \oplus \psi) \multimap \chi))$	
(cp)	$(\varphi \multimap \psi^{\perp}) \multimap (\psi \multimap \varphi^{\perp})$	
(dn)	$\varphi^{\perp\perp} \multimap \varphi$	
(top)	$\varphi \multimap \top$	(top element)
(bot)	$0 \multimap \varphi$	(bottom element)
(1)	1	
$(1\rightarrow)$	$1 \multimap (\varphi \multimap \varphi)$	
(0)	$\perp^{\perp}$	(dualizing element)
(!w)	$\varphi \multimap (!\psi \multimap \varphi)$	(excl. weakening)
(!c)	$(!\varphi \multimap (!\varphi \multimap \psi)) \multimap (!\varphi \multimap \psi)$	(excl. contraction)
(!K)	$!(\varphi \multimap \psi) \multimap (!\varphi \multimap !\psi)$	(excl. modal K)
(!T)	$!\varphi \multimap \varphi$	(excl. modal T)
(!4)	$!\varphi \multimap !!\varphi$	(excl. modal 4)

$$\frac{\varphi \quad \varphi \multimap \psi}{\psi}\ (\mathrm{mp}) \qquad \frac{\varphi \quad \psi}{\varphi \,\&\, \psi}\ (\mathrm{adj}) \qquad \frac{\varphi}{!\varphi}\ (\mathrm{nec})$$

(necessitation)

FIGURE 2.4. Hilbert style linear logic.

Łukasiewicz's logic was arguably[11] the first introduced expressly for the purpose of leaving the two-valued Boolean realm. Consider, says Łukasiewicz after Aristotle, the proposition 'there will be a sea battle tomorrow'. That proposition, to be true, has to describe things the way they really are, so a sea battle has to happen tomorrow. But today no sea battle happened (yet), so our proposition is not true. On the other hand, to be false, the proposition has to describe things the way they really are not, so there has to be no sea battle tomorrow. But, the absence of sea battles today says nothing about sea battles tomorrow, so our proposition is not false either.

[11]But not definitely, as Emil Post's doctoral dissertation introducing what is now known as Post-algebras, another intentional departure from the Boolean realm, also appeared in 1920.

Linear logic	meaning	this book
$\multimap$	implication	$\rightarrow$
&	conjunction	$\wedge$
$\oplus$	disjunction	$\vee$
$\otimes$	fusion	$\cdot$
$\parr$	dual of fusion	$+$
$^{\perp}$	negation	$\neg$
$\perp$	dualizing element, unit for $\parr$	0
1	unit for fusion	1
$\top$	largest element	$\top$
0	smallest element	$\perp$

TABLE 2.1. Translation between linear logic and our notation.

It thus has to have some third value, say, $\frac{1}{2}$. Logical connectives extend to accommodate the third value, resulting in what we now know as the three element algebra $\mathbf{C}_3$. This naturally led to generalizations with values $\frac{1}{n}$ for any natural number n, and also infinite-valued ones.

It was C.C. Chang who coined the name *MV-algebras* for these (MV stands for *many-valued*) and used them in his new proof of completeness of Łukasiewicz's infinite-valued logic (cf. [Cha59]; the first completeness proof is due to Wajsberg but was never published). MV-algebras proved to be of interest far beyond that of logic with more than two truth-values. A recent monograph on MV-algebras is [CDM00]. It defines an MV-algebra as an algebra $(A, \oplus, \neg, O)$ such that $(A, \oplus, O)$ is a commutative monoid and

(1) $\neg\neg x = x$
(2) $x \oplus \neg O = \neg O$
(3) $\neg(\neg x \oplus y) \oplus y = \neg(\neg y \oplus x) \oplus x$.

It can be shown that MV-algebras as defined above are term equivalent to the subvariety MV of $\mathsf{FL_{eo}}$ axiomatized by $(x \rightarrow y) \rightarrow y = x \vee y$ (and $xy = yx$ and $0 \leq x$, relative to FL). The axiomatizing identity is a generalized form of the law of double negation ($\neg\neg x = x$, or $(x \rightarrow 0) \rightarrow 0 = x$) and is known as the *relativized law of double negation.* Exercise 13 asks you to verify that there are two ways in which MV-algebras (as defined in [CDM00]) are term equivalent to the above subvariety of FL-algebras (which will also call MV-algebras from now on).

One is given by

• $x \cdot y = \neg(\neg x \oplus \neg y)$		
• $x \rightarrow y = \neg x \oplus y$		• $O = 0$
• $1 = \neg O$ and $0 = O$	and	• $\neg x = x \rightarrow 0$
• $x \wedge y = x(x \rightarrow y)$		• $x \oplus y = \neg(\neg x \cdot \neg y)$
• $x \vee y = (x \rightarrow y) \rightarrow y$		

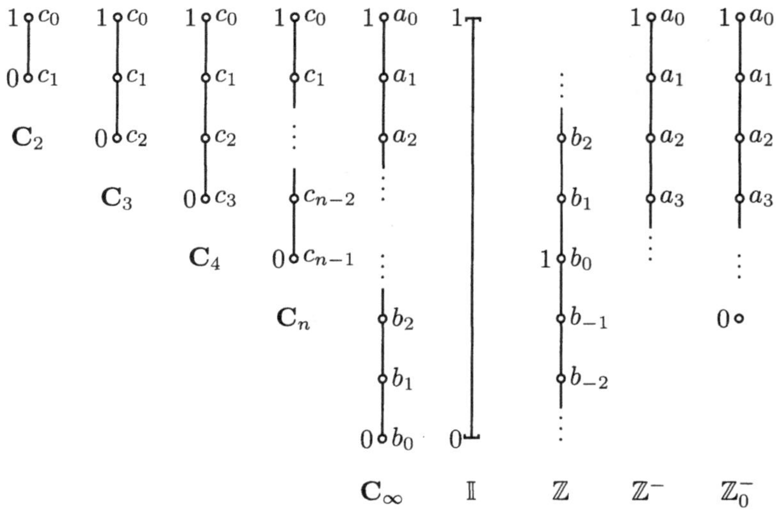

FIGURE 2.5. Some residuated chains.

and the other by

- $x \cdot y = x \oplus y$
- $x \to y = \neg(\neg y \oplus x)$
- $1 = O$ and $0 = \neg O$
- $x \wedge y = (x \to y) \to y$
- $x \vee y = x(x \to y)$

and

- $O = 1$
- $\neg x = x \to 0$
- $x \oplus y = x \cdot y$

This also implies that MV-algebras are term equivalent to themselves in a nontrivial way. We will see that this is a special instance of a term equivalence for involutive FL-algebras and relies on the fact that MV-algebras are self-dual; see Section 3.4.17. Two more varieties that are term equivalent to MV are introduced in Exercises 15 and 16.

Examples of MV-algebras include the algebras $\mathbf{C}_n$ (also known as $\mathbf{Ł_n}$, for Łukasiewicz's n-valued matrix), where $C_n = \{c_{n-1}, \ldots, c_2, c_1, c_0 = 1\}$, $c_i \leq c_j$ iff $i \geq j$ and $c_i \cdot c_j = c_{\min\{i+j,n\}}$; see Figure 2.5. It is easy to verify that $\mathbf{C_n}$ is an FL-algebra. Note that $\mathbf{C}_2$ is isomorphic to the Boolean algebra **2**. The variety generated by $\mathbf{C}_{n+1}$ is denoted by MV_n, hence $\mathsf{BA} = \mathsf{MV}_1$.

Another important example is the unit interval $[0,1]$ with the usual order and $x \cdot y = \max\{0, x+y-1\}$ and $x \to y = \min\{1, 1-x+y\}$ (Łukasiewicz's infinitely valued matrix). Yet another is Chang's infinite chain $\mathbf{C}_\infty$ with universe $C_\infty = \{a_n : n \in \omega\} \cup \{b_n : n \in \omega\}$, where $a_0 = 1$, $b_0 = 0$ and the ordering is as follows: every b_n is below all a_n's, $b_n \leq b_m$ if $n \leq m$, $a_n \leq a_m$ if $n \geq m$. Multiplication in $\mathbf{C}_\infty$ is defined by

$$b_n \cdot b_m = 0, \quad a_n \cdot b_m = b_{m \dot{-} n}, \quad a_n \cdot a_m = a_{n+m},$$

where $\dot{-}$ stands for truncated subtraction. See Figure 2.5.

2.3.6. Fuzzy logics and triangular norms. Fuzziness found its way to logic via the idea that truth of a proposition is a matter of degree, like in 'he is young', which is obviously true about each and every author of this book, but to a rather different degree. From a fuzzy perspective, possible truth values are real numbers from the interval $[0,1]$. Logical connectives are viewed as functions from $[0,1]^n$ to $[0,1]$, where n is the arity of the connective. Another requirement is that connectives should behave classically on classical values 0 and 1, e.g., $0 \to 1 = 1 \to 1 = 1$; yet another is classical behavior with respect to order, so that $x \to y = 1$ for $x \leq y$. Finally, a measure of continuity is required as long as it does not contradict the other requirements.

In particular, for a conjunction-like connective, all these are met by any *triangular norm* (*t-norm*), which is a binary operation on the interval $[0,1]$ that is associative, commutative, monotone and 1 is its unit element. Not all t-norms are continuous, but these that are, give rise to models of P. Hajek's *basic logic*, see [Háj98]. Hajek's logic **BL** is defined as a Hilbert style system in Figure 2.6. Following the original approach of Hajek, we regard conjunction, disjunction, negation and unit as definitional abbreviations.

(sf)	$(\varphi \to \psi) \to ((\psi \to \chi) \to (\varphi \to \chi))$	
(int)	$(\varphi \cdot \psi) \to \varphi$	(integrality)
(com)	$(\varphi \cdot \psi) \to (\psi \cdot \varphi)$	(commutativity)
(conj)	$(\varphi \cdot (\varphi \to \psi)) \to (\psi \cdot (\psi \to \varphi))$	(conjunction)
$(\cdot\!\to)$	$(\varphi \to (\psi \to \chi)) \to ((\varphi \cdot \psi) \to \chi$	
$(\to$pl)	$((\varphi \to \psi) \to \chi) \to (((\psi \to \varphi) \to \chi) \to \chi)$	(arrow prelinearity)
(bot)	$0 \to \varphi$	
	$\varphi \wedge \psi =_{df} \varphi \cdot (\varphi \to \psi)$	(conjunction definition)
	$\varphi \vee \psi =_{df} \neg(\neg\varphi \wedge \neg\psi)$	(disjunction definition)
	$\neg\varphi =_{df} \varphi \to 0$	(negation definition)
	$1 =_{df} 0 \to 0$	(unit definition)

$$\frac{\varphi \quad \varphi \to \psi}{\psi}\ (\mathrm{mp})$$

FIGURE 2.6. Hajek's basic logic.

Basic logic is algebraizable with equivalent quasivariety semantics being (term equivalent to) the variety BL of *BL-algebras*. These are defined as $\mathbf{FL_{eo}}$-algebras that satisfy $x \wedge y = x(x \to y)$ and $x \to y \vee y \to x = 1$. It is easy to see that BL-algebras are integral (set $x = 1$ in the first identity), so the variety BL is a subvariety of $\mathsf{FL_{ew}}$. As with previous examples, we ask the reader to verify the algebraizability and term equivalence of the resulting variety with BL. The reader will also notice that MV is a subvariety of BL and that $\mathsf{HA} \cap \mathsf{BL}$ is precisely the variety of representable Heyting algebras, known as *Gödel algebras*.

Coming back to t-norms, we have said that each continuous t-norm is a model of **BL**. In fact the connection between **BL** and continuous t-norms is stronger than that. Namely, **BL** is complete with respect to continuous t-norms. This is proved algebraically in [CEGT00], where it is shown that the variety of BL-algebras is generated by all continuous t-norms.

Among continuous t-norms, three are of essential importance: Łukasiewicz t-norm $L(x,y) = \max\{x+y-1, 0\}$, product t-norm $\Pi(x,y) = xy$, and Gödel t-norm $G(x,y) = \min\{x,y\}$. They determine three important extensions of **BL**, known as **Ł**, **GL** and **Π**. It turns out that **Ł** is precisely the infinite-valued Łukasiewicz logic (hence the logic of MV), and it is axiomatized relatively to **BL** by adding the single formula

(dn) $$\neg\neg\varphi \to \varphi.$$

Next, **GL** is the Gödel logic, or the superintuitionistic logic defined by adding to **BL** the axiom

$$(\varphi \wedge \psi) \to (\psi \cdot \varphi).$$

Gödel logic is a superintuitionistic logic, in fact the smallest superintuitionistic logic that also is a fuzzy logic. The axiomatization of **GL** relative to **Int** we leave as an exercise to the reader. Finally, **Π**, known also as *product logic*, is defined relatively to **BL** by two axioms.

(**Π**1) $$\neg\neg\chi \to (((\varphi \cdot \chi) \to (\psi \cdot \chi)) \to (\varphi \to \psi)$$

(**Π**2) $$(\varphi \wedge \neg\varphi) \to 0$$

There is also a single formula that can do the job; see [Cin01]. It is

(**Π**) $$\neg\neg\varphi \to (((\varphi \to (\varphi \cdot \psi)) \to (\psi \cdot \neg\neg\psi))$$

Another example of a product logic algebra is the algebra $\mathbb{Z}_0^-$ in Figure 2.5. It consists of a copy of negative integers with an extra bottom element, multiplication defined as the usual addition. This algebra also generates the whole variety corresponding to **Π**.

Because of completeness of $[0,1]$, every continuous t-norm is residuated, but, in fact, only left-continuity (i.e., distribution over arbitrary joins) is required. That is, a t-norm is residuated if and only if it is left-continuous. This naturally leads to a generalization of **BL**, known as **MTL**, or *monoidal t-norm logic*. It can be defined relatively to $\mathbf{FL_{ew}}$ by the linearity axiom.

(pl) $$(\varphi \to \psi) \vee (\psi \to \varphi) \qquad \text{(prelinearity)}$$

Its equivalent quasivariety semantics is the variety $\mathsf{RFL_{ew}}$ of *representable FL_{ew}-algebras*, known also as *MTL-algebras*. It can be shown [JM02] that $\mathsf{RFL_{ew}}$ is generated by all left-continuous t-norms. The logic **MTL** is considered a viable alternative candidate for the smallest fuzzy logic. The recommended reading for fuzzy logic is [Háj98].

2.3.7. Superintuitionistic logics and Heyting algebras. We met intuitionistic logic and Heyting algebras in Chapter 1. Some superintuitionistic logics were also mentioned there. Superintuitionistic logics are all algebraizable, since they are extensions of **FL**. Their equivalent semantics are varieties of Heyting algebras, which we also met in Chapter 1. In particular, the equivalent algebraic semantics for intuitionistic logic is HA.

Recall that one outstanding feature of intuitionistic logic was the absence of the law of excluded middle. The loss of excluded middle is sweetened by the gain of something else, namely, the *disjunction property* which says that a disjunction is a theorem if and only if one of the disjuncts is. This is another manifestation of the constructivist aspect of intuitionism and stands in sharp contrast to classical logic. In Chapter 5 we will come back to the disjunction property in the context of substructural logics.

Despite the contrast between intuitionism and classical logic, it turns out, rather surprisingly, that **Int**, the set of theorems of **HJ**, is not less expressive than **Cl**, the set of theorems of **HK**. Namely, Glivenko proved in [Gli29] that φ is a theorem of **Cl** if and only if $\neg\neg\varphi$ is a theorem of **Int**. This can be viewed as an embedding of classical logic into intuitionism. We will come back to *Glivenko properties* of substructural logics in Chapter 8, where we will see that this translation is not an isolated incident.

Adding excluded middle to intuitionistic logic yields the classical propositional calculus. It was soon realized, however, that there are many theorems of classical logic, which are not theorems of intuitionism but their addition to **Int** does not produce full **Cl**. This realization opened up the realm of *superintuitionistic logics* (then known as *intermediate logics*). They were among the first nonclassical logics widely studied and of the vast literature about them we only mention [CZ97] as a reference book (that, admittedly, studies them mainly by means of non-algebraic tools developed for modal logics) and [ZWC01].

Superintuitionistic logics correspond to varieties of Heyting algebras. Below we give some examples of important superintuitionistic logics and corresponding subvarieties of HA.

Gödel-Dummett logic (**LC**), known also as Gödel logic among fuzzy logicians, defined by prelinearity $(\varphi\to\psi)\vee(\psi\to\varphi)$; the algebras from the variety corresponding to it are called Gödel algebras. Since subdirectly irreducible Gödel algebras are linearly ordered, we use RHA (standing for 'representable Heyting algebras') to denote that variety.

Logic of weak excluded middle (**KC**), defined by $\neg\neg\varphi\vee\neg\varphi$; its corresponding equivalent quasivariety semantics is (of course) the variety defined relatively to HA by $\neg\neg x\vee\neg x = 1$. This variety is characterized as the largest subvariety of HA whose subdirectly irreducible members are precisely the algebras obtained by adding a 'new' bottom element (called 0) to a subdirectly irreducible *Brouwerian algebra* (a Brouwerian algebra is a 0-free subreduct of a Heyting algebra and the paragraph on Johansson's minimal logic below). **KC** and the above construction will be mentioned again in Section 9.6.

Kreisel-Putnam logic (**KP**), defined by $(\neg\varphi \to (\psi \vee \chi)) \to ((\neg\varphi \to \psi) \vee (\neg\varphi \to \chi))$. It was the first example of a logic strictly greater than **Int** but still possessing the disjunction property.

Medvedev logic of finite problems (**ML** or **LM**), which can be defined semantically as (the logic of) the variety generated by all Heyting algebras that arise as sets of downsets of posets $\langle S_n, \leq \rangle$ where $S_n = \wp(\{1, \ldots, n\}) \setminus \{\emptyset\}$ and $\leq$ is reverse inclusion. We ask the reader (Exercise 21) to verify that such a structure really is a Heyting algebra. Thus, **ML** has the finite model property (every non-theorem is refutable in a finite algebraic model) by definition, but Maksimova, Skvortsov and Shehtman show in [MSŠ79] that it cannot be finitely axiomatized. Prucnal shows in [Pru76] that Medvedev's logic is structurally complete (that is every admissible rule of **ML** is derivable in **ML**). It is not known whether Medvedev logic is decidable and this is one of outstanding open problems in superintuitionistic logics.

2.3.8. Minimal logic and Brouwerian algebras. Johansson *minimal logic* **J** is a sublogic of intuitionism **Int** that is obtained by taking its positive fragment $\mathbf{HJ}^+$ (i.e., the theorems of **Int** that do not contain negation) and adding negation to the type without stipulating any conditions for it. Johansson logic can be axiomatized as a Hilbert system by dropping (A9) and (A10) from the axiomatization of intuitionism given in Section 1.3.1 (namely keeping only axioms (A1)–(A8)), but retaining negation in the type. Thus, **J** is exactly $\mathbf{FL_{eci}}$ (which is the same as $\mathbf{FL_{ci}}$). Adding the axiom $0 \to \varphi$ (that is, (o) in our terminology) we obtain **Int** (that is $\mathbf{FL_{cio}} = \mathbf{Int}$, cf. Figure 2.1.3). Johansson logic has also been called *positive logic*, and there is some interest in *negative logics* as well. These are extensions of **J** satisfying $1 \to 0$. The extensions of **J** containing neither (o) nor $1 \to 0$ are sometimes called *paraconsistent logics*. A classification of paraconsistent logics is given by Odintsov in [Odi04] and interpolation and related properties of these logics are discussed by Maksimova in [Mak05].

The logic **J** is algebraizable, as an extension of **FL**, and its equivalent algebraic semantics is the variety $\mathsf{FL_{ci}}$. Another way to these algebras is via the requirement that the multiplication operation $\cdot$ on an FL-algebra residuated lattice $\mathbf{A} = (A, \wedge, \vee, \cdot, \backslash, /, 1)$ be equal to the meet $\wedge$. Obviously, both integrality and $x \leq x^2$ hold for all $x, y \in A$. Conversely, if $\cdot$ is both integral and square-increasing, then $\cdot$ is shown to be equal to $\wedge$, and hence **A** is commutative. In fact, $x \cdot y$ is a lower bound of x and y by integrality, and if z is a lower bound of x and y then $z \leq z^2 \leq x \cdot y$ by the square-increasing property and monotonicity. Therefore, $x \cdot y$ is the greatest lower bound of x and y, and hence is equal to $x \wedge y$. In this case, the law of residuation is nothing but the law of $\wedge$-residuation (see Chapter 1). Such an **A** is called a *Brouwerian algebra*[12]. It is easy to see that Brouwerian algebras are precisely 0-free subreducts of Heyting algebras. A Brouwerian algebra endowed with an additional constant 0, which is not required to satisfy any specific identities, is an FL_{ci}-algebra and a typical model of **J**. Alternatively, Brouwerian algebras may be viewed as a subvariety of $\mathsf{FL_{ci}}$ satisfying $0 = 1$; i.e., as a subvariety of ICRL.

An important related variety is the variety of *Brouwerian semilattices*, that is $\{\to, \wedge, 1\}$-subreducts of Heyting algebras. Unlike Brouwerian algebras, Brouwerian semilattices are locally finite. They play an important role in the theory of varieties with equationally definable principal congruences (EDPC, see Sections 3.6.4 and 11.1 for more on this notion), namely a variety $\mathcal{V}$ has EDPC if and only if for any algebra $\mathbf{A} \in \mathcal{V}$ the join semilattice of compact congruences of **A** is dually isomorphic to a Brouwerian semilattice (see [BKP84]).

2.3.9. Fregean logics and equivalential algebras. At the beginning of 20th Century Frege was puzzled by the difference between *the Evening Star is the Evening Star* and *the Morning Star is the Evening Star*. Clearly the second sentence expresses an important astronomical discovery made a few millenia ago, but the first is a tautological triviality. The problem, Frege said, is that these two sentences have the same *truth value*, but different *meaning*. In the 1960s, Suszko put the same thought in different words: the two sentences are *logically equivalent* but not *identical*. Logical equivalence is thus a weaker notion. It amounts to having exactly the same consequences in the logic in question, or, expressed in our terminology, to generating the same deductive filter. (Deductive filters were defined in Section 1.6.4. For Boolean and Heyting algebras they coincide with nonempty lattice-filters; see Exercise 35 of Chapter 1.) Put differently, the difference lies between mutual provability $\dashv\vdash$ and provability of the equivalence $\leftrightarrow$.

Although generally weaker, in some cases logical equivalence is just as strong as identity. Take, for instance, the Lindenbaum algebra of **Int** and suppose that a and b generate the same deductive filter. Since the deductive filter generated by an element coincides with the upset generated by this element, we have ${\uparrow}a = {\uparrow}b$, and therefore $a = b$. So, in **Int** (and thus in all its extensions) logical equivalence and identity are the same thing. A logic with this property is called *Fregean*. Fregean logics have been subject to quite a lot of investigating effort recently (cf. [CP04] for a survey). It is a feature of every Fregean logic, that it can capture its own logical equivalence relation by means of a definable connective. Calling that connective (surprise, surprise) *equivalence*, we may now study its properties.

[12]Sometimes, the name Brouwerian algebras is used for *duals* of Heyting algebras. Obviously, they are different.

It turns out that they are exactly the properties of the *equivalence fragment* of **Int** (with equivalence defined by $\varphi \leftrightarrow \psi =_{df} (\varphi \to \psi) \wedge (\psi \to \varphi)$).[13]

This logic, known also as intuitionistic *equivalential calculus*[14] was isolated by Tax, who showed in [Tax73] that is can be axiomatized as a Hilbert system by four axioms and two rules (see Figure 2.7).

$$\varphi \leftrightarrow \varphi$$
$$(\varphi \leftrightarrow \psi) \leftrightarrow (\psi \leftrightarrow \varphi)$$
$$(((\varphi \leftrightarrow \psi) \leftrightarrow \chi) \leftrightarrow \chi) \leftrightarrow ((\varphi \leftrightarrow \chi) \leftrightarrow (\psi \leftrightarrow \chi))$$
$$(((\varphi \leftrightarrow \psi) \leftrightarrow ((\varphi \leftrightarrow \chi) \leftrightarrow \chi)) \leftrightarrow ((\varphi \leftrightarrow \chi) \leftrightarrow \chi)) \leftrightarrow (\varphi \leftrightarrow \psi)$$

$$\frac{\varphi \quad \varphi \leftrightarrow \psi}{\psi}\ (\mathrm{mp}) \qquad\qquad \frac{\varphi}{(\varphi \leftrightarrow \psi) \leftrightarrow \psi}\ (\mathrm{t})$$

FIGURE 2.7. Equivalence fragment of **Int**.

An *equivalential algebra* is a groupoid $\mathbf{A} = \langle A, \star \rangle$ satisfying the following identities

(1) $xxy = y$
(2) $xyzz = xy(yz)$
(3) $xy(xzz)(xzz) = xy$

where parentheses are associated to the left and the operation symbol $\star$ is omitted. Equivalential algebras were introduced by Kabziński and Wroński in [KW75]. Overusing the word equivalence a little, we can say that the variety EQ of equivalential algebras[15] is an equivalent algebraic semantics for the equivalence fragment of **Int**, which also is an equivalential logic in the sense of [PW74]. Although introduced as algebraic semantics for Tax's calculus, EQ turned out to be an interesting variety in its own right. First of all, the property of being Fregean applies equally well to varieties of algebras by a congruence property. Namely, a variety $\mathcal{V}$ is *Fregean* if for every algebra $\mathbf{A} \in \mathcal{V}$ and any $a, b, c \in A$ we have

$$\mathrm{Cg}^{\mathbf{A}}(a, c) = \mathrm{Cg}^{\mathbf{A}}(b, c) \text{ implies } a = b.$$

Equivalential algebras are Fregean. Conversely, equivalence subreducts of every Fregean (every Fregean has a definable connective that plays the role of equivalence) variety are a subvariety of EQ. Equivalential algebras are precisely equivalence subreducts of Heyting algebras, as [KW75] shows. They are congruence permutable (with $xyz(xzzx)$ as Mal'cev term) and

[13]We will define $\leftrightarrow$ in a more general setting later.

[14]The name *equivalential logic* was also used by Prucnal and Wroński in [PW74] for a logic that has a definable connective with certain properties of equivalence. In fact their notion is a direct precursor to what we call *algebraizable* logic here.

[15]By yet another unhappy coincidence, there are *equivalence algebras* as well. These are groupoids made out of equivalence relations by putting $xy = x$ if x and y are related and $xy = y$ if they are not (see e.g. [JM01]).

1-regular (with the constant 1 defined by xx). They are locally finite, yet not congruence distributive. These two last properties make equivalential algebras, unlike most other algebras related to logics, accessible to methods of *tame congruence theory* and *commutator theory*. One may say that EQ is the most prominent example of a variety related to logic, to which these two theories are applicable and have been successfully applied. For more on equivalential algebras and their connection with Fregean varieties, see [ISW].

2.3.10. Overview of logics over FL. The class of substructural logics ordered by set inclusion is a complete lattice. Meets are set-theoretical intersection and for a set of logics $\{\mathbf{L}_i : i \in I\}$ their join $\bigvee_{i\in I} \mathbf{L}_i$ is the smallest logic containing $\bigcup_{i\in I} \mathbf{L}_i$. Given axiomatizations for two logics, the join is axiomatized by the union of the two axiomatizations. On the other hand, the meet (intersection) of the logics is hard to compute since it is not, in general, axiomatized by the intersection of their axiomatizations (which might e.g. be disjoint). We will give an axiomatization for meets of logics in Section 9.7. A very schematic picture of the lattice of substructural logics is shown in Figure 2.8. The bottom element is **FL** and the top element is the inconsistent logic. Among consistent logics over **FL**, the classical logic **Cl** is maximal, but as we will see in due course, not the greatest. In fact there are continuum maximal consistent extensions of **FL**, as shown in Chapter 9.

The logic $\mathbf{Ł}_k$, for any positive integer k, is Łukasiewicz's k-valued logic and is defined by the k-element MV-algebra. The logic $\mathbf{Ł}_\infty$ is Łukasiewicz's infinitely valued logic. It is equal to $\bigcap_{n\in\mathbb{N}} \mathbf{Ł}_n$ and its corresponding variety is MV. The interval between $\mathbf{Ł}_\infty$ and **Cl** is dually isomorphic to the lattice of nontrivial subvarieties of MV (Figure 3.4), and that in turn is closely related to the divisibility ordering of natural numbers, cf. Figure 3.3 where MV_n corresponds to $\mathbf{Ł}_{n+1}$. In fact, $\mathbf{Ł}_n$ contains $\mathbf{Ł}_m$ iff $n-1$ divides $m-1$. Thus, $\mathbf{Ł}_{p+1}$, for any prime p, is a maximal non-classical many-valued logic.

The logics $\mathbf{GL}_k$ form a descending chain, with $\mathbf{GL}_n$ containing $\mathbf{GL}_m$ iff $n \leq m$. These are logics of k-element totally ordered Heyting algebras and their corresponding varieties are HA_k. The intersection of this chain is the Gödel-Dummett logic **GL**. In particular $\mathbf{GL}_3$ is the largest superintuitionistic logic properly contained in **Cl**.

The relevant logic **R** extends both **MALL** (without bounds) and $\mathbf{FL_{ec}}$, but unlike them it contains the distributivity axiom $\varphi\wedge(\psi\vee\chi)\to(\varphi\wedge\psi)\vee(\psi\wedge\chi)$. The logic **BL** is Hajek's basic logic and its extension $\mathbf{\Pi}$ is the product logic. Product logic is yet another maximal logic properly contained in **Cl**.

2.4. Parametrized local deduction theorem

Unlike classical and intuitionistic logics, most other substructural logics including **FL** do not have a deduction theorem. For example, if p is a propositional variable, then $p \vdash_{\mathbf{FL_e}} p^2$, but $\nvdash_{\mathbf{FL_e}} p \to p^2$. Also, $p \vdash_{\mathbf{FL}} p^2$, but $\nvdash_{\mathbf{FL}} p\backslash p^2$. The latter can be shown by either an algebraic argument,

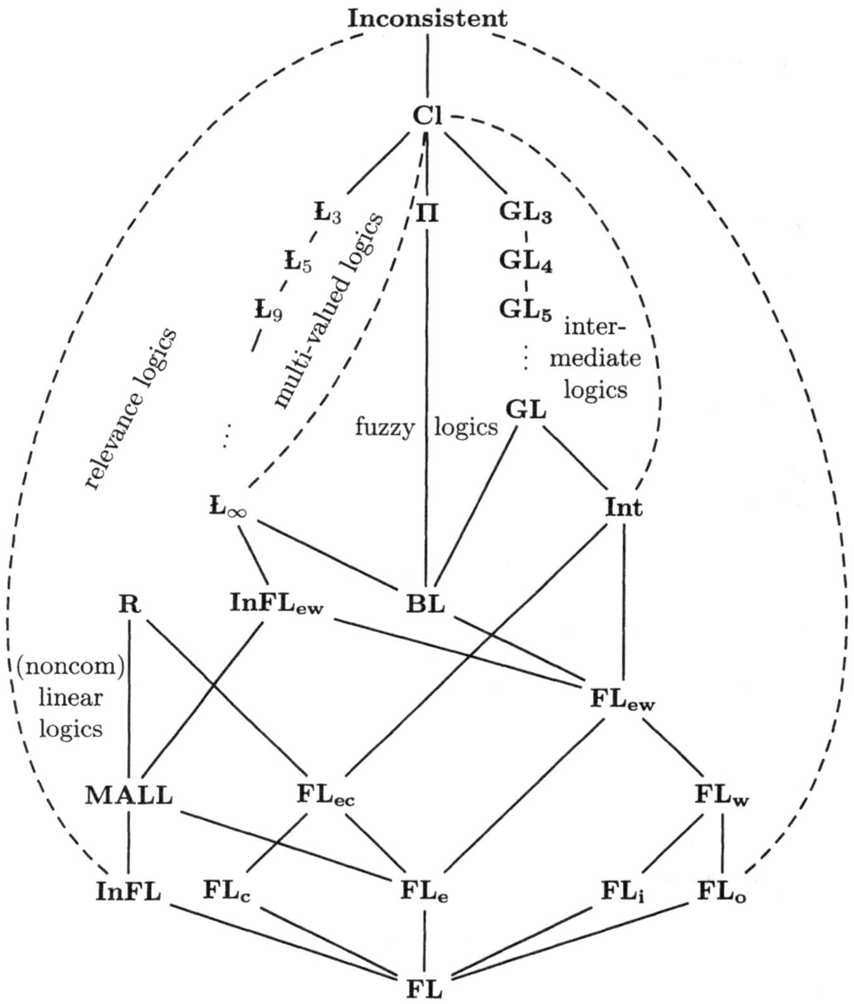

FIGURE 2.8. Some logics extending **FL**.

after obtaining a completeness result for **FL**, or using the fact that **FL** enjoys cut elimination, also to be proved later in the book; see Exercise 4.

Nevertheless, we will show that a weaker form of the deduction theorem called *parametrized local deduction theorem* (PLDT) holds for substructural logics over **FL**. The notion of PLDT was introduced by Czelakowski and Dziobiak in [CD96]; note that they used the different spelling 'parameterized'.

From now on, we write the fusion $\alpha \cdot \beta$ as $\alpha\beta$ when no confusion will occur. For formulas φ, α, we define the *left conjugate*[16] $\lambda_\alpha(\varphi) = (\alpha\backslash(\varphi\alpha))\wedge 1$ and the *right conjugate* $\rho_\alpha(\varphi) = ((\alpha\varphi)/\alpha) \wedge 1$ of φ with respect to α. Hereafter, we use the convention that fusion $\cdot$ binds more strongly than $\backslash$ and $/$, and omit some parentheses. Note that both $\lambda_\alpha(\varphi)$ and $\rho_\alpha(\varphi)$ are provably equivalent to $\varphi \wedge 1$ when α is 1. An *iterated conjugate* of φ is a composition $\gamma_{\alpha_1}(\gamma_{\alpha_2}(\ldots\gamma_{\alpha_n}(\varphi)))$ for some formulas $\alpha_1, \ldots, \alpha_n$, where n is a positive integer and $\gamma_{\alpha_i} \in \{\lambda_{\alpha_i}, \rho_{\alpha_i}\}$ for all i.

LEMMA 2.12. *For all α, β, the following hold.*

(1) $\alpha, \beta \vdash_{\mathbf{FL}} \alpha \wedge \beta$,
(2) $\alpha \vdash_{\mathbf{FL}} \alpha \wedge 1$,
(3) $\alpha \vdash_{\mathbf{FL}} \beta\backslash\alpha\beta$,
(4) $\alpha \vdash_{\mathbf{FL}} \beta\alpha/\beta$,
(5) $\alpha \vdash_{\mathbf{FL}} \gamma(\alpha)$, *where $\gamma(\alpha)$ is any iterated conjugate of α.*

PROOF. The statement (1) is obvious. The statement (2) is obtained from (1) for the case where β is 1. Since $\alpha, \beta \Rightarrow \alpha\beta$ is provable in **FL**, taking a cut with the sequent $\Rightarrow \alpha$ and using $(\Rightarrow \backslash)$ the sequent $\Rightarrow \beta\backslash\alpha\beta$ is provable. Thus, we have (3). Similarly, we can show (4). By using (2), (3) and (4) repeatedly, we obtain (5). □

The following lemma says that weakening and exchange still hold, in a modified form, for iterated conjugates in **FL**.

LEMMA 2.13.

(1) *If a sequent $\Gamma, \Delta \Rightarrow \theta$ is provable in **FL** then so is $\Gamma, \psi \wedge 1, \Delta \Rightarrow \theta$. Thus, in particular, if $\Gamma, \Delta \Rightarrow \theta$ is provable in **FL** then so is $\Gamma, \gamma(\psi), \Delta \Rightarrow \theta$ for any iterated conjugate $\gamma(\psi)$ of any formula ψ.*
(2) *If a sequent $\Gamma, \alpha, \beta, \Delta \Rightarrow \theta$ is provable in **FL** then both $\Gamma, \beta, \lambda_\beta(\alpha), \Delta \Rightarrow \theta$ and $\Gamma, \rho_\alpha(\beta), \alpha, \Delta \Rightarrow \theta$ are also provable in **FL**.*

PROOF. The statement (1) follows by applying (1w) and $(\wedge 2 \Rightarrow)$ to $\Gamma, \Delta \Rightarrow \theta$. The statement (2) is obtained by using the fact that both $\beta, \lambda_\beta(\alpha) \Rightarrow \alpha\beta$ and $\rho_\alpha(\beta), \alpha \Rightarrow \alpha\beta$ are provable in **FL**. □

Now we give a syntactic proof of a *parametrized local deduction theorem* (PLDT) for **FL**. In the theorem, $\prod$ means fusions of finitely many formulas.

THEOREM 2.14. [GO06a] *Suppose that $\Phi \cup \Psi \cup \{\varphi\}$ is a set of formulas and **L** is a substructural logic. Then*

[16]The terminology comes because of the similarity to group theory, where a conjugate is of the form $\alpha^{-1}\varphi\alpha$; we will see the connection with lattice-ordered groups in the next chapter.

$\Phi, \Psi \vdash_{\mathbf{L}} \varphi$ *iff* $\Phi \vdash_{\mathbf{L}} (\prod_{i=1}^{n} \gamma_i(\psi_i))\backslash\varphi$, *for some* n, *where each* $\gamma_i(\psi_i)$ *is an iterated conjugate of a formula* $\psi_i \in \Psi$.

PROOF. In view of the definition of $\vdash_{\mathbf{L}}$, it is enough to show the statement for **FL**. Since $\Psi \vdash_{\mathbf{FL}} \prod_{i=1}^{n} \gamma_i(\psi_i)$ by Lemma 2.12, the if-part follows. To show the converse, take a deduction P of $\Rightarrow \varphi$ from $\Phi \cup \Psi$. Our proof consists of giving a transformation of P into another deduction of $\Rightarrow (\prod_{i=1}^{n} \gamma_i(\psi_i))\backslash\varphi$ from Φ. This is carried out by using induction on the *length* of the (sub)deduction of a given sequent s (in P), and showing that there is a deduction of $\Xi, \Gamma \Rightarrow \delta$ from Φ for some (possibly empty) sequence Ξ of iterated conjugates of some formulas in Ψ, when s is $\Gamma \Rightarrow \delta$. Here, the length of a deduction of s is taken to be the total number of sequents contained in it.

Suppose that s is an initial sequent in P which is of the form $\Rightarrow \psi$ with $\psi \in \Psi$. Then $\psi \wedge 1 \Rightarrow \psi$ is provable in **FL** and moreover $\psi \wedge 1$ is (provably equivalent to) an iterated conjugate of ψ. This yields the required deduction. If s is an initial sequent of a different form, then it is enough to take the empty sequence for Ξ. Next suppose that s is of the form $\Gamma, \alpha\beta, \Lambda \Rightarrow \theta$ and it is obtained from $\Gamma, \alpha, \beta, \Lambda \Rightarrow \theta$ by applying $(\cdot \Rightarrow)$. By the induction hypothesis, for some sequence Ξ of iterated conjugates of some formulas in Ψ there is a deduction of $\Xi, \Gamma, \alpha, \beta, \Lambda \Rightarrow \theta$ from Φ. Then, by applying $(\cdot \Rightarrow)$ to this sequent, we get a deduction of $\Xi, \Gamma, \alpha\beta, \Lambda \Rightarrow \theta$ from Φ. The same argument works for (1w), (0w), $(\Rightarrow \vee)$, $(\wedge \Rightarrow)$ and $(\Rightarrow /)$.

Suppose that s is of the form $\Gamma \Rightarrow \alpha \wedge \beta$ and it is obtained from $\Gamma \Rightarrow \alpha$ and $\Gamma \Rightarrow \beta$ by applying $(\Rightarrow \wedge)$. By the induction hypothesis, for some sequences Ξ and Π of iterated conjugates of some formulas in Ψ there are deductions of $\Xi, \Gamma \Rightarrow \alpha$ and $\Pi, \Gamma \Rightarrow \beta$ from Φ. By using Lemma 2.13 (1), there are also deductions of $\Xi, \Pi, \Gamma \Rightarrow \alpha$ and $\Xi, \Pi, \Gamma \Rightarrow \beta$ from Φ. Then, by applying $(\Rightarrow \wedge)$ to them, we have a deduction of $\Xi, \Pi, \Gamma \Rightarrow \alpha \wedge \beta$ from Φ. The same argument works also for $(\vee \Rightarrow)$.

Suppose that s is of the form $\Gamma \Rightarrow \beta/\alpha$ and it is obtained from $\alpha, \Gamma \Rightarrow \beta$ by applying $(\Rightarrow /)$. By the induction hypothesis, for some sequence Ξ of iterated conjugates of some formulas in Ψ there is a deduction of $\Xi, \alpha, \Gamma \Rightarrow \beta$ from Φ. Then, by using Lemma 2.13 (2) for left-conjugates repeatedly, we get a deduction of $\alpha, \lambda_\alpha(\Xi), \Gamma \Rightarrow \beta$ from Φ. Here, $\lambda_\alpha(\Xi)$ represents a sequence of formulas obtained from Ξ by replacing each η in it by $\lambda_\alpha(\eta)$, and thus is again a sequence of iterated conjugates of some formulas in Ψ. Finally, suppose that s is of the form $\Gamma, \Lambda \Rightarrow \alpha\beta$ and is obtained from both $\Gamma \Rightarrow \alpha$ and $\Lambda \Rightarrow \beta$ by applying $(\Rightarrow \cdot)$. By the induction hypothesis, for some sequences Ξ and Π of iterated conjugates of some formulas in Ψ there are deductions of $\Xi, \Gamma \Rightarrow \alpha$ and $\Pi, \Lambda \Rightarrow \beta$ from Φ. By applying $(\Rightarrow \cdot)$ to them, we have a deduction of $\Xi, \Gamma, \Pi, \Lambda \Rightarrow \alpha\beta$ from Φ. Then, using Lemma 2.13 (2) for right-conjugates repeatedly, we get a deduction of $\Xi, \rho_{\Gamma^{-1}}(\Pi), \Gamma, \Lambda \Rightarrow \alpha\beta$ from Φ. Here, $\rho_{\Gamma^{-1}}(\Pi)$ represents the sequence of formulas obtained from Π by replacing each member by an iterated right-conjugate of it with parameters given by Γ in reverse order, and thus is again a sequence of iterated conjugates of some formulas in Ψ. For example, when Γ is $\varepsilon_1, \varepsilon_2$ and Π is η_1, η_2, η_3, the sequence $\rho_{\Gamma^{-1}}(\Pi)$ becomes $\rho_{\varepsilon_2}(\rho_{\varepsilon_1}(\eta_1)), \rho_{\varepsilon_2}(\rho_{\varepsilon_1}(\eta_2)), \rho_{\varepsilon_2}(\rho_{\varepsilon_1}(\eta_3))$. The same argument works for the remaining rules, including cut.

As a particular case of the above argument, we get a deduction of $\gamma_1(\psi_1), \ldots, \gamma_n(\psi_n) \Rightarrow \varphi$ from Φ for some n such that each $\gamma_i(\psi_i)$ is an iterated conjugate of a formula $\psi_i \in \Psi$. By using both $(\cdot \Rightarrow)$ and $(\Rightarrow \backslash)$, we have a deduction of $\Rightarrow (\prod_{i=1}^{n} \gamma_i(\psi_i))\backslash\varphi$ from Φ. □

The above theorem is said to be *parametrized*, since iterated conjugates in $\prod_{i=1}^{n} \gamma_i(\psi_i)$ contain *parameters*, i.e., formulas other than ψ_i. On the other hand, since both $\alpha \Rightarrow \beta\backslash\alpha\beta$ and $\alpha \Rightarrow \beta\alpha/\beta$ are provable in $\mathbf{FL_e}$, for logics over $\mathbf{FL_e}$ we do not need to use iterated conjugates in the statement of the theorem; hence, we can show the following *local deduction theorem* (LDT). It is called *local* because the number n in the product on the right-hand side depends on Ψ and φ and cannot be taken as fixed.

COROLLARY 2.15. [GO06a] *Suppose that $\Phi \cup \Psi \cup \{\varphi\}$ is a set of formulas.*

(1) *For every substructural logic* $\mathbf{L}$ *over* $\mathbf{FL_e}$,

$$\Phi, \Psi \vdash_{\mathbf{L}} \varphi \quad \text{iff} \quad \begin{array}{l} \Phi \vdash_{\mathbf{L}} (\prod_{i=1}^{n}(\psi_i \wedge 1)) \to \varphi, \text{ for some } n, \\ \text{where } \psi_i \in \Psi \text{ for each } i. \end{array}$$

Thus, if $\mathbf{L}$ *is moreover an extension of* $\mathbf{FL_{ei}}$, *then*

$$\Phi, \Psi \vdash_{\mathbf{L}} \varphi \quad \text{iff} \quad \begin{array}{l} \Phi \vdash_{\mathbf{L}} (\prod_{i=1}^{n} \psi_i) \to \varphi, \text{ for some } n, \\ \text{where } \psi_i \in \Psi \text{ for each } i. \end{array}$$

(2) *In particular, suppose that Ψ consists of a finite number of formulas $\psi_1, \ldots, \psi_m$. Then, for every substructural logic* $\mathbf{L}$ *over* $\mathbf{FL_e}$,

$$\Phi, \psi_1, \ldots, \psi_m \vdash_{\mathbf{L}} \varphi \quad \text{iff} \quad \Phi \vdash_{\mathbf{L}} (\textstyle\prod_{i=1}^{m}(\psi_i \wedge 1))^k \to \varphi, \text{ for some } k.$$

Hence, if $\mathbf{L}$ *is moreover an extension of* $\mathbf{FL_{ec}}$,

$$\Phi, \psi_1, \ldots, \psi_m \vdash_{\mathbf{L}} \quad \text{iff} \quad \Phi \vdash_{\mathbf{L}} (\textstyle\prod_{i=1}^{m}(\psi_i \wedge 1)) \to \varphi.$$

As another corollary, we obtain that the standard *deduction theorem* (DT) holds for all superintuitionistic logics. Using the above results, we have the following theorem.

THEOREM 2.16. [GO06a] *For any set of formulas Φ, the following three conditions are equivalent.*

(1) *Φ is closed under the consequence relation* $\vdash_{\mathbf{FL}}$.
(2) *Φ satisfies the following conditions;*
   - *the set Φ includes all formulas in* $\mathbf{FL}$,
   - *if $\varphi, \varphi\backslash\psi \in \Phi$ then $\psi \in \Phi$,*
   - *if $\varphi \in \Phi$ then $\varphi \wedge 1 \in \Phi$,*
   - *if $\varphi \in \Phi$ and ψ is an arbitrary formula then $\psi\backslash\varphi\psi, \psi\varphi/\psi \in \Phi$.*

(3) Φ *satisfies the following conditions;*
- *the set* Φ *includes all formulas in* **FL**,
- *if* $\varphi, \varphi\backslash\psi \in \Phi$ *then* $\psi \in \Phi$,
- *if* $\varphi \in \Phi$ *then any iterated conjugate of* φ *is in* Φ.

PROOF. Statement (1) implies (2) by Lemma 2.12. It is obvious that (2) implies (3). To show that (3) implies (1), suppose that $\Phi \vdash_{\mathbf{FL}} \varphi$. Then, by Theorem 2.14, for some iterated conjugates $\gamma_i(\psi_i)$ with $\psi_i \in \Phi$, $\vdash_{\mathbf{FL}} (\prod_{i=1}^{n} \gamma_i(\psi_i))\backslash\varphi$, or equivalently $\gamma_n(\psi_n)\backslash(\gamma_{n-1}(\psi_{n-1})\backslash \cdots (\gamma_1(\psi_1)\backslash\varphi)\cdots)$ is in **FL** and hence is in Φ. Since each $\gamma_i(\psi_i)$ is in Φ by our assumption, we have $\varphi \in \Phi$ by using the second condition of (3) repeatedly. □

As a corollary of the above theorem, we have the following result which has already been mentioned.

COROLLARY 2.17. [GO06a] *A set of formulas* **L** *is a substructural logic over* **FL** *iff it satisfies the following conditions;*
- **L** *includes all formulas in* **FL**,
- *if* $\varphi, \varphi\backslash\psi \in \mathbf{L}$ *then* $\psi \in \mathbf{L}$,
- *if* $\varphi \in \mathbf{L}$ *then* $\varphi \wedge 1 \in \mathbf{L}$,
- *if* $\varphi \in \mathbf{L}$ *and* ψ *is an arbitrary formula then* $\psi\backslash\varphi\psi, \psi\varphi/\psi \in \mathbf{L}$,
- **L** *is closed under substitution.*

The above corollary tells us how to get a Hilbert-style calculus for a given substructural logic over **FL**. There are many ways of choosing axiom schemes, but it is enough to take the following (modus ponens, adjunction unit, product normality left and right) as its rules of inference. A set of axiom schemes of a Hilbert-style system for **FL** is given explicitly in the next section.

$$\frac{\varphi\backslash\psi \quad \varphi}{\psi}\ (\mathrm{mp}_\ell) \qquad \frac{\varphi}{\varphi \wedge 1}\ (\mathrm{adj}_u) \qquad \frac{\varphi}{\sigma\backslash\varphi\sigma}\ (\mathrm{pn}_\ell) \qquad \frac{\varphi}{\sigma\varphi/\sigma}\ (\mathrm{pn}_r)$$

2.5. Hilbert systems

The rules (modus ponens, unit adjunction and product normality) of Theorem 2.16 together with the theorems of **FL** can be thought of as an (infinite) Hilbert presentation of $\vdash_{\mathbf{FL}}$. The set of all theorems can be reduced to a finite list of axioms that, together with the rules mentioned above, form a Hilbert-style system that we will call **HFL**. This system simplifies a lot for the case of $\mathbf{FL_e}$ to a system $\mathbf{HFL_e}$, so we start with the latter; see Figure 2.9. The system $\mathbf{HFL_e}$ is a variant of the one given in [vAR04]. There is a lot of choice in the selection of the particular axioms of **HFL**. We do not claim that our presentation is optimal in any sense; actually, $\mathbf{HFL_e}$ and **HFL** unlike their variants in [vAR04] and [GO], respectively, do not enjoy the strong separation property. Nevertheless, we feel that the choice of inference rules facilitates the comparison with the structure theory of the algebraic models, as we will see in the next chapter, and most of the axioms specialize nicely when structural rules are added to the corresponding sequent calculi.

2.5.1. The systems $\mathbf{HFL_e}$ and HFL. See Figures 2.9 and 2.10. Recall that in the presence of exchange the formulas $\alpha\backslash\beta$ and β/α are mutually derivable. Therefore, we collapse the two formulas to $\alpha\to\beta$. In other words we replace the two connectives $\backslash$ and $/$ in our language by a single one. We note that the combination of (adj$_u$) and $(\cdot\wedge)$ can be replaced by (adj) $\{a, b\} \vdash a \wedge b$; see also the remark after Corollary 2.17. None of these rules is needed in the case of intuitionistic logic (or in general for logics with integrality).

The axiom (pf) is the same as the axiom (B) of BCK logic. It is equivalent to (sf) of **E** in the presence of (per), but we write it in the current form because it generalizes easier to (pf$_\ell$) of **HFL**; the axiom (sf) is not valid in **HFL**. On the other hand the generalization of (per) to (a) is more involved. Actually, in $\mathbf{HFL_e}$ the axiom (per) can be replaced by the axiom (as) $\alpha \to [(\alpha\to\beta)\to\beta]$. Nevertheless, in **HFL** both of them are needed, in the forms of (a) and (as$_{\ell\ell}$), while (pn$_\ell$) and (pn$_r$) are equivalent to the generalizations of the rule form $\{\alpha\} \vdash (\alpha\to\beta)\to\beta$ of (as). Also, in the presence of (per) the rules $(1\backslash)$ and $(\backslash 1)$ of **HFL** are equivalent.

It is clear that **LJ** generalizes in a more natural way to **FL** than **HJ** generalizes to **HFL**; this is why we chose to start the chapter with the former.

Due to the lack of commutativity of **FL** and the presence of the two division connectives, the system **HFL** is more complicated than $\mathbf{HFL_e}$. Nevertheless, in many rules the arrow $\to$ simply turns into left division $\backslash$.

2.5.2. Derivable rules. Recall from Section 1.6 that theories of $\vdash_{\mathbf{FL}}$ are deductive filters of $\mathbf{Fm}_{\mathcal{L}}$, so theories are closed under the axioms and the rules (mp$_\ell$), (adj$_u$) and (pn)[=(pn$_\ell$) + (pn$_r$)]. Conversely, to establish that a rule is satisfied by all deductive filters, it is enough to establish that the rule is satisfied by all theories; i.e., it is enough to consider only one algebra of the type of residuated lattices: the absolutely free algebra $\mathbf{Fm}_{\mathcal{L}}$ and its deductive filters (namely the theories). We will refer to the axioms and theorems of **HFL** by (ahfl) and (hfl), respectively.

LEMMA 2.18. [GO06a] [GO] *The following rules are derivable in* **HFL**.

$$\frac{\alpha\backslash\beta \quad \beta\backslash\delta}{\alpha\backslash\delta}\ (\mathrm{tr}_\ell) \qquad \frac{\alpha}{(\alpha\backslash\beta)\backslash\beta}\ (\mathrm{n}_\ell) \qquad \frac{\alpha}{\beta/(\beta/\alpha)}\ (\mathrm{n}_r)$$

$$\frac{\alpha \quad \beta\backslash(\alpha\backslash\delta)}{\beta\backslash\delta}\ (\mathrm{np}_\ell) \qquad \frac{\alpha \quad (\delta/\alpha)/\beta}{\delta\backslash\beta}\ (\mathrm{np}_r) \qquad \frac{\beta\backslash\alpha}{\alpha/\beta}\ (\mathrm{symm})$$

(id)	$\alpha \to \alpha$	(identity)
(pf)	$(\alpha \to \beta) \to [(\delta \to \alpha) \to (\delta \to \beta)]$	(prefixing)
(per)	$[\alpha \to (\beta \to \delta)] \to [\beta \to (\alpha \to \delta)]$	(permutation)
$(\cdot\wedge)$	$[(\alpha \wedge 1)(\beta \wedge 1)] \to (\alpha \wedge \beta)$	(fusion conjunction)
$(\wedge\to)$	$(\alpha \wedge \beta) \to \alpha$	(conjunction implication)
$(\wedge\to)$	$(\alpha \wedge \beta) \to \beta$	(conjunction implication)
$(\to\wedge)$	$[(\alpha \to \beta) \wedge (\alpha \to \delta)] \to [\alpha \to (\beta \wedge \delta)]$	(implication conjunction)
$(\to\vee)$	$\alpha \to (\alpha \vee \beta)$	(implication disjunction)
$(\to\vee)$	$\beta \to (\alpha \vee \beta)$	(implication disjunction)
$(\vee\to)$	$[(\alpha \to \delta) \wedge (\beta \to \delta)] \to [(\alpha \vee \beta) \to \delta]$	(disjunction implication)
$(\to\cdot)$	$\beta \to (\alpha \to \alpha\beta)$	(implication fusion)
$(\cdot\to)$	$[\beta \to (\alpha \to \delta)] \to (\alpha\beta \to \delta)$	(fusion implication)
(1)	1	(unit)
$(1\to)$	$1 \to (\alpha \to \alpha)$	(unit implication)

$$\frac{\alpha \quad \alpha \to \beta}{\beta}\ (\mathrm{mp}) \qquad\qquad \frac{\alpha}{\alpha \wedge 1}\ (\mathrm{adj}_u)$$

(modus ponens) (adjunction unit)

FIGURE 2.9. The system $\mathbf{HFL_e}$.

$$\frac{\alpha \quad \beta}{\alpha\beta}\ (\mathrm{p}) \qquad \frac{\alpha \quad \beta}{\alpha \wedge \beta}\ (\mathrm{adj}) \qquad \frac{\alpha}{\delta\backslash\alpha\delta \wedge 1}\ (\mathrm{pn}_u) \qquad \frac{\alpha}{\delta\alpha/\delta \wedge 1}\ (\mathrm{pn}_u)$$

$$\frac{\alpha \wedge \beta}{\alpha}\ (\mathrm{up}_m) \qquad \frac{\alpha \wedge \beta}{\beta}\ (\mathrm{up}_m) \qquad \frac{\alpha \quad \beta/\alpha}{\beta}\ (\mathrm{mp}_r)$$

$$\frac{\alpha}{(\alpha\backslash\beta)\backslash\beta \wedge 1}\ (\mathrm{n}_u) \qquad \frac{\alpha}{(\beta/(\beta/\alpha) \wedge 1}\ (\mathrm{n}_u)$$

PROOF. In the following proofs we stress the fact that an uppermost formula is an axiom by putting a line over it (and not considering it as a leaf of the proof-tree any more). Leafs of the proof-tree (without a line over them) form the assumptions of the proof.

(tr_ℓ): The following proof shows that (tr_ℓ) is derivable in **HFL**

$$\dfrac{\alpha\backslash\beta \qquad \dfrac{\beta\backslash\delta \quad \overline{(\beta\backslash\delta)\backslash[(\alpha\backslash\beta)\backslash(\alpha\backslash\delta)]}\ (\mathrm{pf}_\ell)}{(\alpha\backslash\beta)\backslash(\alpha\backslash\delta)}\ (\mathrm{mp}_\ell)}{\alpha\backslash\delta}\ (\mathrm{mp}_\ell)$$

(id_ℓ)	$\alpha\backslash\alpha$	(identity)
(pf_ℓ)	$(\alpha\backslash\beta)\backslash[(\delta\backslash\alpha)\backslash(\delta\backslash\beta)]$	(prefixing)
$(\mathrm{as}_{\ell\ell})$	$\alpha\backslash[(\beta/\alpha)\backslash\beta]$	(assertion)
(a)	$[(\beta\backslash\delta)/\alpha]\backslash[\beta\backslash(\delta/\alpha)]$	(associativity)
$(\cdot\backslash/)$	$[(\beta(\beta\backslash\alpha))/\beta]\backslash(\alpha/\beta)$	(fusion divisions)
$(\cdot\wedge)$	$[(\alpha\wedge 1)(\beta\wedge 1)]\backslash(\alpha\wedge\beta)$	(fusion conjunction)
$(\wedge\backslash)$	$(\alpha\wedge\beta)\backslash\alpha$	(conjunction division)
$(\wedge\backslash)$	$(\alpha\wedge\beta)\backslash\beta$	(conjunction division)
$(\backslash\wedge)$	$[(\alpha\backslash\beta)\wedge(\alpha\backslash\delta)]\backslash[\alpha\backslash(\beta\wedge\delta)]$	(division conjunction)
$(\backslash\vee)$	$\alpha\backslash(\alpha\vee\beta)$	(division disjunction)
$(\backslash\vee)$	$\beta\backslash(\alpha\vee\beta)$	(division disjunction)
$(\vee\backslash)$	$[(\alpha\backslash\delta)\wedge(\beta\backslash\delta)]\backslash[(\alpha\vee\beta)\backslash\delta]$	(disjunction division)
$(\backslash\cdot)$	$\beta\backslash(\alpha\backslash\alpha\beta)$	(division fusion)
$(\cdot\backslash)$	$[\beta\backslash(\alpha\backslash\delta)]\backslash(\alpha\beta\backslash\delta)$	(fusion division)
(1)	1	(unit)
$(1\backslash)$	$1\backslash(\alpha\backslash\alpha)$	(unit division)
$(\backslash 1)$	$\alpha\backslash(1\backslash\alpha)$	(division unit)

$$\frac{\alpha \quad \alpha\backslash\beta}{\beta}\ (\mathrm{mp}_\ell) \qquad \frac{\alpha}{\alpha\wedge 1}\ (\mathrm{adj}_u) \qquad \frac{\alpha}{\beta\backslash\alpha\beta}\ (\mathrm{pn}_\ell) \quad \frac{\alpha}{\beta\alpha/\beta}\ (\mathrm{pn}_r)$$

(modus ponens) (adjunction unit) (product normality)

FIGURE 2.10. The system **HFL**.

(n_ℓ): To prove $(\alpha\backslash\beta)\backslash\beta$ from α in **HFL**, we have

$$\dfrac{\dfrac{\alpha}{(\alpha\backslash\beta)\backslash[\alpha(\alpha\backslash\beta)]}\,(\mathrm{pn}_\ell) \qquad \dfrac{\dfrac{\dfrac{}{(\alpha\backslash\beta)\backslash(\alpha\backslash\beta)}\,(\mathrm{id}_\ell) \quad \dfrac{}{[(\alpha\backslash\beta)\backslash(\alpha\backslash\beta)]\backslash([\alpha(\alpha\backslash\beta)]\backslash\beta)}\,(\cdot\backslash)}{[\alpha(\alpha\backslash\beta)]\backslash\beta}\,(\mathrm{mp}_\ell)}{\{(\alpha\backslash\beta)\backslash[\alpha(\alpha\backslash\beta)]\}\backslash[(\alpha\backslash\beta)\backslash\beta]}\,(\mathrm{pf}_\ell),(\mathrm{mp}_\ell)}{(\alpha\backslash\beta)\backslash\beta}\,(\mathrm{mp})$$

Likewise, the opposite formula $\beta/(\beta/\alpha)$ is proved from α

(np_ℓ): It is left as an exercise to check that $[(\alpha\backslash\delta)\backslash\delta]\backslash\{[\beta\backslash(\alpha\backslash\delta)]\backslash(\beta\backslash\delta)\}$ is a theorem of **HFL**.

$$\dfrac{\dfrac{\dfrac{\alpha}{(\alpha\backslash\delta)\backslash\delta}\,(\mathrm{n}_\ell) \quad \dfrac{}{[(\alpha\backslash\delta)\backslash\delta]\backslash\{[\beta\backslash(\alpha\backslash\delta)]\backslash(\beta\backslash\delta)\}}\,(\mathrm{hfl})}{[\beta\backslash(\alpha\backslash\delta)]\backslash(\beta\backslash\delta)}\,(\mathrm{mp}_\ell) \qquad \dfrac{\beta\backslash(\alpha\backslash\delta)}{\{[\beta\backslash(\alpha\backslash\delta)]\backslash(\beta\backslash\delta)\}\backslash(\beta\backslash\delta)}\,(\mathrm{n}_\ell)}{\beta\backslash\delta}\,(\mathrm{mp}_\ell)$$

The proof of (np_r) is left as an exercise.

(symm): For the two directions, we have

$$\dfrac{\alpha\backslash\beta \qquad \dfrac{\dfrac{\dfrac{\dfrac{}{(\alpha\backslash\beta)\backslash(\alpha\backslash\beta)}\,(\mathrm{id}_\ell)}{\alpha\backslash(\beta/(\alpha\backslash\beta))}\,(\mathrm{a})}{[\alpha(\alpha\backslash(\beta/(\alpha\backslash\beta)))]/\alpha}\,(\mathrm{pn}_r)}{(\beta/(\alpha\backslash\beta))/\alpha}\,(\cdot\backslash/)}{\beta/\alpha}\,(\mathrm{np}_r) \qquad\qquad \dfrac{\beta/\alpha \qquad \dfrac{}{\alpha\backslash[(\beta/\alpha)\backslash\beta]}\,(\mathrm{as}_{\ell\ell})}{\alpha\backslash\beta}\,(\mathrm{np}_\ell)$$

For (p) and (adj), we have

$$\dfrac{\beta \qquad \dfrac{\alpha}{\beta\backslash\alpha\beta}\,(\mathrm{pn}_\ell)}{\alpha\beta}\,(\mathrm{mp}_\ell) \qquad \dfrac{\dfrac{\dfrac{\alpha}{\alpha\wedge 1}\,(\mathrm{adj}_u) \quad \dfrac{\beta}{\beta\wedge 1}\,(\mathrm{adj}_u)}{(\alpha\wedge 1)(\beta\wedge 1)}\,(\mathrm{p}) \qquad \dfrac{}{(\alpha\wedge 1)(\beta\wedge 1)\backslash(\alpha\wedge\beta)}\,(\cdot\wedge)}{\alpha\wedge\beta}\,(\mathrm{mp}_\ell)$$

The rule (pn_u) follows from (pn) and (adj_u). Also, the rule (up_m) follows from (mp_ℓ) and $(\wedge\backslash)$; the rule (mp_r) follows from (mp_ℓ) and (sym). Finally, (n_u) follows from (n) and (adj_u). □

We have abbreviated (n_l)+(n_l) to (n), and (pn_l)+(pn_l) to (pn).

LEMMA 2.19. [GO06a] *In the formulation of* **HFL**, *the rule* (pn) *can be replaced by either* (n), *or* (symm), *or both of* (np_ℓ) *and* (np_r). *Moreover, the combination of the rules* (adj_u) *and* (pn) *can be replaced by* (pn_u).

PROOF. In the proof of Lemma 2.18 we showed that, under the assumption of the axioms of **HFL** and the rules (mp_ℓ) and (adj_u), the rule (symm) follows from the combination of (np_ℓ) and (np_r), that the rules (np_ℓ) and (np_r) follow from (n), and that the rule (n) follows from (pn). We will show that, under the same assumptions, (pn) follows from (symm). By $(\backslash\cdot)$, (symm) and (a), we obtain $\alpha\backslash(\alpha\beta/\beta)$; so, from α we deduce $\alpha\beta/\beta$ by (mp_ℓ). By (symm), we have $\beta\backslash\alpha\beta$. Likewise, we show that $\beta\alpha/\beta$ is provable from α.

Assume that conditions (ahfl), (mp_ℓ) and (pn_u) hold. We will show that conditions (adj_u) and (pn) hold as well. From α, we deduce $\delta\backslash\alpha\delta\wedge 1$, by (pn_u). Moreover, $(\delta\backslash\alpha\delta\wedge 1)\backslash(\delta\backslash\alpha\delta))$ by $(\wedge\backslash)$, so we obtain $(\delta\backslash\alpha\delta)$, by (mp_ℓ). Additionally, from α we deduce $1\backslash\alpha 1\wedge 1$, by (pn_u). Since $(1\backslash\alpha 1\wedge 1)\backslash(\alpha\wedge 1)$ is in (hfl), we have $\alpha\wedge 1$, by (mp_ℓ). □

2.5.3. Equality of two consequence relations. For every sequent s of the form $(\alpha_1,\ldots,\alpha_n\Rightarrow\alpha)$, we define the formula $\varphi(s)=((\alpha_1\cdot\cdots\cdot\alpha_n)\backslash\alpha)$.If $n=0$, we define $\varphi(s)=\alpha$. If S is a set of sequents, we set $\varphi[S]=\{\varphi(s): s\in S\}$.

THEOREM 2.20. [GO] *For every set* $S\cup\{s\}$ *of sequents,* $S\vdash^{seq}_{\mathbf{FL}} s$ *implies* $\varphi[S]\vdash_{\mathbf{HFL}}\varphi(s)$.

PROOF. We use induction on the set of rule schemes and axioms of **FL**. Note that every rule scheme of **FL** gives rise to an infinite set of rule schemes to be shown to be derivable in **HFL**; this is because of the freedom in the length of of the sequences of formulas denoted by capital Greek letters. We indicate how this can be done for $(\vee \Rightarrow)$ and leave the rest of the proof as an exercise.

We will verify that the translation of the rule

$$\frac{\Gamma, \alpha, \Delta \Rightarrow \varphi \quad \Gamma, \beta, \Delta \Rightarrow \varphi}{\Gamma, \alpha \vee \beta, \Delta \Rightarrow \varphi}\ (\vee \Rightarrow)$$

is derivable in **HFL**. In other words, since the length n of $\Gamma = (\gamma_1, \ldots, \gamma_n)$ and m of $\Delta = (\delta_1, \ldots, \delta_m)$ can be arbitrary, we will show that for every n, m, and for any formulas $\gamma_1, \ldots, \gamma_n, \delta_1, \ldots, \delta_m$, the rule

$$\frac{\gamma_1 \cdot \cdots \cdot \gamma_n \cdot \alpha \cdot \delta_1 \cdot \cdots \cdot \delta_m \backslash \varphi \quad \gamma_1 \cdot \cdots \cdot \gamma_n \cdot \beta \cdot \delta_1 \cdot \cdots \cdot \delta_m \backslash \varphi}{\gamma_1 \cdot \cdots \cdot \gamma_n \cdot (\alpha \vee \beta) \cdot \delta_1 \cdot \cdots \cdot \delta_m \backslash \varphi}$$

is derivable in **HFL**. We set $\gamma = \gamma_1 \cdot \cdots \cdot \gamma_n$ and $\delta = \delta_1 \cdot \cdots \cdot \delta_m$. We will use the fact that the rules

$$\frac{\varphi\psi\backslash\chi}{\psi\backslash(\psi\backslash\chi)}\ (\mathrm{lr1}) \qquad \frac{\psi\backslash(\psi\backslash\chi)}{\varphi\psi\backslash\chi}\ (\mathrm{lr2}) \qquad \frac{\varphi\psi\backslash\chi}{\varphi\backslash(\chi/\psi)}\ (\mathrm{rr1}) \qquad \frac{\varphi\backslash(\chi/\psi)}{\varphi\psi\backslash\chi}\ (\mathrm{rr2})$$

are derivable in **HFL**, also left as an exercise.

Starting from $(\gamma \cdot \alpha \cdot \delta)\backslash\varphi$, we obtain $(\alpha \cdot \delta)\backslash[\gamma\backslash\varphi]$, using (lr1), and $\alpha\backslash[(\gamma\backslash\varphi)/\delta]$, using (rr1). Likewise, we have $\beta\backslash[(\gamma\backslash\varphi)/\delta]$. Using (adj), $(\backslash\wedge)$ and (mp_ℓ), we obtain $(\alpha \vee b)\backslash[(\gamma\backslash\varphi)/\delta]$. Finally, using the rules (lr2) and (rr2), we have $[\gamma \cdot (\alpha \vee \beta) \cdot \delta]\backslash\varphi$. □

COROLLARY 2.21. [GO] *The relations* $\vdash_{\mathbf{FL}}$ *and* $\vdash_{\mathbf{HFL}}$ *are equal.*

PROOF. By Theorem 2.20, we have $\vdash_{\mathbf{FL}} \subseteq \vdash_{\mathbf{HFL}}$. Also, it is easy to see that modus ponens (mp_ℓ) can be implemented in $\vdash_{\mathbf{FL}}$, as is indicated by the following proof in **FL**,

$$\dfrac{\Rightarrow \varphi\backslash\psi \qquad \dfrac{\varphi \Rightarrow \varphi \quad \psi \Rightarrow \psi}{\varphi, \varphi\backslash\psi \Rightarrow \psi}\ (\backslash \Rightarrow)}{\dfrac{\varphi \Rightarrow \psi \qquad\qquad \Rightarrow \varphi}{\Rightarrow \psi}\ (\mathrm{cut})}\ (\mathrm{cut})$$

and all the axioms of **HFL** are theorems of $\vdash_{\mathbf{FL}}$. For example, for (pf) we have that $\vdash_{\mathbf{FL}} (\beta\backslash\delta)\backslash[(\alpha\backslash\beta)\backslash(\alpha\backslash\delta)]$. Indeed, the following proof in **FL** shows that $\vdash^{seq}_{\mathbf{FL}} \Rightarrow (\beta\backslash\delta)\backslash[(\alpha\backslash\beta)\backslash(\alpha\backslash\delta)]$.

$$\dfrac{\dfrac{\dfrac{\dfrac{\dfrac{\alpha \Rightarrow \alpha \quad \beta \Rightarrow \beta}{\alpha, \alpha\backslash\beta \Rightarrow \beta}\,(\backslash\Rightarrow) \quad \delta \Rightarrow \delta}{\alpha, \alpha\backslash\beta, \beta\backslash\delta \Rightarrow \delta}\,(\backslash\Rightarrow)}{\alpha\backslash\beta, \beta\backslash\delta \Rightarrow \alpha\backslash\delta}\,(\Rightarrow\backslash)}{\beta\backslash\delta \Rightarrow (\alpha\backslash\beta)\backslash(\alpha\backslash\delta)}\,(\Rightarrow\backslash)}{\Rightarrow (\beta\backslash\delta)\backslash[(\alpha\backslash\beta)\backslash(\alpha\backslash\delta)]}\,(\Rightarrow\backslash)$$

So, $\vdash_{\mathbf{FL}}$ contains $\vdash_{\mathbf{HFL}}$. □

Hilbert-style systems that correspond to basic substructural logics can be obtained from **HFL** by adding the corresponding formulas, given in Lemma 2.5, as new axioms. Nevertheless, as for the case of $\mathbf{HFL_e}$ these systems have simpler equivalent formulations. For example, (adj_u) is redundant in $\mathbf{HFL_i}$ (in the presence of integrality).

2.6. Algebraization and deductive filters

As in the case of classical and intuitionistic logic we will prove the completeness and strong completeness of the system **HFL** relative to the variety FL of FL-algebras. In other words we will show that the consequence relations $\vdash_{\mathbf{HFL}}$ and $\models_{\mathsf{FL}}$ are equally expressive. This, together with the fact that the translations are mutually inverse, is usually encoded by the statement "the variety **FL** is the equivalent algebraic semantics for $\vdash_{\mathbf{HFL}}$." The algebraization of $\vdash_{\mathbf{FL}}$ was proved in [GO06a] using results in [BP89] and [BT03]. Here we give a proof from first principles using syntactic arguments. We will also give a description of the deductive filters of the algebraic models.

2.6.1. Algebraization. We begin by defining two translations: ρ from equations to formulas and τ from formulas to equations. Namely we put

$$s = t \quad \overset{\rho}{\mapsto} \quad s\backslash t \wedge t\backslash s$$
$$\varphi \quad \overset{\tau}{\mapsto} \quad 1 = 1 \wedge \varphi$$

An alternative definition with $\rho(s = t)$ translated as $\{s\backslash t, t\backslash s\}$ produces a set of formulas rather than a single formula, but it works equally well and has the advantage of covering weaker signatures without conjunction.

LEMMA 2.22. *If φ is an axiom of* **HFL**, *then $\tau(\varphi)$ is a valid equation in* FL. *Moreover, for a rule of* **HFL** *with assumptions φ (and ψ) and conclusion χ, if $\tau(\varphi)$ (and $\tau(\psi)$) hold(s) in an FL-algebra under a given valuation, then $\tau(\chi)$ holds in the same FL-algebra under the same valuation, as well.*

PROOF. We first note that if φ is of the form $\psi\backslash\chi$, then the resulting equation $\tau(\varphi) = (1 = 1 \wedge \psi\backslash\chi)$ is equivalent, in the theory of FL-algebras, to the inequality $1 \leq \psi\backslash\chi$ and to $\psi \leq \chi$, because of residuation. Therefore, for axioms of the form $\psi\backslash\chi$, we will check the inequality $\psi \leq \chi$. We work out the proofs of only a couple of axioms and leave the rest as an exercise.

Obviously, $\tau(\varphi\backslash\varphi) = (\varphi \leq \varphi)$ holds in FL. For (pf_ℓ), we obtain the equation $\varphi\backslash\psi \leq (\chi\backslash\varphi)\backslash(\chi\backslash\psi)$, which is equivalent to $(\chi\backslash\varphi)(\varphi\backslash\psi) \leq (\chi\backslash\psi)$ and follows from Lemma 2.6(13). For ($\backslash\cdot$), we have $\psi\backslash(\varphi\backslash\chi) \leq \varphi\psi\backslash\chi$, which, by residuation, is equivalent to $(\varphi\psi)[\psi\backslash(\varphi\backslash\chi)] \leq \chi$. The latter follows by the associativity of multiplication and Lemma 2.6(4).

For (mp_ℓ), if $1 \leq \varphi$ and $\varphi \leq \psi$, then $1 \leq \psi$, while for (pn_ℓ) if $1 \leq \varphi$, then $\psi \leq \psi\varphi$, by the fact that multiplication is order preserving. Here we used φ and ψ for evaluations of formulas in an algebra rather than for formulas. □

By induction on the proof of ψ from assumptions Φ, we have the following.

COROLLARY 2.23. *For every set of formulas $\Phi \cup \{\psi\}$, if $\Phi \vdash_{\mathbf{HFL}} \psi$ then $\tau[\Phi] \models_{\mathsf{FL}} \tau(\psi)$.*

Recall from Section 1.6, that if Q is a set of quasiequations, then $\vdash_Q$ is the consequence relation presented by *Con* and the rules of the form (q), for $q \in Q$. In particular, if Q is a set of equations, then the presentation of $\vdash_Q$ consists of *Con* and the equations in Q taken as axioms. Let Ax_{FL} be an equational basis for FL; for example we can take the one given in Theorem 2.7. We will be interested in the relation $\vdash_{Ax_{\mathsf{FL}}}$.

LEMMA 2.24. *If $s = t$ is an equation in Ax_{FL}, then $\rho(s = t)$ is a theorem of* **HFL**. *Also, the ρ images of the axioms and of the rules of Con are derivable in* **HFL**.

PROOF. First note that the formula $\varphi \wedge \psi$ is equivalent in **HFL** to the set of formulas $\{\varphi, \psi\}$. Indeed, from $\varphi \wedge \psi$ we obtain both φ and ψ by ($\wedge\backslash$) and (mp_ℓ). Conversely, from $\{\varphi, \psi\}$, we obtain $\{\varphi \wedge 1, \psi \wedge 1\}$ by (adj_u), $\{\varphi \wedge 1, (\varphi \wedge 1)\backslash(\psi \wedge 1)(\varphi \wedge 1)\}$, by ($\mathrm{pn}_\ell$) and $(\psi \wedge 1)(\varphi \wedge 1)$. Finally, $\psi \wedge \varphi$ follows from ($\cdot\wedge$) and (mp_ℓ).

Moreover, we can show that $\rho(s \leq t) = \rho(s = s \wedge t) = s\backslash(s \wedge t) \wedge (s \wedge t)\backslash s$ is equivalent to $s\backslash t$ by using ($\wedge\backslash$), (mp_ℓ) and (tr_ℓ) for the forward direction and (id_ℓ), ($\backslash\wedge$), (tr_ℓ), ($\wedge\backslash$) and (adj) for the converse.

Therefore, the formula $\rho(s = t) = s\backslash t \wedge t\backslash s$ is equivalent in **HFL** to the set of formulas $\{s\backslash t, t\backslash s\}$. We will go through the axiomatization of FL-algebras and give the proof for selected equations. We leave the remaining ones to the reader.

For $t = t \wedge t$, we need to prove $(t \wedge t)\backslash t$, which is an instance of ($\wedge\backslash$), and $t\backslash(t \wedge t)$. For the latter we start from $t\backslash t$, by (id_ℓ), and apply (adj) to get $(t\backslash t) \wedge (t\backslash t)$. By a combination of ($\backslash\wedge$) and ($\mathrm{mp}_\ell$), we obtain $t\backslash(t \wedge t)$.

For the associativity of meet, we need to verify $[r \wedge (s \wedge t)]\backslash[(r \wedge s) \wedge t]$ (and the inverse of the fraction). From $[r \wedge (s \wedge t)]\backslash(s \wedge t)$ and $(s \wedge t)\backslash s$, we obtain $[r\wedge(s\wedge t)]\backslash s$, which together with $[r\wedge(s\wedge t)]\backslash r$ yields $[r\wedge(s\wedge t)]\backslash(r\wedge s)$. We also have $[r\wedge(s\wedge t)]\backslash t$, from $[r\wedge(s\wedge t)]\backslash(s\wedge t)$ and $(s\wedge t)\backslash t$, so we obtain $[r\wedge(s\wedge t)]\backslash[(r\wedge s)\wedge t]$ by $(\backslash\wedge)$ and (mp_ℓ).

For $r(r\backslash t\wedge s)\leq t$, we need to verify $r(r\backslash t\wedge s)\backslash t$ or simply, in view of $(\backslash\cdot)$ and (mp_ℓ), $(r\backslash t\wedge s)\backslash(r\backslash t)$, which follows from $(\wedge\backslash)$.

For $(s\wedge t/r)r\leq t$, we use $(\mathrm{as}_{\ell\ell})$, $(\backslash\cdot)$, (mp_ℓ) and the rule $(\cdot\backslash\cdot)$, mentioned below.

For the axiom (refl) we use (id_ℓ), the rule (sym) is obvious after we write every equation as two inequalities and (tran) follows from (tr_ℓ). As mentioned in Section 1.6 it is enough to check the rule (rep'_t) for the unary polynomials obtained directly from each fundamental operation. These correspond to the rules:

$$\frac{\alpha\backslash\beta}{(\delta\wedge\alpha)\backslash(\delta\wedge\beta)}\quad\frac{\alpha\backslash\beta}{(\alpha\wedge\delta)\backslash(\beta\wedge\delta)}\quad\frac{\alpha\backslash\beta}{(\delta\vee\alpha)\backslash(\delta\vee\beta)}\quad\frac{\alpha\backslash\beta}{(\alpha\vee\delta)\backslash(\beta\vee\delta)}$$

$$\frac{\alpha\backslash\beta}{(\beta\backslash\delta)\backslash(\alpha\backslash\delta)}\quad\frac{\alpha\backslash\beta}{(\delta\backslash\alpha)\backslash(\delta\backslash\beta)}\quad\frac{\alpha\backslash\beta}{(\delta/\beta)\backslash(\delta/\alpha)}\quad\frac{\alpha\backslash\beta}{(\alpha/\delta)\backslash(\beta/\delta)}$$

$$\frac{\alpha\backslash\beta}{(\delta\cdot\alpha)\backslash(\delta\cdot\beta)}\quad\frac{\alpha\backslash\beta}{(\alpha\cdot\delta)\backslash(\beta\cdot\delta)}\ (\cdot\backslash\cdot)$$

We leave it as an exercise to check that these rules are derivable in **HFL**. □

By induction on the proof of ε from E in Ax_{FL}, we have the following.

COROLLARY 2.25. *For every set of equations $E\cup\{\varepsilon\}$, if $E\vdash_{Ax_{\mathsf{FL}}}\varepsilon$, then $\rho[E]\vdash_{\mathbf{HFL}}\rho(\varepsilon)$.*

COROLLARY 2.26. *For every set of equations $E\cup\{\varepsilon\}$, if $E\models_{\mathsf{FL}}\varepsilon$ then $\rho[E]\vdash_{\mathbf{HFL}}\rho(\varepsilon)$.*

PROOF. Let $E\models_{\mathsf{FL}}\varepsilon$. Without loss of generality, we can assume that E is finite, since FL is an elementary class; we denote by $\mathsf{AND}\ E$ the conjunction of all equations of E. We have $\models_{\mathsf{FL}}(\mathsf{AND}\ E\Rightarrow\varepsilon)$, that is $\mathsf{FL}\models(\mathsf{AND}\ E\Rightarrow\varepsilon)$, so $Ax_{\mathsf{FL}}\vdash_{FOL}(\mathsf{AND}\ E\Rightarrow\varepsilon)$, where Ax_{FL} is an equational basis for FL. By Theorem 1.66, the rule (E,ε) corresponding to the quasiequation $\mathsf{AND}\ E\Rightarrow\varepsilon$ is derivable in $\vdash_{Ax_{\mathsf{FL}}}$; i.e., $E\vdash_{Ax_{\mathsf{FL}}}\varepsilon$. Hence, by Corollary 2.25, $\rho[E]\vdash_{\mathbf{HFL}}\rho(\varepsilon)$. □

It is clear that for every equation $s=t$, we have

$$s=t\ \dashv\models_{\mathsf{FL}}\ 1=1\wedge(s\backslash t\wedge t\backslash s)$$

so using the terminology of Blok and Pigozzi, we have the following theorem. We will refer to this and the following two theorems as the *algebraization theorem* or *result* for FL.

THEOREM 2.27. *The variety FL of FL-algebras is an equivalent algebraic semantics for the system* **HFL**; *the equivalence formula is $\varphi\backslash\psi\wedge\psi\backslash\varphi$ and the defining equation is $1=1\wedge\varphi$.*

We ask the reader as an exercise to verify the details (see page 98 for Blok and Pigozzi's definition of algebraizability). The equivalence formula $\varphi\Delta\psi = \varphi\backslash\psi \wedge \psi\backslash\varphi$ is what we called $\rho(\varphi = \psi)$. The equation $1 \approx 1 \wedge \varphi$ is what we called $\tau(\varphi)$. The definition of algebraization becomes more concise if we write it in terms of τ and ρ. For more on the characterization of these two maps and an equivalent definition the reader is referred to [GT].

For every class $\mathcal{K}$ of pointed residuated lattices and for every set Φ of formulas over $\mathcal{L}$, let $\mathbf{L}(\mathcal{K}) = \{\varphi \in Fm_{\mathcal{L}} \mid \mathcal{K} \models 1 \leq \varphi\}$ and $\mathsf{V}(\Phi) = \mathsf{FL} \cap \mathrm{Mod}(\{1 \leq \varphi \mid \varphi \in \Phi\})$; recall that $1 \leq \varphi$ is short for $1 = 1 \wedge \varphi$. The following two theorems are consequences of Theorem 2.27 and their proofs are left as exercise. We mention the following facts that essentially provide a sketch for the proofs. For a substructural logic $\mathbf{L}$ and for a set $\Phi \cup \{\varphi\}$ of formulas, we have $\Phi \vdash_{\mathbf{L}} \varphi$ iff $\Phi \cup \mathbf{L} \vdash_{\mathbf{FL}} \varphi$ iff $\tau[\Phi] \cup \tau[\mathbf{L}] \models_{\mathsf{FL}} \tau(\varphi)$ iff $\tau[\Phi] \models_{\mathsf{V}(L)} \tau(\varphi)$. Also, for a subvariety $\mathcal{V}$ of FL and for a set $E \cup \{\varepsilon\}$ of equations, we have $E \models_{\mathcal{V}} \varepsilon$ if $E \cup Th_e(\mathcal{V}) \models_{\mathsf{FL}} \varepsilon$ iff $\rho[E \cup Th_e(\mathcal{V})] \vdash_{\mathbf{FL}} \rho(\varepsilon)$ iff $\rho[E] \cup \mathbf{L}(\mathcal{V}) \vdash_{\mathbf{FL}} \rho(\varepsilon)$ iff $\rho[E] \vdash_{\mathbf{L}(\mathcal{V})} \rho(\varepsilon)$. Finally, it is useful to note that $Th_e(\mathsf{V}_1 \cup \mathsf{V}_2) = Th_e(\mathsf{V}_1) \cap Th_e(\mathsf{V}_2)$.

THEOREM 2.28. [GO06a]

(1) *For every $\mathcal{K} \subseteq \mathsf{FL}$, $\mathbf{L}(\mathcal{K})$ is a substructural logic and for every $\Phi \subseteq Fm_{\mathcal{L}}$, $\mathsf{V}(\Phi)$ is a subvariety of* FL.
(2) *The maps $\mathbf{L} : \mathbf{\Lambda}(\mathsf{FL}) \to \mathbf{\Lambda}(\mathbf{FL})$ and $\mathsf{V} : \mathbf{\Lambda}(\mathbf{FL}) \to \mathbf{\Lambda}(\mathsf{FL})$ are mutually inverse, dual lattice isomorphisms.*
(3) *If a substructural logic $\mathbf{L}$ is axiomatized relative to* **HFL** *by a set of formulas Φ, then the variety $\mathsf{V}(\mathbf{L})$ is axiomatized relative to* FL *by the set of equations $\tau[\Phi]$.*
(4) *If a subvariety $\mathcal{V}$ of* FL *is axiomatized relative to* FL *by a set of equations E, then the substructural logic $\mathbf{L}(\mathcal{V})$ is axiomatized relative to* **HFL** *by the set of formulas $\rho[E]$.*
(5) *A substructural logic is commutative, integral or contractive iff the corresponding variety is.*

THEOREM 2.29. [GO06a] *For every substructural logic $\mathbf{L}$ over* **FL**, *its deducibility relation $\vdash_{\mathbf{L}}$ is algebraizable and $\mathsf{V}(\mathbf{L})$ is an equivalent algebraic semantics for it. In particular,*

(1) *if $\Phi \cup \{\varphi\}$ is a subset of $Fm_{\mathcal{L}}$ and $\mathbf{L}$ is a substructural logic, then*
$$\Phi \vdash_{\mathbf{L}} \varphi \text{ iff } \rho[\Phi] \models_{\mathsf{V}(\mathbf{L})} 1 \leq \varphi, \text{ and}$$
(2) *if $E \cup \{t = s\}$ is a set of $\mathcal{L}$-equations and $\mathcal{V}$ is a subvariety of* FL, *then*
$$E \models_{\mathcal{V}} t = s \text{ iff } \tau[E] \vdash_{\mathbf{L}(\mathcal{V})} t\backslash s \wedge s\backslash t.$$

It is worth mentioning that the the relations established between FL-algebras and **FL** can be transferred verbatim to relations between residuated lattices and the positive fragment $\mathbf{FL}^+$, the system obtained from **FL** by removing the constant 0 from the language.

2.6.2. Deductive filters. Recall the definition of a *deductive filter* relative to a consequence relation from Section 1.6. We will focus on deductive filters of FL-algebras relative to $\vdash_{\mathbf{HFL}}$. By definition a subset F of an FL-algebra $\mathbf{A}$ is a deductive filter if it is closed under the axioms and inference rules of **HFL**; i.e., if φ is an axiom, then $f(\varphi) \in F$, for all valuations f in $\mathbf{A}$ ($f : \mathbf{Fm}_{\mathcal{L}} \to \mathbf{A}$) and if (Φ, φ) is an inference rule, then for all valuations f in $\mathbf{A}$, $f[\Phi] \subseteq F$ implies $f(\varphi) \in F$.

Given a subset F of $\mathbf{A}$ that is already closed under the inference rules (mp_ℓ), (adj_u) and (pn) of **HFL**, we have that F is a deductive filter (closed under the axioms of **HFL**) iff it is closed under all theorems of **HFL**; in other words the condition (ahfl) can be replaced by (hfl). We claim, in turn, that this is equivalent to the stipulation that, for all $x \in A$,

(up_u) if $1 \leq x$, then $x \in F$.

To see this, recall that, by the algebraization theorem, t is a theorem of **HFL** iff $1 \leq t$ holds in all FL-algebras. Also, let t be a term such that $1 \leq t$ holds in all FL-algebras, and $\bar{a}$ an element of an appropriate power of A. Then, $1 \leq t(\bar{a})$ is true in $\mathbf{A}$; hence, if we assume (up_u), then $t(\bar{a}) \in F$. Conversely, assume F is closed under all theorems of **HFL** and $1 \leq a$, for some $a \in A$. For $t = (1 \wedge x)\backslash x$, the equation $1 \leq t$ holds in all FL-algebras. So, $a = t(a) \in F$.

Additionally, (up_m) takes the form

(up) $x \in F$ and $x \leq y$, then $y \in F$.

A deductive filter F of an FL-algebra $\mathbf{A}$ is called *integral*, if $x\backslash(1 \wedge x) \in F$, for every $x \in A$; it is called *contractive*, if $x\backslash x^2 \in F$, for every $x \in A$; finally, it is called *commutative*, if $xy\backslash yx \in F$, for every $x, y \in A$. Note that this definition agrees with the one for substructural logics, in view of the correspondence given in the proof of Lemma 2.5.

Note that in every FL-algebra $\mathbf{A}$, its positive part $A^+ = \{a \in A \mid 1 \leq a\}$ is the least deductive filter of $\mathbf{A}$. Moreover, an FL-algebra is commutative, integral or contractive iff its positive part is iff every deductive filter of $\mathbf{A}$ is.

LEMMA 2.30. *If $\mathbf{A}$ is an FL-algebra and F a subset of A, then F is a deductive filter of $\mathbf{A}$ iff it satisfies one of the following sets of conditions.*

(1) *Conditions* (up_u), (mp_ℓ), (adj_u) *and* (n).
(2) *Conditions* (u), (up), (mp_ℓ) *and* (pn_u).
(3) *Conditions* (u), (up), (mp_ℓ), (adj_u) *and* (n).
(4) *Conditions* (u), (up), (p) *and* (pn_u).
(5) *Conditions* (u), (up), (p), (adj_u) *and* (pn).

Among the last list of conditions, (pn) *is redundant, if $\mathbf{A}$ is commutative;* (adj_u) *is redundant, if $\mathbf{A}$ is integral;* (p) *is redundant, if $\mathbf{A}$ is contractive.*

PROOF. To show that (up_u) is equivalent to the combination of (u) and (up), it suffices to show that (up), follows from (up_u), under the assumption of (mp_ℓ). If we assume that $x \in F$ and $x \leq y$, then $1 \leq x\backslash y$; hence $x\backslash y \in F$, by (up_u). By (mp_ℓ), we obtain $y \in F$. Consequently, in the setting of FL-algebras, (hfl) is equivalent to the combination of (u) and (up), under the assumption of (mp_ℓ).

By Lemma 2.19, (pn_u) is equivalent to the combination of (adj_u) and (pn), so a deductive filter is defined by conditions (u), (up), (mp_ℓ) and (pn_u). By the same lemma, (pn_u) is equivalent to the combination of (adj_u) and (n). We will show that under the assumption of (u), (up) and (pn_u), conditions (mp_ℓ) and (p) are equivalent. If $x, x\backslash y \in F$, then $x(x\backslash y) \in F$, by (p). Since $x(x\backslash y) \leq y$, we have $y \in F$, by (up). Conversely, it was shown in the proof of Lemma 2.19, that (p) follows from (mp_ℓ) and (pn). Since (pn) follows from (pn_u), by the same lemma, we obtain the converse implication.

Moreover, under the assumption of (u), (up) and (p), condition (pn_u) is equivalent to the combination of (adj_u) and (pn). Indeed, the backward direction requires no assumptions. Moreover, note that, by Lemma 2.19, the combination of (adj_u) and (pn) follows from (p), under the assumption of (mp_ℓ) and of (hfl). We have seen that, under our hypothesis, (mp_ℓ) follows from (p) and (hfl) follows from (u) and (up).

Note that if **A** is commutative, integral or contractive, then every deductive filter of **A** has these properties. Therefore, the last part of the lemma follows from the corresponding part of Lemma 2.19. □

COROLLARY 2.31. *Let* **A** *be an FL-algebra. Among the conditions in the definition of an deductive filter,* (pn) *is redundant, if F is commutative, and* (adj_u) *is redundant, if F is integral.*

PROOF. If F is commutative, then we have $zx\backslash xz \in F$. So, $x\backslash(z\backslash xz) \in F$ by (mp_ℓ), since $(zx\backslash xz)\backslash[x\backslash(z\backslash xz)] = 1 \subseteq F$. If $x \in F$, then $z\backslash xz \in F$, by (mp_ℓ). Thus F satisfies (pn).

If F is integral, then $x\backslash(x \wedge 1) \in F$. So, $x \wedge 1 \in F$, by (mp_ℓ). Thus, F satisfies (adj_u). □

Exercises

(1) Verify that the algebraic interpretations of the structural rules are what what they are claimed to be on page 81.
(2) Verify that all logics in Figure 2.1.3 are distinct.
(3) Show that the rule (e) is derivable in $\mathbf{FL_{ci}}$.
(4) Show that if p is a propositional variable, then $\nvdash_{\mathbf{FL}} p\backslash p^2$.
(5) Verify the details in the proof of Lemma 2.5.
(6) Give proofs of all identities from Lemma 2.6.

(7) Let $\mathbf{FL_e}^{inv}$ be the sequent calculus obtained from $\mathbf{FL_e}$ by adding all sequents of the form $\neg\neg\beta \Rightarrow \beta$ as initial sequents. Show that for a sequent $\Gamma \Rightarrow \alpha$, where α is a formula or empty, is provable in $\mathbf{FL_e}^{inv}$ iff it is provable in $\mathbf{InFL_e}$.

(8) Show that every FL_e-algebra satisfying the equations $0 \leq x$ and $\sim\sim x \leq x$ is integral.

(9) Show that a substructural logic is commutative, contractive, integral and zero bounded iff it contains the formulas $\varphi\psi\backslash\psi\varphi$, $\varphi\backslash\varphi^2$, $\varphi\backslash(1 \wedge \varphi)$ and $0\backslash\varphi$, respectively.

(10) Show that the four identities from Theorem 2.7 are jointly equivalent to the statement: $x \cdot y \leq z$ iff $y \leq x\backslash z$ iff $x \leq z/y$.

(11) Show that integral, contractive residuated lattices are Brouwerian algebras and hence commutative.

(12) Give proofs of all the sequents that describe derivations of linguistic types on page 2.3.1.

(13) Verify that the two translations of terms for MV-algebras in Section 2.3.5 do indeed provide term equivalences.

(14) Give an example of a Heyting algebra that is not an MV-algebra and an example of an MV-algebra that is not a Heyting algebra.

(15) A *Wajsberg algebra* is an algebra $(A, \rightarrow, \neg, 1)$ such that
- $1 \rightarrow x = x$
- $(x \rightarrow y) \rightarrow ((y \rightarrow z) \rightarrow (x \rightarrow z)) = 1$
- $(x \rightarrow y) \rightarrow y = (y \rightarrow x) \rightarrow x$
- $(\neg x \rightarrow \neg y) \rightarrow (y \rightarrow x) = 1$.

Show that Wajsberg algebras are term equivalent to MV-algebras.

(16) A *bounded commutative BCK-algebra* is an algebra $(A, -, 0, 1)$ such that
- $(x - y) - z = (x - z) - y$
- $x - (x - y) = y - (y - x)$
- $x - x = 0$
- $x - 0 = x$
- $x - 1 = 0$.

Show that bounded commutative BCK-algebras are term equivalent to MV-algebras.

(17) Show that $\mathbf{A}$ is a Brouwerian algebra iff $\mathbf{A}$ is a residuated lattice such that every principal filter is a Heyting algebra.

(18) Show that the following is a quasiequational axiomatization of BCK-algebras.
(a) $(x \rightarrow y) \rightarrow ((z \rightarrow x) \rightarrow (z \rightarrow y)) = 1$,
(b) $(x \rightarrow (y \rightarrow z)) \rightarrow (y \rightarrow (x \rightarrow y)) = 1$,
(c) $x \rightarrow (y \rightarrow x) = 1$,
(d) $x \rightarrow y = 1$ and $y \rightarrow x = 1 \Rightarrow x = y$.

(19) Prove that the algebra $\mathbf{A}$ defined on in the proof of Proposition 2.10 is not a subreduct of any algebra from FL. [Hint: negation and implication

are incompatible, even if you insist that $x \to y = x\backslash y$ and $\neg x = d/x$ for some d.]

(20) Prove that the formula (C) from Section 2.3.2 is derivable in **R**. Hint: show that the following three formulas are theorems of **R**:
- $(((x \to y) \to y) \to z) \to (x \to z)$
- $(((x \to y) \to (z \to y)) \to u) \to ((z \to x) \to u)$
- $((((y \to z) \to z) \to (x \to z)) \to (y \to (x \to z))) \to$
$((x \to (y \to z)) \to (y \to (x \to z)))$

(21) Let S_n be the set of nonempty subsets of $\{1, \ldots, n\}$ ordered by reverse inclusion (it looks like a Boolean algebra with n atoms without the top element). Let **M** be the algebra $(M, \cup, \cap, \to, 0, 1)$, where M is the set of all downsets of S_n ordered by inclusion and $X \to Y$ is the greatest Z with $X \cap Z \subseteq Y$. Show that **M** is a Heyting algebra.

(22) Prove that Medvedev logic **ML** contains the logic of weak excluded middle **KC**. [Hint: See the previous exercise.]

(23) Prove that every subdirectly irreducible algebra in the subvariety of HA defined by $\neg\neg x \vee \neg x = 1$ arises by adjoining a new bottom element 0 to a subdirectly irreducible Brouwerian algebra, and conversely, every algebra obtained that way from a subdirectly irreducible Brouwerian algebra satisfies $\neg\neg x \vee \neg x = 1$.

(24) An algebra $(A, \wedge, \vee, \to, \neg, 1, \bot)$ such that $(A, \wedge, \vee, \neg\bot, \bot)$ is a bounded lattice and the following conditions hold:
(a) $1 \leq a \to a$
(b) $a \to b \leq (b \to c) \to (a \to c)$
(c) $a \to (b \to c) \leq b \to (a \to c)$
(d) $(a \to b) \wedge (a \to c) \leq a \to (b \wedge c)$
(e) $a \to \neg b \leq b \to \neg a$
(f) $\neg\neg a = a$
(g) $\neg(a \vee b) = \neg a \wedge \neg b$
(h) $1 \leq a \to b$ implies $a \leq b$
is called an arabesque (cf. [Agl96]). Prove that arabesques form a variety. [Hint: The quasi-identity can be replaced by the identities $1 \to a = a$ and $a \leq ((a \to b) \wedge 1) \to b$.]

(25) Prove that the variety of arabesques is term-equivalent to $\mathsf{InFL_{e\bot}}$.

(26) Prove using Lambek calculus that *very big mistake* is a noun. [Hint: *Mistake* is a noun (phrase), *big* is an adjective (it takes a noun and produces a noun), and *very* is an adjective modifier (takes an adjective and produces another adjective).]

(27) Prove that the following identities hold in all equivalential algebras (recall that we associate parentheses to the left).
(a) $ab = ba$
(b) $aa = bb$ (this lets you define $1 = xx$)
(c) $abb(baa) = ab$

(d) $abbb = ab$
(e) $abcc = acc(bcc)$
(f) $abbcc = accbb$
(g) $abb(bcc)(bcc) = abb$
(h) $abb(bc)(bc) = abbcc$
(i) $a(bc)b = abcb$
(j) $abbaa = abb$

(28) Prove that equivalential algebras are *congruence orderable*, that is, the binary relation defined on A by putting $a \leq b$ iff $\mathrm{Cg}^{\mathbf{A}}(a,1) \subseteq \mathrm{Cg}^{\mathbf{A}}(b,1)$ is a partial order (recall that 1 is defined by xx).

(29) Using congruence orderability draw a picture of the two-generated free equivalential algebra.

(30) A subset F of an equivalential algebra $\mathbf{A}$ is a filter if
(a) $1 \in F$
(b) $a, ab \in F$ implies $b \in F$
(c) $a \in F$ implies $abb \in F$
hold. Prove that $\theta = \{(a,b) \in A^2 \colon a, b \in F\}$ is a congruence. Conversely, show that if $\theta \in \mathrm{Con}\,\mathbf{A}$, then $[1]_\theta$ is a filter. Conclude that equivalential algebras are 1-regular.

(31) Verify the details in the proofs of Theorem 2.20 and Corollary 2.21.

(32) Verify the details in the proofs of Lemma 2.22 and Lemma 2.24.

(33) Recall that the admissibility of a rule (r) of the form (S, s) in a consequence relation $\vdash$ is equivalent to the stipulation that if all every element of S is a theorem of $\vdash$, then so is s. Show that if $S = \{s_1, \ldots, s_n\}$ and $\mathcal{K}$ is an equivalent algebraic semantics for $\vdash$ under the transformers τ and ρ, then the derivability of (r) is equivalent to the statement that for every substitution σ, if $\mathcal{K} \models \sigma(\tau(s_1))$ and ... and $K \models \sigma(\tau(s_n))$, then $K \models \sigma(\tau(s))$.

Notes

(1) The terms 'system', 'calculus' and 'logic' have been used in the literature with different meanings. As we said in the previous chapter, for the purposes of this book a logic is a set of formulas closed under a fixed consequence relation and under substitution. We do not make a formal distinction between a system and a calculus, but we informally understand a system as simply a set of rules (or sometimes the set of metarules that specify the rules) and a calculus as a system together with a notion of proof in the system that essentially allows for 'calculations'. Nevertheless, a calculus retains the information of its presentation by a system, unlike a 'consequence relation'.
Note also that the term 'system' (Gentzen system and deductive system) has been used in the context of abstract algebraic logic for the

pair consisting of a language and a consequence relation. Also, 'calculus' is taken there as a presentation for a 'system'. This swapping of meanings is the reason that in practice we use the words 'system' and 'calculus' interchangeably in this book.

(2) In proof theory a sequent calculus is called 'classical' if the two sides of the sequents are of the same data type. On the other hand, 'intuitionistic' is reserved for sequents (and systems involving sequents) where the right-hand side is a single formula or empty. For example, for linear logic there is a classical and an intuitionistic calculus, giving rise to different logics. Usually, classical systems contain a single negation connective and are cyclic and involutive; involutive, non-cyclic systems are not discussed in the literature. 'Classical **FL**' has therefore been used for a cyclic and involutive sequent calculus for which the two sides of the sequents are sequences of formulas. Since the term 'classical' seems to include the assumption of cyclicity on top of specifying the form of the sequents, and since we want to discuss involutive, non-cyclic systems like **InFL**, we avoid using the term 'classical' for modifying sequent calculi. We also use $\mathbf{InFl_e}$, since cyclicity follows from exchange.

(3) We have considered only axiomatic extensions of our calculi and focused only on varieties of algebras. Also, we used essentially the relation $\vdash_Q$ only for Q a set of equations. In general, the algebraization theory refers to connections between arbitrary extensions and quasivarieties of algebras; in this case the set Q is a set of quasiequations. This is the reason why in general studies in algebraic logic a 'logic' is taken to be a consequence relation rather than its set of theorems. Such a general exposition is not needed here since the equivalent algebraic semantics for **FL** happens to be a variety, FL.

(4) The proof of the PLDT given in [GO06a] makes use of the structure theory of the algebraic semantics. The proof we give here, see Theorem 2.14, is completely proof-theoretical.

CHAPTER 3

Residuation and structure theory

This chapter forms the core of the book. On the one hand, it closes off the textbook part, giving the essentials of the theory and many examples. On the other, it is an introduction to advanced topics that come later and therefore the examples or constructions discussed here will often reappear, typically, in a more developed form. We begin from the general notion of residuation and its close relative, that of a Galois connection. Galois connections will be central to various constructions of residuated lattices, especially in Chapters 6 and 7.

Then, we introduce residuated structures in a step-by-step fashion, paying attention to which properties follow from residuation alone and which require more structure. Special care is devoted to involutive residuated structures, which are also studied with respect to different signatures and constants are added only in the end. These structures are particularly important from a logical point of view, since they model 'classical' negation by satisfying the double negation law. Nevertheless, we do not assume cyclicity and allow in general for two negation operations, since we work in the noncommutative case with two divisions.

The general definitions are supplemented by a comprehensive list of examples of residuated structures, which are also used to illustrate important notions and basic constructions, introduced along the way. In particular, we define the notion of a nucleus, which is possibly the single most important concept in the study of residuated lattices, in particular through the fundamental construction of a nucleus image of a power set. This is followed by a list of important subvarieties, with concise information about their axiomatization and position in the lattice of subvarieties, accompanied by two related figures. This provides a skeleton of the lattice of subvarieties, based on known and well-studied varieties; general properties of the lattice will be studied in detail in Chapter 9.

The main result for the second part of the chapter is Theorem 3.47, describing congruences in residuated lattices. In particular, they are related algebraically to convex normal subalgebras (very much like in group theory) and logically to deductive filters (corresponding to theories). A discussion of special cases, such as Heyting algebras or ℓ-groups and MV-algebras, in which congruence description has a particularly simple form, provides the necessary intuition and leads to the relatively technical proof of the result mentioned above and known as the structure theory of residuated lattices.

The characterization of congruences provides tools that simplify the study of residuated lattices. As consequences of the structure theory we obtain a description of subdirectly irreducible algebras and we report on the role of central idempotent elements in determining congruences. Additionally, we discuss varieties of residuated lattices with definable principal congruences and ones with the congruence extension property. We conclude the chapter with a construction of adding bounds to a residuated lattice.

3.1. Residuation theory and Galois connections

In this section, we discuss residuated lattices and show some of their basic properties. Before this, we begin with a general discussion on 'residuation;' the subject is called *residuation theory.*

3.1.1. Residuated pairs. We give here a brief survey of basic facts in residuation theory. For more information, see e.g. [BJ72]. Also, see [Blou99].

Let $\mathbf{P}$ and $\mathbf{Q}$ be posets. A map $f : P \to Q$ is *residuated* if there exists a map $f^* : Q \to P$ such that the following holds for any $p \in P$ and any $q \in Q$:

$$f(p) \leq q \Leftrightarrow p \leq f^*(q).$$

In the above case, we say that f and f^* form a *residuated pair*, (or, an *adjunction* in category theory[1]) and that f^* is a (right) *residual* of f. To stress the dependence on the order, we often write $f\colon \mathbf{P} \to \mathbf{Q}$. Note that whether a map is residuated depends on the target set. If the target set is smaller (still containing the image of the map) or larger, the map may fail to be residuated (see Exercise 2).

There is a close relationship between residuated maps and closure operators on posets. In Section 1.6.1 we defined closure operators on powersets; the notion extends to arbitrary posets in a natural way. A *closure operator* γ on a poset $\mathbf{P}$ is a map on P that is expanding, monotone and idempotent, i.e. $x \leq \gamma(x)$, $x \leq y$ implies $\gamma(x) \leq \gamma(y)$, and $\gamma(\gamma(x)) = \gamma(x)$, for all $x, y \in P$. Dually, an *interior operator* δ on $\mathbf{P}$ is a contracting, monotone, idempotent map on $\mathbf{P}$. Any closure operator defines a residuated map $\gamma\colon \mathbf{P} \to \gamma[\mathbf{P}]$ (here we denote by $\gamma[\mathbf{P}]$ the subposet of $\mathbf{P}$ with underlying set $\gamma[P]$, the image of P under γ). The residual of γ is the inclusion map $i\colon \gamma[\mathbf{P}] \to \mathbf{P}$. Actually, in Lemma 3.4 we show that every residuated map factors through a closure operator.

In general, a residuated map f is not a closure operator, but the composition $f^* \circ f$ is.

[1]Every poset can be considered a category with objects the elements of the poset; also, there is a (unique) morphism from p to q iff $p \leq q$.

LEMMA 3.1. *If $f : \mathbf{P} \to \mathbf{Q}$ and $f^* : \mathbf{Q} \to \mathbf{P}$ form a residuated pair, then $f^* \circ f$ is a closure operator and $f \circ f^*$ is an interior operator.*

PROOF. For $x \in P$, $f(x) \leq f(x)$, so $x \leq f^*(f(x))$ by the residuation property. Likewise, $f \circ f^*(x) \leq x$. If $x, y \in P$ and $x \leq y$, then $x \leq y \leq f^*(f(y))$, so $f(x) \leq f(y)$ by residuation. Likewise, f^* is monotone. Moreover, for all $x \in P$, $f \circ f^*(f(x)) \leq f(x)$, since $f \circ f^*$ is contracting; so $f^*(f(f^*(f(x)))) \leq f^*(f(x))$, since f^* is monotone. The reverse inequality follows from the fact that $f^* \circ f$ is expanding, hence $f^* \circ f$ is idempotent. Likewise, $f \circ f^*$ is idempotent. □

In the following for maps $f, g \colon \mathbf{P} \to \mathbf{Q}$, we write $f \leq g$ if $f(x) \leq g(x)$, for all $x \in P$. Also, id_P denoted the identity map on P.

LEMMA 3.2. *The maps $f : \mathbf{P} \to \mathbf{Q}$ and $f^* : \mathbf{Q} \to \mathbf{P}$ form a residuated pair if and only if f and f^* are monotone, $id_P \leq f^* \circ f$ and $f \circ f^* \leq id_Q$.*

PROOF. The forward direction follows from Lemma 3.1. For the converse, assume that $x \in P$, $y \in Q$ and $f(x) \leq y$. Then $x \leq f^*(f(x)) \leq f^*(y)$. Also, if $x \leq f^*(y)$, then $f(x) \leq f(f^*(y)) \leq y$. □

Exercise 1 asks you to show that none of the conditions of Lemma 3.2 can be omitted. Suppose that for a given map $f : \mathbf{P} \to \mathbf{Q}$, both both f^* and f' are right residuals of f. Then, by Lemma 3.2, $f' = id_P \circ f' \leq (f^* \circ f) \circ f' = f^* \circ (f \circ f') \leq f^* \circ id_Q = f^*$, and $f^* \leq f'$. Therefore, the *residual* of a residuated map f is determined uniquely. We will denote it by f^*. In fact, the following lemma gives explicitly how f^* is determined by f.

LEMMA 3.3. *If $f : \mathbf{P} \to \mathbf{Q}$ and $f^* : \mathbf{Q} \to \mathbf{P}$ form a residuated pair, then*
(1) $f^*(q) = \max\{p \in P : f(p) \leq q\}$,
(2) $f(p) = \min\{q \in Q : p \leq f^*(q)\}$.
(3) $f \circ f^* \circ f = f$ *and* $f^* \circ f \circ f^* = f^*$

PROOF. 1) Obviously, $f^*(q)$ is an upper bound of $\{p \in P : f(p) \leq q\}$, since $f(p) \leq q$ implies $p \leq f^*(q)$, and it is an element of the set, because $f(f^*(q)) \leq q$, by Lemma 3.1. The proof of (2) is obtained dually.

The first equality in (3), follows from the fact that $f \circ f^*$ is contracting, by Lemma 3.2, and that $f^* \circ f$ is expanding and f is monotone, by the same lemma. The second equality is obtained dually. □

We now give an important example of a pair of residuated maps. Let A and B be arbitrary sets and let $R \subseteq A \times B$. For every subset X of A, we define

$$R_\exists[X] = \{b \in B \mid R(x, b), \text{ for some } x \in X\}$$

and

$$R_\forall[X] = \{b \in B \mid \text{ for all } a \in A,\ R(a, b) \text{ implies } a \in X\}$$

So, $R_\exists[X]$ contains the elements that are related to *some* element of X, while $R_\forall[X]$ contains the elements that are related *only* to elements of X. We also define the sets $R_\exists^{-1}[Y]$ and $R_\forall^{-1}[Y]$, for all subsets Y of B, where $R^{-1} \subseteq B \times A$ denotes the *inverse* relation of R.

It is easy to show that the maps $\Diamond_{R^{-1}} : \mathcal{P}(A) \to \mathcal{P}(B)$ and $\Box_R : \mathcal{P}(B) \to \mathcal{P}(A)$, where

$$\Diamond_{R^{-1}} X = R_\exists[X] \text{ and } \Box_R Y = R_\forall^{-1}[Y],$$

form a residuated pair. Similarly and dually, the maps $\Diamond_R : \mathcal{P}(B) \to \mathcal{P}(A)$ and $\Box_{R^{-1}} : \mathcal{P}(A) \to \mathcal{P}(B)$, where

$$\Diamond_R Y = R_\exists^{-1}[Y] \text{ and } \Box_{R^{-1}} X = R_\forall[X],$$

form a residuated pair, as well. Clearly, we also have $(\Diamond_{R^{-1}} X)^c = \Box_{R^{-1}} X^c$ and $(\Diamond_R Y)^c = \Box_R Y^c$. The notation of $\Diamond$ and $\Box$ comes from temporal (modal) logic, where the elements of $A = B$ represent moments of time and R is the relation 'is earlier than'. Actually, all residuated maps on powersets arise in this way; see Exercise 5.

For the special case where R is a function $h : A \to B$, we obtain the residuated pair of maps defined as $f(X) = h[X]$ and $f^*[Y] = h^{-1}[Y]$. Recall that a map $f : \mathbf{P} \to \mathbf{Q}$ is said to *reflect the order* if for all $x, y \in P$, $f(x) \leq f(y)$ implies $x \leq y$.

LEMMA 3.4. *Assume that $f : \mathbf{P} \to \mathbf{Q}$ and $f^* : \mathbf{Q} \to \mathbf{P}$ form a residuated pair.*

(1) *f factors through an onto closure operator $f^* \circ f : \mathbf{P} \to \mathbf{P}_{f^* \circ f}$ and an injection $f|_{\mathbf{P}_{f^* \circ f}} : \mathbf{P}_{f^* \circ f} \to \mathbf{Q}$.*
(2) *f^* factors through an onto interior operator $f \circ f^* : \mathbf{Q} \to \mathbf{Q}_{f \circ f^*}$ and an injection $f^*|_{\mathbf{Q}_{f \circ f^*}} : \mathbf{Q}_{f \circ f^*} \to \mathbf{P}$*
(3) *The restrictions of f and f^* to $f^*[\mathbf{Q}]$ and $f[\mathbf{P}]$, respectively, are isomorphisms between those posets.*
(4) *$f[\mathbf{P}] = f[f^*[\mathbf{Q}]]$ and $f^*[\mathbf{Q}] = f^*[f[\mathbf{P}]]$.*

PROOF. 1) The map $f^* \circ f : \mathbf{P} \to \mathbf{P}$ is a closure operator on $\mathbf{P}$ by Lemma 3.1, so it defines an onto map $f^* \circ f : \mathbf{P} \to \mathbf{P}$. We have $f|_{f^* \circ f} \circ f^* \circ f = f \circ f^* \circ f = f$, by Lemma 3.3(3). Statement (2) is proved in the same way as (1).

3) The restriction of f^* on $f[\mathbf{P}]$ is onto $f^*[\mathbf{Q}]$, since for every $x \in f^*[Q]$ there is a $y \in Q$ such that $x = f^*(y) = f^* \circ f \circ f^*(y)$; so $x \in f^*(f[P])$. Moreover, it reflects the order, since if $f^*(x) \leq f^*(y)$, for $x, y \in f[P]$, then $x = f(z), y = f(w)$, for $z, w \in P$ and $x = f(z) = f \circ f^* \circ f(z) = f(f^*(x)) \leq f(f^*(y)) = y$, since f is monotone. Finally, f^* is order reflecting, so its restriction is an order isomorphism from $f[\mathbf{P}]$ is onto $f^*[\mathbf{Q}]$. Its inverse is obviously the restriction of f on $f^*[\mathbf{Q}]$. Statement (4) follows directly from (3). □

Recall that a map $f : \mathbf{P} \to \mathbf{Q}$ is said to *preserve (arbitrary) existing joins* if for all $X \subseteq P$, whenever $\bigvee X$ exists then $\bigvee f[X]$ $(= \bigvee\{f(p) : p \in$

$X\})$ exists and $f(\bigvee X) = \bigvee f[X]$. Maps preserving (arbitrary) existing meets are defined dually.

LEMMA 3.5. *Suppose that $f : \mathbf{P} \to \mathbf{Q}$ and $f^* : \mathbf{Q} \to \mathbf{P}$ form a residuated pair. Then f preserves existing joins and f^* preserves existing meets. Thus, in particular when both $\mathbf{P}$ and $\mathbf{Q}$ are lattices, $f(p \vee p') = f(p) \vee f(p')$ for all $p, p' \in \mathbf{P}$ and $f^*(q \wedge q') = f^*(q) \wedge f^*(q')$ for all $q, q' \in \mathbf{Q}$.*

PROOF. Assume that the join $\bigvee X$ of a subset X of P exists. We will show that $f(\bigvee X)$ is the least upper bound of the set $f[X]$. By the monotonicity of f it is an upper bound. If y is an upper bound of $f[X]$, then $f(x) \leq y$ for all $x \in X$. By the residuation property $x \leq f^*(y)$, for all $x \in X$, so $\bigvee X \leq f^*(y)$. Again by residuation, we have $f(\bigvee X) \leq y$. Likewise, f^* preserves existing meets. □

If both $\mathbf{P}$ and $\mathbf{Q}$ are moreover complete, we have the converse of Lemma 3.5.

LEMMA 3.6. *Suppose that both $\mathbf{P}$ and $\mathbf{Q}$ are complete lattices and that f is a map from $\mathbf{P}$ to $\mathbf{Q}$. Then, f is residuated if and only if f preserves all (possibly infinite) joins. Dually, a map $f^* : \mathbf{Q} \to \mathbf{P}$ is the residual of a map $f : \mathbf{P} \to \mathbf{Q}$ if and only if f^* preserves all (possibly infinite) meets.*

PROOF. One direction follows from Lemma 3.5. For the converse, assume that f preserves all joins. Define $f^*(q) = \bigvee\{p \in P : f(p) \leq q\}$. Then, $f(x) \leq y$ implies $x \leq f^*(y)$. Note that f preserves the order, since if $x \leq y$, then $f(y) = f(x \vee y) = f(x) \vee f(y) \geq f(x)$. So, if $x \leq f^*(y)$, then $f(x) \leq f(f^*(y))$, hence $f(x) \leq f(\bigvee\{p \in P : f(p) \leq y\}) = \bigvee\{f(p) \in P : f(p) \leq y\} = y$. Likewise, we prove the statement for f^*. □

3.1.2. Galois connections. Named after E. Galois, who famously discovered the link between groups and fields, now known as Galois theory, Galois connections are a generalization of that situation. We have seen that both maps in a residuated pair are order preserving. In contrast, both maps in a Galois connection are order reversing. We will see that Galois connections are closely related to residuated pairs. Loosely speaking a Galois connection is a contravariant version of the covariant residuated pair. More precisely, given two posets $\mathbf{P}$ and $\mathbf{Q}$, we say that the maps ${}^{\triangleright} : P \to Q$ and ${}^{\triangleleft} : Q \to P$ form a *Galois connection*, if for all $p \in P$ and $q \in Q$,

$$q \leq p^{\triangleright} \text{ iff } p \leq q^{\triangleleft}.$$

The two maps are called the *polarities* of the Galois connection.

It is immediate that a Galois connection from $\mathbf{P}$ to $\mathbf{Q}$ is just a residuated pair from $\mathbf{P}$ to the dual $\mathbf{Q}^{\partial}$ of $\mathbf{Q}$ (indeed, $p^{\triangleright} \leq_{\mathbf{Q}^{\partial}} q$ iff $p \leq_{\mathbf{P}} q^{\triangleleft}$). This close connection is the reason why sometimes Galois connections in the literature are taken as synonymous to residuated pairs. Unfortunately, such collapsing of terminology creates confusion and does not conform with the historic origins of Galois connections; therefore we chose to distinguish the two notions in this book (see also Note 3).

Of course all the examples of residuated pairs that we mentioned can be converted to examples of Galois connections. Another example of a Galois connection is obtained by taking the map Th_e that gives the equational theory of a class of algebras and the map Mod that produces all algebraic models of a set of equations, both introduced in Chapter 1; see Exercise 8.

We give another natural example without mentioning residuated pairs. Given a relation S between two sets C and D, for $X \subseteq C$ we define

$$S_{\cap}[X] = \{d \in D : x \; S \; d, \text{ for all } x \in X\}.$$

So, $S_{\cap}[X]$ contains the elements of D that are related to *all* elements of X. Note that $S_{\exists}[X] = \bigcup_{x \in X} S_{\exists}[\{x\}]$, while $S_{\cap}[X] = \bigcap_{x \in X} S_{\exists}[\{x\}]$, so this is a different example than the one involving $\Diamond_{S^{-1}}$. We define the maps ${}^{\triangleright} : \mathcal{P}(C) \to \mathcal{P}(D)$ and ${}^{\triangleleft} : \mathcal{P}(D) \to \mathcal{P}(C)$, by $X^{\triangleright} = S_{\cap}[X]$ and $Y^{\triangleleft} = S_{\cap}^{-1}[Y]$. Exercise 9 asks you to verify that these maps form a Galois connection and that all Galois connections on powersets are of this form. The pair $({}^{\triangleright}, {}^{\triangleleft})$ is called the Galois connection *induced* by R. For the relation between this example and the example of the residuated pair $\Diamond_{R^{-1}}$ and $\Box_R$, see Exercise 10.

In the following, we write $X^{\triangleright}$ for $\{x^{\triangleright} : x \in X\}$. The following lemmas about Galois connections is a consequence of the corresponding results about residuated pairs.

LEMMA 3.7. *Assume that the maps* ${}^{\triangleright} : P \to Q$ *and* ${}^{\triangleleft} : Q \to P$ *form a Galois connection between the posets* $\mathbf{P}$ *and* $\mathbf{Q}$. *Then the following properties hold.*

(1) *The maps* ${}^{\triangleright}$ *and* ${}^{\triangleleft}$ *are both order reversing. Moreover, they convert existing joins into meets, i.e. if* $\bigvee X$ *exists in* P *for some* $X \subseteq P$, *then* $\bigwedge X^{\triangleright}$ *exists in* Q *and* $(\bigvee X)^{\triangleright} = \bigwedge X^{\triangleright}$, *and likewise for* ${}^{\triangleleft}$.
(2) *The maps* ${}^{\triangleright\triangleleft} : P \to P$ *and* ${}^{\triangleleft\triangleright} : Q \to Q$ *are both closure operators.*
(3) *We have* ${}^{\triangleright\triangleleft\triangleright} = {}^{\triangleright}$ *and* ${}^{\triangleleft\triangleright\triangleleft} = {}^{\triangleleft}$
(4) *For all* $q \in Q$, $q^{\triangleleft} = max\{p \in P : q \leq p^{\triangleright}\}$ *and for all* $p \in P$, $p^{\triangleright} = max\{q \in Q : p \leq q^{\triangleleft}\}$.
(5) *The restrictions of* ${}^{\triangleright}$ *and* ${}^{\triangleleft}$ *to* $\mathbf{Q}^{\triangleleft}$ *and* $\mathbf{P}^{\triangleright}$ *are mutually inverse, dual isomorphisms between these posets.*
(6) $\mathbf{P}^{\triangleright\triangleleft} = \mathbf{Q}^{\triangleleft}$ *and* $\mathbf{Q}^{\triangleleft\triangleright} = \mathbf{P}^{\triangleright}$.

Given a relation $R \subseteq A \times B$, the induced Galois connection between the powersets $\mathcal{P}(A)$ and $\mathcal{P}(B)$, defined above by $X^{\triangleright} = R_{\cap}[X]$ and $Y^{\triangleleft} = R_{\cap}^{-1}[Y]$, gives rise to the closure operator $\gamma_R : \mathcal{P}(A) \to \mathcal{P}(A)$ associated with R, where $\gamma_R(X) = X^{\triangleright\triangleleft}$, in view of Lemma 3.7(2).

We think of $\_Rd$ as an equation or a *context* with parameter d and say that x satisfies the equation, or fits in the *context*, $\_Rd$ if xRd, and X satisfies the equation if all of its elements do. Then $\gamma_R(X)$ is simply the solution set of all the equations that X satisfies. In algebraic geometry we can take A to be the set of points of some algebraically closed field, B the set of polynomials over the field and R the relation of being a root of a polynomial. The γ-closed elements are algebraic varieties (solution sets of polynomial equations). In universal algebra, we can take A to be a class of algebras over a fixed signature, B the set of equations over the signature and R the relation of modeling (the algebra satisfies the equation), or of the algebra being a 'solution' to the equation. The γ-closed sets are varieties of algebras. This is the historic origin of the term variety in universal algebra.

If γ is a closure operator on a complete lattice $\mathbf{P}$, we say that a subset D of P is a *basis* for γ, if the γ-closed sets are exactly the intersections of elements of D; so in particular $D \subseteq \gamma[P]$. Note that the interior operator in a topological space is a closure operator under the dual order, so the notion of basis here is dual to the one in topology.

LEMMA 3.8. *Let A and B be sets.*

(1) *If R is a relation between A and B, then γ_R is a closure operator on $\mathcal{P}(A)$.*
(2) *If γ is a closure operator on $\mathcal{P}(A)$, then $\gamma = \gamma_R$ for some relation R with domain A.*
(3) *If $(^{\triangleright}, ^{\triangleleft})$ is the Galois connection induced by R, then the collection of the sets $\{b\}^{\triangleleft}$, where $b \in B$, forms a basis for γ_R.*

PROOF. Statement (1) follows from Exercise 9 and Lemma 3.7(2).

For (2), note that $\gamma : \mathbf{P} \to \gamma[\mathbf{P}]^{\partial}$ and $i : \gamma[\mathbf{P}]^{\partial} \to \mathbf{P}$ (the inclusion map) form a Galois connection. So, by Exercise 9, $\gamma = R_{\cap}$ and $i = R_{\cap}^{-1}$, where $x \mathrel{R} y$ iff $y \in \gamma(x)$. Consequently, $\gamma_R(X) = R_{\cap}^{-1}[R_{\cap}[X]] = i \circ \gamma(X) = \gamma(X)$, for all $X \subseteq P$, hence $\gamma = \gamma_R$.

For (3) note that γ-closed elements are of the form $Y^{\triangleleft}$, where $Y \subseteq B$, and that $Y^{\triangleleft} = R_{\cap}^{-1}[Y] = \bigcap_{y \in Y} R_{\exists}^{-1}[\{y\}] = \bigcap_{y \in Y} R_{\cap}^{-1}[\{y\}] = \bigcap_{y \in Y} \{y\}^{\triangleleft}$. □

The following is a consequence of Lemma 3.2.

COROLLARY 3.9. *Two maps form a Galois connection iff they are order-reversing and their compositions are expanding.*

If the maps maps $^{\triangleright} : P \to Q$ and $^{\triangleleft} : Q \to P$ between the posets $\mathbf{P}$ and $\mathbf{Q}$ are order reversing and their compositions are the identity maps (i.e. $x^{\triangleright\triangleleft} = x$ and $y^{\triangleleft\triangleright} = y$ for all $x \in P$ and $y \in Q$, also known as the *law of double negation for* $^{\triangleright}$ *and* $^{\triangleleft}$), then we say that they form an *involutive pair* between $\mathbf{P}$ and $\mathbf{Q}$. If $\mathbf{P} = \mathbf{Q}$, we will say that the maps form an involutive pair on $\mathbf{P}$. According to Corollary 3.9, an involutive pair is a special case of a Galois connection between two posets. Obviously, in the presence of the law of double negation the order reversing stipulation is equivalent to the assumption that the maps form a Galois connection.

We will say that $(P, \leq, ^{\rhd}, ^{\lhd})$ is an *involutive poset*, if $(P, \leq)$ is a poset and $^{\rhd}, ^{\lhd}$ form an involutive pair on $\mathbf{P}$. In addition it is traditionally stipulated that the two maps coincide, but we will not include such an assumption in the definition, since we intend to apply it to the (two distinct) negation operations $\sim$ and $-$ of a (non-commutative) FL-algebra.

3.1.3. Binary residuated maps. Assume that $\mathbf{P}$, $\mathbf{Q}$ and $\mathbf{R}$ are posets and that $\cdot : P \times Q \to R$ is a *binary map.* Here we distinguish between maps and binary maps, by the understanding that the latter comes equipped with a decomposition of its domain into a Cartesian product. We say that $\cdot$ is *(component-wise) residuated* if for all $p \in P$ and $q \in Q$, the maps l_p (left multiplication by p) and r_q (right multiplication by q) are residuated, where $l_p(y) = p \cdot y$ and $r_q(x) = x \cdot q$. Note that this is different than assuming that $\cdot$ viewed as a *unary map* is residuated. Since (component-wise) binary residuated maps are very important we will refer to them simply as *residuated.* In our applications $\cdot$ will be a binary operation on a set, so the use of the term residuated will create no confusion. By setting $x \backslash z = l_x^*(z)$ and $z/y = r_y^*(z)$, we obtain a more concise definition.

A binary map $\cdot : P \times Q \to R$ is *residuated* iff there exist maps $\backslash : P \times R \to Q$ and $/ : R \times Q \to P$ such that all $x \in P$, $y \in Q$ and $z \in R$,

$$\text{(res)} \qquad x \cdot y \leq z \ \text{ iff } \ y \leq x \backslash z \ \text{ iff } \ x \leq z/y.$$

The displayed property (res) is called the *law of residuation.* It follows by Lemma 3.3 that $x \backslash z = max\{y : xy \leq z\}$ and $z/y = max\{x : xy \leq z\}$. Therefore, if $\cdot$ is residuated, the maps $\backslash$ and $/$ are uniquely determined; they are called the *left* and *right division.* We read $x \backslash y$ as 'x under y' and y/x as 'y over x'. In both expressions y is said to be the *numerator* and x the *denominator*.

As an example we can take multiplication on the set of positive reals. Then division (we have $x \backslash y = y/x$) becomes usual division of real numbers.

For another example, let A, B and C be sets and let $R \subseteq A \times B \times C$. We write $R(x, y, z)$ for $(x, y, z) \in R$ and $R[X, Y]$ for $\{z \in P \mid R(x, y, z)$ for some $x \in X, y \in Y\}$. For $X \subseteq A$ and $Y \subseteq B$, we define the binary relations,

$$R_X = \{(y, z) \in B \times C \mid R(x, y, z) \text{ for some } x \in X\}$$
$$R^Y = \{(x, z) \in A \times B \mid R(x, y, z) \text{ for some } y \in Y\}$$

and the binary map $X \cdot Y = R[X, Y]$. It is easy to see that $\cdot$ is residuated and the associated *residuals*, or division operations are $X \backslash Z = \Box^*_{R_X} Z$ and $Z/Y = \Box^*_{R^Y} Z$, where $\Box^*_{R_X}$ denotes the residual of $\Box_{R_X}$.

These are all the residuated (binary) maps on powersets, since it is easy to check that if $\cdot : \mathcal{P}(A) \times \mathcal{P}(B) \to \mathcal{P}(C)$ is a residuated map, then $X \cdot Y = R[X, Y]$, where $R(x, y, z)$ iff $z \in \{x\} \cdot \{y\}$.

Note that if $\cdot : P \times Q \to R$ is a residuated map and $\backslash$, $/$ are the associated division operations, then for every element $r \in R$ the maps $u_r : P \to Q$ and $o_r : Q \to P$, where $u_r(p) = p\backslash r$ and $o_r(q) = r/q$, form a Galois connection.

3.2. Residuated structures

In this section we will use the results on residuation theory to study partially ordered groupoids with residuation. In particular, we will discuss special classes of ordered groupoids which emerge as algebras for substructural logics.

A structure $\mathbf{G} = (G, \cdot, \leq)$ is a *partially ordered groupoid* or *pogroupoid* if $\cdot$ is a binary operation on a poset $(G, \leq)$ that is compatible with the order (or, is *monotone*), i.e.,

$$x \leq x' \text{ and } y \leq y' \text{ implies } x \cdot y \leq x' \cdot y'.$$

A *residuated partially-ordered groupoid* or *residuated pogroupoid* is a structure $\mathbf{G} = (G, \cdot, \backslash, /, \leq)$ such that $(G, \leq)$ is a poset and the law of residuation (res) holds. It follows from Lemma 3.2 that $(G, \cdot, \leq)$ is actually a partially ordered groupoid. When $(G, \cdot)$ forms a semigroup, the structure $\mathbf{G}$ is called a *residuated partially-ordered semigroup*. Similarly, a *residuated partially-ordered monoid* is defined to be a structure $\mathbf{G} = (G, \cdot, \backslash, /, 1, \leq)$ such that $(G, \cdot, \backslash, /, \leq)$ is a residuated groupoid and $(G, \cdot, 1)$ is a monoid.

A *residuated lattice ordered groupoid* (or, a *residuated ℓ-groupoid*) is an algebra $\mathbf{G} = (G, \wedge, \vee, \cdot, \backslash, /)$ such that $(G, \wedge, \vee)$ is a lattice (with associate order $\leq$) and $(G, \cdot, \backslash, /, \leq)$ is a residuated groupoid. Residuated ℓ-semigroups and residuated ℓ-monoids are defined in the obvious way, and the latter are also called *residuated lattices*, as we saw in Chapter 2.

THEOREM 3.10. *Let $\mathbf{G}$ be a residuated groupoid.*

(1) *Multiplication preserves all existing joins in both arguments, i.e., if $\bigvee X$ and $\bigvee Y$ exist for $X, Y \subseteq G$, then so does $\bigvee_{x\in X, y\in Y} xy$, and*

$$\bigvee X \cdot \bigvee Y = \bigvee_{x\in X, y\in Y} xy.$$

(2) *Divisions preserve all existing meets in the numerator, and convert all existing joins in the denominator to meets, i.e., if $\bigvee X$ and $\bigwedge Y$ exist for $X, Y \subseteq G$, then $\bigwedge_{x\in X} x\backslash z$, $\bigwedge_{x\in X} z/x$, $\bigwedge_{y\in Y} z\backslash y$ and $\bigwedge_{y\in Y} y/z$ exist for any $z \in G$, and*

$$z\backslash(\bigwedge Y) = \bigwedge_{y\in Y} z\backslash y, \qquad (\bigwedge Y)/z = \bigwedge_{y\in Y} y/z.$$

$$(\bigvee X)\backslash z = \bigwedge_{x\in X} x\backslash z, \qquad z/(\bigvee X) = \bigwedge_{x\in X} z/x.$$

(3) *$x\backslash z = max\{y \in G : xy \leq z\}$ and $z/y = max\{x \in G : xy \leq z\}$.*

PROOF. (1) follows from the fact that $\cdot$ is residuated in each coordinate and Lemma 3.5. The first two equalities in (2) follow from the same lemma, as l_z^* and r_z^* (see page 148) are the residuals of l_z and r_z. The last two equalities in (2) follow from the Lemma 3.7(1), as u_z and o_z (see page 149) form a Galois connection. Finally, (3) is a restatement of Lemma 3.3(1) for $f = l_x$ and $f = r_x$. □

In particular in a residuated pogroupoid multiplication preserves the order on both sides and the division operations are order-preserving in the numerator and order-reversing in the denominator.

By a *groupoid with order* we mean a structure $(G, \cdot, \leq)$ such that $(G, \leq)$ is a poset and nothing is assumed about the relation of multiplication with the order. The following corollary is a consequence of Lemma 3.6.

COROLLARY 3.11. *A complete groupoid with order is residuated iff multiplication distributes (on both sides) over arbitrary joins.*

A poset $\mathbf{P}$ is called *dually well-ordered* if it has no infinite ascending chain and no infinite antichain.

COROLLARY 3.12. *A dually well ordered groupoid with lattice order is residuated iff multiplication distributes (on both sides) over binary joins.*

PROOF. It is easy to see that every arbitrary join is actually a finite join, so the ordered groupoid is complete. □

COROLLARY 3.13. *A dually well ordered groupoid with total order is residuated iff multiplication is order preserving.*

COROLLARY 3.14. *The following identities hold in every residuated lattice-ordered groupoid.*

(1) $(x \vee y)z = xz \vee yz$ *and* $z(x \vee y) = zx \vee zy$.
(2) $(x \vee y)\backslash z = x\backslash z \wedge y\backslash z$ *and* $z\backslash(x \wedge y) = z\backslash x \wedge z\backslash y$.
(3) $z/(x \vee y) = z/x \wedge z/y$ *and* $(x \wedge y)/z = x/z \wedge y/z$.
(4) $x(x\backslash y) \leq y$ *and* $(y/x)x \leq y$.
(5) $x \leq y/(x\backslash y)$ *and* $x \leq (y/x)\backslash y$.
(6) $(y/(x\backslash y))\backslash y = x\backslash y$ *and* $y/((y/x)\backslash y) = y/x$.

PROOF. The first three conditions follow from Theorem 3.10 and (4) follows from residuation. (5) simply states that $x \leq o_y(u_y(x))$ and follows from Lemma 3.7(2). (6) simply states that $u_y(o_y(u_y(x))) = o_y(x)$ and follows directly from Lemma 3.7(3). □

Associativity of multiplication and the existence of unit yield more.

LEMMA 3.15. *The following identities and their mirror images hold in any residuated pomonoid.*

(1) $(x\backslash y)z \leq x\backslash yz$,
(2) $x\backslash y \leq zx\backslash zy$,

(3) $(x\backslash y)(y\backslash z) \leq x\backslash z$,
(4) $xy\backslash z \leq y\backslash(x\backslash z)$,
(5) $x\backslash(y/z) = (x\backslash y)/z$,
(6) $(x\backslash 1)y \leq x\backslash y$,
(7) $1\backslash x = x$,
(8) $1 \leq x\backslash x$,
(9) $x(x\backslash x) = x$,
(10) $(x\backslash x)^2 = x\backslash x$.
(11) $x\backslash y \leq (z\backslash x)\backslash(z\backslash y)$
(12) $x\backslash y \leq (x\backslash z)/(y\backslash z)$

PROOF. We verify a few of the properties and leave the others to the reader. For (5) we have $w \leq x\backslash(y/z)$ iff $xw \leq y/z$ iff $(xw)z \leq y$ iff $x(wz) \leq y$ iff $w \leq (x\backslash y)/z$, for every w. One inequality in (9) follows from (1), and for the reverse inequality we use (8). One inequality in (10) follows from (3) and the reverse from (8). □

A *pointed po-groupoid* is a structure $\mathbf{G} = (G, \leq \cdot, 0)$, where $(G, \leq \cdot)$ is a po-groupoid and 0 is an arbitrary element of G. Likewise we define *pointed* pogroupoids with unit, ℓ-groupoids, residuated groupoids etc. Note that *pointed residuated lattices* are exactly FL-algebras. As for FL-algebras, for every residuated pointed structure we define the negation operations $\sim$ and $-$ by $\sim x = x\backslash 0$ and $-x = 0/x$. The notions of involutive and cyclic pointed residuated pogroupoids are defined as for FL-algebras.

It follows from Lemma 2.8(1) that the negation operations are order-reversing. In combination with Lemma 2.8(4) and Lemma 3.9 this shows that the negation operations on an FL-algebra form a Galois connection on its poset reduct. Note that an FL-algebra is *involutive* if its poset reduct together with the negation operations is involutive.

3.3. Involutive residuated structures

In this section, we will discuss involutive structures, examples of which are lattice-ordered groups and MV-algebras. Involutive FL-algebras will play an important role in Chapter 8.

Recall that an FL-algebra is called involutive if it satisfies the double negation law $\sim - x = x = - \sim x$, where the negation operations are defined in terms of the constant 0. Note that the condition of involutiveness connects the operations of multiplication and division via the negation operations. Here, we will introduce algebras without the constant 0 but with two negations in the signature, and then relate them to FL-algebras.

In particular, we will add negation operations to various structures (pogroupoids, division posets, pointed residuated groupoids) of different signatures (without multiplication or without divisions). In all of them we will assume the *double negation property*

(DN) $\quad\quad {\sim}{-}x = x = -{\sim}x \quad\quad$ (double negation)

where $\sim$ and $-$ are *new operations* called left and right negation. Moreover, we will assume that these negations are *compatible* with the existing structure. Compatibility with the order will mean that the negations are order reversing, compatibility with multiplication will take the form of (IPG) below and compatibility with divisions will be contraposition (CP). Structures that satisfy these conditions will be called involutive (for their signature). On the other hand we will stick to our definition of an involutive FL-algebra (or more generally of a pointed residuated groupoid) as one that simply satisfies the law of double negation (DN) with respect to the associated negation operations (defined in terms of the constant 0 and the division operations).

The following lemmas will clarify the connections between these structures. We first discuss involutive structures without constants (structures without constants have been considered in relevance logic) and add the constants to the signature at a later stage.

3.3.1. Involutive posets. It was noted on on page 148 that the assumption that the negations are order reversing in an involutive poset (under the stipulation of double negation) can be replaced by the condition that they form a Galois connection

(GN) $\quad\quad x \leq {\sim}y$ iff $y \leq -x. \quad\quad$ (Galois negation)

Indeed, if (GN) is satisfied then $x \leq y = {\sim}{-}y$ implies $-y \leq -x$ and ${\sim}y \leq {\sim}x$. Conversely, if $x \leq {\sim}y$, then $y = -{\sim}y \leq -x$ and vice versa.

Therefore, an involutive poset is simply a structure $(P, \leq, \sim, -)$, where $(P, \leq)$ is a poset and $\sim, -$ satisfy (DN) and (GN). So,

involutive poset = poset + (DN) + (GN).

3.3.2. Involutive pogroupoids. An *involutive po-groupoid* is a structure $(P, \leq, \cdot, \sim, -)$ such that $(P, \leq, \cdot)$ is a po-groupoid (equivalently, a groupoid with order), $(P, \leq, \sim, -)$ is an involutive poset and the involutive pair $(\sim, -)$ is compatible with multiplication

(IPG) $\quad xy \leq z$ iff $y({\sim}z) \leq {\sim}x$ iff $(-z)x \leq -y. \quad\quad$ (inv. pogroupoid)

So,

involutive pogroupoid = pogroupoid + (DN) + (GN) + (IPG),

or

involutive pogroupoid = pogroupoid + inv. poset + (IPG).

Lemma 3.19 shows that an involutive FL-algebra gives rise to an involutive pogroupoid.

As we noted parenthetically above in the definition of an involutive pogroupoid, even if we simply stipulate that we have a groupoid with order, compatibility of multiplication with respect to the order follows: If $x \le z$, then $\sim z \le \sim x$. Since $zy \le zy$, we have $y[\sim(zy)] \le \sim z$, by (IPG), so $y[\sim(zy)] \le \sim x$ and $xy \le zy$, by another application of (IPG). Likewise, we obtain the compatibility of right multiplication. It is interesting to note that the above proof for left multiplication does not use right negation $(-)$ hence it does not use the fact that we have an involutive pair. This naturally leads to the notion of a pogroupoid with left negation, but we do not explore this further.

3.3.3. Involutive division posets. A *division poset* is defined to be a structure $(P, \le, \backslash, /)$ such that $(P, \le)$ is a poset and

(DP) $\quad x \le z/y$ iff $y \le x\backslash z$. $\quad$ (division poset)

An *involutive division poset* is defined as a structure $(P, \le, \backslash, /, \sim, -)$ such that $(P, \le, \sim, -)$ is an involutive poset, $(P, \le, \backslash, /)$ is a division poset and the following compatibility condition holds.

(CP) $\quad y\backslash -x = \sim y/x.$ $\quad$ (contraposition)

So, involutive division poset = division poset + (DN) + (GN) + (CP).

A division poset is called *associative* if it satisfies the identity $x\backslash(y/z) = (x\backslash y)/z$. It is called *commutative* if it satisfies the identity $x\backslash y = y/x$.

Lemma 3.19 shows that an involutive FL-algebra gives rise to an involutive associative division poset.

3.3.4. Term equivalences. We will show that the division operations and the multiplication are interdefinable in the above structures.

LEMMA 3.16. *Involutive po-groupoids and involutive division posets are term equivalent via* $x\backslash y = \sim[(-y)x]$ *and* $y/x = -[x(\sim y)]$, *and* $xy = -(y\backslash\sim x) = \sim(-y/x)$. *Moreover, they are (reducts of) residuated po-groupoids.*

PROOF. Assume that $(P, \le, \cdot, \sim, -)$ is an involutive pogroupoid and define $x\backslash y = \sim[(-y)x]$ and $y/x = -[x(\sim y)]$. We have $z \le x\backslash y$ iff $z \le \sim[(-y)x]$, by definition, iff $(-y)x \le -z$, by (GN), iff $xz \le y$, by (IPG). Likewise, we obtain the opposite condition. Therefore, $(P, \le, \cdot, \backslash, /)$ is a residuated pogroupoid. In particular, (DP) holds. For (CP), we have $z \le y\backslash -x$ iff $yz \le \sim x$, by residuation, iff $zx \le \sim y$, by (IPG), iff $z \le \sim y/z$.

Now assume that $(P, \le, \backslash, /, \sim, -)$ is an involutive division poset. We have,

$$\begin{aligned}
-(y\backslash\sim x) \le z &\Leftrightarrow \sim z \le y\backslash\sim x && \text{(GN), (DN)}\\
&\Leftrightarrow y \le \sim x/\sim z = x\backslash -\sim z = x\backslash z && \text{(DP), (CP)}\\
&\Leftrightarrow x \le z/y = \sim -z/y = -z\backslash -y && \text{(DP), (CP)}\\
&\Leftrightarrow -z \le -y/x && \text{(DP)}\\
&\Leftrightarrow \sim(-y/x) \le z && \text{(GN), (DN)}
\end{aligned}$$

for all z, hence $-(y\backslash\sim x) = \sim(-y/x)$. We define $xy = -(y\backslash\sim x) = \sim(-y/x)$. Note that we have just shown that $xy \le z$ iff $y \le x\backslash z$ iff $x \le z/y$, so multiplication is residuated. For (IPG), we have $xy \leq z$ iff $x \leq z/y = {\sim}{-z}/y = -z\backslash -y$, by (CP), iff $(-z)x \leq -y$. The second equivalence is proved similarly. □

LEMMA 3.17. *In any involutive pogroupoid (equivalently, involutive division poset) the following hold*

(1) $y\backslash x = {\sim}y/{\sim}x$
(2) $x/y = -x\backslash -y$
(3) ${\sim}[(-y)\cdot(-x)] = -[({\sim}y)\cdot({\sim}x)] = x/{\sim}y = -x\backslash y$
(4) ${\sim}[(-y)\backslash(-x)] = {\sim}(y/x) = x({\sim}y)$
(5) $-[({\sim}x)/({\sim}y)] = -(x\backslash y) = (-y)x$

If the pogroupoid is lattice-ordered then the De Morgan laws hold for both negations.

PROOF. Equations (1) and (2) and the last equation of (3) follow from (CP) and (DN). The first two equations of (3) follow from Lemma 3.16. Finally, (4) and (5) follow from (2) and (3).

For the De Morgan laws, we have for all z that $z \leq {\sim}(x \vee y)$ iff $x \vee y \leq -z$, by (GN), iff $x, y \leq z$ iff $z \leq {\sim}x, {\sim}y$, by (GN), iff $z \leq {\sim}x \wedge {\sim}y$, hence ${\sim}(x \vee y) = {\sim}x \wedge {\sim}y$. Moreover, by setting $-x$ for x and $-y$ for y, we have ${\sim}(-x \vee -y) = x \wedge y$ and $-x \vee -y = -(x \wedge y)$, by (DN). □

3.3.5. Constants. An element a in an involutive division poset is called a *negation constant*, if for all x we have ${\sim}x = x\backslash a$ and $-x = a/x$. The element a is called a *division unit* if $a\backslash x = x = x/a$ for all x.

LEMMA 3.18. *An involutive division poset admits a negation constant* 0 *iff it admits a division unit* 1 *iff the term equivalent involutive pogroupoid admits a unit element* 1. *The two constants are connected via the equalities* $1 = {\sim}0 = -0$ *and* $0 = {\sim}1 = -1$.

PROOF. In view of Lemma 3.16, we are free to use operations in both signatures. Assume that 0 is a negation constant. We have $y({\sim}0) \leq {\sim}x$ iff $xy \leq 0$, by (IPG), iff $y \leq {\sim}x$, by (res). By setting $-x$ for x, and because of double negation we have $y({\sim}0) = y$, for all y. Likewise, we have $y = (-0)y$. By setting respectively $y = -0$ and $y = {\sim}0$ in these equations we obtain that ${\sim}0$ and -0 are equal and a unit for multiplication.

Clearly, if 1 is a unit for a residuated pogroupoid, then it is a division unit for the division operations. Indeed, from $1 \cdot x \leq x$, we have $x \leq 1\backslash x$, by residuation, and from $1 \cdot (1\backslash x) \leq x$, we obtain $1\backslash x \leq x$.

Assume, now that 1 is a division unit for an involutive division poset. Then, $x\backslash -1 = {\sim}x/1 = {\sim}x$, by (CP). In particular, $1\backslash -1 = {\sim}1$, so $-1 = {\sim}1$. Also, $-x = 1\backslash -x = {\sim}1/x$, by (CP), so the element $-1 = {\sim}1$ is a negation constant. □

Since the involutive division posets that admit a negation constant (that is the involutive pointed pogroupoids with unit) are residuated by

Lemma 3.16, one might expect that they are term equivalent to involutive pointed residuated pogroupoids (with unit). Nevertheless, the former structures satisfy an additional equation: they are *associating*.

An element a of residuated po-groupoid is called *associating* if for all x, y, we have $x\backslash(a/y) = (x\backslash a)/y$, or equivalently for all x, y, z, we have $(xy)z \leq a$ iff $x(yz) \leq a$. A pointed residuated po-groupoid is called *associating* if 0 is an associating element.

LEMMA 3.19. *Involutive division posets that admit a negation constant (or, equivalently, involutive pointed pogroupoids with unit) are term equivalent to associating involutive pointed residuated pogroupoids (with unit).*

PROOF. The lemma follows from Lemma 3.16 and the observation that (CP), in the presence of a negation constant 0, is a reformulation of the fact that 0 is associating. □

We conjecture that involutive division posets can be embedded into ones with a negation constant.

In view of Lemma 3.19, involutive division posets with a negation constant provide an axiomatization of associating involutive pointed residuated pogroupoids with unit given only in terms of the division operations, the constant 0 and the negation operations, since the existence of multiplication and the fact that it is residuated follow; actually even the negations are definable in terms of divisions and 0. Likewise, involutive pointed pogroupoids with unit provide an axiomatization only in terms of multiplication, negations and 1. We conjecture that there are no such axiomatizations in the restricted signatures for involutive pointed residuated pogroupoids with unit that are not necessarily associating.

Since every associative structure is associating as well, all three structures (associative involutive division posets that admit a unit, associative involutive pogroupoids that admit a unit and involutive residuated pomonoids) are term equivalent. Therefore, involutive FL-algebras can be defined in the restricted signature with lattice operations, negations, multiplication and 1.

We note that in [TW06] involutive cyclic FL-algebras are called *dualizing residuated lattices* and cyclic FL-algebras in the reduced signature mentioned above are called *involutive residuated lattices*. In this book the adjective 'involutive' is used without any assumption of cyclicity. Moreover, we fear that the term 'involutive residuated lattices' is ambiguous, given our identification of RL with the subvariety of FL axiomatized by $0 = 1$. This is because it is not clear if it defines a subvariety of RL (where the negating constant is the unit) or whether we add a new constant in the type. Therefore, we will not use this terminology in the book.

3.3.6. Dual algebras. Given an involutive pogroupoid (equivalently, an involutive division poset), which we chose to display in the full signature of an involutive residuated pogroupoid, $\mathbf{L} = (L, \leq, \cdot, \backslash, /, \sim, -)$, we define its *dual* $\mathbf{L}^\partial = (L, \geq, +, \mathbin{\underset{\cdot}{-}}, \mathbin{\overset{\cdot}{-}}, -, \sim)$ (note the reversing of the order and of the negations), where

$$x + y = {\sim}[(-y) \cdot (-x)] = -[({\sim}y) \cdot ({\sim}x)] = x/{\sim}y = -x\backslash y$$

$$x \mathbin{\overset{\cdot}{-}} y = {\sim}[(-y)\backslash(-x)] = {\sim}(y/x) = x({\sim}y)$$

$$y \mathbin{\underset{\cdot}{-}} x = -[({\sim}x)/({\sim}y)] = -(x\backslash y) = (-y)x$$

By Lemma 3.17, the expressions in each line are equal, so the operations are well defined. The operation $+$ is called addition and the operations $\mathbin{\underset{\cdot}{-}}$ and $\mathbin{\overset{\cdot}{-}}$ are called *left subtraction* and *right subtraction*, respectively.

LEMMA 3.20. *The dual $\mathbf{L}^\partial$ of an involutive pogroupoid (equivalently, an involutive division poset) $\mathbf{L}$ is also an involutive pogroupoid. If $\mathbf{L}$ is associative or lattice-ordered, then so is $\mathbf{L}^\partial$. If $\mathbf{L}$ has unit 1 and negation constant 0, then 0 is a unit and 1 is a negation constant for $\mathbf{L}^\partial$.*

PROOF. We have $x + y \geq z$ iff ${\sim}[(-y) \cdot (-x)] \geq z$ iff $(-y) \cdot (-x) \leq -z$ iff $-y \leq -z/-x$ iff ${\sim}[({\sim}z)/({\sim}x)] \leq y$ iff $y \geq x \mathbin{\underset{\cdot}{-}} z$. The right subtraction is likewise the right residual of addition.

If multiplication is associative, then we have $(x\backslash y)/z = x\backslash(y/z)$, by Lemma 3.15(6). Since $x + y = x/{\sim}y = -x\backslash y$, we obtain $(x + y) + z = (-x\backslash y)/{\sim}z = -x\backslash(y/{\sim}z) = x + (y + z)$.

Clearly, $x + 0 = {\sim}(1 \cdot -x) = x$ and $0 + x = x$. Also, $1 \mathbin{\overset{\cdot}{-}} x = 1({\sim}x) = {\sim}x$ and $x \mathbin{\underset{\cdot}{-}} 1 = -x$. □

COROLLARY 3.21. *Involutive pogroupoids are term equivalent to their duals.*

Clearly, the dual of the dual is the original structure. We will continue the discussion on duals in Sections 3.4.5 and 3.4.2, where we will see examples of involutive FL-algebras and in Section 3.4.17, where we will define the dual with respect to a *dualizing* element.

3.4. Further examples of residuated structures

We now present an extensive list of examples of residuated structures (mostly residuated lattices and FL-algebras) that appear in algebra and logic. En route we also mention some important constructions (negative cone, nucleus image) that produce residuated structures from existing ones.

3.4.1. Boolean algebras and generalized Boolean algebras. We defined Boolean algebras in Chapter 1 as a subvariety of Heyting algebras that satisfies the double negation axiom. Clearly, Heyting algebras are integral commutative FL-algebras, where multiplication is meet. Consequently, Boolean algebras form a subvariety of FL. Nevertheless, Boolean algebras are usually defined as complemented distributive lattices, with complementation and the bounds included in the type. Here we will start from that definition and show how to view Boolean algebras as FL-algebras, justifying in this way the definition we gave in Chapter 1. At the same time we will define generalized Boolean algebras.

A *Boolean lattice* is just the lattice reduct of a Boolean algebra. Boolean lattices and Boolean algebras are obviously definitionally equivalent. We will see that Boolean algebras (defined as complemented distributive lattices) are term equivalent to a subvariety of FL.

A *generalized Boolean lattice* is a lattice such that every principal filter is the lattice reduct of a Boolean algebra. It is clear that a (lower) bounded generalized Boolean lattice is simply a Boolean lattice.

Given a generalized Boolean lattice $(L, \wedge, \vee)$, we denote by 1 its top element and by $x/y = y\backslash x$ the complement of y in the Boolean algebra based on $[x \wedge y, 1]$, where as usual, we set $[a, b] = \{c \in L \mid a \leq c \leq b\}$. *Generalized Boolean algebras* are algebras $(L, \wedge, \vee, \wedge, \backslash, /, 1)$ such that $(L, \wedge, \vee)$ is a generalized Boolean lattice and $1, \backslash, /$ are defined as above. Obviously, generalized Boolean algebras and generalized Boolean lattices are definitionally equivalent.

LEMMA 3.22. [Gal05] *If $(L, \wedge, \vee)$ is a generalized Boolean lattice and 1 is its top element, then the algebra $(L, \wedge, \vee, \wedge, \backslash, /, 1)$ is a residuated lattice.*

PROOF. It is clear that $(L, \wedge, \vee)$ is a lattice and $(L, \wedge, 1)$ is a monoid. Assume that $x \wedge y \leq z$ and let a' denote the complement in $[x \wedge y \wedge z, 1]$ of an element $a \in [x \wedge y \wedge z, 1]$. Then, $z' \leq (x \wedge y)'$, so $z' \leq x' \vee y'$. Hence,

$$z' \wedge y \leq (x' \vee y') \wedge y = (x' \wedge y) \vee (y' \wedge y) = x' \wedge y \leq x',$$

consequently,

$$x = x'' \leq (z' \wedge y)' = z \vee y' = (z \vee y') \wedge (y \vee y') = (z \wedge y) \vee y'.$$

Since y' is the complement of y in $[x \wedge y \wedge z, 1]$, $(z \wedge y) \vee y'$ is the complement of y in $[(z \wedge y) \vee (x \wedge y \wedge z), 1] = [z \wedge y, 1]$, i.e., $(z \wedge y) \vee y' = z/y$. Thus, $x \leq z/y$. Conversely, assume that $x \leq z/y$, where z/y is the complement of y in $[z \wedge y, 1]$. Then, $x \wedge y \leq z/y \wedge y = z \wedge y \leq z$.

For the left division operation, we have $x \wedge y \leq z$ iff $y \wedge x \leq z$ iff $y \leq z/x = x\backslash z$, for all $x, y, z \in L$. □

Thus, a generalized Boolean algebra is a residuated lattice such that every principal lattice filter is the lattice reduct of a Boolean algebra and multiplication coincides with meet. Note that other choices for multiplication can yield residuated lattices as well (a prominent example is relation algebras), so we insist on the last condition. We denote the class of generalized Boolean algebras by GBA.

The algebra $\mathbf{2}_r$ is obviously a generalized Boolean algebra. The following proposition shows that the class GBA is a finitely axiomatized variety and that $\mathsf{GBA} = \mathsf{V}(\mathbf{2}_r)$. Additionally, we provide a number of equational bases for it.

PROPOSITION 3.23. [Gal05] *Let* **L** *be a residuated lattice. The following statements are equivalent.*

(1) **L** *is a generalized Boolean algebra.*
(2) **L** *is in the variety* $\mathsf{V}(\mathbf{2}_r)$.
(3) **L** *satisfies the identities*
   (a) $xy = x \wedge y$, *and*
   (b) $x/(x \vee y) \vee (x \vee y) = 1$.
(4) **L** *satisfies the identities*
   (a) $xy = x \wedge y$, *and*
   (b) $(x \wedge y)/y \vee y = 1$.
(5) **L** *satisfies the identities*
   (a) $xy = x \wedge y$, *and*
   (b) $(y/x)\backslash y = x \vee y$.
(6) **L** *satisfies* $x/(x\backslash y) = x = (y/x)\backslash x$.

PROOF. We will show that (1) $\Rightarrow$ (3) $\Rightarrow$ (4) $\Rightarrow$ (2) $\Rightarrow$ (5) $\Rightarrow$ (1) and that (6) $\Leftrightarrow$ (1).

(1) $\Rightarrow$ (3): It follows directly from the definition of a generalized Boolean algebra.

(3) $\Rightarrow$ (4): The identity (4b) follows from the identity (3b), by substituting $x \wedge y$ for x.

(4) $\Rightarrow$ (2): Assume that **L** satisfies (4). We will show that it is a subdirect product of copies of $\mathbf{2}_r$. Let P be a prime filter of **L** and let $f_P : L \to 2 = \{0, 1\}$ be defined by $f_P(x) = 1$ iff $x \in P$. We will show that f_P is a residuated lattice homomorphism. It is clear that f_P is a lattice homomorphism, thus a monoid homomorphism as well. To prove that it preserves the division operations, given their behavior on $\mathbf{2}_r$, we only need to show that

$$x/y \notin P \text{ iff } x \notin P \text{ and } y \in P.$$

First observe that (4a) implies that $x = x \wedge 1$. Assume that $x/y \notin P$ and $y \notin P$. Since $x/y = (x \wedge y)/y$, by Lemma 2.6(2) and (9), and P is prime, we have $1 = (x \wedge y)/y \vee y \notin P$, a contradiction. Assume that $x/y \notin P$ and $x \in P$. Since $x \leq x/y$ and P is a filter, we have $x/y \in P$, a contradiction. Conversely, if $x \notin P$, $y \in P$ and $x/y \in P$, then $x \geq (x/y) \wedge y \in P$, by Lemma 2.6(4). Hence $x \in P$, which is a contradiction.

Since f_P is a homomorphism, $\ker(f_P)$ is a congruence on **L**. In order to prove that **L** is a subdirect product of copies of $\mathbf{2}_r$, we need only show that the intersection of the congruences above is the diagonal. This follows from the fact that in a distributive lattice any pair of elements (a, b) can be separated by a prime filter; i.e., for all a, b, there exists a prime filter P such that $a \in P$ and $b \notin P$, or $b \in P$ and $a \notin P$. Thus **L** is in $\mathsf{V}(\mathbf{2}_r)$.

(2) $\Rightarrow$ (5): It is trivial to check that $\mathbf{2}_r$ satisfies the identities in (5).

(5) $\Rightarrow$ (1): Assume that x, y are elements of $\mathbf{L}$, such that $x \leq y$. We will show that x/y is the complement of y in $[x, 1]$. We have $x \leq x/y$, since $x \wedge y \leq x$; so $x \leq y \wedge (x/y)$. Moreover, $y \wedge (x/y) \leq x$, by Lemma 2.6(4), hence $y \wedge (x/y) = x$. Additionally,

$$y \vee (x/y) = ((x/y)/y)\backslash(x/y) = (x/(y \wedge y))\backslash(x/y) = (x/y)\backslash(x/y) = 1.$$

It has already been mentioned that Brouwerian algebras have a distributive lattice reduct; so, $\mathbf{L}$ is distributive by (5a). Thus, every principal interval of $\mathbf{L}$ is a Boolean algebra.

(1) $\Leftrightarrow$ (6): Having established the equivalence of (1) and (2), note that the algebra $\mathbf{2}_r$ satisfies the identity (6). Conversely, suppose the equation (6) holds in $\mathbf{L}$. For every element y of L we have $1 = 1/(1\backslash y)$, so $1 \leq 1/y$, i.e., $y \leq 1$. So, $\mathbf{L}$ is an integral residuated lattice. Moreover, for all $x, y \in L$, $x = x/(x\backslash y)$ implies $x\backslash(x\backslash y) = (x/(x\backslash y))\backslash(x\backslash y)$. Thus, by (6) and Lemma 2.6(6), we have $x^2\backslash y = x\backslash y$. By setting $y = x^2$ and $y = x$, we obtain $x^2 = x$, for all $x \in L$. Together with integrality this gives $xy = x \wedge y$, for all $x, y \in L$, since

$$xy \leq x1 \wedge 1y = x \wedge y = (x \wedge y)^2 \leq xy.$$

Thus, $\mathbf{L}$ is a Brouwerian algebra; hence, it has a distributive lattice reduct. Assume now that $y \leq x$. We will show that the complement of x in $[y, 1]$ is $x\backslash y$. Note that $y \leq x\backslash y$, by integrality, so $y \leq x \wedge x\backslash y$. On the other hand we have $x \wedge (x\backslash y) \leq y$, by Lemma 2.6(4), thus $x \wedge x\backslash y = y$. Moreover, using Lemma 2.6(3) and (2), we have

$$\begin{aligned} x \vee x\backslash y &= (x/(x \vee x\backslash y))\backslash(x \vee x\backslash y) \\ &= (x/x \wedge x/(x\backslash y))\backslash(x \vee x\backslash y) \\ &= (x/x \wedge x)\backslash(x \vee x\backslash y) \\ &\geq x\backslash x \geq 1. \end{aligned}$$

So, $x \vee x\backslash y = 1$. □

It follows from Lemma 3.22 that Boolean algebras are term equivalent to a class of FL-algebras; the following corollary characterizes this class. Having justified the definitional equivalence we now go back to using the term equivalent definition of Boolean algebras and view them as FL-algebras. It is clear that generalized Boolean algebras are the 0-free subreducts of Boolean algebras. The next result implies that BA is a finitely axiomatized subvariety of FL generated by the two-element FL-algebra $\mathbf{2}$.

COROLLARY 3.24. *Let* $\mathbf{L}$ *be an FL-algebra. The following statements are equivalent.*

(1) $\mathbf{L}$ *is a Boolean algebra.*
(2) $\mathbf{L}$ *is in the variety* $\mathsf{V}(\mathbf{2})$.
(3) $\mathbf{L}$ *satisfies the identities*
  (a) $xy = x \wedge y$, *and*

(b) $x/(x \vee y) \vee (x \vee y) = 1$.

(4) **L** *satisfies the identities*

(a) $xy = x \wedge y$, *and*

(b) $(x \wedge y)/y \vee y = 1$.

(5) **L** *satisfies the identities*

(a) $xy = x \wedge y$, *and*

(b) $(y/x)\backslash y = x \vee y$.

(6) **L** *satisfies* $x/(x\backslash y) = x = (y/x)\backslash x$.

From Corollary 3.24(5b), for $y = 0$, it follows that $\neg\neg x = x$, since 0 is the least element in Boolean algebras. Consequently, Boolean algebras are (cyclic) involutive. On the other hand, GBA-algebras (for example the collection of all cofinite sets of $\mathbb{N}$; see Exercise 22) do not have a self-dual lattice reduct, so they do not admit involutive negations.

3.4.2. Partially ordered and lattice ordered groups. A *group* is usually defined as an algebra $\mathbf{G} = (G, \cdot, {}^{-1}, 1)$, where $(G, \cdot, 1)$ is a monoid and for all $x \in G$, $xx^{-1} = 1 = x^{-1}x$. Groups are term equivalent (for $x/y = xy^{-1}$, $y\backslash x = y^{-1}x$ and $x^{-1} = 1/x$) to algebras $(G, \cdot, \backslash, /, 1)$, where $(G, \cdot, 1)$ is a monoid and

$$xy = z \text{ iff } y = x\backslash z \text{ iff } x = z/y,$$

or in terms of equations $x(x\backslash y) = x\backslash xy = y = yx/x = (y/x)x$. Therefore, a group is simply a residuated monoid, where the order is trivial (an antichain).

In general, residuated monoids that are groups (not necessarily with a trivial order) are called *partially ordered groups*, and residuated lattices that are groups are called *lattice-ordered groups*, or simply, *ℓ-groups.* Actually, ℓ-groups form a subvariety LG of RL that is axiomatized by the identity $x(x\backslash 1) = 1$; see Lemma 3.25. The integers $\mathbb{Z}$, the rationals $\mathbb{Q}$ and the reals $\mathbb{R}$ under addition and the usual order form commutative ℓ-groups; actually the variety of commutative (or abelian) ℓ-groups is generated by each of those algebras, due to a celebrated result of Weinberg [Wei63], so $\mathsf{CLG} = \mathsf{V}(\mathbb{Z})$. An example of a non-commutative ℓ-group is obtained by taking the order automorphisms of a totally ordered set; actually every ℓ-group can be embedded in such an ℓ-group due to a result of C. Holland [Hol63].

We show here that every nontrivial ℓ-group $\mathbf{G}$ is unbounded, i.e. if there exists an element $x \in G$ different from the unit element 1 then $\mathbf{G}$ has neither a greatest nor a least element. This is in contrast with groups, since there are no finite non-trivial ℓ-groups (every unbounded lattice is infinite). Suppose that $x \neq 1$ for an element $x \in G$. If $x < 1$ then $1 < x^{-1}$. If $x \not< 1$ then $1 < 1 \vee x$. Therefore, G contains an element y such that $1 < y$. Hence, G cannot have a greatest element $\top$ in it, since then $1 < y \leq y \leq \top < \top^2$, contradicting the maximality of $\top$. If G has a least element $\bot$ then it also has a greatest element $\bot^{-1}$, so G cannot have a least element, either.

LEMMA 3.25. [Gal03] *Each of the following sets of equations forms an equational basis for* LG.

(1) $(1/x)x = 1$
(2) $x = 1/(x\backslash 1)$ *and* $x/x = 1$
(3) $x = y/(x\backslash y)$
(4) $x/(y\backslash 1) = xy$ *and* $1/x = x\backslash 1$
(5) $(y/x)x = y$
(6) $x/(y\backslash z) = (z/x)\backslash y)$

PROOF. Recall that an ℓ-group has a group and a lattice reduct and multiplication distributes over joins. Obviously all the equations are valid in ℓ-groups, if we define $x/y = xy^{-1}$ and $y\backslash x = y^{-1}x$.

A residuated lattice that satisfies the first identity is a monoid such that every element has a right inverse; so it is a group. Multiplication distributes over joins, by Lemma 2.6, so we obtain an ℓ-group.

Using the two identities of (2) and Lemma 2.6 we get

$$\begin{aligned}(1/x)/(1/x) = 1 \quad &\Rightarrow 1/(1/x)x = 1\\ &\Rightarrow [1/(1/x)x]\backslash 1 = 1\backslash 1\\ &\Rightarrow (1/x)x = 1.\end{aligned}$$

So, (2) implies (1). Setting $y = 1$ in (3), we obtain $x = 1/(x\backslash 1)$; setting $x = 1$, we get $1 = y/y$. So (3) implies (2). Setting $x = 1$ in (4), we have $1/(y\backslash 1) = y$; setting $y = 1/x$, we get $x/[(1/x)\backslash 1] = x(1/x)$, so $x/x = x(x\backslash 1)$. It follows from Lemma 2.6(9) and (4) that x/x is an element of the positive cone and that $x(x\backslash 1)$ is an element of the negative cone, so $x/x = x(x\backslash 1) = 1$. Identity (5) yields (1) for $y = 1$. Finally, for $x = z = 1$ in (6), we get $y = 1/(y\backslash 1)$ and for $z = x, y = 1$ we have $1 \leq x/x = (x/x)\backslash 1$, so $x/x \leq 1$, so $x/x = 1$. Thus, (6) implies (1). □

An important construction on ℓ-groups is that of a lexicographic product, which we define for two factors. Given two ℓ-groups $\mathbf{G}$ and $\mathbf{H}$, we define an algebra of the type of residuated lattices on their cartesian product $G \times H$. The order is *lexicographic*, namely $(g_1, h_1) \leq_{lex} (g_2, h_2)$ iff $g_1 < g_2$, or $g_1 = g_2$ and $h_1 \leq h_2$, and multiplication is defined coordinatewise; the divisions are defined as the residuals (to be shown to exist) of multiplication. Exercise 26 asks you to prove that this algebra forms an ℓ-group called the *lexicographic product* of $\mathbf{G}$ and $\mathbf{H}$.

In Exercise 27 you are asked to verify that the dual of an ℓ-group (viewed as an FL-algebra with $0 = 1$) is also an ℓ-group; see also Lemma 3.20. Note that we actually have $x + y = xy$ and $\sim x = -x = x^{-1}$. In fact, for any choice of negating constant 0 in an ℓ-group $\mathbf{G}$ the dual is also an ℓ-group with $x + y = x \cdot 0 \cdot y$ (and $\sim x = x^{-1}0$, $-x = 0x^{-1}$). We will see in

Section 3.4.17 that this is the *translation* $\mathbf{G}^a$ of $\mathbf{G}$ by the element $a = 0^{-1}$; see also Proposition 3.45.

THEOREM 3.26. *Every involutive residuated pomonoid that is a residuated and dually residuated with the same multiplication (in other words the operations of addition and multiplication coincide* $\cdot = +$*) and is either cyclic or cancellative (or even satisfies either* $x\backslash x = 1$ *or* $x/x = 1$*) is an* ℓ*-group.*

PROOF. Since $xy = x + y = {\sim}[(-y)\cdot(-x)]$, we get $x \cdot 0 = {\sim}[(-0)(-x)] = {\sim}(1 \cdot -x) = x$ and $0 \cdot x = x$, so 0 is the identity element 1. We can show that $x/y = x(-y)$, see Exercise 21. If the algebra is cyclic, then $1 \leq x/x = x(-x) = x(0/x) = x(x\backslash 0) \leq 0 = 1$, so $x(-x) = 1$, for every x; hence the algebra is an ℓ-group, because of associativity. If $x/x = 1$ the same argument works. □

Exercise 30 asks you to show that the preceding theorem is not true without the assumption of cyclicity or cancellativity.

3.4.3. The negative cone of a residuated lattice. Obviously, Boolean algebras, generalized Boolean algebras, Heyting algebras and Brouwerian algebras are all integral. On the other hand there is no non-trivial integral ℓ-group. We describe a construction that produces an integral residuated lattice from an arbitrary one. Integral residuated lattices remain unchanged by the construction.

An element a of a residuated lattice $\mathbf{L}$ is called *positive* if $a \geq 1$ and *negative* if $a \leq 1$. The *positive part* of $\mathbf{L}$ is defined as the set $L^+ = \{x \in L \mid x \geq 1\}$ of all negative elements of $\mathbf{L}$; likewise we define the *negative part* $L^- = \{x \in L \mid x \leq 1\}$ of $\mathbf{L}$.

The *negative cone* of $\mathbf{L}$ is the algebra $\mathbf{L}^- = (L^-, \wedge, \vee, \cdot, \backslash_{\mathbf{L}^-}, /_{\mathbf{L}^-}, 1)$, where $x\backslash_{\mathbf{L}^-} y = x\backslash y \wedge 1$ and $x/_{\mathbf{L}^-} y = x/y \wedge 1$. It is easy to check that $\mathbf{L}^-$ is also a residuated lattice, which is obviously integral. If $\mathcal{K}$ is a class of residuated lattices, we denote the class of negative cones of elements of $\mathcal{K}$ by $\mathcal{K}^-$. Clearly, $\mathsf{BA}^- = \mathsf{BA}$ and $\mathsf{GBA}^- = \mathsf{GBA}$, but $\mathsf{LG}^- \neq \mathsf{LG}$. Actually, Theorem 3.30 shows that LG^- is a variety. Moreover, it is shown in [BCG+03] that the negative cones of abelian ℓ-groups form a variety that is generated by the negative cone of the ℓ-group of integers, $\mathsf{CLG}^- = \mathsf{V}(\mathbb{Z}^-)$.

3.4.4. Cancellative residuated lattices. A residuated lattice (or FL-algebra) is called *cancellative* if multiplication is cancellative; i.e., if $xz = yz$ implies $x = y$, and $zx = zy$ implies $x = y$. The class of all cancellative semigroups (or monoids) forms a proper quasivariety. Nevertheless, because of the expressive power of residuated lattices due to the presence of the division operations, cancellative residuated lattices and FL-algebras form varieties, which we denote by CanRL and CanFL. In general, for any class $\mathcal{K}$, we will denote by $\mathsf{Can}\mathcal{K}$ the class of cancellative members of $\mathcal{K}$.

LEMMA 3.27. [BCG+03] *A residuated lattice is cancellative iff it satisfies the identities*

$$\text{(Can)} \qquad xy/y = x = y\backslash yx.$$

PROOF. The identity $(xy/y)y = xy$ holds in any residuated lattice since $xy/y \leq xy/y$ implies $(xy/y)y \leq xy$, and $xy \leq xy$ implies $x \leq xy/y$, hence $xy \leq (xy/y)y$. By right cancellativity, we have $xy/y = x$. Conversely, suppose $xy/y = x$ holds, and consider elements a, b, c such that $ac = bc$. Then $a = ac/c = bc/c = b$, so right cancellativity is satisfied. □

Note that ℓ-groups and their negative cones are examples of cancellative residuated lattices. On the other hand, the only cancellative (generalized) Boolean or Heyting algebras are trivial.

We now give an example, taken from [BCG+03], of an integral, cancellative, commutative residuated chain that is not the negative cone of an ℓ-group, even though its semigroup reduct is isomorphic to the semigroup reduct of the negative cone of an ℓ-group.

Let F be the universe of the free 2-generated commutative monoid on a and b. We denote the empty word by e and order F by dual shortlex order, i.e. for words $u, v \in F$ we have $u \leq v$ iff $|u| > |v|$, or $|u| = |v|$ and $u <_{\text{lex}} v$ in the lexicographic order generated by $b < a$; here $|u|$ denotes the length of u. For example,

$$1 > a > b > a^2 > ab > b^2 > a^3 > a^2b > ab^2 > b^3 > \cdots$$

Then $\mathbf{F} = (F, \wedge, \vee, \cdot, \backslash, /, 1)$ is a cancellative, commutative, integral residuated chain. However it is not the negative cone of an ℓ-group. Indeed, it is easy to see that the negative cone of an ℓ-group satisfies the law $(x/y)y \approx x \wedge y$ (see Theorem 3.30), whereas we have $(b/a)a = a^2 \neq b = a \wedge b$ in $\mathbf{F}$.

THEOREM 3.28. [BCG+03] *Any lattice is a sublattice of the lattice reduct of some cancellative, integral residuated lattice.*

PROOF. Let $\mathbf{L}$ be a lattice. Since any lattice can be embedded in a lattice with a top element, we may assume that $\mathbf{L}$ has a top element, say 1. Let $\mathbf{L}^*$ be the ordinal sum of $\mathbf{L}^n$ (the cartesian power, ordered pointwise) for $n = 0, 1, 2, \ldots$, with $\mathbf{L}^n$ above $\mathbf{L}^{n+1}$ (see Figure 3.1), and define the multiplication by concatenation of sequences (where L^0 is the set containing the empty sequence). Then the monoid reduct of $\mathbf{L}^*$ is the free monoid generated by the elements in L^1, so $\mathbf{L}^*$ is cancellative. It is residuated since each block has a largest element, namely the constant sequence $\mathbf{1}_k = (1, 1, \ldots, 1) \in L^k$. The left residual can be calculated explicitly:

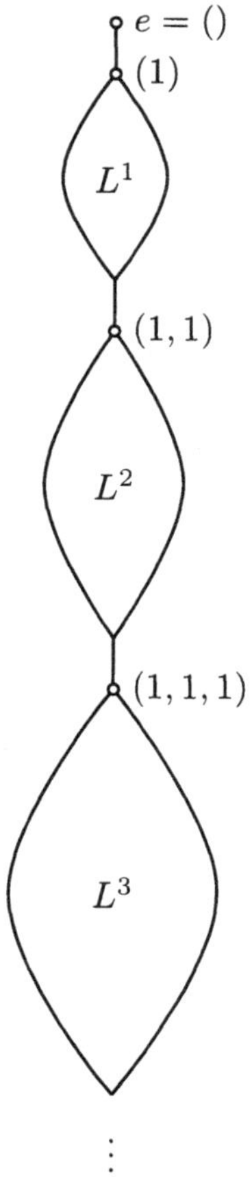

FIGURE 3.1. $\mathbf{L}^*$, a cancellative expansion of a given lattice $\mathbf{L}$.

$$
\begin{aligned}
&(x_1, \ldots, x_m)\backslash(y_1, \ldots, y_n) \\
&\quad = \begin{cases} (y_{m+1}, \ldots, y_n) & \text{if } m \le n \text{ and } x_i \le y_i \text{ for } 1 \le i \le m \\ \mathbf{1}_k & \text{otherwise, where } k = \max(n - m + 1, 0), \end{cases}
\end{aligned}
$$

and the right residual is similar. □

Note that the lattice reduct of $\mathbf{L}^*$ is in the variety generated by $\mathbf{L}$, since lattice varieties are closed under the operation of ordinal sum and adding a top element. If we take $\mathbf{L}$ to be $\mathbf{M}_3$, the 5-element modular lattice, then $\mathbf{L}^*$ is a 3-generated modular, nondistributive, cancellative, integral residuated lattice. If we take $\mathbf{L}$ to be $\mathbf{N}_5$, the 5-element nonmodular lattice, then $\mathbf{L}^*$

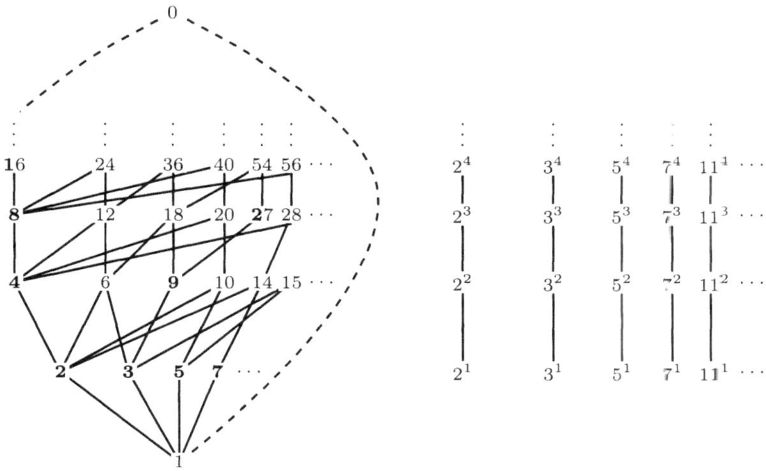

FIGURE 3.2. The distributive lattice $(\mathbb{N}, \gcd, \mathrm{lcm})$ and its poset of completely join-irreducibles.

is a nonmodular, cancellative, integral residuated lattice. In fact, given any lattice variety $\mathcal{V}$, we can define a subvariety of CanRL by

$$\hat{\mathcal{V}} = \mathcal{V}(\{\mathbf{L}^* : \mathbf{L} \in \mathcal{V}\}).$$

COROLLARY 3.29. [BCG+03] *The map $\mathcal{V} \mapsto \hat{\mathcal{V}}$ is an order-preserving injection of the lattice of all lattice varieties into $\mathbf{\Lambda}(\mathsf{CanRL})$.*

3.4.5. MV-algebras and generalized MV-algebras. In Section 2.3.5 we saw that MV-algebras form a subvariety MV of $\mathsf{FL_{eo}}$ axiomatized by the identity $(x \to y) \to y = x \vee y$. They include Boolean algebras, by Lemma 3.24(5b), and they are involutive (set $y = 0$). We know from Lemma 3.20 that the dual of an MV-algebra is an FL-algebra; Exercise 32 asks you to show that it actually is an MV-algebra, as well.

Figure 3.4 shows the bottom part of the lattice $\mathbf{\Lambda}(\mathsf{MV})$ of subvarieties of MV. The structure of this lattice was analyzed by Komori in [Kom81] where, among other results, it is shown that the lattice is countable. Here $\mathsf{MV_n}$ is the variety generated by the $n+1$-element MV-chain $\mathbf{C}_{n+1}$, defined in Section 2.3.5. The variety $\mathsf{MV_n}^{\varepsilon}$ is generated by the chain $\mathbf{C}^{\varepsilon}_{n+1}$ which is defined as $\mathbf{C}_{n+1}$ with a chain of infinitesimal elements (isomorphic to the integers) above and below each element. More precisely, $\mathbf{C}^{\varepsilon}_{n+1}$ is defined as follows: Let

$$C^{\varepsilon}_{n+1} = \{a_{0,j} : j \leq 0\} \cup \{a_{i,j} : i = 1, \ldots, n-1,\ j \in \mathbb{Z}\} \cup \{a_{n,j} : j \geq 0\}$$

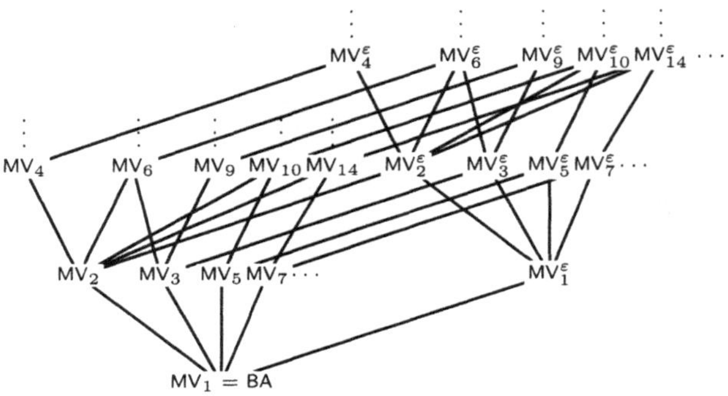

FIGURE 3.3. The completely join-irreducible MV-varieties.

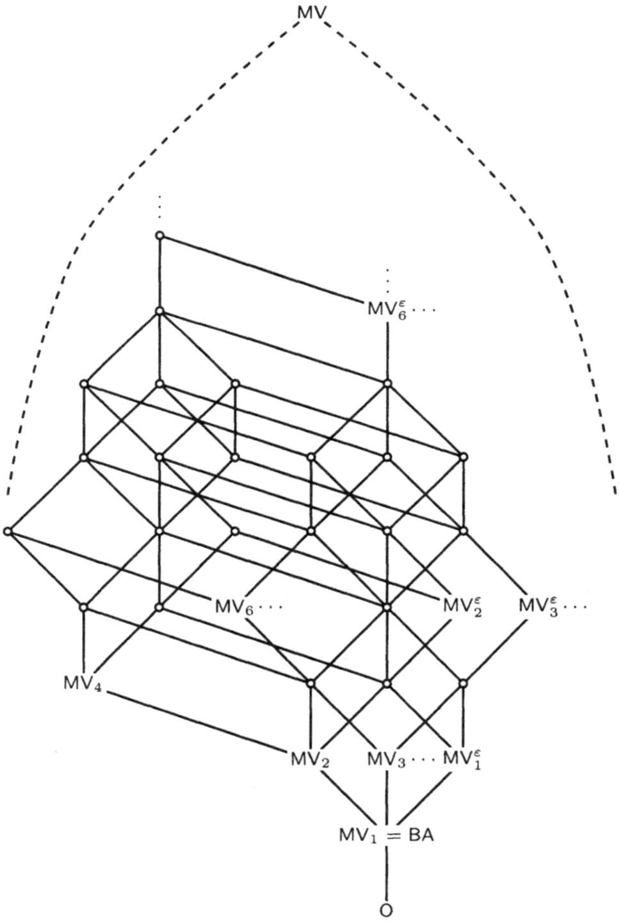

FIGURE 3.4. A glimpse of the lattice of MV-varieties.

and order this set linearly by the order given by $a_{i,j} < a_{k,l}$ iff $i > k$ or ($i = k$ and $j < l$). Multiplication is defined by

$$a_{i,j} \cdot a_{k,l} = \begin{cases} a_{i+k,j+l} & \text{if } i + k < n \\ a_{n,\max(j+l,0)} & \text{if } i + k \geq n. \end{cases}$$

It is easy to check that this defines an associative residuated operation on the chain.

The proper subvarieties of MV that are generated by a single chain are completely join-irreducible elements in the distributive lattice $\mathbf{\Lambda}(\mathsf{MV})$ and every MV-variety is a join of such subvarieties. Hence the structure of $\mathbf{\Lambda}(\mathsf{MV})$ is determined by the poset of completely join-irreducible elements given in Figure 3.3. This poset is itself a lattice, given by the direct product of the two-element lattice and the divisibility lattice of the natural numbers in Figure 3.2. The divisibility lattice plays a role here since $\mathbf{C}_m$ is a subalgebra of $\mathbf{C}_n$ iff $m|n$, while the direct product with the two-element lattice follows from the observation that $\mathbf{C}_n$ is a subalgebra of $\mathbf{C}_n^\varepsilon$.

MV-algebras are closely related to (abelian) ℓ-groups. Given an abelian ℓ-group $\mathbf{G} = (\wedge, \vee, \cdot, \to, 1)$ and a positive ($a \geq 1$) element $a \in G$, we define $\Gamma(\mathbf{G}, a) = ([1, a], \wedge, \vee, \circ^a, \to^a, a, 1)$, where $x \circ^a y = (a \to xy) \vee 1$ and $x \to^a y = (x \to ya) \wedge a$; then $\Gamma(\mathbf{G}, a)$ is an MV-algebra. Here we use the interval notation $[a, b] = \{x : a \leq x \leq b\}$. C. C. Chang [Cha59] showed that every totally ordered MV-algebra is of the form $\Gamma(\mathbf{G}, a)$ for some abelian ℓ-group G and a positive element $a \in G$. D. Mundici [Mun86] generalized this result to all MV-algebras.

Alternatively, one can chose a negative ($a \leq 1$) element $a \in G$ and define $\Gamma(\mathbf{G}, a) = ([a, 1], \wedge, \vee, \circ_a, \to_a, a, 1)$, where $x \circ_a y = xy \vee a$ and $x \to_a y = (x \to y) \wedge 1$; in this way the operations are related in a more natural way. It can be shown that $\Gamma(\mathbf{G}, a)$ is an MV-algebra, as well; this construction is a special case of Lemma 3.40. The two different ways of obtaining an MV-algebra from an abelian ℓ-group are related to the two ways of defining a term equivalence for MV-algebras that we gave in Section 2.3.5. It can be shown that if $a \in G$, then $\Gamma(\mathbf{G}, a^{-1})$ is isomorphic to what we have called in Section 3.3.6 the dual of $\Gamma(\mathbf{G}, a)$; see also Section 3.4.17. Note that $\mathbf{C}_n$, defined in Section 2.3.5, is isomorphic to $\Gamma(\mathbb{Z}, n)$ and to $\Gamma(\mathbb{Z}, -n)$, and the example on the unit interval $[0, 1]$ is isomorphic to $\Gamma(\mathbb{R}, 1)$ and to $\Gamma(\mathbb{R}, -1)$. Also, observe that $\mathbf{C}_n^\varepsilon = \Gamma(\mathbb{Z} \overrightarrow{\times} \mathbb{Z}, (-n, 0))$, where $\mathbb{Z} \overrightarrow{\times} \mathbb{Z}$ is the lexicographic product of the integers under addition with itself.

It is interesting to note that, although this is not stipulated explicitly in their axiomatization as special FL-algebras, MV-algebras are integral. Indeed, for $y = 1$ in the axiomatizing identity, we get $(x \to 1) \to 1 = x \vee 1$, so $1 \leq (x \to 1) \to 1$ and $[(x \to 1) \to 1] \to 1 \leq 1 \to 1 = 1$, since $\to$ is order reversing in the denominator (first argument). By Lemma 2.6(11), we get $x \to 1 \leq 1$.

For $x = z \to 1$ this yields, together with Lemma 2.6(10), $z \leq (z \to 1) \to 1 \leq 1$; hence $z \leq 1$, for all z. Therefore, $\mathsf{MV} \subseteq \mathsf{FL_{ew}}$.

By removing the constant 0 from the type of MV-algebras and the axiom $0 \leq x$, we can define a subvariety of RL (which we will denote by ICGMV). This step is analogous to moving from Boolean algebras to generalized Boolean algebras or from Heyting algebras to Brouwerian algebras and defines the 0-free subreducts of MV. Algebras of this kind are such that every principal filter forms an MV-algebra.

We would like to generalize MV-algebras, but not only in this way. In particular, the connection between MV-algebras and abelian ℓ-groups leads to the question of what kind of algebras correspond to arbitrary ℓ-groups and suggests that the commutativity assumption in MV-algebras is dropped. Finally, the fact that MV-algebras turn out to be integral raises the question on whether there is a natural non-integral generalization. The culminating point of this search leads to what we will call *generalized MV-algebras.*

The removal of commutativity for MV-algebras would imply that the new algebras are FL-algebras (not FL_e-algebras), but also the axiom $x \vee y = (x \to y) \to y$ would have to be written using the two division operations. It turns out that the proper writing is the one where both division operations are used successively

(IGMV) $$(y/x)\backslash y = x \vee y = y/(x\backslash y).$$

This is justified a posteriori and also by the fact that the map $x \mapsto (x \to y) \to y$ is a closure operator in FL_e-algebras, but of all possibilities for the division operations only $x \mapsto (y/x)\backslash y$ and $x \mapsto y/(x\backslash y)$ are closure operators in FL-algebras. FL_o-algebras that satisfy the identity $(y/x)\backslash y = x \vee y = y/(x\backslash y)$ are known in the literature as *pseudo MV-algebras*; we will denote their variety by $\mathsf{IGMV_o}$ and also by psMV. Their 0-free subreducts form a subvariety of RL that we will denote by IGMV. Pseudo MV-algebras are intervals of arbitrary ℓ-groups in the same way as MV-algebras are intervals of abelian ℓ-groups; see [Dvu02] and [GT05]. On the other hand, IGMV includes LG^-.

Finally, for dropping integrality, we first note that in the context of IRL, the identity $(y/x)\backslash y = x \vee y = y/(x\backslash y)$ is equivalent to the quasi-identity $x \leq y \Rightarrow (y/x)\backslash y = y = y/(x\backslash y)$; see Exercise 33. This, in its equational form for RL, is exactly the axiom we take for our algebras. Consequently, a *generalized MV-algebra* or *GMV-algebra* is a residuated lattice that satisfies

(GMV) $$[y/(x \vee y)]\backslash y = x \vee y = y/[(x \vee y)\backslash y].$$

For a study of GMV-algebras the reader is referred to [GT05].

We know that the 0-free reducts of MV-algebras are GMV-algebras; actually, they are exactly the bounded (integral) commutative GMV-algebras (integrality follows from the existence of bounds and (GMV)). It is interesting that ℓ-groups are also GMV-algebras. Indeed, in ℓ-groups we have

$[y/(x \vee y)]\backslash y = [y/(x \vee y)]^{-1}y = [y(x \vee y)^{-1}]^{-1}y = (x \vee y)y^{-1}y = x \vee y$ and $y/[(x \vee y)\backslash y] = x \vee y$. Moreover, the negative cones of ℓ-groups are (integral) GMV-algebras; see Exercise 35. Actually, the subvariety GMV^0 of FL that is axiomatized by the equation (GMV)—we will refer to its members as *pointed GMV-algebras*—has all three MV, LG and LG^- as subvarieties.

Even though (pointed) GMV-algebras are very general they are completely understood relative to ℓ-groups. In particular, every GMV-algebra is a direct product of an ℓ-group and an integral GMV-algebra. Further, every integral GMV-algebra, even though not always a negative interval of an ℓ-group unless it is bounded, is a filter of the negative cone of ℓ-group. These filters are characterized as exactly the images of a nucleus (see Section 3.4.11) on the negative cone of the ℓ-group. Alternatively, every GMV-algebra is an image of an ℓ-group under the composition of a kernel (see Section 3.4.16) and a nucleus. For more details and the proofs of these results, see [GT05].

Moreover, integral GMV-algebras are exactly the 0-free subreducts of pseudo MV-algebras and are characterized as residuated lattices such that every principal filter forms a pseudo MV-algebra.

Although MV-algebras are representable (as subdirect products of totally ordered algebras), GMV-algebras are not always, since they include, for example, non-representable ℓ-groups. Nevertheless, commutative GMV-algebras are representable; see Exercise 37.

It is shown in [BCG$^+$03] that the cancellative integral GMV-algebras are exactly the negative cones of ℓ-groups, $\mathsf{CanIGMV} = \mathsf{LG}^-$. This provides an axiomatization of LG^-, relative to RL, by the axioms (Can) and (IGMV). Furthermore, in [GT05] it is shown that cancellative GMV-algebras are exactly the cartesian products of ℓ-groups with negative cones of ℓ-groups, $\mathsf{CanGMV} = \mathsf{LG} \times \mathsf{LG}^-$; in the same paper they are also characterized as kernel retractions of ℓ-groups. These results, based on the decidability of ℓ-groups, lead to the decidability (of the equational theories) of the aforementioned varieties; see Section 4.3 for the details.

3.4.6. BL-algebras and generalized BL-algebras. Recall from Section 2.3.6, that BL-algebras were defined as FL_{eo}-algebras that satisfy $x \wedge y = x(x \to y)$ and $(x \to y) \vee (y \to x) = 1$. *Pseudo BL-algebras* are non-commutative generalizations of BL-algebras and are defined as FL_{io}-algebras that satisfy

(Div) $\qquad x(x\backslash y) = x \wedge y = (y/x)x \qquad$ (divisibility)

and

(PL) $\qquad x\backslash y \vee y\backslash x = 1. \qquad$ (prelinearity)

Clearly, BL-algebras are exactly the commutative pseudo BL-algebras. Pseudo BL-algebras were defined originally by Georgescu and Iorgulescu

[GI01] in a term equivalent way that generalizes the definition of MV-algebras given in [CDM00].

We will generalize BL-algebras in the same way we generalized MV-algebras. The second defining equation for BL-algebras (also known as *prelinearity*) is equivalent to their representability. Note that representability for MV-algebras was not explicitly stipulated and it followed from the other axioms. Moreover, as noted above, GMV-algebras are not representable, but the addition of commutativity guarantees their representability. Therefore, it makes sense to remove prelinearity, together with commutativity, in the definition of a GBL-algebra.

For dropping integrality, we first note that in the context of IRL, the identity $x(x\backslash y) = x \wedge y = (y/x)x$ is equivalent to the quasi-identity $y \leq x \Rightarrow x(x\backslash y) = y = (y/x)x$; see Exercise 34. This, in its equational form for RL, is exactly the axiom we take for our algebras. Consequently, a *generalized BL-algebra* or *GBL-algebra* is a residuated lattice that satisfies

$$\text{(GBL)} \qquad x[x\backslash(x \wedge y)] = x \wedge y = [(x \wedge y)/x]x.$$

GMV-algebras and Brouwerian algebras are examples of GBL-algebras; see [BCG$^+$03]. *Pointed GBL-algebras* are FL-algebras with a GBL-algebra residuated lattice reduct. We will see in Lemma 8.39 that involutive pointed GBL-algebras are pointed GMV-algebras. We denote by GBL and GBL^0 the varieties of GBL-algebras and pointed GBL-algebras. BL-algebras are exactly the representable, integral, commutative pointed GBL-algebras; alternatively, $\mathsf{BL} = \mathsf{RFL_{eo}} \cap \mathsf{GBL}^0$, as integrality follows from the other axioms. Moreover, every bounded GBL-algebra can be shown to be integral; see [GT05]. In the same paper it is shown that, as in the case of GMV-algebras, every GBL-algebra is the direct product of an ℓ-group and an integral GBL-algebra. Also, in [BCG$^+$03] it is shown that every GBL-algebra is distributive and that cancellative integral GBL-algebras are exactly the negative cones of ℓ-groups, $\mathsf{CanIGBL} = \mathsf{LG}^-$. This provides an alternative axiomatization for LG^- by (Can) and (IGBL). Furthermore, it follows from [GT05] that $\mathsf{CanGBL} = \mathsf{LG} \times \mathsf{LG}^-$.

THEOREM 3.30. [BCG$^+$03] *The negative cones of ℓ-groups from a variety LG^- that is axiomatized by (Can) and either one of (IGMV) or (IGBL).*

3.4.7. Hoops. A *hoop*, originally introduced in by Büchi and Owens under an equivalent definition, can be defined as an algebra $\mathbf{A} = (A, \cdot, \rightarrow, 1)$, where $(A, \cdot, 1)$ is a commutative monoid and the following identities hold:

$$x \rightarrow x = 1 \quad x(x \rightarrow y) = y(y \rightarrow x) \quad (xy) \rightarrow z = y \rightarrow (x \rightarrow z)$$

It is easy to see that the relation defined by $a \leq b$ iff $1 = a \rightarrow b$ is a partial order and that $\mathbf{A}$ is a hoop iff $(A, \cdot, \rightarrow, 1, \leq)$ is an integral residuated pomonoid that satisfies $x(x \rightarrow y) = y(y \rightarrow x)$. Actually, if $\mathbf{A}$ is a hoop, then

$(A, \leq)$ admits a meet operation defined by $x \wedge y = x(x \to y)$. Consequently, every hoop satisfies the divisibility condition (Div). It turns out that not all hoops have a lattice reduct; the ones that do are exactly the join-free reducts of commutative, integral GBL-algebras. Also, among those, the ones that satisfy prelinearity (PL) are exactly the reducts of BL-algebras and are known as *basic hoops*; the corresponding subvariety of RL is denoted by BH. If a hoop satisfies

$$(x \to y) \to y = (y \to x) \to x$$

then it admits a join given by $x \vee y = (x \to y) \to y$. Such hoops are known as *Łukasiewicz hoops* and as *Wasjberg hoops* and they are term equivalent to commutative integral GMV-algebras; we will use WH as another name for ICGMV.

For more on hoops and Wasjberg hoops the reader is advised to look at [AFM99], [Fer92], [BF93], [BP94a], [Kom81], [FRT84] and their references.

3.4.8. Relation algebras. The set $\mathcal{P}(X \times X)$ of all binary relations on a set X is of course a Boolean algebra. We can also define composition of relations and the converse of a relation and consider the empty, the universal and the identity relation. This is the primary example of a relation algebra.

A *relation algebra* is an algebra $\mathbf{A} = (A, \wedge, \vee, {}^-, \bot, \top, \cdot, 1, {}^\smile)$, such that $(A, \wedge, \vee, {}^-, \bot, \top)$ is a Boolean algebra, $(A, \cdot, 1)$ is a monoid and for all $a, b, c \in A$

(i) $(a^\smile)^\smile = a$, $(ab)^\smile = b^\smile a^\smile$;
(ii) $a(b \vee c) = ab \vee ac$, $(b \vee c)a = ba \vee ca$, $(a \vee b)^\smile = a^\smile \vee b^\smile$; and
(iii) $a^\smile(ab)^- \leq b^-$.

In the example of all binary relations on a set X, $\bot$ is the empty relation, $\top$ is $X \times X$ and 1 is the identity relation on X. Our notation is not standard, since these are usually denoted by 0, 1 and $1'$, respectively, while composition is denoted by semicolon ; and sometimes the lattice operations are written as $\cdot$ and $+$. Most of these symbols are reserved in our case, so we use the above notation.

Given a relation algebra $\mathbf{A}$, the structure $\mathcal{R}(\mathbf{A}) = (A, \wedge, \vee, \cdot, \backslash, /, 1)$, where $a \backslash b = (a^\smile b^-)^-$ and $b/a = (b^- a^\smile)^-$ is a residuated lattice. The only thing to be checked is that the division operations are the residuals of multiplication, i.e., the last condition in the definition of a residuated lattice. If $ab \leq c$ then $c^- \leq (ab)^-$. So, $a^\smile c^- \leq a^\smile(ab)^- \leq b^-$, by (iii); hence $b \leq (a^\smile c^-)^-$. On the other hand, if $b \leq (a^\smile c^-)^-$, then $ab \leq a(a^\smile c^-)^- \leq c$, by (iii), and the idempotency of $^\smile$ and $^-$.

Note that the operations of converse $^\smile$ and complement $^-$ commute. Indeed, first observe that for every element x in a relation algebra, we have $x \leq \top$, so $\top = x \vee \top$ and $\top^\smile = x^\smile \vee \top^\smile$, by (ii), hence $x^\smile \leq \top^\smile$ for all x. Therefore, $\top = \top^{\smile\smile} \leq \top^\smile$; i.e. $\top^\smile = \top$. Now, for all x, we have $x \vee x^- = \top$, so $x^\smile \vee x^{-\smile} = \top^\smile = \top$, by (ii) and the above calculation, so

$x^{\smile -} \leq x^{-\smile}$, because we are in a Boolean algebra (say, by taking meet of both sides by $x^{\smile -}$). For the converse inequality apply the already proved one to $x^{\smile}$ to obtain $x^{-} \leq x^{\smile - \smile}$ and $x^{-\smile} \leq x^{\smile -}$.

This observation allows us to show that we can define an involutive cyclic negation on any relation algebra. Indeed, define $\neg x = x^{\smile -} = x^{-\smile}$. Since the operations of converse and complementation commute and are idempotent, we have $\neg\neg x = x$. Moreover, the negation admits a negation constant which is the complement 1^{-} of the identity, namely $1^{-} = \neg 1$ and $\neg x = 1^{-}/x = x\backslash 1^{-}$. So, $\mathcal{R}(\mathbf{A})$ can be expanded to an FL-algebra.

A relation algebra is called *symmetric* if it satisfies the identity $x^{\smile} = x$. In this case we obtain $\neg x = x^{-}$. Since in every relation algebra $1 \wedge 1^{-}$ is the least element and $1 \vee 1^{-}$ is the greatest element, every symmetric relation algebra satisfies $1 \wedge \neg 1 \leq x \leq 1 \vee \neg 1$. In other words, if $\mathbf{A}$ is a symmetric relation algebra the bounds are definable in $\mathcal{R}(\mathbf{A})$. Consequently, symmetric relation algebras are term equivalent to the subvariety SRA of CyInFL that satisfies $x \vee \neg x = 1 \vee \neg 1$ (and $x \wedge \neg x = 1 \wedge \neg 1$, a fact that follows from the de Morgan laws). Note that SRA is actually commutative, since $xy = (xy)^{\smile} = y^{\smile}x^{\smile} = yx$, and of course distributive.

The following partial result is known.

THEOREM 3.31. [JL94] *The subvariety lattice of* SRA *has at least 13 atoms.*

3.4.9. Ideals of a ring. Assume that $\mathbf{R}$ is a ring with unit and let $\mathcal{I}(\mathbf{R})$ be the collection of all two-sided ideals of $\mathbf{R}$. This set, ordered by inclusion, is a lattice. The meet of two ideals is their intersection and their join is the ideal generated by the union. We define multiplication of two ideals I, J in the usual way

$$I \cdot J = \{\textstyle\sum_{x \in X, y \in Y} x \cdot y : X, Y \text{ are finite subsets of } I, J\}.$$

Then $\mathcal{I}(\mathbf{R})$ forms a residuated lattice with unit the ring $\mathbf{R}$ itself and divisions given by

$$I\backslash J = \{k \in R : I \cdot \{k\} \subseteq J\} \text{ and } J/I = \{k \in R : \{k\} \cdot I \subseteq J\}.$$

This example is integral and, if $\mathbf{R}$ is commutative, it is commutative as well. It was in this setting that (commutative, integral) residuated lattices were first defined by Ward and Dilworth [WD39].

3.4.10. Powerset of a monoid. Let $\mathbf{M} = (M, \cdot, 1)$ be a monoid. For any two elements X, Y of the power set $\mathcal{P}(M)$ of M, we denote their intersection, union and complex product respectively, by $X \cap Y$, $X \cup Y$ and $X \cdot Y = \{x \cdot y \mid x \in X, y \in Y\}$. Also, we define the sets $X/Y = \{z \mid \{z\} \cdot Y \subseteq X\}$ and $Y\backslash X = \{z \mid Y \cdot \{z\} \subseteq X\}$. It is easy to see that the algebra $\mathcal{P}(\mathbf{M}) = (\mathcal{P}(M), \cap, \cup, \cdot, \backslash, /, \{1\})$ is a residuated lattice.

Actually, we obtain a residuated lattice even if we start from a *partial monoid*, i.e., a set with a partial operation and an element 1 such that $x1 = 1x = x$, for all x and if one of $x(yz)$ and $(xy)z$ is defined then they are both defined and equal. See Exercise 42 for details and a generalization of this result. Therefore, the powerset of a monoid and the algebra of all relations on a set are examples of the residuated lattice obtained as a powerset of a partial monoid.

The particular case of Exercise 42 that we will also use is the powerset $\mathcal{P}(\mathbf{A})$ of a groupoid $\mathbf{A}$ (with unit). The operations are defined in the same way as above.

THEOREM 3.32. *If* $\mathbf{A}$ *is a groupoid (with unit) then* $\mathcal{P}(\mathbf{A})$ *is a residuated lattice-ordered groupoid (with unit).*

3.4.11. The nucleus image of a residuated lattice. We now define an important notion in the context of residuated lattices.

A *nucleus* on a pogroupoid $\mathbf{L}$ is a closure operator γ on (the poset reduct of) $\mathbf{L}$ such that $\gamma(a)\gamma(b) \leq \gamma(ab)$, for all $a, b \in L$.

The concept of a nucleus was originally defined in the context of Brouwerian algebras [ST77] and of quantales [Ros90b]. Recall that L_γ is the image $\gamma[L]$ of L under γ; also recall the definition of a basis for a closure operator. A proof of the first three statements of the following lemma can be found in [GT05]; the last statement is taken from [GJ].

LEMMA 3.33. [GT05] [GJ] *If* γ *is a closure operator on a pogroupoid* $\mathbf{L}$, *then the following statements are equivalent:*

(1) γ *is a nucleus.*

(2) $\gamma(\gamma(x)\gamma(y)) = \gamma(xy)$, *for all* $x, y \in L$.

If $\mathbf{L}$ *is residuated, then the above conditions are also equivalent to:*

(3) x/y, $y\backslash x \in L_\gamma$, *for all* $x \in L_\gamma$, $y \in L$.

If, additionally, $\mathbf{L} = \mathcal{P}(\mathbf{A})$, *for some groupoid* $\mathbf{A}$, *and* D *is a basis for* γ, *then each of the above conditions are also equivalent to:*

(4) $d/\{a\}$, $\{a\}\backslash d \in L_\gamma$, *for all* $d \in D$ *and* $a \in A$.

PROOF. (1) $\Rightarrow$ (2): Let $x \in L_\gamma$ and $y \in L$. Since γ is extensive and monotone, we have $\gamma(xy) \leq \gamma(\gamma(x)\gamma(y))$. On the other hand, by the defining property of a nucleus and monotonicity, we have $\gamma(\gamma(x)\gamma(y)) \leq \gamma(\gamma(xy))$. So, $\gamma(\gamma(x)\gamma(y)) \leq \gamma(xy)$, since γ is idempotent.

(2) $\Rightarrow$ (3): Since $x \in L_\gamma$, we get $\gamma(x) = x$. So,

$$\begin{aligned} \gamma(x/y)\cdot y &\leq \gamma(\gamma(x/y)\cdot\gamma(y)) && (\gamma \text{ is extensive}) \\ &= \gamma((x/y)\cdot y) && (2) \\ &\leq \gamma(x) && (\text{Lemma 2.6(4) and monotonicity}) \\ &= x. \end{aligned}$$

So, $\gamma(x/y) \leq x/y$, by the defining property of residuated lattices. Since the reverse inequality follows by the extensivity of γ, we have $x/y = \gamma(x/y) \in L_\gamma$. Similarly, we get the result for the other division operation.

(3) $\Rightarrow$ (1): Since γ is extensive, $xy \leq \gamma(xy)$, so $x \leq \gamma(xy)/y$. By the monotonicity of γ and the hypothesis, we have $\gamma(x) \leq \gamma(xy)/y$. Using the

defining property of residuated lattices, we get $y \leq \gamma(x)\backslash\gamma(xy)$. Invoking the monotonicity of γ and the hypothesis, once more, we obtain $\gamma(y) \leq \gamma(x)\backslash\gamma(xy)$, namely $\gamma(x)\gamma(y) \leq \gamma(xy)$.

Obviously, (3) implies (4). For the converse, let $x \in L_\gamma$ and $y \in L$. Then, there exists a $D_x \subseteq D$, such that $x = \bigcap D_x$. We have $x/y = (\bigcap D_x)/y = \bigcap_{d\in D_x} d/y = \bigcap_{d\in D_x} d/(\bigcup_{a\in y}\{a\}) = \bigcap_{d\in D_x}\bigcap_{a\in y} d/\{a\}$, by Theorem 3.10 for the residuated lattice $\mathbf{L} = \mathcal{P}(\mathbf{A})$. If $d/\{a\}$ is a γ-closed element, for all $d \in D$ and $a \in A$, then $\bigcap_{d\in D_x}\bigcap_{a\in y} d/\{a\}$ is γ-closed, since the intersection of closed elements of a closure operator on a complete lattice is also closed. Indeed, if C is a set of closed elements, then $\gamma(\bigcap C) \leq \bigcap\{\gamma(c) : c \in C\} \leq \bigcap C$, since $\gamma(c) \leq c$ for all $c \in C$. □

Note that condition (3) of Lemma 3.33 allows for the definition of nuclei on division posets.

Actually, it can be shown that an arbitrary map γ on a residuated lattice $\mathbf{L}$ is a nucleus if and only if $\gamma(a)/b = \gamma(a)/\gamma(b)$ and $b\backslash\gamma(a) = \gamma(b)\backslash\gamma(a)$, for all $a, b \in L$.

If $\mathbf{L} = (L, \leq, \cdot)$ is a residuated groupoid and γ a nucleus on $\mathbf{L}$, then the structure $\mathbf{L}_\gamma = (L_\gamma, \leq, \circ_\gamma)$, where $x \circ_\gamma y = \gamma(x \cdot y)$ is called the *γ-retraction* or *γ-image* of $\mathbf{L}$. If $\mathbf{L}$ has a unit, is lattice ordered and/or is residuated, then the γ-retraction is defined to have the operations $\gamma(1)$, $\wedge$, $\vee_\gamma$ (where $x \vee_\gamma y = \gamma(x \vee y)$), and $\backslash$, $/$, respectively. So for example if $\mathbf{L} = (L, \wedge, \vee, \cdot, \backslash, /, 1)$ is a residuated lattice and γ a nucleus on $\mathbf{L}$, then the γ-retraction of $\mathbf{L}$ is the algebra $\mathbf{L}_\gamma = (L_\gamma, \wedge, \vee_\gamma, \circ_\gamma, \backslash, /, \gamma(1))$, where $x \circ_\gamma y = \gamma(x \cdot y)$ and $x \vee_\gamma y = \gamma(x \vee y)$.

THEOREM 3.34.

(1) *The nucleus retraction $\mathbf{L}_\gamma$ of a pogroupoid $\mathbf{L}$ is a pogroupoid and the properties of lattice-ordering, lattice-completeness, being residuated and having a unit are preserved.*
(2) *In the above cases, the nucleus γ is a $\{\cdot, \vee, 1\}$-homomorphism from $\mathbf{L}$ to $\mathbf{L}_\gamma$ (if $\vee$ and 1 exist); also it is order preserving as such a map. In particular, if t is a $\{\cdot, \vee, 1\}$-formula, then $\gamma(t^{\mathbf{L}}(\bar{x})) = t^{\mathbf{L}_\gamma}(\gamma(\bar{x}))$, for all sequences $\bar{x}$ of elements in K.*
(3) *All equations and inequations involving $\{\cdot, \vee, 1\}$ are preserved. For example, if $\mathbf{L}$ is associative, commutative, integral or contracting, then so is $\mathbf{L}_\gamma$.*
(4) *In particular, if $\mathbf{L}$ is a (complete) residuated lattice and γ a nucleus on it, then the γ-retraction $\mathbf{L}_\gamma$ of $\mathbf{L}$ is a (complete) residuated lattice.*

PROOF. (1) Obviously, $\gamma(1)$ is the multiplicative identity of $\mathbf{L}_\gamma$ and $\mathbf{L}_\gamma$ is closed under $\circ_\gamma$ and $\wedge_\gamma$. By Lemma 3.33, it is also closed under the division operations. To prove that $\mathbf{L}_\gamma$ is closed under meets, note that for $x, y \in L_\gamma$, $\gamma(x \wedge y) \leq \gamma(x) \wedge \gamma(y) = x \wedge y$. The reverse inequality follows by the fact that γ is extensive, so $x \wedge y \in L_\gamma$. Thus, $\mathbf{L}_\gamma$ is closed under all operations, and it is a meet-subsemilattice of $\mathbf{L}$.

To show that $\mathbf{L}_\gamma$ is a lattice note that for elements $x, y, z \in L_\gamma$, $x, y \leq z$ is equivalent to $x \vee y \leq z$. Since $\gamma(z) = z$ and γ is extensive, this is, in turn, equivalent to $\gamma(x \vee y) \leq z$, namely to $x \vee_\gamma y \leq z$. Thus, $\vee_\gamma$ is the join in $\mathbf{L}_\gamma$. Completeness, follows from Exercise 7.

We next show that multiplication is associative. Let $x, y, z \in L_\gamma$. Using Lemma 3.33 and the definition of multiplication, we get

$$\begin{aligned}(x \circ_\gamma y) \circ_\gamma z &= \gamma(x \cdot y) \circ_\gamma z \\ &= \gamma(\gamma(x \cdot y) \cdot z) \\ &= \gamma(\gamma(x \cdot y) \cdot \gamma(z)) \\ &= \gamma((x \cdot y) \cdot z) \\ &= \gamma(x \cdot y \cdot z).\end{aligned}$$

Similarly,

$$x \circ_\gamma (y \circ_\gamma z) = \gamma(x \cdot y \cdot z).$$

Hence, multiplication in $\mathbf{L}_\gamma$ is associative and $(L, \circ_\gamma, \gamma(e))$ is a monoid.

Finally, to check that $\circ_\gamma$ is residuated, consider $x, y, z \in L_\gamma$. We have

$$\begin{aligned}x \circ_\gamma y \leq z &\Leftrightarrow \gamma(x \cdot y) \leq z \\ &\Leftrightarrow x \cdot y \leq z \qquad (x \cdot y \leq \gamma(x \cdot y) \text{ and } z = \gamma(z)) \\ &\Leftrightarrow y \leq x\backslash z.\end{aligned}$$

Likewise, $x \circ_\gamma y \leq z \Leftrightarrow x \leq z/y$.

(2) Note that for all $x, y \in L$, $\gamma(x), \gamma(y) \leq \gamma(x \vee y)$, by the monotonicity of γ. So, $\gamma(x) \vee \gamma(y) \leq \gamma(x \vee y)$ and $\gamma(\gamma(x) \vee \gamma(y)) \leq \gamma(x \vee y)$. The converse inequality follows from the monotonicity of γ and $x \leq \gamma(x)$, $y \leq \gamma(y)$, so $\gamma(x \vee y) = \gamma(\gamma(x) \vee \gamma(y)) = \gamma(x) \vee_\gamma \gamma(x)$. Also, by Lemma 3.33(2), $\gamma(xy) = \gamma(\gamma(x)\gamma(y)) = \gamma(x) \cdot_\gamma \gamma(x)$. Finally, $\gamma(1_\mathbf{L}) = 1_{\mathbf{L}_\gamma}$. So, $\gamma : \mathbf{L} \to \mathbf{L}_\gamma$ it is a $\{\cdot, \vee, 1\}$-homomorphism; also, it is clearly order preserving since it is order preserving in $\mathbf{L}$ and the order on $\mathbf{L}_\gamma$ is the induced one.

Properties (3) and (4) follow directly from (1), (2) and Exercise 7. □

The preceding construction is quite general as it can be seen in Theorem 3.38.

Consider the FL-algebra terms $\lambda(x) = -{\sim}x$ and $\rho(x) = {\sim}{-}x$; we will use the same symbols for the term operations that these terms define on particular FL-algebras.

LEMMA 3.35. *Let* $\mathbf{A}$ *be an FL-algebra.*

(1) *The maps* λ *and* ρ *are closure operators on* $\mathbf{A}$.
(2) *If for all* $x \in \mathbf{A}$, $(x\backslash 0)\backslash 0 = 0/(x\backslash 0)$, *then* λ *is a nucleus on* $\mathbf{A}$. *The same holds for* ρ, *if* $(x\backslash 0)\backslash 0 = 0/(x\backslash 0)$, *for all* x.
(3) *In particular, if* $\mathbf{A}$ *is cyclic, then* λ *and* ρ *are nuclei on* $\mathbf{A}$.

(4) *If, additionally,* λ *(or* ρ*) is a nucleus on* $\mathbf{A}$*, then* $\lambda(0) = 0$ *(*$\rho(0) = 0$*) and* $\mathbf{A}_\lambda$ *(*$\mathbf{A}_\rho$*) is left involutive (right involutive, respectively).*

PROOF. The fact that λ is a closure operator follows from (2), (3) and (4) of Lemma 2.8. Using Lemma 2.6(4) and (13), we have

$$\begin{aligned}[y\backslash(x\backslash 0)]\lambda(x)\lambda(y) &= [y\backslash(x\backslash 0)][0/(x\backslash 0)][0/(y\backslash 0)] \\ &= [y\backslash(x\backslash 0)][(x\backslash 0)\backslash 0][(y\backslash 0)\backslash 0] \\ &\leq [y\backslash 0][(y\backslash 0)\backslash 0] \leq 0.\end{aligned}$$

So, we have $\lambda(x)\lambda(y) \leq [y\backslash(x\backslash 0)]\backslash 0 = (xy\backslash 0)\backslash 0 = 0/(xy\backslash 0) = \lambda(xy)$, by Lemma 2.6(6) and the assumption. We get $\lambda(0) = 0$, by Lemma 2.6(12). □

Recall that if $R \subseteq A \times B$, we can define a Galois connection between the powersets $\mathcal{P}(A)$ and $\mathcal{P}(B)$ by $X^{\triangleright} = R_{\cap}[X]$ and $Y^{\triangleleft} = R_{\cap}^{-1}[Y]$ (see page 146). By Lemma 3.7(2), we obtain a closure operator $\gamma_R : \mathcal{P}(A) \to \mathcal{P}(A)$, where $\gamma_R(X) = X^{\triangleright\triangleleft}$. Conversely, every closure operator on $\mathcal{P}(A)$ is of the form γ_R, for some relation R with domain A.

Moreover, we have seen that if $\mathbf{A}$ is a groupoid, then $\mathcal{P}(\mathbf{A})$ is a residuated groupoid (associativity and unit are preserved). We would like to know for which relations $R \subseteq A \times B$ the map γ_R is a nucleus on $\mathcal{P}(\mathbf{A})$. The following condition characterizes the relations that give rise to nuclei.

A relation $N \subseteq A \times B$ is called *nuclear* on a groupoid $\mathbf{A}$ if for every $a_1, a_1 \in A, b \in B$, there exist subsets $a_1 \mathbin{\backslash\!\backslash} b$ and $b \mathbin{/\!/} a_2$ of B such that

$$a_1 \cdot a_2 \; N \; b \text{ iff } a_1 \; N \; b \mathbin{/\!/} a_2 \text{ iff } a_2 \; N \; a_1 \mathbin{\backslash\!\backslash} b.$$

Here $x \; N \; y$ for all $y \in Y$ was abbreviated to $x \; N \; Y$. We allow ourselves to write $b^{\triangleleft}$ for $\{b\}^{\triangleleft}$.

LEMMA 3.36. *If* $\mathbf{A}$ *is a groupoid and* $N \subseteq A \times B$*, then* γ_N *is a nucleus on* $\mathcal{P}(\mathbf{A})$ *iff* N *is a nuclear relation.*

PROOF. By Lemma 3.8(3), the collection D of sets $\{b\}^{\triangleleft}$, where $b \in B$, form a basis for γ_N. So, by Lemma 3.33(4) γ_N is a nucleus iff $\{b\}^{\triangleleft}/\{a\}$ and $\{a\}\backslash\{b\}^{\triangleleft}$ are γ-closed, for all $a \in A$ and $b \in B$. Since D is a basis, $\{a_1\}\backslash\{b\}^{\triangleleft}$ is closed iff $\{a_1\}\backslash\{b\}^{\triangleleft} = \cap D_x$, for some $D_x \subseteq D$ iff $\{a_1\}\backslash\{b\}^{\triangleleft} = \bigcap_{c \in a_1 \backslash\!\backslash b}\{c\}^{\triangleleft}$, for some $a_1 \mathbin{\backslash\!\backslash} b \subseteq B$. This is equivalent to the statement that for all $a_2 \in A$,

$$a_2 \in \{a_1\}\backslash\{b\}^{\triangleleft} \text{ iff } a_2 \in \textstyle\bigcap_{c \in a_1 \backslash\!\backslash b}\{c\}^{\triangleleft}.$$

Transforming this statement further we obtain for all $a_2 \in A$,

$$a_1 \cdot a_2 \in \{b\}^{\triangleleft} \text{ iff } a_2 \in \{c\}^{\triangleleft} \text{ for all } c \in a_1 \mathbin{\backslash\!\backslash} b.$$

or, equivalently, $a_1 \cdot a_2 \; N \; b$ iff $a_2 \; N \; c$ for all $c \in a_1 \mathbin{\backslash\!\backslash} b$. So, i.e. $\{a_1\}\backslash\{b\}^{\triangleleft}$ is closed iff there exists $a_1 \mathbin{\backslash\!\backslash} b \subseteq B$ such that $a_1 \cdot a_2 \; N \; b$ iff $a_2 \; N \; a_1 \mathbin{\backslash\!\backslash} b$. Likewise, we obtain the second equivalence of a nuclear relation. □

In most of the applications of this result the sets $a_1 \backslash\!\backslash b$ and $b /\!/ a_2$ are singletons.

3.4.12. The Dedekind-MacNeille completion of a residuated lattice. Let $\mathbf{L}$ be a residuated lattice. Exercise 11 shows that the maps ^{u} and ^{l} on $\mathcal{P}(L)$, where $X^u = \{a \in L \mid a \geq x \text{ for all } x \in X\}$ and $X^l = \{a \in L \mid a \leq x$ for all $x \in X\}$, for all $X \subseteq L$, form a Galois connection on $\mathcal{P}(L)$. By Lemma 3.7(2), the map ^{ul}, is a closure operator on $\mathcal{P}(L)$. Alternatively, note directly that $X^u = X^{\rhd} = \leq_{\cap}[X]$ and $X^l = X^{\lhd} = \geq_{\cap}[X]$, so $\gamma_{\leq} = {}^{ul}$ is a closure operator on the lattice $\mathcal{P}(L)$; here $\gamma_{\leq}(X) = X^{ul}$, for all $X \subseteq A$. The poset $\mathcal{P}(L)^{ul}$ of closed elements is called the *Dedekind-MacNeille completion* of the poset reduct of $\mathbf{L}$. This completion was originally introduced in [Mac37] as a generalization of Dedekind's completion of rational numbers.

Moreover, $\leq$ is a nuclear relation on L, for $a \backslash\!\backslash b = \{a \backslash b\}$ and $b /\!/ a = \{b \backslash a\}$ by the residuation condition, so $\gamma_{\leq} = {}^{ul}$ is a nucleus on the residuated lattice $\mathcal{P}(\mathbf{L})$. By Theorem 3.32, the image $\mathcal{P}(\mathbf{L})^{ul}$ of $\mathcal{P}(\mathbf{L})$ under ^{ul} is a residuated lattice, which we call the *Dedekind-MacNeille completion* of the residuated lattice $\mathbf{L}$.

We know from Exercise 12 that the lattice reduct of $\mathbf{L}$ embeds into the lattice reduct of $\mathcal{P}(\mathbf{L})^{ul}$, but we can show more.

LEMMA 3.37. *The map* $\downarrow : \mathbf{L} \to \mathcal{P}(\mathbf{L})^{ul}$ *is a residuated lattice embedding.*

PROOF. We need only check that $\downarrow$ preserves the unit, multiplication and the division operations.

It is clear that multiplication by $\downarrow 1$ of any downset fixes that downset. Since every element of $\mathcal{P}(\mathbf{L})^{ul}$ is a downset, $\downarrow 1$ is the unit.

If X is any subset of L with a maximum element m, then $X^{ul} = \downarrow m$. Obviously, $\downarrow x \cdot \downarrow y$ has maximum element xy, so $\downarrow x \cdot_{ul} \downarrow y = (\downarrow x \cdot \downarrow y)^{ul} = \downarrow (xy)$.

Finally, for left division (right division is handled in the opposite way) we have $\downarrow (x \backslash y) = \{z \in L : xz \leq y\}$ and $\downarrow x \backslash_{\mathcal{P}(\mathbf{L})^{ul}} \downarrow y = \downarrow x \backslash_{\mathcal{P}(\mathbf{L})} \downarrow y = \{w \in L : (\downarrow x)w \subseteq \downarrow y\}$. Clearly, the two sets are equal, since multiplication is order preserving. □

Therefore, every residuated lattice can be embedded into a complete one. Moreover, by Exercise 12, every complete residuated lattice is the nucleus image of the power set of a monoid. So, we obtain the following *representation theorem for residuated lattices.*

THEOREM 3.38. cf. [Ros90b], [Blou99] *Every residuated lattice is a subalgebra of the nucleus image of the power set of a monoid.*

The same result holds for residuated lattice-ordered groupoids (with unit). Furthermore, it follows from the proof of this result that for representing commutative residuated lattices we can take commutative monoids. More generally, residuated lattices defined by monoid equations that are preserved under the powerset construction (by Theorem 3.34(3), all monoid identities are preserved by nuclei) are represented by the class of monoids that satisfy these identities. It is natural to ask which classes of residuated lattices are represented by various special classes of monoids.

3.4.13. Order ideals of a partially ordered monoid. For a pomonoid $\mathbf{M} = (M, \cdot, 1, \leq)$, let $\mathcal{O}$ be the set of all order ideals (downsets) of the underlying partially ordered set. For every $X, Y \in \mathcal{O}$, we define $X \bullet Y = {\downarrow}(X \cdot Y)$, the downset of their complex product.

LEMMA 3.39. [Gal03] *The algebra* $\mathcal{O}(\mathbf{M}) = (\mathcal{O}, \cap, \cup, \bullet, \backslash, /, {\downarrow}\{1\})$ *is a residuated lattice.*

PROOF. To prove this we show that the map γ on $\mathcal{P}(\mathbf{M})$ defined by $\gamma(X) = {\downarrow}X$ is a nucleus. Indeed, if $z \in \gamma(X)\gamma(Y) = ({\downarrow}X)({\downarrow}Y)$, then $z = ab$, $a \leq x$ and $b \leq y$, for some $x \in X$ and $y \in Y$. So, $z \leq xy$, namely $z \in {\downarrow}XY$. Finally notice that for any two order ideals X, Y, $\gamma(X \cup Y) = {\downarrow}(X \cup Y) = X \cup Y$. Thus, by Proposition 3.34, $\mathcal{O}(\mathbf{M}) = \mathcal{P}(\mathbf{M})_\gamma$. □

If $\mathbf{M}$ has the discrete order, this reduces to the case of Section 3.4.10. Note that we could have taken order filters instead of order ideals. Exercises 43–51 expand on this idea by considering ideals of various other structures.

3.4.14. Quantales. Quantales were introduced by Mulvey in [Mul86] and discussed extensively in [Ros90b]. An algebra $\mathbf{A} = (A, \bigvee, \cdot)$ is a *quantale* if

(1) $(A, \bigvee)$ is a complete lattice (and hence bounded),
(2) $(A, \cdot)$ is a semigroup,
(3) $(\bigvee X) \cdot y = \bigvee(X \cdot y)$ and $y \cdot (\bigvee X) = \bigvee(y \cdot X)$ for $X \subseteq A$ and $y \in A$.

In the above, $X \cdot y$ and $y \cdot X$ denote the sets $\{xy : x \in X\}$ and $\{yx : x \in X\}$, respectively. Thus, each quantale is a complete lattice ordered semigroup satisfying condition (3) of distributivity of the semigroup operation over infinite joins. Then, by Corollary 3.11, $\cdot$ is residuated, and hence $\mathbf{A}^* = (A, \wedge, \vee, \cdot, \backslash, /)$ is a residuated ℓ-semigroup satisfying

$$x \backslash y = \bigvee\{z : x \cdot z \leq y\} \text{ and } y/x = \bigvee\{w : w \cdot x \leq y\}.$$

Conversely, every complete residuated ℓ-semigroup, namely a residuated ℓ-semigroup whose lattice reduct is complete as a lattice, satisfies the condition (3) by Theorem 3.10. Thus, quantales are essentially equivalent to complete residuated ℓ-semigroups. A quantale $\mathbf{A}$ is *unital* if $(A, \cdot)$ has a unit element. In this case, $\mathbf{A}^*$ augmented by an additional constant for the unit becomes a residuated lattice.

A quantale in which $\cdot$ is equal to $\wedge$ is called a *locale* (or a *frame*). It is easy to see that locales are essentially equivalent to complete Heyting algebras.

3.4.15. Retraction to an interval. For a residuated lattice $\mathbf{L}$ and a negative element a of $\mathbf{L}$, we consider the structure $\mathbf{L}_a = ([a, 1], \wedge, \vee, \circ_a, \backslash_a, /_a, 1)$, where $x \circ_a y = xy \vee a$, $x \backslash_a y = (x \backslash y) \wedge 1$ and $y /_a x = (y/x) \wedge 1$.

LEMMA 3.40. [Gal03] *If $\mathbf{L}$ is a residuated lattice and a a negative element, then $\mathbf{L}_a$ is a residuated lattice.*

PROOF. The map γ on $\mathbf{L}^-$, defined by $\gamma(x) = x \vee a$ is obviously a closure operator. Moreover, if $x, y \leq a$, then $xa, ya, a^2 \leq a$, so

$$\gamma(x) \cdot \gamma(y) = (x \vee a)(y \vee a) = xy \vee xa \vee ay \vee a^2 \leq xy \vee a = \gamma(xy).$$

Therefore, γ is a nucleus and $\mathbf{L}_a = (\mathbf{L}^-)_\gamma$, is a residuated lattice, by Theorem 3.34. □

We have already mentioned that if $\mathbf{L}$ is a commutative ℓ-group and a is a negative element of $\mathbf{L}$, then $\mathbf{L}_a = \Gamma(\mathbf{L}, a)$ is an MV-algebra. Note that if $a \in L^-$, then $\mathbf{L}_a = (\mathbf{L}^-)_a$.

3.4.16. Conuclei and kernel contractions. A *conucleus* δ on a residuated lattice $\mathbf{L}$ is an interior operator such that for all x, y in L

$$\delta(x)\delta(y) \leq \delta(xy) \text{ and } \delta(1)\delta(x) = \delta(x) = \delta(x)\delta(1).$$

Clearly the first condition is equivalent to $\delta(\delta(x)\delta(y)) = \delta(x)\delta(y)$.

Let $\mathbf{L}$ be a residuated lattice and δ a conucleus on it. The *δ-contraction* of $\mathbf{L}$ is the algebra $\mathbf{L}_\delta = (L_\delta, \wedge_\delta, \vee, \cdot, \backslash_\delta, /_\delta, \delta(1))$, where $x \wedge_\delta y = \delta(x \wedge y)$, $x /_\delta y = \delta(x/y)$ and $x \backslash_\delta y = \delta(x \backslash y)$.

PROPOSITION 3.41. [Gal03] *The δ-contraction $\mathbf{L}_\delta$ of a residuated lattice $\mathbf{L}$ under a conucleus δ on $\mathbf{L}$ is a residuated lattice.*

PROOF. Note that L_δ is closed under join, since δ is an interior operator, and under multiplication, by the first property of a conucleus. Moreover, $\delta(1)$ is the unit element, by the second property of a conucleus, and it is obviously closed under $\backslash_\delta$, $/_\delta$ and $\wedge_\delta$.

Finally, $\mathbf{L}_\delta$ is residuated. Indeed, for all $x, y, z \in L_\delta$, $x \leq z /_\delta y$ is equivalent to $x \leq \delta(x/y)$, which in turn is equivalent to $x \leq z/y$, since δ is contracting and $x = \delta(x)$. □

A *kernel* δ is a conucleus that additionally satisfies $\delta(1) = 1$ and $\delta(x) \wedge y = \delta(\delta(x) \wedge y)$.

PROPOSITION 3.42. [Gal03] *If $\mathbf{L}$ is a residuated lattice and δ is a kernel on $\mathbf{L}$, then the δ-contraction $\mathbf{L}_\delta$ is a lattice-ideal of $\mathbf{L}$.*

PROOF. By the second property of a kernel and the fact that it is closed under joins, L_δ is an ideal of $\mathbf{L}$. So, $\mathbf{L}_\delta$ is closed under all the operations. □

The δ-contraction construction, where δ is a kernel, is a generalization of the negative cone construction, defined before, for $\delta(x) = x \wedge 1$.

3.4.17. The dual of a residuated lattice with respect to an element. Let $\mathbf{L} = (L, \wedge, \vee, \cdot, \backslash, /, 1)$ be a residuated lattice and $a \in L$ a *dualizing element*, i.e., an element of L such that

$$x = a/(x\backslash a) = (a/x)\backslash a,$$

for all $x \in L$. Then, the *dual* of $\mathbf{L}$ with respect to the element a is the algebra $\mathbf{L}^{\partial_a} = (L, \vee, \wedge, +, \stackrel{\backslash}{-}, \stackrel{/}{-}, a)$, where $x + y = x/(y\backslash a)$, $x \stackrel{/}{-} y = x(y\backslash a)$ and $x \stackrel{\backslash}{-} y = (a/x)y$. Notice that the underlying lattice of $\mathbf{L}^{\partial_a}$ is the dual of the lattice reduct of $\mathbf{L}$.

PROPOSITION 3.43. [Gal03] *The dual $\mathbf{L}^{\partial_a}$ of a residuated lattice $\mathbf{L}$ with respect to a dualizing element a of $\mathbf{L}$ is also a residuated lattice.*

PROOF. First observe that

$$\begin{aligned} x + y &= x/(y\backslash a) \\ &= ((a/x)\backslash a)/((y\backslash a) \\ &= (a/x)\backslash(a/(y\backslash a)) \\ &= (a/x)\backslash y \end{aligned}$$

and that $1 = a/(1\backslash a) = (a/1)\backslash a$, i.e., $1 = a/a = a\backslash a$.

It is obvious that $(L, \vee, \wedge)$ is a lattice. Multiplication is associative because

$$\begin{aligned} (x + y) + z &= [(a/x)\backslash y]/(z\backslash a) \\ &= (a/x)\backslash[y/(z\backslash a)] \\ &= x + (y + z); \end{aligned}$$

and a is the additive identity since

$$x + a = x/(a\backslash a) = x/1 = x$$

and

$$a + x = (a/a)\backslash x = 1\backslash x = x.$$

Finally multiplication is residuated, since

$$\begin{aligned} x + y \leq_{\mathbf{L}^{\partial_a}} z &\Leftrightarrow x + y \geq_{\mathbf{L}} z \\ &\Leftrightarrow x/(y\backslash a) \geq_{\mathbf{L}} z \\ &\Leftrightarrow x \geq_{\mathbf{L}} z(y\backslash a) \\ &\Leftrightarrow x \geq_{\mathbf{L}} z \stackrel{/}{-} y \\ &\Leftrightarrow x \leq_{\mathbf{L}^{\partial_a}} z \stackrel{/}{-} y \end{aligned}$$

and similarly for $\stackrel{\backslash}{-}$. □

Passing to the dual is a generalization of a construction for MV-algebras. The dual of an MV-algebra with respect to its least element is known to be an MV-algebra.

Moreover, it is easy to see that the dual $\mathbf{L}^{\partial_0}$ of a residuated lattice $\mathbf{L} = (L, \wedge, \vee, \cdot, \backslash, /, 1)$ with respect to the element $0 \in L$ is simply the dual $(\mathbf{L}')^{\partial}$ of the FL-algebra $\mathbf{L}' = (L, \wedge, \vee, \cdot, \backslash, /, 1, 0)$, viewed as an involutive pogoupoid, in the sense of Section 3.3.6.

3.4.18. Translations with respect to an invertible element. An element a in a residuated lattice $\mathbf{L}$ is called *invertible*, if there exists an element a^{-1} such that $aa^{-1} = 1 = a^{-1}a$. It is easy to see that a is invertible iff $a(a\backslash 1) = 1 = (1/a)a$; in this case, $a^{-1} = 1/a = a\backslash 1$.

To establish an equality between two elements a, b of a residuated lattice, we will frequently prove that $x \leq a \Leftrightarrow x \leq b$, for every element x. By setting $x = a$, we have $a \leq b$. On the other hand, by setting $x = b$, we obtain $b \leq a$.

LEMMA 3.44. [Gal03] *If a is invertible, then for all x, y we have*
(1) $x/a = xa^{-1}$ *and* $a\backslash x = a^{-1}x$;
(2) $a(x \wedge y) = ax \wedge ay$ *and* $(x \wedge y)a = xa \wedge ya$;
(3) $a\backslash a = 1$ *and* $a/a = 1$;
(4) $(x/a)y = x(a\backslash y)$; *and*
(5) $a(a^{-1}x/a^{-1}y) = (x/y)a$ *and* $a(a^{-1}y\backslash a^{-1}x) = a(y\backslash x)$.
Moreover, (4) *implies that a is invertible.*

PROOF. 1) For every element z we have

$$z \leq x/a \Leftrightarrow za \leq x \Leftrightarrow z \leq xa^{-1},$$

so $x/a = xa^{-1}$. Similarly, we get the opposite equality $a\backslash x = a^{-1}x$.

2) We have $a(x \wedge y) \leq ax, ay$, so $a(x \wedge y) \leq ax \wedge ay$. For the reverse inequality, note that

$$a^{-1}(ax \wedge ay) \leq a^{-1}ax \wedge a^{-1}ay = x \wedge y,$$

hence $ax \wedge ay \leq a(x \wedge y)$. Similarly we get the opposite equality.

3) This is a direct consequence of (1).

4) If a is invertible, then $(x/a)y = xa^{-1}y = x(a\backslash y)$. Conversely, if we set $x = 1$ and $y = a$ in $(x/a)y = x(a\backslash y)$, we get $(1/a)1 = a\backslash a$ Since, by Lemma 2.6(4) and (9), $(1/a)a \leq 1$ and $a\backslash a \geq 1$, we obtain $(1/a)a = a\backslash a = 1$. Similarly, $(a\backslash 1)a = 1$.

5) For every z, we have

$$\begin{aligned} z \leq a(a^{-1}x/a^{-1}y) &\Leftrightarrow a^{-1}z \leq a^{-1}x/a^{-1}y \\ &\Leftrightarrow a^{-1}za^{-1}y \leq a^{-1}x \\ &\Leftrightarrow za^{-1}y \leq x \\ &\Leftrightarrow za^{-1} \leq x/y \\ &\Leftrightarrow z \leq (x/y)a \end{aligned}$$

The opposite equation follows, since the definition of an invertible element is self-opposite. □

Let $\mathbf{L}$ be a residuated lattice, a an invertible element of L and f_a the map on L defined by $f_a(x) = ax$. Note that the map f_a is invertible and $f_a^{-1}(x) = a^{-1}x$. Consider the structure $\mathbf{L}^a = (L, \wedge^a, \vee^a, \cdot^a, \backslash^a, /^a, 1^a)$, where $1^a = a$ and for every binary operation $\star \in \{\wedge, \vee, \cdot, \backslash, /\}$,

$$x \star^a y = f(f^{-1}(x) \star f^{-1}(y)).$$

By Lemma 3.44, we have

$$x \wedge^a y = a(a^{-1}x \wedge a^{-1}y) = aa^{-1}(x \wedge y) = x \wedge y.$$

Similarly, $\vee^a = \vee$. Moreover, by Lemma 3.44,

$$x \cdot^a y = a(a^{-1}xa^{-1}y) = xa^{-1}y,$$

$$x/^a y = a(a^{-1}x/a^{-1}y) = (x/y)a$$

and

$$y\backslash x = a(a^{-1}y\backslash a^{-1}x) = a(y\backslash x).$$

Note that if we take $g_a(x) = xa$, then we obtain the same structure, so $\mathbf{L}^a$ does not depend on the choice of left or right multiplication by a. The algebra $\mathbf{L}^a = (L, \wedge, \vee, \cdot^a, \backslash^a, /^a, a)$ is called the *translation of* $\mathbf{L}$ *with respect to* a. We remark that we could have defined the operations as follows: $x \cdot^a y = (x/a)y$, $y\backslash^a x = (y/a)\backslash x$ and $x/^a y = x/(a\backslash y)$.

PROPOSITION 3.45. [Gal03] *The translation* $\mathbf{L}^a$ *of a residuated lattice* $\mathbf{L}$ *with respect to an invertible element* a *is a residuated lattice.*

PROOF. It is trivial to check that multiplication is associative and a is the multiplicative identity. To show that multiplication is residuated, let $x, y, z \in L$. We have

$$x \cdot^a y \leq z \Leftrightarrow xa^{-1}y \leq z \Leftrightarrow a^{-1}y \leq x\backslash z \Leftrightarrow y \leq a(x\backslash z) \Leftrightarrow y \leq x\backslash^a z$$

and similarly for the other division. □

Note that the translation by an invertible element and the negative cone constructions on a residuated lattice $\mathbf{L}$ commute, i.e., $(\mathbf{L}^a)^- = (\mathbf{L}^-)^a$.

3.5. Subvariety lattices

In this section we list some important varieties together with their axiomatizations for easy reference (an earlier version of this information appeared in [Jip03]). The page references to these varieties in the index at the end of the book point to this list, as well as to the first place in the text where the variety was first defined. The list is accompanied by diagrams (Figures 3.5 and 3.6) of the subvariety lattices of FL and RL, and by Tables 3.2 and 3.1.

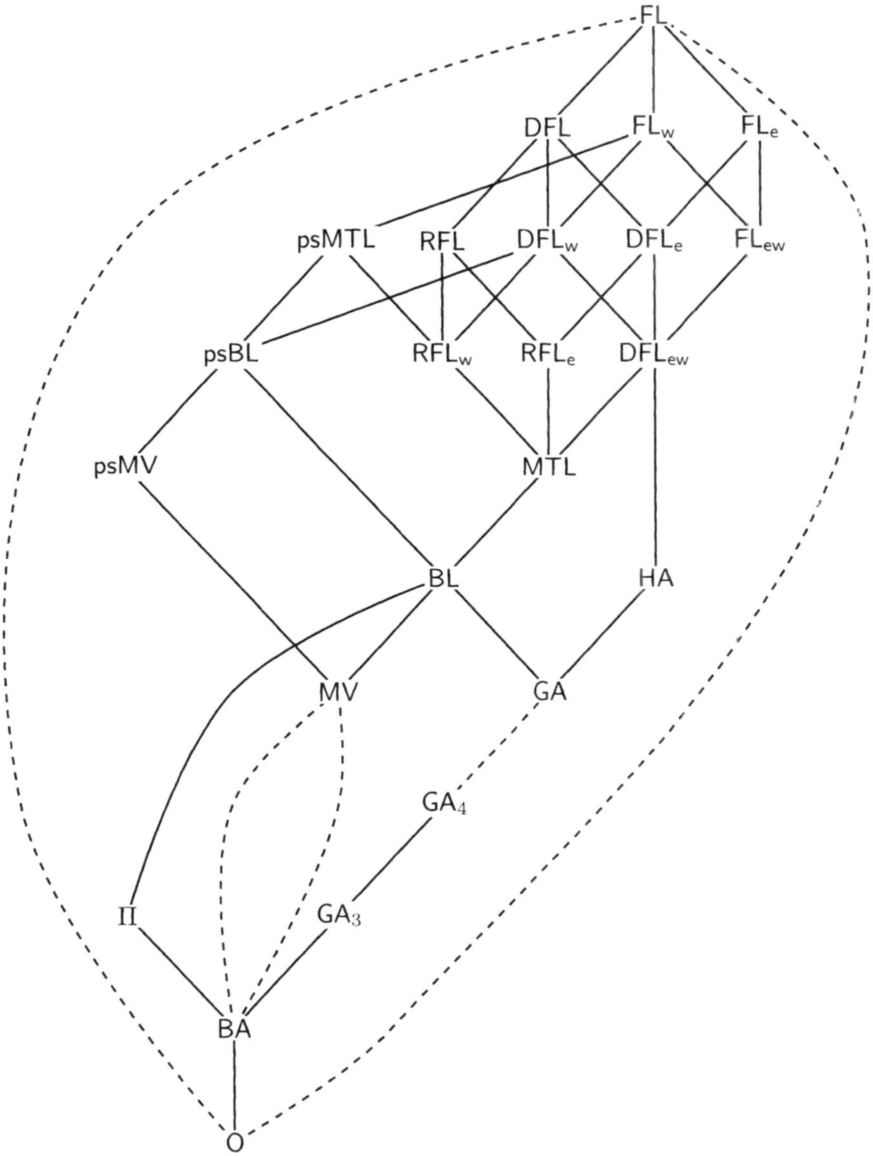

FIGURE 3.5. Some subvarieties of FL ordered by inclusion.

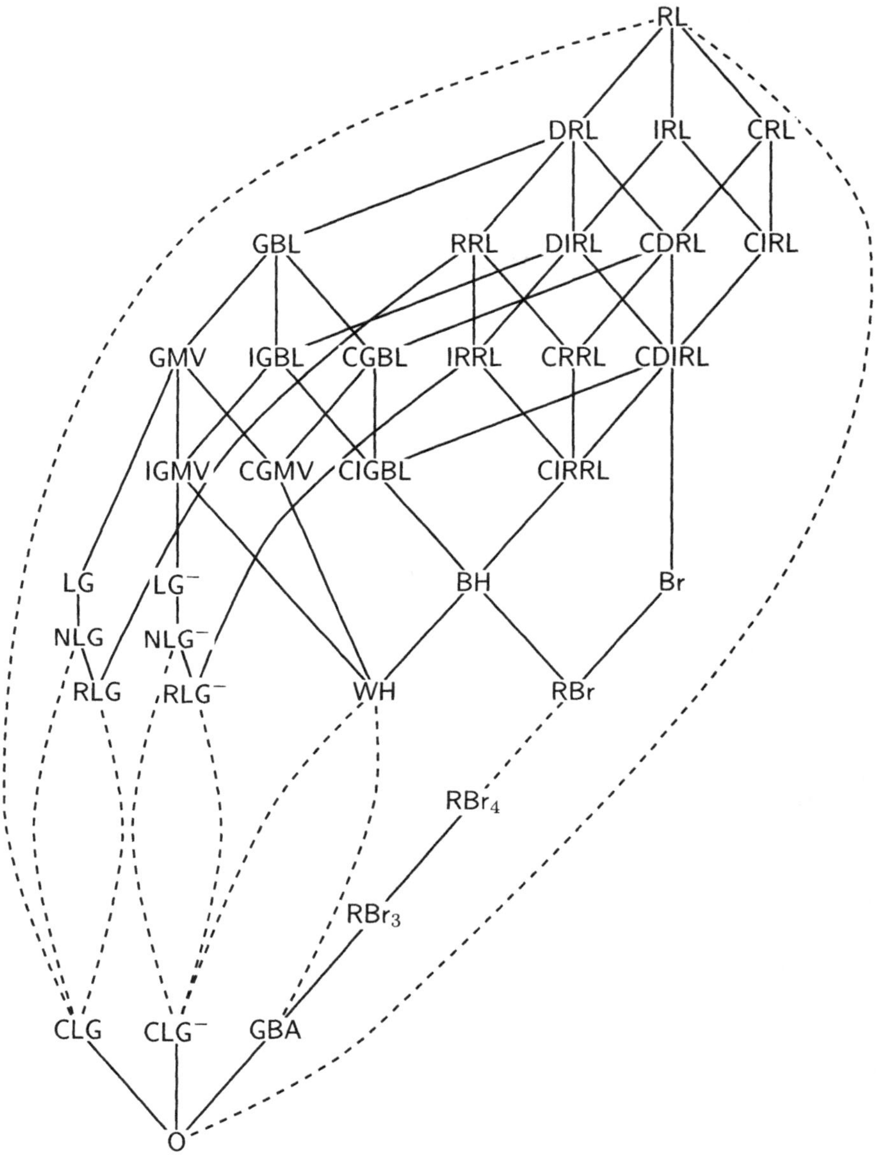

FIGURE 3.6. Some subvarieties of RL ordered by inclusion.

3.5.1. Some subvarieties of FL. We list here some important subvarieties of FL and RL, together with their various names.

- RL = *residuated lattices* [JT02, BT03]: FL-algebras that satisfy $0 = 1$.
- $\mathsf{FL_e}$ = *FL_e-algebras* [Ono03a]: FL-algebras that satisfy $x \cdot y = y \cdot x$ (or equivalently $x\backslash y = y/x$). In this case one usually writes $x \to y$ instead of $x\backslash y$ or y/x.
- $\mathsf{FL_i}$ = *FL_i-algebras* [Ono03a]: integral FL-algebras, i.e. FL-algebras that satisfy $x \leq 1$.
- $\mathsf{FL_o}$ = *FL_o-algebras* [Ono03a]: FL-algebras that satisfy $0 \leq x$.
- $\mathsf{FL_w}$ = *FL_w-algebras* [Ono03a]: FL-algebras that satisfy weakening, i.e. both $0 \leq x$ and $x \leq 1$.
- $\mathsf{FL_c}$ = *FL_c-algebras* [Ono03a]: FL-algebras that satisfy contraction, i.e. $x \leq xx$, hence they are also called square increasing.
- InFL = *involutive FL-algebras*: FL-algebras that satisfy $\sim -x = x = -\sim x$ where $\sim x = x\backslash 0$ and $-x = 0/x$.
- CyFL = *cyclic FL-algebras*: FL-algebras that satisfy $\sim x = -x$, i.e. $x\backslash 0 = 0/x$; this unary operation is usually denoted by $\neg x$.
- $\mathsf{P_nFL}$ = *n-potent FL-algebras*: FL-algebras that satisfy $x^n = x^{n+1}$.
- DFL = *distributive FL-algebras*: FL-algebras that satisfy $x \wedge (y \vee z) = (x \wedge y) \vee (x \wedge z)$.
- RFL = *representable FL-algebras*: FL-algebras that are subdirect products of linearly ordered FL-algebras, or equivalently satisfy the identity $1 \leq u\backslash((x \vee y)\backslash x)u \vee v((x \vee y)\backslash y)/v$; see Corollary 9.75.
- SRA = *symmetric relation algebras*: involutive distributive FL_e-algebras that satisfy $x \wedge \neg x \leq y$.

3.5.2. Some subvarieties of $\mathsf{FL_w}$.

- psMTL = *pseudo monoidal t-norm algebras, or weak-pseudo-BL algebras* [FGI01]: FL_w-algebras that satisfy prelinearity ($x\backslash y \vee y\backslash x = 1$ and $x/y \vee y/x = 1$).
- $\mathsf{FL_{ew}}$ = *FL_{ew}-algebras* [KO01]: algebras that are both FL_e-algebras and FL_w-algebras.
- MTL = *monoidal t-norm algebras* [EG01]: FL_{ew}-algebras that satisfy prelinearity.
- psBL = *pseudo BL-algebras* [FGI01], [DNGI02]: FL_w-algebras that satisfy divisibility ($x \wedge y = x(x\backslash y) = (y/x)x$).
- BL = *basic logic algebras* [Háj98]: MTL-algebras that satisfy divisibility, or equivalently, commutative prelinear pseudo BL-algebras.
- HA = *Heyting algebras* [BD74], [Day72]: FL_o-algebras that satisfy $x \wedge y = xy$, or equivalently FL_w-algebras that are idempotent ($xx = x$).
- psMV = *pseudo MV-algebras* [GI01]: pseudo BL-algebras that satisfy $x \vee y = x/(y\backslash x) = (x/y)\backslash x$.

- MV = *multi-valued logic algebras, or Łukasiewicz algebras* [CDM00], cf. [FRT84]: BL-algebras that satisfy $\neg\neg x = x$ or equivalently, commutative pseudo MV-algebras.
- GA = *Gödel logic algebras, or linear Heyting algebras* [Háj98], [Dum59]: BL-algebras that satisfy $x \cdot x = x$ or equivalently, Heyting algebras that satisfy prelinearity.
- GA_n = *Gödel logic algebras of degree n*: subdirect products of the linearly ordered n-element Heyting algebra.
- Π = *product logic algebras* [Háj98] [Cin01]: BL-algebras that satisfy $\neg\neg x \leq (x \to xy) \to y(\neg\neg y)$.
- BA = *Boolean algebras*: Heyting algebras that satisfy $\neg\neg x = x$, or equivalently, MV-algebras that are idempotent ($xx = x$).

3.5.3. Some subvarieties of RL.

- DRL = *distributive residuated lattices*: Residuated lattices that satisfy $x \wedge (y \vee z) = (x \wedge y) \vee (x \wedge z)$.
- IRL = *integral residuated lattices*: Residuated lattices that satisfy $x \leq 1$.
- CRL = *commutative residuated lattices* [HRT02]: Residuated lattices that satisfy $x \cdot y = y \cdot x$ (or equivalently $x\backslash y = y/x$). In this case one usually writes $x \to y$ instead of $x\backslash y$ or y/x.
- RRL = *representable residuated lattices* [BT03]: Residuated lattices that are subdirect products of residuated chains, or equivalently satisfy the identity $1 \leq u\backslash((x \vee y)\backslash x)u \vee v((x \vee y)\backslash y)/v$.
- GBL = *generalized BL-algebras* [JT02]: Residuated lattices that satisfy $x \wedge y = x(x\backslash(x \wedge y)) = ((x \wedge y)/x)x$.
- GMV = *generalized MV-algebras* [JT02] [GT05]: Residuated lattices that satisfy $x \vee y = x/((x \vee y)\backslash x) = (x/(x \vee y))\backslash x$.
- Fleas [Háj05]: Integral residuated lattices that satisfy prelinearity ($x\backslash y \vee y\backslash x = 1$ and $x/y \vee y/x = 1$).
- GBH = IGBL = *generalized basic hoops, pseudo hoops*: Residuated lattices that satisfy $x \wedge y = x(x\backslash y) = (y/x)x$. It follows that they are integral.
- BH = *basic hoops* [AFM99]: Commutative representable generalized basic hoops, i.e., commutative representable residuated lattices that satisfy divisibility ($x \wedge y = x(x\backslash y)$).
- WH = *Wajsberg hoops*: Commutative integral generalized MV-algebras.
- CanRL = *cancellative residuated lattices* [BCG$^+$03]: Residuated lattices that satisfy $x = (x/y)y = y(y\backslash x)$. These identities are equivalent to cancellativity of fusion.

- LG = *lattice-ordered groups or ℓ-groups* [AF88] [GH89]: Residuated lattices that satisfy $1 = x(x\backslash 1)$. NLG = *normal-valued ℓ-groups*, defined by $(x \wedge 1)^2(y \wedge 1)^2 \leq (y \wedge 1)(x \wedge 1)$. RLG = *representable ℓ-groups*, defined by $1 \leq (e\backslash x)yx \vee 1\backslash y$. CLG = *commutative ℓ-groups*
- LG^- = *negative cones of lattice-ordered groups* [JT02]: Cancellative integral generalized BL-algebras. NLG^- = *negative cones of normal-valued ℓ-groups*, defined by $x^2y^2 \leq yx$ relative to LG^-. RLG^- = *negative cones of representable ℓ-groups*, defined as cancellative integral representable generalized BL-algebras. CLG^- = *negative cones of commutative ℓ-groups*, defined as cancellative basic hoops.
- Br = *Brouwerian algebras*: Residuated lattices that satisfy $x \wedge y = xy$.
- RBr = *representable Brouwerian algebras*: Brouwerian algebras that satisfy prelinearity, or equivalently, basic hoops that are idempotent ($xx = x$).
- $\mathsf{RBr_n}$ = *representable Brouwerian algebras of degree n*, defined as subdirect products of the linearly ordered n-element Brouwerian algebra.
- GBA = *generalized Boolean algebras*: Brouwerian algebras that satisfy $x \vee y = (x \to y) \to y$, or equivalently, Wajsberg hoops that are idempotent ($xx = x$).

Many further varieties can be obtained from these by combining some of the identities mentioned above. For example the prefixes C, D, I, are used to denote the commutative, distributive and integral identities respectively. The relationships between various subvarieties of RL and FL are shown in Figures 1 and 2. Note that joins in these figures do not in general agree with joins in the lattice of subvarieties.

3.6. Structure theory

In this section we will identify certain subsets of an FL-algebra that correspond to congruences of the algebra. This is based on the fact that FL-algebras are 1-regular. Moreover, we will discuss the connection with deductive filters; see also Note 2.

3.6.1. Structure theory for special cases. Recall from Section 1.6 that if $\vdash$ is a consequence relation on the set of $\mathcal{L}$-formulas, for some language $\mathcal{L}$, then a matrix model of $\vdash$ is a pair $(\mathbf{A}, F)$, where $\mathbf{A}$ is an $\mathcal{L}$-algebra and F is a subset of A such that for every assignment f into $\mathbf{A}$, and for every set $\Phi \cup \{\psi\}$ of $\mathcal{L}$-formulas, such that $\Phi \vdash \psi$, if $f[\Phi] \in F$ then $f(\psi) \in F$. In particular, if $\vdash$ is presented by a Hilbert-style system, then it is enough to check the above condition for the inference rules (Φ, ψ) and the axioms $(\emptyset, \psi)$ of the system. In this case F is called a $\vdash$-deductive filter of $\mathbf{A}$, or a deductive filter of $\mathbf{A}$ with respect to $\vdash$.

A subset F of a Boolean algebra $\mathbf{A}$ is a deductive filter of $\mathbf{A}$ with respect to $\vdash_{\mathbf{HK}}$ iff

FL	RL	Defining identities
$\mathsf{FL_e}$	CRL	$xy = yx$
DFL	DRL	$x \wedge (y \vee z) = (x \wedge y) \vee (x \wedge z)$
RFL	RRL	$1 \leq u\backslash((x \vee y)\backslash x)u \vee v((x \vee y)\backslash y)/v$
		Below add $0 \leq x$ for subvarieties of FL
$\mathsf{FL_w}$	IRL	$x \leq 1$
$\mathsf{FL_{ew}}$	CIRL	$xy = yx$, $x \leq 1$
psMTL	Fleas	$x\backslash y \vee y\backslash x = 1 = x/y \vee y/x$
MTL	CIRRL	$xy = yx$, $(x \to y) \vee (y \to x) = 1$
	GBL	$x \wedge y = x(x\backslash(x \wedge y)) = ((x \wedge y)/x)x$
psBL	IGBL	$x \wedge y = x(x\backslash y) = (y/x)x$
BL	BH	$xy = yx$, $x \wedge y = x(x \to y)$, $(x \to y) \vee (y \to x) = 1$
	GMV	$x \vee y = x/((x \vee y)\backslash x) = (x/(x \vee y))\backslash x$
psMV	IGMV	$x \vee y = x/(y\backslash x) = (x/y)\backslash x$
MV	WH	$xy = yx$, $x \vee y = (x \to y) \to y$
HA	Br	$x \wedge y = xy$
GA	RBr	$x \wedge y = xy$, $(x \to y) \vee (y \to x) = 1$
Π		BL and $\neg\neg x \leq (x \to xy) \to y(\neg\neg y)$
BA	GBA	$x \wedge y = xy$, $x \vee y = (x \to y) \to y$

TABLE 3.1. Definition and correspondence of subvarieties of FL and RL.

FL	RL	Generated by
RFL_w	IRRL	all residuated pseudo-t-norms
MTL	CIRRL	all residuated t-norms [JM02]
BL	BH	all continuous t-norms [Háj98] [CEGT00]
MV	WH	Łukasiewicz $xy = \max\{0, x + y - 1\}$ [Cha59]
GA	RBr	Gödel $xy = \min\{x, y\}$ [Háj98]
Π		Product xy = multiplication on $[0, 1]$ [Háj98]

TABLE 3.2. Some varieties generated by t-norms.

- $1 \in F$ and
- $a, a \to b \in F$ implies $b \in F$.

This follows from the fact that in a Boolean algebra all axioms of **HK** evaluate to 1 under all assignments and (mp) is the only inference rule.

It is not hard to see that the above conditions can be replaced by

- $1 \in F$,
- $a \in F$ and $a \leq b$ implies $b \in F$ and
- $a, b \in F$ implies $a \wedge b \in F$.

therefore the deductive filters of Boolean algebras are exactly the nonempty lattice filters of its lattice reduct.

Every congruence θ on a Boolean algebra is completely determined by the block $[1]_\theta$ of 1, which is a filter of the algebra. Indeed, $a\ \theta\ b$ iff $(a \leftrightarrow b)\ \theta\ 1$; also, $[1]_\theta$ contains 1, is convex and it is closed under meets. Here $a \leftrightarrow b$ is short for $(a \to b) \wedge (b \to a)$ and it will be defined in a more general setting on page 191. Conversely, if F is a deductive filter on a Boolean algebra $\mathbf{A}$, then the relation $\Theta(F)$, defined by $a\ \Theta(F)\ b$ iff $a \leftrightarrow b \in F$, is a congruence relation on $\mathbf{A}$. Finally, we can see that the above define a lattice isomorphism between congruences and deductive filters of $\mathbf{A}$. Exercise 52 asks you to check the details of the above claims. Exactly the same situation holds for Heyting algebras; see also Section 1.2.4.

Although the deductive filters of MV-algebras do not coincide with lattice filters we still get a lattice isomorphism between congruences and deductive filters given by the maps $\theta \mapsto [1]_\theta$ and $F \mapsto \Theta(F)$. It can be shown that the deductive filters of an MV-algebra $\mathbf{A}$ are the subsets F of A that satisfy the conditions

- $1 \in F$,
- $a \in F$ and $a \leq b$ implies $b \in F$, and
- $a, b \in F$ implies $a \cdot b \in F$.

Note that deductive filters are lattice filters, since $xy \leq x \wedge y$ in MV-algebras, but the converse is not true. Nevertheless, in view of the fact that every Boolean algebra is an MV-algebra, the conditions for a deductive filter specialize exactly to those for Boolean algebras (since $xy = x \wedge y$).

The above indicate that there is a connection between the deductive filters (with respect to $\vdash_{\mathbf{FL}}$) and the congruences of an FL-algebra. In this section, we will show that this is the case, even though the deductive filters are more general; they have to satisfy normality and adjunction. This is not surprising in view of Theroem 2.16 and algebraization.

On a seemingly different direction, recall that congruences on groups are also determined by the block $[1]_\theta$ of 1, which is a normal subgroup of the algebra. Note that if $\mathbf{A}$ is a Boolean algebra, then $(A, \leftrightarrow, 1)$ is an abelian group of order 2 (every element is its own inverse). The normal subgroups of $(A, \leftrightarrow, 1)$ are exactly the nonempty lattice ideals of $\mathbf{A}$.

Furthermore, congruences on ℓ-groups are determined by the block $[1]_\theta$ of 1 ($a\ \theta\ b$ iff $ab^{-1} \in [1]_\theta$), which is a convex (relative to the order) normal subgroup of the algebra. In view of the correspondence between abelian ℓ-groups and MV-algebras, see Section 2.3.5, the convex (normal) subgroups become filters of the corresponding MV-algebras. This fact seems to indicate a connection between deductive filters and certain subalgebras.

Note that the deductive filters in MV-algebras (and also in Heyting and Boolean algebras) are simply convex subalgebras of the 0-free reducts. This should not lead to the conjecture that for FL-algebras convex subalgebras of their residuated lattice (or 0-free) reducts are the same as deductive filters, as this is true only for FL_{ei}-algebras. It is not true (without modifications) if we lack integrality, since deductive $\vdash_{\mathbf{FL}}$-filters of ℓ-groups are not convex subalgebras. For example, the set of all positive elements is a deductive filter, but it is not a subalgebra (since $1/a$ is not positive, if $a \neq 1$ is positive). As for the lack of commutativity, we mention that we will need the condition of normality (closure under iterated conjugates) for convex subalgebras.

We will see that the upward closure of a convex normal subalgebra is a deductive filter. Also, the set of all elements of a deductive filter F that are between $1/a$ and a, for $a \in F$, form a convex normal subalgebra. This will yield a lattice isomorphism between deductive filters and convex normal subalgebras.

Yet another type of subset can be defined by the observation that every deductive filter (and every convex normal subalgebra) is determined by its negative cone. We will identify these subsets as convex normal (in the whole algebra) submonoids of the negative cone of the algebra and show that these form a fourth lattice, isomorphic to the ones mentioned above.

3.6.2. Convex normal subalgebras and submonoids, congruences and deductive filters. Recall the definition of a conjugate of a formula by another formula from page 121. This definition applies to the absolutely free algebra of formulas, but it can be generalized to any algebra over the same type. In particular we will be interested in the case where iterated conjugates denote the following homonymous elements or polynomials.

Let $\mathbf{A}$ be a residuated lattice or FL-algebra. For $a, x \in A$, we define the *left conjugate* $\lambda_a(x) = a\backslash xa \wedge 1$ and the *right conjugate* $\rho_a(x) = ax/a \wedge 1$ of x with respect to a. An *iterated conjugate* of x is a composition $\gamma_{a_1}(\gamma_{a_2}(\dots\gamma_{a_n}(x)))$, where n is a positive integer, $a_1, a_2, \dots, a_n \in A$ and $\gamma_{a_i} \in \{\lambda_{a_i}, \rho_{a_i}\}$, for all $i \in \{1, 2, \dots, n\}$. We denote the set of all iterated conjugates of elements of $X \subseteq A$ by $\Gamma(X)$. We will also allow X to be a set of variables, in which setting we will consider *conjugate polynomials*. So, an iterated conjugate (polynomial) is a composition of polynomials of the form $\lambda_a(x) = a\backslash xa \wedge 1$ and $\rho_b(x) = bx/b \wedge 1$, for various values of a and b. For example, $\gamma(x) = a\backslash(b\backslash(cx/c \wedge 1)b \wedge 1)a \wedge 1$ is an iterated conjugate (polynomial). In analogy with groups, a subset X of A is called *normal*, if for all $x \in X$ and $a \in A$, $\lambda_a(x), \rho_a(x) \in X$.

LEMMA 3.46. [BT03] *If $\mathbf{A}$ is a residuated lattice or an FL-algebra and $u, a_1, \dots, a_n \in A$, then*

$$\prod \lambda_u(a_i) \leq \lambda_u(\prod a_i) \quad \textit{and} \quad \prod \rho_u(a_i) \leq \rho_u(\prod a_i).$$

PROOF. For $n = 2$, we have

$$\lambda_u(a_1)\lambda_u(a_2) = (u\backslash a_1 u \wedge 1)(u\backslash a_2 u \wedge 1) \leq (u\backslash a_1 u)(u\backslash a_2 u) \wedge 1$$

$$\leq u\backslash a_1u(u\backslash a_2u) \wedge 1 \leq u\backslash a_1a_2u \wedge 1 = \lambda_u(a_1a_2).$$

The rest of the proof is a simple induction; likewise we prove the statement for ρ_u. □

A term $t(u_1, \ldots, u_m, x_1, \ldots, x_n)$ in the language of a class $\mathcal{K}$ of similar algebras with a constant e is called an *ideal term of* $\mathcal{K}$ *in* $x_1, \ldots, x_n$ if $\mathcal{K}$ satisfies the identity $t(u_1, \ldots, u_m, e, \ldots, e) = e$. We also write the term as $t_{u_1,\ldots,u_m}(x_1, \ldots, x_n)$ to indicate the distinction between the two types of variables.

Examples of ideal terms for FL and RL are $\lambda_u(x) = (u\backslash xu) \wedge 1$, $\rho_u(x) = (ux/u) \wedge 1$, as well as $\kappa_u(x, y) = (u \wedge x) \vee y$, and the fundamental operations $x \diamond y$ for $\diamond \in \{\vee, \wedge, \cdot, \backslash, /\}$.

A subset H of $\mathbf{A} \in \mathcal{K}$ is a $\mathcal{K}$*-ideal of* $\mathbf{A}$ if for all ideal terms t of $\mathcal{K}$, and all $a_1, \ldots, a_m \in A$, $b_1, \ldots, b_n \in H$ we have $t^{\mathbf{A}}(a_1, \ldots, a_m, b_1, \ldots, b_n) \in H$. Note that we use the superscript $\mathbf{A}$ to distinguish the term function $t^{\mathbf{A}}$ from the (syntactic) term t that defines it.

Clearly any e-congruence class is a $\mathcal{K}$-ideal. A class $\mathcal{K}$ is called an *ideal class* if in every member of $\mathcal{K}$ every ideal is an e-congruence class. We will prove below that this is the case for residuated lattices, and that the ideals of a residuated lattice are characterized as those subalgebras that are closed under the ideal terms λ, ρ and κ.

The closed interval $\{u \in A : x \leq u \leq y\}$ is denoted by $[x, y]$. As for posets, we call S *convex* if $[x, y] \subseteq S$ for all $x, y \in S$. Note that for a sub*lattice* S the property of being convex is equivalent to $\kappa_u(x, y) \in S$ for all $u \in A$ and $x, y \in S$. Thus a convex normal subalgebra is precisely a subalgebra of $\mathbf{A}$ that is closed under the RL-ideal terms λ, ρ and κ.

Now it follows immediately that every RL-ideal is a convex normal subalgebra, and since we observed earlier that every e-congruence class is an RL-ideal, we have shown that every e-congruence class is a convex normal subalgebra. As we will see below, the converse requires a bit more work.

For $a, b \in A$, we define $a \leftrightarrow b = a\backslash b \wedge b\backslash a \wedge 1$ and $a \leftrightarrow' b = b/a \wedge a/b \wedge 1$; clearly $a \leftrightarrow 1 = a\backslash 1 \wedge a \wedge 1$. Moreover, for every subset X of A, we define the sets

$$\begin{aligned}
&X \wedge 1 = \{x \wedge 1 \mid x \in X\},\\
&\Delta(X) = \{x \leftrightarrow 1 \mid x \in X\},\\
&\Pi(X) = \{x_1x_2 \cdots x_n \mid n \geq 1, x_i \in X\} \cup \{1\},\\
&\Xi(X) = \{a \in A \mid x \leq a \leq x\backslash 1, \text{ for some } x \in X\} \text{ and}\\
&\Xi^-(X) = \{a \in A \mid x \leq a \leq 1, \text{ for some } x \in X\}.
\end{aligned}$$

Note that the negative part $A^- = \{a \in A \mid a \leq 1\}$ of A is closed under multiplication and it contains 1, so it is a submonoid of $\mathbf{A}$. If $\mathbf{A}$ is an FL-algebra, we denote its 0-free residuated lattice reduct by $\mathbf{A}_r$. If $\mathbf{A}$ is a residuated lattice, we set $\mathbf{A}_r = \mathbf{A}$.

THEOREM 3.47. [BT03] [JT02] [GO06a] *For every residuated lattice or FL-algebra* $\mathbf{A}$, *the following properties hold.*

(1) *If S is a convex normal subalgebra of $\mathbf{A}_r$, M a convex normal in $\mathbf{A}$ submonoid of A^-, θ a congruence on $\mathbf{A}$ and F a deductive filter of $\mathbf{A}$, then*
  (a) $M_s(S) = S^-$, $M_c(\theta) = [1]_\theta^-$ *and* $M_f(F) = F^-$ *are convex, normal in* $\mathbf{A}$ *submonoids of* $\mathbf{A}^-$,
  (b) $S_m(M) = \Xi(M)$, $S_c(\theta) = [1]_\theta$ *and* $S_f(F) = \Xi(F^-)$ *are convex normal subalgebras of* $\mathbf{A}_r$,
  (c) $F_s(S) = \uparrow S$, $F_m(M) = \uparrow M$, *and* $F_c(\theta) = \uparrow[1]_\theta$ *are deductive filters of* $\mathbf{A}$.
  (d) $\Theta_s(S) = \{(a,b) \mid a \leftrightarrow b \in S\}$, $\Theta_m(M) = \{(a,b) \mid a \leftrightarrow b \in M\}$ *and* $\Theta_f(F) = \{(a,b) \mid a \leftrightarrow b \in F\} = \{(a,b) \mid a\backslash b, b\backslash a \in F\}$ *are congruences on* $\mathbf{A}$.

Moreover, the deductive filters, as well as the congruence relations and the convex normal submonoids, of $\mathbf{A}$ *and* $\mathbf{A}_r$ *are identical.*

(2) (a) *The convex, normal subalgebras of* $\mathbf{A}_r$, *the convex, normal in* $\mathbf{A}$ *submonoids of* A^- *and the deductive filters of* $\mathbf{A}$ *form lattices, denoted by* $\mathbf{CNS}(\mathbf{A}_r)$, $\mathbf{CNM}(\mathbf{A})$ *and* $\mathbf{Fil}(\mathbf{A})$, *respectively.*
  (b) *All the above lattices are isomorphic to the congruence lattice* $\mathbf{Con}(\mathbf{A})$ *of* $\mathbf{A}$ *via the appropriate pairs of maps defined above.*
  (c) *The composition (whenever defined) of any two of the above maps gives the corresponding map; e.g.,* $M_s(S_c(\theta)) = M_c(\theta)$.

(3) *If X is a subset of A^- and Y is a subset of A, then*
  (a) *the convex, normal in* $\mathbf{A}$ *submonoid* $M(X)$ *of* A^- *generated by* X *is equal to* $\Xi^-\Pi\Gamma(X)$.
  (b) *The convex, normal subalgebra* $S(Y)$ *of* $\mathbf{A}$ *generated by* Y *is equal to* $\Xi\Pi\Gamma\Delta(Y)$.
  (c) *The deductive filter* $F(Y)$ *of* $\mathbf{A}$ *generated by* $Y \subseteq A$ *is equal to* $\uparrow\Pi\Gamma(Y) = \uparrow\Pi\Gamma(Y \wedge 1)$.
  (d) *The congruence* $\Theta(P)$ *on* $\mathbf{A}$ *generated by a set of pairs* $P \subseteq A^2$ *is equal to* $\Theta_m(M(P'))$, *where* $P' = \{a \leftrightarrow b \mid (a,b) \in P\}$.

PROOF. Throughout the proof we will use properties of residuated lattices and FL-algebras and the characterization of deductive filters freely, without any particular reference to Theorem 2.6 or Theorem 2.30.

(1a) Clearly, F^- and S^- are convex submonoids of A^-. Moreover, F^- is normal, since for $u \in A$ and $a \in F$, we have $u\backslash au, ua/u \in F$, by (pn), and $u\backslash au \wedge 1, ua/u \wedge 1 \in F$, by (adj$_u$). Also, S^- is normal since the conjugates are negative elements. Consequently, $[1]_\theta^-$ is also a convex normal in $\mathbf{A}$ submonoid of $\mathbf{A}-$.

(1b) It is easy to see that, $[1]_\theta$ is a convex normal subalgebra; it also follows from the fact that it is an RL-ideal. We will show that $\Xi(M)$ is a convex normal subalgebra. The same will follow for $\Xi(F^-)$ by (1a).

Observe that $a \in \Xi(M)$ iff there are $y, z \in M$ such that $y \leq a \leq z\backslash 1$. For the non trivial direction, note that $yz \leq y \leq a \leq z\backslash 1 \leq yz\backslash 1$ and $yz \in M$.

If $a, b \in \Xi(M)$, then there exist $x, y \in M$ such that $x \leq a \leq x\backslash 1$ and $y \leq b \leq y\backslash 1$. If $a \leq c \leq b$, for some $c \in A$, then $x \leq a \leq c \leq b \leq y\backslash 1$, so $c \in \Xi(M)$; hence $\Xi(M)$ is convex. Moreover,

$$x \wedge y \leq a \wedge b \leq x\backslash 1 \wedge y\backslash 1 = (x \vee y)\backslash 1,$$

$$x \vee y \leq a \vee b \leq (x \wedge y)\backslash 1,$$

$$xy \leq ab \leq (x\backslash 1)(y\backslash 1) \leq x\backslash(y\backslash 1) = (yx)\backslash 1$$

$$\lambda_a(yx) \leq a\backslash yxa \leq a\backslash[y/(x\backslash 1)]a \leq a\backslash[b/a]a \leq a\backslash b \leq x\backslash(y\backslash 1) = yx\backslash 1$$

and

$$xy \leq x/(y\backslash 1) \leq a/b \leq (x\backslash 1)/y \leq [x\rho_{(x\backslash 1)/y}(y)]\backslash 1$$

since for $u = (x\backslash 1)/y$ we have $x\rho_u(y)u \leq x\{uy/u\}u \leq xuy \leq 1$. Therefore, $\Xi(M)$ is a subalgebra.

To show that $\Xi(M)$ is normal, we consider an element u in A; we will show that $\lambda_u(x), \rho_u(x) \in \Xi(M)$ for all $a \in \Xi(M)$. Indeed,

$$\lambda_u(x) \leq \lambda_u(a) \leq \lambda_u(x\backslash 1) \leq \lambda_u(x)\backslash 1,$$

since $\lambda_u(x)\lambda_u(x\backslash 1) \leq u\backslash x(x\backslash 1)u \wedge 1 \leq u\backslash u \wedge 1 = 1$, by Lemma 3.46, and likewise for $\rho_u(a)$.

(1c) We will show that $\uparrow M$ is a deductive filter. The remaining two claims are shown in the same way. Note that $\uparrow M$ contains 1 and is upward closed. If $a, b \in \uparrow M$, then there exist $x, y \in M$ such that $x \leq a$ and $y \leq b$. So, $xy \leq ab$ and $xy \in M$. Therefore, $\uparrow M$ is closed under (p). Finally, for $u \in A$, we have $\lambda_u(x) \leq \lambda_u(a) \leq u\backslash au$ and $\lambda_u(x) \in M$, so $u\backslash au \in \uparrow M$; likewise we have $ua/u \in \uparrow M$, so $\uparrow M$ is closed under (pn).

(1d) Clearly $\Theta_s(S)$ is reflexive and symmetric. Assuming $a \leftrightarrow b, b \leftrightarrow c \in S$, we have

$$(a \leftrightarrow b)(b \leftrightarrow c) \wedge (b \leftrightarrow c)(a \leftrightarrow b) \leq (a\backslash b)(b\backslash c) \wedge (c\backslash b)(b\backslash a) \wedge 1 \leq (a \leftrightarrow c) \leq 1.$$

Since S is convex, $a \leftrightarrow b \in S$, and therefore $\Theta_s(S)$ is an equivalence relation; we will show that it is a congruence.

Assuming $a \leftrightarrow b \in S$ and $c \in A$, it remains to show that $c \diamond a \leftrightarrow c \diamond b, a \diamond c \leftrightarrow b \diamond c \in S$ for $\diamond \in \{\cdot, \wedge, \vee, \backslash, /\}$. Since S is convex and $x \leftrightarrow y \leq 1 \in S$, it suffices to construct elements of S that are below these two expressions.

We have $a\backslash b \leq ca\backslash cb$, hence $a \leftrightarrow b \leq ca \leftrightarrow cb$. Also, $\lambda_c(a \leftrightarrow b) \leq c\backslash(a\backslash b)c \wedge c\backslash(b\backslash a)c \wedge e \leq ac \leftrightarrow bc$.

Since $(a \wedge c) \cdot (a \leftrightarrow b) \leq a(a \leftrightarrow b) \wedge c(a \leftrightarrow b) \leq b \wedge c$, we have $a \leftrightarrow b \leq (a \wedge c)\backslash(b \wedge c)$, and similarly $a \leftrightarrow b \leq (b \wedge c)\backslash(a \wedge c)$. Therefore $a \leftrightarrow b \leq (a \wedge c) \leftrightarrow (b \wedge c) \leq 1$. The computation for $\vee$ is the same.

We have $a\backslash b \leq (c\backslash a)\backslash(c\backslash b)$ and $b\backslash a \leq (c\backslash b)\backslash(c\backslash a)$, whence $a \leftrightarrow b \leq (c\backslash a) \leftrightarrow (c\backslash b)$. Likewise, we get $a\backslash b \leq (a\backslash c)/(b\backslash c)$ and $b\backslash a \leq (b\backslash c)/(a\backslash c)$ therefore $a \leftrightarrow b \leq (a\backslash c) \leftrightarrow' (b\backslash c)$. It follows that $(a\backslash c) \leftrightarrow' (b\backslash c)$ is in S, so by Exercise 55 we have $(a\backslash c) \leftrightarrow (b\backslash c) \in S$. The computation for $/$ is a mirror image of the above.

Note that $\Theta_s(S) = \{(a, b) \mid a \leftrightarrow b \in S^-\}$, since $a \leftrightarrow b$ is a negative element. Using the results proved in (2b), we have $\Theta_m(M) = \{(a, b) | a \leftrightarrow b \in (\Xi(M))^-\} = \Theta_s(S_m(M))$. Likewise, we get $\Theta_f(F) = \Theta_s(S_f(F))$, so all the expressions in (1d) define congruences.

(2a) It is clear that the three sets form closure families, so they form lattices, where meet is intersection.

(2b) It is clear that all the maps in (1) are order preserving. The lattice isomorphisms will follow after we show that the three following pairs of maps define bijections.

We will show that S_c and Θ_s are inverse maps. Note that $S = [1]_{\Theta_s(S)}$ holds because S is a convex subalgebra. Indeed, $a \in S$ implies $a \leftrightarrow 1 \in S$, and the reverse implication holds since $(a \leftrightarrow 1) \leq a \leq (a \leftrightarrow 1)\backslash 1$.

For the other composition, note that if $(a, b) \in \Theta_s(S_c(\theta))$ then $a \leftrightarrow b \in [1]_\theta$ so $a \leftrightarrow b \; \theta \; 1$. Therefore, $a \; \theta \; a(a \leftrightarrow b) \leq a(a\backslash b) \leq b$, so $a \vee b \; \theta \; b$. Likewise, $a \vee b \; \theta \; a$, so $a \; \theta \; b$. Conversely, if $a \; \theta \; b$, then $1 = (a\backslash a \wedge b\backslash b \wedge 1) \; \theta \; (a\backslash b \wedge b\backslash a \wedge 1) = a \leftrightarrow b$.

We will show that S_m and M_s are inverse maps. Note that $a \in \Xi(S^-)$ iff there exists $x \in S^-$ such that $x \leq a \leq x\backslash 1$. Since S contains 1 is closed under division and convex, we have $a \in S$, so $\Xi(S^-) \subseteq S$. The other inclusion is obvious, so $S_m(M_s(S)) = S$. Moreover, $a \in \Xi(M)^-$ iff $a \leq 1$ and there exists $x \in M$ such that $x \leq a \leq x\backslash 1$; hence $x \leq a \leq 1$. Since M contains 1 and is convex, we have $a \in M$, so $\Xi(M)^- \subseteq M$. The other inclusion is obvious, so $M_s(S_m(M) = M$.

We will show that M_f and F_m are inverse maps. If $a \in ({\uparrow}M)^-$, then there is an $x \in M$ such that $x \leq a \leq 1$; since M contains 1 and is convex, we have $a \in M$. The converse direction is trivial, so $({\uparrow}M)^- = M$.

Clearly, ${\uparrow}(F^-) \subseteq {\uparrow}F = F$. Also, if $a \in F$, then $a \wedge 1 \in F$, by (adj$_u$) and $a \wedge 1 \leq a$, so $F \subseteq {\uparrow}(F^-)$.

(2c) We will verify some of the compositions; Exercise 56 asks you to prove the remaining ones.

We will first show that $S_f(F) = S_c(\Theta_f(F))$. If $a \in S_c(\Theta_f(F))$, then $a \; \Theta_f(F) \; 1$, so $a\backslash 1, 1\backslash a \in F$. Hence $a, 1/a \in F$, by (symm). Since $1 \in F$, we get $x = a \wedge 1/a \wedge 1 \in F^-$, by (adj). Obviously, $x \leq a$; also $a \leq (1/a)\backslash 1 \leq x\backslash 1$. Thus, $a \in S_f(F)$. Conversely, if $a \in S_f(F)$, then $x \leq a \leq x\backslash 1$, for some

$x \in F^-$. So, $a \in F$, by (up), and $1/(x\backslash 1) \leq 1/a$. Since, $x \leq 1/(x\backslash 1)$, we have $x \leq 1/a$ and $1/a \in F$, by (up). Thus both $a\backslash 1$ and $1\backslash a$ are in F. Hence, $a \in [1]_{\Theta_f(F)}$.

$M_f(F) = M_s(S_f(F))$ follows from the fact that $F^- = (\Xi(F^-))^-$. Indeed, $a \in (\Xi(F^-))^-$, iff $a \leq 1$ and $x \leq a \leq 1/x$ for some $x \in F^-$, iff $a \in F^-$.

To show that $F_s(S) = F_c(\Theta_s(S))$, note that $a \in F_c(\Theta_s(S))$ iff $1 \; \Theta_s(S) \; 1 \wedge a$ iff $1\backslash(1 \wedge a) \wedge 1, (1 \wedge a)\backslash 1 \wedge 1 \in S$ iff $1 \wedge a \in S$ iff $a \in \uparrow S$. An identical argument shows that $F_m(M) = F_c(\Theta_m(M))$.

Also, note that congruence relations do not change under the expansion of the language by a constant. So, the congruences of a pointed residuated lattice coincide with the congruences of its 0-free residuated-lattice reduct.

(3a) Clearly, if M is a convex, normal in $\mathbf{A}$ submonoid of $\mathbf{A}^-$ that contains X, then it contains $\Gamma(X)$, by normality, $\Pi\Gamma(X)$, since M is closed under product, and $\Xi^-\Pi\Gamma(X)$, since M is convex and contains 1. We will now show that $\Xi^-\Pi\Gamma(X)$ itself is a convex, normal in $\mathbf{A}$ submonoid of A^-; it obviously contains X. It is clearly convex and a submonoid of $\mathbf{A}^-$. To show that it is convex, consider $a \in \Xi^-\Pi\Gamma(X)$ and $u \in A$. There are $x_1, \ldots, x_n \in X$ and iterated conjugates $\gamma_1, \ldots, \gamma_n$ such that $\gamma_1(x_1)\cdots\gamma_n(x_n) \leq a \leq 1$. Using Lemma 3.46 we obtain

$$\prod \lambda_u(\gamma_i(x_i)) \leq \lambda_u(\prod \gamma_i(x_i)) \leq \lambda_u(a) \leq 1.$$

Also, $\lambda_u(\gamma_i(x_i)) \in \Gamma(X)$ and $\prod \lambda_u(\gamma_i(x_i)) \in \Pi\Gamma(X)$, so $\lambda_u(a) \in \Xi^-\Pi\Gamma(X)$. Likewise, we have $\rho_u(a) \in \Xi^-\Pi\Gamma(X)$.

(3b) Clearly, if a convex, normal subalgebra of $\mathbf{A}$ contains Y, then it also contains $\Xi\Pi\Gamma\Delta(Y)$. By (3a) and (1b) $\Xi\Xi^-\Pi\Gamma\Delta(Y) = \Xi\Pi\Gamma\Delta(Y)$ is a convex, normal subalgebra of $\mathbf{A}$ and it obviously contains Y.

(3c) We will first show that $F(Y) = F(Y \wedge 1)$. If $y \in Y$, then $y \wedge 1 \leq y$, so $y \in \uparrow(Y \wedge 1) \subseteq F(Y \wedge 1)$. Hence, $F(Y) \subseteq F(Y \wedge 1)$. Conversely, if $y \in Y$, then $y \wedge 1 \in F(Y)$, so $Y \wedge 1 \subseteq F(Y)$.

Thus, $F(Y \wedge 1) \subseteq F(Y)$.

By (1b), $(F(Y \wedge 1))^-$ is a convex, normal in $\mathbf{A}$ submonoid of A^- and contains $Y \wedge 1$. So, it contains $M(Y \wedge 1)$; hence $M(Y \wedge 1) \subseteq F(Y \wedge 1)$. Since $F(Y \wedge 1)$ is increasing, we have $\uparrow M(Y \wedge 1) \subseteq F(Y \wedge 1)$. On the other hand, $\uparrow M(Y \wedge 1)$ is a deductive filter, by (1d), and it contains $Y \wedge 1$, so it contains $F(Y \wedge 1)$. Thus, $F(Y \wedge 1) = \uparrow M(Y \wedge 1) = \uparrow \Xi^-\Pi\Gamma(Y \wedge 1) = \uparrow\Pi\Gamma(Y \wedge 1)$, by (3a). Consequently, $F(Y) = \uparrow\Pi\Gamma(Y \wedge 1)$.

If $x \in \uparrow\Pi\Gamma(Y)$, then it is greater or equal to a product of conjugates of elements y_i of Y. This product is in turn greater or equal to the product of the same conjugates of the elements $y_i \wedge 1$; so, $x \in \uparrow\Pi\Gamma(Y \wedge 1)$. Thus, $\Pi\Gamma(Y) \subseteq \uparrow\Pi\Gamma(Y \wedge 1)$. Conversely, if $x \in \uparrow\Pi\Gamma(Y \wedge 1)$, then it is greater or equal to a product of conjugates of elements $y_i \wedge 1$, where $y_i \in Y$. Note that $y_i \wedge 1$ is the left conjugate of y_i by 1, so x is greater or equal to a product of conjugates of the elements y_i; i.e., $x \in {\uparrow}\Pi\Gamma(Y)$. Consequently, ${\uparrow}\Pi\Gamma(Y) = {\uparrow}\Pi\Gamma(Y \wedge 1)$.

(3d) First we will show that $\Theta(P) = \Theta(P' \times \{1\})$. For every congruence θ on $\mathbf{A}$, and every $a, b \in A$, if $a \; \theta \; b$, then $(a\backslash b \wedge 1) \; \theta \; (b\backslash b \wedge 1) = 1$. Likewise, $b\backslash a \wedge 1 \; \theta \; 1$, so $a\backslash b \wedge b\backslash a \wedge 1 \; \theta \; 1$. Conversely, if $a\backslash b \wedge b\backslash a \wedge 1 \; \theta \; 1$, then since $a\backslash b \wedge b\backslash a \wedge 1 \leq a\backslash b \wedge 1 \leq 1$ and the congruence blocks of θ are convex (θ is a lattice congruence), we have $a\backslash b \wedge 1 \; \theta \; 1$; so, $a(a\backslash b \wedge 1) \; \theta \; a$. Since $a(a\backslash b \wedge 1) \leq a(a\backslash b) \wedge a \leq b \wedge a \leq a$, we get $a \wedge b \; \theta \; a$. Likewise, we have $a \wedge b \; \theta \; b$, so $a \; \theta \; b$. Thus, for every congruence θ on $\mathbf{A}$, and every $a, b \in A$, $(a, b) \in \theta$ iff $(a\backslash b \wedge b\backslash a \wedge 1, 1) \in \theta$. Consequently, for every $(a, b) \in P$, we have $(a\backslash b \wedge b\backslash a \wedge 1, 1) \in \Theta(P)$. So, $P' \times \{1\} \subseteq \Theta(P)$; hence $\Theta(P' \times \{1\}) \subseteq \Theta(P)$. Conversely, for every $(a, b) \in P$, we have $(a, b) \in \Theta(P' \times \{1\})$, since $(a\backslash b \wedge b\backslash a \wedge 1, 1) \in \Theta(P' \times \{1\})$. So, $P \subseteq \Theta(P' \times \{1\})$; hence $\Theta(P) \subseteq \Theta(P' \times \{1\})$.

Finally, we will prove that for every subset X of A^-, $\Theta(X \times \{1\}) = \Theta_m(M(X))$. If $x \in X$, then $x/1 \wedge 1 = x, 1/x \wedge 1 = 1 \in M(X)$, so $(x, 1) \in \Theta_m(M(X))$. Consequently, $X \times \{1\} \subseteq \Theta_m(M(X))$, hence $\Theta(X \times \{1\}) \subseteq \Theta_m(M(X))$. Conversely, if $(a, b) \in \Theta_m(M(X))$, then $a/b \wedge 1, b/a \wedge 1 \in M(X) = \Xi^-\Pi\Gamma(X)$, so $p \leq a/b$ and $q \leq b/a$, for some $p, q \in \Pi\Gamma(X)$. Thus, $pb \leq a$, $qa \leq b$, so $pqa \leq pb \leq a$. On the other hand since every element of X is congruent to 1 modulo $\Theta(X \times \{1\})$, every conjugate and every product of conjugates of elements of X is congruent to 1. In particular, p, q and pq are congruent to 1. So, $(pqa, a), (pb, b) \in \Theta(X \times \{1\})$. Since $pqa \leq pb \leq a$, we have $(a, b) \in \Theta(X \times \{1\})$. □

COROLLARY 3.48. *The variety* RL *is* 1*-regular. In particular, convex normal subalgebras of residuated lattices coincide with* RL*-ideals.*

Congruences of commutative residuated lattices, as we have seen quite a few times already, behave particularly well. We will see another example of that in the next lemma, which shows that in the commutative case all information about congruences is contained in the negative cone.

LEMMA 3.49. *Let* $\mathbf{A}$ *be a* FL_e*-algebra. For every* $\theta \in \operatorname{Con}\mathbf{A}$*, the restriction* $\theta|_{A^-}$ *belongs to* $\operatorname{Con}\mathbf{A}^-$ *and every* $\psi \in \operatorname{Con}\mathbf{A}^-$ *extends uniquely to a congruence* $\varphi \in \operatorname{Con}\mathbf{A}$ *such that* $\psi = \varphi|_{A^-}$*. Thus, the extension and restriction maps are mutually inverse and establish an isomorphism between* $\mathbf{Con\,A}$ *and* $\mathbf{Con\,A^-}$*.*

PROOF. It is immediate that $\theta|_{A^-}$ is a congruence on $\mathbf{A}^-$. Take a $\psi \in \operatorname{Con}\mathbf{A}^-$ and let $C = \mathrm{cn}^{\mathbf{A}}([1]_\psi)$. By Theorem 3.47 and commutativity of the negative cone, we obtain that $C = \Xi\Pi\Delta([1]_\psi)$. Since $[1]_\psi$ is the contained in the negative cone, we further get that $\Delta([1]_\psi) = [1]_\psi$. Thus, $C = \Xi\Pi([1]_\psi)$ and therefore $b \in C$ iff $a \leq b \leq a \to 1$ for some $a \in [1]_\psi$. Let φ be the congruence on $\mathbf{A}$ with $[1]_\varphi = C$. We will show $\psi = \varphi|_{A^-}$. By 1-regularity, it suffices to establish $[1]_\psi = [1]_\varphi \cap A^-$. Clearly, any $a \in [1]_\psi$ belongs to $[1]_\varphi$, so it remains to show that $[1]_\psi \supseteq [1]_\varphi \cap A^-$. Take any $b \in [1]_\varphi \cap A^- = C \cap A^-$. Then $a \leq b \leq a \to 1$ for some $a \in [1]_\psi$. Since $[1]_\psi$ is upward closed and b is negative, we get $b \in [1]_\psi$.

To close the circle, take $\psi = \theta|_{A^-}$ and extend it to $\varphi \in \mathrm{Con}\,\mathbf{A}$ as above. By previous paragraph, $[1]_\psi = [1]_\theta \cap A^- = [1]_\varphi \cap A^-$, so by Corollary 3.48 we get $\varphi = \theta$ as required. □

The above statement is not true in the non-commutative case.

Before moving on to some special cases and applications of the structure theory, we state and prove a technical lemma. The lemma allows for a certain degree of commutativity when dealing with inequalities, the right-hand side of which is negated. Note that we cannot use Lemma 2.9, as we do not assume cyclicity.

LEMMA 3.50. *Let x_i and y be elements of an FL-algebra, and γ_i iterated conjugates, for all $i \in \{1, 2, \ldots, n\}$, where n is a non-negative integer. If $\prod_{i=1}^n \gamma_i(x_i) \leq -y$, then there exist iterated conjugates $\overline{\gamma}_i$, for $i \in \{1, 2, \ldots, n\}$, such that $\prod_{i=n}^1 -{\sim}\overline{\gamma}_i(x_i) \leq -y$. [Note the change in the order of the product.]*

PROOF. If $\gamma_1(x_1)\gamma_2(x_2)\gamma_3(x_3)\cdots\gamma_n(x_n) \leq -y$, then we have

$$[-{\sim}\gamma_1(x_1)]\gamma_2(x_2)\gamma_3(x_3)\cdots\gamma_n(x_n) \leq -y,$$

by Lemma 2.8(13). Recalling that $\rho_a(b)a \leq ab$, by Lemma 2.6, we have successively

$$\rho_{[-{\sim}\gamma_1(x_1)]}(\gamma_2(x_2))[-{\sim}\gamma_1(x_1)]\gamma_3(x_3)\cdots\gamma_n(x_n) \leq -y,$$

$$\rho_{[-{\sim}\gamma_1(x_1)]}(\gamma_2(x_2))\rho_{[-{\sim}\gamma_1(x_1)]}(\gamma_3(x_3))[-{\sim}\gamma_1(x_1)]\cdots\gamma_n(x_n) \leq -y,$$

and finally, for $a = [-{\sim}\gamma_1(x_1)]$,

$$\rho_a(\gamma_2(x_2))\rho_a(\gamma_3(x_3))\cdots\rho_a(\gamma_n(x_n))[-{\sim}\gamma_1(x_1)] \leq -y,$$

or simply, for $\gamma_i'(x_i) = \rho_a(\gamma_i(x_i))$,

$$\gamma_2'(x_2)\gamma_3'(x_3)\cdots\gamma_n'(x_n)[-{\sim}\gamma_1(x_1)] \leq -y$$

By another application of Lemma 2.8(13), we have

$$[-{\sim}\gamma_2'(x_2)]\gamma_3'(x_3)\cdots\gamma_n'(x_n)[-{\sim}\gamma_1(x_1)] \leq -y$$

and proceeding in the same spirit as above, we have, for $\gamma_i''(x_i) = \rho_b(\gamma_i'(x_i))$ and $b = [-{\sim}\gamma_2'(x_2)]$,

$$\gamma_3''(x_3)\cdots\gamma_n''(x_n)[-{\sim}\gamma_2'(x_2)][-{\sim}\gamma_1(x_1)] \leq -y.$$

Proceeding inductively, we obtain

$$[-{\sim}\overline{\gamma}_n(x_n)]\cdots[-{\sim}\overline{\gamma}_2(x_2)][-{\sim}\overline{\gamma}_1(x_1)] \leq -y,$$

for some iterated conjugates $\overline{\gamma}_i$. □

3.6.3. Central negative idempotents. Let $\mathbf{L}$ be a residuated lattice and $S \subseteq L$. We denote the set of *idempotent* elements of S by $E(S) = \{a \in S \mid a^2 = a\}$ and the set of *central* elements, or the *center* of S, by $Z(S) = \{a \in S \mid ax = xa, \text{ for all } x \in L\}$. The set of *central idempotents* of S is $CE(S) = \{a \in S \mid a^2 = a \text{ and } ax = xa, \text{ for all } x \in L\}$.

LEMMA 3.51. [Gal03] *Let $\mathbf{L}$ be a residuated lattice. If $a \in CE(L^-)$, then $[a, 1/a]$ is a convex normal subalgebra of $\mathbf{L}$. Conversely, if N is a convex normal subalgebra of $\mathbf{L}$ with a least element a, then $N = [a, 1/a]$ and $a \in CE(L^-)$.*

PROOF. Let $a \in CE(L^-)$. Note that $a(1/a) = (1/a)a \leq 1$, so $1/a \leq a\backslash 1$. Similarly we get $a\backslash 1 \leq 1/a$, hence, $1/a = a\backslash 1$. Moreover, since a is negative, $1 \leq 1/a$, so by Lemma 2.6,

$$1/a \leq (1/a)(1/a) \leq (1/a)1/a \leq (1/a)/a = 1/a^2 = 1/a,$$

hence $1/a \in E(L)$. For every $x, y \in [a, 1/a]$, we have

$$a = a^2 \leq xy \leq (1/a)(1/a) = 1/a,$$

thus, $xy \in [a, 1/a]$. Moreover,

$$a = a^2 \leq a/(1/a) \leq x/y \leq (1/a)/a = 1/a^2 = 1/a,$$

that is $x/y \in [a, 1/a]$; likewise, we have $y\backslash x \in [a, 1/a]$. Since, $x \vee y, x \wedge y, 1 \in [a, 1/a]$, the interval $[a, 1/a]$ is a subalgebra, which is obviously convex. To prove that $[a, 1/a]$ is normal, let $x \in [a, 1/a]$ and $z \in L$. We have,

$$a = a \wedge 1 \leq az/z \wedge 1 = za/z \wedge 1 \leq zx/z \wedge 1 \leq 1,$$

that is $\rho_z(x) \in [a, 1/a]$. Likewise, we show that $\lambda_z(x) \in [a, 1/a]$.

Conversely, assume that N is a convex normal subalgebra with a least element a. The element a is in the negative cone, so $a^2 \leq a$. Since $a^2 \in N$, we get $a = a^2$, i.e., $a \in E(L)$. By the normality of N, for all $z \in L$, $za/z \wedge 1$ is an element of N, hence $a \leq za/z \wedge 1$. Since a is already negative, this is equivalent to $a \leq za/z$, thus $az \leq za$ for all $z \in L$. Symmetrically, we get $za \leq az$ for all $z \in L$, so $a \in CE(L^-)$. Moreover, since N is a convex subalgebra $[a, 1/a] \subseteq N$. On the other hand, for every $b \in N$, we have $1/b \in N$, so $a \leq 1/b$, i.e., $ab \leq 1$. By the centrality of a we get $ba \leq 1$, i.e., $b \leq 1/a$, hence $b \in [a, 1/a]$. Consequently, $[a, 1/a] = N$. □

The next theorem shows that the congruence lattice of a finite residuated lattice $\mathbf{L}$ is dually isomorphic to a join-subsemilattice of $\mathbf{L}$.

THEOREM 3.52. [Gal03] *Let $\mathbf{L}$ be a finite residuated lattice. Then the structure $\mathbf{CE}(L^-) = (CE(L^-), \cdot, \vee)$ is a lattice and the map $a \mapsto \Theta(a, 1)$ is a dual lattice embedding of $\mathbf{CE}(L^-)$ into $\mathbf{Con}\mathbf{L}$. In particular, if $\mathbf{L}$ is finite, then $\mathbf{Con}\mathbf{L} \cong (\mathbf{CE}(L^-))^\partial$.*

PROOF. It is easy to see that $\mathbf{CE}(L^-)$ is a lattice and that for all $a, b \in CE(L^-)$,

$$a = ab \Leftrightarrow a \leq b \Leftrightarrow a \vee b = b.$$

We define the map $\varphi : CE(L^-) \to CNS(\mathbf{L})$, by $\varphi(a) = [a, 1/a]$. If follows from the previous lemma that φ is well defined. If $\varphi(a) = \varphi(b)$ for some $a, b \in CE(L^-)$, then $[a, 1/a] = [b, 1/b]$, so $a = b$; hence φ is one-to-one. If $N \in CNS(L)$, then, by the Lemma 3.51, $N = [a, 1/a]$, for some $a \in CE(L^-)$, so φ is onto. The map φ reverses the order, since if $a \leq b$, then $[b, 1/b] \subseteq [a, 1/a]$. Moreover, if $[a, 1/a] \subseteq [b, 1/b]$ then $b \leq a$. Therefore, φ is a dual embedding of $\mathbf{CE}(L^-)$ in $\mathbf{CNS(L)}$. Using the isomorphism between **ConL** and $\mathbf{CNS(L)}$ provided in Theorem 3.47, we get a dual embedding of $\mathbf{CE}(L^-)$ in **ConL**.

If $\mathbf{L}$ is finite, then every convex sublattice, hence every convex normal subalgebra, is an interval. By Lemma 3.51, convex normal subalgebra is of the form $[a, 1/a]$, so φ is onto. □

In the commutative case we do not need the centrality assumption.

COROLLARY 3.53. [Gal03] *Let* $\mathbf{L}$ *be a finite commutative residuated lattice. Then* $\mathbf{E}(L^-)$ *is a lattice, with multiplication as meet and* $\mathbf{ConL} \cong (\mathbf{E}(L^-))^\partial$.

Note that the statement is false without the assumption of finiteness. For example, $|\mathbf{Con}\mathbb{Z}^-| = 2$, but $|CE(\mathbb{Z}^-)| = 1$.

3.6.4. Varieties with (equationally) definable principal congruences. We use the description of congruence relations to characterize commutative varieties of residuated lattices with definable principal congruences, along the lines of Theorem 5.4 of [BP94a].

LEMMA 3.54. [Gal03] *Let* $\mathcal{V}$ *be a variety that satisfies the identity* $(x \wedge 1)^k y = y(x \wedge 1)^k$, *for some* $k \in \mathbb{N}^+$. *Then for every* $\mathbf{L} \in \mathcal{V}$ *and for all* $a, b, c, d \in L$, $(a, b) \in \mathrm{Cg}(c, d)$ *is equivalent to* $(c \leftrightarrow d)^l \leq a \leftrightarrow b$, *for some* $l \in \mathbb{N}$.

PROOF. Let $\mathbf{L}$ be a residuated lattice and $a, b \in L$. It follows from (the proof of) Theorem 3.47(2b) that $a \; \theta \; b$ iff $(a \leftrightarrow b) \; \theta \; e$. Consequently, $\mathrm{Cg}(a, b) = \mathrm{Cg}(a \leftrightarrow b, e)$; moreover, $(a, b) \in \mathrm{Cg}(c, d)$ iff $a \leftrightarrow b \in [1]_{\mathrm{Cg}(c \leftrightarrow d, 1)}$. Since $a \leftrightarrow b$ is negative, $(a, b) \in \mathrm{Cg}(c, d)$ is equivalent to $a \leftrightarrow b \in M(c \leftrightarrow d)$, by Theorem 3.47. Using the description of the convex, normal submonoid $M(s)$ generated by a negative element s given Theorem 3.47(3), we see that this is in turn equivalent to $\prod_{i=1}^m \gamma_i(c \leftrightarrow d) \leq a \leftrightarrow b$, for some $m \in \mathbb{N}$ and some iterated conjugates $\gamma_1, \ldots, \gamma_n$. Recall that $f \leq \gamma(f)$, for every negative element $f \in L$ and for every iterated conjugate γ, so,

$$(c \leftrightarrow d)^{km} = ((c \leftrightarrow d)^k)^m \leq \prod_{i=1}^m \gamma_i((c \leftrightarrow d)^k) \leq \prod_{i=1}^m \gamma_i(c \leftrightarrow d),$$

thus $(a, b) \in \mathrm{Cg}(c, d)$ is equivalent to $(c \leftrightarrow d)^l \leq a \leftrightarrow b$, for some $l \in \mathbb{N}$. □

Recall that a variety $\mathcal{V}$ has *definable principal congruences* or *DPC* if there is a formula $\varphi(x, y, z, w)$ in the first order language of $\mathcal{V}$ such that for every algebra $\mathbf{A} \in \mathcal{V}$ and for all elements $a, b, c, d \in A$ we have that (a, b) is in the congruence generated by (c, d) iff $\varphi(a, b, c, d)$ holds. If $\varphi(a, b, c, d)$ can be taken to be a conjunction of equations, then $\mathcal{V}$ is said to have *equationally definable principal congruences* or *EDPC*.

THEOREM 3.55. [Gal03] *Let $\mathcal{V}$ be a variety that satisfies $(x \wedge 1)^k y = y(x \wedge 1)^k$, for some positive integer k. Then the following conditions are equivalent:*

(1) *$\mathcal{V}$ satisfies $(x \wedge 1)^n = (x \wedge 1)^{n+1}$, for some $n \in \mathbb{N}$.*
(2) *$\mathcal{V}$ has EDPC.*
(3) *$\mathcal{V}$ has DPC.*

PROOF. Assume that $\mathcal{V}$ satisfies $(x \wedge 1)^n = (x \wedge 1)^{n+1}$, for some $n \in \mathbb{N}$ and let $\mathbf{L} \in \mathcal{V}$ and $a, b, c, d \in L$. Since, $(c \leftrightarrow d)^n \leq (c \leftrightarrow d)^l$, for every l, by Lemma 3.54 we get

$$(a, b) \in \mathrm{Cg}(c, d) \Leftrightarrow (c \leftrightarrow d)^n \leq a \leftrightarrow b$$

which proves EDPC with $\varphi(x, y, z, t)$ being (some equational rendering of) $(z \leftrightarrow t)^n \leq x \leftrightarrow y$. EDPC, in turn, trivially implies DPC.

It remains to demonstrate that (1) follows from (3). By DPC, there is a formula $\varphi(x, y, z, w)$ in the first-order language of $\mathcal{V}$ such that for all $a, b, c, d \in \mathbf{A} \in \mathcal{V}$ we have $(c, d) \in \Theta(a, b)$ iff $\mathbf{A} \models \varphi(a, b, c, d)$. Suppose for contradiction that $(x \wedge 1)^n = (x \wedge 1)^{n+1}$ fails for every natural number n. Then for every n there exist $A_n \in \mathcal{V}$ and $a_n \in A_n$, $a_n < 1$, such that $a_n^{n+1} < a_n^n$. Let $A = \prod_{i=1}^n A_n$, $a = (a_n)_{n \in \mathbb{N}}$ and $b = (a_n^{n+1})_{n \in \mathbb{N}}$. Now, at every coordinate n the pair $(a_n^{n+1}, 1_n)$ belongs to the congruence generated by $(a_n, 1_n)$. By DPC, $\mathbf{A}_n \models \varphi(a_n^{n+1}, 1_n, a_n, 1_n)$, and thus $\mathbf{A} \models \varphi(b, 1, a, 1)$. Therefore, $(b, 1) \in \mathrm{Cg}(a, 1)$ and, since a and b are negative, this is equivalent to $a^k \leq b$, for some $k \in \mathbb{N}$. Thus, Thus, $a_n^k \leq a_n^{n+1}$ for all n; in particular, $a_k^k \leq a_k^{k+1}$, a contradiction. □

An element a in a residuated lattice $\mathbf{A}$ is called *n-potent* for some natural number n, if $a^{n+1} = a^n$. Recall that residuated lattice is called *n-potent*, if every element in it is n-potent; i.e. it satisfies $x^{n+1} = x^n$.

COROLLARY 3.56. [Gal03] *A variety of commutative residuated lattices has DPC iff it has EDPC iff the negative cones of the algebras in the variety are n-potent, for some natural number n.*

It follows that the only subvarieties of $\mathsf{FL_{ew}}$ that have EDPC (DPC) are the ones of the form $\mathsf{E_n}$, for some n.

3.6.5. The congruence extension property. Recall that a variety has the *congruence extension property* or *CEP* if for every algebra $\mathbf{A}$ in the variety, for any subalgebra $\mathbf{B}$ of $\mathbf{A}$ and for any congruence θ on $\mathbf{B}$, there exists a congruence $\bar{\theta}$ on $\mathbf{A}$, such that $\bar{\theta} \cap B^2 = \theta$.

Note that in view of Theorem 3.47 congruences of subalgebras can be extended to the whole algebra iff convex normal subalgebras can be extended.

LEMMA 3.57. [Gal03] *If a variety satisfies* $(x \wedge 1)^k y = y(x \wedge 1)^k$, *for some* k, *then it enjoys the congruence extension property. In particular* CRL *has the CEP.*

PROOF. Recall that by Theorem 3.47, congruences on a residuated lattice are in one-to-one correspondence with convex normal (in the whole residuated lattice) submonoids of the negative cone. Let $\mathbf{A}$ be a residuated lattice, $\mathbf{B}$ a subalgebra of it and N a convex normal submonoid of $\mathbf{B}$. If N' is the convex normal submonoid of $\mathbf{A}$ generated by N, it suffices to show that $N = N' \cap B$. For the non-obvious inclusion, let $b \in N' \cap B$. Then $\prod_{i=1}^{n} \gamma_i(a_i) \leq b$, for some $a_1, \ldots, a_n \in N$ and some iterated conjugates $\gamma_1, \ldots, \gamma_n$. Since, k-powers of the negative cone are in the center, $a_i^k \leq \gamma_i(a_i^k)$. Moreover, $\gamma_i(a_i^k) \leq \gamma_i(a_i)$, because a_i are in the negative cone. Thus, $\prod_{i=1}^{n} a_i^k \leq b$. Since, $a_i \in N$ and $b \in B$, we get $b \in N$. □

Not every residuated lattice satisfies the CEP. Let $A = \{0, c, b, a, e\}$ and $0 < c < b < a < 1$. Define $a^2 = a, b^2 = ba = ab = b, ac = bc = c$, and let all other non-trivial products be 0. It is easy to see that $\mathbf{A}$ is a residuated lattice and $B = \{1, a, b\}$ defines a subalgebra of it. $\mathbf{B}$ has the non-trivial congruence $\{\{1, a\}, \{b\}\}$, while $\mathbf{A}$ is simple. To see that, let θ be a non-trivial congruence and $a \,\theta\, 1$; then $(ca/c) \,\theta\, c1/c$, namely $c \,\theta\, 1$. So, $0 \,\theta\, 1$, hence $\theta = A \times A$.

The paper [vA05b] contains information about the EDPC and CEP for suvarieties of RL and some examples.

3.6.6. Subdirectly irreducible algebras. Our first result on subdirectly irreducible FL-algebras is a consequence of Theorem 3.47.

LEMMA 3.58. *An FL-algebra* $\mathbf{A}$ *is subdirectly irreducible if and only if there exists a strictly negative* $a \in A$ *such that for every strictly negative* $b \in A$ *there is a positive integer* n *and iterated conjugates* γ_{u_i} *with* $i \in \{1, \ldots, n\}$ *such that* $\gamma_{u_1}(b) \cdot \gamma_{u_2}(b) \cdot \ldots \cdot \gamma_{u_n}(b) \leq a$.

PROOF. By Theorem 3.47, $\mathbf{A}$ is subdirectly irreducible if and only if $\mathbf{A}_r$ has a smallest deductive filter $\mathrm{F_c}(\mu)$, where μ is the monolith congruence. Therefore, $\mathrm{F_c}(\mu)$ is nontrivial, i.e., different from $\uparrow 1$, and generated by a single strictly negative element a. So, by Theorem 3.47, $\mathrm{F_c}(\mu) = \mathrm{Fg}(a) = \uparrow \Pi\Gamma(\{a\})$. Let b be strictly negative. Then, $\mathrm{Fg}(b)$, the deductive filter generated by b, contains $\mathrm{F_c}(\mu)$. Hence, by Theorem 3.47 again, $\mathrm{Fg}(a) \subseteq \uparrow \Pi\Gamma(\{b\})$. In particular, $a \in \uparrow \Pi\Gamma(\{b\})$ proving that the required property holds. Conversely, if there exists a strictly negative a with the required property, then each nontrivial deductive filter of $\mathbf{A}$ must contain it; hence, $\mathrm{Fg}(a)$ is the smallest nontrivial deductive filter of $\mathbf{A}$. □

Examples of subdirectly irreducible residuated lattices and FL-algebras are strewn across this book, so we feel no need to present any particular list here. The next lemma provides a sufficient condition for subdirect irreducibility. For finite, commutative FL-algebras this condition is also necessary.

LEMMA 3.59. *If the unit element of* $\mathbf{A}$ *has a unique subcover, then* $\mathbf{A}$ *is subdirectly irreducible. If* $\mathbf{A}$ *is a finite, subdirectly irreducible* FL_e*-algebra, then its unit element has a unique subcover.*

PROOF. Let $c \prec 1$ be the unique subcover of 1 in A. Since for any nontrivial congruence θ, the class $[1]_\theta$ contains a strictly negative element, $\mathrm{Cg}^{\mathbf{A}}(\{c, 1\})$ is the smallest nontrivial congruence on $\mathbf{A}$, so the first statement follows. For the second, suppose $\mathbf{A}$ is a finite subdirectly irreducible FL_e-algebra with two strictly negative elements a and b such that $a \vee b = 1$. Without loss of generality we can assume that b is in the convex normal subalgebra $S(\{a\})$ generated by a. Since conjugates are not needed in the commutative case, we get that that there is a positive integer n such that $a^n \leq b$, from Lemma 3.58. On the other hand, for every k we have $(a \vee b)^k = 1$. Using commutativity, we obtain $(a \vee b)^n = a^n \vee a^{n-1}b \vee \cdots \vee ab^{n-1} \vee b^n = 1$. But as a and b are strictly negative, every factor in that join is below b. Thus, $1 \leq b < 1$ a contradiction. □

We will generalize a little the result above, with an eye on an application in Chapter 9. The origins of the next lemma are in [Kow95].

LEMMA 3.60. *Let* $\mathbf{A}$ *be a subdirectly irreducible residuated lattice with monolith* μ *and set* $M = [1]_\mu \cap A^-$. *If* M *has a least element* b, *there is an* $n \in \mathbb{N}$ *with* $x^n = b$ *for all* $a \in M \setminus \{1\}$ *and* $xy = yx$ *for all* $x, y \in M$, *then* 1 *is completely join irreducible in* $\mathbf{A}$; *hence, it has a unique subcover.*

PROOF. Suppose the contrary and let D be a strictly ascending chain of elements of $M \setminus \{0\}$ such that $\bigvee D = 1$. Clearly, we have $(\bigvee D)^n = 1$. On the other, by distributivity of multiplication over infinite joins, we get $(\bigvee D)^n = \bigvee\{d_{\varphi(1)} \cdot \cdots \cdot d_{\varphi(n)} : \varphi \in \Phi\}$, where Φ is the set of choice functions for D^n. By commutativity, we can assume that $d_{\varphi(1)} \leq \cdots \leq d_{\varphi(n)}$ for each φ, so every factor in the join is smaller that $d^n_{\varphi(1)} = b$. Thus $(\bigvee D)^n = b$, a contradiction. □

In particular, in n-potent, commutative varieties all subdirectly irreducible algebras are characterized by having a unique subcover of 1. This characterization fails in the noncommutative case, even for finite subdirectly irreducibles. Figure 3.7 shows the smallest example of a subdirectly irreducible (even simple) FL_w-algebra $\mathbf{S}$ with two subcovers of 1. Notice that both $A = \{1, a, 0\}$ and $B = \{1, b, 0\}$ are subuniverses of S and the the algebras $\mathbf{A}$ and $\mathbf{B}$ are both isomorphic to the three-element Heyting algebra. Thus our example also shows (again) that $\mathsf{FL_w}$ lacks CEP.

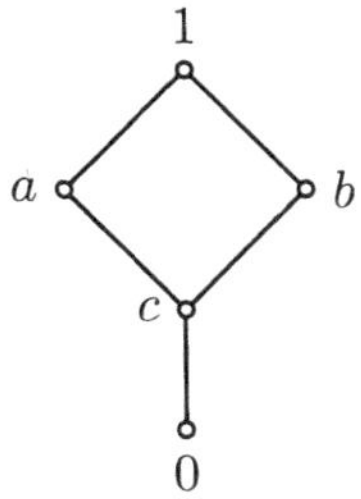

$\cdot$	1	a	b	c	0
1	1	a	b	c	0
a	a	a	0	0	0
b	b	c	b	c	0
c	c	c	0	0	0
0	0	0	0	0	0

FIGURE 3.7. A simple FL-algebra with two subcovers of 1.

3.6.7. Constants. Certain FL-algebras or residuated lattices, for instance all finite ones, are bounded. In applications to some logics it is further required that the bounds be constants in the type. Moreover, as it happens for example in linear logic, the bounds in general do not coincide with 1 or 0. To deal with these situations we define $\mathrm{FL}_\bot$-algebras to be structures $(L, \vee, \wedge, \cdot, \backslash, /, 1, 0, \top, \bot)$ such that $(L, \vee, \wedge, \cdot, \backslash, /, 1, 0)$ is an FL-algebra and $\top$ and $\bot$ are constants satisfying $\bot \leq x \leq \top$.

LEMMA 3.61. [GR04] *Every FL-algebra can be embedded in an $FL_\bot$-algebra, and the embedding is functorial, i.e, it extends to homomorphisms.*

PROOF. Let $\mathbf{A}$ be an FL-algebra. Let $A_\bot$ be $A \cup \{\top, \bot\}$, with $\{\top, \bot\} \cap A = \emptyset$. The order on $A_\bot$ is defined by extending the order on A setting $\bot \leq a \leq \top$ for each $a \in A$. The lattice operations also extend to $A_\bot$ in the obvious way, so it remains to extend the multiplication and residuals. We put:

- $\top \cdot x = x \cdot \top = \begin{cases} \top & \text{if } x \in A \cup \{\top\} \\ \bot & \text{if } x = \bot \end{cases}$
- $\bot \cdot x = x \cdot \bot = \bot$
- $\top \backslash x = x / \top = \begin{cases} \bot & \text{if } x \in A \cup \{\bot\} \\ \top & \text{if } x = \top \end{cases}$
- $x \backslash \top = \top / x = \top$
- $x \backslash \bot = \bot / x = \begin{cases} \bot & \text{if } x \in A \cup \{\top\} \\ \top & \text{if } x = \bot \end{cases}$
- $\bot \backslash x = x / \bot = \top$

It is not difficult to verify that under these definitions $A_\bot$ becomes an $\mathrm{FL}_\bot$-algebra. Now, let $\mathbf{A}$ and $\mathbf{B}$ be FL-algebras and e_A and e_B their respective embeddings into $\mathbf{A}_\bot$ and $\mathbf{B}_\bot$. Suppose h is a homomorphism from $\mathbf{A}$ to $\mathbf{B}$. We define a map $h_\bot : A_\bot \longrightarrow B_\bot$ by putting $h_\bot(x) = e_B(h(e_A^{-1}(x)))$ for $x \in e_A(A)$ and completing it with $h_\bot(\bot) = \bot$ and $h_\bot(\top) = \top$. This map makes the appropriate diagram commute by definition, so we only need to verify that it is a homomorphism of $\mathrm{FL}_\bot$-algebras. Of a number of cases to go through, we present one. Let $x \in e_A(A)$; we have $h_\perp(\top) \cdot h_\perp(x) = \top \cdot e_B(h(e_A^{-1}(x)))$ by definition. Since $e_A^{-1}(x) \in A$ we get $h(e_A^{-1}(x)) \in B$. Therefore, $e_B(h(e_A^{-1}(x))) \in B_\perp \backslash \{\top, \perp\}$ and thus, by definition of operations in $B_\perp$ we obtain $\top \cdot e_B(h(e_A^{-1}(x))) = \top$ On the other hand, $h_\perp(\top \cdot x) = \top$ as well, so we have $h_\perp(\top) \cdot h_\perp(x) = h_\perp(\top \cdot x)$ as needed. □

Despite being functorial, the embedding above is slightly unsatisfactory in that it does preserve existing bounds. This is particularly inconvenient if the bounds are defined by constant terms in the variety of **FL** algebras we are working with. Note however, that if the bottom is defined by a constant term b, then we have $b \leq b/x$ for any x, and thus $bx \leq b$. It follows that $x \leq b \backslash b$ for any x, hence, $b \backslash b$ defines the top. If both top and bottom are defined by constant terms, we can declare $\top = b \backslash b$ and $\perp = b$ extending the type trivially. So the only really bothersome case arises when the top element is defined by a constant term, but the bottom element is not.

Exercises

(1) Give examples of pairs of maps $f : \mathbf{P} \to \mathbf{Q}$ and $g : \mathbf{Q} \to \mathbf{P}$ that do not form a residuated pair and fail exactly one of the following conditions
- f is monotone,
- f^* is monotone,
- $id_P \leq f^* \circ f$,
- $f \circ f^* \leq id_Q$.

(2) Give an example of posets $\mathbf{P}, \mathbf{Q}, \mathbf{R}, \mathbf{S}$, where $\mathbf{R} \leq \mathbf{Q} \leq \mathbf{S}$, and of a map $f\colon P \to R$ such that $f\colon P \to Q$ is residuated, but $f\colon P \to R$ and $f\colon P \to S$ are not; here we use the same symbol for maps with the same graph.

(3) Show that the map $f_m : \mathbb{N} \to \mathbb{N}$, defined by $f_m(n) = mn$, is residuated for every non-zero natural number m. What is its residual?

(4) Show that the converse of Lemma 3.3 is not true. Can you think of a condition to add to (1) or (2) of the lemma, so that the converse is also true?

(5) Show that if $f\colon \mathcal{P}(A) \to \mathcal{P}(B)$ is residuated, then $f = \Diamond_{R^{-1}}$ and $f^* = \Box_R$, where $x \mathrel{R} y$ iff $y \in f(\{x\})$.

(6) Give a counterexample to the converse of Lemma 3.5 by defining a map $f : \mathbf{P} \to \mathbf{Q}$ that preserves existing joins but is not residuated.

(7) Show that if γ is a closure operator on a poset $\mathbf{P}$ and $\bigwedge X$ exists in $\mathbf{P}$, for some $X \subseteq \gamma[P]$, then $\bigwedge X$ is the meet of X in $\mathbf{P}_\gamma$. Moreover, if $\bigvee X$ exists in $\mathbf{P}$, for some $X \subseteq \gamma[P]$, then $\gamma(\bigvee X)$ is the join of X in $\mathbf{P}_\gamma$. Consequently, if γ is a closure operator on a complete lattice $\mathbf{L}$, then $\mathbf{L}_\gamma$ forms a complete lattice on $\gamma[L]$, where (arbitrary) meets agree with meets in $\mathbf{L}$ and (arbitrary) joins are the γ-closure of joins in $\mathbf{L}$.

(8) Recall from Section 1.1 that, given a class $\mathcal{K}$ of algebras, $Th_e(\mathcal{K})$ denotes the equational theory of $\mathcal{K}$ (the set of all true identities of $\mathcal{K}$) and that,

given a set E of equations, Mod(E) is the class of all algebras that satisfy all the equations from E. Prove that Th_e and Mod form a Galois connection on classes.

(9) Prove that the pair of maps defined relative to a relation S on page 146 form indeed a Galois connection. Also, show that if ${}^{\triangleright} : \mathcal{P}(A) \to \mathcal{P}(B)$ and ${}^{\triangleleft} : \mathcal{P}(B) \to \mathcal{P}(A)$ form a Galois connection, then ${}^{\triangleright} = S_{\cap}$ and ${}^{\triangleleft} = S_{\cap}^{-1}$, where $x \; S \; y$ iff $y \in \{x\}^{\triangleright}$.

(10) The example of a Galois connection on page 146 involves a relation $S \subseteq C \times D$ and maps ${}^{\triangleright} : \mathcal{P}(C) \to \mathcal{P}(D)$ and ${}^{\triangleleft} : \mathcal{P}(D) \to \mathcal{P}(C)$. Consequently, the maps $f : \mathcal{P}(C) \to \mathcal{P}(D)^{\partial}$ and $g : \mathcal{P}(D)^{\partial} \to \mathcal{P}(C)$, which are set-theoretically the same as ${}^{\triangleright}$ and ${}^{\triangleleft}$, form a residuated pair. Although the target set of f is not a powerset, hence it does not fall into the scope of Exercise 5, the poset $\mathcal{P}(D)^{\partial}$ can be thought of as the powerset of a the set D' consisting of the elements $d' = \{d\}^c$, for $d \in D$. With this convention in mind find a relation $R \subseteq C \times D'$ such that $f = \Diamond_{R^{-1}}$ and prove your claim. [Hint: $c \; R \; d'$ iff $c \; S^c \; d$.]

(11) Let $\mathbf{P}$ be a poset. For every $X \subseteq P$, we define $X^u = \{y \in P : x \leq y$, for all $x \in X\}$ and $X^l = \{y \in P : y \leq x$, for all $x \in X\}$. Show that the maps ^{u} and ^{l} form a Galois connection. [Hint: $X^u = \leq_{\cap}[X]$ and $X^u = \geq_{\cap}[X]$ (see page 146), where $\leq$ is the order relation of $\mathbf{P}$.] The poset $\mathbf{P}^{ul}$ is called the *Dedekind-MacNeille completion* of $\mathbf{P}$.

(12) Show that the Dedekind-MacNeille completion $\mathbf{P}^{ul}$ of $\mathbf{P}$ is a complete lattice and that $\varphi(x) = {\downarrow}x$ defines an order embedding $\varphi : \mathbf{P} \to \mathbf{P}^{ul}$ that preserves existing joins and meets. Also, verify that the Dedekind-MacNeille completion of a complete lattice is isomorphic to it. [Even stronger, if $\mathbf{Q}$ is a subposet of a complete lattice $\mathbf{P}$ and every element of P is a join and a meet (in $\mathbf{P}$) of elements from Q, then the Dedekind-MacNeille completion of $\mathbf{Q}$ is isomorphic to $\mathbf{P}$.]

(13) Give direct proofs of Lemma 3.7 and Lemma 3.9.

(14) Show that if $\cdot : \mathcal{P}(A) \times \mathcal{P}(B) \to \mathcal{P}(C)$ is a residuated map, then $X \cdot Y = R[X, Y]$, where $R(x, y, z)$ iff $z \in \{x\} \cdot \{y\}$; see page 148.

(15) Verify the remaining properties in Lemma 3.15.

(16) Given a monoid $\mathbf{K}$, define a lattice on the set $M_K = K \cup \{\bot, \top\}$ by putting the trivial order on K and adding bounds $\bot, \top \notin K$. Also, extend the multiplication of $\mathbf{K}$ to M_K by stipulating that $\top$ is an absorbing element for $K \cup \{\top\}$, and that $\bot$ is an absorbing element for M_K. Show that this pomonoid on M_K extends to a residuated lattice $\mathbf{M_K}$ iff $\mathbf{K}$ is cancellative.

(17) Show that every bounded lattice with at least one completely join-irreducible element 1 is the reduct of a residuated lattice with unit 1. In particular every finite lattice can be residuated. [Hint. Define the product of two elements to be the (i) bottom, if they are both strictly negative, (ii) the element 1, if none of them is negative and (iii) the strictly negative one, if exactly one of them is strictly negative.]

(18) Let **L** be the lattice obtained by adding a least element to a downward branching tree (a root system). Show that **L** cannot be the lattice reduct of a residuated lattice.

(19) Show that an FL-algebra **A** is cyclic if and only if $yz = 0$ implies $zy = 0$ for any $y, z \in A$.

(20) Show that the same (associative) pogroupoid may be involutive in two different ways (with two different pairs of negation operations).

(21) Show that in every involutive residuated pomonoid, where multiplication and addition coincide, we have $x/y = x(-y)$, $y\backslash x = ({\sim}y)x$, $y \mathbin{\dot{-}} x = (-y)x$, $x \mathbin{-\!\!\cdot} y = x({\sim}y)$.

(22) Show that the collection of all cofinite sets of $\mathbb{N}$ form a generalized Boolean algebra.

(23) Verify that the different formulation of a group in terms of divisions is actually term equivalent to the traditional one.

(24) Prove that ℓ-groups form a subvariety of RL that is axiomatized by the identity $x(x\backslash 1) = 1$. Also, show that they can also be characterized as structures with a group and a lattice reduct in which multiplication preserves order (or equivalently it distributes over meets, or over joins).

(25) Show that in ℓ-groups left multiplication l_x and right multiplication r_x by any element x is an automorphism of the underlying lattice reduct. Also, observe that l_x and r_x preserve and reflect strict order.

(26) Verify that the lexicographic product of two ℓ-groups, defined in Section 3.4.2, is indeed an ℓ-group.

(27) Show that the dual of an ℓ-group (viewed as an FL-algebras with $0 = 1$) is also an ℓ-group.

(28) Verify the claim that under any choice of a negating element the dual of an ℓ-group is also an ℓ-group.

(29) For a poset **P** and a map f on **P**, we denote its residual by ${\sim}f$ and its dual residual by $-f$ if they exist. We also use ${\sim}^n f$ and $-^n f$ for iterated residuals and dual residuals; so ${\sim}^2 f = {\sim}({\sim}f)$. Let $RDR^\infty(\mathbf{P})$ be the set of all order preserving maps f on **P** such that ${\sim}^n f$ and $-^n$ exist for all n. Show that $RDR^\infty(\mathbf{P})$, under composition and pointwise order, is an involutive pomonoid such that addition coincides with multiplication. If **P** is a chain, then $RDR^\infty(\mathbf{P})$ is lattice-ordered. Show that if **P** is a finite chain then $RDR^\infty(\mathbf{P})$ is the set of all order automorphisms of **P**. What is $RDR^\infty(\mathbf{P})$ if **P** is a finite poset or the real numbers?

(30) Find an example of a pomonoid that is residuated and dually residuated (with the same multiplication) and is not an ℓ-group. Note that by Theorem 3.26 your example should not be cyclic or cancellative (and not even satisfy $x\backslash x = 1$ or $x/x = 1$). [Hint: Show that in $RDR^\infty(\mathbb{Z})$

$$-f(y) = \min f^{-1}(\min\{\uparrow y \cap f[\mathbb{Z}\})$$

$${\sim}f(y) = \max f^{-1}(\max\{\downarrow y \cap f[\mathbb{Z}\})$$

and prove that $RDR^{\infty}(\mathbb{Z})$ is the set of all order-preserving maps such that the pre-image of every point is bounded.]

(31) Verify that the negative cone $\mathbf{L}^-$ of a residuated lattice $\mathbf{L}$ is a residuated lattice, as well.

(32) Prove that the dual of an MV-algebra is also an MV-algebra. Verify that the MV-algebras that are obtained by the two different term equivalences of page 111 are dual.

(33) Show that for integral residuated lattices $(y/x)\backslash y = x \vee y = y/(x\backslash y)$ is equivalent to the quasi-identity $x \leq y \Rightarrow (y/x)\backslash y = y = y/(x\backslash y)$.

(34) Show that for integral residuated lattices $x(x\backslash y) = x \wedge y = (y/x)x$ is equivalent to the quasi-identity $y \leq x \Rightarrow x(x\backslash y) = y = (y/x)x$.

(35) Show that the negative cones of ℓ-groups are integral GMV-algebras.

(36) Show that GMV-algebras are GBL-algebras.

(37) Show that commutative GMV-algebras are representable.

(38) Derive the distributivity identity from (IGBL).

(39) Verify all the claims made about hoops (and Wasjberg hoops) in Section 3.4.7.

(40) Show that if $\mathbf{G}$ is an abelian ℓ-group, then $\Gamma(\mathbf{G}, a)$ (Section 3.4.5) is an MV-algebra in both cases where a is positive or negative. Also show that if a is positive or negative, then $\Gamma(\mathbf{G}, a^{-1})$ is isomorphic to the dual with respect to a of $\Gamma(\mathbf{G}, a)$; see Section 3.4.17.

(41) Verify that the ideals of a ring form a residuated lattice.

(42) Let $\mathbf{M} = (M, R, E)$ be a structure, where M is a set, $R \subseteq M^3$ a ternary relation on M and E is a subset of M. Define the following operations on the power set $\mathcal{P}(M)$ of M: $X \circ Y = R[X, Y]$, $X/Y = \{z \in M \mid \{z\} \circ Y \subseteq X\}$ and $Y\backslash X = \{z \in M \mid Y \circ \{z\} \subseteq X\}$. Show that the algebra $\mathcal{P}(\mathbf{M}) = (M, \cap, \cup, \circ, \backslash, /, E)$ is a residuated lattice iff for all $x, y, z, w \in M$,
   (a) $R(x, e, y)$ for some $e \in E$ iff $x = y$ iff $R(e, x, y)$ for some $e \in E$; and
   (b) $R(x, y, u)$ and $R(u, z, w)$, for some $u \in M$ iff $R(x, v, w)$ and $R(y, z, v)$, for some $v \in M$. (In short $(x \star_R y) \star_R z = x \star_R (y \star_R z)$, where $x \star_R y = \{m \in M \colon (x, y, m) \in R\}$.)

   In this case, the residuated lattice $\mathcal{P}(\mathbf{M})$ is called the power set of $\mathbf{M}$. If R is a partial operation, then E is a singleton and $\mathbf{M}$ is a partial monoid.

(43) Given a *partial semigroup* $(S, *)$ (a partial algebra such that if any of the two sides of the associativity condition is defined, then the other side is also defined and they are equal), and two subsets X, Y of S, we define $X * Y = \{x * y : x * y \in S,\ x \in X,\ y \in Y\}$, the *complex product* of X and Y, $\langle X \rangle_* = \{x_1 * x_2 * \cdots * x_n \in S : n \in \mathbb{N},\ x_1, \ldots x_n \in X\}$, the *subsemigroup* generated by X, and $[X]_* = X \cup (S * X) \cup (X * S) \cup (S * X * S)$, the *semigroup ideal* generated by X. Show that if $\mathbf{S}$ is commutative, then the set of its semigroup ideals forms a residuated

lattice. [Hint: Show that γ, defined by $\gamma(X) = [X]_* = X \cup (S * X)$ is a nucleus on $\mathcal{P}(\mathbf{S})$.]

(44) Show that the subsemilattices of a partial lower-bounded join-semilattice $\mathbf{L} = (L, \vee)$ form a residuated lattice; see Exercise 43. [Hint: Show that γ, defined by $\gamma(X) = \langle X \rangle_\vee$ is a nucleus on $\mathcal{P}(\mathbf{L})$.]

(45) A *partial semiring* is a structure $\mathbf{S} = (S, *, e, +)$ such that $(S, *, e)$ is a monoid, $(S, +)$ is a partial semigroup and $*$ distributes over existing binary sums, namely if $x + y \in S$, then $x * z + x * y = (x + y) * z$ and $z * x + z * y = z * (x + y)$. Note, that it follows that $*$ distributes over finite existing sums. Show that the $+$-subsemigroups of a partial semiring $\mathbf{S}$ form a residuated lattice; see Exercise 43. [Hint: Show that γ, defined by $\gamma(X) = \langle X \rangle_+$ is a nucleus on $\mathcal{P}(\mathbf{L})$.]

(46) Show that the semigroup-ideals of a partial $+$-commutative semiring $\mathbf{S} = (S, *, e, +)$ form a residuated lattice $\mathcal{I}_S(\mathbf{M})$; see Exercise 45. [Hint: Show that γ, defined by $\gamma(X) = [X]_+$ is a nucleus on $\mathcal{P}(\mathbf{S})$.]
Note that a partially ordered monoid $\mathbf{M} = (M, \cdot, 1, \leq)$ can be identified with a partial semiring $\mathbf{M}' = (M, \cdot, 1, \wedge)$ such that $x \wedge x = x$ for all $x \in M$ and if $x \wedge y \in M$ then $x \wedge y = y \wedge x$. The definitional equivalence is given by $x \wedge y = x$ iff $x \leq y$. Moreover, ${\downarrow}X = [X]_\wedge$, namely the notions of order-ideal of $\mathbf{M}$ and semigroup-ideal of $\mathbf{M}'$ coincide. So, the order-ideals $\mathcal{O}(\mathbf{M})$ of $\mathbf{M}$ (Section 3.4.9) is a special case of the semigroup-ideals $\mathcal{I}_S(\mathbf{M})$ of $\mathbf{M}'$.

(47) Given a partial semiring $\mathbf{S} = (S, *, e, +)$, it is easy to see that the collection $\mathcal{I}(\mathbf{S}) = \{X \subseteq S : X = [\langle X \rangle_+]_*\}$ of all *ideals* of $\mathbf{S}$ is a subalgebra of the residuated lattice of $+$-subsemigroups of $\mathbf{S}$, given in Exercise 45. Show that if $\mathbf{S}$ is $*$-commutative, $\mathcal{I}(\mathbf{S})$ can be realized as the image of the power set of $\mathbf{S}$ under the nucleus defined by $\gamma(X) = [\langle X \rangle_+]_*$, the composition of the nuclei given in Exercises 45 and 43. If $\mathbf{S}$ is a ring with unit, then this reduces to the case of Section 3.4.9.

(48) A *join-semilattice-ordered monoid* $\mathbf{M} = (M, *, 1, \vee)$ is an algebra, such that $(M, *, 1)$ is a monoid, $(M, \vee)$ is a join-semilattice and multiplication distributes over binary joins. Show that the join-closed subsets of $\mathbf{M}$ form a residuated lattice. [Hint: Use Exercise 47.]

(49) Show that the join-ideals of a join-semilattice-ordered monoid $\mathbf{M} = (M, *, 1, \vee)$ form a residuated lattice. [Hint: Show that the composition of the maps, which in our case commute, given by $\gamma_1(X) = {\downarrow}X$ and $\gamma_2(X) = \langle X \rangle_\vee$ (see Section 3.4.13 and Exercise 44) is a nucleus.]
In view of the remark in Exercise 46, a join-semilattice-ordered monoid can be viewed as a structure $\mathbf{M} = (M, *, e, \vee, \wedge)$, where $(M, *, e, \vee)$ and $(M, *, e, \wedge)$ are partial semirings. Clearly, the join-ideals of a join-semilattice-ordered monoid are the $\vee$-subsemigroups that happen to be $\wedge$-ideals.

(50) Show that the ideals of an ℓ-monoid form a residuated lattice. [Hint: See Exercise 49.]

(51) Show that the ideals of a (bounded) distributive lattice form a residuated lattice, actually a Brouwerian algebra. [Hint: For the bounded case, see Exercise 49. For the general case show that $\gamma(X) = \downarrow(X)_\vee = [(X)_\vee]_\wedge$ is a nucleus.]

(52) Verify the characterization of deductive filters for Boolean algebras and the details of their connections to congruences mentioned in the beginning of Section 3.6.1.

(53) Verify the characterization of deductive filters for MV-algebras, mentioned in the beginning of Section 3.6.1, and show that they do not coincide with lattice filters.

(54) Show that if $\mathbf{A}$ is a Boolean algebra, then $(A, \leftrightarrow, 1)$ is an abelian group of order 2 (every element is its own inverse).

(55) Show that if a, b are elements of a residuated lattice, then $\lambda_a(a \leftrightarrow' b) \cdot \lambda_b(a \leftrightarrow' b) \leq a \leftrightarrow b$ and $\rho_a(a \leftrightarrow b) \cdot \rho_b(a \leftrightarrow b) \leq a \leftrightarrow' b$.

(56) Verify the remaining compositions of Theorem 3.47(2c).

(57) Prove the form of PLDT given in Theorem 2.14 algebraically, by using Theorem 3.47.

(58) Show that if S is a convex normal subalgebra of a residuated lattice $\mathbf{A}$, then $\Theta_s(S) = \{(a, b) : (\exists s \in S)(sa \leq b \text{ and } sb \leq a)\}$.

(59) Show that the join of two deductive filters is their (element-wise) product. Characterize the join of two convex normal subalgebras and of two convex normal submonoids.

(60) For two elements a, b of a residuated lattice $\mathbf{A}$, their *left* and *right commutator* are defined to be $[a, b]_l = ba \backslash ab \wedge 1$ and $[a, b]_r = ab/ba \wedge 1$, respectively. A subset X of A is *closed under commutators* if for all $a \in A$ and $x \in X$, $[x, a]_l, [a, x]_r \in X$. Show that a convex subalgebra of $\mathbf{A}$ (or a convex submonoid of $\mathbf{A}^-$) is normal iff it is closed under commutators.

(61) *Open problem:* Prove or disprove that involutive division posets can be embedded into ones with a negation constant.

(62) *Open problem:* Prove or disprove that there is no axiomatization in the restricted signatures without division or without multiplication for involutive pointed residuated pogroupoids with unit that are not necessarily associating.

(63) *Open problem*: Find an example of a non-associative involutive pogroupoid (with unit), where addition coincides with multiplication.

(64) *Project*: Come up with a presentation for a consequence relation with respect to which the deductive filters of residuated lattices are their convex normal subalgebras. Do the same for the convex normal submonoids.

Notes

(1) The parts of (1), (2a) and (2b) in Theorem 3.47 that do not refer to deductive filters, as well as (3a) and (3b) essentially follow from results in [BT03], as they have been reformulated in [JT02]. The remaining parts are essentially from [GO06a]. The connections between congruences and deductive filters hold in general for algebraizable logics. In particular, it follows from Theorem 2.29 and from Theorem 5.1 and Lemma 5.2 of [BP89] that for every residuated lattice or FL-algebra $\mathbf{A}$, the maps $F \mapsto \Theta_f(F)$ and $\theta \mapsto F_\theta = \{a \in A \mid 1\,\theta\,1 \wedge a\}$ are mutually inverse lattice isomorphisms between the lattices $\mathbf{Fil}(\mathbf{A})$ and $\mathbf{Con}(\mathbf{A})$. It is immediate that $F_\theta = F_c(\theta)$. It appears that the specifics of the above correspondence for residuated lattices was, to a large extend, known to W. Blok. Finally, the connection between congruence and ideals hold in general for ideal determined varieties.

(2) We have seen that for residuated lattices deductive filters are in bijective correspondence with convex normal subalgebras. Both of them correspond to congruences of the residuated lattice and therefore to RL-congruences of the absolutely free algebra $\mathbf{Fm}_{\mathcal{L}}$. Although theories (the deductive filters of $\mathbf{Fm}_{\mathcal{L}}$) do not correspond to convex normal subalgebras of $\mathbf{Fm}_{\mathcal{L}}$ (after all even the definition of convexity is not clear) they do so in the free residuated lattice. Therefore, we have an intuitive algebraic understanding of the theories (modulo equivalence) as convex normal subalgebras of the free residuated lattice. This can lead, for example, to an alternative proof of the PLDT; see [GO06a].

(3) The notation for the polarities of a Galois connection is essentially taken from [DP02]. We would like to mention, though, that our notation conforms more with the first edition of the book, which made the distinction between Galois connections and residuated maps, and also used black triangles ($^{\blacktriangleleft}$ and $^{\blacktriangleright}$) for the polarities. The second edition [DP02] does not make the distinction and uses white triangles ($^{\triangleleft}$ and $^{\triangleright}$). We use white triangles, but in the same way that black triangles were used in the first edition.

(4) Section 3.3 was inspired by ideas communicated to the first author by James Raftery.

CHAPTER 4

Decidability

In this chapter we discuss the decidability of various substructural logics and varieties of residuated lattices. The first part of the chapter uses a proof-theoretic analysis on the sequent calculus systems we have discussed. In the second half of the chapter we will use mainly semantical arguments. The exposition of the first part is based on [Ono98b] while the second is based on [Gal02]. A lot of the results in this chapter are summarized in Table 4.3.

Recall that a substructural logic is said to be *decidable*, if there is an algorithm that decides whether a formula is a theorem of the logic or not. A substructural logic $\mathbf{L}$ has a *decidable deducibility relation* if there is an algorithm that decides whether $\Phi \vdash_{\mathbf{L}} \{\psi\}$, for all sets $\Phi \cup \psi$ of formulas. A class of algebras has a *decidable* (quasi)equational theory if there is an algorithm that decides whether a (quasi)equation holds in the class or not. Note that decidability problems for varieties of FL-algebras axiomatized by 0-free set of equations reduce to the corresponding problems for the varieties of residuated lattices axiomatized by the same equations.

Because of the algebraization result for substructural logics, a substructural logic is decidable iff the corresponding variety of FL-algebras has a decidable equational theory and vice versa. Furthermore, a substructural logic has a decidable deducibility relation iff the corresponding variety of FL-algebras has a decidable quasiequational theory and vice versa. Therefore, in many cases we will state the results only for logics or varieties without explicit reference to the direct implications due to the algebraization theorem.

4.1. Syntactic proof of cut elimination

We show the following theorem in this section, which is shown in [OK85] and [Ono90]. For more information on historical remarks about cut elimination for substructural logics, see the notes at the end of the present chapter.

THEOREM 4.1. *Cut elimination holds for the Gentzen calculus* $\mathbf{FL_R}$ *with any* $R \subseteq \{e, c, i, o\}$ *except* $R = \{c\}$, *and for also* $\mathbf{InFL_Q}$ *with any* $Q \in \{e, ec, ei, eo, ew\}$.

Here, we will give an outline of a syntactic proof of cut elimination for the calculus **FL** first, which is in fact much simpler than the proof of cut elimination for **LJ**. Then, we will show how to get a proof of cut elimination for other calculi of basic substructural logics by modifying the proof, and why the proof becomes more complicated when a calculus contains the contraction rule. In Chapter 7, we give a proof of cut elimination in an algebraic way. It will be interesting to compare syntactic proofs with algebraic ones.

4.1.1. Basic idea of cut elimination. Suppose that a sequent s is provable in **FL** and P is a proof of s. We will try to eliminate all applications of the cut rule in P without changing its endsequent s. We note that the cut rule is not derivable from other rules of **FL**. More precisely, it is impossible to replace the cut rule in a uniform way by repeated applications of other rules. Therefore, we have to eliminate each application of the cut rule in P depending on how it appears in the proof.

We will eliminate applications of the cut rule inductively. We first consider an application of the cut rule of the following form;

$$\frac{\Gamma \Rightarrow \alpha \quad \Theta, \alpha, \Pi \Rightarrow \delta}{\Theta, \Gamma, \Pi \Rightarrow \delta} \text{ (cut)}$$

such that this is one of uppermost applications of the cut rule in P, namely one such that both proofs of its upper sequents (as subproofs of P) contain no cut. Recall here that the formula α in the above cut is called the *cut formula* of this application of the cut rule. The *grade* of this cut is defined as the length of the cut formula; i.e., the total number of symbols appearing in α. The *size* of this cut is the sum $k_1 + k_2$, where k_1 and k_2 are the total numbers of sequents appearing in the proof of the left upper sequent $\Gamma \Rightarrow \alpha$ and the right upper sequent $\Theta, \alpha, \Pi \Rightarrow \delta$, respectively.

Our strategy of eliminating this cut is either *pushing the cut up* or *replacing the cut formula by a simpler one.* More precisely, cut elimination is proved by the double induction on the grade and the size, i.e. either the grade becomes smaller, or the size becomes smaller while the grade remains the same. We explain this below, by taking some typical examples.

1. Pushing the cut up.

Consider the following cut.

$$\frac{\dfrac{\Sigma, \beta, \Gamma \Rightarrow \alpha}{\Sigma, \beta \wedge \gamma, \Gamma \Rightarrow \alpha} \quad \Theta, \alpha, \Pi \Rightarrow \delta}{\Theta, \Sigma, \beta \wedge \gamma, \Gamma, \Pi \Rightarrow \delta} \text{ (cut)}$$

This cut can be replaced by the following, which has the same endsequent, but its size becomes smaller by one:

$$\dfrac{\dfrac{\Sigma, \beta, \Gamma \Rightarrow \alpha \quad \Theta, \alpha, \Pi \Rightarrow \delta}{\Theta, \Sigma, \beta, \Gamma, \Pi \Rightarrow \delta} \text{(cut)}}{\Theta, \Sigma, \beta \wedge \gamma, \Gamma, \Pi \Rightarrow \delta}$$

The same argument holds also when either the right upper sequent or the left upper sequent is a lower sequent of a rule in which the cut formula is a side formula.

2. Replacing the cut formula by a simpler one.

We now assume that both the right upper sequent and the left upper sequent of the cut are lower sequents of rules such that the cut formula φ is the main formula. For instance, suppose that φ is of the form $\beta \wedge \gamma$, and consider the following cut.

$$\dfrac{\dfrac{\Gamma \Rightarrow \beta \quad \Gamma \Rightarrow \gamma}{\Gamma \Rightarrow \beta \wedge \gamma} \quad \dfrac{\Theta, \beta, \Pi \Rightarrow \delta}{\Theta, \beta \wedge \gamma, \Pi \Rightarrow \delta}}{\Theta, \Gamma, \Pi \Rightarrow \delta} \text{(cut)}$$

This cut can be replaced by the following, which has the same endsequent, but clearly the cut formula below is simpler (i.e. its grade is smaller) than that in the above:

$$\dfrac{\Gamma \Rightarrow \beta \quad \Theta, \beta, \Pi \Rightarrow \delta}{\Theta, \Gamma, \Pi \Rightarrow \delta} \text{(cut)}$$

Now, by applying these two kinds of replacement (which we call *reductions*) repeatedly, we will eventually come to such an application of the cut rule that at least one of the upper sequents is one of initial sequents. For instance, if the cut rule is of the form

$$\dfrac{\Gamma \Rightarrow \alpha \quad \alpha \Rightarrow \alpha}{\Gamma \Rightarrow \alpha} \text{(cut)}$$

then we can replace this proof by the proof of the left upper sequent $\Gamma \Rightarrow \alpha$, which is cut-free, by our assumption. Other cases are treated in a similar way.

Reasoning as above, we establish the cut elimination property for **FL**. This proof works well also for the sequent calculi for other (involutive) basic substructural logics that have weakening rule(s) but do not have the contraction rule. In fact, in this case, it is enough to consider also the cases where the cut formula is introduced by the weakening rule just above the cut rule. For example, let us consider the following case.

$$\dfrac{\Gamma \Rightarrow \alpha \quad \dfrac{\Theta, \Pi \Rightarrow \delta}{\Theta, \alpha, \Pi \Rightarrow \delta}}{\Theta, \Gamma, \Pi \Rightarrow \delta} \text{(cut)}$$

Then, $\Theta, \Gamma, \Pi \Rightarrow \delta$ is obtained from $\Theta, \Pi \Rightarrow \delta$ by applying the weakening rule several times, and thus it is provable without using cut.

4.1.2. Contraction rule and mix rule. On the other hand, the contraction rule causes certain difficulties in eliminating the cut rule. Let us consider the following cut:

$$\dfrac{\Gamma \Rightarrow \alpha \quad \dfrac{\alpha, \alpha, \Pi \Rightarrow \delta}{\alpha, \Pi \Rightarrow \delta}\,(\mathrm{c})}{\Gamma, \Pi \Rightarrow \delta}\,(\mathrm{cut})$$

One may reduce this to the following proof containing two cuts with the same grade:

$$\dfrac{\dfrac{\Gamma \Rightarrow \alpha \quad \dfrac{\Gamma \Rightarrow \alpha \quad \alpha, \alpha, \Pi \Rightarrow \delta}{\alpha, \Gamma, \Pi \Rightarrow \delta}\,(cut)}{\Gamma, \Gamma, \Pi \Rightarrow \delta}\,(\mathrm{cut})}{\Gamma, \Pi \Rightarrow \delta}\ \text{some (e)(c)}$$

Here, while the size of the upper application of cut becomes smaller, the size of the lower cut will not. Thus, we need to modify the proof for calculi, like **LK** and **LJ**, which have contraction.

To overcome the above difficulty, Gentzen [Gen35] introduced the following substitute, called the *mix rule*, for the cut rule.

$$\dfrac{\Gamma \Rightarrow \alpha \quad \Pi \Rightarrow \delta}{\Gamma, \Pi_{\alpha} \Rightarrow \delta}\,(\mathrm{mix})$$

Here Π has at least one occurrence of α, and Π_{α} is the sequence of formulas obtained from Π by deleting *all* occurrences of α. We call the formula α, the *mix formula* of the above application of the mix rule. The idea in the special case of the last example is that the two consecutive cuts are now replaced by one application of the mix rule.

We will give the details only for the sequent calculus **LJ**, but similar arguments works also for **LK**. First, we show that the mix rule is interchangeable with the cut rule.

Let $\mathbf{LJ}^{\dagger}$ be the sequent calculus obtained from **LJ** by replacing the cut rule by the mix rule. We can show the following.

LEMMA 4.2. *A sequent is provable in* $\mathbf{LJ}^{\dagger}$ *if and only if it is provable in* **LJ**.

PROOF. It is enough to show that the cut rule is derivable in $\mathbf{LJ}^{\dagger}$ and conversely that the mix rule is derivable in **LJ**. First, consider the following cut;

$$\frac{\Gamma \Rightarrow \alpha \quad \alpha, \Pi \Rightarrow \delta}{\Gamma, \Pi \Rightarrow \delta}$$

Then, the following figure shows that this cut is derivable in $\mathbf{LJ}^\dagger$:

$$\cfrac{\cfrac{\Gamma \Rightarrow \alpha \quad \alpha, \Pi \Rightarrow \delta}{\Gamma, \Pi_\alpha \Rightarrow \delta}\ (\text{mix})}{\Gamma, \Pi \Rightarrow \delta}\ \text{some (w)}$$

Conversely, consider the following mix;

$$\frac{\Gamma \Rightarrow \alpha \quad \Pi \Rightarrow \delta}{\Gamma, \Pi_\alpha \Rightarrow \delta}$$

Then, this mix is derivable in **LJ**, as the following figure shows.

$$\cfrac{\Gamma \Rightarrow \alpha \quad \cfrac{\Pi \Rightarrow \delta}{\alpha, \Pi_\alpha \Rightarrow \delta}\ \text{some (c)(e)}}{\Gamma, \Pi_\alpha \Rightarrow \delta}\ (\text{cut})$$

□

Notice that in the above lemma the weakening rule is used in an essential way in showing that the cut rule can be replaced by the mix rule. Now, we return to cut elimination of **LJ**. Suppose that a proof P of a sequent s in **LJ** is given. As shown in the proof of the above lemma, we can get a proof Q of the sequent s in $\mathbf{LJ}^\dagger$, by replacing all applications of the cut rule by the mix rule. Now it remains to show the *mix elimination* for $\mathbf{LJ}^\dagger$. This will produce a mix-free proof of the sequent s in $\mathbf{LJ}^\dagger$ and, by another use of the above lemma, we will obtain a cut-free proof of s in **LJ**, yielding, in this way, the cut elimination of **LJ**.

We eliminate applications of the mix rule by induction, in a similar way to the procedure of cut elimination, by considering one of uppermost applications of the mix rule in Q. Now, for the case where the right upper sequent of the mix rule is introduced by contraction rule, the application

$$\cfrac{\Gamma \Rightarrow \alpha \quad \cfrac{\alpha, \alpha, \Pi \Rightarrow \delta}{\alpha, \Pi \Rightarrow \delta}\ (\text{c})}{\Gamma, \Pi_\alpha \Rightarrow \delta}\ (\text{mix})$$

will be reduced to:

$$\frac{\Gamma \Rightarrow \alpha \quad \alpha, \alpha, \Pi \Rightarrow \delta}{\Gamma, \Pi_\alpha \Rightarrow \delta}\ (\text{mix})$$

Obviously, the size of the latter proof becomes smaller than that of the former.

Every reduction process introduced in the previous subsection seems to go well for mix elimination. Nevertheless, our definition of size is seriously affected by the replacement of the cut rule by the mix rule, and it does not work anymore, as it is shown in the following example. Here, we assume that $\alpha \wedge \beta$ does not occur in Γ.

$$\dfrac{\Gamma \Rightarrow \alpha \wedge \beta \quad \dfrac{\dfrac{\alpha, \beta \Rightarrow \delta}{\alpha, \alpha \wedge \beta \Rightarrow \delta}}{\alpha \wedge \beta, \alpha \wedge \beta \Rightarrow \delta}}{\Gamma \Rightarrow \delta}\ \text{(mix)}$$

The above application of the mix rule can be replaced by the following;

$$\dfrac{\dfrac{\Gamma \Rightarrow \alpha \wedge \beta \quad \dfrac{\dfrac{\Gamma \Rightarrow \alpha \wedge \beta \quad \dfrac{\alpha, \beta \Rightarrow \delta}{\alpha, \alpha \wedge \beta \Rightarrow \delta}}{\alpha, \Gamma \Rightarrow \delta}\ \text{(mix)}}{\alpha \wedge \beta, \Gamma \Rightarrow \delta}}{\Gamma, \Gamma \Rightarrow \delta}\ \text{(mix)}}{\Gamma \Rightarrow \delta}\ \text{some (c)(e)}$$

The size of the lower application of the mix rule in the second proof becomes bigger than the size of the mix rule in the first one.

Hence, we need to introduce another measure called the *rank* of a given application of the mix rule, instead of the size. The rank of an application of the mix rule in a given proof Q with the mix formula γ is defined as follows. When we regard a proof as a finite labeled tree, a *branch* in Q is defined to be a sequence $(n_1, \ldots, n_m)$ of nodes of Q such that for each $i < m$ n_i is a cover of n_{i+1}. Note that we move down the branch as we move to the right of the sequence. When a sequent s_i is attached to a node n_i for each i, we say also that the sequence $(s_1, \ldots, s_m)$ of sequents is a branch in Q, by abuse of language. Thus, in this case s_i is (one of) its upper sequent(s) and s_{i+1} is its lower sequent for each $i < m$.

Now, consider any branch consisting of sequents with at least one occurrence of the mix formula γ in the succedent, whose last element is the left upper sequent of the mix under consideration. Then the *left rank* r_1 of this mix rule is the maximum among the lengths of these branches. Similarly, we define *right rank* r_2 to be the maximum among lengths of branches consisting of sequents with at least one occurrence of γ in the antecedent, whose last element is the right upper sequent of the mix. Now, the *rank* of this mix rule is the sum $r_1 + r_2$.

In the above example, the right rank r_2 of the mix rule is 2 in the original proof, as $\alpha \wedge \beta$ appears in both $\alpha \wedge \beta, \alpha \wedge \beta \Rightarrow \delta$ and $\alpha, \alpha \wedge \beta \Rightarrow \delta$. On the other hand, both of the right ranks of the two applications of the mix rule are 1 in the second proof, while all of the left ranks remain the same.

Once we take the rank instead of the sizes, we can show by using the double induction on the grade and the rank that each uppermost application of the mix is eliminable. This completes the proof of the mix elimination of $\mathbf{LJ}^\dagger$.

By using a similar idea, we can show the cut elimination for $\mathbf{FL_{ec}}$ and $\mathbf{InFL_{ec}}$, either of which has both the exchange and contraction rules, but does not have the weakening rule; see [KO91]. For $\mathbf{FL_{ec}}$, we need to take the following *generalized mix rule*, since cut rule cannot be replaced by the mix rule in the original form without the help of the weakening rule.

$$\frac{\Gamma \Rightarrow \alpha \quad \Pi \Rightarrow \gamma}{\Gamma, \tilde{\Pi}_\alpha \Rightarrow \gamma} \text{ (gmix)}$$

where Π has at least one occurrence of α, and $\tilde{\Pi}_\alpha$ is a sequence of formulas obtained from Π by deleting *at least one* occurrence of α. The same idea can be applied also to $\mathbf{InFL_{ec}}$.

This replacement of the cut rule by the generalized mix rule works well only when we have the exchange rule, as it can be seen in the following result.

LEMMA 4.3. [BO96] *Cut elimination does not hold for* $\mathbf{FL_c}$. *In fact, it does not hold even for the implicational fragment of* $\mathbf{FL_c}$.

In fact, the following sequent is provable in $\mathbf{FL_c}$ but is not provable in $\mathbf{FL_c}$ without cut, where p, q, r, s and t are propositional variables.

$$q\backslash t, t\backslash r, (q\backslash r)\backslash p, p\backslash((q\backslash r)\backslash s) \Rightarrow s$$

Let us consider the following rule, called the *global contraction rule*, which is a stronger form of the contraction rule, where Σ is any finite sequence of formulas.

$$\frac{\Gamma, \Sigma, \Sigma, \Delta \Rightarrow \varphi}{\Gamma, \Sigma, \Delta \Rightarrow \varphi}$$

Let $\mathbf{FL_{gc}}$ be the sequence calculus $\mathbf{FL}$ with the global contraction rule. By using rules for fusions, we can show that a sequent is provable in $\mathbf{FL}_{gc}$ if and only if it s provable in $\mathbf{FL_c}$. Moreover, we can show the following.

LEMMA 4.4. *Cut elimination holds for* $\mathbf{FL_{gc}}$.

We will discuss sequent calculi for involutive logics in Section 7.3.1.

4.2. Decidability as a consequence of cut elimination

As we have mentioned already in Chapter 1 for the case of **LK**, cut elimination has some important consequences, one of them being the subformula property. This follows from the fact that except for the cut rule, in every other rule in a sequent calculus for a basic substructural logic every formula appearing in the upper sequent(s) is a subformula of some formula in the lower sequent.

THEOREM 4.5. *Let R be any subset of $\{e, c, i, o\}$ different than $\{c\}$. If a sequent s is provable in* $\mathbf{FL_R}$ *then there exists a proof of s such that every formula appearing in it is a subformula of some formula in s. In particular, any cut-free proof of s has this property. This holds also for any* $\mathbf{InFL_Q}$ *with $Q \in \{e, ec, ei, eo, ew\}$.*

Similarly to **LK**, we can infer the *conservativity* of $\mathbf{FL_R}$ from the subformula property. Recall from Section 1.6.3 that if $\mathcal{K}$ is a sublanguage of $\mathcal{L}$, then the *$\mathcal{K}$-fragment* of $\mathbf{FL_R}$ is the sequent calculus obtained from $\mathbf{FL_R}$ by restricting the applications of the rules (and axioms) to $\mathcal{K}$-formulas. Also, recall that a system (set of rules) is a conservative extension of its $\mathcal{K}$-fragment, if the two systems prove exactly the same $\mathcal{K}$-formulas.

COROLLARY 4.6. *Assume that $\mathcal{K}$ is a sublanguage of $\mathcal{L}$ and R is a subset of $\{e, c, i, o\}$ different than $\{c\}$. Then,* $\mathbf{FL_R}$ *is a conservative extension of its $\mathcal{K}$-fragment. This holds also for any* $\mathbf{InFL_Q}$ *with $Q \in \{e, ec, ei, eo, ew\}$.*

4.2.1. Decidability of basic substructural logics without contraction rule. Another important consequence of cut elimination is the decidability of basic substructural logics. The contraction rule causes complications not only in the proof of cut elimination, but also in proving decidability. We, therefore, start our discussion from the case for substructural logics without the contraction rule.

THEOREM 4.7. *For any subset R of $\{e, i, o\}$, the substructural logic* $\mathbf{FL_R}$ *is decidable. For any $Q \in \{e, ei, eo, ew\}$, the substructural logic* $\mathbf{InFL_Q}$ *is decidable.*

PROOF. We give a proof of the decidability for such a sequent calculus $\mathbf{FL_R}$, as the decidability of $\mathbf{InFL_Q}$ is shown similarly. Since the cut elimination holds for $\mathbf{FL_R}$, it is enough to search for a proof of a given sequent s among cut-free proofs. Such proofs, if they exist, have the subformula property. Moreover, in every rule appearing in such a cut-free proof except the contraction rule and the exchange rule, (both of) its upper sequent(s) is strictly simpler than the lower sequent. But the contraction rule never appears by our assumption. For the exchange rule, if exists, it is enough to consider a sequent calculus without the exchange rule, but whose sequents are of the form $\Gamma \Rightarrow \delta$ with a multiset Γ of formulas. Therefore, if s is provable in $\mathbf{FL_R}$, it has a cut-free proof such that every sequent s' in it consists of subformulas of a formula in s and moreover the length of s' is less than or equal to the length of s. Obviously, the number of such sequents is finite. Since there is a finite number of rules in the calculus, we can make a thorough finite search of a possible cut-free proof of s. If we can find one then s is provable, and it is not provable otherwise. This yields the decidability of the calculus. □

In view of the algebraization of **FL** and its extensions, we have the following result.

COROLLARY 4.8. *For any subset R of $\{e, i, o\}$, variety* $\mathsf{FL_R}$ *has a decidable equational theory. Also, the varieties* RL, CRL, IRL *and* CIRL *have decidable equational theories.*

In [Tam74], the decidability of RL and IRL is shown, by using cut elimination for the corresponding sequent calculi. Most of the decidability results in Theorem 4.7 are shown in [Kom86], [Došs88] and [Ono90]. Došen [Doš88] has proved also the cut elimination and decidability of several extensions of *non-associative* substructural logics. For further results on non-associative logics, see [GO].

As we have mentioned already in Chapter 2, the cut elimination of **InFL** is shown algebraically in Chapter 7. Therefore, by using the same argument as above we have the decidability of **InFL**. Related to this, we notice that the following result was shown by Yetter [Yet90] about **CyInFL** using proof-theoretic methods and algebraically by Wille [Wil05] about CyInFL.

THEOREM 4.9. [Yet90] [Wil05] *The logic* **CyInFL** *is decidable and cyclic involutive FL-algebras have a decidable equational theory.*

4.2.2. Decidability of intuitionistic logic — Gentzen's idea. Gentzen gave in [Gen35] a decision procedure of the sequent calculi **LK** and **LJ**. In the following we show this for **LJ**. As above, it suffices to search for a proof of a given sequent s among cut-free proofs, and thus proofs with the subformula property. On the other hand, because of the presence of the contraction rule, the condition that upper sequents are strictly simpler than the lower sequent does not always hold. Thus, there may be infinitely many possible proofs. In other words, the proof search as it stands is infinite and it will not terminate if s is not provable. Therefore, it is necessary to impose certain restrictions on proofs, so that provability of sequents will remain the same, but the proof search will terminate.

Proofs with no redundancies: We say that a proof Q of a sequent s has *redundancies*, if the same sequent appears twice in a branch of Q. Clearly, any proof having some redundancies can be transformed into one with no redundancies. Indeed, suppose that Q has redundant occurrence of $\Sigma \Rightarrow \delta$, as shown below.

$$\begin{array}{c} \mathsf{P}_0 \\ \Sigma \Rightarrow \delta \\ \vdots \\ \Sigma \Rightarrow \delta \end{array}$$

$$\begin{array}{c}\mathsf{P}_1 \\ s\end{array}$$

Then, this redundancy can be eliminated as follows.

$$\begin{array}{c}\mathsf{P}_0 \\ \Sigma \Rightarrow \delta \\ \mathsf{P}_1 \\ s\end{array}$$

This restriction is still not enough to reduce the number of all possible proofs to finitely many. To see this, suppose that a given sequent $\beta, \Theta \Rightarrow \gamma$ consists only of subformulas of some formulas in s. Then, for every natural number n, a sequent of the form $\underbrace{\beta, \ldots, \beta}_{n}, \Theta \Rightarrow \gamma$ would appear in a possible proof.

Proofs consisting only of reduced sequents: We introduce a further restriction. A sequent $\Sigma \Rightarrow \delta$ is called *reduced*, if every formula in it occurs at most three times in its antecedent Σ. Also, we say that a sequent $\Sigma \Rightarrow \delta$ is *1-reduced* if every formula in the antecedent Σ occurs exactly once, i.e. every formula in Σ is distinct.

Now, a sequent $\Gamma^\dagger \Rightarrow \delta$ is called a *contraction* of a sequent $\Gamma \Rightarrow \delta$, if it is obtained from $\Gamma \Rightarrow \delta$ by applications of the rules of contraction and exchange. For example, the sequent $\alpha, \alpha, \gamma \Rightarrow \beta$ is a contraction of a sequent $\alpha, \gamma, \alpha, \alpha, \gamma \Rightarrow \beta$. Then, we can show that for any given sequent $\Gamma \Rightarrow \delta$, there exists a reduced, or even a 1-reduced sequent $\Gamma^* \Rightarrow \delta$ which is a contraction of $\Gamma \Rightarrow \delta$, and moreover that $\Gamma \Rightarrow \delta$ is provable in **LJ** if and only if $\Gamma^* \Rightarrow \delta$ is provable in **LJ**. In fact, if a formula appears more than three times in Γ then we can reduce the number of occurrences of the formula to three (or, even to one) by applying the contraction rule (and the exchange rule, if necessary) repeatedly. In this way, we can get a (1-)reduced sequent $\Gamma^* \Rightarrow \delta$. Conversely, by applying the weakening rule to $\Gamma^* \Rightarrow \delta$ as many times as necessary, we can recover $\Gamma \Rightarrow \delta$.

Thus, there exists an *effective* way of getting a (1-)reduced sequent s' for a given sequent s, whose provability is the same as that of s. So, it suffices to give an algorithm which can decide whether a given *reduced sequent* is provable or not. We show the following.

LEMMA 4.10. *Suppose that $\Gamma \Rightarrow \delta$ is a sequent which is provable in* **LJ** *and that $\Gamma^* \Rightarrow \delta$ is any 1-reduced contraction of $\Gamma \Rightarrow \delta$. Then, there exists a cut-free proof of $\Gamma^* \Rightarrow \delta$ in* **LJ** *such that every sequent appearing in it is reduced.*

PROOF. Let us take a cut-free proof P of $\Gamma \Rightarrow \delta$. We will prove our lemma by induction on the *length* of P, i.e. the maximum length of branches in P which start from an initial sequent and end at the endsequent of P. Our lemma is trivial when $\Gamma \Rightarrow \delta$ is an initial sequent. Suppose that $\Gamma \Rightarrow \delta$ is the lower sequent of an application of a rule (r). When (r) has a single upper sequent, it must be of the following form:

$$\frac{\Lambda \Rightarrow \sigma}{\Gamma \Rightarrow \delta}\ (r)$$

Let $\Lambda^* \Rightarrow \sigma$ be any 1-reduced contraction of $\Lambda \Rightarrow \sigma$. Then, by the hypothesis of induction, there exists a cut-free proof of $\Lambda^* \Rightarrow \sigma$ such that every sequent appearing in it is reduced. Let $\Gamma' \Rightarrow \delta$ be the sequent obtained by applying the same rule (r) to $\Lambda^* \Rightarrow \sigma$. (When (r) is the contraction rule (c), we cannot apply it to $\Lambda^* \Rightarrow \sigma$. In this case, it is enough to take $\Lambda^* \Rightarrow \sigma$ for $\Gamma' \Rightarrow \delta$.) Then, $\Gamma' \Rightarrow \delta$ is always reduced (but not necessarily, 1-reduced) for every rule (r) of **LJ**. By applying the contraction rule (and the exchange rule if necessary), we can get a 1-reduced (in fact, any 1-reduced) contraction of $\Gamma \Rightarrow \delta$. Clearly, the proof thus obtained consists only of reduced sequents. The same argument holds also in the case where (r) has two upper sequents. □

Here we will add an explanation of the reason why the number "three" appears in the definition of reduced sequents. Let us consider the case when (r) (in the above proof) is $(\rightarrow \Rightarrow)$;

$$\frac{\Lambda \Rightarrow \alpha \qquad \beta, \Sigma \Rightarrow \delta}{\alpha \rightarrow \beta, \Lambda, \Sigma \Rightarrow \delta}\ (r)$$

When $\alpha \rightarrow \beta$ occurs once in Λ and once in Σ, the antecedent of the lower sequent contain three occurrences of $\alpha \rightarrow \beta$. Thus, we have to admit three occurrences of the same formula to make the lower sequent reduced.

Now, we are ready for showing the decidability of **LJ**. Take any sequent s. We can assume from the beginning that it is reduced. We need to check whether s is provable or not. By the above observations, if s is provable, there exists a proof Q consisting only of reduced sequents such that every formula in it is a subformula of a formula in s. Clearly, the number of such reduced sequents is finite. As we can moreover assume that Q contains no redundancy, the number of *possible* proofs of s without redundancy is finite. Thus, our decision algorithm is to make a thorough search of possible proofs and to check whether there is a correct proof of s among them or not.

In practice, making a thorough search of possible proofs is quite inefficient. What we should do instead is to construct a possible proof from the endsequent *upward*. The last inference must be either one of structural rules or one of rules for logical connectives. In the latter case, the logical connective for which a rule is applied must be an outermost logical connective of a formula in the endsequent. Therefore, there are finitely many possibilities. We repeat this process again for sequents which are obtained already in this way. Note that a possible proof will be branching upward, and therefore by choosing one of these branches we need to continue the above construction until we come to the top of the branch, i.e. until we reach a sequent which does not produce any new sequent. Then we check whether this sequent is an initial sequent or not. If it is, we do the same for other branches. If not, we have to abandon this choice and to backtrack, and try to find other choices. Such a procedure is called a *proof-search algorithm.*

Here are two examples of possible proofs of $\Rightarrow p \vee (p \to q)$. The left one fails to be a proof, but the right one is a correct proof.

$$\dfrac{\dfrac{p \Rightarrow q}{p \Rightarrow q \vee r}\,(\Rightarrow \vee 1)}{p \wedge (p \to q) \Rightarrow q \vee r}\,(\wedge 1 \Rightarrow) \qquad \dfrac{\dfrac{\dfrac{\dfrac{\dfrac{p \Rightarrow p \quad q \Rightarrow q}{p, p \to q \Rightarrow q}\,(\to \Rightarrow)}{p, p \wedge (p \to q) \Rightarrow q}\,(\wedge 2 \Rightarrow)}{p \wedge (p \to q), p \wedge (p \to q) \Rightarrow q}\,(\wedge 1 \Rightarrow)}{p \wedge (p \to q) \Rightarrow q}\,(c \Rightarrow)}{p \wedge (p \to q) \Rightarrow q \vee r}\,(\Rightarrow \vee 1)$$

THEOREM 4.11. *Both classical and intuitionistic logics are decidable.*

COROLLARY 4.12. *The varieties* BA *and* HA *have decidable equational theories.*

4.2.3. Decidability of basic substructural logics with the contraction rule. Decision problems for substructural logics with the contraction rule but without the weakening rule, like $\mathbf{FL_{ec}}$ and $\mathbf{InFL_{ec}}$, are even more complicated. In the following, we discuss the decision problem for $\mathbf{FL_{ec}}$ in detail. The proof is based on the idea introduced by Kripke [Kri59] for the decidability of the implicational fragment of $\mathbf{FL_{ec}}$, and developed by Meyer [Mey66] for $\mathbf{InFL_{ec}}$ in his dissertation. See also [BJW65] and [Dun86]. It will be more technical than the case of substructural logics without the contraction rule. Therefore readers may skip the details of the proofs in their first reading.

As we have mentioned already, main difficulties are caused by the presence of the contraction rule. That is, in a proof of a given sequent s, there may occur sequents which are much more complicated than s. Unlike in intuitionistic logic, we cannot rely on the idea of reduced sequents, since $\mathbf{FL_{ec}}$ does not have the weakening rule.

Auxiliary calculus $\mathbf{FL_{ec}}'$ *without explicit contraction rule.* To overcome this difficulty, we remove the contraction rule from $\mathbf{FL_{ec}}$ and incorporate it into each logical rule as *implicit contractions.* The resulting calculus is called $\mathbf{FL_{ec}}'$. To avoid an overload of technicalities, we will omit the full definition $\mathbf{FL_{ec}}'$, and illustrate the idea by discussing only its implicational fragment. According to our original definition, sequents of $\mathbf{FL_{ec}}$ are expressions of the form $\Gamma \Rightarrow \delta$ with a (possibly empty) sequence Γ of formulas and a (possibly empty) formula δ. On the other hand, sequents of $\mathbf{FL_{ec}}'$ are expressions of the form $\Gamma \Rightarrow \delta$ with a (possibly empty) *finite multiset* Γ of formulas and a (possibly empty) formula δ. (This is only for the sake of brevity. In this way, we can dispense with the exchange rule.) The system $\mathbf{FL_{ec}}'$ does not have the contraction rule, but its logical rules are slightly modified. In the following, by $\sharp_\Pi(\beta)$ we mean the multiplicity of a formula β in a given multiset Π, i.e. the number of occurrences of β in Π.

Recall that the rules for implication in $\mathbf{FL_{ec}}$ are

$$\frac{\Gamma \Rightarrow \alpha \quad \beta, \Sigma \Rightarrow \delta}{\alpha \to \beta, \Gamma, \Sigma \Rightarrow \delta}\ (\to \Rightarrow) \qquad\qquad \frac{\alpha, \Gamma \Rightarrow \beta}{\Gamma \Rightarrow \alpha \to \beta}\ (\Rightarrow \to)$$

On the other hand, the rules for the implication in $\mathbf{FL_{ec}}'$ are taken to be

$$\frac{\Gamma \Rightarrow \alpha \quad \beta, \Sigma \Rightarrow \delta}{\alpha \to \beta, \Pi \Rightarrow \delta}\ (\to \Rightarrow) \qquad\qquad \frac{\alpha, \Gamma \Rightarrow \beta}{\Gamma \Rightarrow \alpha \to \beta}\ (\Rightarrow \to)$$

Here $\alpha \to \beta, \Pi$ is any multiset which is a *contraction* of the multiset $\alpha \to \beta, \Gamma, \Sigma$ (i.e. $\alpha \to \beta, \Pi$ is any multiset obtained from the multiset $\alpha \to \beta, \Gamma, \Sigma$ by deleting some of duplicated formulas in it) that satisfies the following requirements:

(1) $\sharp_\Pi(\alpha \to \beta) \geq \sharp_{(\Gamma,\Sigma)}(\alpha \to \beta) - 2$, when $\alpha \to \beta$ belongs to both Γ and Σ. Otherwise, $\sharp_\Pi(\alpha \to \beta) \geq \sharp_{(\Gamma,\Sigma)}(\alpha \to \beta) - 1$.

(2) For any formula γ in Γ, Σ except $\alpha \to \beta$, $\sharp_\Pi(\gamma) \geq \sharp_{(\Gamma,\Sigma)}(\gamma) - 1$, when γ belongs to both Γ and Σ, and $\sharp_\Pi(\gamma) = \sharp_{(\Gamma,\Sigma)}(\gamma)$ otherwise.

Note that there is some freedom in the choice of Π. Here is an example of an application of $(\to \Rightarrow)$ of $\mathbf{FL_{ec}}'$:

$$\frac{\alpha \to \beta, \gamma, \delta \Rightarrow \alpha \quad \beta, \alpha \to \beta, \gamma, \varphi \Rightarrow \psi}{\alpha \to \beta, \gamma, \delta, \varphi \Rightarrow \psi}$$

LEMMA 4.13. (1) *Cut elimination holds for* $\mathbf{FL_{ec}}'$.

(2) *A formula is provable in* $\mathbf{FL_{ec}}'$ *if and only if it is provable in* $\mathbf{FL_{ec}}$.

A proof P in $\mathbf{FL_{ec}}'$ has *redundant contractions* if there exist sequents s and s' in a branch of P such that s' is a contraction of s and appears below s.

$$\begin{array}{c} \vdots \\ s \\ \vdots \\ s' \\ \vdots \end{array}$$

Obviously, proofs without redundant contractions are proofs without redundancies. The following lemma can be proved by using induction on the length of the proof (i.e. the maximum length of branches in the proof) of s.

LEMMA 4.14. (Curry's Lemma) *Suppose that a given sequent s has a cut-free proof of length m in* $\mathbf{FL_{ec}}'$ *and that s' is a contraction of s. Then, s' has a cut-free proof in* $\mathbf{FL_{ec}}'$ *whose length is not greater than m.*

By using this lemma, we have the following result.

COROLLARY 4.15. *If a sequent is provable in* $\mathbf{FL_{ec}}'$*, then it has a cut-free proof in* $\mathbf{FL_{ec}}'$ *which has no redundant contractions.*

PROOF. Let P a cut-free proof of s in $\mathbf{FL_{ec}}'$. Our corollary can be shown by using induction on the length n of P. Suppose moreover that P has a redundant contraction shown in the following figure, where s_2 is a contraction of s_1, and the lengths of the subproofs of s_1 and s_2 (in P) are k and m, respectively, with $k < m$.

$$\begin{array}{c} \vdots \\ s_1 \\ \vdots \\ s_2 \\ \vdots \\ s. \end{array}$$

Then, by Curry's Lemma, there exists a cut-free proof Q of s_2 with the length k' such that $k' \leq k$. So, we can get another proof P' of s by replacing the subproof of s_2 in P by Q:

$$\begin{array}{c} \mathsf{Q} \\ s_2 \\ \vdots \\ s \end{array}$$

with the length n' such that $n' < n$, since $k' \leq k < m$. Hence, by the hypothesis of induction we can get a proof of s with no redundant contractions. □

Termination of the proof-search algorithm. There is no hope of having a finite number of sequents in all possible proofs in $\mathbf{FL_{ec}}'$. Instead, we can restrict the possible proofs to those with *the subformula property* and with *no redundant contractions*. Are these two restrictions enough to make the total number of possible proofs finite? This is not obvious. Let us consider the following example. Here, we will write the sequence $\alpha, \ldots, \alpha$ with n occurrences of α as α^n. Let s be the sequent $\alpha^3, \beta^2, \gamma^5 \Rightarrow \delta$, and consider the sequents $\alpha^4, \beta, \gamma^4 \Rightarrow \delta$, $\alpha^2, \beta^3, \gamma^4 \Rightarrow \delta$ and $\alpha^5, \beta^2, \gamma^3 \Rightarrow \delta$. Clearly, s is not a contraction of any of thm. Moreover, among the three sequents, none is a contraction of an other. Thus, all three sequents can appear in the same branch of a possible proof of s with no redundant contractions. We are, thus, facing the following combinatorial problem.

Suppose that formulas $\alpha_1, \ldots, \alpha_m, \delta$ are given. Consider a sequence $(s_1, s_2, \ldots)$ of sequents such that each member is always of the form ${\alpha_1}^{k_1}, \ldots, {\alpha_m}^{k_m} \Rightarrow \delta$, where each k_i is positive. (Any sequence of this kind is called a sequence of *cognate* sequents.) Moreover, assume that s_i is not a contraction of s_j whenever $i < j$. Our question is; *can such a sequence be of infinite length*? Or, is the length of such a sequence always finite? (Note that the replacement of the condition "whenever $i < j$" by "whenever $i \neq j$" does not affect the answer, since the number of sequents which are contractions of a given sequent s_i is obviously finite.)

This problem is mathematically the same as the following. Let $\mathbb{N}$ be the set of all positive integers, and $\mathbb{N}^m$ be the set of all n-tuples $(k_1, \ldots, k_m)$ such that $k_i \in \mathbb{N}$ for each i. Define a binary relation $\leq^*$ on $\mathbb{N}^m$ by $(k_1, \ldots, k_m) \leq^* (h_1, \ldots, h_m)$ if and only if $k_i \leq h_i$ for each i. Clearly, $\leq^*$ is a partial order. A subset W of $\mathbb{N}^m$ is called an *antichain* if for any distinct $\mathbf{u}, \mathbf{v} \in W$ neither $\mathbf{u} \leq^* \mathbf{v}$ nor $\mathbf{v} \leq^* \mathbf{u}$ hold. Then, the above question is restated as; *are there infinite antichains of* $\mathbb{N}^m$? The answer is negative.

LEMMA 4.16. *Any antichain of* $\mathbb{N}^m$ *is finite.*

This result can be proved by various ways. One way is to show the following lemma, from which the above lemma follows immediately. A partial order $\preceq$ on a given set Y is *well-founded* if there are no descending chains in Y, i.e. there are no elements $\{s_i\}_{i\in\mathbb{N}}$ in Y such that $s_{i+1} \prec s_i$ for each i. A partial order $\preceq$ on a given set Y is a *well partial order* if it is well-founded and moreover Y contains no infinite antichain with respect to $\preceq$.

LEMMA 4.17.

(1) *The natural order on* $\mathbb{N}$ *is a well partial order.*
(2) *If* $\leq_1$ *and* $\leq_2$ *are well partial orders on the sets* U *and* V*, respectively, then their direct product order* $\leq$ *is also a well partial order on* $U \times V$*. Here,* $\leq$ *is defined by* $(u, v) \leq (u', v')$ *if and only if* $u \leq_1 u'$ *and* $v \leq_2 v'$*.*

COROLLARY 4.18. (Kripke's Lemma) *Every sequence* $(s_1, s_2, \ldots)$ *of cognate sequents, such that* s_i *is not a contraction of* s_j *whenever* $i < j$*, is necessarily finite.*

Using these results, we can give the following proof-search algorithm for $\mathbf{FL_{ec}}$ which is similar to the proof-search algorithm for **LJ** but is more complicated.

Take any sequent (or any pair of sequents) from which a given sequent s follows by a single application of a rule of inference except the cut rule. Write any of them just over s as long as it consists only of subformulas of formulas in s. We regard each of them as an *immediate predecessor* of s

in our proof-search. Then, repeat this process again for each of immediate predecessors, and continue this as long as possible. But at any stage of generating immediate predecessors if a newly generated sequent has a contraction below it, we must abandon the possibility of adding it to a proof-search tree. When there are no rules which generate immediate predecessors, we check whether this sequent is one of initial sequents or not. If not, we must abandon this possibility.

We can see that if s is provable then there exists a proof which forms a subtree of the tree constructed by our proof-search of s. We can see that the number of the branching at each "node" of the tree of our proof-search is finite, since each sequent has only finitely many sequents that can be its upper sequents. Furthermore, the finiteness of the length of each branch in the tree of our proof-search is assured by Kripke's Lemma. The following lemma says that the tree of our proof-search is in fact finite.

LEMMA 4.19. (König's Lemma) *A tree is finite if and only if the number of the branching at each node of the tree is finite and the length of each branch in the tree is finite.*

Thus, it turns out that a given sequent s is provable if and only if there exists a proof of s which is a subtree of the tree of the proof-search. Thus we have the following, due to [Mey66] for $\mathbf{InFL_{ec}}$ and to [KO91] for $\mathbf{FL_{ec}}$.

THEOREM 4.20. *The substructural logics* $\mathbf{FL_{ec}}$ *and* $\mathbf{InFL_{ec}}$ *are decidable.*

COROLLARY 4.21. *The varieties* $\mathsf{FL_{ec}}$, $\mathsf{InFL_{ec}}$ *and* CKRL *have decidable equational theories.*

4.3. Further results

In this section, we will give a short survey and make some historical remarks about (un)decidability problems related to substructural logics and residuated lattices.

In the beginning of the sixties, H. Wang [Wan63] observed that (even) the predicate logic obtained from **LK** be deleting the contraction rule is decidable. For more information on (un)decidable substructural predicate logics, see [Kom86] and [KO91]. On the other hand, as we have seen in the previous section, the proof of the decidability of substructural (propositional) logics with contraction is rather involved. The decision algorithm for such logics relies on Kripke's Lemma. Due to the non-constructive character of the latter we do not have a good bound on the computational complexity of the decision algorithm. In fact, Urquhart showed the following.

THEOREM 4.22. [Urq99b] *There is no primitive recursive decision procedure for the positive fragment of* $\mathbf{InFL_{ec}}$.

Obviously, adding the distributive law to logics with the contraction rule makes the situation worse. In fact, in 1984 A. Urquhart showed the following. Note that the relevant logic **R**, is equivalent to **InFL**$_{\mathbf{ec}}$ plus distributivity, has a cut-free sequent calculus, but this does not guarantee decidability. Recall that the positive fragment $\mathbf{L}^+$ of a logic **L** is the fragment without negation and without the constant 0.[1]

THEOREM 4.23. [Urq84] *The positive fragments* $\mathbf{T}^+, \mathbf{E}^+$ *and* $\mathbf{R}^+$ *of the relevant logics* $\mathbf{T}, \mathbf{E}$ *and* **R**, *respectively, are undecidable.*

In contrast, [Mey66] shows that, by an effective translation, the decision problem of the positive fragment of **InFL**$_{\mathbf{ec}}$ can be translated to that of the $\{\to, \wedge\}$-fragment of **R**. The computational complexity is still relatively high, as shown in [Urq99b].

THEOREM 4.24. *The* $\{\to, \wedge\}$*-fragments of* **T**, **E** *and* **R** *are decidable, but there is no primitive recursive decision procedure for either of them.*

As the combination of contraction and distributivity cause serious complications, we consider contraction-less versions of the relevant logics **T**, **E** and **R**, and denote them **TW**, **EW** and **RW**, respectively. Giambrone [Gia85] showed that the constant-free fragments (namely, the 1-free fragments of the positive fragments of) **TW** and of **RW** are decidable. Additionally, Brady [Bra90], [Bra91] showed that **TW**, **EW** and **RW** are all decidable. Recall from Section 2.3.3 that the algebraic semantics of **RW** is the variety $\mathsf{InDFL_e}$ (up to term equivalence between negation and 0). The decidability of $\mathbf{RW}^+$ follows from results in [GR04]. As a consequence of the above we get the following result.

COROLLARY 4.25. *The equational theories of* $\mathsf{InDFL_e}$ *and* CDRL *are decidable.*

We also mention that Restall [Res98] used proof theoretic methods based on display calculi to obtain the decidability of some substructural logics.

Even though there is no natural sequent calculus for the Abelian logic of R. Meyer, it can be shown [HM79] that the variety of ℓ-groups has a decidable equational theory. Furthermore, the universal theory of abelian ℓ-groups is decidable [His66] and actually co-NP-complete [Weis86]. Based on the decidability of ℓ-groups, [BCG$^+$03] proved the decidability of the equational theories of LG^- and CLG^- and [GT05] proved the decidability of the equational theories of GMV, IGMV and CGMV.

Another important way of obtaining decidability is through the *finite model property* (FMP). A substructural logic **L** is said to have the FMP, if for any non-provable formula α there exists a *finite* model of **L** (namely an

[1]Often in the study of relevant logics, versions of logics with or without constants have been considered and often this choice makes a difference. For more information on relevant logics, the reader is referred to the two volumes [AB75] and [ABD92].

FL-algebra $\mathbf{A}$ such that $\mathbf{A} \models \mathbf{L}$ of $\mathbf{A} \in \mathsf{V}(\mathbf{L})$) such that $\mathbf{A} \not\models \alpha$. Harrop's theorem specializes to the following.

THEOREM 4.26. [Har56] *If a substructural logic* $\mathbf{L}$ *is finitely axiomatizable and has the finite model property, then it is decidable.*

In fact, the finite axiomatizability of $\mathbf{L}$ implies that $\mathbf{L}$ is recursively enumerable as a set of formulas, and the finite model property implies that the complement of $\mathbf{L}$ is also recursively enumerable. Therefore, it is a recursive set by Post's theorem.

As this proof shows, although the theorem guarantees that the logic $\mathbf{L}$ is decidable, the direct implementation of the decision algorithm is not practical. In modal logic many decidability results are obtained by showing the finite model property (using Kripke frames). On the other hand, relatively little is known about the finite model property of substructural logics in general, except through the *finite embeddability property* (FEP). The FEP is a stronger property than FMP and will be discussed in detail in Chapter 6.

We also mention the following result by Komori on MV-algebras and Łukasiewicz logic.

THEOREM 4.27. [Kom81]

(1) *For each proper extension* $\mathbf{L}$ *of Łukasiewicz logic,* $\mathbf{L}$ *has the FMP iff the implicational fragment of* $\mathbf{L}$ *differs from that of Łukasiewicz logic.*
(2) *All extensions of Łukasiewicz logic are finitely axiomatizable.*
(3) *All extensions of Łukasiewicz logic are decidable.*

Thus, some extensions of Łukasiewicz logic are decidable without having the FMP. Another result on the FMP of superintuitionistic logics is given by Komori.

THEOREM 4.28. [Kom75] *Every superintuitionistic logic which is not weaker than the Gödel-Dummett logic has the FMP.*

We also have the following.

THEOREM 4.29. [Ued00]

(1) *There exist uncountably many extensions of* $\mathbf{InFL_{ew}}$ *with the FMP, and uncountably many extensions of* $\mathbf{InFL_{ew}}$ *without the FMP.*
(2) *There exist countably many finitely axiomatizable extensions of* $\mathbf{InFL_{ew}}$ *without the FMP.*

We do not mention here in detail results on the FMP of fragments of substructural logics. Such results have been obtained by using Kripke-type semantics, and the methods seem to be related to those used to prove the FEP. For more on these topics, see [Mey73], [MO94], [vAR99], [Bus96], and [Bus02].

Name	Eq. Theory	Word prob.	Univ. Th.
FL	FMP Thm. 7.18	Und. Thm. 4.35	Und. Thm. 4.30
FL_e	FMP Thm. 7.19		Und.Thm. 4.32
FL_i	FMP Thm. 7.19		FEP Thm. 6.46
FL_w	FMP Thm. 7.19		FEP Thm. 6.46
FL_{ew}	FMP Thm. 7.19		FEP Thm. 6.46
DFL		Und. Cor. 4.48	Und. Cor. 4.48
DFL_e	Dec. Cor. 4.25	Und. Cor. 4.48	Und. Cor. 4 48
MFL	Und. Thm. 4.33	Und. Thm. 4.33	Und. Thm. 4.33
BL	FMP	Decidable	FEP Thm. 6.54
MV	Dec. Thm. 4.27		FEP Thm. 6.55
HA	FMP	Decidable	FEP Cor. 6.52
GA	FMP	Decidable	FEP Thm. 6.51
BA	FMP	Decidable	FEP
InFL	Dec. Cor. 7.15		
CyInFL	Dec. Thm. 4.9	Und. Thm. 4.35	Und. Thm. 4.35
$InDFL_e$	Dec. Cor. 4.25	Und. Cor. 4.49	Und. Cor. 4.49

TABLE 4.1. (Un)decidability of some subvarieties of F_-.

In Chapter 7, we discuss the FMP of basic substructural logics. Interestingly enough, the FMP for those logics is shown via the cut elimination result and resulting decision procedure.

Some of the results about decidability mentioned throughout the book are organized in Table 4.1 in three columns, referring to the decidability of the equational theory, the word problem and the universal theory of a variety. In the table Dec. and Und. stand for decidability and undecidability. Often a stronger result, like FMP in the first column and FEP in the last one are mentioned. Note that since all listed classes are finitely axiomatized, those with FEP have a decidable universal theory, and those with FMP have a decidable equational theory.

The table is far from comprehensive and is meant to serve only as an aid. In particular, we did not always put the reference to the result obtained historically first, as it was easier to refer to theorems of the book, which in turn refer to original sources. Many of the decidability results followed from cut elimination in the book are not mentioned in the table and the reader is advised to look at the statements of the theorem referenced there.

4.4. Undecidability

In this section we will mention some undecidability results. In particular we will show that the quasiequational theory of RL and DRL, as well as the equational theory of modular residuated lattices are undecidable. In fact, we will prove that the quasiequational theory (actually the word problem) for a range of subvarieties DRL, including the subvariety CDRL, is undecidable.

4.4.1. The quasiequational theory of residuated lattices. Recall that residuated lattices have a decidable equational theory. Nevertheless, the same does not hold for their quasiequational theory.

THEOREM 4.30. [JT02] *The quasi-equational theory of residuated lattices is undecidable. The same holds for any subvariety that contains all powersets of finite monoids.*

PROOF. Consider the class $\mathcal{K}$ of all monoid reducts of RL. Note that a quasi-identity that uses only $\cdot$, holds in all residuated lattices iff it holds in $\mathcal{K}$ iff it holds in all subalgebras of $\mathcal{K}$. Any semigroup $\mathbf{S}$ can be embedded in some member of $\mathcal{K}$ follows. Embed $\mathbf{S}$ in a monoid $\mathbf{S}_1$, where $S_1 = S \cup \{1\}$, and construct the residuated lattice $\mathcal{P}(\mathbf{S}_1)$ on the powerset of S_1. The collection of singletons is closed under multiplication in $\mathcal{P}(\mathbf{S}_1)$ and isomorphic to $\mathbf{S}_1$.

On the other hand, $\mathcal{K}$ is contained in the variety SG of all semigroups, hence SG coincides with the class of all subalgebras of $\mathcal{K}$. But the quasi-equational theory of semigroups is undecidable, hence the same is true for residuated lattices.

For the second part of the theorem, we use the result of [GL84] that the quasi-equational theory of semigroups is recursively inseparable from the quasi-equational theory of finite semigroups. □

Since $\mathcal{P}(\mathbf{S}_1)$ is distributive, it follows that the variety of distributive residuated lattices has an undecidable quasi-equational theory.

COROLLARY 4.31. *The deducibility relations of* **FL** *and* **DFL** *are undecidable.*

The proof of the undecidability of linear logic (with exponentials) given by Lincoln, Mitchell, Scedrov and Shankar in [LMSS92] implies the following. (See also [Tro92].)

THEOREM 4.32. cf. [LMSS92] *The deducibility relation of* $\mathbf{FL_e}$ *is undecidable.*

The next result shows that there are varieties of residuated lattices with even undecidable equational theory.

THEOREM 4.33. [JT02] *The variety of modular residuated lattices has an undecidable equational theory.*

PROOF. Let $\mathbf{M}$ be a modular lattice and define $A = M \cup \{0, 1, T\}$ where $0 < 1 < x < T$ for all $x \in M$. Then $\mathbf{A}$ is still modular (in fact in the same variety as M).

We define $\cdot$ on $\mathbf{A}$ by $x0 = 0 = 0x$, $x1 = x = 1x$ and $xy = 1$ if $x, y \neq 0, 1$. It is easy to check that $\cdot$ is associative and residuated, hence the $\vee, \wedge$-equational theory of modular residuated lattices coincides with the equational theory of modular lattices. Since modular lattices have an undecidable equational theory ([Fre80]), the same is true for modular residuated lattices. □

COROLLARY 4.34. *The logic obtained from* **FL** *by adding the modularity axiom is undecidable.*

4.4.2. The word problem. Let $\mathcal{V}$ be a variety, X a set of variables and R a set of equations over X. Starting from the free algebra in $\mathcal{V}$ over X, we factor the smallest equivalence relation containing R (up to identification of the equivalence classes) and we obtain an algebra that we denote by $\mathbf{A} = (X|\, R)$.[2] We say that X and R form a *presentation* for $\mathbf{A}$. If both X and R are finite, then we say that $\mathbf{A}$ is *finitely presented.*

Let $\mathbf{A} = (X|\, R)$ be a finitely presented algebra in a variety $\mathcal{V}$. The algebra $\mathbf{A}$ is said to have an *undecidable (local) word problem* if there is no algorithm that decides whether or not any two given words (or rather terms) in the absolutely free term algebra represent the same element of $\mathbf{A}$. If $\mathcal{V}$ contains such an algebra $\mathbf{A}$, we say that it has an undecidable word problem.

Note that two terms s, t over X represent the same element of $\mathbf{A}$ iff the quasiequation $\mathsf{AND}\ R \Rightarrow s = t$ holds in $\mathcal{V}$, see Lemma 4.44. Therefore, $\mathcal{V}$ has a decidable word problem if for every conjunction $\mathsf{AND}\ R$ of a equations, there is an algorithm that decides all quasiequations with left-hand side $\mathsf{AND}\ R$. A stronger demand is that there exists a (uniform or global) algorithm that decides all quasiequations; this is simply the decidability of the quasiequational theory, also referred to as the decidability of the *global word problem.* Clearly, if the quasiequational theory of a variety is decidable, then so is its word problem.

It is well known that the word problem, hence also the quasiequational theory, for the varieties of semigroups and groups is undecidable. The same holds for the variety of ℓ-groups; see [GG83]. Moreover, the first-order theory of abelian ℓ-groups is hereditarily undecidable [Gur67].

Wille [Wil07] proved that the word problem is undecidable for a number of subvarieties of FL. The undecidability of FL is not stated in the paper, but is follows directly from the result.

---

[2]To be more precise, we start from the absolutely free algebra over X and factor both the congruence generated by R and the fully invariant congruence, factoring by which produces the free algebra in $\mathcal{V}$ over X.

THEOREM 4.35. [Wil07] *The word problem is undecidable for the varieties* FL, RL, CyInFL, CyFL *and for any of their subvarieties axiomatized by non-trivial lattice equations.*

The main result of the following two sections, which we present in detail, is the undecidability of the word problem (hence also of the quasiequational theory) for a range of varieties including the variety of distributive residuated lattices and the variety of commutative distributive ones. The result for a subrange, including the latter variety, is a consequence of a theorem by Urqhart [Urq84]. The proof here is based on the undecidability of the word problem for the variety of semigroups and makes use of the concept of an n-frame, introduced by von Neumann. The methods in the proof extend ideas used by Lipshitz and Urquhart to establish undecidability results for the varieties of modular lattices and distributive lattice-ordered semigroups, respectively.

In the next section we give the definition of an n-frame and the results for modular lattices from [Lip74], some of which will be used later on, before presenting the modified definition for residuated lattices together with the corresponding theorem.

4.4.3. Modular lattices.

We begin with an equivalent version of the original definition due to von Neumann.

DEFINITION 4.1. A *modular n-frame* in a lattice $\mathbf{L}$ is an $n \times n$ matrix, $C = [c_{ij}]$, $c_{ij} \in \mathbf{L}$, (set $a_i = c_{ii}$ and $e = \bigwedge\{a_i \mid i \in \mathbb{N}_n\}$; $\mathbb{N}_n = \{1, ..., n\}$), such that:

i) $\bigvee A_1 \wedge \bigvee A_2 = \bigvee(A_1 \cap A_2)$, for all $A_1, A_2 \subseteq \{a_1, a_2, ..., a_n\}$, where $\bigvee \emptyset = e$;
ii) $c_{ij} = c_{ji}$, for all $i, j \in \mathbb{N}_n$;
iii) $a_i \vee a_j = a_i \vee c_{ij}$, for all $i, j \in \mathbb{N}_n$;
iv) $a_i \wedge c_{ij} = e$, for all distinct $i, j \in \mathbb{N}_n$;
v) $(c_{ij} \vee c_{jk}) \wedge (a_i \vee a_k) = c_{ik}$, for all distinct triples $i, j, k \in \mathbb{N}_n$.

The following examples, taken from [AGN97], give some idea of the motivation for the definition.

Consider the real projective plane P. The lattice $\mathbf{L}$ of subspaces of P contains points, projective lines, P and $\emptyset$, ordered under inclusion. Meet is intersection of subsets of P, while the join of two projective subspaces is the least subspace containing both of them. Modularity of $\mathbf{L}$ is well known and easy to establish. A modular 3-frame, see Figure 4.1, will consist of essentially six points: $a_1, a_2, a_3, c_{12}, c_{13}, c_{23}$, because of (ii). The points a_1, a_2, a_3 are not collinear, by condition (i); c_{ij} has to be on the line $a_i \vee a_j$, by (iii), while c_{12}, c_{13}, c_{23} are collinear, by condition (v); actually, c_{ik} is the point of intersection of the lines $a_i \vee a_j$ and $c_{ij} \vee c_{jk}$.

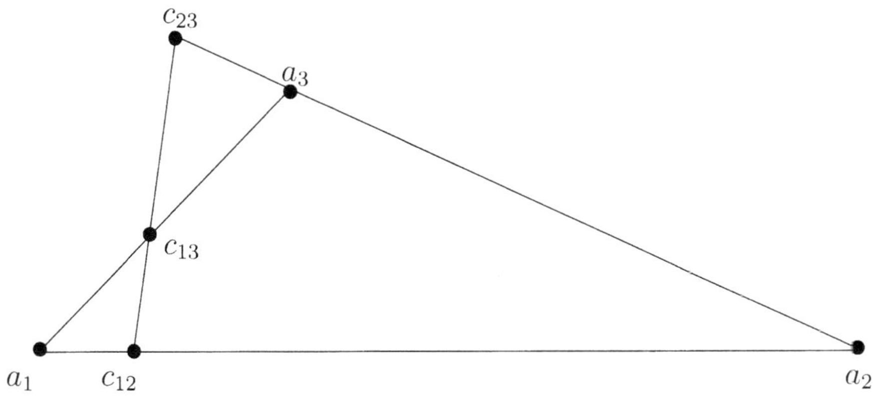

FIGURE 4.1. The geometric meaning of a modular 3-frame.

Let $\mathbf{V}$ be an n-dimensional real inner product space, $\{\mathbf{e}_i \mid i \in \mathbb{N}_n\}$ an orthonormal base of $\mathbf{V}$, $a_i = \langle \mathbf{e}_i \rangle$, the subspace generated by $\mathbf{e}_i$, and $c_{ij} = \langle \mathbf{e}_i - \mathbf{e}_j \rangle$. Then $[c_{ij}]$, $i, j \in \mathbb{N}_n$, is a modular n-frame in the lattice $\mathbf{L}$ of subspaces of $\mathbf{V}$.

Given a modular n-frame one can define operations of multiplication and addition on certain elements of the lattice. If $[c_{ij}]$ is a modular n-frame in a modular lattice $\mathbf{L}$, we define

- $L_{ij} = \{x \in \mathbf{L} \mid x \vee a_j = a_i \vee a_j \text{ and } x \wedge a_j = e\}$, for all distinct $i, j \in \mathbb{N}_n$;
- $b \otimes_{ijk} d = (b \vee d) \wedge (a_i \vee a_k)$, for all $b \in L_{ij}, d \in L_{jk}$;
- $b \odot_{ij} d = (b \otimes_{ijk} c_{jk}) \otimes_{ikj} (c_{ki} \otimes_{kij} d)$, for all $b, d \in L_{ij}$;
- $b \oplus_{ij} d = [((b \vee c_{ik}) \wedge (a_j \vee a_k)) \vee ((d \vee a_k) \wedge (a_j \vee c_{ik}))] \wedge (a_i \vee a_j)$, for all $b, d \in L_{ij}$.

In [Lip74] it is shown that the definitions of $\odot_{ij}$ and $\oplus_{ij}$ are independent of the choice of $k \in \mathbb{N}_n$, for $k \neq i, k \neq j$.

The following theorem of Von Neumann justifies the terminology of multiplication and addition, and validates the connection between projective geometry and modular lattices.

THEOREM 4.36. [vN60] *Let $C = [c_{ij}]$ be a modular n-frame in a modular lattice $\mathbf{L}$, where $n \geq 4$. Then $\mathbf{R}_{ij} = (L_{ij}, \oplus_{ij}, \odot_{ij}, a_i, c_{ij})$ is a ring for all distinct $i, j \in \mathbb{N}_n$. Moreover, all rings $\mathbf{R}_{ij}$ are isomorphic.*

In view of the last statement of the previous theorem the choice of indices i, j in $\mathbf{R}_{ij}$ is inessential. So, $\mathbf{R}_{12} = (L_{12}, \oplus_{12}, \odot_{12}, a_1, c_{12})$ is called the *ring associated with the modular n-frame C of* $\mathbf{L}$.

For a vector space, $\mathbf{V}$, denote by $L(\mathbf{V})$ the set of all subspaces of $\mathbf{V}$. It is well known that $\mathbf{L}(\mathbf{V}) = (L(\mathbf{V}), \wedge, \vee)$ is a modular lattice, where meet

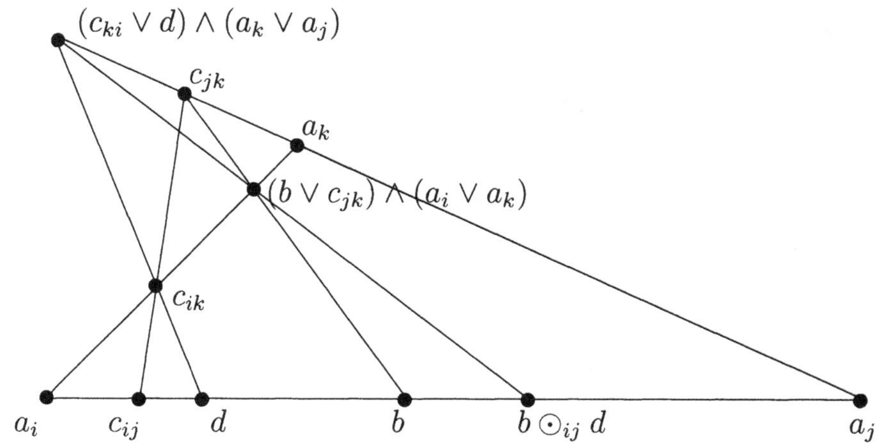

FIGURE 4.2. The geometric meaning of $\odot_{ij}$.

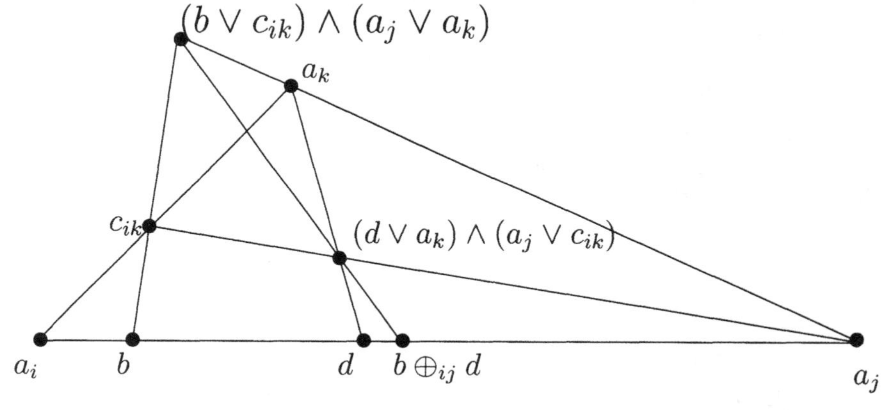

FIGURE 4.3. The geometric meaning of $\oplus_{ij}$.

is intersection and the join of two subspaces is the subspace generated by their union. The following results of Lipshitz make use of the definition of a modular n-frame.

LEMMA 4.37. [Lip74] *Let* $\mathbf{V}$ *be an infinite-dimensional vector space. Then,*

i) $\mathbf{L}(\mathbf{V})$ *contains a 4-frame,* C*, where* e *is the least element of* $\mathbf{L}(\mathbf{V})$ *and*
ii) *Any countable semigroup is a subsemigroup of the multiplicative semigroup of the ring associated with* C.

THEOREM 4.38. [Lip74] *The word problem for modular lattices is undecidable.*

4.4.4. Distributive residuated lattices. We modify the definition of a modular n-frame, in order to be applicable in the study of distributive residuated lattices.

DEFINITION 4.2. A *residuated-lattice n-frame* (or just *n-frame*) in a residuated lattice $\mathbf{L}$ is an $n \times n$ matrix, $C = [c_{ij}]$, $c_{ij} \in \mathbf{L}$, (set $a_i = c_{ii}$), such that:

i) $a_i a_j = a_j a_i$, for all $i, j \in \mathbb{N}_n$;
ii) $\prod A_1 \wedge \prod A_2 = \prod(A_1 \cap A_2)$, for all $A_1, A_2 \subseteq \{a_1, a_2, ..., a_n\}$, where $\prod \emptyset = 1$;
iii) $a_i^2 = a_i$, for all $i \in \mathbb{N}_n$;
iv) $c_{ij}c_{jk} \wedge a_i a_k = c_{ik}$, for all distinct triples $i, j, k \in \mathbb{N}_n$;
v) $c_{ij} = c_{ji}$, for all $i, j \in \mathbb{N}_n$;
vi) $c_{ij}a_j = a_i a_j$, for all $i, j \in \mathbb{N}_n$;
vii) $c_{ij} \wedge a_j = 1$, for all distinct $i, j \in \mathbb{N}_n$.

It is clear that if multiplication is replaced by join, the conditions in the definition reduce to the ones in Definition 4.1.

An element a of a residuated lattice $\mathbf{L}$ is called *modular* if $c(b \wedge a) = cb \wedge a$ and $(a \wedge b)c = a \wedge bc$ for all elements b, c of L, such that $c \leq a$. An n-frame of a residuated lattice is called *modular* if $\prod A$ is modular, for all $A \subseteq \{a_1, a_2, ..., a_n\}$.

If $[c_{ij}]$ is an n-frame in a residuated lattice $\mathbf{L}$, we define

- $L_{ij} = \{x \in \mathbf{L} \mid xa_j = a_i a_j \text{ and } x \wedge a_j = 1\}$, for all distinct $i, j \in \mathbb{N}_n$;
- $b \otimes_{ijk} d = bd \wedge a_i a_k$, for all $b \in L_{ij}, d \in L_{jk}$;
- $b \odot_{ij} d = (b \otimes_{ijk} c_{jk}) \otimes_{ikj} (c_{ki} \otimes_{kij} d)$, for all $b, d \in L_{ij}$ and for all distinct triples i, j, k.

The definition of $\odot_{ij}$ does not depend on the choice of k, as is shown in the lemma below.

LEMMA 4.39. [Gal02] *Let $c = [c_{ij}]$ be a modular 4-frame in a residuated lattice $\mathbf{L}$.*

i) *If $b \in L_{ij}$, then $b \leq a_i a_j$;*
ii) *If $b \in L_{ij}$ and $d \in L_{jk}$, then $b \otimes_{ijk} d \in L_{ik}$, for all distinct triples $i, j, k \in \mathbb{N}_n$;*
iii) *If $b \in L_{ij}$, $d \in L_{jk}$ and $f \in L_{kl}$, then $(b \otimes_{ijk} d) \otimes_{ikl} f = b \otimes_{ijl} (d \otimes_{jkl} f)$, for all distinct quadruples $i, j, k, l \in \mathbb{N}_n$;*
iv) *If $b, d \in L_{ij}$, then $(b \otimes_{ijk} c_{jk}) \otimes_{ikj} (c_{ki} \otimes_{kij} d) = (b \otimes_{ijl} c_{jl}) \otimes_{ilj} (c_{li} \otimes_{lij} d)$, for all distinct quadruples $i, j, k, l \in \mathbb{N}_4$.*

PROOF. i) $b = be \leq ba_j = a_i a_j$.

ii) We first show that $(b \otimes_{ijk} d)a_k = a_i a_k$.

$$\begin{aligned}
(b \otimes_{ijk} d)a_k &= (a_i a_k \wedge bd)a_k, & \\
&= bda_k \wedge a_i a_k, & a_i a_k \text{ is modular and } a_k \leq a_i a_k \\
&= ba_j a_k \wedge a_i a_k, & da_k = a_j a_k, \text{ since } d \in L_{jk} \\
&= a_i a_j a_k \wedge a_i a_k, & ba_j = a_i a_j, \text{ since } b \in L_{ij} \\
&= a_i a_k, & \text{(ii) of Def. 4.2}
\end{aligned}$$

To prove $b \otimes_{ijk} d \in L_{ij}$ we also need to show that $(b \otimes_{ijk} d) \wedge a_k = 1$.

$$\begin{aligned}
& (b \otimes_{ijk} d) \wedge a_k & \\
= \ & bd \wedge a_i a_k \wedge a_k & \\
= \ & bd \wedge a_k, & a_k \leq a_i a_k, \text{ since } 1 \leq a_i \\
\leq \ & bd \wedge a_j a_k, & a_k \leq a_j a_k, \text{ since } 1 \leq a_i \\
= \ & (b \wedge a_j a_k)d, & a_j a_k \text{ is modular and } d \leq a_j a_k, \text{ by (i)} \\
= \ & (b \wedge a_j a_k \wedge a_i a_j)d, & b \wedge a_i a_j = b, \text{ by (i)} \\
= \ & (b \wedge a_j)d, & a_j a_k \wedge a_i a_j = a_j, \text{ by (ii) of Def. 4.2} \\
= \ & d, & b \wedge a_j = 1, \text{ since } b \in L_{ij}
\end{aligned}$$

So, $(b \otimes_{ijk} d) \wedge a_k = b \otimes_{ijk} d \wedge a_k \wedge a_k \leq d \wedge a_k = 1$, since $d \in L_{jk}$. Moreover,

$$1 = 1 \cdot 1 \wedge 1 \cdot 1 \wedge 1 \leq bd \wedge a_i a_k \wedge a_k = (b \otimes_{ijk} d) \wedge a_k.$$

Thus, $(b \otimes_{ijk} d) \wedge a_k = 1$.

iii) Since $b \in L_{ij}$, $d \in L_{jk}$ and $f \in L_{kl}$, by (ii) we get, $b \otimes_{ijk} d \in L_{ik}$ and $d \otimes_{jkl} f \in L_{jl}$; thus, $(b \otimes_{ijk} d) \otimes_{ikl} f$, $b \otimes_{ijl} (d \otimes_{jkl} f) \in L_{il}$.

$$\begin{aligned}
& (b \otimes_{ijk} d) \otimes_{ikl} f & \\
= \ & (bd \wedge a_i a_k)f \wedge a_i a_l & \\
= \ & (bd \wedge a_i a_j a_k \wedge a_i a_k a_l)f \wedge a_i a_l, & \text{(ii) of Def. 4.2} \\
= \ & (bd \wedge a_i a_k a_l)f \wedge a_i a_l, & bd \leq a_i a_j a_j a_k = a_i a_j a_k, \\
& & \text{by (i), since } b \in L_{ij}, d \in L_{jk} \\
& & \text{and } a_j^2 = a_j \\
= \ & bdf \wedge a_i a_k a_l \wedge a_i a_l, & a_i a_k a_l \text{ is modular and} \\
& & f \leq a_k a_l \leq a_i a_k a_l, \\
& & \text{since } f \in L_{kl} \\
= \ & bdf \wedge a_i a_l, & \text{(ii) of Def. 4.2}
\end{aligned}$$

Similarly, $b \otimes_{ijl} (d \otimes_{jkl} f) = bdf \wedge a_i a_l$, so

$$(b \otimes_{ijk} d) \otimes_{ikl} f = b \otimes_{ijl} (d \otimes_{jkl} f).$$

iv) First note that condition (iv) of the definition of an n-frame can be written as $c_{rs} = c_{rt} \otimes_{rts} c_{ts}$, for all distinct triples $r, t, s \in \mathbb{N}_n$.

$$\begin{aligned}
(a \otimes_{ijk} c_{jk}) \otimes_{ikj} (c_{ki} \otimes_{kij} b) &= (a \otimes_{ijk} (c_{jl} \otimes_{jlk} c_{lk})) \otimes_{ikj} (c_{ki} \otimes_{kij} b) \\
&= ((a \otimes_{ijl} c_{jl}) \otimes_{ilk} c_{lk}) \otimes_{ikj} (c_{ki} \otimes_{kij} b) \\
&= (a \otimes_{ijl} c_{jl}) \otimes_{ilj} (c_{lk} \otimes_{lkj} (c_{ki} \otimes_{kij} b)) \\
&= (a \otimes_{ijl} c_{jl}) \otimes_{ilj} ((c_{lk} \otimes_{lki} c_{ki}) \otimes_{lij} b) \\
&= (a \otimes_{ijl} c_{jl}) \otimes_{ilj} (c_{li} \otimes_{lij} b)
\end{aligned}$$

Thus the definition of $\odot_{ij}$ is independent of k. ☐

LEMMA 4.40. [Gal02] *Let $c = [c_{ij}]$ be a modular 4-frame in a residuated lattice* $\mathbf{L}$.

i) *If $b, d \in L_{ij}$, then $b \odot_{ij} d \in L_{ij}$, for all distinct $i, j \in \mathbb{N}_n$.*
ii) *If $b, d, f \in L_{12}$, then $(b \odot_{12} d) \odot_{12} f = b \odot_{12} (d \odot_{12} f)$.*

PROOF. i) Since $b \in L_{ij}$, $c_{jk} \in L_{jk}$ and $c_{ki} \in L_{ki}$, $d \in L_{ij}$, we have $b \otimes_{ijk} c_{jk} \in L_{ik}$ and $c_{ki} \otimes_{kij} d \in L_{kj}$. So,

$$b \odot_{ij} d = (b \otimes_{ijk} c_{jk}) \otimes_{ikj} (c_{ki} \otimes_{kij} d) \in L_{ij}.$$

ii)
$$\begin{aligned}
& (b \odot_{12} d) \odot_{12} f \\
= & \{[(b \otimes_{123} c_{23}) \otimes_{132} (c_{31} \otimes_{312} d)] \otimes_{123} c_{23}\} \otimes_{132} (c_{31} \otimes_{312} f) \\
= & \{[(b \otimes_{124} c_{24}) \otimes_{142} (c_{41} \otimes_{412} d)] \otimes_{123} c_{23}\} \otimes_{132} (c_{31} \otimes_{312} f) \\
= & \{(b \otimes_{124} c_{24}) \otimes_{143} [(c_{41} \otimes_{412} d) \otimes_{423} c_{23}]\} \otimes_{132} (c_{31} \otimes_{312} f) \\
= & (b \otimes_{124} c_{24}) \otimes_{142} \{[(c_{41} \otimes_{412} d) \otimes_{423} c_{23}] \otimes_{432} (c_{31} \otimes_{312} f)\} \\
= & (b \otimes_{124} c_{24}) \otimes_{142} \{[c_{41} \otimes_{413} (d \otimes_{123} c_{23})] \otimes_{432} (c_{31} \otimes_{312} f)\} \\
= & (b \otimes_{124} c_{24}) \otimes_{142} \{c_{41} \otimes_{412} [(d \otimes_{123} c_{23}) \otimes_{132} (c_{31} \otimes_{312} f)]\} \\
= & (b \otimes_{123} c_{23}) \otimes_{132} \{c_{31} \otimes_{312} [(d \otimes_{123} c_{23}) \otimes_{132} (c_{31} \otimes_{312} f)]\} \\
= & b \odot_{12} (d \odot_{12} f)
\end{aligned}$$

A fact that establishes the associativity of $\odot_{12}$. ☐

COROLLARY 4.41. [Gal02] *Let $C = [c_{ij}]$ be a modular residuated-lattice 4-frame in a residuated lattice* $\mathbf{L}$. *Then, $\mathbf{S}_{12} = (L_{12}, \odot_{12})$ is a semigroup, called the semigroup associated with the 4-frame C.*

LEMMA 4.42. [Gal02] *Let $\mathbf{L}$ be a distributive residuated lattice, with a top element, T, and a bottom element, B. If $a, \tilde{a}, \in L$, $a^2 \leq a$, $a\tilde{a} \leq \tilde{a}$, $\tilde{a}a \leq \tilde{a}$, $a \wedge \tilde{a} = B$ and $a \vee \tilde{a} = T$, then, a is modular.*

PROOF. Let $b, c \in L, c \leq a$. Then, $(a \wedge b)c \leq ac \leq a^2 \leq a$ and $(a \wedge b)c \leq bc$; thus, $(a \wedge b)c \leq a \wedge bc$. On the other hand,

$$\begin{aligned}
a \wedge bc &= a \wedge (b \wedge T)c = a \wedge (b \wedge (a \vee \tilde{a}))c \\
&= a \wedge ((b \wedge a) \vee (b \wedge \tilde{a}))c = a \wedge ((b \wedge a)c \vee (b \wedge \tilde{a})c) \\
&\leq a \wedge ((b \wedge a)c \vee \tilde{a}a) \leq (a \wedge (b \wedge a)c) \vee (a \wedge \tilde{a}) \\
&\leq (b \wedge a)c \vee B = (b \wedge a)c
\end{aligned}$$

Thus, $a \wedge bc = (b \wedge a)c$. Similarly, we get the other condition $a \wedge cb = c(a \wedge b)$. ☐

Let $\mathbf{V}$ be a vector space and let A, B be elements of the power set $L_{\mathbf{V}} = \mathcal{P}(V)$ of $\mathbf{V}$. We define $A \wedge B = A \cap B$, $A \vee B = A \cup B$, $AB = \{a + b \mid a \in A, y \in B\}$, $A \backslash B = B/A = \{c \mid \{c\}A \subseteq B\}$ and $1 = \{0_{\mathbf{V}}\}$. It's easy to see that $\mathbf{L_V} = (L_{\mathbf{V}}, \wedge, \vee, \cdot, \backslash, /, 1)$ is a distributive residuated lattice. Moreover, $L(\mathbf{V})$ is a subset of $L_{\mathbf{V}}$, but $\mathbf{L}(\mathbf{V})$ is not a sublattice of the lattice reduct of $\mathbf{L_V}$. Nevertheless, a subset A of V is in $L(\mathbf{V})$ if and only if $1 \leq A$ and $AA = A$. Additionally, $\wedge_{\mathbf{L}(\mathbf{V})} = \wedge_{\mathbf{L_V}}$ and $\vee_{\mathbf{L}(\mathbf{V})} = {}_{\mathbf{L_V}}$.

If $\mathbf{S} = (S, \bullet)$, $S = (x_1, x_2, ..., x_n |\, r_1^\bullet(\overline{x}) = s_1^\bullet(\overline{x}), ..., r_k^\bullet(\overline{x}) = s_k^\bullet(\overline{x}))$, is a finitely presented semigroup and $\mathcal{V}$ is a variety of residuated lattices, let $\mathbf{L}(\mathbf{S}, \mathcal{V})$ be the residuated lattice in $\mathcal{V}$ with the presentation described below: (we define $\mathcal{A}(C) = \mathcal{P}(\{a_1, a_2, a_3, a_4\})$)

Generators:

$$x_1', x_2', ..., x_n',\ c_{ij}\ (i, j \in \mathbb{N}_4),\ \top, \bot \ \text{ and } \widetilde{\prod A}\ (A \in \mathcal{A}(C)).$$

Relations:

i) Equations (i)-(vii) of Definition 4.2 (for $n = 4$);
ii) $x_i' a_2 = a_1 a_2$ and $x_1' \wedge a_2 = 1$, for all $i \in \mathbb{N}_n$;
iii) $r_i^{\odot_{12}}(\overline{x}') = s_i^{\odot_{12}}(\overline{x}')$, for all $i \in \mathbb{N}_k$, where $t^{\odot_{12}}$ denotes the evaluation of t in the semigroup associated with the 4-frame $[c_{ij}]$;
iv) $\bot^2 = \bot$, $\top^2 = \top = \top/\bot = \bot/\bot = \bot\backslash\top = \bot\backslash\bot$, $\bot \leq 1 \leq \top$ and $\bot \leq x \leq \top$, $\bot x = x\bot = \bot$, for every generator x;
v) $x^2 \leq x$, $x\tilde{x} \leq \tilde{x}$, $\tilde{x}x \leq \tilde{x}$, $x \wedge \tilde{x} = \bot$ and $x \vee \tilde{x} = \top$, for all x of the form $\prod A$, $A \in \mathcal{A}(C)$.

Let $R(\overline{x})$ denote the conjunction $\bigwedge_{i \in \mathbb{N}_k} r_i(\overline{x}) = s_i(\overline{x})$ of the relations of $\mathbf{S}$ and $R'(\overline{x}', C, \overline{\mathcal{A}}(C), \bot, \top)$ the conjunction of the relations of $\mathbf{L}(\mathbf{S}, \mathcal{V})$; $\overline{x}$ abbreviates $(x_1, x_2, ..., x_n)$.

LEMMA 4.43. [Gal02] *For every semigroup $\mathbf{S}$, $\mathbf{L}(\mathbf{S}, \mathcal{V})$ has a bounded lattice reduct and $\bot$, $\top$ are the bottom and top elements.*

PROOF. We will prove that $\bot \leq w \leq \top$, for every word w in the generators. We first prove that "$\bot \leq w$ and $\bot w = w\bot = \bot$" for every word, w, using induction on the complexity of w. The statement is true for the generators and for 1, by (iv) in the relations of $\mathbf{L}(\mathbf{S}, \mathcal{V})$.
If the statement is true for words u, v, i.e. $\bot \leq u, v$ and $\bot u = u\bot = \bot v = v\bot = \bot$ then:

- $\bot(u \vee v) = \bot u \vee \bot v = \bot$ and $(u \vee v)\bot = \bot$. Also, $\bot \leq u \vee v$.
- $\bot \leq u \wedge v$ and $\bot = \bot\bot \leq \bot(u \wedge v) \leq \bot u \wedge \bot v = \bot$, while $(u \wedge v)\bot = \bot$ is proven in a similar way.
- $\bot \leq \bot\bot \leq uv$ and $\bot uv = \bot v = \bot$, while the other products equal $\bot$, also.
- Since $\bot v = \bot \leq u$, we have $\bot \leq u/v$; thus $\bot = \bot\bot \leq \bot(u/v)$ and $\bot \leq (u/v)\bot$.

Moreover, $\bot(u/v) \leq (\bot u)/v = \bot/v \leq \bot/\bot = \top$, so $u/v \leq \bot\backslash\top = \bot\backslash\bot = \bot/\bot$; hence $\bot(u/v) \leq \bot$ and $(u/v)\bot \leq \bot$. Thus, $\bot(u/v) = \bot$ and $(u/v)\bot = \bot$. For left division we work analogously. Consequently, $\bot$ is the bottom element of $\mathbf{L}$ and $\top = \bot/\bot$ is the top element of $\mathbf{L}$. □

The following lemma is a well known fact from the theory of Universal Algebra.

LEMMA 4.44. *Let* $A = (\overline{x} \mid R(\overline{x}))$, *be a finite presentation of an algebra* $\mathbf{A}$ *in a variety* $\mathcal{V}$ *where* $\overline{x} = (x_1, ..., x_n)$, $n \in \mathbb{N}$ *is the sequence of generators and* $R(\overline{x})$ *the conjunction of the relations. Also, let* r, s *be* n*-ary semigroup terms; then the following are equivalent:*

i) $\mathbf{A}$ *satisfies* $r^{\mathbf{A}}(\overline{x}) = s^{\mathbf{A}}(\overline{x})$.
ii) *For every algebra* $\mathbf{B}$ *in* $\mathcal{V}$, *if there exist elements* $y_1, ..., y_n \in B$, *such that* $R(\overline{y})$ *holds in* $\mathbf{B}$, *then* $\mathbf{B}$ *satisfies* $r^{\mathbf{B}}(\overline{y}) = s^{\mathbf{B}}(\overline{y})$.

PROOF. For the non-trivial direction, note that the natural epimorphism $F_{\mathcal{V}}(\overline{x}) \to Sg_{\mathbf{B}}(\overline{y})$ from the free algebra of $\mathcal{V}$ on $\overline{x}$ to the subalgebra of $\mathbf{B}$ generated by $\overline{y}$, given by $x_i \mapsto y_i$, factors through $\mathbf{F}_{\mathcal{V}}(\overline{x})/R(\overline{x}) \cong \mathbf{A}$. So, $f: A \to Sg_{\mathbf{B}}(\overline{y}) \subseteq B$, $x_i \mapsto y_i$ is a homomorphism. Since $r^{\mathbf{A}}(\overline{x}) = s^{\mathbf{A}}(\overline{x})$, we get

$$r^{\mathbf{B}}(\overline{y}) = r^{\mathbf{B}}(f(\overline{x})) = f(r^{\mathbf{A}}(\overline{x})) = f(s^{\mathbf{A}}(\overline{x})) = s^{\mathbf{B}}(f(\overline{x})) = s^{\mathbf{B}}(\overline{y})$$

in $\mathbf{B}$. □

LEMMA 4.45. [Gal02] *Assume that* $\mathbf{S}$ *is a semigroup,* r, s *are semigroup terms and* $\mathcal{V}$ *is a variety of distributive residuated lattices. If* $\mathbf{S}$ *satisfies* $r^{\bullet}(\overline{x}) = s^{\bullet}(\overline{x})$ *then* $\mathbf{L}(\mathbf{S}, \mathcal{V})$ *satisfies* $r^{\odot_{12}}(\overline{x}') = s^{\odot_{12}}(\overline{x}')$.

PROOF. $C = [c_{ij}]$ is a 4-frame in $\mathbf{L}(\mathbf{S}, \mathcal{V})$, by (i) of $R'(\overline{x}', C, \widetilde{\mathcal{A}}(C), \bot, \top)$ and $x_i' \in L_{12}$, by (ii). Moreover, by (v) and Lemma 4.42, $\prod A$ is modular, for all $A \in \mathcal{A}(C)$, hence C is modular. By Corollary 4.41, $(L_{12}, \odot_{12})$ is a semigroup and, by (iii), it satisfies $R(\overline{x}')$; thus, by Lemma 4.44, it also satisfies $r^{\odot_{12}}(\overline{x}') = s^{\odot_{12}}(\overline{x}')$. □

We can now prove the main theorem of the section.

THEOREM 4.46. [Gal02] *Let* $\mathcal{V}$ *be a variety of distributive residuated lattices, containing* $\mathbf{L_V}$, *for some infinite-dimensional vector space* $\mathbf{V}$. *Then, there is a finitely presented residuated lattice in* $\mathcal{V}$, *with undecidable word problem.*

PROOF. Let $\mathbf{S} = (S, \bullet)$, $S = (x_1, x_2, ..., x_n \mid r_1^{\bullet}(\overline{x}) = s_1^{\bullet}(\overline{x}), ..., r_k^{\bullet}(\overline{x}) = s_k^{\bullet}(\overline{x}))$, be a finitely presented semigroup with undecidable word problem (see [RS94]) and consider $\mathbf{L}(\mathbf{S}, \mathcal{V})$.

We will show that, for every pair r, s of semigroup words, $\mathbf{S}$ satisfies $r^{\bullet}(\overline{x}) = s^{\bullet}(\overline{x})$ if and only if $\mathbf{L}(\mathbf{S}, \mathcal{V})$ satisfies $r^{\odot_{12}}(\overline{x}') = s^{\odot_{12}}(\overline{x}')$. Since one direction follows from Lemma 4.45, suppose that $\mathbf{S}$ does not satisfy $r(\overline{x}) = s(\overline{x})$. By Lemma 4.37, $\mathbf{S}$ is embeddable, via f, say, into the multiplicative semigroup of the ring, $\mathbf{R}$, associated with a modular 4-frame $\hat{C}$ in the modular lattice $\mathbf{L}(\mathbf{V})$. So, $r^{\mathbf{R}}(f(\overline{x})) = s^{\mathbf{R}}(f(\overline{x}))$ is false in $\mathbf{R}$, where $f(\overline{x}) = (f(x_1), ..., f(x_n))$, thus also false in $\mathbf{L}(\mathbf{V})$, if viewed as a lattice equation. Since, as noted before, $\wedge_{\mathbf{L(V)}} = \wedge_{\mathbf{L_V}}$ and $\vee_{\mathbf{L(V)}} = \cdot_{\mathbf{L_V}}$, $\mathbf{L_V}$ fails $r^{\mathbf{R}}(f(\overline{x})) = s^{\mathbf{R}}(f(\overline{x}))$, where the latter is considered a residuated lattice equation ($r^{\mathbf{L_V}}(f(\overline{x})) = s^{\mathbf{L_V}}(f(\overline{x}))$). On the other hand, if we view the above mentioned modular 4-frame as a residuated lattice 4-frame and take $\emptyset$ as $\bot$, V as $\top$ and $V - x$ as $\overline{x}$, for all $\overline{x} \in \widetilde{\mathcal{A}}(\hat{C})$, it follows that $\mathbf{L_V}$ satisfies $R'(f(\overline{x}), \hat{C}, \widetilde{\mathcal{A}}(\hat{C}), \bot, \top)$. Indeed, (i) and (iv) of $R'(f(\overline{x}), \hat{C}, \widetilde{\mathcal{A}}(\hat{C}), \bot, \top)$ are obvious, while (ii) is true, since $f(x_i)$ is a member of the multiplicative semigroup of $\mathbf{R}$ and this semigroup plays the role of L_{12}. Condition (iii) holds because it holds in $\mathbf{S}$ for $\overline{x}$ and (v) is very easy to check. So, for $\overline{y} = f(\overline{x})$, $\mathbf{L_V}$ satisfies $R'(\overline{y}, C, \widetilde{\mathcal{A}}(C), \bot, \top)$, but not $r^{\mathbf{L_V}}(\overline{y}) = s^{\mathbf{L_V}}(\overline{y})$; hence, by Lemma 4.44, $\mathbf{L}(\mathbf{S}, \mathcal{V})$ fails $r^{\odot_{12}}(\overline{x}') = s^{\odot_{12}}(\overline{x}')$.

If the word problem for $\mathbf{L}(\mathbf{S}, \mathcal{V})$ were decidable then the one for $\mathbf{S}$ would be decidable, too. Thus, $\mathcal{V}$ has an undecidable word problem. □

Corollary 4.47. [Gal02] *If $\mathcal{V}$ is a variety such that* $\mathsf{HSP}(\mathbf{L_V}) \subseteq \mathcal{V} \subseteq \mathsf{DRL}$, *for some infinite-dimensional vector space* $\mathbf{V}$, *then $\mathcal{V}$ has an undecidable quasi-equational theory, namely, there is no algorithm that decides whether a quasi-equation in the language of residuated lattices is valid in $\mathcal{V}$ or not.*

Corollary 4.48. [Gal02] *The word problem and quasi-equational theory for (commutative) distributive residuated lattices is undecidable.*

We note that results in [Urq84] imply the consequence of Theorem 4.46 in the commutative case. Moreover, the result for DRL alone can be proved in a much more simple and direct way.

We have seen that the equational theory of RL is known to be decidable. It is an open problem, though, whether the same is true for DRL.

The algebraization of $\mathbf{RW}^+$ and the embedding results into involutive structures of [GR04], in combination with the above result, yield the following.

Corollary 4.49. [GR04] *The word problem for* $\mathsf{InDFL_e}$ *and* CDRL *is undecidable.*

Exercises

(1) Let us consider the sequent calculus $\mathbf{LJ}'$ whose sequents are of the form $\Gamma \Rightarrow \Delta$ with finite sequences Γ and Δ. The rules of $\mathbf{LJ}'$ are the same as those of $\mathbf{LK}$ except $(\Rightarrow \neg)$ and $(\Rightarrow \to)$, and only $(\Rightarrow \neg)$ and $(\Rightarrow \to)$ are the same as those of $\mathbf{LJ}$. Therefore, the lower sequent of either of $(\Rightarrow \neg)$ and $(\Rightarrow \to)$ must be a sequent which has a single formula in the right-hand side. Show that for each sequent of the form $\Sigma \Rightarrow \alpha$, $\Sigma \Rightarrow \alpha$ is provable in $\mathbf{LJ}'$ iff it is provable in $\mathbf{LJ}$. (The calculus $\mathbf{LJ}'$ is discussed by [Mae54] and [Ume55].)
(2) Show the cut elimination for $\mathbf{LJ}'$. (See [Hos74].)
(3) Show that a sequent is provable in $\mathbf{FL_{gc}}$ if and only if it s provable in $\mathbf{FL_c}$.
(4) Give a proof of cut elimination for $\mathbf{FL_{gc}}$ (see Lemma 4.4).

(5) Consider the calculus $\mathbf{FL_{ec}}'$ in Lemma 4.13. Show that a formula is provable in $\mathbf{FL_{ec}}'$ if and only if it is provable in $\mathbf{FL_{ec}}$.
(6) Let α be an arbitrary formula which does not contain the constant 0 (and therefore no negation either). Show that α is provable in $\mathbf{InFL_e}$ iff it is provable in $\mathbf{FL_e}$. Show also that this relation holds between $\mathbf{InFL_{ew}}$ and $\mathbf{FL_{ew}}$, and also between $\mathbf{InFL_{ec}}$ and $\mathbf{FL_{ec}}$. (This fact is mentioned in [Ono90].)
(7) Obviously, the relation in the above exercise does not hold between **LK** and **LJ**. (For example, the formula $((p \to q) \to p) \to p$, which does not contain 0, is provable in **LK**, but not in **LJ**.) Point out where your proof for the above exercise does not work for the present case.
(8) For each pair (n, k) of natural numbers, let $\mathbf{FL_e}^{n \rightsquigarrow k}$ be the sequent calculus obtained from $\mathbf{FL_e}$ by adding the following rule, the $(n \rightsquigarrow k)$-rule, where α^m denotes a sequence of m copies of a formula α:

$$\frac{\alpha^n, \Gamma \Rightarrow \delta}{\alpha^k, \Gamma \Rightarrow \delta}$$

Show the cut elimination for $\mathbf{FL_e}^{n \rightsquigarrow 1}$ for $n > 1$.
(9) Show that $\mathbf{FL_e}^{n \rightsquigarrow 1}$ is decidable for $n > 1$.
(10) For each $k > 1$, the *k-mingle* rule is a rule of the following form:

$$\frac{\Gamma_1, \Sigma \Rightarrow \delta \quad \dots \quad \Gamma_k, \Sigma \Rightarrow \delta}{\Gamma_1, \dots, \Gamma_k, \Sigma \Rightarrow \delta}$$

Show that a sequent is provable in $\mathbf{FL_e}^{1 \rightsquigarrow k}$ if and only if it is provable in the calculus $\mathbf{FL_e}$ with the k-mingle rule.
(11) Show the cut elimination for $\mathbf{FL_e}$ with the k-mingle rule. (Note that cut elimination does not hold in $\mathbf{FL_e}^{1 \rightsquigarrow k}$, as shown in [HOS94].)
(12) Verify that the two examples given in Section 4.4.3 are indeed modular n-frames.
(13) Show that the lattice all subspaces of a vector space $\mathbf{V}$ is a modular.
(14) *Open problem* : Is $\mathbf{FL_c}$ decidable? How about the implicational fragment of $\mathbf{FL_c}$?

Notes

(1) For Lambek calculus and cut elimination, and its motivation, see [Lam58]. Cut elimination for **FL**, $\mathbf{FL_e}$, $\mathbf{FL_w}$, $\mathbf{FL_{ew}}$ and related calculi was shown also by Tamura [Tam74], Dardžaniá [Dar77], Idziak [Idz84a] and also [Gri82] for $\mathbf{InFL_{ew}}$. For more results on extensions of **FL** by structural rules that enjoy cut elimination, see [GO].
(2) A survey of decision problems of substructural logics is given in [Ono98a] and [Ono98b]. The former contains some information of results on (un)decidability of substructural logics.
(3) In Section 4.2, we have explained a way of incorporating the contraction rule into rules for logical connectives and as a result, of eliminating

the *explicit* contraction rule. A similar idea can be applied also to intuitionistic propositional logic. But, the system still contains some circularities in the proof-search, caused by the rule ($\to\Rightarrow$). To remove these circularities, we need to divide the rule ($\to\Rightarrow$) into several rules. Then we get an efficient decision algorithm for intuitionistic propositional logic. This was done independently by Dyckhoff [Dyc92] and Hudelmaier [Hud93].

(4) Some decision algorithms for nonclassical logics based on cut elimination are effective to a certain extent, and therefore they are implemented as automated reasoning of logics. An example of such an attempt in an early stage is seen in [TMM88].

(5) Section 4.4.4 is heavily influenced by work on similar problems. Lipshitz [Lip74] established the undecidability of the word problem for modular lattices and Urquhart for DL-semigroups [Urq95] and models of relevance logic [Urq84]. Moreover, [AGN97] contains undecidability results about relation algebras, while Freese [Fre80] proved that the word problem for the free modular lattice on five generators is undecidable. The proofs of all the above make use of the notion of an *n-frame*, introduced by von Neumann in [vN60]. It is a geometric concept that was originally used in the definition of the von Staudt product of two points on a projective line. Taking advantage of the intrinsic connections between projective geometry and modular lattices, von Neumann defined this product in the latter. In other words, the notion of an n-frame can be used to define a semigroup structure in a modular lattice. Lipshitz used this fact to reduce the decidability of the word problem for modular lattices to the one for semigroups. Going one step further and using a modified version of an n-frame, Urquhart applied similar ideas to DL-semigroups.

(6) The definitions of $b \odot_{ij} d$ and $b \oplus_{ij} d$ given in Section 4.4.3 differ from [vN60] and [Lip74]. There multiplication and addition are not defined for elements of L_{ij}, but for L-numbers. An L-number α in a modular n-frame C is a set of lattice elements indexed by $\{(i,j) | \, i,j \in \mathbb{N}_n, \, i \neq j\}$, such that $(\alpha)_{kh} = [P(i,j,k,h)]((\alpha)_{ij})$, where $(\alpha)_{ij}$ symbolizes the (i,j)-coordinate of α and $P(i,j,k,h)$ is the composition of the two perspective isomorphisms with axes c_{jh} and c_{ik}. Lemma 6.1 of [vN60] guarantees that one can work with the fixed (i,j)-coordinates of L-numbers instead of them, since given $i,j \in \mathbb{N}_n$ the correspondence between α and $(\alpha)_{ij}$ is a bijection. Moreover, this bijection between L-numbers under the multiplication and addition defined in [vN60] and L_{ij} under $\odot_{ij}$ and $\oplus_{ij}$ is a ring isomorphism, as it can be deduced from Lemmas 6.2, 6.3, Theorem 6.1 and the appendix to Chapter 6, Part II of [vN60]. Freese, in [Fre80], is the first one to use $\odot_{ij}$ and $\oplus_{ij}$, instead of multiplication and addition of L-numbers, and essentially the definition of an n-frame presented here.

In the context of the first example of Section 4.4.3, L_{ij} is the set of all points x on the line $a_i \vee a_j$, $(x \vee a_i = a_i \vee a_j)$, different from a_i, $(x \wedge a_j = e)$, $b \otimes_{ijk} d$ is by definition the intersection of the lines $b \vee d$ and $a_i \vee a_j$, for $b \in L_{ij}, d \in L_{jk}$, while $b \odot_{ij} d$ and $b \ominus_{ij} d$, $b, d \in L_{ij}$ are the (von Staudt) product, see Figure 4.2, and sum of b and d on the line $a_i \vee a_j$, where a_i plays the role of zero, c_{ij} is the unit and a_j is infinity. Some projective geometry is required to verify this assertion.

CHAPTER 5

Logical and algebraic properties

This chapter is devoted to discussing some basic logical properties of substructural logics and their algebraic characterizations. These properties include the disjunction property, Halldén completeness, Maksimova's variable separation property and several versions of the interpolation property. We show that each of them can be characterized by an equivalent algebraic property, typically a form of an embedding property. Many of these will turn out to be generalizations of well-known algebraic properties such as the amalgamation property or the joint-embedding property.

The first two sections develop a syntactic approach, based on cut elimination, while the algebraic characterizations are discussed in later sections. Such two-sided approach usually brings about a deeper understanding.

5.1. Syntactic approach to logical properties

5.1.1. Disjunction property. A logic **L** has the *disjunction property* (DP) when for any formulas α and β, if $\alpha \vee \beta$ is provable in **L** then either α or β is provable in it. Classical logic does not have the disjunction property, as $p \vee \neg p$ is provable but neither of p and $\neg p$ are provable, for a propositional variable p. On the other hand, the DP for intuitionistic logic follows as an easy consequence of cut elimination of **LJ**.

THEOREM 5.1. *Intuitionistic logic has the disjunction property.*

PROOF. Suppose that the sequent $\Rightarrow \alpha \vee \beta$ is provable in **LJ**; we will show that $\Rightarrow \alpha$ or $\Rightarrow \beta$ is provable in it. The last inference in any cut-free proof P of $\Rightarrow \alpha \vee \beta$ in P will be either ($\Rightarrow$ w) (namely (o)) or ($\Rightarrow \vee$). If it is ($\Rightarrow$ w) then the upper sequent should be $\Rightarrow$. But this is impossible, since the subformula property implies that every formula in the initial sequents of P must appear in the endsequent as a subformula. Thus, it must be ($\Rightarrow \vee$) and the upper sequent must be either $\Rightarrow \alpha$ or $\Rightarrow \beta$. □

Using a similar argument, we have the following theorem. Note that the disjunction property of $\mathbf{FL_c}$ follows from the cut elimination of $\mathbf{FL_{gc}}$, i.e., **FL** with the global contraction rule (cf. Section 4.1).

THEOREM 5.2. *Each of* **FL**, $\mathbf{FL_e}$, $\mathbf{FL_w}$, $\mathbf{FL_c}$, $\mathbf{FL_{ew}}$ *and* $\mathbf{FL_{ec}}$ *has the disjunction property.*

The reason why the above argument does not work for **LK** is due to the fact that ($\Rightarrow$ c) can be also the last inference in **LK**. In fact, ($\Rightarrow$ c) is the last inference of any cut-free proof of $p \vee \neg p$. On the other hand, this means that the above argument remains valid even in involutive substructural logics, as long as they do not have right contraction ($\Rightarrow$ c).

THEOREM 5.3. *Both* $\mathbf{InFL_e}$ *and* $\mathbf{InFL_{ew}}$ *have the disjunction property.*

An immediate consequence of the above theorem is that the sequent $\Rightarrow p \vee \neg p$ is not provable in $\mathbf{InFL_{ew}}$. Note that $\mathbf{InFL_{ew}}$ is equivalent to the sequent calculus obtained from $\mathbf{FL_{ew}}$ by adding sequents of the form $\neg\neg\alpha \Rightarrow \alpha$ as additional initial sequents. On the other hand, the system obtained from $\mathbf{FL_{ew}}$ by adding $\Rightarrow \alpha \vee \neg\alpha$ as initial sequents can be shown to be classical logic.

In the same way as in **LK**, we can show that $\mathbf{InFL_{ec}}$ does not have the disjunction property. On the other hand, the presence of the right contraction rule ($\Rightarrow$ c) does not always imply the failure of the disjunction property. For instance, consider the sequent calculus **LJ′**, introduced in Exercises of Chapter 4, which has the same rules as **LK**, except that both ($\Rightarrow \neg$) and ($\Rightarrow \rightarrow$) can be applied only when there are at most one formula in the right-hand side of the arrow. Note that **LJ′** has ($\Rightarrow$ c), but it can be shown that **LJ′** is a cut-free sequent calculus for intuitionistic logic, and therefore it has the disjunction property.

5.1.2. Craig interpolation property. Craig proved in [Cra57b] the following result on classical logic, which is now called *Craig's interpolation theorem* for classical logic. In the following, $\mathrm{var}(\gamma)$ denotes the set of all propositional variables in a formula γ. When Γ is a sequence or a set of formulas $\gamma_1, \ldots, \gamma_m$, we define $\mathrm{var}(\Gamma) = \mathrm{var}(\gamma_1) \cup \ldots \cup \mathrm{var}(\gamma_m)$.[1]

THEOREM 5.4. *If a formula* $\alpha \rightarrow \beta$ *is provable in classical logic then there exists a formula* γ *such that both* $\alpha \rightarrow \gamma$ *and* $\gamma \rightarrow \beta$ *are provable, and that* $\mathrm{var}(\gamma) \subseteq \mathrm{var}(\alpha) \cap \mathrm{var}(\beta)$.

Any formula γ satisfying the conditions in the above theorem is called an *interpolant* of $\alpha \rightarrow \beta$. We note that whether interpolation theorem holds or not depends highly on the language $\mathcal{L}$, i.e., on both connectives and constants. For instance, the above interpolation theorem does not always imply Craig's interpolation theorem for fragments of classical logic.

For classical predicate logic, Craig's interpolation theorem is known to be equivalent to *Beth's definability theorem* and also to *Robinson's consistency theorem.* (See e.g., [CK90] for the details, in which the equivalence is shown by using model-theoretical methods.)

---

[1]Here we assume that our language $\mathcal{L}$ contains some constants. Thus $\mathrm{var}(\gamma) = \emptyset$ means that γ is a formula consisting only of constants. The form of interpolation property when $\mathcal{L}$ contains no constants will be discussed later in this section.

Hereafter, we assume that our language contains at least 0 and 1, if we do not mention otherwise. We say that a substructural logic **L** has the *Craig interpolation property* (CIP) if the following holds for **L**:

> for all formulas α and β, if $\alpha\backslash\beta$ is in **L** then there exists a formula γ such that both $\alpha\backslash\gamma$ and $\gamma\backslash\beta$ are in **L**, and that $\mathrm{var}(\gamma) \subseteq \mathrm{var}(\alpha) \cap \mathrm{var}(\beta)$.

Note that for any logic **L** under consideration, $\alpha\backslash\beta$ is in **L** iff β/α is in **L**, for all formulas α and β. Thus, we may as well take the right residual instead of the left residual in the definition above.

Using the fact that the Craig interpolation property of a superintuitionistic logic **L** is equivalent to an algebraic property, called *amalgamation property* (AP), of the subvariety of Heyting algebras corresponding to **L**, L.L. Maksimova showed in [Mak77] the following striking result. Recall that there exist uncountably many superintuitionistic logics, as mentioned in Theorem 1.59. For the Craig interpolation property of modal logics, see [Mak79b].

THEOREM 5.5. *Only seven consistent superintuitionistic logics have the Craig interpolation property.*

Relations between the CIP and the AP in substructural logics will be discussed in Section 5.5. Extending the idea by Krzystek and Zachorowski [KZ77], Komori showed the following.

THEOREM 5.6. [Kom81] *Classical logic is the single consistent extension of Łukasiewicz logic which has the Craig interpolation property.*

Also, A. Urquhart showed the failure of the Craig interpolation property for a large number of relevant logics (see [Urq93, Urq98, Urq99b]). These results say that the cases where the CIP holds are rather rare. Let **T** be the logic of ticket entailment, **TW** the logic obtained form **T** by removing contraction, and $\mathbf{TW}^+$ the positive fragment of **TW** (see [Dun86] for the details).

THEOREM 5.7. *Craig interpolation property fails for all positive relevant logics between* $\mathbf{TW}^+$ *and* **R**.

5.1.3. Maehara's method.

Various methods have been used to prove the CIP. Craig used a semantical method in [Cra57a], [Cra57b] to show the theorem. In 1960, S. Maehara [Mae60] introduced a syntactic proof of Craig's interpolation theorem for classical logic, which is obtained as a consequence of the cut elimination theorem of **LK**. Unlike semantical methods, the proof by Maehara's method gives a concrete form of an interpolant of a formula $\alpha \to \beta$, once a cut-free proof of $\alpha \to \beta$ is given (see also e.g., [Tak87]). The following exposition is based on the paper [Ono98b].

Craig interpolation property of intuitionistic logic. In the following, we explain how Maehara's method works for intuitionistic logic and other basic substructural logics. First, we consider the CIP for intuitionistic logic. For any given finite sequence Γ of formulas, a pair $\langle\Gamma_1;\Gamma_2\rangle$ of (possibly empty) sequences Γ_1 and Γ_2 of formulas is a *partition* of Γ, if the multiset union of Γ_1 and Γ_2 is equal to Γ if we regard Γ, Γ_1 and Γ_2 as multisets of formulas. We show the following, which is a generalization of the original CIP.

THEOREM 5.8. *Suppose that a sequent $\Gamma \Rightarrow \delta$ is provable in* **LJ**. *Then, for any partition $\langle\Gamma_1;\Gamma_2\rangle$ of Γ, there exists a formula φ such that both $\Gamma_1 \Rightarrow \varphi$ and $\varphi, \Gamma_2 \Rightarrow \delta$ are provable in* **LJ** *and that $var(\varphi) \subseteq var(\Gamma_1) \cap var(\Gamma_2, \delta)$.*

PROOF. For a given partition $\langle\Gamma_1;\Gamma_2\rangle$ of Γ, we call a formula φ satisfying the conditions of the above theorem, an *interpolant* of $\Gamma \Rightarrow \delta$ (with respect to the partition $\langle\Gamma_1;\Gamma_2\rangle$). By the cut elimination of **LJ**, there exists a cut-free proof P of $\Gamma \Rightarrow \delta$. We prove the theorem by induction on the length n of P. If $n = 1$ then $\Gamma \Rightarrow \delta$ must be an initial sequent. Let it be $\alpha \Rightarrow \alpha$. We need to consider two partitions, i.e., $\langle\alpha;\emptyset\rangle$ and $\langle\emptyset;\alpha\rangle$. For the former case, we can take α itself for an interpolant. For the latter case, we can take 1 for an interpolant, as both sequents $\Rightarrow 1$ and $1, \alpha \Rightarrow \alpha$ are provable. Similarly, we can show the existence of an interpolant also when $\Gamma \Rightarrow \delta$ is one of other initial sequents.

Next suppose that $n > 1$. Let (r) be the last rule of inference in P. By the hypothesis of induction, we can assume that there exists an interpolant of (each of) the upper sequent(s) of (r) with respect to any partition. We need to show that the lower sequent $\Gamma \Rightarrow \delta$ has also an interpolant with respect to any partition. It is necessary to check this when (r) is an arbitrary rule of inference of **LJ**. In the following, we show the claim when (r) is either $(\vee \Rightarrow)$ or $(\Rightarrow \vee)$.

1) Suppose that Γ is $\alpha \vee \beta, \Sigma$ and that (r) is an application of $(\vee \Rightarrow)$ as follows;

$$\frac{\alpha, \Sigma \Rightarrow \delta \quad \beta, \Sigma \Rightarrow \delta}{\alpha \vee \beta, \Sigma \Rightarrow \delta}$$

1.1) Consider a partition $\langle\alpha \vee \beta, \Sigma_1;\Sigma_2\rangle$ of $\alpha \vee \beta, \Sigma$. Taking a partition $\langle\alpha, \Sigma_1;\Sigma_2\rangle$ of α, Σ and a partition $\langle\beta, \Sigma_1;\Sigma_2\rangle$ of β, Σ and using the hypothesis of induction, we can assume that there exist formulas θ and ψ such that

- both $\alpha, \Sigma_1 \Rightarrow \theta$ and $\theta, \Sigma_2 \Rightarrow \delta$ are provable,
- both $\beta, \Sigma_1 \Rightarrow \psi$ and $\psi, \Sigma_2 \Rightarrow \delta$ are provable,
- $\mathrm{var}(\theta) \subseteq \mathrm{var}(\alpha, \Sigma_1) \cap \mathrm{var}(\Sigma_2, \delta)$,
- $\mathrm{var}(\psi) \subseteq \mathrm{var}(\beta, \Sigma_1) \cap \mathrm{var}(\Sigma_2, \delta)$

Then, from the first two we can derive that both $\alpha \vee \beta, \Sigma_1 \Rightarrow \theta \vee \psi$ and $\theta \vee \psi, \Sigma_2 \Rightarrow \delta$ are provable. Moreover by the third and fourth conditions,

$\text{var}(\theta \vee \psi) \subseteq \text{var}(\alpha \vee \beta, \Sigma_1) \cap \text{var}(\Sigma_2, \delta)$. It means that the formula $\theta \vee \psi$ is a required interpolant.

1.2) Next, consider a partition $\langle \Sigma_1; \alpha \vee \beta, \Sigma_2 \rangle$ of $\alpha \vee \beta, \Sigma$. This time, we take a partition $\langle \Sigma_1; \alpha, \Sigma_2 \rangle$ of α, Σ and a partition $\langle \Sigma_1; \beta, \Sigma_2 \rangle$ of β, Σ. Then, by the hypothesis of induction, we can assume that there exist formulas θ and ψ such that

- both $\Sigma_1 \Rightarrow \theta$ and $\theta, \alpha, \Sigma_2 \Rightarrow \delta$ are provable,
- both $\Sigma_1 \Rightarrow \psi$ and $\psi, \beta, \Sigma_2 \Rightarrow \delta$ are provable,
- $\text{var}(\theta) \subseteq \text{var}(\Sigma_1) \cap \text{var}(\alpha, \Sigma_2, \delta)$,
- $\text{var}(\psi) \subseteq \text{var}(\Sigma_1) \cap \text{var}(\beta, \Sigma_2, \delta)$

It follows that both $\Sigma_1 \Rightarrow \theta \wedge \psi$ and $\theta \wedge \psi, \alpha \vee \beta, \Sigma_2 \Rightarrow \delta$ are provable, and that $\text{var}(\theta \wedge \psi) \subseteq \text{var}(\Sigma_1) \cap \text{var}(\alpha \vee \beta, \Sigma_2, \delta)$. Therefore, the formula $\theta \wedge \psi$ is a required interpolant in this case.

2) Suppose next that (r) is an application of $(\Rightarrow \vee 1)$ as follows:

$$\frac{\Gamma \Rightarrow \alpha}{\Gamma \Rightarrow \alpha \vee \beta}$$

Let $\langle \Gamma_1; \Gamma_2 \rangle$ be any partition of Γ. By the hypothesis of induction, there exists a formula ψ such that

- both $\Gamma_1 \Rightarrow \psi$ and $\psi, \Gamma_2 \Rightarrow \alpha$ are provable,
- $\text{var}(\psi) \subseteq \text{var}(\Gamma_1) \cap \text{var}(\Gamma_2, \alpha)$.

Then, clearly both $\Gamma_1 \Rightarrow \psi$ and $\psi, \Gamma_2 \Rightarrow \alpha \vee \beta$ are provable, and also it holds that $\text{var}(\psi) \subseteq \text{var}(\Gamma_1) \cap \text{var}(\Gamma_2, \alpha \vee \beta)$. Hence, the formula ψ is an interpolant. The case $(\Rightarrow \vee 2)$ can be treated in the same way. □

If in Theorem 5.8 we take $\alpha \Rightarrow \beta$ for $\Gamma \Rightarrow \delta$ and consider the partition $\langle \alpha; \emptyset \rangle$ of α, then we obtain Craig's interpolation theorem for intuitionistic logic.

THEOREM 5.9. *If a sequent $\alpha \Rightarrow \beta$ is provable in* **LJ** *then there exists a formula φ such that both $\alpha \Rightarrow \varphi$ and $\varphi \Rightarrow \beta$ are provable in* **LJ**, *and moreover* $\text{var}(\varphi) \subseteq \text{var}(\alpha) \cap \text{var}(\beta)$.

Maehara's method was originally introduced, to show the CIP of classical logic. In this case, we need to modify the above definition of partitions as follows. Suppose that any sequent $\Gamma \Rightarrow \Delta$ of **LK** is given. We say that $\langle (\Gamma_1 : \Delta_1); (\Gamma_2 : \Delta_2) \rangle$ is a *partition* of $\Gamma \Rightarrow \Delta$, if the multiset union of Γ_1 and Γ_2 (Δ_1 and Δ_2) is equal to Γ (Δ, respectively) as multisets of formulas. Then, instead of Theorem 5.8, we can show the following.

THEOREM 5.10.

(1) *Suppose that $\Gamma \Rightarrow \Delta$ is provable in* **LK** *and that $\langle (\Gamma_1 : \Delta_1); (\Gamma_2 : \Delta_2) \rangle$ is an arbitrary partition of $\Gamma \Rightarrow \Delta$. Then, there exists a formula φ*

such that both $\Gamma_1 \Rightarrow \Delta_1, \varphi$ *and* $\varphi, \Gamma_2 \Rightarrow \Delta_2$ *are provable in* **LK**, *and moreover that* $var(\varphi) \subseteq var(\Gamma_1, \Delta_1) \cap var(\Gamma_2, \Delta_2)$.

(2) *In particular, Craig interpolation property holds for classical logic.*

In order to show the Craig interpolation property in Theorems 5.8 and 5.10, we needed to consider arbitrary partitions. This is because the idea of Maehara's method is to construct an interpolant of the lower sequent by using interpolants of upper sequents. Consider for example the following application of $(\Rightarrow \to)$:

$$\frac{p \wedge r, p \to q \Rightarrow q \vee s}{p \wedge r \Rightarrow (p \to q) \to (q \vee s)}$$

In this case, one of interpolants of the upper sequent is q, and any other interpolant must be a formula containing only q as a variable. On the other hand, one of interpolants of the lower one is p, and any other interpolant is a formula containing only p as variables. Thus, there seems to be no way of finding an interpolant of the lower sequent from these of the upper one. This difficulty is resolved by considering partitions. In fact, an interpolant of the upper sequent with respect to the partition $\langle p \wedge r; p \to q \rangle$ is p. Thus, we need to consider partitions as long as the sequent calculus under consideration has a rule whose main formula is composed of formulas that come from the opposite side of $\Rightarrow$ in the upper sequent, like rules for implication and negation.

Interpolation property for languages without constants. In the above proof we assumed that our language contains propositional constants. We consider here the CIP when our language does not contains any constant. Then, there exists no formula φ such that $\mathrm{var}(\varphi) \subseteq \mathrm{var}(\Gamma_1) \cap \mathrm{var}(\Gamma_2, \delta)$, if $\mathrm{var}(\Gamma_1) \cap \mathrm{var}(\Gamma_2, \delta)$ is empty. In this case, the statement in Theorem 5.8 becomes meaningless. But we can recover the CIP by modifying the statement slightly as shown below.

To distinguish **LJ** in the language with propositional constants from that without constants, we call temporarily the latter $\mathbf{LJ}^-$. By using cut elimination for **LJ**, we can see that **LJ** is a conservative extension of $\mathbf{LJ}^-$. Let α and β be formulas without constants such that $\alpha \Rightarrow \beta$ is provable in $\mathbf{LJ}^-$. Obviously, this is provable also in **LJ**. So there exists an interpolant φ, by Theorem 5.9. When φ contains propositional constants 0 and 1, we can simplify the formula φ by using the following logical equivalence. Here, $\gamma \equiv \delta$ means that γ is logically equivalent to δ in **LJ**.

- $1 \to \gamma \equiv \gamma$, $0 \to \gamma \equiv \gamma \to 1 \equiv 1$, $\gamma \to 0 \equiv \neg\gamma$,
- $1 \wedge \gamma \equiv \gamma \wedge 1 \equiv \gamma$, $0 \wedge \gamma \equiv \gamma \wedge 0 \equiv 0$,
- $1 \vee \gamma \equiv \gamma \vee 1 \equiv 1$, $0 \vee \gamma \equiv \gamma \vee 0 \equiv \gamma$,
- $\neg 1 \equiv 0$, $\neg 0 \equiv 1$.

By repeating this simplification, φ will be transformed into a formula ψ which is either a formula with no propositional constants, or a propositional constant 0 or 1. In the former case, both $\alpha \Rightarrow \psi$ and $\psi \Rightarrow \beta$ are provable also in $\mathbf{LJ}^-$ since $\mathbf{LJ}$ is a conservative extension of $\mathbf{LJ}^-$. Thus, we get an interpolant ψ with no propositional constants.

Let us consider the latter case. First, suppose that the set $\mathrm{var}(\alpha) \cap \mathrm{var}(\beta)$ is nonempty in Theorem 5.9. Since ψ is either 0 or 1, it is logically equivalent to either $p \wedge \neg p$ or $\neg(p \wedge \neg p)$, respectively, where p is any propositional variable in $\mathrm{var}(\alpha) \cap \mathrm{var}(\beta)$. By replacing 0 (and 1) by $p \wedge \neg p$ (and $\neg(p \wedge \neg p)$, respectively), we get an interpolant with no propositional constants.

Next suppose that the set $\mathrm{var}(\alpha) \cap \mathrm{var}(\beta)$ is empty. When an interpolant ψ is 1, it means that both $\alpha \Rightarrow 1$ and $1 \Rightarrow \beta$ are provable. But, the first sequent is always provable, and the second is provable if and only if β is provable. When ψ is 0, both $\alpha \Rightarrow 0$ and $0 \Rightarrow \beta$ must be provable. The second sequent is always provable, and the first is provable if and only if $\alpha \Rightarrow$ is provable, or equivalently $\neg\alpha$ is provable. Thus, we have the following.

THEOREM 5.11. *Suppose that a formula $\alpha \to \beta$ is provable in* $\mathbf{LJ}$ *with the language $\mathcal{L}$ having no propositional constants. If the set $var(\alpha) \cap var(\beta)$ is nonempty, there exists a formula φ (without constants) of $\mathcal{L}$ such that both $\alpha \to \varphi$ and $\varphi \to \beta$ are provable in* $\mathbf{LJ}$ *and $var(\varphi) \subseteq var(\alpha) \cap var(\beta)$. If the set $var(\alpha) \cap var(\beta)$ is empty, then either $\neg\alpha$ or β is provable in* $\mathbf{LJ}$. *This holds also for* $\mathbf{LK}$.

We mentioned already that Craig's interpolation theorem for classical and intuitionistic logic does not necessarily imply the theorem for fragments of them. For example, our proof of Theorem 5.8 shows that a conjunctive formula $\theta \wedge \psi$ is needed in the case 1.2) where the rule of inference is for the disjunction.

We noticed before that interpolants of a given formula $\alpha \to \beta$ are not always determined uniquely (up to logical equivalence). For instance, consider the following formula which is provable in $\mathbf{LJ}$:

$$(p \wedge (p \to (r \wedge s))) \to (q \to (r \vee s)).$$

We can see that each of $r \wedge s$, r, s and $r \vee s$ is an interpolant for the above formula. For classical and intuitionistic logic, it can be shown that there exists both a *least* φ_1 and a *greatest* φ_2 among interpolants of a given $\alpha \to \beta$. That is, both $\varphi_1 \to \psi$ and $\psi \to \varphi_2$ are provable for any interpolant ψ of $\alpha \to \beta$. For intuitionistic logic, this was proved by A.M. Pitts in [Pit92].

Craig interpolation property of basic substructural logics. Next, we discuss the Craig interpolation property of basic substructural logics. First, suppose that our language contains propositional constants. Maehara's method essentially works well for them. But when a given substructural logic, like $\mathbf{FL}$ and $\mathbf{FL_w}$, lacks the exchange rule we need to modify the definition of partition in the following way. Let $\Gamma \Rightarrow \delta$ be an arbitrary sequent (of $\mathbf{FL}$

or $\mathbf{FL_w}$). Then, a triple $\langle\Gamma_1;\Gamma_2;\Gamma_3\rangle$ is a partition of Γ, if the sequence $\Gamma_1,\Gamma_2,\Gamma_3$ is exactly equal to Γ (without changing the order of formulas). Then, by an inductive argument similar to that from the proof of Theorem 5.8, we can show the next lemma, which holds also with **FL** replaced by $\mathbf{FL_w}$.

LEMMA 5.12. *Suppose that a sequent $\Gamma \Rightarrow \delta$ is provable in* **FL** *and that $\langle\Gamma_1;\Gamma_2;\Gamma_3\rangle$ is any partition of Γ. Then, there exists a formula φ such that both $\Gamma_2 \Rightarrow \varphi$ and $\Gamma_1,\varphi,\Gamma_3 \Rightarrow \delta$ are provable in* **FL**, *and moreover $var(\varphi) \subseteq var(\Gamma_2) \cap var(\Gamma_1,\Gamma_3,\delta)$.*

Using this lemma, we obtain the following result. (See [OK85], [Ono90]. For the Craig interpolation property of fragments of $\mathbf{FL_{ew}}$, see [Wro84a] and [Idz84a].)

THEOREM 5.13. *The substructural logics* **FL**, $\mathbf{FL_w}$, $\mathbf{FL_e}$, $\mathbf{FL_{ew}}$, $\mathbf{FL_{ec}}$, $\mathbf{InFL_e}$, $\mathbf{InFL_{ew}}$ *and* $\mathbf{InFL_{ec}}$ *have the Craig interpolation property (in the language with propositional constants).*

When a given calculus has both left- and right-weakening rules, $1 = \top$ and $0 = \bot$ hold, and moreover the following logical equivalences hold. For any formula γ,

$$1\cdot\gamma \equiv \gamma\cdot 1 \equiv \gamma,\ 0\cdot\gamma \equiv \gamma\cdot 0 \equiv 0.$$

Then we can use the same argument as that in the proof of Theorem 5.11. That is, when $\mathrm{var}(\alpha) \cap \mathrm{var}(\beta)$ is nonempty and the interpolant ψ is equal to either 0 or 1, it is enough to replace it by either $p\cdot{\sim}p$ or ${\sim}(p\cdot{\sim}p)$, where p is any propositional variable in $\mathrm{var}(\alpha)\cap\mathrm{var}(\beta)$. Then suppose that $\mathrm{var}(\alpha) \cap \mathrm{var}(\beta)$ is empty. If ψ is 1, both $\alpha \Rightarrow 1$ and $1 \Rightarrow \beta$ are provable. The first sequent is always provable because of the left-weakening rule, and the second is provable if and only if β is provable. On the other hand, if ψ is 0, then $\alpha \Rightarrow 0$ and $0 \Rightarrow \beta$ are provable. The second sequent is always provable because of the right-weakening rule, and the first is provable if and only if $\alpha \Rightarrow$ is provable, or equivalently ${\sim}\alpha$ is provable.

COROLLARY 5.14. *Suppose that a formula $\alpha\backslash\beta$ is provable in $\mathbf{FL_w}$ with the language having no propositional constants. If the set $var(\alpha) \cap var(\beta)$ is nonempty, there exists a formula φ (without constants) of $\mathcal{L}$ such that both $\alpha\backslash\varphi$ and $\varphi\backslash\beta$ are provable in $\mathbf{FL_w}$ and $var(\varphi) \subseteq var(\alpha)\cap var(\beta)$. If the set $var(\alpha) \cap var(\beta)$ is empty, then either ${\sim}\alpha$ or β is provable in $\mathbf{FL_w}$. This holds also for both calculi $\mathbf{FL_{ew}}$ and $\mathbf{InFL_{ew}}$ with the language having no propositional constants.*

On the other hand, we cannot eliminate propositional constants in general when a logic lacks either of weakening rules, as both 1 and 0 can take an arbitrary value. Nevertheless, this is not an essential problem as we will see below.

5.1.4. Variable sharing property of logics without the weakening rules. We will call a formula *constant-free* if it contains no occurrences of propositional constants. The notion extends naturally to sequents and proofs. We say that a logic **L** has the *variable sharing property* (VSP) when for all constant-free formulas α and β, if $\alpha\backslash\beta$ is provable in **L** then the formulas α and β share some propositional variables, i.e., $\mathrm{var}(\alpha)\cap\mathrm{var}(\beta)$ is nonempty. This is equivalent to saying that whenever $\mathrm{var}(\alpha) \cap \mathrm{var}(\beta)$ is empty, and α, β are constant-free, then $\alpha\backslash\beta$ is not provable in **L**. It is clear that the VSP does not hold for a logic with weakening rules. For instance, $p\backslash(q\backslash q)$ is provable in $\mathbf{FL_w}$ but obviously the VSP fails. The VSP is also known as the *relevance principle*, especially in the context of relevant logics.

It is clear by the definition that if a logic stronger than **L** has the VSP, then **L** has it as well. In [BJ60], N. Belnap showed using algebraic methods that the relevant logic **R** has the VSP. W. Dziobiak [Dzi83] showed that there are uncountably many logics with the VSP between the relevant logics **R** and **RM**. Since **R** is an extension of $\mathbf{InFL_{ec}}$, the following holds as a corollary.

COROLLARY 5.15. *The substructural logic* $\mathbf{InFL_{ec}}$ *has the variable sharing property, and a fortiori* $\mathbf{FL}$, $\mathbf{FL_e}$, $\mathbf{FL_c}$, $\mathbf{FL_{ec}}$ *and* $\mathbf{InFL_e}$ *have the variable sharing property.*

We will give here a proof of this fact for $\mathbf{FL_{ec}}$, using a syntactic method similar to Maehara's. The proof for $\mathbf{InFL_{ec}}$ can be obtained by a simple modification. The proof of Theorem 5.16 given below was originally obtained by H. Naruse in his master's thesis in 1996. Also T. Seki [Sek01] gave a syntactic proof which yields the CIP and the VSP at the same time.

THEOREM 5.16. *Suppose that* $\Gamma \Rightarrow \alpha$ *is a constant-free sequent such that* Γ *is nonempty. If* $\Gamma \Rightarrow \alpha$ *is provable in* $\mathbf{FL_{ec}}$ *then* $var(\Gamma)\cap var(\alpha)$ *is nonempty.*

PROOF. Suppose that P is a cut-free proof of a constant-free sequent $\Gamma \Rightarrow \alpha$ with nonempty Γ. By subformula property, P is a constant-free proof. A sequent $\Sigma \Rightarrow \delta$ in P is *good* if it contains at least two formulas, i.e., either both Σ and δ are nonempty, or δ is empty but Σ contains at least two formulas. By our assumption, $\Gamma \Rightarrow \alpha$ is good. A branch in P is called a *good branch* if it begins with an initial sequent and contains only good sequents. Note that since P is constant-free, initial sequents are of the form $\varphi \Rightarrow \varphi$ and thus are always good. We show that for any good sequent $\Sigma \Rightarrow \delta$ in P there exists a good branch in P which ends with $\Sigma \Rightarrow \delta$. This is verified by checking that for every rule of inference of $\mathbf{FL_{ec}}$ if the lower sequent is good then at least one of the upper sequent(s) is good. Observe that for rules $(\to\ \Rightarrow)$ and $(\Rightarrow\ \cdot)$ one of the upper sequents must be good while the other may not be. (On the other hand, the weakening rules can have a good lower sequent, but a not good upper one.) For any occurrence of good sequent s, define the *depth* of s to be the maximum length of good branches which end with this occurrence of s.

For any good sequent $\Sigma \Rightarrow \delta$ in P, we say that $\langle\Sigma_1; \Sigma_2, \delta\rangle$ is a *partition* of $\Sigma \Rightarrow \delta$ if (1) the multiset union of Σ_1 and Σ_2 is equal to Σ, and (2) both Σ_1 and Σ_2, δ are nonempty. Since $\Sigma \Rightarrow \delta$ is good, there exists at least one partition of $\Sigma \Rightarrow \delta$. A partition $\langle\Sigma_1; \Sigma_2, \delta\rangle$ *shares variables* if $\mathrm{var}(\Sigma_1) \cap \mathrm{var}(\Sigma_2, \delta)$ is nonempty. We show the following by using induction on the depth of a given good sequent.

For any good sequent $\Sigma \Rightarrow \delta$ in P, every partition of $\Sigma \Rightarrow \delta$ shares variables.

In particular, taking $\Gamma \Rightarrow \alpha$ for $\Sigma \Rightarrow \delta$, we have that the partition $\langle\Gamma; \alpha\rangle$ shares variables. This completes our proof. □

5.2. Maksimova's variable separation property

Using semantical methods, Maksimova [Mak76] showed that some relevant logics, including **R** and **E**, have the following property.

Suppose that $\alpha_1 \to \alpha_2$ and $\beta_1 \to \beta_2$ have no propositional variables in common. If a formula $\alpha_1 \wedge \beta_1 \to \alpha_2 \vee \beta_2$ is provable, then either $\alpha_1 \to \alpha_2$ or $\beta_1 \to \beta_2$ is provable.

We call the above property for a given logic **L**, *Maksimova's variable separation property* (MVP)[2] (or *Maksimova's principle of variable separation*) for **L**. (In [CZ97], **L** is said to be *Maksimova complete* in this case.) Maksimova noted in [Mak79a] that the MVP holds for every extension of Gödel logic.

In this section, we will consider a proof-theoretical approach to the MVP for substructural logics.

THEOREM 5.17. *Maksimova's variable separation property holds for* **FL**, $\mathbf{FL_e}$, $\mathbf{FL_{ec}}$, $\mathbf{FL_w}$, $\mathbf{FL_{ew}}$, $\mathbf{InFL_e}$, $\mathbf{InFL_{ec}}$ *and* $\mathbf{InFL_{ew}}$.[3]

Detailed proofs of the theorem are given in the paper [NBSO98]. The method is applied to some relevant logics, and the MVP for positive fragments of **R**, **RW** and **TW** is shown also in the paper. As it is pointed out in [NBSO98] there seem to be some relations between the MVP and the CIP at least when a logic has the weakening rule. For example, Maksimova showed that if a superintuitionistic logic **L** has the CIP then it has the MVP.

To see this, let us consider the MVP of classical logic. Suppose that formulas $\alpha_1 \to \alpha_2$ and $\beta_1 \to \beta_2$ have no propositional variables in common, and that the sequent $\alpha_1 \wedge \beta_1 \Rightarrow \alpha_2 \vee \beta_2$ is provable in **LK**. So $\alpha_1, \beta_1 \Rightarrow \alpha_2, \beta_2$ is provable, and hence $\alpha_1, \neg\alpha_2 \Rightarrow \neg\beta_1, \beta_2$ is provable. By Theorem 5.11,

[2]We use the acronym MVP to avoid confusion between variable *sharing* property (VSP) and variable *separation* property (MVP).

[3]To be precise, what is shown below is the MVP restricted to constant-free formulas.

either $\alpha_1, \neg\alpha_2 \Rightarrow$ or $\Rightarrow \neg\beta_1, \beta_2$ is provable. Thus, either $\alpha_1 \Rightarrow \alpha_2$ or $\beta_1 \Rightarrow \beta_2$ is provable in **LK**.

In the following, we give a brief outline of the MVP (restricted to constant-free formulas) for $\mathbf{FL_{ec}}$. For the sake of brevity, we assume that our language does not contain any propositional constants. In the following, for a formula δ, $S(\delta)$ denotes the set of all subformulas of δ and for a sequence $\gamma_1, \ldots, \gamma_m$ of formulas, $S(\gamma_1, \ldots, \gamma_m)$ denotes $S(\gamma_1) \cup \ldots \cup S(\gamma_m)$. Also, for a sequent $\Gamma \Rightarrow \Delta$, $S(\Gamma \Rightarrow \Delta)$ denotes $S(\Gamma) \cup S(\Delta)$.

LEMMA 5.18. *Let $\alpha_1, \alpha_2, \beta_1, \beta_2$ be constant-free formulas. Suppose $\alpha_1 \to \alpha_2$ and $\beta_1 \to \beta_2$ have no propositional variables in common. If $\Pi \Rightarrow \delta$ is a constant-free sequent satisfying the following three conditions:*

- $S(\Pi \Rightarrow \delta) \subseteq S(\alpha_1 \wedge \beta_1) \cup S(\alpha_2 \vee \beta_2)$,
- $S(\Pi \Rightarrow \delta) \cap (S(\alpha_1) \cup S(\alpha_2)) \neq \emptyset$,
- $S(\Pi \Rightarrow \delta) \cap (S(\beta_1) \cup S(\beta_2)) \neq \emptyset$,

then $\Pi \Rightarrow \delta$ is not provable in $\mathbf{FL_{ec}}$.

PROOF. Let us suppose that $\Pi \Rightarrow \delta$ is provable. Then there exists a cut-free proof P of $\Pi \Rightarrow \delta$. By checking every application of the rules of $\mathbf{FL_{ec}}$ in P, we can show that if the lower sequent satisfies the above three conditions then at least one of the upper sequents satisfies these three conditions. Thus, at least one of the initial sequents in P must also satisfy them. Note that by our assumption P is a constant-free proof. Thus, each initial sequent is of the form $\gamma \Rightarrow \gamma$. Then, the second and third conditions together with the condition $\mathrm{var}(\alpha_1 \to \alpha_2) \cap \mathrm{var}(\beta_1 \to \beta_2) = \emptyset$ give a contradiction. □

COROLLARY 5.19. *Let $\alpha_1, \alpha_2, \beta_1, \beta_2$ be constant-free formulas. Suppose that $\alpha_1 \to \alpha_2$ and $\beta_1 \to \beta_2$ have no propositional variables in common. Moreover, suppose that P is a cut-free proof of a given constant-free sequent $\Pi \Rightarrow \delta$ in $\mathbf{FL_{ec}}$ such that $S(\Pi \Rightarrow \delta) \subseteq S(\alpha_1 \wedge \beta_1) \cup S(\alpha_2 \vee \beta_2)$ and $S(\Pi \Rightarrow \delta) \cap (S(\alpha_1) \cup S(\alpha_2)) \neq \emptyset$. Then every sequent $\Theta \Rightarrow \psi$ appearing in P satisfies the following condition*

(∗) $S(\Theta \Rightarrow \psi) \subseteq S(\alpha_1 \wedge \beta_1) \cup S(\alpha_2 \vee \beta_2)$ *and*
$S(\Theta \Rightarrow \psi) \cap (S(\alpha_1) \cup S(\alpha_2)) \neq \emptyset$.

Hence, there are no applications of the following rules of inference in P:

$$\frac{\Gamma, \beta_1, \Sigma \Rightarrow \varphi}{\Gamma, \alpha_1 \wedge \beta_1, \Sigma \Rightarrow \varphi}\ (\wedge 2 \Rightarrow) \qquad \frac{\Gamma \Rightarrow \alpha_1 \quad \Gamma \Rightarrow \beta_1}{\Gamma \Rightarrow \alpha_1 \wedge \beta_1}\ (\Rightarrow \wedge)$$

$$\frac{\Gamma, \alpha_2, \Sigma \Rightarrow \varphi \quad \Gamma, \beta_2, \Sigma \Rightarrow \varphi}{\Gamma, \alpha_2 \vee \beta_2, \Sigma \Rightarrow \varphi}\ (\vee \Rightarrow) \qquad \frac{\Gamma \Rightarrow \beta_2}{\Gamma \Rightarrow \alpha_2 \vee \beta_2}\ (\Rightarrow \vee 2)$$

PROOF. We can prove that for each application of a rule (r) of $\mathbf{FL_{ec}}$ in P if the lower sequent of (r) satisfies the condition (∗) then (both of) the upper sequent(s) must satisfy the condition (∗). Since $\Pi \Rightarrow \delta$ satisfies (∗), every sequent in P must satisfy $(*)$. Next suppose that one of the four rules above is applied in P. Then, in each case, at least one of the upper sequent(s) s contains either β_1 or β_2. At the same time, the sequent s must satisfy also $(*)$. Thus, s satisfies all of the three conditions in Lemma 5.18. But this is a contradiction since s is provable. □

THEOREM 5.20. *Maksimova's variable separation property holds for* $\mathbf{FL_{ec}}$. *More precisely, let* $\alpha_1, \alpha_2, \beta_1, \beta_2$ *be constant-free formulas. Suppose that formulas* $\alpha_1 \to \alpha_2$ *and* $\beta_1 \to \beta_2$ *have no propositional variables in common. Then the following hold.*

(1) *If the sequent* $\alpha_1 \wedge \beta_1 \Rightarrow \alpha_2 \vee \beta_2$ *is provable, then either* $\alpha_1 \Rightarrow \alpha_2$ *or* $\beta_1 \Rightarrow \beta_2$ *is provable.*
(2) *If the sequent* $\alpha_1 \wedge \beta_1 \Rightarrow \alpha_2$ *is provable, then* $\alpha_1 \Rightarrow \alpha_2$ *is provable.*
(3) *If the sequent* $\alpha_1 \wedge \beta_1 \Rightarrow \beta_2$ *is provable, then* $\beta_1 \Rightarrow \beta_2$ *is provable.*

PROOF. Suppose that the sequent $\alpha_1 \wedge \beta_1 \Rightarrow \alpha_2 \vee \beta_2$ is provable in $\mathbf{FL_{ec}}$. Clearly, it is not an initial sequent. We can see that the lowest part of any cut-free proof P of this sequent must be of the following form such that (r) is a rule of inference other than the exchange and contraction rules. (Of course, such (r) must exist.)

$$\dfrac{\dfrac{\vdots}{\alpha_1 \wedge \beta_1, \ldots, \alpha_1 \wedge \beta_1 \Rightarrow \alpha_2 \vee \beta_2}\,(r)}{\alpha_1 \wedge \beta_1 \Rightarrow \alpha_2 \vee \beta_2}\ \text{some } (\mathrm{c} \Rightarrow), (\mathrm{e} \Rightarrow)$$

So, (r) must be a logical rule and therefore it should be one of the following rules: $(\wedge 1 \Rightarrow), (\wedge 2 \Rightarrow), (\Rightarrow \vee 1)$, and $(\Rightarrow \vee 2)$. Let us suppose first that (r) is $(\wedge 1 \Rightarrow)$. Then:

$$\dfrac{\alpha_1 \wedge \beta_1, \ldots, \alpha_1, \ldots, \alpha_1 \wedge \beta_1 \Rightarrow \alpha_2 \vee \beta_2}{\alpha_1 \wedge \beta_1, \ldots, \alpha_1 \wedge \beta_1, \ldots, \alpha_1 \wedge \beta_1 \Rightarrow \alpha_2 \vee \beta_2}\ (\wedge 1 \Rightarrow)$$

Here, the antecedent of the upper sequent of (r) contains only one α_1 and others are $\alpha_1 \wedge \beta_1$. Then by Corollary 5.19, the proof of the upper sequent and hence P cannot contain any application of rules of inference mentioned in Corollary 5.19. This means that whenever any (occurrence of) one of formulas $\alpha_1 \wedge \beta_1$ and $\alpha_2 \vee \beta_2$ is introduced in the proof P it must be introduced only by rules of the following form:

$$\dfrac{\Gamma, \alpha_1, \Sigma \Rightarrow \varphi}{\Gamma, \alpha_1 \wedge \beta_1, \Sigma \Rightarrow \varphi}\ (\wedge 1 \Rightarrow) \qquad \dfrac{\Gamma \Rightarrow \alpha_2}{\Gamma \Rightarrow \alpha_2 \vee \beta_2}\ (\Rightarrow \vee 1)$$

Now, we replace all occurrences of $\alpha_1 \wedge \beta_1$ by α_1 and of $\alpha_2 \vee \beta_2$ by α_2 in P and remove redundancies which are caused by this replacement. (Note that the formulas $\alpha_1 \wedge \beta_1$ and $\alpha_2 \vee \beta_2$ may be introduced in several places in P.) This gives us a proof of $\alpha_1 \Rightarrow \alpha_2$ in $\mathbf{FL_{ec}}$. Other cases where (r) is any one of other rules mentioned in the above can be dealt with in a similar way. In each case, we get a proof of either $\alpha_1 \Rightarrow \alpha_2$ or $\beta_1 \Rightarrow \beta_2$. □

5.3. Algebraic characterizations

In the rest of the present chapter, we discuss algebraic characterizations of several logical properties, including the disjunction property, Halldén completeness (HC), Maksimova's variable separation property (MVP) and its deductive form, Craig interpolation property (CIP) and deductive interpolation property (DIP), which is the deductive form of the CIP. Furthermore,we show how structural rules affect these algebraic characterizations.

These topics have been studied already for modal and superintuitionistic logics, e.g., in Wroński [Wro76] and Maksimova [Mak77], [Mak95], [Mak99].

5.3.1. Disjunction property. In [Mak86], Maksimova gave an algebraic characterization of the disjunction property for superintuitionistic logics, by using *well-connected* Heyting algebras. Here a Heyting algebra $\mathbf{A}$ is well-connected iff for all $x, y \in A$, $x \vee y = 1$ implies either $x = 1$ or $y = 1$. The result can be easily extended to all substructural logics. We say that an FL-algebra $\mathbf{A}$ is *well-connected* iff for all $x, y \in A$, $x \vee y \geq 1$ implies either $x \geq 1$ or $y \geq 1$.

THEOREM 5.21. *Suppose that* $\mathbf{L}$ *is a substructural logic over* $\mathbf{FL}$ *and that* $\mathcal{K}$ *is a class of FL-algebras such that* $\mathbf{L} = \mathbf{L}(\mathcal{K})$. *Then, the following are equivalent:*

(1) $\mathbf{L}$ *has the disjunction property,*
(2) *for all* $\mathbf{A}, \mathbf{B} \in \mathcal{K}$ *there exist a well-connected FL-algebra* $\mathbf{C} \in \mathsf{V}(\mathbf{L})$ *and a surjective homomorphism from* $\mathbf{C}$ *to the direct product* $\mathbf{A} \times \mathbf{B}$ *of* $\mathbf{A}$ *and* $\mathbf{B}$.

PROOF. Suppose first that $\mathbf{L}$ has the DP. Take arbitrary algebras $\mathbf{A}, \mathbf{B} \in \mathcal{K}$. Clearly, the product $\mathbf{A} \times \mathbf{B}$ is in $\mathsf{V}(\mathbf{L})$. Let $\mathbf{C}$ be the Lindenbaum-Tarski algebra of $\mathbf{L}$ (over a suitable set of variables). Then by the universal mapping property, there exists a surjective homomorphism from $\mathbf{C}$ to $\mathbf{A}\times\mathbf{B}$. The well-connectedness of $\mathbf{C}$ follows from the DP of $\mathbf{L}$.

Conversely, suppose that the second condition holds but that $\mathbf{L}$ does not have the DP. Then, for some φ, ψ, while $\varphi \vee \psi$ is provable in $\mathbf{L}$, neither of φ and ψ are provable. By our assumption on $\mathcal{K}$, there are $\mathbf{A}, \mathbf{B} \in \mathcal{K}$ and assignments f and g, respectively on each of them, such that both $\mathbf{A}, f \models 1 \not\leq \varphi$ and $\mathbf{B}, g \models 1 \not\leq \psi$ hold. Now, by our assumption, there exist a well-connected FL-algebra $\mathbf{C} \in \mathsf{V}(\mathbf{L})$ and a surjective homomorphism h from $\mathbf{C}$ to $\mathbf{A}\times\mathbf{B}$. Then, both $\pi_1 \circ h$ and $\pi_2 \circ h$ are also surjective homomorphisms from $\mathbf{C}$ to $\mathbf{A}$ and from $\mathbf{C}$ to $\mathbf{B}$, respectively, where π_i is the projection to i-th component for $i = 1, 2$. Since $\varphi \vee \psi$ is provable in $\mathbf{L}$, $\mathbf{C} \models 1 \leq \varphi \vee \psi$. By the well-connectedness, either $\mathbf{C} \models 1 \leq \varphi$ or $\mathbf{C} \models 1 \leq \psi$ holds. Since equations are preserved by surjective homomorphisms, it follows from this that either $\mathbf{A} \models 1 \leq \varphi$ or $\mathbf{B} \models 1 \leq \psi$. In either case, this leads us to a contradiction. □

Theorem 5.21 is useful for showing the DP of logics for which the syntactic method discussed in the previous section does not work. Here we give some examples to show how our algebraic method works, following [Sou07]. First, we consider the following axiom scheme e_n^m for $m \geq 0$ and $n \geq 0$.

(e_n^m) $p^m \backslash p^n$

Note that e_n^m is equivalent to the contraction rule when $m = 1$ and $n = 2$, and is equivalent to the left-weakening rule when $m = 1$ and $n = 0$. Let $\mathbf{FL} + \{e_n^m\}$ and $\mathbf{FL_e} + \{e_n^m\}$ be the extensions of $\mathbf{FL}$ and $\mathbf{FL_e}$, respectively, by adding the axiom scheme e_n^m. When either (1) $n \geq 1$ and $m = 1$ or (2) $n = 1$ and $m \geq 1$, there exists a cut-free sequent calculus for $\mathbf{FL} + \{e_n^m\}$ (see [HOS94]), but we do not know if this holds for other cases.

THEOREM 5.22. *Both* $\mathbf{FL} + \{e_n^m\}$ *and* $\mathbf{FL_e} + \{e_n^m\}$ *have the disjunction property for every* m, n.

PROOF. By Theorem 5.21, it suffices to produce an appropriate FL-algebra $\mathbf{C}$ for given FL-algebras $\mathbf{A}$ and $\mathbf{B}$ in both of which e_n^m is valid. Suppose that $\mathbf{A}$ and $\mathbf{B}$ are given as follows.

- $\mathbf{A} = \langle A, \wedge^A, \vee^A, \cdot^A, \backslash^A, /^A, 1_A, 0_A, \rangle$,
- $\mathbf{B} = \langle B, \wedge^B, \vee^B, \cdot^B, \backslash^B, /^B, 1_B, 0_B \rangle$.

We will define $\mathbf{C} = (C, \wedge, \vee, \cdot, \backslash, /, 1, 0)$ step by step below.

- $C = (A \times B \times \{l\}) \cup \{\langle a, b, u \rangle |\ 1_A \leq a \in A, 1_B \leq b \in B\}$, with $1 = \langle 1_A, 1_B, u\rangle$ and $0 = \langle 0_A, 0_B, l\rangle$ (read l and u as *lower* and *upper*, respectively, assuming that the order relation $l{<^*}u$ holds; also note that we omitted superscripts ^{A} and ^{B}, as they are clear from context),
- a binary relation $\leq$ on C is defined by:
  $\langle a, b, i\rangle \leq \langle a', b', j\rangle \Leftrightarrow a \leq a', b \leq b', i{\leq^*}j.$
- a binary function $\cdot$ on C is then defined by:
  $\langle a, b, i\rangle \cdot \langle a', b', j\rangle = \langle a \cdot a', b \cdot b', i{\cdot^*}j\rangle$
  where $i{\cdot^*}j = \min\{i, j\}$ for $i, j \in \{l, u\}$.

It is easy to show that $\leq$ is a partial order on C, and moreover that $\langle C, \leq\rangle$ forms a lattice with lattice operations $\wedge$ and $\vee$. Also, $\langle C, \cdot, 1\rangle$ forms a monoid, and in fact $\langle C, \wedge, \vee, \cdot, 1\rangle$ is a lattice ordered monoid. Now, define $\backslash$ and $/$ on C as follows. First, we define $\backslash^*$ and $/^*$ on $\{l, u\}$, by letting $i\backslash^*j = j/^*i = u$ when $i{\leq^*}j$, and otherwise $i\backslash^*j = j/^*i = l$. Then,

- if either $1_A \not\leq a\backslash a'$ or $1_B \not\leq b\backslash b'$, and if $i\backslash' j = u$, then $\langle a, b, i\rangle\backslash\langle a', b', j\rangle = \langle a\backslash a', b\backslash b', l\rangle$,
- otherwise, $\langle a, b, i\rangle\backslash\langle a', b', j\rangle = \langle a\backslash a', b\backslash b', i\backslash^*j\rangle$.

The right-division / is defined just in the same way as \, but replacing all left-divisions by right-divisions. Then, it can be shown that **C** is an FL-algebra, and also it is easily seen that it is an $\mathrm{FL_e}$-algebra when both **A** and **B** are so. Also, inspecting the definition of operations on **C** one can see that the first and second components always preserve operations on **A** and **B**, respectively. Therefore, the mapping h defined by $h(\langle a, b, i\rangle) = \langle a, b\rangle$ for $a \in A, b \in B, j \in \{l, u\}$ is a surjective homomorphism from **C** to $\mathbf{A} \times \mathbf{B}$.

So, it remains to show that e_n^m is valid in **C** whenever it is valid in both **A** and **B**. For all $a \in A$ and $b \in B$, $1_A \le a^m\backslash a^n$, $1_B \le b^m\backslash b^n$. Thus, for all $\langle a, b, i\rangle \in C$,

$$\begin{aligned} &\langle a, b, i\rangle^m\backslash\langle a, b, i\rangle^n \;=\; \langle a^m, b^m, i^m\rangle\backslash\langle a^n, b^n, i^n\rangle \\ &= \langle a^m\backslash a^n, b^m\backslash b^n, i^m\backslash^* i^n\rangle \;\ge\; \langle 1_A, 1_B, u\rangle \;= 1. \end{aligned}$$

Hence e_n^m is valid in **C**. □

Since e_2^1 corresponds to the contraction rule, the DP of $\mathbf{FL_{ec}}$ follows. Also, it is easily seen that if we assume moreover that the lattice reducts of both **A** and **B** are distributive, then so is **C**. Now, let (dis) be the axiom scheme defined by:

$$(dis) \quad ((p \wedge (q \vee r))\backslash((p \wedge q) \vee (p \wedge r))).$$

we have the following.

COROLLARY 5.23. *For all m, n, $\mathbf{FL}+\{e_n^m, dis\}$ and $\mathbf{FL_e}+\{e_n^m, dis\}$ have the disjunction property. In particular $\mathbf{FL}+\{dis\}$ has the disjunction property.*

As the existence of the zero element 0 of **C** does not play any essential role in the proof of Theorem 5.22, we can derive that the positive fragment of each of these logics also has the disjunction property. Since the positive relevant logic $\mathbf{R}^+$ is equal to the positive fragment of $\mathbf{FL_e} + \{e_2^1, dis\}$, we get an alternative proof of the disjunction property of $\mathbf{R}^+$, which was first proved by R. K. Meyer in [Mey76].

The algebra **C** in the proof of Theorem 5.22 is not involutive, even if both **A** and **B** are so. Thus, some modification of the construction of **C** becomes necessary to show the DP of **InFL**. This can be done as follows.

Suppose that **A** and **B** are given involutive FL-algebras. Define **D** by $\langle D, \wedge, \vee, \cdot, \backslash, /, 1, 0\rangle$ by:

- $D = (A \times B \times \{m\}) \cup \{\langle a, b, u\rangle \mid 1_A \le a \in A, 1_B \le b \in B\} \cup \{\langle a, b, l\rangle \mid a \le 0_A, b \le 0_B, a \in A, b \in B\}$, with $1 = \langle 1_A, 1_B, u\rangle$ and $0 = \langle 0_A, 0_B, l\rangle$ (read m as *middle*, assuming this time that $l <^* m <^* u$),
- a binary relation $\le$ on C is defined by:
  $\langle a, b, i\rangle \le \langle a', b', j\rangle \Leftrightarrow a \le a', b \le b', i \le^* j$.
- a binary function $\cdot$ on C is then defined as follows (as before $i \cdot^* j = \min\{i, j\}$ for $i, j \in \{l, m, u\}$):

(1) Suppose that $a \cdot a' \leq 0_A, b \cdot b' \leq 0_B$, and neither of i and j are equal to u. Define $\langle a, b, i\rangle \cdot \langle a', b', j\rangle = \langle a \cdot a', b \cdot b', l\rangle$.
(2) Suppose that either $a \cdot a' \not\leq 0_A$ or $b \cdot b' \not\leq 0_B$, and $i \cdot^* j = l$. Define $\langle a, b, i\rangle \cdot \langle a', b', j\rangle = \langle a \cdot a', b \cdot b', m\rangle$.
(3) Suppose otherwise. Define $\langle a, b, i\rangle \cdot \langle a', b', j\rangle = \langle a \cdot a', b \cdot b', i \cdot^* j\rangle$.

- define $\backslash$ and $/$ on C as follows. First, we define $\backslash^*$ and $/^*$ on $\{l, u\}$, by letting $i\backslash^* j = j/^* i = u$ when $i \leq^* j$, $i\backslash^* j = j/^* i = m$ when $i = u$ and $j = m$, and otherwise $i\backslash^* j = j/^* i = l$. Then:
  (1) Suppose that $i\backslash^* j = u$ and either $1_A \not\leq a\backslash a'$ or $1_B \not\leq b\backslash b'$. Define $\langle a, b, i\rangle\backslash\langle a', b', j\rangle = \langle a\backslash a', b\backslash b', m\rangle$.
  (2) Suppose that $i = m$ and $\langle a', b', j\rangle \leq \langle 0_A, 0_B, l\rangle$. Define $\langle a, b, i\rangle\backslash\langle a', b', j\rangle = \langle a\backslash a', b\backslash b', m\rangle$.
  (3) Suppose otherwise. Define $\langle a, b, i\rangle\backslash\langle a', b', j\rangle = \langle a\backslash a', b\backslash b', i\backslash^* j\rangle$.

  Right division $/$ is defined similarly.

Then, we can see that **D** is an FL-algebra, and moreover it is an FL_e-algebra when both **A** and **B** are. Also, as before, the mapping h defined by $h(\langle a, b, i\rangle) = \langle a, b\rangle$ for $a \in A, b \in B, j \in \{l, m, u\}$ is a surjective homomorphism. It remains to show that **D** is involutive. This can be derived immediately from the fact that the following equations for $\sim$ and the corresponding equations for $-$ hold.

$$\sim\langle a, b, l\rangle = \langle\sim a, \sim b, u\rangle,\ \sim\langle a, b, m\rangle = \langle\sim a, \sim b, m\rangle,$$
$$\sim\langle a, b, u\rangle = \langle\sim a, \sim b, l\rangle.$$

Also in this case, if the lattice reducts of **A** and **B** are distributive, so is **D**. Thus, we have the following. Note that the constant-free fragment of $\mathbf{InFL_e} + \{dis\}$ (expanded by negation) is equivalent to the contraction-less relevant logic **RW**, whose disjunction property is shown in [Sla84] by using *metavaluations*.

THEOREM 5.24. *All of* **InFL**, $\mathbf{InFL_e}$, $\mathbf{InFL} + \{dis\}$ *and* $\mathbf{InFL_e} + \{dis\}$ *have the disjunction property.*

5.3.2. Halldén Completeness. A substructural logic **L** is *Halldén complete*, if for all formulas φ and ψ with no variables in common, if $\varphi \vee \psi$ is in **L** then φ or ψ is in **L**. The notion was introduced by S. Halldén in [Hal51]. Obviously the disjunction property implies Halldén completeness (HC). But it is easy to see that while classical logic is Halldén complete, it does not have the disjunction property. Galanter [Gal88] showed that in fact there exist uncountably many Halldén complete superintuitionistic logics without the DP. The HC can be regarded as a special case of Maksimova's variable separation property.

Theorem 5.28 is a generalization of results on superintuitionistic logics by both Lemmon [Lem66] and Wroński [Wro76]. Recall that a logic **L** over $\mathbf{FL_{ew}}$ is *meet irreducible* (in the lattice of all substructural logics over **FL**) if it is not an intersection of two incomparable logics, or equivalently, if it is not the finite intersection of strictly bigger logics (see Subsection 1.1.3). First, we show two technical lemmas. The first one is on an axiomatization of the meet of logics over $\mathbf{FL_{ew}}$. This is generalized in Section 9.7, where an axiomatization of the meet of logics over **FL** is discussed.

LEMMA 5.25. *Suppose that* $\mathbf{L}$ *is a logic over* $\mathbf{FL_{ew}}$ *and that* $\mathbf{L}_i$ *is the logic obtained from* $\mathbf{L}$ *by adding an axiom* φ_i *for* $i = 1, 2$. *Without loss of generality, we can assume moreover that formulas* φ_1 *and* φ_2 *have no variables in common. Then, the meet* $\mathbf{L}_1 \cap \mathbf{L}_2$ *is axiomatized over* $\mathbf{L}$ *by a formula* $\varphi_1 \vee \varphi_2$.

PROOF. Clearly, $\varphi_1 \vee \varphi_2$ belongs to the meet $\mathbf{L}_1 \cap \mathbf{L}_2$. So, it suffices to show that for any formula $\psi \in \mathbf{L}_1 \cap \mathbf{L}_2$, ψ follows in $\mathbf{L}$ from some substitution instances of $\varphi_1 \vee \varphi_2$. Since $\psi \in \mathbf{L}_1 \cap \mathbf{L}_2$, by using the local deduction theorem for $\mathbf{FL_{ew}}$ in Chapter 2 there are substitution instances δ_i with $i = 1, \dots, n$ of φ_1, and σ_j with $j = 1, \dots, m$ of φ_2, respectively, such that both formulas

$$\prod_{i=1}^n \delta_i \to \psi \quad \text{and} \quad \prod_{j=1}^m \sigma_j \to \psi$$

are in $\mathbf{L}$, and hence the formula

$$\left(\prod_{i=1}^n \delta_i \vee \prod_{j=1}^m \sigma_j\right) \to \psi.$$

is also provable in it. On the other hand, it is easily seen that for all formulas α_i $(i = 1, \dots, k)$ and β_j $(j = 1, \dots, l)$;

$$\prod_{i=1}^k \prod_{j=1}^l (\alpha_i \vee \beta_j) \to \left(\prod_{i=1}^k \alpha_i \vee \prod_{j=1}^l \beta_j\right)$$

is always provable in $\mathbf{FL_{ew}}$, by using the distributivity of $\cdot$ over disjunction. Thus, we have that

$$\prod_{i=1}^n \prod_{j=1}^m (\delta_i \vee \sigma_j) \to \psi$$

is also provable in it. Since φ_1 and φ_2 have no variables in common, every formula $\delta_i \vee \sigma_j$ is a substitution instance of $\varphi_1 \vee \varphi_2$. Thus $\psi \in \mathbf{L} + \{\varphi_1 \vee \varphi_2\}$. □

The second lemma is on prime deductive filters of FL_{ew}-algebras. In the following, filters mean always deductive filters. Recall that for an $\mathbf{FL_{ew}}$-algebra $\mathbf{A}$, the *deductive* filter $\mathrm{Fg}^{\mathbf{A}}(S)$ generated by a subset S of A is expressed as:

$$\{x \in A : a_1 \cdots a_m \leq x \text{ for some } a_i \in S\}.$$

See Section 2.6. In particular, the deductive filter $\mathrm{Fg}^{\mathbf{A}}(\{c\})$ generated by an element c is $\{x \in A : c^n \leq x \text{ for some } n\}$. Thus, we can show that an FL_{ew}-algebra $\mathbf{A}$ is subdirectly irreducible iff there is an element $z < 1$ such that for any $x \in A$ if $x < 1$ then there exists a natural number n such that $x^n \leq z$ (see Lemma 3.58).

We will show that each subdirectly irreducible FL_{ew}-algebra is well-connected. Suppose that an FL_{ew}-algebra $\mathbf{A}$ is subdirectly irreducible and both $x, y \in A$ are smaller than 1. Then, for some $z < 1$ there exist natural numbers m and n such that $x^m \leq z$ and $y^n \leq z$. Let $t = m + n - 1$. Then, by the distributivity of $\cdot$ over $\vee$, $(x \vee y)^t = \bigvee_{i=0}^{t} x^i \cdot y^{t-i}$. If $i \geq m$ then $x^i \cdot y^{t-i} \leq x^i \leq x^m \leq z$. Similarly, if $t - i \geq n$, $x^i \cdot y^{t-i} \leq y^{t-i} \leq y^n \leq z$. But, for each i either $i \geq m$ or $t - i \geq n$ holds, and hence $(x \vee y)^t \leq z$. Thus, $x \vee y < 1$.

LEMMA 5.26. *Let G be a proper deductive filter of an FL_{ew}-algebra $\mathbf{A}$ such that $a \notin G$. Then there exists a prime deductive filter F_a of $\mathbf{A}$ such that $a \notin F_a$ and $G \subseteq F_a$.*

PROOF. The lemma can be proved similarly to the prime filter theorem of distributive lattices (Theorem 1.21). Let K be the set of all deductive filters F of $\mathbf{A}$ such that $a \notin F$ and $G \subseteq F$. By Zorn's lemma, K has a maximal element F_a. We now show that F_a is prime. Assume that both $x \notin F_a$ and $y \notin F_a$. By the maximality of F_a in K, there exist some natural numbers m and n, and $u, v \in F_a$ such that $x^m \cdot u \leq a$ and $y^n \cdot v \leq a$. By using the same argument as the above, we show $x \vee y \notin F_a$. Let $t = m+n-1$. Then,

$$(x \vee y)^t \cdot u \cdot v = \bigvee_{i=0}^{t} x^i \cdot y^{t-i} \cdot u \cdot v.$$

If $i \geq m$ then $x^i \cdot y^{t-i} \cdot u \cdot v \leq x^m \cdot u \leq a$. Similarly, if $t-i \geq n$, $x^i \cdot y^{t-i} \cdot u \leq a$. But, for each i either $i \geq m$ or $t-i \geq n$ holds. Therefore, $(x \vee y)^t \cdot u \cdot v \leq a$, and hence $x \vee y \notin F_a$. Thus, F_a is a prime deductive filter. □

COROLLARY 5.27. *For each element $a < 1$ in an FL_{ew}-algebra $\mathbf{A}$, there exists a prime deductive filter F of $\mathbf{A}$ such that $a \notin F$ and the quotient algebra $\mathbf{A}/F$ is subdirectly irreducible.*

PROOF. Taking the singleton set $\{1\}$ for G in the proof of Lemma 5.26, there exists a prime deductive filter F which is maximal among deductive filters such that $a \notin F$. It means that every deductive filter of $\mathbf{A}$, which properly includes F, contains the deductive filter $\mathrm{Fg}^{\mathbf{A}}(F \cup \{a\})$ generated by the set $F \cup \{a\}$. Thus, the lattice of deductive filters of $\mathbf{A}$ which includes F has a second smallest deductive filter $\mathrm{Fg}^{\mathbf{A}}(F \cup \{a\})$. Thus, the quotient algebra $\mathbf{A}/F$ is subdirectly irreducible. □

THEOREM 5.28. *The following conditions are equivalent for every substructural logic $\mathbf{L}$ over $\mathbf{FL_{ew}}$.*

1. *$\mathbf{L}$ is Halldén complete,*
2. *$\mathbf{L} = \mathbf{L}(\mathbf{A})$ for some well-connected FL_{ew}-algebra $\mathbf{A}$,*
3. *$\mathbf{L}$ is meet irreducible.*

PROOF. First we show that (2) implies (3). Let $\mathbf{A}$ be a well-connected FL_{ew}-algebra and $\mathbf{L} = \mathbf{L}(\mathbf{A})$. Suppose that $\mathbf{L} = \mathbf{L}_1 \cap \mathbf{L}_2$ for some incomparable logics $\mathbf{L}_1$ and $\mathbf{L}_2$. Then there exist some formulas φ and ψ such that $\varphi \in \mathbf{L}_1 \setminus \mathbf{L}_2$ and $\psi \in \mathbf{L}_2 \setminus \mathbf{L}_1$. Obviously, neither of them belong to $\mathbf{L}$. We can assume here that φ and ψ have no variables in common, since we may replace variables by other distinct variables when necessary. Thus there exists an assignment f on $\mathbf{A}$ such that $f(\varphi) < 1$ and $f(\psi) < 1$. Since we assume that $\mathbf{A}$ is well-connected, $f(\varphi \vee \psi) < 1$. On the other hand, as $\varphi \vee \psi$ belongs both to $\mathbf{L}_1$ and $\mathbf{L}_2$, it must belong to $\mathbf{L}$. This is a contradiction. Thus, $\mathbf{L}$ is meet irreducible.

Next, we show that (3) implies (1). Suppose that (1) does not hold. Then, there exist formulas φ_1 and φ_2 with no variables in common such that $\varphi_1 \vee \varphi_2$ is in $\mathbf{L}$ while neither φ_1 nor φ_2 is in it. Define logics $\mathbf{L}_i = \mathbf{L} + \{\varphi_i\}$ for $i = 1, 2$. Clearly, both $\mathbf{L}_1$ and $\mathbf{L}_2$ are strictly stronger than $\mathbf{L}$. By Lemma 5.25, $\mathbf{L}_1 \cap \mathbf{L}_2$ is axiomatized over $\mathbf{L}$ by the formula $\varphi_1 \vee \varphi_2$. But $\varphi_1 \vee \varphi_2$ is in $\mathbf{L}$ by our assumption, and hence $\mathbf{L}_1 \cap \mathbf{L}_2 = \mathbf{L}$. Thus, $\mathbf{L}$ is not meet irreducible, which is a contradiction.

Lastly, we show that (1) implies (2). Take any FL_{ew}-algebra $\mathbf{C}$ such that $\mathbf{L} = \mathbf{L}(\mathbf{C})$. (For instance, take the Lindenbaum-Tarski algebra of $\mathbf{L}$ for $\mathbf{C}$.) From this $\mathbf{C}$, we construct a well-connected FL_{ew}-algebra $\mathbf{A}$ such that $\mathbf{L} = \mathbf{L}(\mathbf{A})$.

Let $\{G_i : i \in I\}$ be an enumeration of all prime deductive filters of $\mathbf{C}$ such that the quotient algebra $\mathbf{C}/G_j$ is subdirectly irreducible. For each formula φ, define a subset $R(\varphi)$ of (r) by

$$R(\varphi) = \{j \in I : \mathbf{C}/G_j \not\models \varphi\}.$$

Note that each algebra $\mathbf{C}/G_i$ is well-connected. When φ is not provable in $\mathbf{L}$, $R(\varphi)$ is nonempty. In fact, if φ is not provable in $\mathbf{L}$, $f(\varphi) < 1$ for an assignment f in C. Then there exists a prime deductive filter G_k of $\mathbf{C}$ such that $f(\varphi) \notin G_k$ by Corollary 5.27 such that $\mathbf{C}/G_k$ is subdirectly irreducible. Define an assignment f^* in $\mathbf{C}/G_k$ by $f^*(p) = h(f(p))$ for each propositional variable p, where h is a natural homomorphism induced by the congruence determined by the prime deductive filter G_k. Then, $f^*(\varphi) < 1$ in $\mathbf{C}/G_k$. Thus, $k \in R(\varphi)$.

We show next that the set $E = \{R(\varphi) | \varphi \notin \mathbf{L}\}$ has the finite intersection property, i.e., every intersection of finitely many members from E is nonempty. To see this, it suffices to show that for each $R(\varphi)$ and $R(\psi)$ in E there exists a formula γ such that $R(\gamma) \subseteq R(\varphi) \cap R(\psi)$ and $R(\gamma) \in E$. It may happen that φ and ψ have some variables in common. Then we take a formula φ' which is obtained from φ by *renaming* propositional variables so that φ' and ψ have no variables in common. Here, a renaming means a substitution of distinct variables by other distinct variables. Clearly, $\varphi' \notin \mathbf{L}$ and $R(\varphi') = R(\varphi)$. Thus we can assume from the beginning that φ and ψ have no variables in common. Since we assume that $\mathbf{L}$ is Halldén complete,

$R(\varphi \vee \psi) \in E$. Now, we show that $R(\varphi \vee \psi) \subseteq R(\varphi) \cap R(\psi)$. Suppose that $j \in R(\varphi \vee \psi)$. Then $\mathbf{C}/G_j \not\models \varphi \vee \psi$. Then for an assignment g in $\mathbf{C}/G_j$, $g(\varphi \vee \psi) = g(\varphi) \vee g(\psi) < 1$. Therefore, $\mathbf{C}/G_j \not\models \varphi$ and $\mathbf{C}/G_j \not\models \psi$. Hence $j \in R(\varphi) \cap R(\psi)$.

By the finite intersection property of E, there exists an ultrafilter U over (r) such that $E \subseteq U$. Let $\mathbf{A}$ be the ultraproduct $(\prod_{i \in I}(\mathbf{C}/G_i))/U$. Since each algebra $\mathbf{C}/G_i$ is well-connected and moreover the well-connectedness can be expressed as a first-order sentence, $\mathbf{A}$ is also well-connected by Corollary 1.29. Obviously, $\mathbf{A} \in \mathsf{V}(\mathbf{L})$. If a formula φ is not provable in $\mathbf{L}$ then $R(\varphi) \in U$ and hence $\mathbf{A} \not\models \varphi$. Thus, $\mathbf{L} = \mathbf{L}(\mathbf{A})$. □

Complete meet-irreducibility is a stronger form of meet-irreducibility. That is, a logic $\mathbf{L}$ is completely meet-irreducible iff it is not the intersection of all logics strictly bigger than $\mathbf{L}$ (see Section 1.1.3). Then, we can easily show the following. The second result in the following lemma can be shown by using Theorem 5.28 and the fact that every subdirectly irreducible FL_{ew}-algebra is well-connected.

LEMMA 5.29.

(1) *For each logic* $\mathbf{L}$ *over* $\mathbf{FL}$, *if* $\mathbf{L}$ *is completely meet-irreducible then* $\mathbf{L} = \mathbf{L}(\mathbf{A})$ *for some subdirectly irreducible FL-algebra* $\mathbf{A}$,

(2) *For each logic* $\mathbf{L}$ *over* $\mathbf{FL_{ew}}$, $\mathbf{L} = \mathbf{L}(\mathbf{A})$ *for some subdirectly irreducible* FL_{ew}*-algebra* $\mathbf{A}$, *then* $\mathbf{L}$ *is meet-irreducible.*

In [Wro76], Wroński proved that the converse of the above (2) holds for superintuitionistic logics. That is, a superintuitionistic logic $\mathbf{L}$ is meet-irreducible, or equivalently, it is Halldén complete iff:

$\mathbf{L} = \mathbf{L}(\mathbf{A})$ for a *subdirectly irreducible* Heyting algebra $\mathbf{A}$.

This result follows from the proof of our Theorem 5.28, since every algebra $\mathbf{C}/G_i$ in the proof is subdirectly irreducible and moreover the subdirect irreducibility of Heyting algebras can be expressed by the following first-order sentence:

$$\exists z < 1, \forall x < 1, \ x \leq z.$$

On the other hand, the subdirect irreducibility of FL_{ew}-algebras is written as :

$$\exists z < 1, \forall x < 1 \text{ for some } n \in \mathbb{N} \ x^n \leq z,$$

which is not a first-order sentence.

In the proof of Theorem 5.28, we can take an enumeration of all prime deductive filters of $\mathbf{C}$ for $\{G_i : i \in I\}$, instead of an enumeration of a subset of all prime deductive filters with a stronger condition on $\{G_i : i \in I\}$. But, we chose such a subset, in order to clarify these differences between Heyting algebras and FL_{ew}-algebras. When a logic over $\mathbf{FL_{ew}}$ satisfies the axiom of m-potency, i.e., $\alpha^m \to \alpha^{m+1}$, then we can take a fixed number m for n in the above statement and hence it becomes a first-order sentence. Thus, we have the following.

COROLLARY 5.30. *The following two conditions are equivalent for every substructural logic* $\mathbf{L}$ *over* $\mathbf{FL_{ew}}$ *satisfying the axiom of* m*-potency for some* m.

(1) $\mathbf{L}$ *is Halldén complete,*
(2) $\mathbf{L} = \mathbf{L}(\mathbf{A})$ *for some subdirectly irreducible* FL_{ew}*-algebra* $\mathbf{A}$.

We do not know whether the converses of the statements in Lemma 5.29 hold. The meet-primeness of a given logic, which implies the complete meet-irreducibility, has a close relation to splittings discussed in Chapter 10. For more information on meet-primeness and complete meet-irreducibility, see [Kra99].

A similar result for logics over $\mathbf{FL_e}$ can be shown, but some modifications are necessary, as the unit 1 is not assumed to be a greatest element.

THEOREM 5.31. *The following conditions are equivalent for every substructural logic* $\mathbf{L}$ *over* $\mathbf{FL_e}$.

(1) $\mathbf{L}$ *is weakly Halldén complete, i.e., for all formulas* φ *and* ψ *which have no variables in common, if* $(\varphi \wedge 1) \vee (\psi \wedge 1)$ *is in* $\mathbf{L}$ *then either* φ *or* ψ *is in* $\mathbf{L}$,
(2) $\mathbf{L} = \mathbf{L}(\mathbf{A})$ *for some* FL_e*-algebra* $\mathbf{A}$ *satisfying that* $x \vee y = 1$ *implies* $x = 1$ *or* $y = 1$ *for all* $x, y \in A$,
(3) $\mathbf{L}$ *is meet irreducible.*

This result shows us a characterization of the meet-irreducibility, but not Halldén completeness for logics over $\mathbf{FL_e}$.

5.4. Maksimova's property and well-connected pairs

We consider algebraic characterizations of two forms of Maksimova's variable separation property and derive an algebraic characterization of Halldén completeness as a particular case. Results in the rest of this chapter, including results on Halldén completeness in the previous subsection, came out from Kihara's dissertation [Kih06] and will be included in the paper [KO07] in full detail.

A substructural logic $\mathbf{L}$ is said to have the Maksimova's variable separation property (MVP), when for all formulas $\alpha_1\backslash\alpha_2$ and $\beta_1\backslash\beta_2$ with no propositional variables in common if the formula $(\alpha_1 \wedge \beta_1)\backslash(\alpha_2 \vee \beta_2)$ is provable in $\mathbf{L}$, then $\alpha_1\backslash\alpha_2$ or $\beta_1\backslash\beta_2$ is provable in it.

We can define also the *deductive form* of Maksimova's property as follows. A substructural logic $\mathbf{L}$ has the *deductive Maksimova's variable separation property* (DMVP), when for all formulas $\alpha_1\backslash\alpha_2$ and $\beta_1\backslash\beta_2$ with no variables in common, $\alpha_1 \wedge \beta_1 \vdash_{\mathbf{L}} \alpha_2 \vee \beta_2$ implies $\alpha_1 \vdash_{\mathbf{L}} \alpha_2$ or $\beta_1 \vdash_{\mathbf{L}} \beta_2$.

Since the condition $\gamma, \sigma \vdash_{\mathbf{L}} \psi$ is equivalent to the condition $\gamma \wedge \sigma \vdash_{\mathbf{L}} \psi$ in general, we can state it also in the following way (in view of compactness).

> A substructural logic $\mathbf{L}$ has the DMVP, if for all sets of formulas $\Gamma \cup \{\varphi\}$ and $\Sigma \cup \{\psi\}$ with no variables in common, $\Gamma, \Sigma \vdash_{\mathbf{L}} \varphi \vee \psi$ implies ($\Gamma \vdash_{\mathbf{L}} \varphi$ or $\Sigma \vdash_{\mathbf{L}} \psi$).

In this section, we give algebraic characterizations of both the MVP and the DMVP for substructural logics. Two subalgebras $\mathbf{B}$ and $\mathbf{C}$ of an FL-algebra $\mathbf{A}$ form a *strongly well-connected pair* if for all elements $b_1, b_2 \in B$ and $c_1, c_2 \in C$, $b_1 \wedge_{\mathbf{A}} c_1 \leq b_2 \vee_{\mathbf{A}} c_2$ implies $b_1 \leq b_2$ or $c_1 \leq c_2$. When $b_1 = c_1 = 1$, the above becomes the condition that for any elements $b \in B$ and $c \in C$, $1 \leq b \vee_{\mathbf{A}} c$ implies $1 \leq b$ or $1 \leq c$. When this restricted form holds for $\mathbf{B}$ and $\mathbf{C}$, we say that $\mathbf{B}$ and $\mathbf{C}$ form a *well-connected pair* of $\mathbf{A}$. It is obvious that an algebra $\mathbf{A}$ is well-connected iff $\mathbf{A}$ and $\mathbf{A}$ itself form a well-connected pair, and also iff every pair of subalgebras of $\mathbf{A}$ form a well-connected pair.

THEOREM 5.32. *Let* $\mathbf{L}$ *be a logic over* $\mathbf{FL}$. *Then the following two conditions are equivalent;*

(1) $\mathbf{L}$ *has the MVP,*
(2) *for every two non-degenerate FL-algebras* $\mathbf{A}, \mathbf{B}$ *in* $\mathsf{V}(\mathbf{L})$, *there exist an FL-algebra* $\mathbf{C}$ *and subalgebras* $\mathbf{C}_1, \mathbf{C}_2$ *of* $\mathbf{C}$ *in* $\mathsf{V}(\mathbf{L})$ *such that* $\mathbf{C}_1$ *and* $\mathbf{C}_2$ *form a strongly well-connected pair, and moreover that* $\mathbf{A}$ *and* $\mathbf{B}$ *are homomorphic images of* $\mathbf{C}_1$ *and* $\mathbf{C}_2$, *respectively.*

PROOF. Suppose first that $\mathbf{L}$ has the MVP, and let $\mathbf{A}$ and $\mathbf{B}$ be non-degenerate FL-algebras in $\mathsf{V}(\mathbf{L})$. Take disjoint sets of variables Y and Z that are enough big to ensure the surjective maps from Y to A and Z to B, respectively. Let X be the union of Y and Z, and let $\mathbf{C}$, $\mathbf{C}_1$ and $\mathbf{C}_2$ be free algebras in $\mathsf{V}(\mathbf{L})$, generated by X, Y and Z, respectively. By the universal mapping property, $\mathbf{A}$ and $\mathbf{B}$ are homomorphic images of $\mathbf{C}_1$ and $\mathbf{C}_2$. Also both $\mathbf{C}_1$ and $\mathbf{C}_2$ are regarded as subalgebras of $\mathbf{C}$. So, it remains to show that $\mathbf{C}_1$ and $\mathbf{C}_2$ form a strongly well-connected pair.

Take arbitrary elements $a_1, a_2 \in C_1$ and $b_1, b_2 \in C_2$. Then there exist terms s_1, s_2 over the set Y and terms t_1, t_2 over Z such that

$$a_1 = s_1/{\equiv_{\mathbf{L}}}, a_2 = s_2/{\equiv_{\mathbf{L}}}, b_1 = t_1/{\equiv_{\mathbf{L}}} \text{ and } b_2 = t_2/{\equiv_{\mathbf{L}}}.$$

Recall that $\equiv_{\mathbf{L}}$ is a congruence relation satisfying that $s \equiv_{\mathbf{L}} t$ iff $(s\backslash t) \wedge (t\backslash s)$ is provable in $\mathbf{L}$ for all terms s and t. Now suppose that $a_1 \not\leq a_2$ and $b_1 \not\leq b_2$. Then, neither $s_1\backslash s_2$ nor $t_1\backslash t_2$ are provable in $\mathbf{L}$. Since $s_1\backslash s_2$ and $t_1\backslash t_2$ have no variables in common, $(s_1 \wedge t_1)\backslash(s_2 \vee t_2)$ is neither provable in $\mathbf{L}$ by our assumption that $\mathbf{L}$ has the MVP. This means that $a_1 \wedge b_1 \not\leq a_2 \vee b_2$. Thus, $\mathbf{C}_1$ and $\mathbf{C}_2$ form a strongly well-connected pair.

We show next that the condition (2) implies the MVP of a logic $\mathbf{L}$. Suppose that for given formulas $\varphi_1\backslash\varphi_2$ and $\psi_1\backslash\psi_2$ which have no variables in common, neither $\varphi_1\backslash\varphi_2$ nor $\psi_1\backslash\psi_2$ are provable in $\mathbf{L}$. Let Y and Z be the sets of variables appearing in $\varphi_1\backslash\varphi_2$ and $\psi_1\backslash\psi_2$, respectively. Then there exist non-degenerate FL-algebras $\mathbf{A}, \mathbf{B}$ in $\mathsf{V}(\mathbf{L})$ and valuations f on $\mathbf{A}$ and g on $\mathbf{B}$ such that

$$f(\varphi_1) \not\leq_{\mathbf{A}} f(\varphi_2) \text{ and } g(\psi_1) \not\leq_{\mathbf{B}} g(\psi_2).$$

By our assumption, for some FL-algebra $\mathbf{C}$ and some subalgebras $\mathbf{C}_1, \mathbf{C}_2$ of $\mathbf{C}$ in $\mathsf{V}(\mathbf{L})$, there exist surjective homomorphisms h from $\mathbf{C}_1$ to $\mathbf{A}$ and j from $\mathbf{C}_2$ to $\mathbf{B}$ and $\mathbf{C}_1$ and $\mathbf{C}_2$ form a strongly well-connected pair. We define a valuation k in $\mathbf{C}$ for formulas over the set of variables $Y \cup Z$ so that $k(p) \in (h^{-1} \circ f)(p)$ for each $p \in Y$ and $k(q) \in (j^{-1} \circ g)(q)$ for each $q \in Z$. Such a map k exists since both h and j are surjective, and is well-defined since Y and Z are disjoint. Then, for each formula φ over Y, $f(\varphi) = (h \circ k)(\varphi)$ holds and similarly for each formula ψ over Z, $g(\psi) = (j \circ k)(\psi)$ holds. Therefore,

$$(h \circ k)(\varphi_1) \not\leq_{\mathbf{A}} (h \circ k)(\varphi_2) \text{ and } (j \circ k)(\psi_1) \not\leq_{\mathbf{B}} (j \circ k)(\psi_2).$$

Hence,

$$k(\varphi_1) \not\leq k(\varphi_2) \text{ and } k(\psi_1) \not\leq k(\psi_2).$$

Since $k(\varphi_1), k(\varphi_2) \in C_1$ and $k(\psi_1), k(\psi_2) \in C_2$,

$$k(\varphi_1 \wedge \psi_1) = k(\varphi_1) \wedge k(\psi_1) \not\leq k(\varphi_2) \vee k(\psi_2) = k(\varphi_2 \vee \psi_2)$$

by the strong well-connectedness of $\mathbf{C}_1$ and $\mathbf{C}_2$. Thus, $(\varphi_1 \wedge \psi_1)\backslash(\varphi_2 \vee \psi_2)$ is not provable in $\mathbf{L}$. Hence, $\mathbf{L}$ has the MVP. ☐

We assume that both $\mathbf{A}$ and $\mathbf{B}$ are non-degenerate in the condition (2) of Theorem 5.32. But by checking the proof, we can see that this assumption is replaced by a stronger assumption that *both* $\mathbf{A}$ *and* $\mathbf{B}$ *are subdirectly irreducible.*

Halldén completeness can be regarded as a special case of the MVP. On the other hand, it is pointed out in [CZ93] that there exist uncountably many superintuitionistic logics with the HC but without the MVP. Similarly to the above Theorem 5.32, we can give another characterization of the HC by replacing "strong well-connectedness" by "well-connectedness".

THEOREM 5.33. *Let* $\mathbf{L}$ *be a logic over* $\mathbf{FL}$. *Then the following two conditions are equivalent;*

(1) $\mathbf{L}$ *is Halldén complete,*
(2) *for every two non-degenerate* $\mathbf{FL}$*-algebras* $\mathbf{A}, \mathbf{B}$ *in* $\mathsf{V}(\mathbf{L})$, *there exist an FL-algebra* $\mathbf{C}$ *and subalgebras* $\mathbf{C}_1, \mathbf{C}_2$ *of* $\mathbf{C}$ *in* $\mathsf{V}(\mathbf{L})$ *such that* $\mathbf{C}_1$ *and* $\mathbf{C}_2$ *form a well-connected pair and moreover that* $\mathbf{A}$ *and* $\mathbf{B}$ *are homomorphic images of* $\mathbf{C}_1$ *and* $\mathbf{C}_2$, *respectively.*

It will be interesting to compare this with an algebraic characterization of the disjunction property in Theorem 5.21, since from a syntactic point of view the Halldén completeness follows immediately from the disjunction property. Theorem 5.21 says that the necessary and sufficient condition for a logic $\mathbf{L}$ to have the DP is that for all $\mathbf{A}, \mathbf{B}$ there exist a well-connected algebra $\mathbf{C} \in \mathsf{V}(\mathbf{L})$ and a surjective homomorphism h from $\mathbf{C}$ to the direct product $\mathbf{A} \times \mathbf{B}$ of $\mathbf{A}$ and $\mathbf{B}$. But, when $\mathbf{C}$ is well-connected, $\mathbf{C}$ with itself form a well-connected pair. Moreover, the composition of h and each projection map gives a surjective homomorphism from $\mathbf{C}$ to $\mathbf{A}$ or from $\mathbf{C}$ to $\mathbf{B}$. This shows how the disjunction property implies the Halldén completeness semantically.

An algebraic characterization of the DMVP given below shows that the strong well-connectedness and the existence of homomorphisms in Theorem 5.32 can be replaced by the well-connectedness and the existence of isomorphisms, respectively.

THEOREM 5.34. *Let* $\mathbf{L}$ *be a logic over* $\mathbf{FL_e}$. *Then the following two conditions are equivalent;*

(1) $\mathbf{L}$ *has the DMVP,*
(2) *for every two non-degenerate* FL_e*-algebras* $\mathbf{A}, \mathbf{B}$ *in* $\mathsf{V}(\mathbf{L})$*, there exist an FL-algebra* $\mathbf{D}$ *and subalgebras* $\mathbf{D}_1, \mathbf{D}_2$ *of* $\mathbf{D}$ *in* $\mathsf{V}(\mathbf{L})$ *such that* $\mathbf{D}_1$ *and* $\mathbf{D}_2$ *form a well-connected pair and moreover that* $\mathbf{A}$ *and* $\mathbf{B}$ *are isomorphic to* $\mathbf{D}_1$ *and* $\mathbf{D}_2$*, respectively. In other words, every two non-degenerate FL-algebras* $\mathbf{A}, \mathbf{B}$ *in* $\mathsf{V}(\mathbf{L})$ *can be jointly embedded into an FL-algebra* $\mathbf{D} \in \mathsf{V}(\mathbf{L})$ *and their images form a well-connected pair.*

PROOF. The proof goes similarly to that of Theorem 5.32. Suppose that $\mathbf{L}$ has the DMVP, and let $\mathbf{A}$ and $\mathbf{B}$ be non-degenerate FL-algebras in $\mathsf{V}(\mathbf{L})$. Like before, we take disjoint sets of variables Y and Z that are enough big to ensure the surjective maps from Y to A and Z to B, respectively. Let X be the union of Y and Z, and let $\mathbf{C}$, $\mathbf{C}_1$ and $\mathbf{C}_2$ be free algebras in $\mathsf{V}(\mathbf{L})$ generated by X, Y and Z, respectively. Algebras $\mathbf{C}_1$ and $\mathbf{C}_2$ can be regarded as subalgebras of $\mathbf{C}$. By the universal mapping property, there exist surjective homomorphisms $h : \mathbf{C}_1 \to \mathbf{A}$ and $k : \mathbf{C}_2 \to \mathbf{B}$. Let $F_1 = h^{-1}(\uparrow 1_{\mathbf{A}})$ and $F_2 = k^{-1}(\uparrow 1_{\mathbf{B}})$. Then both F_1 and F_2 are proper deductive filters of $\mathbf{C}_1$ and $\mathbf{C}_2$, respectively, since both $\mathbf{A}$ and $\mathbf{B}$ are non-degenerate. Moreover, by the homomorphism theorem, $\mathbf{A} \cong \mathbf{C}/F_1$ and $\mathbf{B} \cong \mathbf{C}/F_2$.

Now, we show that there exists a deductive filter G of $\mathbf{C}$ such that

(a) $F_1 = C_1 \cap G$ and $F_2 = C_2 \cap G$,
(b) for any $b \in C_1$ and any $c \in C_2$, $b \vee c \in G$ implies either $b \in G$ or $c \in G$.

Let G be the deductive filter of $\mathbf{C}$ generated by the set $F_1 \cup F_2$. Obviously, G is written as follows:

$G = \{x \in C : \Pi_{i=1}^n \gamma_i(a_i) \leq x$ for some $a_i \in F_1 \cup F_2$ and some iterated conjugates γ_i on $\mathbf{C}$ with $1 \leq i \leq n\}$.

We show that $F_1 = C_1 \cap G$. It is easy to see that $F_1 \subseteq C_1 \cap G$. Suppose that $d \in C_1 \cap G$. Then, there exist some $a_i \in F_1 \cup F_2$ and some iterated conjugates γ_i on $\mathbf{C}$ with $1 \leq i \leq m$ such that $\Pi_{i=1}^m \gamma_i(a_i) \leq d$. Since d belongs to C_1, there exists a formula s^* over Y such that $d = s^*/\equiv_{\mathbf{L}}$. Similarly, if a_j belongs to F_1 (F_2) there exists a formula u_j over Y (over Z, respectively) such that $a_j = u_j/\equiv_{\mathbf{L}}$. Then, $\Pi_{i=1}^m \gamma_i(a_i) \leq d$ holds in $\mathbf{C}$ iff $1_{\mathbf{C}} \leq (\Pi_{i=1}^m \gamma_i(a_i))\backslash d$ holds in $\mathbf{C}$. The latter implies that $(\Pi_{i=1}^m \sigma_i(u_i))\backslash s^*$ is provable in $\mathbf{L}$, where each $\sigma_i(u_i)$ is a formula (over X) corresponding to $\gamma_i(a_i)$ for each i. Then, this in turn implies $\{u_i : 1 \leq i \leq m\} \vdash_{\mathbf{L}} s^*$. Now let us take a formula t^* over Z such that $t^*/\equiv_{\mathbf{L}} \notin F_2$, as F_2 is a proper deductive filter. Obviously, $\{u_i : 1 \leq i \leq m\} \vdash_{\mathbf{L}} s^* \vee t^*$. Since Y and Z are disjoint, by our assumption on the DMVP of $\mathbf{L}$ either $\{s_j\} \vdash_{\mathbf{L}} s^*$ or $\{t_l\} \vdash_{\mathbf{L}} t^*$ holds, where each s_j (and t_l) is a formula in $\{u_i : 1 \leq i \leq m\}$ over Y (and Z, respectively). Suppose that $\{t_l\} \vdash_{\mathbf{L}} t^*$. Let k' be a valuation on $\mathbf{B}$ defined by $k'(z) = k(z/\equiv_{\mathbf{L}})$ for $z \in Z$. Then, we have $\mathbf{B}, k' \models t_l$ by the definition of F_2, and also $k(t_l/\equiv_{\mathbf{L}}) \geq 1_{\mathbf{B}}$. Hence, $\mathbf{B}, k' \models t^*$, which implies $k'(t^*) = k(t^*/\equiv_{\mathbf{L}}) \geq 1_{\mathbf{B}}$. Hence, $t^*/\equiv_{\mathbf{L}} \in F_2$. This contradicts to the choice of t^*. Thus, $\{s_j\} \vdash_{\mathbf{L}} s'$ must hold. Using the same argument as the above, this implies that $d = s'/\equiv_{\mathbf{L}} \in F_1$. Therefore, $F_1 = C_1 \cap G$. Similarly, we can show $F_2 = C_2 \cap G$.

It remains to show that G satisfies the condition (b). Suppose that $u \vee v \in G$ for $u \in C_1$ and $v \in C_2$. Then, by the definition of G, there exist some $a_i \in F_1 \cup F_2$ and iterated conjugates γ_i on $\mathbf{C}$ with $1 \leq i \leq m$ such that $\Pi_{i=1}^m \gamma_i(a_i) \leq u \vee v$. Then, there exist formulas s' over Y and t' over Z such that $u = s'/ \equiv_{\mathbf{L}}$ and $v = t'/ \equiv_{\mathbf{L}}$. Also, there exists a formula u_j over Y (Z) if a_j belongs to F_1 (F_2, respectively) such that $a_j = u_j/\equiv_{\mathbf{L}}$. Then, $\Pi_{i=1}^m \gamma_i(a_i) \leq u \vee v$ iff $1_{\mathbf{C}} \leq (\Pi_{i=1}^m \gamma_i(a_i))\backslash(u \vee v)$, which implies $(\Pi_{i=1}^m \sigma_i(\varphi_i))\backslash(s' \vee t')$ is provable in $\mathbf{L}$, where each $\sigma_i(u_i)$ is a suitable formula (over X) corresponding to $\gamma_i(a_i)$ for each i. Thus, we have $\{u_i : 1 \leq i \leq m\} \vdash_{\mathbf{L}} s' \vee t$.

By the DMVP of $\mathbf{L}$, either $\{s_j\} \vdash_{\mathbf{L}} s'$ or $\{t_l\} \vdash_{\mathbf{L}} t'$ holds, where each s_j (and t_l) is a formula in $\{u_i : 1 \leq i \leq m\}$ over Y (and Z, respectively). Suppose that the latter holds. Taking the same valuation k' on $\mathbf{B}$ introduced in the above, we can show that $v = t'/ \equiv_{\mathbf{L}} \in F_2$. Since $F_2 = C_2 \cap G$ holds, we have $v \in G$. Similarly, $\{s_j\} \vdash_{\mathbf{L}} s'$ implies $u \in G$. Therefore, either $u \in G$ or $v \in G$.

We continue the proof. We show next that both $\mathbf{C}_1/F_1$ and $\mathbf{C}_2/F_2$ are embedded into $\mathbf{C}/G$. Define mappings $f : \mathbf{C}_1/F_1 \to \mathbf{C}/G$ and $g : \mathbf{C}_2/F_2 \to \mathbf{C}/G$ by $f(x/F_1) = x/G$ and $g(y/F_2) = y/G$, respectively. Since F_1 and F_2 are subsets of G, these mappings f and g are well-defined homomorphisms. For $x, x' \in C_1$, $f(x/F_1) = f(x'/F_1)$ implies $x/G = x'/G$, and by the property of G shown above, $x/F_1 = x'/F_1$. Thus, f is injective. Similarly, g is also injective. Now, let us denote $\mathbf{C}/G$, $f(\mathbf{C}_1/F_1)$ and $g(\mathbf{C}_2/F_2)$ by $\mathbf{D}$, $\mathbf{D}_1$ and $\mathbf{D}_2$, respectively. Then, $\mathbf{D}_1$ and $\mathbf{D}_2$ are subalgebras of $\mathbf{D}$, and $\mathbf{A}$ and $\mathbf{B}$ are isomorphic to $\mathbf{D}_1$ and $\mathbf{D}_2$, respectively. It remains to show that $\mathbf{D}_1$ and $\mathbf{D}_2$ form a well-connected pair. Suppose that $a \vee b \geq 1_{\mathbf{D}}$ for $a \in D_1$ and $b \in D_2$. Then there exist a formula s over Y and a formula t over Z such that $a = f(s/F_1)$ and $b = g(t/F_2)$. Then, $a \vee b = s/G \vee t/G = (s \vee t)/G \geq 1_{\mathbf{D}}$ and hence, $s \vee t \in G$. By using the property (b) of G shown above, either $s \in G$ or $t \in G$. Thus, either $a = s/G \geq 1$ or $b = t/G \geq 1$ holds. Therefore, $\mathbf{D}_1$ and $\mathbf{D}_2$ form a well-connected pair.

Conversely, we assume the second condition (2) in our theorem and show that $\mathbf{L}$ has the DMP. Suppose that for formulas $\varphi_1, \varphi_2, \psi_1$ and ψ_2 such that any variable appearing either of φ_1 and φ_2 appears neither of ψ_1 and ψ_2, neither $\varphi_1 \vdash_{\mathbf{L}} \varphi_2$ nor $\psi_1 \vdash_{\mathbf{L}} \psi_2$ hold. Then, there are FL-algebras $\mathbf{A}$ and $\mathbf{B}$ in $\mathsf{V}(\mathbf{L})$ and valuations f and g on them such that $\mathbf{A}, f \models \varphi_1$ but $\mathbf{A}, f \not\models \varphi_2$, and $\mathbf{B}, g \models \psi_1$ but $\mathbf{B}, g \not\models \psi_2$. By our assumption, there exist an FL-algebra $\mathbf{D}$ and subalgebras $\mathbf{D}_1, \mathbf{D}_2$ of $\mathbf{D}$ in $\mathsf{V}(\mathbf{L})$ such that $\mathbf{D}_1$ and $\mathbf{D}_2$ form a well-connected pair and moreover that $\mathbf{A}$ and $\mathbf{B}$ are isomorphic to $\mathbf{D}_1$ and $\mathbf{D}_2$, respectively. For the sake of simplicity, we identify $\mathbf{A}$ with $\mathbf{D}_1$ and $\mathbf{B}$ with $\mathbf{D}_2$, since each of these pairs is isomorphic. By our assumption on the disjointness of variables, we can define a valuation h on $\mathbf{D}$ such that $h(p) = f(p)$ for each variable p appearing in either of φ_1 and φ_2, and $h(q) = g(q)$ for each variable q appearing in either of ψ_1 and ψ_2. Clearly $h(\varphi_i) = f(\varphi_i)$ and $h(\psi_i) = g(\psi_i)$ hold for $i = 1, 2$. Then $h(\varphi_1 \wedge \psi_1) = f(\varphi_1) \wedge g(\psi_1) \geq 1$, but $h(\varphi_2) = f(\varphi_2) \not\geq 1$ and $h(\psi_2) = g(\psi_2) \not\geq 1$. Therefore, $h(\varphi_2 \vee \psi_2) = h(\varphi_2) \vee h(\psi_2) \not\geq 1$ since $\mathbf{A}$ and $\mathbf{B}$ form a well-connected pair. That is, $\mathbf{D}, h \models \varphi_1 \wedge \psi_1$ but $\mathbf{D}, h \not\models \varphi_2 \vee \psi_2$. Thus, $\varphi_1 \wedge \psi_1 \vdash_{\mathbf{L}} \varphi_2 \vee \psi_2$ does not hold. This completes the proof of the DMVP of $\mathbf{L}$. □

We note again that also in the condition (2) of Theorem 5.34 the assumption that $\mathbf{A}$ and $\mathbf{B}$ are non-degenerate can be replaced by the stronger one that they are subdirectly irreducible.

If we restrict our attention to logics over $\mathbf{FL_{ew}}$, we can give simpler algebraic characterizations. The following is an extension of a result in [Mak95] for superintuitionistic logics.

THEOREM 5.35. *The following conditions are equivalent for every substructural logic* $\mathbf{L}$ *over* $\mathbf{FL_{ew}}$.

(1) $\mathbf{L}$ *has the DMVP,*
(2) *all pairs of subdirectly irreducible algebras in* $\mathsf{V}(\mathbf{L})$ *are jointly embeddable into a well-connected (or even a subdirectly irreducible) algebra in* $\mathsf{V}(\mathbf{L})$.

The characterizations given in this subsection are summarized in Figure 5.1. Each of them involves a condition on two subalgebras of an algebra and a condition on mappings between them.

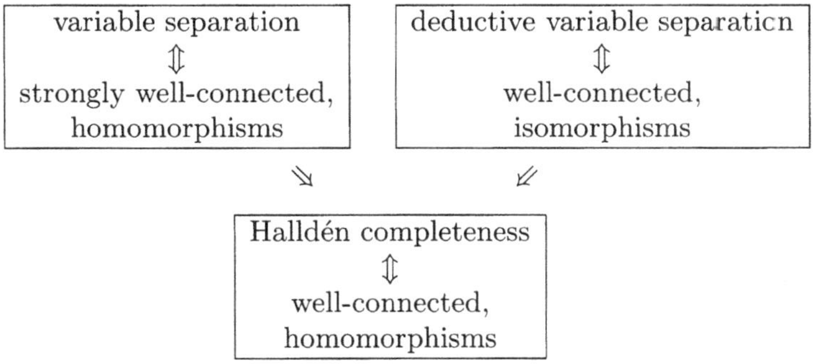

FIGURE 5.1. Relations among MVP, DMVP and HC.

5.5. Deductive interpolation properties

In the present and next sections, we provide algebraic characterizations of several forms of interpolation property for substructural logics. They include characterizations of Craig interpolation property and *deductive interpolation property* of a logic, in terms of *super-amalgamation property* and *amalgamation property* of the corresponding variety. In this section, we begin with two kinds of deductive interpolation properties and with the Robinson property. We show that Robinson property corresponds to the amalgamation property.

5.5.1. Strong deductive interpolation property. Recall that a substructural logic **L** over **FL** has the Craig interpolation property (CIP), if for all formulas φ, ψ, whenever $\varphi\backslash\psi$ is provable in **L**, there exists a formula δ such that

(1) both $\varphi\backslash\delta$ and $\delta\backslash\psi$ are provable in **L**,
(2) $\mathrm{var}(\delta) \subseteq \mathrm{var}(\varphi) \cap \mathrm{var}(\psi)$,

where $\mathrm{var}(\gamma)$ denotes the set of propositional variables in a formula γ.

By replacing provability of implications in the definition of the CIP by deducibility, we obtain other interpolation properties. A substructural logic **L** has the *strong deductive interpolation property* (SDIP), if for any set of formulas $\Gamma \cup \Sigma \cup \{\psi\}$, if $\Gamma, \Sigma \vdash_{\mathbf{L}} \psi$, then there exists a set of formulas Δ such that

(1) $\Gamma \vdash_{\mathbf{L}} \delta$, for all $\delta \in \Delta$ and $\Delta, \Sigma \vdash_{\mathbf{L}} \psi$,
(2) $\mathrm{var}(\Delta) \subseteq \mathrm{var}(\Gamma) \cap \mathrm{var}(\Sigma \cup \{\psi\})$.

The SDIP is called *Maehara interpolation property* in [CP99] and GINT in [Ono86]. Since $\vdash_{\mathbf{L}}$ is finitary and conjunctive for each substructural logic **L**, the modifications of the SDIP obtained by stipulating that each of the sets Γ, Σ and Δ is finite or a single formula, is equivalent to the SDIP. This holds also for all of interpolation properties discussed below. For the SDIP, we usually take a single formula δ for an interpolant, instead of a set Δ, and thus δ is assumed to satisfy that

(1) $\Gamma \vdash_{\mathbf{L}} \delta$, and $\delta, \Sigma \vdash_{\mathbf{L}} \psi$,
(2) $\mathrm{var}(\delta) \subseteq \mathrm{var}(\Gamma) \cap \mathrm{var}(\Sigma \cup \{\psi\})$.

In [Wro84b], Wroński introduced the notion of the equational interpolation property for a class of algebras. In the present context, this is stated as follows. A subvariety $\mathcal{V}$ of FL has the *equational interpolation property* (EqIP), if for every set of equations $G \cup E \cup \{\varepsilon\}$, whenever $G, E \models_{\mathcal{V}} \varepsilon$, there exists a set of equations D such that

(1) $G \models_{\mathcal{V}} \delta$, for all $\delta \in D$ and $D, E \models_{\mathcal{V}} \varepsilon$, and
(2) $\mathrm{var}(D) \subseteq \mathrm{var}(G) \cap \mathrm{var}(E \cup \{\varepsilon\})$.

This property with the assumption that E is empty is introduced and discussed also in [Jón65] and [Bac75]. By the algebraization theorem, the EqIP of a variety $\mathsf{V}(\mathbf{L})$ can be translated into the SDIP of a substructural logic $\mathbf{L}$. More precisely,

LEMMA 5.36. *A substructural logic* $\mathbf{L}$ *has the SDIP iff* $\mathsf{V}(\mathbf{L})$ *has the EqIP.*

Taking the empty set for Σ in the definition of the SDIP, we get the definition of the deductive interpolation property. A substructural logic $\mathbf{L}$ has the *deductive interpolation property* (DIP), if for any set of formulas $\Gamma \cup \{\psi\}$, if $\Gamma \vdash_{\mathbf{L}} \psi$, then there exists a formula δ such that

(1) $\Gamma \vdash_{\mathbf{L}} \delta$, and $\delta \vdash_{\mathbf{L}} \psi$,
(2) $\mathrm{var}(\delta) \subseteq \mathrm{var}(\Gamma) \cap \mathrm{var}(\psi)$.

The DIP is called also *turnstile interpolation property* by Madarász [Mad98], and *interpolation property for deducibility* by Maksimova [Mak95].

5.5.2. Robinson property. In Ono [Ono86] two properties for classes of algebras, ROB* and limited GINT are introduced, both of which are weaker than the EqIP. In view of the algebraization theorem, they are translated into the logical properties RP and ExIP, introduced below. Both of them are also discussed in [CP99], where RP is called *ordinary interpolation property.*

A substructural logic $\mathbf{L}$ has the *Robinson property* (RP) if for every set of formulas $\Gamma \cup \Sigma \cup \{\psi\}$, $\Gamma, \Sigma \vdash_{\mathbf{L}} \psi$ implies $\Sigma \vdash_{\mathbf{L}} \psi$, whenever $\Gamma \vdash_{\mathbf{L}} \alpha$ iff $\Sigma \vdash_{\mathbf{L}} \alpha$ for every formula α such that $\mathrm{var}(\alpha) \subseteq \mathrm{var}(\Gamma) \cap \mathrm{var}(\Sigma \cup \{\psi\})$.

A substructural logic $\mathbf{L}$ has the *extension interpolation property* (ExIP),[4] if for every set of formulas $\Gamma \cup \Sigma \cup \{\psi\}$, if $\Gamma, \Sigma \vdash_{\mathbf{L}} \psi$, then there exists a formula δ such that

(1) $\Gamma \vdash_{\mathbf{L}} \delta$ and $\delta, \Sigma \vdash_{\mathbf{L}} \psi$,
(2) $\mathrm{var}(\delta) \subseteq \mathrm{var}(\Sigma \cup \{\psi\})$.

---

[4]Note that in the definition of the limited GINT given in [Ono86] requires that $\mathrm{var}(\Sigma \cup \{\psi\}) \subseteq \mathrm{var}(\Gamma)$. But it is easy to see that this assumption can be omitted.

THEOREM 5.37. *For any substructural logic* **L**, **L** *has the SDIP iff it has both the RP and the ExIP.*

PROOF. Suppose first that **L** has the SDIP. The ExIP follows immediately from the SDIP. To show the RP, we assume that $\Gamma, \Sigma \vdash_{\mathbf{L}} \psi$ holds, and moreover that for every formula α such that $\mathrm{var}(\alpha) \subseteq X$, $\Gamma \vdash_{\mathbf{L}} \alpha$ iff $\Sigma \vdash_{\mathbf{L}} \alpha$, where $X = \mathrm{var}(\Gamma) \cap \mathrm{var}(\Sigma \cup \{\psi\})$. By applying the SDIP to the assumption that $\Gamma, \Sigma \vdash_{\mathbf{L}} \psi$, we obtain a formula δ such that

(1) $\Gamma \vdash_{\mathbf{L}} \delta$ and $\delta, \Sigma \vdash_{\mathbf{L}} \psi$,
(2) $\mathrm{var}(\delta) \subseteq X$.

Then by our assumption, $\Gamma \vdash_{\mathbf{L}} \delta$ implies $\Sigma \vdash_{\mathbf{L}} \delta$. Combining this with $\delta, \Sigma \vdash_{\mathbf{L}} \varphi$, we have $\Sigma \vdash_{\mathbf{L}} \varphi$. Thus, the RP holds.

Suppose conversely that **L** has both the RP and the ExIP hold and assume that $\Gamma, \Sigma \vdash_{\mathbf{L}} \psi$. Let $X = \mathrm{var}(\Gamma) \cap \mathrm{var}(\Sigma \cup \{\psi\})$. Define the sets of formulas $\Gamma^\dagger$ and $\Sigma^\dagger$, by $\Gamma^\dagger = \{\gamma : \Gamma \vdash_{\mathbf{L}} \gamma \text{ and } \mathrm{var}(\gamma) \subseteq X\}$ and $\Sigma^\dagger = \{\gamma : \Sigma \vdash_{\mathbf{L}} \gamma \text{ and } \mathrm{var}(\gamma) \subseteq X\}$, respectively. We show that for every formula α such that $\mathrm{var}(\alpha) \subseteq X$, we have $\Gamma, \Sigma^\dagger \vdash_{\mathbf{L}} \alpha$ iff $\Gamma^\dagger, \Sigma \vdash_{\mathbf{L}} \alpha$. To prove this, suppose that $\Gamma, \Sigma^\dagger \vdash_{\mathbf{L}} \alpha$ for a formula α with $\mathrm{var}(\alpha) \subseteq X$. Then, by the ExIP, there exists a formula δ such that $\mathrm{var}(\delta) \subseteq \mathrm{var}(\Sigma^\dagger \cup \{\alpha\})$, $\Gamma \vdash_{\mathbf{L}} \delta$, and $\delta, \Sigma^\dagger \vdash_{\mathbf{L}} \alpha$. Since $\mathrm{var}(\delta) \subseteq X$, we have $\delta \in \Gamma^\dagger$, and hence $\Gamma^\dagger, \Sigma \vdash_{\mathbf{L}} \alpha$. Similarly, the converse implication holds.

Obviously, our assumption $\Gamma, \Sigma \vdash_{\mathbf{L}} \psi$ implies that $\Gamma, \Sigma^\dagger, \Gamma^\dagger, \Sigma \vdash_{\mathbf{L}} \psi$. Thus, by the RP, we have $\Gamma^\dagger, \Sigma \vdash_{\mathbf{L}} \psi$. Now, by compactness, if we take a suitable finite conjunction δ^* of $\Gamma^\dagger$, both $\Gamma \vdash_{\mathbf{L}} \delta^*$ and $\delta^*, \Sigma \vdash_{\mathbf{L}} \psi$ hold, and moreover $\mathrm{var}(\delta^*) \subseteq X$. Thus, the SDIP holds in **L**. □

The above theorem tells us that the SDIP implies the RP. Also, the RP implies the DIP as shown in the next lemma. To illustrate the differences among these notions, we introduce alternative definitions of the SDIP and the DIP. We use these definitions to obtain the corresponding algebraic characterizations.

A substructural logic **L** has the SDIP* if for every set of formulas $\Gamma \cup \Sigma \cup \{\psi\}$, we have that $\Gamma, \Sigma \vdash_{\mathbf{L}} \psi$ implies $\Sigma \vdash_{\mathbf{L}} \psi$, whenever $\Gamma \vdash_{\mathbf{L}} \alpha$ *implies* $\Sigma \vdash_{\mathbf{L}} \alpha$ for every formula α with $\mathrm{var}(\alpha) \subseteq \mathrm{var}(\Gamma) \cap \mathrm{var}(\Sigma \cup \{\psi\})$. Also, **L** has the DIP* if for every set of formulas $\Gamma \cup \Sigma \cup \{\psi\}$, $\Gamma \vdash_{\mathbf{L}} \psi$ implies $\Sigma \vdash_{\mathbf{L}} \psi$, whenever $\Gamma \vdash_{\mathbf{L}} \alpha$ iff $\Sigma \vdash_{\mathbf{L}} \alpha$ for every formula α with $\mathrm{var}(\alpha) \subseteq \mathrm{var}(\Gamma) \cap \mathrm{var}(\Sigma \cup \{\psi\})$.

LEMMA 5.38.

(1) *For every substructural logic, the SDIP holds iff the* SDIP* *holds.*
(2) *For every substructural logic, the DIP holds iff the* DIP* *holds.*

Thus, the SDIP implies the RP, and the RP implies the DIP.

PROOF. First suppose that the SDIP holds for a logic **L**. We assume that $\Gamma \vdash_{\mathbf{L}} \alpha$ implies $\Sigma \vdash_{\mathbf{L}} \alpha$ for every formula α with $\mathrm{var}(\alpha) \subseteq X$, and that

$\Gamma, \Sigma \vdash_{\mathbf{L}} \psi$. Here, X is $\mathrm{var}(\Gamma) \cap \mathrm{var}(\Sigma \cup \{\psi\})$. By the SDIP, there exists a formula

δ with $\mathrm{var}(\delta) \subseteq X$ such that $\Gamma \vdash_{\mathbf{L}} \delta$, and $\delta, \Sigma \vdash_{\mathbf{L}} \psi$. By our assumption, we have $\Sigma \vdash_{\mathbf{L}} \delta$, so $\Sigma \vdash_{\mathbf{L}} \psi$ and the SDIP* holds. Similarly, we can show that the DIP implies the DIP*.

Conversely, suppose that the SDIP* holds for a logic $\mathbf{L}$, and that $\Gamma, \Sigma \vdash_{\mathbf{L}} \psi$. Define $\Gamma^{\dagger} = \{\gamma : \Gamma \vdash_{\mathbf{L}} \gamma \text{ and } \mathrm{var}(\gamma) \subseteq X\}$, where X is the set of variables defined as above. Clearly, for any formula α with $\mathrm{var}(\alpha) \subseteq X$, $\Gamma \vdash_{\mathbf{L}} \alpha$ implies $\Gamma^{\dagger}, \Sigma \vdash_{\mathbf{L}} \alpha$. Now, $\Gamma, \Sigma \vdash_{\mathbf{L}} \psi$ implies always $\Gamma, \Gamma^{\dagger}, \Sigma \vdash_{\mathbf{L}} \psi$, from which $\Gamma^{\dagger}, \Sigma \vdash_{\mathbf{L}} \psi$ follows by the SDIP*. Since $\vdash_{\mathbf{L}}$ is finitary and conjunctive, there exists a formula δ, which is a conjunction of finitely many formulas in $\Gamma^{\dagger}$, such that $\delta, \Sigma \vdash_{\mathbf{L}} \psi$. By the definition of $\Gamma^{\dagger}$ it is clear that $\Gamma \vdash_{\mathbf{L}} \delta$ and $\mathrm{var}(\delta) \subseteq X$. Thus, the SDIP holds.

To show that the DIP follows from the DIP*, we need to modify the above proof slightly. Suppose that the DIP* holds for a logic $\mathbf{L}$, and that $\Gamma \vdash_{\mathbf{L}} \psi$. Define $\Gamma^{\dagger}$ just in the same way as above, but taking the set $\mathrm{var}(\Gamma) \cap \mathrm{var}(\psi)$ for X. Then it is obvious that $\Gamma \vdash_{\mathbf{L}} \alpha$ iff $\Gamma^{\dagger} \vdash_{\mathbf{L}} \alpha$ for every formula α with $\mathrm{var}(\alpha) \subseteq X$. Thus, by applying the DIP*, $\Gamma^{\dagger} \vdash_{\mathbf{L}} \psi$ follows from $\Gamma \vdash_{\mathbf{L}} \psi$. Then, by taking a suitable finite conjunction δ of $\Gamma^{\dagger}$, we can show that $\Gamma \vdash_{\mathbf{L}} \delta$ and $\delta \vdash_{\mathbf{L}} \psi$. As $\mathrm{var}(\delta) \subseteq X$, the DIP follows. Clearly, the SDIP* implies the RP, which in turn implies DIP*. □

LEMMA 5.39.

(1) *The ExIP holds for every substructural logic over* $\mathbf{FL_e}$.
(2) *For every substructural logic over* $\mathbf{FL_e}$, *the DIP is equivalent to the SDIP.*

PROOF. We show first that the ExIP holds always for each substructural logic over $\mathbf{FL_e}$. Suppose that $\gamma, \sigma \vdash_{\mathbf{L}} \varphi$. By the local deduction theorem, $\gamma \vdash_{\mathbf{L}} (\sigma \wedge 1)^n \to \psi$ for some n. For δ equal to $(\sigma \wedge 1)^n \to \psi$, both $\gamma \vdash_{\mathbf{L}} \delta$ and $\delta, \sigma \vdash_{\mathbf{L}} \psi$ hold. Moreover, $\mathrm{var}(\delta) \subseteq \mathrm{var}(\sigma \cup \{\psi\})$. Thus, the ExIP follows.

Next, suppose that a logic over $\mathbf{FL_e}$ has the DIP and that $\gamma, \sigma \vdash_{\mathbf{L}} \psi$. By the local deduction theorem, we have $\gamma \vdash_{\mathbf{L}} (\sigma \wedge 1)^n \to \psi$ for some n. Then by the DIP, there exists a formula δ such that (1) $\gamma \vdash_{\mathbf{L}} \delta$, (2) $\delta \vdash_{\mathbf{L}} (\sigma \wedge 1)^n \to \psi$ and (3) $\mathrm{var}(\delta) \subseteq \mathrm{var}(\gamma) \cap \mathrm{var}((\sigma \wedge 1)^n \to \psi)$. So, both $\gamma \vdash_{\mathbf{L}} \delta$ and $\delta, \sigma \vdash_{\mathbf{L}} \psi$ hold for a formula δ with $\mathrm{var}(\delta) \subseteq \mathrm{var}(\gamma) \cap \mathrm{var}(\sigma, \psi)$. Thus, the SDIP holds. □

Thus, the SDIP, the RP and the DIP are mutually equivalent for every logic over $\mathbf{FL_e}$. These equivalences hold in fact whenever a logic satisfies the following *local deduction property*:

> for any Γ, σ, ψ, if $\Gamma, \sigma \vdash_{\mathbf{L}} \psi$ then there exists a formula $\star(\sigma)$ with $\mathrm{var}(\star(\sigma)) \subseteq \mathrm{var}(\sigma)$ such that $\Gamma \vdash_{\mathbf{L}} \star(\sigma)\backslash\psi$ and $\sigma \vdash_{\mathbf{L}} \star(\sigma)$.

5.5.3. Amalgamation property and Robinson property. Next we show the following theorem, which is originally shown in a more general setting in [Ono86]. A variety $\mathcal{V}$ has the *amalgamation property* (AP), if for all $\mathbf{A}, \mathbf{B}, \mathbf{C}$ in $\mathcal{V}$ and for all embeddings $f : \mathbf{A} \to \mathbf{B}$ and $g : \mathbf{A} \to \mathbf{C}$, there exists an algebra $\mathbf{D}$ in $\mathcal{V}$ and embeddings $h : \mathbf{B} \to \mathbf{D}$, $k : \mathbf{C} \to \mathbf{D}$ such that $h \circ f = k \circ g$. The next result says that the amalgamation property for $\mathsf{V}(\mathbf{L})$ is equivalent to the Robinson property for $\mathbf{L}$.

THEOREM 5.40. *For each substructural logic* $\mathbf{L}$, $\mathbf{L}$ *has the RP iff* $\mathsf{V}(\mathbf{L})$ *has the AP.*

PROOF. Suppose that $\mathbf{L}$ has the RP. We assume moreover that there exist embeddings $f : \mathbf{A} \to \mathbf{B}$ and $g : \mathbf{A} \to \mathbf{C}$ for FL-algebras $\mathbf{A}, \mathbf{B}$ and $\mathbf{C}$ in $\mathsf{V}(\mathbf{L})$. We take a set of variables X which is big enough to assure the existence of a surjective homomorphism $\eta_{\mathbf{A}} : \mathbf{Fm}_{\mathcal{L}}(X) \to \mathbf{A}$. Next, take sets of variables Y and Z with $Y \cap Z = X$ such that

> there are surjective homomorphisms $\eta_{\mathbf{B}} : \mathbf{Fm}_{\mathcal{L}}(Y) \to \mathbf{B}$ and $\eta_{\mathbf{C}} : \mathbf{Fm}_{\mathcal{L}}(Z) \to \mathbf{C}$ that satisfy $\eta_{\mathbf{B}}(\alpha) = f(\eta_{\mathbf{A}}(\alpha))$ and $\eta_{\mathbf{C}}(\alpha) = g(\eta_{\mathbf{A}}(\alpha))$ for every $\alpha \in \mathbf{Fm}_{\mathcal{L}}(X)$.

Define sets of formulas $\Gamma_{\mathbf{B}}$ and $\Gamma_{\mathbf{C}}$ by

> $\Gamma_{\mathbf{B}} = \{\varphi \in \mathbf{Fm}_{\mathcal{L}}(Y) : \eta_{\mathbf{B}}(\varphi) \geq 1_{\mathbf{B}}\}$ and $\Gamma_{\mathbf{C}} = \{\psi \in \mathbf{Fm}_{\mathcal{L}}(Z) : \eta_{\mathbf{C}}(\psi) \geq 1_{\mathbf{C}}\}$,

respectively. We introduce a binary relation $\equiv$ on $\mathbf{Fm}_{\mathcal{L}}(Y \cup Z)$ by

> $\beta \equiv \gamma$ iff $\Gamma_{\mathbf{B}}, \Gamma_{\mathbf{C}} \vdash (\beta \backslash \gamma) \wedge (\gamma \backslash \beta)$,

where $\vdash$ denotes $\vdash_{\mathbf{L}}$. Then, it is easily seen that $\equiv$ is a congruence relation on $\mathbf{Fm}_{\mathcal{L}}(Y \cup Z)$ and that the quotient algebra $\mathbf{Fm}_{\mathcal{L}}(Y \cup Z)/{\equiv}$ is a member of $\mathsf{V}(\mathbf{L})$. Let us call this algebra, $\mathbf{D}$. We will show that this $\mathbf{D}$ is a required algebra satisfying the conditions for the AP.

To define required embeddings $h : \mathbf{B} \to \mathbf{D}$ and $k : \mathbf{C} \to \mathbf{D}$, we show that for each $\alpha \in \mathbf{Fm}_{\mathcal{L}}(X)$, $\Gamma_{\mathbf{B}} \vdash \alpha$ iff $\Gamma_{\mathbf{C}} \vdash \alpha$. Suppose that $\Gamma_{\mathbf{B}} \vdash \alpha$. Then, $f(1_{\mathbf{A}}) = 1_{\mathbf{B}} \leq \eta_{\mathbf{B}}(\alpha) = f(\eta_{\mathbf{A}}(\alpha))$ by using the definition of $\Gamma_{\mathbf{B}}$. Since f is injective, $1_{\mathbf{A}} \leq \eta_{\mathbf{A}}(\alpha)$ and hence $1_{\mathbf{C}} = g(1_{\mathbf{A}}) \leq g(\eta_{\mathbf{A}}(\alpha)) = \eta_{\mathbf{C}}(\alpha)$. Thus, $\Gamma_{\mathbf{C}} \vdash \alpha$. The converse implication can be shown in the same way by using the fact that g is injective. Now define mappings $h : \mathbf{B} \to \mathbf{D}$ and $k : \mathbf{C} \to \mathbf{D}$ by

- $h(b) = (\varphi/{\equiv})$ when $b = \eta_{\mathbf{B}}(\varphi)$ for a formula $\varphi \in \mathbf{Fm}_{\mathcal{L}}(Y)$,
- $k(c) = (\psi/{\equiv})$ when $c = \eta_{\mathbf{C}}(\psi)$ for a formula $\psi \in \mathbf{Fm}_{\mathcal{L}}(Z)$.

We prove that both h and k are well-defined embeddings. To show that h is well-defined, suppose that $\eta_{\mathbf{B}}(\varphi) = \eta_{\mathbf{B}}(\varphi')$ for $\varphi, \varphi' \in \mathbf{Fm}_{\mathcal{L}}(Y)$. Then, $\Gamma_{\mathbf{B}} \vdash (\varphi \backslash \varphi') \wedge (\varphi' \backslash \varphi)$, and hence $\varphi \equiv \varphi'$. It is easy to see that h is a homomorphism. To show that h is injective, suppose that $h(b) = h(b')$, where $b = \eta_{\mathbf{B}}(\varphi)$ and $b' = \eta_{\mathbf{B}}(\varphi')$ for $\varphi, \varphi' \in \mathbf{Fm}_{\mathcal{L}}(Y)$. Then, $\varphi \equiv \varphi'$, and thus $\Gamma_{\mathbf{B}}, \Gamma_{\mathbf{C}} \vdash (\varphi \backslash \varphi') \wedge (\varphi' \backslash \varphi)$ by the definition of $\equiv$. From the RP, it follows that $\Gamma_{\mathbf{B}} \vdash (\varphi\backslash\varphi') \wedge (\varphi'\backslash\varphi)$, which implies that $b = \eta_{\mathbf{B}}(\varphi) = \eta_{\mathbf{B}}(\varphi') = b'$. Similarly, k is shown to be a well-defined embedding. Note that in this case it is necessary to interchange the role of $\Gamma_{\mathbf{B}}$ and $\Gamma_{\mathbf{C}}$ in applying the RP, as the RP is of the symmetric form.

It remains to see that $h \circ f = k \circ g$. Take an arbitrary element $a \in A$. Then there exists a formula $\alpha \in \mathbf{Fm}_{\mathcal{L}}(X)$ such that $a = \eta_{\mathbf{A}}(\alpha)$. Then, $(h \circ f)(a) = h(f(\eta_{\mathbf{A}}(\alpha))) = h(\eta_{\mathbf{B}}(\alpha)) = (\alpha/\equiv)$. Similarly, $(k \circ g)(a) = (\alpha/\equiv)$. Thus, $h \circ f = k \circ g$.

We show next that the AP implies the RP. Let us assume that Γ and $\Sigma \cup \{\psi\}$ are sets of formulas such that $\Gamma \subseteq \mathbf{Fm}_{\mathcal{L}}(Y)$ and $\Sigma \cup \{\psi\} \subseteq \mathbf{Fm}_{\mathcal{L}}(Z)$ for some sets of variables Y and Z. Moreover, we assume that

- $\Gamma \vdash \alpha$ iff $\Sigma \vdash \alpha$ for every formula $\alpha \in \mathbf{Fm}_{\mathcal{L}}(X)$, where $X = Y \cap Z$,
- $\Gamma, \Sigma \vdash \psi$.

Define $\Delta = \{\alpha \in \mathbf{Fm}_{\mathcal{L}}(X) : \Gamma \vdash \alpha\}$, which is obviously equal to $\{\alpha \in \mathbf{Fm}_{\mathcal{L}}(X) : \Sigma \vdash \alpha\}$. The set Δ determines a binary relation $\equiv_\Delta$ on $\mathbf{Fm}_{\mathcal{L}}(X)$ by

$$\alpha \equiv_\Delta \beta \text{ iff } \Delta \vdash (\alpha\backslash\beta) \wedge (\beta\backslash\alpha),$$

which is in fact a congruence relation. We denote the quotient algebra of $\mathbf{Fm}_{\mathcal{L}}(X)$ determined by this $\equiv_\Delta$ as $\mathbf{Fm}_{\mathcal{L}}(X)/\Delta$. Similarly, we can introduce quotient algebras $\mathbf{Fm}_{\mathcal{L}}(Y)/\Gamma$ of $\mathbf{Fm}_{\mathcal{L}}(Y)$, and $\mathbf{Fm}_{\mathcal{L}}(Z)/\Sigma$ of $\mathbf{Fm}_{\mathcal{L}}(Z)$, by taking sets of formulas Γ and Σ, respectively, in the place of Δ. It is clear that all of these algebras $\mathbf{Fm}_{\mathcal{L}}(X)/\Delta$, $\mathbf{Fm}_{\mathcal{L}}(Y)/\Gamma$ and $\mathbf{Fm}_{\mathcal{L}}(Z)/\Sigma$ are members of $\mathsf{V}(\mathbf{L})$. We define mappings $f : \mathbf{Fm}_{\mathcal{L}}(X)/\Delta \to \mathbf{Fm}_{\mathcal{L}}(Y)/\Gamma$ and $g : \mathbf{Fm}_{\mathcal{L}}(X)/\Delta \to \mathbf{Fm}_{\mathcal{L}}(Z)/\Sigma$ by

$$f(\alpha/\equiv_\Delta) = \alpha/\equiv_\Gamma \text{ and } g(\alpha/\equiv_\Delta) = \alpha/\equiv_\Sigma.$$

They are shown to be well-defined embeddings, by using the definition of Δ.

Since we assume that the AP holds for $\mathsf{V}(\mathbf{L})$, there exist an algebra $\mathbf{D} \in \mathsf{V}(\mathbf{L})$, and two embeddings $h : \mathbf{Fm}_{\mathcal{L}}(Y)/\Gamma \to \mathbf{D}$ and $k : \mathbf{Fm}_{\mathcal{L}}(Z)/\Sigma \to \mathbf{D}$ satisfying $h \circ f = k \circ g$. Now, consider a valuation w over $\mathbf{D}$ for formulas in $\mathbf{Fm}_{\mathcal{L}}(Y \cup Z)$ defined as follows: For every $x \in Y \cup Z$,

$$w(x) = \begin{cases} h(x/\equiv_\Gamma) & \text{if } x \in Y \\ k(x/\equiv_\Sigma) & \text{if } x \in Z \end{cases}$$

The mapping w is well-defined, since if $x \in X$, $h(x/\equiv_\Gamma) = (h \circ f)(x/\equiv_\Delta) = (k \circ g)(x/\equiv_\Delta) = k(x/\equiv_\Sigma)$. As usual, w is extended to a mapping from $\mathbf{Fm}_{\mathcal{L}}(Y \cup Z)$ to $\mathbf{D}$, which satisfies that

$$w(\gamma) = \begin{cases} h(\gamma/\equiv_\Gamma) & \text{if } \gamma \in \mathbf{Fm}_{\mathcal{L}}(Y) \\ k(\gamma/\equiv_\Sigma) & \text{if } \gamma \in \mathbf{Fm}_{\mathcal{L}}(Z) \end{cases}$$

For each formula $\gamma \in \Gamma$, $(\gamma/\equiv_\Gamma) \geq 1_{\mathbf{Fm}_{\mathcal{L}}(Y)/\Gamma}$ and hence $w(\gamma) = h(\gamma/\equiv_\Gamma) \geq 1_{\mathbf{D}}$. Similarly, $w(\gamma) \geq 1_{\mathbf{D}}$ for each formula $\gamma \in \Sigma$. Thus, $\mathbf{D}, w \models \gamma$ for every

$\gamma \in \Gamma \cup \Sigma$. Since $\Gamma, \Sigma \vdash \psi$ by our assumption and $\mathbf{D} \in \mathsf{V}(\mathbf{L})$, $\mathbf{D}, w \models \psi$ must hold. That is, $k(\psi/\equiv_\Sigma) \geq 1_{\mathbf{D}}$, which implies $\psi/\equiv_\Sigma \geq 1_{\mathbf{Fm}_\mathcal{L}(Z)/\Sigma}$. This means that $\Sigma \vdash \psi$. This completes our proof. □

The following is an immediate corollary of Theorem 5.40, combining with Lemma 5.39.

COROLLARY 5.41. *For each substructural logic* $\mathbf{L}$ *over* $\mathbf{FL_e}$, $\mathbf{L}$ *has the DIP iff* $\mathsf{V}(\mathbf{L})$ *has the AP.*

We note that the following holds. This was originally proved in [Ono86], where the equivalence between the equational form of the ExIP and the congruence extension property is shown.

LEMMA 5.42. *For each substructural logic* $\mathbf{L}$, $\mathbf{L}$ *has the ExIP iff* $\mathsf{V}(\mathbf{L})$ *has the CEP.*

5.5.4. Algebraic characterization of the deductive interpolation property. We give next algebraic characterization of both the SDIP and the DIP. The algebraic characterization of the SDIP is essentially due to Wroński [Wro84b]. An algebraic characterization of the DIP by the *flat amalgamation property* is given in [Jón65], [Pig72], [Bac75] and [CP99]. We show another characterization due to by Kihara (see [KO07]).

A variety $\mathcal{V}$ has *transferable injections* (TI), if for all $\mathbf{A}, \mathbf{B}, \mathbf{C}$ in $\mathcal{V}$ and for any embedding $f : \mathbf{A} \to \mathbf{B}$ and any homomorphism $g : \mathbf{A} \to \mathbf{C}$, there exists an algebra $\mathbf{D}$ in $\mathcal{V}$, a homomorphism $h : \mathbf{B} \to \mathbf{D}$, and an embedding $k : \mathbf{C} \to \mathbf{D}$ such that $h \circ f = k \circ g$.

THEOREM 5.43. *A substructural logic* $\mathbf{L}$ *has the SDIP iff* $\mathsf{V}(\mathbf{L})$ *has TI.*

PROOF. We give here an outline of the proof, referring the proof of Theorem 5.40, since the proof goes almost in the same way. First, we show that $\mathsf{V}(\mathbf{L})$ has the TI, by assuming the SDIP* of $\mathbf{L}$. For given $\mathbf{A}, \mathbf{B}$ and $\mathbf{C}$ we construct an algebra $\mathbf{D}$ just in the same way as in the proof of Theorem 5.40. Since f is injective by our assumption, we can only show that for each $\alpha \in \mathbf{Fm}_\mathcal{L}(X)$, $\Gamma_\mathbf{B} \vdash \alpha$ implies $\Gamma_\mathbf{C} \vdash \alpha$. But this is enough for applying the SDIP*. Now, we define mappings $h : \mathbf{B} \to \mathbf{D}$ and $k : \mathbf{C} \to \mathbf{D}$ in the same way. We prove that both h and k are well-defined homomorphisms. Then we show that k is injective, by using the fact that for all $\psi, \psi' \in \mathbf{Fm}_\mathcal{L}(Z)$, $\Gamma_\mathbf{B}, \Gamma_\mathbf{C} \vdash (\psi\backslash\psi') \wedge (\psi'\backslash\psi)$ implies $\Gamma_\mathbf{C} \vdash (\psi\backslash\psi') \wedge (\psi'\backslash\psi)$, by using SDIP* of $\mathbf{L}$.

The converse direction is also shown in the same way, but this time, we assume that

- $\Gamma \vdash \alpha$ implies $\Sigma \vdash \alpha$ for every formula $\alpha \in \mathbf{Fm}_\mathcal{L}(X)$, where $X = Y \cap Z$,
- $\Gamma, \Sigma \vdash \psi$.

As before, we define $\Delta = \{\alpha \in \mathbf{Fm}_\mathcal{L}(X) : \Gamma \vdash \alpha\}$. Also, we define mappings f and g in the same way as before. Both maps are homomorphisms, but this time we can only show that f is an embedding. Then by TI, there exist an algebra $\mathbf{D}$, a homomorphism h and an embedding k. As before, we define a valuation w and conclude that $k(\psi/\equiv_\Sigma) \geq 1_\mathbf{D}$. Since k is injective, we can derive that $\psi/\equiv_\Sigma \geq 1_{\mathbf{Fm}_\mathcal{L}(Z)/\Sigma}$. Therefore $\Sigma \vdash \psi$. Thus, SDIP* holds for $\mathbf{L}$. □

Combining this Theorem 5.43 with Theorem 5.40 and Lemma 5.42, the following corollary is derived from Theorem 5.37. See [Bac75] for a related result.

COROLLARY 5.44. *For any subvariety $\mathcal{V}$ of* FL, *$\mathcal{V}$ has the TI iff it has both the AP and the CEP.*

The amalgamation property says that if for all $\mathbf{A}, \mathbf{B}, \mathbf{C}$ in a variety $\mathcal{V}$ and for all embeddings $f : \mathbf{A} \to \mathbf{B}$ and $g : \mathbf{A} \to \mathbf{C}$, there exists an algebra $\mathbf{D}$ in $\mathcal{V}$ and embeddings $h : \mathbf{B} \to \mathbf{D}$, $k : \mathbf{C} \to \mathbf{D}$ satisfying certain conditions. Obviously, h (k) gives an isomorphism between $\mathbf{B}$ and a subalgebra $h(\mathbf{B})$ of $\mathbf{D}$ (between $\mathbf{C}$ and a subalgebra $k(\mathbf{C})$ of $\mathbf{D}$, respectively). Equivalently, this can be expressed in such a way that there exist subalgebras $\mathbf{D}_1$ and $\mathbf{D}_2$ of $\mathbf{D}$ and isomorphisms from $\mathbf{D}_1$ to $\mathbf{B}$ and from $\mathbf{D}_2$ to $\mathbf{C}$ (in fact, they are h^{-1} and k^{-1}, respectively) satisfying certain conditions. By taking homomorphisms instead of isomorphisms, we obtain the definition of the generalized amalgamation property as follows. A variety $\mathcal{V}$ has the *generalized amalgamation property* (GAP) if for all $\mathbf{A}, \mathbf{B}, \mathbf{C}$ in $\mathcal{V}$ and for all embeddings $f : \mathbf{A} \to \mathbf{B}$ and $g : \mathbf{A} \to \mathbf{C}$, there exist an algebra $\mathbf{D}$ in $\mathcal{V}$, subalgebras $\mathbf{D}_1$ and $\mathbf{D}_2$ of $\mathbf{D}$, and surjective homomorphisms $i : \mathbf{D}_1 \to \mathbf{B}$ and $j : \mathbf{D}_2 \to \mathbf{C}$ such that for all $a \in A$ there exists $d \in D_1 \cap D_2$ such that $f(a) = i(d)$ and $g(a) = j(d)$. When i or j is injective, we call it the *generalized amalgamation property with injections*, and write it as IGAP. Of course, when both of them are injective, we get the AP.

THEOREM 5.45. *A substructural logic* $\mathbf{L}$ *has the DIP iff* $\mathsf{V}(\mathbf{L})$ *has the IGAP.*

PROOF. By Lemma 5.38, it is enough to show that $\mathbf{L}$ has the DIP* iff $\mathsf{V}(\mathbf{L})$ has the IGAP. Once again, the proof proceeds like the proof of Theorem 5.40. So, we point out only the necessary modifications. First, we show that $\mathsf{V}(\mathbf{L})$ has the IGAP, by assuming the DIP* of $\mathbf{L}$. We define $\Gamma_\mathbf{B}$ and $\Gamma_\mathbf{C}$ in the same way, but unlike in the proof of Theorem 5.40, we define a binary relation $\equiv$ on $\mathbf{Fm}_\mathcal{L}(Y \cup Z)$ by

$$\beta \equiv \gamma \text{ iff } \Gamma_\mathbf{B} \vdash (\beta\backslash\gamma) \wedge (\gamma\backslash\beta).$$

Now, let $\mathbf{D}$ be the quotient algebra $\mathbf{Fm}_\mathcal{L}(Y \cup Z)/\equiv$ which is in $\mathsf{V}(\mathbf{L})$. Also, let $\mathbf{D}_1$ and $\mathbf{D}_2$ be $\mathbf{Fm}_\mathcal{L}(Y)/\equiv$ and $\mathbf{Fm}_\mathcal{L}(Z)/\equiv$, respectively. Clearly they are subalgebras of $\mathbf{D}$. Now define mappings $i : \mathbf{D}_1 \to \mathbf{B}$ and $j : \mathbf{D}_2 \to \mathbf{C}$ by

- $i(\varphi/\equiv) = \eta_\mathbf{B}(\varphi)$ for a formula $\varphi \in \mathbf{Fm}_\mathcal{L}(Y)$,
- $j(\psi/\equiv) = \eta_\mathbf{C}(\psi)$ for a formula $\psi \in \mathbf{Fm}_\mathcal{L}(Z)$.

For all formulas $\varphi, \varphi' \in \mathbf{Fm}_\mathcal{L}(Y)$,

$\varphi \equiv \varphi'$ iff $\Gamma_{\mathbf{B}} \vdash (\varphi\backslash\varphi') \wedge (\varphi'\backslash\varphi)$ iff $\eta_{\mathbf{B}}(\varphi) = \eta_{\mathbf{B}}(\varphi')$.

Thus, i is a well-defined isomorphism. On the other hand, since for each $\alpha \in \mathbf{Fm}_{\mathcal{L}}(X)$, $\Gamma_{\mathbf{B}} \vdash \alpha$ iff $\Gamma_{\mathbf{C}} \vdash \alpha$, $\Gamma_{\mathbf{B}} \vdash \beta$ implies $\Gamma_{\mathbf{C}} \vdash \beta$ for any $\beta \in \mathbf{Fm}_{\mathcal{L}}(Z)$, by the DIP*. Therefore, for all formulas $\psi, \psi' \in \mathbf{Fm}_{\mathcal{L}}(Z)$,

$\psi \equiv \psi'$ iff $\Gamma_{\mathbf{B}} \vdash (\psi\backslash\psi') \wedge (\psi'\backslash\psi)$, which implies
$\Gamma_{\mathbf{C}} \vdash (\psi\backslash\psi') \wedge (\psi'\backslash\psi)$, iff $\eta_{\mathbf{C}}(\varphi) = \eta_{\mathbf{C}}(\varphi')$.

Thus, j is a well-defined isomorphism. Moreover, for each $a \in A$ there exists a formula $\alpha \in \mathbf{Fm}_{\mathcal{L}}(X)$ such that $\eta_{\mathbf{A}}(\alpha) = a$. It is easy to see that d, taken to be $\alpha/\equiv$, satisfies the required condition for the GAP. Thus the IGAP holds.

Conversely, suppose that $\mathsf{V}(\mathbf{L})$ has the IGAP. The proof also goes like the proof of Theorem 5.40. So, we introduce algebras $\mathbf{Fm}_{\mathcal{L}}(X)/\Delta$, $\mathbf{Fm}_{\mathcal{L}}(Y)/\Gamma$ and $\mathbf{Fm}_{\mathcal{L}}(Z)/\Sigma$, and mappings $f : \mathbf{Fm}_{\mathcal{L}}(X)/\Delta \to \mathbf{Fm}_{\mathcal{L}}(Y)/\Gamma$ and $g : \mathbf{Fm}_{\mathcal{L}}(X)/\Delta \to \mathbf{Fm}_{\mathcal{L}}(Z)/\Sigma$ by

$f(\alpha/\equiv_\Delta) = \alpha/\equiv_\Gamma$ and $g(\alpha/\equiv_\Delta) = \alpha/\equiv_\Sigma$.

As before, they are shown to be well-defined embeddings. By using the IGAP of $\mathsf{V}(\mathbf{L})$, we have that there exist an algebra $\mathbf{D} \in \mathsf{V}(\mathbf{L})$, subalgebras $\mathbf{D}_1$ and $\mathbf{D}_2$ of $\mathbf{D}$, and an isomorphism $i : \mathbf{D}_1 \to \mathbf{B}$ and a surjective homomorphism $j : \mathbf{D}_2 \to \mathbf{C}$ such that for all $a \in A$ there exists $d \in D_1 \cap D_2$ such that $f(a) = i(d)$ and $g(a) = j(d)$. Note that such an element d is determined uniquely since i is injective.

Now, we define a valuation w over $\mathbf{D}$ for formulas in $\mathbf{Fm}_{\mathcal{L}}(Y \cup Z)$ so as to satisfy the following: For every $x \in Y \cup Z$,

$$w(x) = \begin{cases} i^{-1}(x/\equiv_\Gamma) & \text{if } x \in Y \\ \text{an element in } j^{-1}(x/\equiv_\Sigma) & \text{if } x \in Z \setminus Y \end{cases}$$

Thus, $(i \circ w)(x) = x/\equiv_\Gamma$ for $x \in Y$ and $(j \circ w)(x) = x/\equiv_\Sigma$ for $x \in Z$. In fact, when $x \in X$, $w(x) = i^{-1}(x/\equiv_\Gamma)$ by the definition of w. Thus, $f(x/\equiv_\Delta) = x/\equiv_\Gamma = i(w(x))$. By the condition of the IGAP, we have $x/\equiv_\Sigma = g(x/\equiv_\Delta) = j(w(x)) = (j \circ w)(x)$. The mapping w is extended to a homomorphism from $\mathbf{Fm}_{\mathcal{L}}(Y \cup Z)$ to $\mathbf{D}$, which satisfies that:

- $(i \circ w)(\gamma) = \gamma/\equiv_\Gamma$ for $\gamma \in \mathbf{Fm}_{\mathcal{L}}(Y)$,
- $(j \circ w)(\beta) = \beta/\equiv_\Sigma$ for $\beta \in \mathbf{Fm}_{\mathcal{L}}(Z)$

For each formula $\gamma \in \Gamma$, $(i \circ w)(\gamma) = (\gamma/\equiv_\Gamma) \geq 1_{\mathbf{Fm}_{\mathcal{L}}(Y)/\Gamma}$ and hence $w(\gamma) \geq 1_{\mathbf{D}}$ by the injectivity of i. Thus, $\mathbf{D}, w \models \gamma$ for every $\gamma \in \Gamma$. So, if $\Gamma \vdash \psi$ for a formula $\psi \in \mathbf{Fm}_{\mathcal{L}}(Z)$, then $\mathbf{D}, w \models \psi$. Thus, $(j \circ w)(\psi) = (\psi/\equiv_\Sigma) \geq 1_{\mathbf{C}}$. Since $\mathbf{C} = \mathbf{Fm}_{\mathcal{L}}(Z)/\Sigma$, it implies $\Sigma \vdash \psi$. This completes our proof. □

5.6. Craig interpolation property

In this section, we discuss a similar algebraic characterization of the Craig interpolation property: the *super-amalgamation property* (superAP).

5.6.1. Extensions of Craig interpolation property. We introduce two extensions of the CIP. A substructural logic $\mathbf{L}$ has the *strong Craig interpolation property* (SCIP), if for any set of formulas $\Gamma \cup \Sigma \cup \{\varphi, \psi\}$, if $\Gamma, \Sigma \vdash_{\mathbf{L}} \varphi\backslash\psi$, then there exists a formula δ such that

(1) $\Gamma \vdash_{\mathbf{L}} \varphi\backslash\delta$, and $\Sigma \vdash_{\mathbf{L}} \delta\backslash\psi$,
(2) $\mathrm{var}(\delta) \subseteq \mathrm{var}(\Gamma \cup \{\varphi\}) \cap \mathrm{var}(\Sigma \cup \{\psi\})$.

Next, a substructural logic $\mathbf{L}$ has the *strong Robinson property* (SRP), provided that for any set of formulas $\Gamma \cup \Sigma \cup \{\varphi, \psi\}$ such that $\Gamma \vdash_{\mathbf{L}} \sigma$ iff $\Sigma \vdash_{\mathbf{L}} \sigma$ for every formula $\sigma \in \mathbf{Fm}_{\mathcal{L}}(X)$ with $X = \mathrm{var}(\Gamma \cup \{\varphi\}) \cap \mathrm{var}(\Sigma \cup \{\psi\})$, the following holds:

(1) $\Gamma, \Sigma \vdash_{\mathbf{L}} \alpha$ implies $\Gamma \vdash_{\mathbf{L}} \alpha$ for any α with $\mathrm{var}(\alpha) \subseteq \mathrm{var}(\Gamma \cup \{\varphi\})$,
(2) $\Gamma, \Sigma \vdash_{\mathbf{L}} \beta$ implies $\Sigma \vdash_{\mathbf{L}} \beta$ for any β with $\mathrm{var}(\beta) \subseteq \mathrm{var}(\Sigma \cup \{\psi\})$,
(3) if $\Gamma, \Sigma \vdash_{\mathbf{L}} \varphi\backslash\psi$ then there exists a formula $\delta \in \mathbf{Fm}_{\mathcal{L}}(X)$ such that $\Gamma \vdash_{\mathbf{L}} \varphi\backslash\delta$, and $\Sigma \vdash_{\mathbf{L}} \delta\backslash\psi$.

Similarly to the arguments in the previous section, we can introduce an alternative definition of the SCIP, which we call the SCIP*. A logic $\mathbf{L}$ has the SCIP*, provided that for any set of formulas $\Gamma \cup \Sigma \cup \{\varphi, \psi\}$ such that $\Gamma \vdash_{\mathbf{L}} \sigma$ *implies* $\Sigma \vdash_{\mathbf{L}} \sigma$ for every formula $\sigma \in \mathbf{Fm}_{\mathcal{L}}(X)$ with $X = \mathrm{var}(\Gamma \cup \{\varphi\}) \cap \mathrm{var}(\Sigma \cup \{\psi\})$, the following holds:

(1) $\Gamma, \Sigma \vdash_{\mathbf{L}} \alpha$ implies $\Gamma \vdash_{\mathbf{L}} \alpha$ for any α with $\mathrm{var}(\alpha) \subseteq \mathrm{var}(\Gamma \cup \{\varphi\})$,
(2) $\Gamma, \Sigma \vdash_{\mathbf{L}} \beta$ implies $\Sigma \vdash_{\mathbf{L}} \beta$ for any β with $\mathrm{var}(\beta) \subseteq \mathrm{var}(\Sigma \cup \{\psi\})$,
(3) if $\Gamma, \Sigma \vdash_{\mathbf{L}} \varphi\backslash\psi$ then there exists a formula $\delta \in \mathbf{Fm}_{\mathcal{L}}(X)$ such that $\Gamma \vdash_{\mathbf{L}} \varphi\backslash\delta$, and $\Sigma \vdash_{\mathbf{L}} \delta\backslash\psi$.

Then, the following lemma is shown in the same way as Lemma 5.38.

LEMMA 5.46. *For each substructural logic, the SCIP holds iff the* SCIP* *holds.*

LEMMA 5.47. *For every substructural logic, the following hold.*

(1) *The SCIP implies the SRP and the SRP implies the CIP.*
(2) *The SCIP implies the SDIP.*
(3) *The SRP implies the RP.*

PROOF. Since the SCIP is equivalent to the SCIP* by Lemma 5.46, the first implication of (1) follows immediately. The CIP follows from the SRP, by taking the empty set for both Γ and Σ. To show (2), suppose that $\Gamma, \Sigma \vdash_{\mathbf{L}} \varphi$. Then, $\Gamma, \Sigma \vdash_{\mathbf{L}} 1\backslash\varphi$ holds. Thus, by the SCIP, there exists a formula δ with $\mathrm{var}(\delta) \subseteq \mathrm{var}(\Gamma) \cap \mathrm{var}(\Sigma \cup \{\psi\})$ such that $\Gamma \vdash_{\mathbf{L}} 1\backslash\delta$ and $\Sigma \vdash_{\mathbf{L}} \delta\backslash\psi$. From them, both $\Gamma \vdash_{\mathbf{L}} \delta$ and $\delta, \Sigma \vdash_{\mathbf{L}} \psi$ follow. It is obvious that the RP follows from the SRP. □

Thus we summarize the relations among these interpolation properties in Figure 5.2.

On the other hand, if a given logic is over $\mathbf{FL_e}$, we have the following.

$$\begin{array}{ccccc} \text{SCIP} & \Longrightarrow & \text{SRP} & \Longrightarrow & \text{CIP} \\ \Downarrow & & \Downarrow & & \\ \text{SDIP} & \Longrightarrow & \text{RP} & \Longrightarrow & \text{DIP} \end{array}$$

FIGURE 5.2. Relations among interpolation properties.

LEMMA 5.48. *For every substructural logic over* $\mathbf{FL_e}$,

(1) *the CIP implies the SCIP, and hence all of the SCIP, the SRP and the CIP are equivalent,*
(2) *the CIP implies the DIP.*

PROOF. Let $\mathbf{L}$ be a substructural logic over $\mathbf{FL_e}$ with the CIP. Suppose that $\gamma, \sigma \vdash_{\mathbf{L}} \varphi \to \psi$. Then by the local deduction theorem a formula $((\gamma \wedge 1)^m \cdot \varphi) \to ((\sigma \wedge 1)^n \to \psi)$ is provable in $\mathbf{L}$ for some m and n. By the CIP, there exists a formula δ such that both $((\gamma \wedge 1)^m \cdot \varphi) \to \delta$ and

$\delta \to ((\sigma \wedge 1)^n \to \psi)$ are provable in $\mathbf{L}$ and $\mathrm{var}(\delta) \subseteq \mathrm{var}((\gamma \wedge 1)^m \cdot \varphi) \cap \mathrm{var}((\sigma \wedge 1)^n \to \psi)$. Then, both $\gamma \vdash_{\mathbf{L}} \varphi \to \delta$ and $\sigma \vdash_{\mathbf{L}} \delta \to \psi$ hold for a formula δ which satisfies $\mathrm{var}(\delta) \subseteq \mathrm{var}(\gamma, \varphi) \cap \mathrm{var}(\sigma, \psi)$. Thus, the SCIP holds for $\mathbf{L}$. (2) follows immediately from (1) with Lemmas 5.39 and 5.47. □

Thus, we have:

$$\boxed{\text{SCIP} \Leftrightarrow \text{SRP} \Leftrightarrow \text{CIP}}$$
$$\Downarrow$$
$$\boxed{\text{SDIP} \Leftrightarrow \text{RP} \Leftrightarrow \text{DIP}}$$

FIGURE 5.3. Relations among interpolation properties in the commutative case.

The above figure shows that the CIP always implies the SDIP for every substructural logic over $\mathbf{FL_e}$. But they are in general independent, as the following theorem shows.

LEMMA 5.49. *The properties CIP and SDIP are independent.*

PROOF. We first show that the CIP does not imply the SDIP. It follows from Theorem 5.13 that $\mathbf{FL_w}$ has the CIP. On the other hand, the variety $\mathsf{FL_w}$ of FL_w-algebras does not have the CEP. In fact, in Section 3.6.5 we saw a 5-element integral FL-algebra without the CEP. By Lemma 5.42 and Theorem 5.37, $\mathbf{FL_w}$ cannot have the SDIP.

To see that SDIP does not imply the CIP, let $\mathbf{MV}$ be Łukasiewicz' infinite-valued logic. Recall that the corresponding variety MV is generated by the FL_{ew}-algebra whose underlying set is the unit interval of real numbers

$[0,1]$ with the usual order, where $ab = max\{0, a+b-1\}$ and $a \to b = min\{1, 1-a+b\}$. It is easy to see that $(p \wedge \neg p) \to (q \vee \neg q)$ is provable in **MV**, where p, q are distinct propositional variables, but the formula has no interpolants. On the other hand, since MV has the AP as shown in [Mun86] and **MV** is over $\mathbf{FL_e}$, **MV** has the SDIP. In particular, for the example under discussion, i.e., for $p \wedge \neg p \vdash_{\mathbf{MV}} q \vee \neg q$ we have $p \wedge \neg p \vdash_{\mathbf{MV}} 0$, because $\vdash_{\mathbf{MV}} (p \wedge \neg p)^2 \to 0$, and $0 \vdash_{\mathbf{MV}} q \vee \neg q$. □

5.6.2. Super-amalgamation property and strong Robinson property. We extend Theorem 5.40 to a characterization result of the strong Robinson property. A variety $\mathcal{V}$ has the *super-amalgamation property* (superAP), if whenever $\mathbf{A}, \mathbf{B}, \mathbf{C}$ are in $\mathcal{V}$ and $f : \mathbf{A} \to \mathbf{B}$, $g : \mathbf{A} \to \mathbf{C}$ are embeddings, then there exists an algebra $\mathbf{D}$ in $\mathcal{V}$ and embeddings $h : \mathbf{B} \to \mathbf{D}$, $k : \mathbf{C} \to \mathbf{D}$ such that $h \circ f = k \circ g$, and moreover that for all $b \in B$ and $c \in C$ if $h(b) \leq k(c)$ there there exists $a \in A$ for which both $b \leq f(a)$ and $g(a) \leq c$ hold. Clearly, the superAP implies the AP.

THEOREM 5.50. *A substructural logic* $\mathbf{L}$ *has the SRP iff* $\mathsf{V}(\mathbf{L})$ *has the superAP.*

PROOF. We show our theorem using the proof of Theorem 5.40. Suppose first that $\mathbf{L}$ has the SRP. Since it has the RP, we can construct an algebra $\mathbf{D}$ and embeddings h and k as we have done in the proof of Theorem 5.40. So, it remains to show that the last condition for the superAP holds. Let $b \in B$ and $c \in C$ such that $h(b) \leq k(c)$. Then, for some formulas $\varphi \in \mathbf{Fm}_{\mathcal{L}}(Y)$ and $\psi \in \mathbf{Fm}_{\mathcal{L}}(Z)$, $b = \eta_{\mathbf{B}}(\varphi)$ and $c = \eta_{\mathbf{C}}(\psi)$. From the definitions of h, k and $\equiv$, we can see that the condition $h(b) \leq k(c)$ is equivalent to $\Gamma_{\mathbf{B}}, \Gamma_{\mathbf{C}} \vdash \varphi\backslash\psi$. Since the required condition for applying the SRP is satisfied by $\Gamma_{\mathbf{B}}$ and $\Gamma_{\mathbf{C}}$, as shown in the proof of Theorem 5.40, there exists a formula $\delta \in \mathbf{Fm}_{\mathcal{L}}(X)$ such that $\Gamma_{\mathbf{B}} \vdash \varphi\backslash\delta$ and $\Gamma_{\mathbf{C}} \vdash \delta\backslash\psi$. Let $a = \eta_{\mathbf{A}}(\delta)$. Then, we have $b = \eta_{\mathbf{B}}(\varphi) \leq \eta_{\mathbf{B}}(\delta) = f(\eta_{\mathbf{A}}(\delta)) = f(a)$ and $g(a) = g(\eta_{\mathbf{A}}(\delta)) = \eta_{\mathbf{C}}(\delta) \leq \eta_{\mathbf{C}}(\psi) = c$. Thus, the superAP holds.

Conversely, suppose that $\mathsf{V}(\mathbf{L})$ has the superAP. As we have done in the proof of Theorem 5.40, for given embeddings $f : \mathbf{Fm}_{\mathcal{L}}(X)/\Delta \to \mathbf{Fm}_{\mathcal{L}}(Y)/\Gamma$ and $g : \mathbf{Fm}_{\mathcal{L}}(X)/\Delta \to \mathbf{Fm}_{\mathcal{L}}(Z)/\Sigma$, there exist an algebra $\mathbf{D}$ and embeddings $h : \mathbf{Fm}_{\mathcal{L}}(Y)/\Gamma \to \mathbf{D}$ and $k : \mathbf{Fm}_{\mathcal{L}}(Z)/\Sigma \to \mathbf{D}$ satisfying $h \circ f = k \circ g$. Then a valuation w over $\mathbf{D}$ is introduced, which satisfies that

$$w(\gamma) = \begin{cases} h(\gamma/{\equiv_\Gamma}) & \text{if } \gamma \in \mathbf{Fm}_{\mathcal{L}}(Y) \\ k(\gamma/{\equiv_\Sigma}) & \text{if } \gamma \in \mathbf{Fm}_{\mathcal{L}}(Z) \end{cases}$$

To complete the proof of the SRP of $\mathbf{L}$, we assume that $\Gamma, \Sigma \vdash \alpha\backslash\beta$ for $\alpha \in \mathbf{Fm}_{\mathcal{L}}(Y)$ and $\beta \in \mathbf{Fm}_{\mathcal{L}}(Z)$. Then, $\mathbf{D}, w \models \alpha\backslash\beta$ holds, i.e., $w(\alpha) \leq w(\beta)$, or equivalently, $h(\alpha/{\equiv_\Gamma}) \leq k(\beta/{\equiv_\Sigma})$. By the superAP, there exists an element $d \in \mathbf{Fm}_{\mathcal{L}}(X)/\Delta$ such that $(\alpha/{\equiv_\Gamma}) \leq f(d)$ and $g(d) \leq (\beta/{\equiv_\Sigma})$. This implies that there exists a formula $\delta \in \mathbf{Fm}_{\mathcal{L}}(X)$ such that $d = (\delta/{\equiv_\Delta})$,

$(\alpha/{\equiv_\Gamma}) \leq (\delta/{\equiv_\Gamma})$ and $(\delta/{\equiv_\Sigma}) \leq (\beta/{\equiv_\Sigma})$. Therefore, $\Gamma \vdash \alpha\backslash\delta$ and $\Sigma \vdash \delta\backslash\beta$ for a formula $\delta \in \mathbf{Fm}_{\mathcal{L}}(X)$. Thus, the SRP holds. □

Thus, we have the following as a corollary.

COROLLARY 5.51. *A substructural logic* $\mathbf{L}$ *over* $\mathbf{FL_e}$ *has the CIP iff* $\mathsf{V}(\mathbf{L})$ *has the superAP.*

5.6.3. Algebraic characterization of Craig interpolation property. Like in Theorem 5.50, we can give algebraic characterizations of the SCIP and the CIP. The equivalence of the SCIP to the SCIP* suggests the following definition. A variety $\mathcal{V}$ has *super transferable injections* (superTI), if whenever $\mathbf{A}, \mathbf{B}, \mathbf{C}$ are in $\mathcal{V}$, $f : \mathbf{A} \to \mathbf{B}$ is an embedding and $g : \mathbf{A} \to \mathbf{C}$ is a homomorphism, then there exists an algebra $\mathbf{D}$ in $\mathcal{V}$, a homomorphism $h : \mathbf{B} \to \mathbf{D}$, and an embedding $k : \mathbf{C} \to \mathbf{D}$ such that $h \circ f = k \circ g$, and moreover for all $b \in B$ and $c \in C$ if $h(b) \leq k(c)$ then there exists $a \in A$ for which both $b \leq f(a)$ and $g(a) \leq c$ hold. By combining the proof of Theorem 5.50 with that of Theorem 5.43, we have the following.

THEOREM 5.52. *A substructural logic* $\mathbf{L}$ *has the SCIP iff* $\mathsf{V}(\mathbf{L})$ *has the superTI.*

Finally, we give an algebraic characterization of the CIP, by using the following notion of the superGAP. A variety $\mathcal{V}$ has the *super generalized amalgamation property* (superGAP), if for all $\mathbf{A}, \mathbf{B}, \mathbf{C}$ in $\mathcal{V}$ and for all embeddings $f : \mathbf{A} \to \mathbf{B}$ and $g : \mathbf{A} \to \mathbf{C}$, there exist an algebra $\mathbf{D}$ in $\mathcal{V}$, subalgebras $\mathbf{D}_1$ and $\mathbf{D}_2$ of $\mathbf{D}$, and surjective homomorphisms $i : \mathbf{D}_1 \to \mathbf{B}$ and $j : \mathbf{D}_2 \to \mathbf{C}$ that satisfy the following:

(1) for all $a \in A$ there exists $d \in D_1 \cap D_2$ such that $f(a) = i(d)$ and $g(a) = j(d)$,
(2) for all $d_1 \in D_1$, $d_2 \in D_2$ such that $d_1 \leq d_2$, there exists $a \in A$ such that $i(d_1) \leq f(a)$ and $g(a) \leq j(d_2)$.

THEOREM 5.53. *A substructural logic* $\mathbf{L}$ *has the CIP iff* $\mathsf{V}(\mathbf{L})$ *has the superGAP.*

PROOF. The proof goes like the proofs of Theorems 5.50 and 5.45. First, we show that the superGAP holds, by assuming the CIP. But this time we define a binary relation $\equiv$ on $\mathbf{Fm}_{\mathcal{L}}(Y \cup Z)$ simply by

$$\beta \equiv \gamma \text{ iff } \vdash (\beta\backslash\gamma) \wedge (\gamma\backslash\beta).$$

Then, let $\mathbf{D}$ be the quotient algebra $\mathbf{Fm}_{\mathcal{L}}(Y \cup Z)/{\equiv}$ which is in $\mathsf{V}(\mathbf{L})$, and let $\mathbf{D}_1$ and $\mathbf{D}_2$ be $\mathbf{Fm}_{\mathcal{L}}(Y)/{\equiv}$ and $\mathbf{Fm}_{\mathcal{L}}(Z)/{\equiv}$, respectively. We define mappings $i : \mathbf{D}_1 \to \mathbf{B}$ and $j : \mathbf{D}_2 \to \mathbf{C}$ by

- $i(\varphi/{\equiv}) = \eta_{\mathbf{B}}(\varphi)$ for a formula $\varphi \in \mathbf{Fm}_{\mathcal{L}}(Y)$,
- $j(\psi/{\equiv}) = \eta_{\mathbf{C}}(\psi)$ for a formula $\psi \in \mathbf{Fm}_{\mathcal{L}}(Z)$.

Then both i and j are well-defined and are surjective homomorphisms. It is easily shown that they satisfy the first condition of the superGAP, like in the proof of Theorem 5.45. To show the second condition, suppose that $d_1 \in D_1$, $d_2 \in D_2$ such that $d_1 \leq d_2$. Then, there exist formulas $\varphi \in \mathbf{Fm}_{\mathcal{L}}(Y)$ and $\psi \in \mathbf{Fm}_{\mathcal{L}}(Z)$ such that $d_1 = \varphi/{\equiv}, d_2 = \psi/{\equiv}$ and $\vdash \varphi\backslash\psi$. By the CIP, there exists a formula $\delta \in \mathbf{Fm}_{\mathcal{L}}(X)$ such that both $\vdash \varphi\backslash\delta$ and $\vdash \delta\backslash\psi$ hold. Thus, $(\varphi/{\equiv}) \leq (\delta/{\equiv})$ holds in $\mathbf{D}_1$ and $(\delta/{\equiv}) \leq (\psi/{\equiv})$ holds in $\mathbf{D}_2$. Let $a = \eta_{\mathbf{A}}(\delta)$. Then, by using the first condition of the superGAP, we have $i(d_1) = i(\varphi/{\equiv}) \leq i(\delta/{\equiv}) = \eta_{\mathbf{B}}(\delta) = f(\eta_{\mathbf{A}}(\delta)) = f(a)$, and similarly $g(a) \leq j(d_2)$. Thus, the superGAP holds.

Conversely, suppose that the superGAP holds for $\mathsf{V}(\mathbf{L})$. By using the same congruence relation $\equiv$ as in the above, define $\mathbf{A}$, $\mathbf{B}$ and $\mathbf{C}$ by $\mathbf{Fm}_{\mathcal{L}}(X)/{\equiv}$, $\mathbf{Fm}_{\mathcal{L}}(Y)/{\equiv}$ and $\mathbf{Fm}_{\mathcal{L}}(Z)/{\equiv}$. Then there exist inclusion maps $f : \mathbf{A} \to \mathbf{B}$ and $g : \mathbf{A} \to \mathbf{C}$. By the superGAP, there exist an algebra $\mathbf{D}$ in $\mathsf{V}(\mathbf{L})$, subalgebras $\mathbf{D}_1$ and $\mathbf{D}_2$ of $\mathbf{D}$, and surjective homomorphisms $i : \mathbf{D}_1 \to \mathbf{B}$ and $j : \mathbf{D}_2 \to \mathbf{C}$ that satisfy the following:

(1) for all $a \in A$ there exists $d \in D_1 \cap D_2$ such that $f(a) = i(d)$ and $g(a) = j(d)$,
(2) for all $d_1 \in D_1$, $d_2 \in D_2$ such that $d_1 \leq d_2$, there exists $a \in A$ such that $i(d_1) \leq f(a)$ and $g(a) \leq j(d_2)$.

Now, we define a valuation w over $\mathbf{D}$ for formulas in $\mathbf{Fm}_{\mathcal{L}}(Y \cup Z)$. By the above condition (1), for each variable $x \in X$, there exists $d \in D_1 \cap D_2$ such that $f(x/{\equiv}) = i(d)$ and $g(x/{\equiv}) = j(d)$. Let d_x be one of such elements d. Now, define w as follows: For every $x \in Y \cup Z$,

$$w(x) = \begin{cases} d_x & \text{if } x \in X \\ \text{an element in } i^{-1}(x/{\equiv}) & \text{if } x \in Y \setminus X \\ \text{an element in } j^{-1}(x/{\equiv}) & \text{if } x \in Z \setminus X \end{cases}$$

Thus, we can show that $(i \circ w)(x) = x/{\equiv}$ for $x \in Y$ and $(j \circ w)(x) = x/{\equiv}$ for $x \in Z$. As usual, the mapping w is extended to a mapping from $\mathbf{Fm}_{\mathcal{L}}(Y \cup Z)$ to $\mathbf{D}$, for which the following holds.

- $(i \circ w)(\gamma) = (\gamma/{\equiv})$ for $\gamma \in \mathbf{Fm}_{\mathcal{L}}(Y)$,
- $(j \circ w)(\beta) = (\beta/{\equiv})$ for $\beta \in \mathbf{Fm}_{\mathcal{L}}(Z)$

Suppose that $\vdash \varphi\backslash\psi$ for formulas $\varphi \in \mathbf{Fm}_{\mathcal{L}}(Y)$ and $\psi \in \mathbf{Fm}_{\mathcal{L}}(Z)$. Then, $w(\varphi\backslash\psi) \geq 1_{\mathbf{D}}$. Thus, $w(\varphi) \leq w(\psi)$. By the second condition of the superGAP, there exists $a \in A$ such that $i(w(\varphi)) \leq f(a) = a$ and $a = g(a) \leq j(w(\psi))$. Let $a = \delta/{\equiv}$ for $\delta \in \mathbf{Fm}_{\mathcal{L}}(X)$. Then, $(\varphi/{\equiv}) \leq (\delta/{\equiv})$ and $(\delta/{\equiv}) \leq (\varphi/{\equiv})$. That is, $\vdash \varphi\backslash\delta$ and $\vdash \delta\backslash\psi$. Thus, the CIP holds. □

Though we do not mention the details here, these algebraic characterizations of interpolation properties hold for many fragments of substructural logics. If we take a sublanguage $\mathcal{L}'$ of our language $\mathcal{L}$, interpolants must be formulas of $\mathcal{L}'$ in interpolation properties, and homomorphisms in their algebraic characterizations must be mappings which preserve operations and constants in $\mathcal{L}'$.

We summarize our results on algebraic characterizations of interpolation properties in Figure 5.4. It is an interesting problem whether the superGAP implies the IGAP or not.

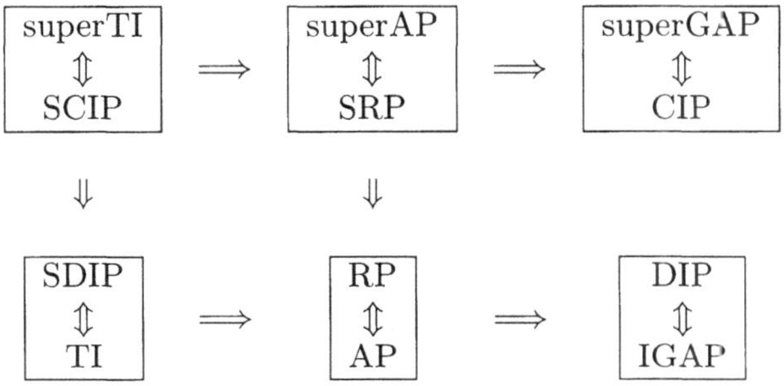

FIGURE 5.4. Algebraic characterizations of interpolation properties.

5.6.4. Joint embedding property. In these deductive interpolation properties and (extended) Craig interpolation properties, it may happen that the two sets of variables under consideration are disjoint and therefore the interpolants are formulas without variables. Obvious modifications of the results shown above lead us to algebraic characterizations of these cases. To see this, let us consider the relation between the RP and the AP, for example.

We say that a substructural logic $\mathbf{L}$ has the *$\sharp$-Robinson property* ($RP^\sharp$), if for every set of formulas $\Gamma \cup \Sigma \cup \{\psi\}$ such that $\mathrm{var}(\Gamma) \cap \mathrm{var}(\Sigma \cup \{\psi\})$ is empty and moreover that $\Gamma \vdash_{\mathbf{L}} \delta$ iff $\Sigma \vdash_{\mathbf{L}} \delta$ for every formula δ without variables, $\Gamma, \Sigma \vdash_{\mathbf{L}} \psi$ implies $\Sigma \vdash_{\mathbf{L}} \psi$. Corresponding to this property, we introduce the following notion of *$\sharp$-amalgamation property*. A variety $\mathcal{V}$ has the $\sharp$-amalgamation property ($AP^\sharp$), if for all $\mathbf{B}, \mathbf{C}$ in $\mathcal{V}$ such that their 0-generated subalgebras are isomorphic, there exists an algebra $\mathbf{D}$ in $\mathcal{V}$ into which both of these algebras are embedded. Almost in the same way as the proof of Theorem 5.40, we can show the following.

THEOREM 5.54. *A substructural logic* $\mathbf{L}$ *has the* $RP^\sharp$ *iff* $\mathsf{V}(\mathbf{L})$ *has the* $AP^\sharp$.

We say that a substructural logic $\mathbf{L}$ has the *$\sharp$-deductive interpolation property* ($DIP^\sharp$), if for any set of formulas $\Gamma \cup \{\psi\}$ such that $\mathrm{var}(\Gamma) \cap \mathrm{var}(\psi)$ is empty, if $\Gamma \vdash_{\mathbf{L}} \psi$, then there exists a formula δ without variables such that $\Gamma \vdash_{\mathbf{L}} \delta$ and $\delta \vdash_{\mathbf{L}} \psi$. Then, as a corollary of Theorem 5.54, we have the following.

COROLLARY 5.55. *A substructural logic* $\mathbf{L}$ *over* $\mathbf{FL_e}$ *has the* $DIP^\sharp$ *iff* $\mathsf{V}(\mathbf{L})$ *has the* $AP^\sharp$.

We say that a variety $\mathcal{V}$ has the *joint embedding property* (JEP), if for all non-degenerate algebras $\mathbf{B}, \mathbf{C}$ in $\mathcal{V}$, there exists an algebra $\mathbf{D}$ in $\mathcal{V}$ into which both of these algebras are embedded. The $\sharp$-amalgamation property can be regarded as a restricted form of the JEP. On the other hand, for FL_w-algebras, every 0-generated subalgebra of a given non-degenerate algebra is isomorphic to an FL_w-algebra with two elements $\{0, 1\}$, the JEP is equivalent to the $AP^\sharp$.

5.6.5. Interpolation property and pseudo-relevance property. A substructural logic $\mathbf{L}$ has the *pseudo-relevance property* (PRP), if for all pairs of formulas α, β with $\mathrm{var}(\alpha) \cap \mathrm{var}(\beta) = \emptyset$, if $\alpha\backslash\beta$ is provable in $\mathbf{L}$ then either $\alpha\backslash\bot$ or β is provable in it. The notion was introduced by Suzuki [Suz89].

Similarly, we can introduce its deductive forms as follows: A substructural logic $\mathbf{L}$ has the *strong deductive pseudo-relevance property* (SDPRP), if for all sets of formulas $\Gamma \cup \Sigma \cup \{\psi\}$ with $\mathrm{var}(\Gamma) \cap \mathrm{var}(\Sigma \cup \{\psi\}) = \emptyset$, $\Gamma, \Sigma \vdash_{\mathbf{L}} \psi$ implies either $\Gamma \vdash_{\mathbf{L}} \bot$ or $\Sigma \vdash_{\mathbf{L}} \psi$. When Σ is to the empty set in this definition, $\mathbf{L}$ is said to have the *deductive pseudo-relevance property* (DPRP).

Two algebras $\mathbf{B}, \mathbf{C}$ are *jointly embeddable* into an algebra $\mathbf{D}$, if there exist embeddings $h : \mathbf{B} \to \mathbf{D}$ and $j : \mathbf{C} \to \mathbf{D}$. Then, we have the following. (See Maksimova [Mak95] for related results.)

THEOREM 5.56. *Let $\mathbf{L}$ be a substructural logic.*

(1) *$\mathbf{L}$ has the SDPRP iff every pair of subdirectly irreducible algebras in $\mathsf{V}(\mathbf{L})$ are jointly embeddable into an algebra in $\mathsf{V}(\mathbf{L})$.*
(2) *SDPRP implies DPRP for every $\mathbf{L}$, and the converse holds also when $\mathsf{V}(\mathbf{L})$ has the CEP.*
(3) *If a subvariety $\mathcal{V}$ of $\mathsf{FL_{ew}}$ has the AP, then all pairs of subdirectly irreducible algebras in $\mathcal{V}$ are jointly embeddable into an algebra in $\mathcal{V}$. Thus, the DIP implies the DPRP for every substructural logic over $\mathbf{FL_{ew}}$.*

The implication in the above 3 does not hold, if we drop either of the conditions $1 = \top$ and $0 = \bot$. For instance, $\mathbf{FL_e}$ has the CIP and hence the DIP, but it does not have the DPRP. We can show the following in the same way as a result by Komori [Kom78], by using an extension of Glivenko's theorem.

THEOREM 5.57. *Every extension of the logic $\mathbf{FL_{ew}}$ with the axiom $\neg(\varphi\wedge\neg\varphi)$ has the PRP.*

PROOF. Let $\mathbf{L}$ be an extension of $\mathbf{FL_{ew}}$ with $\neg(\varphi \wedge \neg\varphi)$. Suppose that neither $\neg\alpha$ nor β is provable in $\mathbf{L}$. Since β is not provable in $\mathbf{L}$, there exists an FL_{ew}-algebra $\mathbf{A}$ which validates $\mathbf{L}$ but $f(\beta) < 1_{\mathbf{A}}$ for some valuation f. On the other hand, since $\neg\alpha$ is not provable in $\mathbf{L}$, it is neither provable in classical logic by Glivenko theorem and by the fact that **L** is an extension of $\mathbf{FL_{ew}}$ with $\neg(\varphi \wedge \neg\varphi)$ (see Corollary 8.36). Therefore, there exists a two valued valuation g such that $g(\neg\alpha) = 0$, i.e., $g(\alpha) = 1$. The algebra **A** has a subalgebra isomorphic to two-valued Boolean algebra, and hence g can be regarded also as a valuation of **A**. As there are no common variables appearing in α and β, we can define another valuation h of **A** such that $h(p) = f(p)$ for every variable $p \in \mathrm{var}(\alpha)$ and $h(q) = g(q)$ for every variable $q \in \mathrm{var}(\beta)$. Clearly, $h(\alpha) = g(\alpha) = 1$ and $h(\beta) = f(\beta) < 1_{\mathbf{A}}$. Thus, $h(\alpha \to \beta) < 1_{\mathbf{A}}$. Hence, $\alpha \to \beta$ is not provable in **L**. □

Exercises

(1) Show that the sequent calculus obtained from $\mathbf{FL_{ew}}$ by adding sequents of the form $\Rightarrow \alpha \vee \neg\alpha$ is equivalent to classical logic.
(2) Show that classical logic is Halldén complete.
(3) In Exercises of Chapter 4, the $(n \rightsquigarrow k)$-rule is introduced. Quite similarly to the case for $\mathbf{FL_{e}}^{n\rightsquigarrow 1}$, we can show that the cut elimination holds also for $\mathbf{FL_{ew}}^{n\rightsquigarrow 1}$. Now, using Maehara's method, show that $\mathbf{FL_{ew}}^{n\rightsquigarrow 1}$ has the Craig interpolation property. (Thus, there are infinitely many extensions of $\mathbf{FL_{ew}}$ that have the Craig interpolation property. This makes an interesting contrast with Theorem 5.5 by Maksimova.)
(4) Give a direct proof of the fact that the TI implies the AP for any variety $\mathcal{V}$. [Hint. In fact, to show this it is enough to assume that $\mathcal{V}$ is closed under products.]
(5) Show that the AP implies the TI for every variety which has the congruence extension property.
(6) Is the AP followed from the IGAP by assuming the congruence extension property?
(7) Show that a substructural logic **L** has the ExIP iff the corresponding variety $\mathsf{V}(\mathbf{L})$ has the CEP (Lemma 5.42).
(8) *Open problem*: Is there a logic over $\mathbf{FL_{ew}}$ which is characterized by a single well-connected algebra but is not characterized by any single subdirectly irreducible algebra?
(9) *Open problem*: Find examples of substructural logics with MVP but without DMVP, and of logics with DMVP but without MVP (if any).
(10) *Open problem*: Does the superGAP imply the IGAP?

Notes

The main source of proof-theoretic arguments developed in this chapter is a survey paper [Ono98b]. Algebraic characterizations given in this chapter came mostly from [Sou07] and [KO07].

Kreisel-Putnam logic in Section 2.3 is the first example of a superintuitionistic logic with the DP stronger than intuitionistic logic, and Medvedev logic is known to be one of maximal superintuitionistic logics with the DP. There are uncountably many superintuitionistic logics with the DP [Wro73], and moreover there are uncountably many, maximal superintuitionistic logics with the DP [Cha92]. Since all superintuitionistic logic which are not included by Gödel-Dummett logic have the DP ([Ono72]), and there are uncountably many such logics ([Kuz75]), uncountably many superintuitionistic logics do not have the DP. The papers [CZ91],[CZ93] and the book [CZ97] by Chagrov and Zakharyaschev contain a lot of information on the DP of superintuitionistic logics and related problems.

For more information on interpolation properties and various types of amalgamation properties, see [CP99] and [GM05]. The paper [Mon06] by Montagna shows the AP and the *strong* AP of some subvarieties of $\mathsf{B}L$, and consequently derives the DIP and *deductive Beth property*, respectively, of corresponding extensions of **BL**.

CHAPTER 6

Completions and finite embeddability

Intuitively, a completion is a way of filling gaps in a given structure in order to obtain a smoother one. The use of Dedekind cuts to define real numbers out of the rationals is a paramount example. Embedding a partial algebra into a total one is another example, in fact underlying proofs of finite model property and finite embeddability. In this chapter we present ways of completing a (possibly partial) residuated groupoid to a residuated lattice. Among completions of algebras with lattice reduct, two have achieved prominent places: Dedekind-MacNeille completions and canonical extensions. Both are applicable in the context of residuated lattices, yet both view the residuated monoid structure as superimposed on a lattice, in what appears to be somewhat artificial way. At the very least, the intrinsic connection between the monoid structure and lattice structure is not apparent. Our preferred completion – completion by nucleus – does treat both the lattice and the monoid structure in a natural way, and we will show that both the Dedekind-MacNeille completion and the canonical extension of a residuated lattice can be obtained as its instances. We will begin by presenting Dedekind-MacNeille completions and canonical extensions in a unified framework and then move on to completions by nuclei.

6.1. Completions of posets

A *completion* of a poset $\mathbf{P}$ is a pair $(e, \mathbf{C})$ with $\mathbf{C}$ a complete lattice and e an order embedding preserving the existing finite meets and joins of $\mathbf{P}$. An element $a \in C$ is called *open* if $a = \bigvee e[X]$ for some subset X of P, where the join is taken in C; note that in this case X can be taken to be the set $\{x \in P : a \leq e(x)\}$. Dually, $a \in C$ is *closed* if $a = \bigwedge e[X]$ for some $X \subseteq P$. We will use $K^{\mathbf{P}}$ and $O^{\mathbf{P}}$. to denote sets of closed and open elements of $\mathbf{P}$. When $\mathbf{P}$ is clear from the context we will just use K and O. A completion $(e, \mathbf{C})$ is called *join dense* if every element of C is open; it is called *meet dense* if every element of C is closed. A completion that is both join and meet dense is often called dense, but following [GH01] we will use that name for a weaker notion. Namely, a completion $(e, \mathbf{C})$ will be called *dense* if every element of C can be expressed both as a join of meets and as a meet of joins of elements of $e[P]$.

Let now $\mathcal{F}$ and $\mathcal{I}$ be families of (possibly empty) upsets and downsets of P such that

- each member of $\mathcal{F}$ ($\mathcal{I}$) is closed under existing finite meets (joins)
- $\mathcal{F}$ ($\mathcal{I}$) contains all principal upsets (downsets)
- $\mathcal{F}$ ($\mathcal{I}$) contains the empty set iff $\mathbf{P}$ has no upper (lower) bound

For want of a better term we will call families satisfying the conditions above *rich enough.* Given some rich enough $\mathcal{F}$ and $\mathcal{I}$ we define a completion $(e, \mathbf{C})$ to be *internally compact with respect to* $(\mathcal{F}, \mathcal{I})$ if, for any $S, T \subseteq P$, it satisfies

- $\bigwedge e[S] \leq \bigvee e[T]$ iff $f \cap i \neq \emptyset$ for each $f \in \mathcal{F}$ and $i \in \mathcal{I}$ such that $S \subseteq f$ and $T \subseteq i$.

Internal compactness is thus a form of compactness for subsets of the image of P that is also relative to families $\mathcal{F}$ and $\mathcal{I}$. Often, the pair $(\mathcal{F}, \mathcal{I})$ will be clear from context; in such cases we will call the relevant completion just *internally compact.* A completion is *compact* if for any set A of closed elements and any set B of open elements, we have $\bigwedge A \leq \bigvee B$ if and only if there are finite subsets A_0 of A and B_0 of B with $\bigwedge A_0 \leq \bigvee B_0$. Notice that even this stronger notion is a property of $(e, \mathbf{C})$ and does not force compactness of the lattice $\mathbf{C}$ itself.

Let R be the set of all pairs $(f, i) \in \mathcal{F} \times \mathcal{I}$ such that $f \cap i \neq \emptyset$. The polarities of R establish a Galois connection

$$^{\rhd} : \mathcal{P}(\mathcal{F}) \leftrightarrows \mathcal{P}(\mathcal{I}) : {}^{\lhd}$$

with $^{\rhd} : \mathcal{P}(\mathcal{F}) \to \mathcal{P}(\mathcal{I})$ and $^{\lhd} : \mathcal{P}(\mathcal{I}) \to \mathcal{P}(\mathcal{F})$ given by

$$X^{\rhd} = \{i \in \mathcal{I} \colon i \cap f \neq \emptyset \text{ for each } f \in X\}$$
$$Y^{\lhd} = \{f \in \mathcal{F} \colon f \cap i \neq \emptyset \text{ for each } i \in Y\}$$

for any $X \subseteq \mathcal{F}$ and $Y \subseteq \mathcal{I}$. As usual, $U \in \mathcal{P}(\mathcal{F})$ is Galois closed if $U = S^{\lhd}$ for some $S \in \mathcal{P}(\mathcal{I})$, and similarly for $S \in \mathcal{P}(\mathcal{I})$.

Since $\mathcal{P}(\mathcal{F})$ and $\mathcal{P}(\mathcal{I})$ form complete lattices with set theoretic operations, the sets $\mathcal{P}(\mathcal{F})^{\rhd\lhd}$ and $\mathcal{P}(\mathcal{I})^{\lhd\rhd}$ of Galois closed elements form complete lattices where meet is intersection and join is the Galois closure of the union, by Exercise 7 of Chapter 3. Also, recall that the maps $^{\rhd}|_C$ and $^{\lhd}|_D$ are mutually inverse complete dual isomorphisms between $\mathcal{P}(\mathcal{F})^{\rhd\lhd}$ and $\mathcal{P}(\mathcal{I})^{\lhd\rhd}$, by Lemma 3.7.

It is clear that

$$\{\uparrow x\}^{\lhd\rhd} = \{f \in \mathcal{F} \colon x \in f\} \text{ and } \{\downarrow x\}^{\rhd\lhd} = \{i \in \mathcal{I} \colon x \in i\},$$

so we can define the maps $e \colon P \to \mathcal{P}(\mathcal{F})^{\rhd\lhd}$ and $d \colon P \to \mathcal{P}(\mathcal{I})^{\lhd\rhd}$ by $e(x) = \{f \in \mathcal{F} \colon x \in f\}$ and $d(x) = \{i \in \mathcal{I} \colon x \in i\}$.

LEMMA 6.1. *The map e is an embedding and the map d a dual embedding. Moreover, for any $A \subseteq L$, any $X \subseteq \mathcal{F}$ and any $Y \subseteq \mathcal{I}$ the following hold:*

(1) $\bigwedge e[A] = \{f \in \mathcal{F} \colon f \supseteq A\}$,
(2) $\bigwedge d[A] = \{i \in \mathcal{I} \colon i \supseteq A\}$,
(3) $e[A]^{\rhd} = d[A]$ *and* $d[A]^{\lhd} = e[A]$,
(4) $\bigvee e[A] = \{f \in \mathcal{F} \colon f \cap i \neq \emptyset$ *for all* $i \supseteq A\}$,
(5) $\bigvee d[A] = \{i \in \mathcal{I} \colon i \cap f \neq \emptyset$ *for all* $f \supseteq A\}$,
(6) $X^{\lhd\rhd} = \bigvee\{\bigwedge e[f] \colon f \in X\}$,
(7) $X^{\rhd\lhd} = \bigvee\{\bigwedge d[i] \colon i \in Y\}$.

PROOF. Exercise 1 asks you to verify that e and d are indeed embeddings. For (1), note that for every $f \in \mathcal{F}$, we have $f \in \bigwedge e[A] = \bigwedge\{e(x) : x \in A\}$ iff $f \in e(x)$, for all $x \in A$, that is $x \in f$, for all $x \in A$, or $A \subseteq f$; likewise, we verify (2). Note that for all $i \in \mathcal{I}$, we have $i \in e(x)^{\rhd}$ iff $i \cap f \neq \emptyset$, for all f with $x \in f \in \mathcal{F}$, iff $x \in i$, since $x \in \uparrow x \in \mathcal{F}$; thus $e(x)^{\rhd} = d(x)$, for all $x \in P$, and $e[A]^{\rhd} = d[A]$. Likewise, $d[A]^{\lhd} = e[A]$. =

For (4) observe that since $\bigvee e[A]$ is Galois closed we have $\bigvee e[A] = (\bigvee e[A])^{\lhd\rhd}$. This equals $(\bigwedge e[A]^{\rhd})^{\lhd}$ because a Galois connection gives rise to complete dual isomorphism between the lattices of closed sets. Then we obtain $(\bigwedge e[A]^{\rhd})^{\lhd} = (\bigwedge d[A])^{\lhd} = \{i \in \mathcal{I} \colon A \subseteq i\}^{\lhd}$ and that equals $\{f \in \mathcal{F} \colon f \cap i \neq \emptyset$ for all $A \subseteq i\}$ by the definition of $^{\lhd}$. Claim (5) is dual. To prove (6), we first observe that $\bigvee\{\uparrow_{\mathcal{F}} f \colon f \in X\} \subseteq \bigvee\{\uparrow_{\mathcal{F}} f \colon f \in X^{\lhd\rhd}\} = (\uparrow_{\mathcal{F}}(X^{\lhd\rhd}))^{\lhd\rhd} = X^{\lhd\rhd}$. Note that the first set contains X and, since it is closed, it is equal to $X^{\lhd\rhd}$. Now (6) follows from the fact that $\bigwedge e[f] = \uparrow_{\mathcal{F}} f$, due to (1). Claim (7) is obtained dually. ☐

Consider now a poset $\mathbf{P}$, and let $\mathcal{M}$ and $\mathcal{J}$ be sets of subsets of P, $\mathcal{M}, \mathcal{J} \subseteq \mathcal{P}(P)$, such that $\bigwedge M$ and $\bigvee J$ exist for each $M \in \mathcal{M}$ and $J \in \mathcal{J}$. We say that the family $\mathcal{F}$ of upsets of $\mathbf{P}$ is *closed under meets of the members of* $\mathcal{M}$, if

- for every $f \in \mathcal{F}$ and $M \in \mathcal{M}$, if $M \subseteq f$ then $\bigwedge M \in f$.

Likewise, we say that the family $\mathcal{I}$ of downsets of $\mathbf{P}$ is *closed under joins of the members of* $\mathcal{J}$, if

- for every $i \in \mathcal{I}$ and $J \in \mathcal{J}$, if $J \subseteq i$ then $\bigwedge M \in i$.

Then, $(e, \mathcal{P}(\mathcal{F})^{\rhd\lhd})$ is the unique completion of $\mathbf{P}$ that is dense, internally compact with respect to $\mathcal{F}$ and $\mathcal{I}$, and preserves meets from $\mathcal{M}$ and joins from $\mathcal{J}$.

THEOREM 6.2. *Let* $\mathbf{P}$ *be a poset, let* $\mathcal{M}, \mathcal{J}$ *be subsets of* $\mathcal{P}(P)$, *such that* $\bigwedge M$ *and* $\bigvee J$ *exist for each for each* $M \in \mathcal{M}$ *and* $J \in \mathcal{J}$, *and let* $\mathcal{F}, \mathcal{I}$ *be rich enough families of downsets and upsets of* $\mathbf{P}$ *that are closed under meets of the members of* $\mathcal{M}$ *and joins of the members of* $\mathcal{J}$, *respectively.*

(1) *The completion* $(e, \mathcal{P}(\mathcal{F})^{\rhd\lhd})$ *of* $\mathbf{P}$ *is dense, internally compact with respect to* $\mathcal{F}$ *and* $\mathcal{I}$, *and preserves meets from* $\mathcal{M}$ *and joins from* $\mathcal{J}$.
(2) *Let* $(g, \mathbf{D})$ *be a completion of* $\mathbf{P}$ *with the following properties:*

(a) *every element of D is a meet of joins and a join of meets of $g[P]$ (density),*
(b) *for each $M \in \mathcal{M}$, we have $g(\bigwedge M) = \bigwedge g[M]$ (preservation of meets).*
(c) *for each $J \in \mathcal{J}$, we have $g(\bigvee J) = \bigvee g[J]$ (preservation of joins),*
(d) *for any $S, T \subseteq P$ we have $\bigwedge g[S] \leq \bigvee g[T]$ iff $f \cap i \neq \emptyset$ for each $f \in \mathcal{F}$ and $i \in \mathcal{I}$ such that $S \subseteq f$ and $T \subseteq i$ (internal compactness).*

Then $(g, \mathbf{D})$ is isomorphic to $(e, \mathbf{C})$.

PROOF. From Lemma 6.1 it follows that $(e, \mathcal{P}(\mathcal{F})^{\rhd\lhd})$ has all the required properties. For instance, to show preservation of meets from $\mathcal{M}$, we calculate $\bigwedge e[M] = \{f \in \mathcal{F}\colon M \subseteq f\} \subseteq \{f \in \mathcal{F}\colon \bigwedge M \in f\} = e(\bigwedge M)$, where the first equality follows by Lemma 6.1, the second by preservation properties of $\mathcal{F}$ and the last is definitional. The reverse inclusion follows from the monotonicity of e. The other properties follow in similarly straightforward way.

Consider now a completion $(g, \mathbf{D})$ that has these properties. Then, as every closed element d of D can be expressed as $\bigwedge\{g(a)\colon a \in P,\ g(a) \geq d\}$, we can map $K^{\mathbf{D}}$ to $K^{\mathcal{P}(\mathcal{F})^{\rhd\lhd}}$ by extending the natural map $g(a) \mapsto e(a)$. Namely, we put $h(d) = \bigwedge e\big[\{a \in P\colon g(a) \geq d\}\big]$. This is clearly an order preserving map. Take any $c \in K^{\mathcal{P}(\mathcal{F})^{\rhd\lhd}}$ and let $A = \{a \in P\colon e(a) \geq c\}$. Then, $\bigwedge g[A]$ is a closed element of D, so by definition of h we get $h(\bigwedge g[A]) = \bigwedge e\big[\{a \in P\colon g(a) \geq \bigwedge g[A]\}\big]$. Then, by internal compactness of $(g, \mathbf{D})$ we have $g(a) \geq \bigwedge g[A]$ iff $f \cap i \neq \emptyset$ for each $f \in \mathcal{F}$ and $i \in \mathcal{I}$ with $A \subseteq f$ and $a \in i$. By internal compactness of $(e, \mathbf{C})$ now, this holds if and only if $e(a) \geq \bigwedge e[A]$. Thus, $h(\bigwedge g[A]) = \bigwedge e\big[\{a \in P\colon e(a) \geq \bigwedge e[A]\}\big] = \bigwedge e\big[\{a \in P\colon e(a) \geq c\}\big] = c$. So h is onto.

Take closed $d, d' \in D$ with $h(d) = h(d')$. Letting $A = \{a \in P\colon g(a) \geq d\}$ and $A' = \{a \in P\colon g(a) \geq d'\}$ we obtain $\bigwedge e[A] = \bigwedge e[A']$. Therefore, by Lemma 6.1 $\{f \in \mathcal{F}\colon f \supseteq A\} = \{f \in \mathcal{F}\colon f \supseteq A'\}$, and thus, $d = \bigwedge g[A] = \bigwedge g[A'] = d'$. So h is injective.

Hence, internal compactness determines the poset of closed elements. Dually, it also determines the poset of open elements. Then, density ensures that h is a lattice isomorphism. □

Many completions can be rendered that way, by varying the choice of $\mathcal{F}$ and $\mathcal{I}$. We obtain the Dedekind-MacNeille completion by taking $\mathcal{F}$ and $\mathcal{I}$ to be, respectively, the set of all principal filters (augmented by $\emptyset$, if P is unbounded from above) and the set of all principal ideals (augmented by $\emptyset$, if P is unbounded from below). This completion is tightest, in the sense that it preserves all of the infinitary structure of the lattice. By expanding $\mathcal{F}$ and $\mathcal{I}$ we obtain completions that forget some of it. The mechanism of this is quite simple: if a set S has a supremum in the original structure, but there is a filter in $\mathcal{F}$ that contains S but not its supremum, then $\bigvee S$ will not be preserved by the completion. At the extreme end, no infinitary structure will be preserved. This happens if we take $\mathcal{F}$ and $\mathcal{I}$ to be the largest rich enough families. The completion we obtain then is the *canonical extension* of $\mathbf{P}$. Alternatively, the canonical extension can be characterized among dense extensions as the unique one for which internal compactness implies compactness, as the next lemma shows.

LEMMA 6.3. *Let $(e, \mathbf{C})$ be a completion of $\mathbf{P}$ and $\mathcal{F}$, $\mathcal{I}$ the largest rich enough families. The following are equivalent:*

(1) *the completion $(e, \mathbf{C})$ is compact,*
(2) *for any $S, T \subseteq P$ we have $\bigwedge e[S] \leq \bigvee e[T]$ iff $\bigwedge S_0 \leq \bigvee T_0$, for some finite $S_0 \subseteq S$ and $T_0 \subseteq T$.*
(3) *for any $S, T \subseteq P$ we have $\bigwedge e[S] \leq \bigvee e[T]$ iff $f \cap i \neq \emptyset$ for each $f \in \mathcal{F}$ and $i \in \mathcal{I}$ such that $S \subseteq f$ and $T \subseteq i$.*

PROOF. Clearly compactness implies (2), and (2) implies (3). To show that (3) implies (1), take $A \subseteq K$ and $B \subseteq O$ with $\bigwedge A \leq \bigvee B$. Then $\bigwedge A = \bigwedge(\bigcup_{n \in N} S_n)$ and $\bigvee B = \bigvee(\bigcup_{m \in M} T_m)$ for some $S_n, T_m \subseteq P$, such that for each $a \in A$ we have $a = \bigwedge S_n$ for some $n \in N$ and dually for B. By (3) we obtain $f \cap i \neq \emptyset$ for all $f \in \mathcal{F}$ and $i \in \mathcal{I}$ such that $\bigcup_{n \in N} S_n \subseteq f$ and $\bigcup_{m \in M} T_m \subseteq i$. Suppose that for all finite $S' \subseteq \bigcup_{n \in N} S_n$ and $T' \subseteq \bigcup_{m \in M} T_m$ we have $\bigwedge S' \not\leq \bigvee T'$. Then, let $g = \{\uparrow(\bigwedge S') \colon S' \subseteq_{\text{fin}} S\} \cap P$ and $j = \{\downarrow(\bigvee T') \colon T' \subseteq_{\text{fin}} T\} \cap P$. Since $\mathcal{F}$ and $\mathcal{I}$ are the largest rich enough families, $g \in \mathcal{F}$ and $j \in \mathcal{I}$. Moreover, $\bigcup_{n \in N} S_n \subseteq g$ and $\bigcup_{m \in M} T_m \subseteq j$, but $g \cap j = \emptyset$. Now by the choice of S_n and T_m, for any finite $A' \subseteq A$ there is a finite $S' \subseteq \bigcup_{n \in N} S_n$ with $\bigwedge S' = \bigwedge A'$ and, similarly, for any finite $B' \subseteq B$ there is a finite $T' \subseteq \bigcup_{m \in M} T_m$ with $\bigvee T' = \bigvee B'$. But $\bigwedge S' \not\leq \bigvee T'$. Thus, $\bigwedge A' \not\leq \bigvee B'$ for all $A' \subseteq_{\text{fin}} A$ and $B' \subseteq_{\text{fin}} B$ contradicting compactness. □

6.1.1. Some properties of canonical extensions. It follows from Theorem 6.2 that every poset has a unique canonical extension. We will write $\mathbf{P}^\sigma$ for the canonical extension of $\mathbf{P}$. If $\mathbf{P}^\partial$ is the dual of $\mathbf{P}$, then $(\mathbf{P}^\partial)^\sigma$ is isomorphic to $(\mathbf{P}^\sigma)^\partial$.

For a nonempty family of sets X, we will denote by Φ_X the family of choice functions for X. Let $\mathbf{P}$ be a poset and X and Y be families of subsets of P^σ such that each $A \in X$ is a downward directed set of closed elements, and each $B \in Y$ is an upward directed set of open elements.

LEMMA 6.4. *The following infinite distributive laws hold in $\mathbf{P}^\sigma$:*

(1) $\bigvee\{\bigwedge A : A \in X\} = \bigwedge\{\bigvee \operatorname{ran}\varphi : \varphi \in \Phi_X\}$,
(2) $\bigwedge\{\bigvee B : B \in Y\} = \bigvee\{\bigwedge \operatorname{ran}\varphi : \varphi \in \Phi_Y\}$,

PROOF. Since $\bigwedge A \leq \bigvee \operatorname{ran}\varphi$ holds for every $A \in X$ and every $\varphi \in \Phi_X$, we have $\bigvee\{\bigwedge A : A \in X\} \leq \bigwedge\{\bigvee \operatorname{ran}\varphi : \varphi \in \Phi_X\}$. For the converse, we will show that every element majorizing $\bigvee\{\bigwedge A : A \in X\}$ majorizes

$\bigwedge\{\bigvee \operatorname{ran}\varphi : \varphi \in \Phi_X\}$ as well. As every element is a meet of open elements, it suffices to show that for an $o \in O$. Then $o \geq \bigwedge A$ for every $A \in X$ and therefore by compactness of the completion there is a finite $A_0 \subseteq A$ with $\bigwedge A_0 \leq o$, for each $A \in X$. Since A is downward directed, we get $\bigwedge A_0 \in A$ and thus there is a choice function φ with $\varphi(A) = \bigwedge A_0$. Thus $\bigvee \operatorname{ran}\varphi \leq o$ and therefore $\bigwedge\{\bigvee \operatorname{ran}\varphi\colon \varphi \in \Phi_X\}$ as needed. The second statement is dual. □

Because the poset **P** may be a nondistributive lattice itself, full distributivity cannot, of course, hold. The equalities above are referred to as *infinite restricted distributivity* laws.

Let now **L** be a lattice. The next lemma and the theorem following it provide a link between the canonical extension of **L** and its representation in the fashion of Urquhart's [Urq78].

LEMMA 6.5. *The poset O of all open elements of $\mathbf{L}^\sigma$ is isomorphic to the lattice $\mathcal{I}_\mathbf{L}$ of ideals of* **L**. *The poset K of all closed elements of $\mathbf{L}^\sigma$ is isomorphic to the lattice $\mathcal{F}_\mathbf{L}$ of filters of* **L**.

PROOF. Consider the map $\mu\colon \mathcal{I} \to O$ defined for each $i \in \mathcal{I}$ by $\mu(i) = \bigvee e[i]$. This map is clearly order preserving and since for each $S \subseteq L$ we have $\bigvee S = \bigvee i_S$, where i_S is the ideal generated by S, it is onto. Suppose $\mu(i) = \mu(j)$. Since every element of the canonical extension is the meet of all closed elements above it, we get $\bigvee e[i] = \bigwedge\{p \in K\colon p \geq \bigvee e[i]\}$ and similarly for $\bigvee e[j]$. By compactness then, there are finite $i_0 \subseteq i$ and $K_0 \subseteq \{p \in K\colon p \geq \bigvee e[i]\}$ such that $\bigvee e[i_0] \geq \bigwedge K_0 \geq \bigvee e[i]$. Hence, $\bigvee e[i_0] = \bigvee e[i]$. Using similar argument for j, we obtain $\bigvee e[i] = \bigvee e[i_0] = \bigvee e[j_0] = \bigvee e[j]$. Now since e is an embedding, we get $\bigvee e[i] = e[\bigvee i_0] = e[\bigvee j_0] = \bigvee e[j]$ and therefore $i = {\downarrow}\{\bigvee i_0)\} = {\downarrow}\{\bigvee j_0\} = j$, proving that μ is injective.

Since finite joins of open elements are clearly open, it remains to show that $p \wedge q$ is open for all $p, q \in O$. By definition, $p = \bigvee e[S]$ and $q = \bigvee e[T]$ for some $S, T \subseteq L$. Let $i, j \in \mathcal{I}$ be, respectively, the ideal generated by S and the ideal generated by T. Then, $p = \bigvee e[i]$ and $q = \bigvee e[j]$. Let Y be the family $\{e[i], e[j]\}$. We have $p \wedge q = \bigwedge\{\bigvee B\colon B \in Y\}$. Therefore, since Y satisfies the conditions of Lemma 6.4 we get $p \wedge q = \bigvee\{\bigwedge \operatorname{ran}\varphi : \varphi \in \Phi_Y\} = \bigvee\{e(a) \wedge e(b)\colon a \in i, b \in j\} = \bigvee\{e(a \wedge b)\colon a \in i, b \in j\}$. This is an open element. □

Recall that an element x of a lattice **L** is completely join irreducible if $x = \bigvee X$ for some $X \subseteq L$ implies $x \in X$. Completely meet irreducible elements are defined dually. Let $J(\mathbf{L})$ be the set of completely join irreducible elements of **L** and $M(\mathbf{L})$ the set of its completely meet irreducible elements. A pair $(f, i) \in \mathcal{F}_\mathbf{L} \times \mathcal{I}_\mathbf{L}$ is called *maximal disjoint* if f is maximal among all filters disjoint from i and i is maximal among all ideals disjoint from f.

THEOREM 6.6. *The following hold in $\mathbf{L}^\sigma$:*

(1) $x \in J(\mathbf{L}^\sigma)$ *iff* $x = \bigwedge e[f]$ *for some maximal disjoint pair* (f,i).
(2) $x \in M(\mathbf{L}^\sigma)$ *iff* $x = \bigvee e[i]$ *for some maximal disjoint pair* (f,i).

Moreover, each element of $\mathbf{L}^\sigma$ *is a join of completely join irreducible elements and a meet of completely meet irreducible elements.*

PROOF. Let (f,i) be a maximal pair. We will show that $u = \bigwedge e[f]$ is completely join irreducible. Suppose for contradiction that $u = \bigvee C$ but $u \notin C$. Without loss of generality we can assume that C is a set of closed elements. For each $c \in C$ consider the set $\uparrow c \cap e[L]$ and let f_c be its inverse image in L. Thus f_c is a filter properly containing f and therefore by maximality of (f,i) we get that $f_c \cap i \neq \emptyset$ for each $c \in C$. For each $c \in C$ choose an element $a_c \in f_c \cap i$. Since each a_c belongs to C, it follows that $\bigwedge e[f] = u = \bigvee C \leq \bigvee\{e(a_c) : c \in C\} \leq \bigvee e[i]$. By internal compactness, we obtain $f \cap i \neq \emptyset$, which contradicts the assumption that (f,i) is a maximal pair. Thus, u is completely join irreducible, which proves the 'if' direction of (1).

Next we will show that every closed element p is a join of completely join irreducible elements arising from maximal pairs. As p is the meet of all open elements above it, it suffices to show that for every open element o such that $p \not\leq o$ there is a maximal pair (f,i) such that $u = \bigwedge e[f]$ is below p but not below o. Since p is closed and o is open, there is a filter g and an ideal j with $p = \bigwedge e[g]$ and $o = \bigvee e[j]$. Moreover, $g \cap j = \emptyset$ because $p \not\leq o$. Extend the disjoint pair (g,j) to a maximal one, say (f,i) (which exists by Zorn's Lemma), and put $u = \bigwedge e[f]$. We have $u \leq p$, and, by internal compactness, $u \not\leq o$. This proves the 'moreover' part for completely join irreducible elements.

Finally, we show that every completely join irreducible element arises from a maximal pair. Since every element of L^σ is a join of closed elements, we have that every element of L^σ is a join of completely join irreducible elements arising from maximal pairs. In particular every completely join irreducible element c is such a join, hence c itself arises from a maximal pair. That proves the 'only if' part of (1). The proof for completely meet irreducible elements is dual. □

6.1.2. Canonical extensions of maps. Let $f : P \to Q$ be any map between posets. Since every element of P^σ is expressible as a join of meets and a meet of joins of elements of P, we have two natural ways of extending f to P^σ. We define maps $f^\sigma, f^\pi : P^\sigma \to Q^\sigma$ putting:

$$f^\sigma(x) = \bigvee\{\bigwedge\{f(a) : a \in P, p \leq a \leq q\} : p \in K, q \in O, p \leq x \leq q\}$$

$$f^\pi(x) = \bigwedge\{\bigvee\{f(a) : a \in P, p \leq a \leq q\} : p \in K, q \in O, p \leq x \leq q\}$$

If $f^\sigma = f^\pi$, we say that f is *smooth*. The following two lemmas are easy consequences of definitions (see Exercise 6).

LEMMA 6.7. *Both f^σ and f^π extend f. Moreover, $f^\sigma \leq f^\pi$ under the pointwise ordering.*

LEMMA 6.8. *Let $f : P \to Q$ be an order preserving map. Then,*
(1) $f^\sigma(p) = \bigwedge\{f(a) : a \in P, p \leq a\}$, *for all* $p \in K$;
(2) $f^\pi(q) = \bigvee\{f(a) : a \in P, q \geq a\}$, *for all* $q \in O$;
(3) $f^\sigma(x) = \bigvee\{f^\sigma(p) : p \in K, p \leq x\}$, *for all* $x \in P^\sigma$;
(4) $f^\pi(x) = \bigwedge\{f^\pi(q) : q \in O, q \geq x\}$, *for all* $x \in P^\sigma$;
(5) f^σ *and* f^π *are equal on* $K \cup O$.

The extensions f^σ and f^π are sometimes called the lower and upper envelope of f. The next corollary shows why.

COROLLARY 6.9. *The map $f^\sigma|_K$ is the greatest order preserving extension of f to K, and f^σ is the smallest order preserving extension of $f^\sigma|_K$ to P^σ. Dual statements hold for f^π.*

LEMMA 6.10. *Extensions σ and π commute with homomorphic images, subalgebras and finite direct products.*

PROOF. We show this for σ extensions on finite products leaving the other cases as an exercise (Exercise 7). Consider the i-th projection map $f_i\colon (P_1 \times \cdots \times P_n) \to P_i$. This map is order preserving, so by Lemma 6.8

$$f^\sigma(\overline{x}) = \bigvee\{f_i^\sigma(\overline{p})\colon \overline{p} \in K^{\mathbf{P}_1 \times \cdots \times \mathbf{P}_n}, \overline{p} \leq \overline{x}\}.$$

Further,

$$\begin{aligned} f^\sigma(p_1, \dots, p_n) &= \bigwedge\{f_i(a_1, \dots, a_n)\colon (a_1, \dots, a_n) \geq (p_1, \dots, p_n)\} \\ &= \bigwedge\{a_i\colon (a_1, \dots, a_n) \geq (p_1, \dots, p_n)\} \\ &= \bigwedge\{a_i\colon a_i \geq p_i\} = p_i. \end{aligned}$$

Therefore, $f^\sigma(x_1, \dots, x_n) = \bigvee\{p_i\colon p_i \leq x_i\} = x_i$, so f_i^σ is the i-th projection on $\mathbf{P}_1^\sigma \times \cdots \times \mathbf{P}_n^\sigma$. Thus, the map $(f_1^\sigma, \dots, f_n^\sigma)$ establishes an isomorphism between $(\mathbf{P}_1 \times \cdots \times \mathbf{P}_n)^\sigma$ and $\mathbf{P}_1^\sigma \times \cdots \times \mathbf{P}_n^\sigma$. □

Since an n-ary operation f on a poset $\mathbf{P}$ is a map $f : P^n \to P$, we naturally obtain extensions $f^\sigma : (P^n)^\sigma \to P^\sigma$ and $f^\pi : (P^n)^\sigma \to P^\sigma$. But $(\mathbf{P}^n)^\sigma = (\mathbf{P}^\sigma)^n$, so we obtain natural extensions of operations.

LEMMA 6.11. *For a lattice $\mathbf{L}$, the extensions $\wedge^\sigma, \wedge^\pi$ are both equal to the meet in $\mathbf{L}^\sigma$. Similarly, the extensions $\vee^\sigma, \vee^\pi$ are both the join in $\mathbf{L}^\sigma$.*

PROOF. Firstly, we have $x \wedge^\sigma y = \bigvee\{p \wedge^\sigma q\colon (x, y) \geq (p, q) \in K \times K\} = \bigvee\{p \wedge^\sigma q\colon x \geq p \in K, y \geq q \in K\}$. Then, for closed p and q, we get $p \wedge^\sigma q = \bigwedge\{a \wedge b\colon a \geq p, b \geq q\} = \bigwedge\{a\colon a \geq p\} \wedge \bigwedge\{b\colon b \geq q\} = p \wedge q$. Thus, $x \wedge^\sigma y = \bigvee\{p \wedge q\colon (x, y) \geq (p, q) \in K \times K\} = x \wedge y$. The proof for $\wedge^\pi$ is completely analogous and the dual argument works for join. □

6.1.3. Operators and preservation of identities. A map f on a poset $\mathbf{P}$ is an *operator* if it preserves existing finite joins in each coordinate. A map g is a *dual operator* if it preserves existing finite meets in each coordinate. Operators and dual operators behave nicely with respect to canonical extensions.

LEMMA 6.12. *Let $\mathbf{P}$ be a poset, f, f' operators on P and g, g' dual operators on P. Then the following hold:*

(1) *f^σ preserves arbitrary non-empty joins in each coordinate.*
(2) *g^π preserves arbitrary non-empty meets in each coordinate.*
(3) *f^σ preserves upward directed joins.*
(4) *g^π preserves downward directed meets.*
(5) *$(f'f)^\sigma = f'^\sigma f^\sigma$.*
(6) *$(g'g)^\pi = g'^\pi g^\pi$.*

PROOF. For (1) we can assume f is unary. Since f is order preserving, so is f^σ and thus $f^\sigma(\bigvee X) \geq \bigvee\{f^\sigma(x) \colon x \in X\}$. We need to show the reverse inequality, for which it suffices to prove that $f^\sigma(\bigvee X) \leq \bigvee\{f^\sigma(q) \colon q \in {\downarrow}X \cap K\}$. On the other hand, we also know that $f^\sigma(\bigvee X) = \bigvee\{f^\sigma(p) \colon p \in K, p \leq \bigvee X\}$, so the task reduces to showing $f^\sigma(p) \leq \bigvee\{f^\sigma(q) \colon q \in {\downarrow}X \cap K\}$, for a closed $p \leq \bigvee X$. Let $Z = {\downarrow}X \cap K$. Then, $f^\sigma(p) \leq f^\sigma(\bigvee Z)$, so to complete the proof we need to show that $f^\sigma(\bigvee Z) \leq \bigvee\{f^\sigma(z) \colon z \in Z\} = \bigvee\{\bigwedge\{f(a) \colon a \geq z\} \colon z \in Z\} = \bigvee\{\bigwedge\{F_a \colon a \geq z\} \colon z \in Z\}$, where F_a is the filter generated in P^σ by $\{f(a) \colon a \geq z\}$. Let G be the family $\{\{F_a \colon a \geq z\} \colon z \in Z\}$. Since each F_a is downward directed, restricted distributivity applies yielding $\bigvee\{\bigwedge\{f(a) \colon a \geq z\} \colon z \in Z\} = \bigwedge\{\bigvee \operatorname{ran}\varphi : \varphi \in \Phi_G\}$. Hence, the task reduces to showing that $f^\sigma(\bigvee Z) \leq \bigvee \operatorname{ran}\varphi$ for any choice function $\varphi \in \Phi_G$. But this clearly holds, as every member of $\operatorname{ran}\varphi$ is above some member of Z.

The proof that (1) implies (3), we leave to the reader (Exercise 9). For (5), observe that from Lemma 6.8 it follows that $(f'f)^\sigma \leq f'^\sigma f^\sigma$ always holds, and on $K \cup O$ we have $(f'f)^\sigma \leq f'^\sigma f^\sigma$. We need to show the reverse inequality. Also, $(f'^\sigma f^\sigma)$ and $f'^\sigma f^\sigma$ agree on closed elements, so we obtain $(f'f)^\sigma(x) = \bigvee\{f'^\sigma f^\sigma(p) \colon p \in K, p \leq x\}$. Letting $U = \{f^\sigma(p) \colon p \in K, p \leq x\}$, we can rewrite $(f'f)^\sigma(x)$ as $\bigvee\{f'^\sigma(u) \colon u \in U\}$. But this join is upward directed and thus f' preserves it. Therefore, $(f'f)^\sigma(x) = f'^\sigma(\bigvee U)$. On the other hand, $\bigvee U = \bigvee\{f^\sigma(p) \colon p \in K, p \leq x\} = f^\sigma(x)$, so $(f'f)^\sigma(x) = f'^\sigma f^\sigma(x)$ as claimed.

The proofs of (2), (4) and (6) are dual. □

LEMMA 6.13. *Let $\mathbf{P} = (P, f_i(i \in I))$ be a poset with additional operations and $s(x_1, \ldots, x_k)$ a term in the appropriate language. If all operations comprising s interpret to operators, then $(s^{\mathbf{P}}(x_1, \ldots, x_k))^\sigma = s^{\mathbf{P}^\sigma}(x_1, \ldots, x_k)$. Similarly, if all operations comprising s interpret to dual operators, then $(s^{\mathbf{P}}(x_1, \ldots, x_k))^\pi = s^{\mathbf{P}^\pi}(x_1, \ldots, x_k)$.*

PROOF. Induction on complexity of s. We only sketch the argument, asking the reader to fill the missing details (Exercise 8). To be able to use Lemma 6.12 in the inductive step, we need to treat sequences $(f_1, \ldots, f_k)$ of (say, n-ary) operations on P as single maps $(f_1, \ldots, f_k)\colon P^n \to P^k$. It is not difficult to show that $(f_1, \ldots, f_k)^\sigma = (f_1^\sigma, \ldots, f_k^\sigma)$.

Having done that, the base case is with s being a projection or a constant. If s is a projection, then all we need is in already in the proof of Lemma 6.10. If s is a constant, then $s^{\mathbf{P}^\sigma} = s^{\mathbf{P}}$ since $\mathbf{P}$ is a subalgebra of $\mathbf{P}^\sigma$, and $(s^{\mathbf{P}})^\sigma = \bigwedge\{a \in P\colon a \geq s^{\mathbf{P}}\} = s^{\mathbf{P}}$, since $s^{\mathbf{P}}$ is closed.

For the inductive case, assume $s = f(g_1, \ldots, g_k)$, that is $f * (g_1, \ldots, g_k)$, with $*$ standing for map composition. Then an application of Lemma 6.12 yields $(f * (g_1, \ldots, g_k))^\sigma = f^\sigma * (g_1, \ldots, g_k)^\sigma$ and the inductive hypothesis can be accessed. □

THEOREM 6.14. *Let $s = t$ be an identity satisfied by $\mathbf{P}$. If all operations comprising s and t interpret to operators, then $\mathbf{P}^\sigma$ satisfies $s = t$. Similarly, if all operations comprising s and t interpret to dual operators, then $\mathbf{P}^\pi$ satisfies $s = t$.*

PROOF. By Lemma 6.13. □

6.2. Canonical extensions of residuated groupoids

Consider now a residuated pogroupoid $\mathbf{P} = (P, \cdot, \backslash, /, \leq)$. Since multiplication distributes over existing joins, it follows by Corollary 6.14 that $\cdot^\sigma$ is residuated. It turns out that its residuals are precisely $\backslash^\pi$ and $/^\pi$.

LEMMA 6.15. *The operations $\backslash^\pi$ and $/^\pi$ are the right and left residuals of the multiplication $\cdot^\sigma$ on P^σ.*

PROOF. We use Lemma 6.7 to facilitate calculations. To this end, we will regard $\cdot$ as a map from $P \times P$ into P and the residuals as maps from $P^\partial \times P$ into P so that all the three be order preserving. We need to prove that

$$y \leq x \backslash^\pi z \quad \text{iff} \quad x \cdot^\sigma y \leq z \quad \text{iff} \quad x \leq z /^\pi y$$

where the ordering is the natural lattice ordering on $\mathbf{P}^\sigma$. Notice that the maps $\backslash^\pi$ and $/^\pi$ remain the same (as set maps) regardless of whether we view them as maps from $P^\partial \times P$ to P or as maps from $P \times P$ to P. We have

$$x \cdot^\sigma y = \bigvee\{p \cdot^\sigma q : (p, q) \in K^{P \times P}, (p, q) \leq (x, y)\},$$

which is smaller than z if and only if $p \cdot^\sigma q \leq z$ for all $(p, q) \in K^{P \times P}$ with $(p, q) \leq (x, y)$. Moreover, z is the value of the identity function on z, so $z = \bigwedge\{r \in O : r \geq z\}$ and we finally obtain that $x \cdot^\sigma y \leq z$ if and only if $p \cdot^\sigma q \leq r$ for all $(p, q) \in K^{P \times P}$ with $(p, q) \leq (x, y)$ and for all $r \in O$ with $r \geq z$. Further still, we get

$$p \cdot^\sigma q = \bigwedge\{a \cdot b : (a, b) \in P \times P, (p, q) \leq (a, b)\} \leq \bigvee\{c \in P : c \leq r\} = r$$

and to subsets of P (and $P \times P$) internal compactness applies, so we get

$$a_1 \cdot b_1 \wedge \cdots \wedge a_k \cdot b_k \leq c_1 \vee \cdots \vee c_n$$

for some finite subsets of P such that p and q are respective lower bounds of $\{a_1, \ldots, a_k\}$ and $\{b_1, \ldots, b_k\}$, and r is an upper bound of $\{c_1, \ldots, c_k\}$. Thus, of course, $p \leq a_1 \wedge \cdots \wedge a_k$ as well as $q \leq b_1 \wedge \cdots \wedge b_k$, and $c_1 \vee \cdots \vee c_n \leq r$. By monotonicity of $\cdot$ it now follows that $(a_1 \wedge \cdots \wedge a_k) \quad (b_1 \wedge \cdots \wedge b_k) \leq a_1 \cdot b_1 \wedge \cdots \wedge a_k \cdot b_k$ and so our finite sets may be taken to be singletons. Taking into account that $(p, q) \in K^{P \times P}$ iff $p \in K$ and $q \in K$, we can state that $x \cdot^\sigma y \leq z$ if and only if the following condition holds:

$$\begin{array}{ll} \text{for all } p, q \in K \text{ and } r \in O \text{ such that } p \leq x,\ q \leq y \text{ and } r \geq z, & \\ \text{there are } a, b, c \in P \text{ such that } p \leq a,\ q \leq b,\ r \geq c \text{ and } a \cdot b \leq c. & (1) \end{array}$$

On the other hand,

$$x \backslash^\pi z = \bigwedge \{s \backslash^\pi t : (s, t) \in O^{P^\partial \times P}, (s, t) \geq (x, z)\},$$

which is greater than y if and only if $s \backslash^\pi t \geq y$ for all $(s, t) \in O^{P^\partial \times P}$ with $(s, t) \geq (x, z)$, which in turn unwinds to $s \leq x$ and $t \geq z$ because of the dual ordering on the first coordinate. As before, we can express y as $\bigvee \{u \in K : u \leq y\}$ obtaining $x \backslash^\pi z \geq y$ if and only if for all $(s, t) \in O^{P^\partial \times P}$ with $s \leq x$ and $t \geq z$, and for all $u \in K$ with $u \leq y$ we have

$$s \backslash^\pi t = \bigvee \{d \backslash e : (d, e) \in P^\partial \times P, (d, e) \leq (s, t)\} \geq \bigwedge \{f \in P : u \leq f\} = u.$$

By internal compactness again, we obtain

$$d_1 \backslash e_1 \vee \cdots \vee d_k \backslash e_k \geq f_1 \wedge \cdots \wedge f_n$$

where $d_1 \wedge \cdots \wedge d_k \geq s$ (notice the dual ordering), $e_1 \vee \cdots \vee e_k \leq t$ and $f_1 \wedge \cdots \wedge f_n \geq u$. By residuated lattice properties it is straightforward that $(d_1 \wedge \cdots \wedge d_k) \backslash (e_1 \vee \cdots \vee e_k) \geq d_1 \backslash e_1 \vee \cdots \vee d_k \backslash e_k$, so again we may take single elements d, e and f with $d \backslash e \geq f$. Also, open elements in the dual ordering become closed, so $(s, t) \in O^{P^\partial \times P}$ if and only if $s \in K$ and $t \in O$. Therefore, $x \backslash^\pi z \geq y$ if and only if we have

$$\begin{array}{ll} \text{for all } s, u \in K \text{ and } t \in O \text{ such that } s \leq x,\ u \leq y \text{ and } t \geq z, & \\ \text{there are } d, e, f \in P \text{ such that } s \leq d,\ u \leq f,\ t \geq e \text{ and } f \leq d \backslash e. & (2) \end{array}$$

By residuation for $\mathbf{P}$ it immediately follows that (1) and (2) are equivalent. The proof for the other residual is completely symmetric. □

We will call the algebra $\mathbf{P}^\sigma = (P^\sigma, \wedge^\sigma, \vee^\sigma, \cdot^\sigma, \backslash^\pi, /^\pi)$ the *canonical extension* of $\mathbf{P}$. By Lemma 6.2 it follows that $\mathbf{P}^\sigma$ is unique. Since multiplication and the constant 1 are operators, by Lemma 6.12 we get that $\mathbf{P}^\sigma$ is a residuated lattice whenever $\mathbf{P}$ is a residuated semigroup. We already mentioned (see page 293) that canonical extensions preserve no existing infinitary meets and joins of $\mathbf{P}$. In fact, by fine-tuning the families $\mathcal{F}$ and $\mathcal{I}$ we can obtain extensions that preserve exactly the meets and joins we want preserved. Namely, let $\mathcal{M}$ and $\mathcal{J}$ be families of subsets of P such that meets from $\mathcal{M}$ (joins from $\mathcal{J}$) exist in $\mathbf{P}$ and are preserved by $\mathcal{F}$ (by $\mathcal{I}$).

THEOREM 6.16. *There exist a dense and internally compact extension* $(e, \mathbf{C})$ *of* $\mathbf{P}$ *such that* e *preserves precisely the meets from* $\mathcal{M}$ *and the joins from* $\mathcal{J}$. *This extension is unique up to isomorphism.*

PROOF. We claim that $\mathbf{P}^\sigma$ is the desired extension. For the lattice reduct $(P^\sigma, \vee^\sigma, \wedge^\sigma)$ of $\mathbf{P}^\sigma$ the claim follows by Theorem 6.2. To conclude the argument we need to show that $\mathbf{P}^\sigma$ is residuated. But this follows from Lemma 6.15 by noticing that only internal compactness was used in the proof. □

THEOREM 6.17. *Let* $\mathbf{P}$ *be a residuated groupoid. If* $\mathbf{P}$ *is unital, so is* $\mathbf{P}^\sigma$. *If* $\mathbf{P}$ *is associative and/or commutative, so is* $\mathbf{P}^\sigma$. *In particular, if* $\mathbf{P}$ *is a residuated monoid, then* $\mathbf{P}^\sigma$ *is a residuated lattice.*

PROOF. Constants are operators and associativity and commutativity involve only multiplication, which also is an operator. Thus, the claim follows by Lemma 6.14. □

6.2.1. Canonicity. A variety $\mathcal{V}$ of FL-algebras is *closed under canonical extensions* or *canonical* if $\mathbf{A}^\sigma \in \mathcal{V}$ for every $\mathbf{A} \in \mathcal{V}$. Clearly, the canonical extension of a residuated lattice is a residuated lattice and the canonical extension of an FL-algebra is an FL-algebra. Thus, RL and FL are canonical varieties. The next lemma yields more.

LEMMA 6.18. *Let* $\mathbf{A}$ *be an FL-algebra. If* $s = t$ *is an identity involving only constants, multiplication and join, then the following are equivalent:*

(1) $\mathbf{A} \models s = t$,
(2) $\mathbf{A}^\sigma \models s = t$,

PROOF. Immediate from Lemma 6.14. □

We could also state the above lemma in a slightly different manner, saying that the relevant identities are *preserved under canonical extensions.* Following the usage in modal logic (and theory of Boolean algebras with operators) we will call such identities *canonical.* Obviously, if Σ is a set of canonical identities, then $\operatorname{Mod}\Sigma$ is a canonical variety. The converse does not hold for certain varieties of Boolean algebras with operators. Indeed, it does not hold for representable relation algebras, as Hodkinson and Venema showed (see [HV05]). It will be interesting to see whether there exist a canonical variety of FL-algebras without canonical axiomatization, and a good candidate to look at would be the variety of *symmetric* representable relation algebras.

COROLLARY 6.19. *Let* Σ *be a set of identities involving only constants, multiplication and join or meet but not both. Then,* Σ *is canonical. In particular, for any* $S \subseteq \{e, c, i, o, w\}$, *the variety* FL_S *is canonical.*

Other examples of canonical varieties include HA and n-potent varieties. Notice that since the defining identity of HA is $x \wedge y = xy$, we can use distributivity of multiplication over join to conclude that in this case meet *is* an operator, and so the conclusion follows by Corollary 6.19.

LEMMA 6.20. *Let $\mathcal{V}$ be a canonical subvariety of* FL *and $\mathcal{V}_D$ be the subvariety of $\mathcal{V}$ satisfying distributivity. Then $\mathcal{V}_D$ is canonical.*

PROOF. It follows from Lemma 6.4 that canonical extensions preserve distributivity. □

LEMMA 6.21. *Let* **A** *be a DFL-algebra. If $s = t$ is an identity not involving divisions. Then the following are equivalent:*

(1) $\mathbf{A} \models s = t$,
(2) $\mathbf{A}^\sigma \models s = t$,

PROOF. Since **A** is distributive as a lattice, meet is an operator. Then the claim follows by Lemma 6.14. □

COROLLARY 6.22. *Let Σ be a set of identities not involving divisions. Then the variety* DFL $\cap$ Mod(Σ) *is canonical.*

Even when divisions are involved all is not lost, provided they are not mixed with multiplication in one and the same identity. This will be the content of our next result, encompassing fairly large class of canonical varieties of FL-algebras.

THEOREM 6.23. *Let $\mathcal{V}$ be a canonical subvariety of* FL *and Σ a set of identities such that the basic operations occurring in any member of Σ are among:*

(1) *constants, multiplication, meet, or*
(2) *constants, multiplication, join, or*
(3) *constants, divisions, meet, or*
(4) *constants, divisions, join.*

Then $\mathcal{V} \cap \mathrm{Mod}(\Sigma)$ is canonical. Further, if $\mathcal{V}$ is a subvariety of DFL *and the basic operations occurring in any member of Σ are among:*

(1) *constants, multiplication, meet, join, or*
(2) *constants, divisions, meet, join.*

Then $\mathcal{V} \cap \mathrm{Mod}(\Sigma)$ is canonical.

PROOF. Take an algebra $\mathbf{A} \in \mathcal{V} \cap \mathrm{Mod}(\Sigma)$. We need to show that $\mathbf{A}^\sigma \in \mathcal{V} \cap \mathrm{Mod}(\Sigma)$ as well. By canonicity of $\mathcal{V}$ we have immediately that $\mathbf{A}^\sigma \in \mathcal{V}$, so it remains to show that $\mathbf{A}^\sigma \in \mathrm{Mod}(\Sigma)$. It follows from Lemma 6.11 that the algebras $(A^\sigma, \wedge^\sigma, \vee^\sigma, \backslash^\pi, /^\pi)$ and $(A^\pi, \wedge^\pi, \vee^\pi, \backslash^\pi, /^\pi)$ are isomorphic. Moreover, $1^\sigma = 1 = 1^\pi$ and $0^\sigma = 0 = 0^\pi$. Thus, we can use either one of the signatures above for the extension $\mathbf{A}^\sigma$. Let $s = t$ be an identity in Σ. If $s = t$ does involve divisions then by Theorem 6.14 we get

$(A^\pi, \wedge^\pi, \vee^\pi, \cdot^\sigma, \backslash^\pi, /^\pi, 1^\pi, 0^\pi) \models s = t$. If $s = t$ does not involve divisions, then $(A^\sigma, \wedge^\sigma, \vee^\sigma, \cdot^\sigma, \backslash^\pi, /^\pi, 1^\sigma, 0^\sigma) \models s = t$. □

Canonicity of many important subvarieties can be derived from the result above. The reader is encouraged to check which varieties on the list in Section 3.5 pass the test. Many do. As soon as we begin mixing multiplication with divisions, however, things go wrong. We will see a wide class of examples in Section 6.4.

An important class of canonical varieties not derived from the result above is covered by the next lemma, whose proof can be found in [GH01], but will be omitted here as it would take us too far afield. For the curious: it uses the fact that canonical extensions commute with Boolean products and that every direct product can be rendered as a Boolean product of ultraproducts.

LEMMA 6.24. *A finitely generated variety is canonical.*

6.2.2. A counterexample for canonical extensions. We will show here that the extension $(A^\sigma, \wedge^\sigma, \vee^\sigma, \cdot^\sigma, \backslash^\sigma, /^\sigma, 1)$ could only work for a finite residuated unital groupoid $\mathbf{A}$, so the choice $\mathbf{A}^\sigma = (A^\sigma, \wedge^\sigma, \vee^\sigma, \cdot^\sigma, \backslash^\pi, /^\pi, 1)$ as the canonical extensions is the right one. Our example will also demonstrate a somewhat curious fact that the right notion of canonicity depends on the choice of basic operations.

THEOREM 6.25. *Let $\mathbf{A}$ be any infinite residuated unital groupoid. Then, the operations $\backslash^\sigma$ and $/^\sigma$ are not residuals of $\cdot^\sigma$ in $(A^\sigma, \wedge^\sigma, \vee^\sigma, \cdot^\sigma, 1)$.*

PROOF. We will argue by contradiction. Suppose $\backslash^\sigma$ is the right residual of $\cdot^\sigma$. Since A is infinite, there exists an $S \subseteq A$ such that $s = \bigwedge S \in A^\sigma$ but $s \notin A$. Consider the element $s\backslash^\sigma s$. We obtain $s\backslash^\sigma s = \bigvee\{p\backslash^\sigma q : (s, s) \geq (p, q) \in K\} = \bigvee\{p\backslash^\sigma q : s \leq p \in O, s \geq q \in K\}$. Since s is closed, this further equals $\bigvee\{p\backslash^\sigma s : s \leq p \in O\}$. Further still, (p, s) is closed in $(A^\partial)^\sigma \times A^\sigma$ and therefore $p\backslash^\sigma s = \bigwedge\{a\backslash b : (p, s) \leq (a, b)\} = \bigwedge\{a\backslash b : p \geq a, s \leq b\}$. As $s = \bigwedge S$, $s \leq p$ and $p = \bigvee P$ for some $P \subseteq A$ (because p is open), by compactness there exist at least two distinct elements $a, b \in A$ such that $a, b \in \{x \in A\colon s < x \leq p\}$ and $a \not\leq b$. Thus, $p\backslash^\sigma s \not\geq 1$, for every open p.

Now, by properties of residuation we have $s\backslash^\sigma s \geq 1$, that is, $\bigvee\{p\backslash^\sigma s\colon p \in O\} \geq 1$. By restricted infinite distributivity we then get $\bigwedge\{\bigvee \operatorname{ran}\varphi\colon \varphi \in \Phi_X\}$, where $X = \big\{\{a\backslash b\colon p \geq a, s \leq b\}\colon s \leq p \in O\big\} \geq 1$. So, for every choice function φ picking out one $a\backslash b$ for each open p, we should have $\bigvee \operatorname{ran}\varphi \geq 1$. Then, by compactness, for every such choice function there should be elements $a_1, \ldots, a_n$ and $b_1, \ldots, b_n$ such that $(a_1\backslash b_1) \vee \cdots \vee (a_n\backslash b_n) \geq 1$.

But we will construct a choice function φ such that for every finite $U \subseteq \operatorname{ran}\varphi$ we have $\bigvee U \not\geq 1$. Namely, as we have seen, for each open $p \geq s$ we can choose $a_p\backslash b_p$ with $a_p \not\leq b_p$. We can arrange the choices so that $\{a_p\colon s \leq p \in O\}$ is a nondecreasing chain and $\{b_p\colon p \in O\}$ is a nonincreasing chain. Then each finite subset U of $\operatorname{ran}\varphi$ has the largest element $a_q\backslash b_q$, for some open q greater or equal to s, with $a_q \not\leq b_q$. Thus $\bigvee U = a_q\backslash b_q \not\geq 1$ as claimed. It follows that $s\backslash^\sigma s \not\geq 1$ and that contradicts properties of residuation. □

Since a Boolean algebra is a residuated groupoid with meet being the groupoid operation, the counterexample above applies also to any infinite Boolean algebra $\mathbf{A}$ showing that $\mathbf{A}^\sigma$ fails to be residuated. Precisely speaking, it shows that $\rightarrow^\sigma$ is not the residual of $\wedge^\sigma$. This does not contradict the well-known fact that operators on Boolean algebras are smooth. The apparent contradiction is due to the fact that in the standard presentation of Boolean algebras. $\rightarrow$ is not a basic operation. Therefore, $x \rightarrow y$ is given in the extension by $\neg^\sigma x \vee^\sigma y = \neg^\sigma x \vee y$ and not by $(\neg x \vee y)^\sigma = x \rightarrow^\sigma y$. On the other hand, by Lemmas 6.12 and 6.11 we have $(\neg x \vee y)^\pi = \neg^\pi x \vee^\pi y = \neg^\pi x \vee y$. Since negation in Boolean algebras is smooth (Exercise 10 asks the reader to verify this), we obtain $x \rightarrow^\pi y = (\neg x \vee y)^\pi = \neg x \vee y$ exactly as required.

COROLLARY 6.26. *Canonicity depends on the choice of basic operations.*

6.3. Nuclear completions of residuated groupoids

Having dealt with canonical extensions defined in a somewhat abstract way, we now turn to what will be our standard way of completing a residuated groupoid. We begin with recalling the following basic construction underlying the completions; see Section 3.4.10. Given a groupoid $\mathbf{G} = (G, \cdot)$, we first define the operations $\circ$, $\backslash$ and $/$ for every $X, Y \subseteq G$ by

$$X \circ Y = \{a \cdot b \in G : a \in X \text{ and } b \in Y\},$$
$$X\backslash Y = \{a \in G : X \circ \{a\} \subseteq Y\},$$
$$Y/X = \{a \in G : \{a\} \circ X \subseteq Y\}.$$

Then, we put $\mathcal{P}(\mathbf{G}) = (\mathcal{P}(G), \cap, \cup, \circ, \backslash, /)$. It follows by Theorem 3.32 that $\mathcal{P}(\mathbf{G})$ is a residuated ℓ-groupoid. The following lemma follows immediately from Exercise 42 of Chapter 3.

LEMMA 6.27. *The algebra $\mathcal{P}(\mathbf{G})$ is a complete distributive residuated ℓ-groupoid. If $\mathbf{G}$ is a semigroup, then $\mathcal{P}(\mathbf{G})$ is a complete distributive residuated ℓ-semigroup. If $\mathbf{G}$ is a monoid with unit 1, then $\mathcal{P}(\mathbf{G})$ is a complete distributive residuated lattice with unit $\{1\}$.*

Again recall from Chapter 3 that a closure operator γ on a pogroupoid $\mathbf{L}$ is a nucleus if $\gamma(x) \cdot \gamma(y) \leq \gamma(x \cdot y)$ holds for all $x, y \in L$. Further, if $\mathbf{L}$ is a residuated lattice, the nucleus retraction $\mathbf{L}_\gamma$ of $\mathbf{L}$ is the algebra $\mathbf{L}_\gamma = (L_\gamma, \wedge, \vee_\gamma, \cdot_\gamma, \backslash, /, 1_\gamma)$, with meet and divisions coming from $\mathbf{L}$ and the other operations defined by

$x \cdot_\gamma y = \gamma(x \cdot y)$, $x \vee_\gamma y = \gamma(x \vee y)$, and $1_\gamma = \gamma(1)$.

Recall that Theorem 3.34 states that if $\mathbf{L}$ is a residuated ℓ-groupoid and γ a nucleus on G, then $\mathbf{L}_\gamma$ is a residuated ℓ-groupoid as well. Furthermore, the properties of associativity, commutativity, integrality, being square increasing, lattice-ordering, lattice-completeness and having a unit ($\gamma(1)$) are preserved.

Combining Theorem 3.34 and Lemma 6.27 we obtain the following.

COROLLARY 6.28. *If* $\mathbf{M}$ *is a monoid and* γ *is a nucleus on* $\mathcal{P}(\mathbf{M})$, *then* $\mathcal{P}(\mathbf{M})_\gamma = (\gamma[\mathcal{P}(M)], \cap, \cup_\gamma, \circ_\gamma, \backslash, /, \gamma(\{1\}))$ *is a complete residuated lattice.*

The construction described above, and its certain refinements, will be our preferred way of constructing complete residuated lattices; it will be instantiated in this section to obtain the Dedekind-MacNeille completion and the canonical extension of a residuated lattice, and it will also be used in Section 6.5.1 and in various places in Chapter 7.

As we saw in the representation theorem Theorem 3.38 for residuated lattices and in Exercise 12 of Chapter 3 any residuated lattice can be embedded into one of the form $\mathcal{P}(\mathbf{M})_\gamma$ and any complete residuated lattice is isomorphic to one of this form. If we apply this construction to a residuated lattice $\mathbf{L}$, the pomonoid reduct $(L, \leq, \cdot, 1)$ of $\mathbf{L}$ determines another residuated lattice $\mathcal{P}(\mathbf{L})_\gamma$. It is then natural to ask to what extent $\mathcal{P}(\mathbf{L})_\gamma$ reflects the structure of $\mathbf{L}$, in particular its lattice operations and residuals. The next result shows that if the nucleus γ can be chosen in a certain way, we can completely preserve the structure of any subset B of L. The set B may be thought of as containing some important *base* that we want preserved. In Section 6.5 it will be a finite partial subalgebra.

For a pogroupoid $\mathbf{G}$ and a nucleus γ on $\mathcal{P}(\mathbf{G})$, we say that γ is a *downward* nucleus if all γ-closed sets are downward closed.

THEOREM 6.29. cf. [Ono03c] *Suppose that* $\mathbf{G}$ *is a pogroupoid and* γ *is a downward nucleus on* $\mathcal{P}(\mathbf{G})$. *Let* B *be a subset of* G *such that*

(1) *for all* $b \in B$ *the principal downset* ${\downarrow}b$ *is closed, and*
(2) *for all* $S \subseteq B$ *if* $\bigvee S \in B$ *then* $\bigvee S \in \gamma(S)$.

Then the map $h : B \to \gamma(\mathcal{P}(G))$ *defined by* $h(b) = {\downarrow}b$ *is a complete embedding, i.e., an order-preserving map that also preserves all existing products, residuals, (arbitrary) joins and meets in* B.

PROOF. Clearly the map h is order-preserving, as $\mathcal{P}(\mathbf{G})_\gamma$ is ordered by inclusion. We need to show that h preserves existing products, residuals, joins and meets in B.

Suppose that $a \cdot b \in B$ for $a, b \in B$. Since γ is a downward nucleus, ${\downarrow}(a \cdot b)$ is the least γ-closed set which includes the set ${\downarrow}a \circ {\downarrow}b$. Therefore, $h(a) \circ_\gamma h(b) = \gamma({\downarrow}a \circ {\downarrow}b) = {\downarrow}(a \cdot b) = h(a \cdot b)$.

Next suppose that $a\backslash b \in B$ for $a, b \in B$. Since ${\downarrow}a \circ {\downarrow}(a\backslash b) = \{x \cdot y \in G : x \leq a \text{ and } y \leq a\backslash b\} \subseteq {\downarrow}b \in \gamma(G)$, we have ${\downarrow}a \circ_\gamma {\downarrow}(a\backslash b) \subseteq {\downarrow}b$. If ${\downarrow}a \circ_\gamma X \subseteq {\downarrow}b$

for a γ-closed set X, then, for any $c \in X$, $a \cdot c \leq b$ and therefore $c \leq a\backslash b$. Hence $X \subseteq {\downarrow}(a\backslash b)$. This means that ${\downarrow}(a\backslash b)$ is maximal among such γ-closed sets X that satisfy ${\downarrow}a \circ_\gamma X \subseteq {\downarrow}b$. That is, ${\downarrow}a\backslash{\downarrow}b = {\downarrow}(a\backslash b)$. Therefore, $h(a)\backslash h(b) = {\downarrow}(a\backslash b) = h(a\backslash b)$. The argument for $/$ is a mirror image.

It is easily seen that h preserves all existing meets, since $\bigwedge_i {\downarrow}a_i = {\downarrow}(\bigwedge_i a_i)$ holds. It remains to show that h preserves all existing joins. Suppose that $\bigvee_i a_i$ exists and belongs to B where $\{a_i : i \in I\} \subseteq B$. Since ${\downarrow}a_j \subseteq {\downarrow}(\bigvee_i a_i)$ for each $j \in I$, we obtain that $\gamma(\bigvee_i {\downarrow}a_i) \subseteq {\downarrow}(\bigvee_i a_i)$. For a given γ-closed set X, suppose that ${\downarrow}a_j \subseteq X$ for each $j \in I$. Then $a_j \in X$ for each $j \in I$. By our assumption that γ is join-closed, $\bigvee_i a_i$ must belong to X. Hence ${\downarrow}(\bigvee_i a_i) \subseteq X$, and therefore $\gamma(\bigvee_i h(a_i)) = \gamma(\bigvee_i {\downarrow}a_i) = {\downarrow}(\bigvee_i a_i) = h(\bigvee_i a_i)$. □

6.3.1. Canonical extensions as nuclear completions. Let $\mathbf{A}$ be a residuated lattice and $\mathcal{F}_\mathbf{A}$ and $\mathcal{I}_\mathbf{A}$ respectively the sets of filters and the set of ideals of $\mathbf{A}$. For $f, g \in \mathcal{F}_\mathbf{A}$ we will employ the usual coset notation $fg = \{xy\colon x \in f, y \in g\}$. We define multiplication on $\mathcal{F}_\mathbf{A}$, by putting $f \circ g = {\uparrow}(fg)$. Under this definition $(\mathcal{F}_\mathbf{A}, \circ, {\uparrow}1)$ is a monoid (see Exercise 11). In the usual fashion, setting $A \circ B = \{{\uparrow}(fg)\colon f \in A, g \in B\}$, we can lift multiplication to the powerset (see Chapter 3, in particular Exercise 42 for details of that lifting).

LEMMA 6.30. *The structure $(\mathcal{P}(\mathcal{F}_\mathbf{A}), \subseteq, \circ, \{{\uparrow}1\})$ is a pomonoid.*

Define $N \subseteq \mathcal{F}_\mathbf{A} \times \mathcal{I}_\mathbf{A}$ by putting $(f, i) \in N$ iff $f \cap i \neq \emptyset$ (notice that N is exactly the relation defined on page 290). Further, for $f \in \mathcal{F}_\mathbf{A}$ and $i \in \mathcal{I}_\mathbf{A}$ define

$$f \mathbin{\backslash\!\backslash} i = \{j \in \mathcal{I}_\mathbf{A} : x\backslash y \in j \text{ for all } x \in f, y \in i\}$$
$$i \mathbin{/\!/} f = \{j \in \mathcal{I}_\mathbf{A} : y/x \in j \text{ for all } x \in f, y \in i\}$$

LEMMA 6.31. *The relation N is nuclear, i.e., the following equivalences hold*

$$g\, N\, f \mathbin{\backslash\!\backslash} i \quad \textit{iff} \quad f \circ g\, N\, i \quad \textit{iff} \quad f\, N\, g \mathbin{/\!/} i$$

for all $f, g \in \mathcal{F}_\mathbf{A}$ and all $i \in \mathcal{I}_\mathbf{A}$.

PROOF. First we show that the sets ${\downarrow}\{u\backslash v\colon u \in f, v \in i\}$ and ${\downarrow}\{u/v\colon u \in f, v \in i\}$ are ideals, for any $f \in \mathcal{F}_\mathbf{A}$ and $i \in \mathcal{I}_\mathbf{A}$. For take $z, z' \in \{u\backslash v\colon u \in f, v \in i\}$. Then $z = u\backslash v$ and $z' = u'\backslash v'$ for $u, u' \in f$ and $v, v' \in i$. Since $u \wedge u' \in f$ and $v \vee v' \in i$ we get that $(u \wedge u')\backslash(v \vee v') \in \{u\backslash v\colon u \in f, v \in i\}$. But as $(u \wedge u')\backslash(v \vee v') \geq (u\backslash v) \vee (u'\backslash v') = z \vee z'$ we get that $z \vee z' \in {\downarrow}\{u\backslash v\colon u \in f, v \in i\}$.

Now suppose $g\ N\ f \mathbin{\backslash\!\backslash} i$ holds. By definition of N and the observation above, we have that $g \cap {\downarrow}\{x\backslash y\colon x \in f, y \in i\} \neq \emptyset$. Since g is a filter, it follows that for some $x \in f$ and $y \in i$ we have $x\backslash y \in g$. Thus, $x(x\backslash y) \in fg$. But, $x(x\backslash y) \leq y$ and thus $y \in h$ for every filter h containing fg. Therefore $h \cap i \neq \emptyset$ as needed. Now suppose $f \circ g\ N\ i$ holds. It follows that for some $x \in f$ and $z \in g$ we have $xz \in i$. Consider $x \backslash xz$. Since $z \leq x \backslash xz$ we obtain that $z \in j$ for every ideal j containing $\{u \backslash v : u \in f, v \in i\}$. Therefore $g \cap j \neq \emptyset$ as needed. The other equivalence is proved similarly. □

It now follows by Lemma 3.36 that the map $\gamma = \gamma_N : \mathcal{P}(\mathcal{F}_{\mathbf{A}}) \to \mathcal{P}(\mathcal{F}_{\mathbf{A}})$ sending a set of filters to its Galois closure is a nucleus on $(\mathcal{P}(\mathcal{F}_{\mathbf{A}}), \subseteq, \circ, \{\uparrow 1\})$. Consequently, the map $e : \mathbf{A} \to \mathcal{P}(\mathcal{F}_{\mathbf{A}})_\gamma$ is an embedding.

LEMMA 6.32. *Residuated lattices $\mathcal{P}(\mathcal{F}_{\mathbf{A}})_\gamma$ and $\mathbf{A}^\sigma$ are isomorphic.*

PROOF. It follows from Theorem 6.2 that $\mathcal{P}(\mathcal{F}_{\mathbf{A}})_\gamma$ and $\mathbf{A}^\sigma$ are isomorphic as lattices. Thus, it suffices to prove that multiplication $\circ^\gamma$ on $\mathcal{P}(\mathcal{F}_{\mathbf{A}})_\gamma$ coincides with $\cdot^\sigma$ as the rest will follow by uniqueness of residuals. We will do it piecemeal, beginning with closed elements. Let p, q be closed elements of $\mathcal{P}(\mathcal{F}_{\mathbf{A}})_\gamma$. Then $p = \bigwedge e[S] = \{f \in \mathcal{F}_{\mathbf{A}} : f \supseteq S\}$ and $q = \bigwedge e[T] = \{g \in \mathcal{F}_{\mathbf{A}} : g \supseteq T\}$ for some $S, T \subseteq A$. We have

$$\begin{aligned} p \circ^\gamma q &= \gamma(\{f \in \mathcal{F}_{\mathbf{A}} : f \supseteq T\} \circ \{g \in \mathcal{F}_{\mathbf{A}} : g \supseteq T\}) \\ &= \gamma(\{\uparrow(fg) : f \supseteq S, g \supseteq T\}) \\ &= \{h \in \mathcal{F}_{\mathbf{A}} : h \supseteq ST\} \\ &= \bigwedge e[ST] \\ &= \bigwedge \{e(st) : s \in S, t \in T\} \\ &= \bigwedge \{e(s) \cdot e(t) : s \in S, t \in T\} \\ &= \bigwedge \{e(a) \cdot e(b) : a, b \in A, a \geq p, b \geq q\} \\ &= p \cdot^\sigma q \end{aligned}$$

proving that multiplications coincide on closed elements. Now, since every element is the join of the set of closed elements below it, and both multiplications give rise to residuated lattices, we have

$$\begin{aligned} x \circ^\gamma y &= \bigvee \{p \in K^A : p \leq x\} \circ^\gamma \bigvee \{q \in K^A : q \leq y\} \\ &= \bigvee \{p \circ^\gamma q : p, q \in K^A, p \leq x, q \leq y\} \\ &= \bigvee \{p \cdot^\sigma q : p, q \in K^A, p \leq x, q \leq y\} \\ &= \bigvee \{p \in K^A : p \leq x\} \cdot^\sigma \bigvee \{q \in K^A : q \leq y\} \\ &= x \cdot^\sigma y \end{aligned}$$

as claimed. □

6.4. Negative results for completions

Although, as we have seen, every residuated lattice $\mathbf{D}$ has a completion, there are residuated lattices whose every completion lies outside the variety the original algebra generates. This section is based on [KL07]. To make the setting precise, we will say that a class $\mathcal{K}$ *admits completions* if for every $\mathbf{D} \in \mathcal{K}$ there is a complete algebra $\mathbf{M} \in \mathcal{K}$ such that $\mathbf{D}$ is a subalgebra of $\mathcal{M}$. Our aim in this section is to present a rather large class of varieties of residuated lattices not admitting completions.

One example of a variety not admitting completions is the variety LG of ℓ-groups. To see that it suffices to notice that no non-trivial ℓ-group is bounded as a lattice (see Exercise 14). Thus, LG does not admit completions, but in an admittedly trivial way. For such cases, a more interesting notion is that of *local completeness*. A (residuated) lattice $\mathbf{D}$ is locally complete if every interval of D is complete. Reals or natural numbers with standard order are locally complete but not complete. For bounded lattices, local completeness is equivalent to completeness.

A class of algebras K *admits local completions* if every algebra from K can be embedded into a (locally) complete algebra from $\mathcal{K}$. Let us stress once again: we *do not* require this embedding to preserve infinite meets and joins, only finitary operations. Canonical extensions, for example, do not preserve infinitary operations. Thus, when proving that closure under completions fails, we cannot use too much information about suprema and infima in an algebra witnessing this failure. As was mentioned in the introduction: we only use (possibly infinitary) laws which hold in the whole variety.

Let $\mathbf{D}$ be a GBL-algebra. Suppose there are subsets A and B of D with the following properties:

(1) B is a nonprincipal ideal without supremum in D.
(2) $B < A$, i.e., for each $a \in A$ and $b \in B$ we have $b < a$.
(3) there are elements $w, u \in D$ with $w < u$ such that $b\backslash u \in A$ for all $b \in B$, and
  (a) for each $a \in A$ there is $b \in B$ with $b\backslash u \leq a$,
  (b) for each $b \in B$ there is $a \in A$ with $ba \leq w$.

We will then call (B, A) a *residuation discontinuity*, and $\mathbf{D}$ a *discontinuous GBL-algebra.*

The choice of the name will become clear from the proof of the next theorem, where indeed it turns out that the function $x(x\backslash u)$ is not continuous at a certain point presumed to exist (namely b_∞).

THEOREM 6.33. [KL07] *No variety of GBL-algebras containing a discontinuous GBL-algebra is closed under sectional completions.*

PROOF. Let $\mathbf{D}$ be a discontinuous GBL-algebra with a residuation discontinuity (B, A). Suppose for contradiction that $\mathbf{D}$ is embedded into a complete residuated lattice $\mathbf{E}$. Define $b_\infty = \bigvee B$ and $a_\infty = \bigwedge A$. Clearly a_∞ and b_∞ are elements of E and by definition of discontinuity $b_\infty \notin D$, but we make no similar assumption on a_∞. It may well belong to D or even to A.

Consider $b_\infty(b_\infty\backslash u)$. As $b_\infty > u$ we obtain $b_\infty(b_\infty\backslash u) = u$. On the other hand, $\bigvee B \cdot (\bigvee B\backslash u) = \bigvee B \cdot (\bigwedge\{b\backslash u : b \in B\})$. Since every $a \in A$ is minorized by $b\backslash u$ for some $b \in B$ and each $b\backslash u$ belongs to A, we have $\bigwedge\{b\backslash u : b \in B\} = a_\infty$. Therefore, $\bigvee B \cdot (\bigwedge\{b\backslash u : b \in B\}) = \bigvee B \cdot a_\infty = \bigvee\{b \cdot a_\infty : b \in B\}$. Now, for each $b \in B$ choose an $a_b \in A$ with $b \cdot a_b \leq w$. Since $a_\infty \leq a$ for all $a \in A$, we get $b \cdot a_\infty \leq b \cdot a_b \leq w < u$. Thus, $b_\infty(b_\infty\backslash u) < u$, a contradiction. □

We propose three applications of Theorem 6.33. Taken together, they show that the class of varieties that do not admit (local) completions is quite large.

6.4.1. MV-algebras. Recall from Chapter 2 the MV-algebra $\mathbf{C}_\infty$, known as Chang's chain, with the universe is $C_\infty = \{a_n\}_{n\in\mathbb{N}} \cup \{b_n\}_{n\in\mathbb{N}}$ with $a_0 = 1$ and $b_0 = 0$. The lattice order is defined as follows: every b_n is below all a_n's, $b_n \leq b_m$ if $n \leq m$, $a_n \leq a_m$ if $n \geq m$. Multiplication is defined by

$$b_n \cdot b_m = 0, \quad a_n \cdot b_m = b_{m \dot{-} n}, \quad a_n \cdot a_m = a_{n+m},$$

where $\dot{-}$ stands for truncated subtraction. See Figure 2.5.

LEMMA 6.34. [KL07] *Chang's chain $\mathbf{C}_\infty$ is a discontinuous GBL-algebra.*

PROOF. Assume $\mathbf{A}$ is a complete MV-algebra having $\mathbf{C}_\infty$ as a subalgebra. Define $A = \{a_n\}_{n\in\mathbb{N}}$, $B = \{b_n\}_{n\in\mathbb{N}}$, $w = b_0$, $u = b_1$. Conditions (1) and (2) of the definition of residuation discontinuity are immediate. For (3), observe that for every i, $\neg b_i = a_i$ and $b_{i+1} \to b_1 = a_i$. □

COROLLARY 6.35. [KL07] *No variety of GBL-algebras containing $\mathbf{C}_\infty$ is closed under completions. In particular, neither BL-algebras nor MV-algebras are closed under completions.*

For varieties of MV-algebras, closure under completions is equivalent to canonicity and both in turn are equivalent to being finitely generated. Let us prove it in more detail.

LEMMA 6.36. [KL07] *A variety $\mathcal{V}$ of MV-algebras does not contain $\mathbf{C}_\infty$ if and only if it is finitely generated.*

PROOF. Since each finite MV-algebra satisfies the identity $x^{n+1} = x^n$ for some $n \in \mathbb{N}$, we get that finitely generated $\mathcal{V}$ does not contain $\mathbf{C}_\infty$. For the converse, suppose $\mathcal{V}$ is not finitely generated. Then for every $k \in \mathbb{N}$ there is a subdirectly irreducible $\mathbf{A}_k \in \mathcal{V}$ falsifying $x^{k+1} = x^k$. Choose a suitable $a_k \in A_k$ witnessing that fact. Let $\mathbf{A}$ be the ultraproduct $\prod_{k\in\mathbb{N}} \mathbf{A}_k/U$ for some nonprincipal U. Then, the element $a = (a_k : k \in \mathbb{N})/U$ has $a^{n+1} < a^n$ for every $n \in \mathbb{N}$. Also, $\mathbf{A}$ is linearly ordered, since all $\mathbf{A}_k$ are. It is now easy to verify that the subalgebra of $\mathbf{A}$ generated by a is isomorphic to $\mathbf{C}_\infty$. □

Combining the lemma above with Lemma 6.24 we obtain the next result.

THEOREM 6.37. [KL07] *For a variety $\mathcal{V}$ of MV-algebras the following are equivalent:*

(1) *$\mathcal{V}$ is canonical,*
(2) *$\mathcal{V}$ is closed under completions,*
(3) *$\mathcal{V}$ is finitely generated,*
(4) *$\mathcal{V}$ is n-potent for some $n \in \mathbb{N}$.*

6.4.2. Lattice-ordered groups. In Section 3.4.5 we saw that Abelian ℓ-groups can be obtained in a certain way from MV-algebras. The ℓ-group obtained this way from Chang's chain is, as should be expected, a good example that ℓ-groups are not closed under local completions. This algebra is isomorphic to the lexicographic product of integers with themselves. To be precise, we define the algebra $\mathbf{I} = (\mathbb{Z} \times \mathbb{Z}, \wedge, \vee, \cdot, ^{-1}, (0,0))$, where $\wedge$ and $\vee$ are respectively the minimum and maximum with respect to the lexicographic order, and multiplication and inverse are defined pointwise, as addition and unary minus.

THEOREM 6.38. [KL07] *The algebra $\mathbf{I}$ is a discontinuous GBL-algebra. Thus, no nontrivial variety of ℓ-groups is closed under (local) completions.*

PROOF. Define $A = \{(k,n) \in I : k \geq 1\}$ and $B = \{(\ell, m) \in I : \ell \leq -1\}$. So defined A and B clearly satisfy conditions 1 and 2. Now putting $u = (0,1)$ and $w = (0,0)$ it is easy to verify that conditions 3a and 3b hold as well. To satisfy 3a it suffices to take for each $(k,n) \in A$, the element $(-k, 1-n)$. To satisfy 3b, we just take the inverse of every $(\ell, m) \in B$.

For the second statement it suffices to notice that $\mathbf{I}$ is Abelian and the variety of Abelian ℓ-groups is minimal, so every other nontrivial variety of ℓ-groups is closed under local completions. □

6.4.3. Product algebras. The variety of product algebras is generated by the real interval $[0,1]$ as a residuated lattice under natural order and natural multiplication. In particular, the algebra $\mathbb{Z}_0^-$, best described as the upper half $A = \{a_n : n \in \mathbb{N}\}$ of Chang's chain with an additional zero element at the bottom (cf. Figure 2.5 again), is a product algebra. Notice that for any $i \in \mathbb{N}$ we have $\neg a_i = 0$. The algebra $\mathbb{Z}_0^-$ is subdirectly irreducible with A being its monolith deductive filter. In fact, $\mathbb{Z}_0^-$ also generates the whole variety of product algebras. Unlike $\mathbf{C}_\infty$, the algebra $\mathbb{Z}_0^-$ is complete itself. It will be our example of a complete algebra whose ultrapower is not embeddable in any complete algebra from the variety it generates. Take an ultrapower $\mathbf{R} = (\mathbb{Z}_0^-)^I/U$ for some infinite I and a nonprincipal U.

THEOREM 6.39. [KL07] *The algebra $\mathbf{R}$ is a discontinuous GBL-algebra. Thus, the variety of product algebras is not closed under completions. The same holds for every larger variety of GBL-algebras.*

PROOF. The algebra $\mathbf{R}$ is subdirectly irreducible, with the monolith deductive filter isomorphic to A. Pick elements $w, u \in R \setminus A$ with $0 < w \prec u$. This is always possible by properties of ultraproducts, moreover, every element $e \in R \setminus A$ has an immediate successor e' (i.e., a cover with respect to the natural ordering). Define inductively $u_0 = u$ and $u_{n+1} = u'_n$. Let B be the downward closure of $\{u_k : k \in \mathbb{N}\}$. Clearly, B is an ideal without supremum in R. Notice that $\mathbb{Z}_0^-$ has the following property:

- for every x and y, if $x > 0$ and y is the k-th successor of x, then $y \to x$ is the k-th predecessor of 1, i.e., equals a_k.

The property above is first-order expressible, and thus carries over to $\mathbf{R}$. Therefore, $u_k \to u = a_k \in A$. Since every $b \in B$ is either smaller than u or equal to u_k for some $k \in \mathbb{N}$, we have $b \to u \in A$ for every $b \in B$. Further, for each $j \in \mathbb{N}$ we get $a_j = u_j \to u$. This takes care of all the required properties up to 3(a). To show that 3(b) holds as well, note first that if $b \leq w$, then $xb \leq w$ for any x whatsoever, so it remains to show that for each $j \in \mathbb{N}$ there is an $a \in A$ with $au_j \leq w$. To this end, observe that $\mathbb{Z}_0^-$ has the following, first-order expressible, property.

- for every x and y, if $x > 0$ and y is the k-th predecessor of x, then the $k+1$-th successor of 1, i.e., a_{k+1} has $a_{k+1} \cdot y \prec x$.

We conclude that $a_{k+1} \cdot u_k = w \prec u$ holds in $\mathbf{R}$. This completes the list of requirements. The conclusion now follows by Theorem 6.33. □

6.5. Finite embeddability property

We say that the class $\mathcal{K}$ of algebras has the *finite embeddability property* when for any given finite *partial* subalgebra $\mathbf{B}$ of an algebra $\mathbf{A}$ in $\mathcal{K}$, there exists a finite algebra $\mathbf{D}$ in $\mathcal{K}$ into which $\mathbf{B}$ can be embedded. It clearly follows that $\mathcal{K}$ is generated by its finite members (i.e., it has the finite model property), but in fact FEP is stronger than that. We say that $\mathcal{K}$ has the *strong finite model property* (SFMP) if every quasi-identity that fails in $\mathcal{K}$ is falsified on a finite member of $\mathcal{K}$. It turns out that SFMP and FEP are equivalent in quasivarieties of finite type.

LEMMA 6.40. *For any quasivariety $\mathcal{K}$ of finite type the following are equivalent:*

(1) *$\mathcal{K}$ has FEP,*
(2) *$\mathcal{K}$ has SFMP,*
(3) *$\mathcal{K}$ is generated as a quasivariety by its finite members.*

PROOF. Suppose $\mathcal{K}$ has FEP and q is a quasi-identity that fails in $\mathcal{K}$. Take an algebra $\mathbf{A} \in \mathcal{K}$ witnessing that. So, we have a sequence of elements $\overline{a}$ of K such that $\mathbf{K} \not\models q(\overline{a})$. Let $\mathbf{P}$ be the finite partial subalgebra of $\mathbf{K}$ defined as $\{t(\overline{a}) : \psi \in T\}$, where T is the set of subterms of q. The partial operations on P are defined naturally by the structure of q. Then the finite algebra $\mathbf{D}$ into which P is embedded falsifies q. Thus every quasi-identity that fails in $\mathcal{K}$ fails on a finite algebra from $\mathcal{K}$, hence $\mathcal{K}$ has SFMP.

Now suppose $\mathcal{K}$ has SFMP. Then $\mathrm{Th}_q \mathcal{K} = \mathrm{Th}_q \mathcal{K}_{fin}$, where $\mathcal{K}_{fin}$ denotes the class of finite member of $\mathcal{K}$, and thus $\mathcal{K} = \mathrm{Mod}(\mathrm{Th}_q \mathcal{K}_{fin}) = \mathsf{ISPP_U}(\mathcal{K}_{fin})$.

Finally, suppose $\mathcal{K} = \mathsf{ISPP_U}(\mathcal{K}_{fin})$. Take an algebra $\mathbf{A}$ from $\mathcal{K}$ and a finite partial subalgebra $\mathbf{P}$ of $\mathbf{A}$ with universe $\{p_1, \ldots, p_n\}$. For each p_i fix a variable x_{p_i} and for each basic operation f of the language consider the identity $f(x_{p_j}, \ldots, x_{p_k}) = x_{p_m}$ if $f^{\mathbf{P}}(p_j, \ldots, p_k) = p_m$. Let now $\delta_{\mathbf{P}}$ be the conjunction of all identities $f(x_{p_j}, \ldots, x_{p_k}) = x_{p_m}$. For $1 \leq s < t \leq n$ consider the quasi-identity $q_{s,t}$ defined as $\delta_{\mathbf{P}} \Rightarrow (x_s = x_t)$. Obviously $\mathbf{P}$ falsifies it and hence so does $\mathbf{A}$. Since $\mathcal{K}$ is generated as a quasivariety by its finite members, there is a finite $\mathbf{C}_{s,t} \in \mathcal{V}$ falsifying $\delta_{\mathbf{P}} \Rightarrow (x_s = x_t)$. So, $C_{s,t}$ contains elements $c_1, \ldots, c_n$, not necessarily pairwise distinct, but such that $\delta_{\mathbf{P}}(c_1, \ldots, c_n)$ and also $c_s \neq c_t$. Then, taking $\mathbf{D} = \prod_{1\leq s<t\leq n} \mathbf{C}_{s,t}$ we define the map $e \colon P \to D$ putting $e(p_i) = \overline{c}_i$ where $\overline{c}_i$ is the sequence $(c_i)_{s,t}$. Since $\mathbf{D}$ satisfies $\delta_{\mathbf{P}}(\overline{c}_1, \ldots, \overline{c}_n)$, we get that e is a (partial) homomorphism. But, by construction, $\mathbf{D}$ separates points of $\mathbf{P}$ and thus e is an embedding. □

Another consequence of FEP has to do with decidability. Namely, if a class $\mathcal{K}$ has FEP then every universal sentence that fails in $\mathcal{K}$ fails also in a finite member of $\mathcal{K}$. To see that, let φ be a universal sentence (in normal form with all quantifiers in front) and $\mathbf{A} \in \mathcal{K}$ an algebra falsifying φ under a valuation v. Let P be the set of subterms of φ. Clearly, the image $v[P] \subseteq A$ is finite. Moreover, it gives rise to a partial subalgebra of $\mathbf{A}$ with $v(s) \star v(t)$ defined, as $v(s \star t)$, if $s \star t$ is a subterm of φ, where $\star$ ranges over the operations occurring in φ. By FEP, $v[P]$ can be embedded in a finite algebra $\mathbf{D} \in \mathcal{K}$ and it is easy to see that the valuation defined by sending each variable x occurring in φ to the image of $v(x)$ in D (and arbitrary for other variables) falsifies φ in $\mathbf{D}$. Thus, if $\mathcal{K}$ is finitely axiomatizable, then its universal theory is decidable.

A typical example of a class with finite embeddability property is the variety HA of Heyting algebras. Let $\mathbf{B}$ be a finite partial subalgebra of a Heyting algebra $\mathbf{A}$. The Heyting subalgebra generated by the domain of $\mathbf{B}$ is not always finite, since the class of Heyting algebras is not locally finite. Nevertheless, the sublattice $\mathbf{D}$ generated by $B \cup \{0, 1\}$ *is* finite, by the local finiteness of distributive lattices, and thus can be expanded to a Heyting algebra. Exercise 15 asks the reader to verify that such an expansion preserves the existing residuals from $\mathbf{B}$. Thus, $\mathbf{B}$ can be embedded into (the expansion of) $\mathbf{D}$.

This example, although it nicely illustrates the idea, uses two crucial facts that are not available to us in the general case of residuated lattices: local finiteness of the lattice reduct and fusion coinciding with meet. Below we present a construction that uses neither.

6.5.1. An embedding construction. Let $\mathbf{B}$ be a partial subalgebra of a residuated pomonoid $\mathbf{A}$. We will describe how to embed $\mathbf{B}$ into a complete algebra $\mathbf{D}$, in such a way that in certain cases $\mathbf{D}$ will be finite whenever $\mathbf{B}$ is. The construction is taken from [BvA02], but we modify the presentation in certain parts in order to show the connections of the construction to the ones used in the next chapter. Let $\mathbf{M} = (M, \cdot, 1, \leq)$ be the subpomonoid generated by the domain B of $\mathbf{B}$. The algebra $\mathbf{D}$ will be a nucleus image on the powerset $\mathcal{P}(\mathbf{M})$; the nucleus itself will be of the form γ_N, for some nuclear relation N.

For each pair $u, w \in M$ and $c, b \in B$, we define

$$c\ N\ (u, w, b) \text{ iff } ucw \leq b;$$

note that N is a relation between M and $M^2 \times B$. Moreover, it is nuclear, since for all $c, d \in M$, we have

$$c \cdot d\ N\ (u, w, b) \text{ iff } ucdw \leq b \text{ iff } c\ N\ (u, dw, b) \text{ iff } d\ N\ (uc, w, b).$$

Clearly, $(u, w, b)^{\lhd} = \{c \in M : ucw \leq b\}$; if $u\backslash b/w$ happens to be an element of M, then $(u, w, b)^{\lhd} = \{c \in M : c \leq u\backslash b/w\} = {\downarrow_M}(u\backslash b/w)$. Since ${\downarrow_M}(u\backslash b/w)$ represents in $\mathcal{P}(M)$ the element $u\backslash b/w$, intuitively speaking $(u, w, b)^{\lhd}$ plays the role of $u\backslash b/w$; if it exists then the two elements coincide. Note that ${\downarrow_M}(u\backslash b/w)$ is actually an element in $\mathcal{P}(\mathbf{M})_{\gamma_N}$, since it is a Galois closed (basic) element. Therefore, by this construction we create in $\mathbf{D} = \mathcal{P}(\mathbf{M})_{\gamma_N}$ all the missing (two sided) residuals of $\mathbf{B}$ of the form $u\backslash b/w$. Furthermore, since the elements of the form $(u, w, b)^{\lhd}$, for $u, w \in M$ and $b \in B$, form a basis (which we will denote by $\overline{D}$), all other elements in $\mathbf{D}$ will be intersections of them. Consequently, loosely speaking, $\mathbf{D}$ consists of all formal meets of residuals of the form $u\backslash b/w$. It is interesting that all we need to add to $\mathbf{B}$ are these meets of residuals in order to obtain an algebra $\mathbf{D}$ into which $\mathbf{B}$ embeds.

It follows directly from the fact that $\overline{D}$ is a basis that for all $x \subseteq M$, we have $\gamma_N(X) = \bigcap\{Z \in \overline{D} : X \subseteq Z\}$; from now on we will write simply γ for γ_N.

LEMMA 6.41. *The map $\gamma : \mathcal{P}(M) \to \mathcal{P}(M)$ is a downward nucleus on $\mathcal{P}(\mathbf{M})$. Moreover, if $\bigvee X \in M$, for some $X \subseteq M$, then $\bigvee X \in \gamma(X)$.*

PROOF. The map γ is a closure operator because N is a nuclear relation. To show that any closed set is downward closed, it suffices to check only the basic closed sets, since downward closed sets are closed under intersections. Indeed, if $c \in (u, w, b)^{\lhd}$ and $d \leq d$, then $udw \leq ucw \leq b$, so $d \in (u, w, b)^{\lhd}$. Finally, suppose that $\bigvee X \in M$, for some $X \subseteq M$. To prove that $\bigvee X \in \gamma(X)$ is suffices to show that if $X \subseteq (u, w, b)^{\lhd}$, then $\bigvee X \in (u, w, b)^{\lhd}$. So, if $x \in (u, w, b)^{\lhd}$, for all $x \in X$, then $uxw \leq b$ and $x \leq u\backslash b/w$, for all $x \in X$. Consequently, $\bigvee X \leq u\backslash b/w$ and $u \cdot \bigvee X \cdot w \leq b$; hence $\bigvee X \in (u, w, b)^{\lhd}$. □

By Theorem 6.29 it follows that $\mathbf{B}$ is completely embedded into the complete residuated lattice $\mathcal{P}(\mathbf{M})_\gamma$ via the map $b \mapsto {\downarrow_M} b = (1, 1, b)^{\lhd}$. Moreover the embedding preserves integrality, commutativity and the property of being square-increasing. However, in general neither M nor $\overline{D}$ are finite even if B is. The next step in the construction deals with this problem.

Let $\mathbf{F}(k)$ be the free monoid on k generators $x_1, \ldots, x_k$. As usual, we will think of the generators as letters and the elements of $F(k)$ as words on these letters, with the empty word acting as identity element. Define a relation $\sqsubseteq$ on $F(k)$ by putting $u \sqsubseteq w$ if w is a *deletion instance* of u, that is if w can be obtained from u by deleting some subwords of u. Notice that being a deletion instance generalizes the usual subword relation. It is not difficult to verify (cf. Exercise 16) that the structure $(F(k), \cdot, \sqsubseteq)$ is an integral pomonoid. Moreover, since $\mathbf{F}(k)$ is free, we have that if $s, t, u, v \in F(k)$ are nonempty words and $st = uv$, then $s = u$ and $t = v$.

LEMMA 6.42. [BvA02] *Let $s, t, u, v \in F(k)$ and t, v be nonempty words. If $su \sqsubseteq tv$ then either $s \sqsubseteq tv$ or $u \sqsubseteq tv$ or both $s \sqsubseteq t$ and $u \sqsubseteq v$.*

PROOF. Since $su \sqsubseteq tv$, we can delete some subwords of su to obtain tv. Let s' and u' be the resulting deletion instances of s and u, so that $s'u' = tv$. If s' is the empty word, then $u' = tv$ and thus $u \sqsubseteq tv$. Analogously, if u' is empty then $s \sqsubseteq tv$. Suppose both s' and u' are nonempty. Then by freeness $s' = t$ and $u' = v$, and so $s \sqsubseteq t$ and $u \sqsubseteq v$ as claimed. □

LEMMA 6.43. [BvA02] *The pomonoid $\langle F(k), \cdot, \sqsubseteq\rangle$ is residuated. For $s, t \in F(k)$ the residuals are characterized as follows:*

(1) *$s\backslash t$ is the shortest final subword t_2 of t such that $t_1t_2 = t$ and $s \sqsubseteq t_1$,*
(2) *t/s is the shortest initial subword t_1 of t such that $t_1t_2 = t$ and $s \sqsubseteq t_2$.*

PROOF. We will prove it only for the right residual, proceeding by induction with step 2 on the length of t. If t is the empty word, i.e., $t = 1$, then, as the empty word is a deletion instance of every word, we obtain $s\backslash t = 1$. If t is a single letter, say $t = x_1$, then $s\backslash x_1 = 1$ if x_1 occurs in s and $s\backslash x_1 = x_1$ if x_1 does not occur in s. So the claim holds for the base case.

We can now assume that $t = uw$, with both u and w nonempty. Consider the set $R = \{v \in F(k) \colon sv \sqsubseteq t\}$. Notice that R is nonempty, for example $t \in R$. We will prove that R has a largest element. For each $v \in R$, there is a deletion of subwords of sv such that the resulting $s'v'$ is equal to t. It follows that sv' also can be reduced to t and therefore $v' \in R$ and $v \sqsubseteq v'$. Without loss of generality we can thus assume $v' = v$, thereby considering only maximal elements of R. If v is empty, then $s \sqsubseteq t$ and $s\backslash t = 1$. If v is nonempty but s' is empty, then $v \sqsubseteq t$ and thus by maximality assumption $v = t$. Then, $s\backslash t = t$. The remaining case is $s'v = t = uw$ with both s' and v nonempty. Then, we have $sv \sqsubseteq uw$ and by Lemma 6.42 we get $s \sqsubseteq u$ and $v \sqsubseteq w$. By maximality of v we obtain $v = w$ and therefore $s\backslash t = w$. □

LEMMA 6.44. [BvA02] *The pomonoid $\langle F(k), \cdot, \sqsubseteq\rangle$ contains neither infinite antichains nor infinite ascending chains.*

PROOF. The lack of infinite ascending chains is clear. The lack of infinite antichains we will prove by induction on the number of generators k. For $k = 1$ the claim trivially holds as there are no antichains at all. Suppose it holds for all $n < k$, and let A be an antichain in $F(k)$. If the set of words in A in which all the generators occur is finite, then, setting A_i to be the subantichain of A in which x_i does not occur, we obtain by inductive hypothesis that A_i is finite for every $i \in \{1, \ldots, k\}$; hence A is finite. Therefore the subantichain A' of A where all generators occur in every word is infinite. Pick an $a \in A'$, say $a = v_1 \cdots v_m$ with $\{v_1, \ldots, v_m\} = \{x_1, \ldots, x_k\}$. Now A' contains words of arbitrary length and there are only k letters, so in a sufficiently long word they must occur in exactly the order $v_1, \ldots, v_m$. Therefore, there is a $b \in A'$ such that $b = s_1 v_1 s_2 \cdots s_m v_m s_{m+1}$. But then $b \sqsubseteq a$, so A is not an antichain contradicting the assumption. □

Let now G be a finite subset of $\mathbf{F}(k)$ and put $H = \{p\backslash g/q : p, q \in F(k) \text{ and } g \in G\}$. By Lemma 6.43 we get that H is contained in $\uparrow G$ and thus by Lemma 6.44 it follows that H is finite.

LEMMA 6.45. [BvA02] *If* $\mathbf{A}$ *is integral and* B *is finite, then* $\overline{D}$ *is finite. Therefore,* $\mathcal{P}(\mathbf{M})$ *is also finite.*

PROOF. Assume $|B| = k$ and map the generators of $\mathbf{F}(k)$ onto the elements of B. Since $\mathbf{M}$ is integral, this map extends to a pomonoid homomorphism $h : \mathbf{F}(k) \to \mathbf{M}$. For each $b \in B$ let $\mathrm{Crit}(b)$ be the set of maximal elements in $h^{-1}(b)$ and define $G = \bigcup_{b \in B} \mathrm{Crit}(b)$. By Lemma 6.44, G is finite and therefore so is H defined as above. We will show that for every $u, w \in M$ and every $b \in B$ the set $(u, w, b)^{\lhd}$ is the image of $\downarrow Y$ for some $Y \subseteq H$. Pick a $p \in h^{-1}(u)$ and a $q \in h^{-1}(w)$. Define $Y = \{p\backslash r/q : r \in \mathrm{Crit}(b)\}$. We have:

$$\begin{aligned} z \in h^{-1}((u, w, b)^{\lhd}) &\text{ iff } h(z) \in (u, w, b)^{\lhd} \\ &\text{ iff } u \cdot h(z) \cdot w \le b \\ &\text{ iff } h(p) \cdot h(z) \cdot h(q) \le b \\ &\text{ iff } h(pzq) \le b \\ &\text{ iff } pzq \in h^{-1}(b) \\ &\text{ iff } pzq \le r \text{ for some } r \in \mathrm{Crit}(b) \\ &\text{ iff } z \le p\backslash r/q \text{ for some } r \in \mathrm{Crit}(b) \\ &\text{ iff } z \in \downarrow Y. \end{aligned}$$

Since $\overline{D}$ is the set of all $(u, w, b)^{\lhd}$ with $u, w \in M$ and $b \in B$, by finiteness of H we obtain that $\overline{D}$ is finite. □

The next theorem is now straightforward to obtain. In fact, it was proved in [BvA02] even for nonassociative case, i.e., for the quasivariety of integral residuated *groupoids*; see also Note 2 in the end of the chapter.

THEOREM 6.46. [BvA02] *The quasivariety of integral residuated pomonoids has FEP. The varieties* $\mathsf{FL_i}$ *and* IRL *have FEP as well. Moreover, any subvariety of* $\mathsf{FL_i}$ *defined by a combination of commutativity, zero-boundedness and idempotency has FEP. Thus, any subvariety of* IRL *defined by a combination of commutativity and idempotency has FEP as well.*

PROOF. The part that is not an immediate corollary of the series of lemmas above is concerned with the constant 0. Observe first that if $\mathbf{A}$ has this constant and $\mathbf{B}$ is a finite partial subalgebra of $\mathbf{A}$, then $\mathbf{B}_0$, with $B_0 = B \cup \{0\}$, is a finite partial subalgebra of $\mathbf{A}$, too, and it contains $\mathbf{B}$. Carrying out the construction for B_0 we get that $(1,1,0)^{\lhd} \in \overline{D}$. In particular, if 0 is the bottom element of $\mathbf{A}$, then $(1,1,0)^{\lhd} = \{0\}$ and that is the bottom element of the algebra $\mathcal{P}(\mathbf{M})_\gamma$. □

6.5.2. FEP for some subvarieties of FL**.** We have already seen that the variety HA of Heyting algebras is an example. By precisely the same reasoning the variety of RHA of linear Heyting algebras has FEP. The apparent triviality of the argument should not lure the reader into thinking that it will work for any variety of Heyting algebras. Indeed, there are uncountably many subvarieties of HA not generated by their finite members (see Chapter 6 of [CZ91] for particularly simple examples, under the guise of superintuitionistic logics that are not *finitely approximable*).

The next few results employ the construction described above, due to Blok-van Alten, and apply it to some subvarieties of $\mathsf{FL_i}$. To smoothen the presentation it will be helpful to recall what the residuals look like in the algebra $\mathcal{P}(\mathbf{M})_\gamma$. Namely, we have for $X, Y \in D$

$$X\backslash Y = \{a \in M \colon Xa \subseteq Y\} \qquad X/Y = \{a \in M \colon aX \subseteq Y\}$$

in particular

$$\sim X = \{a \in M \colon Xa \subseteq {\downarrow}0\} \qquad -X = \{a \in M \colon aX \subseteq {\downarrow}0\}$$

We will also need the following sufficient condition for FEP (see e.g. [BF00]). Recall that $\mathcal{V}_{\mathsf{SI}}$ stands for the class of all subdirectly irreducible members of a variety $\mathcal{V}$.

LEMMA 6.47. *For a variety* $\mathcal{V}$*, if* $\mathcal{V}_{\mathsf{SI}}$ *has FEP, then so does* $\mathcal{V}$.

PROOF. Let $\mathbf{A} \in \mathcal{V}$ have a finite partial subalgebra $\mathbf{B}$. Take a subdirect representation $\prod_{i\in I} \mathbf{A}_i$ of $\mathbf{A}$. Let $\mathbf{B}_i$ be the projection of $\mathbf{B}$ on the coordinate i. Clearly each $\mathbf{B}$ is a finite partial subalgebra of $\mathbf{A}_i$ and thus is embeddable into a finite algebra $\mathbf{C}_i$. Since $\mathbf{B}$ is finite, finitely many coordinates suffice to separate points of B. Letting J index these coordinates we get that $\mathbf{B}$ embeds into $\prod_{j\in J} \mathbf{C}_j$, which is a finite algebra from $\mathcal{V}$. □

The following lemma extends the proof in [KO06], given with the assumption of commutativity and least element.

LEMMA 6.48. *The variety* $\mathsf{RFL_i}$ *of representable* FL_i*-algebras has FEP.*

PROOF. By Lemma 6.47 it suffices to show that the class of subdirectly irreducible members of $\mathsf{RFL_i}$ has FEP. Now, any algebra $\mathbf{A}$ in that class is linearly ordered, and thus so is $\mathbf{M}$. Since each member of $\overline{D}$ is a downset of M, the set $\overline{D}$ is also linearly ordered by set inclusion. Further, each member of $\mathcal{P}(\mathbf{M})_\gamma$ is a finite intersection of members of $\overline{D}$ and thus the universe of $\mathcal{P}(\mathbf{M})_\gamma$ is exactly $\overline{D}$. Now, take arbitrary members X and Y of $\mathcal{P}(M)_\gamma = \overline{D}$. Then, either $X \subseteq Y$ or $Y \subseteq X$ holds. Thus, $\mathcal{P}(\mathbf{M})_\gamma$ is linearly ordered and being finite it has

a unique subcover of 1. Therefore, $\mathcal{P}(\mathbf{M})_\gamma$ is subdirectly irreducible and thus belongs to $\mathsf{RFL_i}$. □

It was shown in [KO06], that $\mathsf{InFL_{ew}}$ has the FEP. The following result extends the proof given there.

LEMMA 6.49. *The varieties* $\mathsf{CyFL_i}$ *and* $\mathsf{CyInFL_i}$ *have FEP.*

PROOF. We will modify slightly the construction at the very beginning. Namely, if $\mathbf{B}$ is finite partial subalgebra of an algebra $\mathbf{A}$ in any of the varieties that this lemma deals with, we define $B^{\sim} = \{{\sim}b \colon b \in B\}$, $B^{-} = \{-b \colon b \in B\}$ and $B^{\star} = B^{\sim} \cup B \cup B^{-}$. Then we consider a finite partial subalgebra $\mathbf{B}^{\star}$ of $\mathbf{A}$ with universe $B^{\star}$. Let $\mathbf{D}$ be a finite algebra into which $\mathbf{B}^{\star}$ is embedded. Since $\mathbf{B} \subseteq \mathbf{B}^{\star}$ the original partial algebra $\mathbf{B}$ is also embedded into $\mathbf{D}$. It remains to show that this embedding preserves the relevant properties.

Cyclicity is straightforward. Let $z \in {\sim}X$. This means in our coset notation that $Xz \leq 0$. Thus, $X \leq 0/z = -z$ and by cyclicity $-z = {\sim}z$. Therefore $X \leq z\backslash 0$ and so $zX \leq 0$; hence $z \in -X$ as required.

To show involutiveness, given cyclicity, let $z \in -{\sim}X$. Recalling the characterization of residuals in $\mathbf{D}$ it is easy to see that it amounts to the following property

$$\text{for all } a \in M \colon Xa \leq 0 \text{ implies } za \leq 0$$

Let $X = \bigcap_i (u_i, w_i, c_i)^{\lhd}$, where $u_i, w_i \in M$, $c_i \in B^{\star}$ and i ranges over some unspecified index set. Take any $x \in X$. We have $x \leq u_i\backslash c_i/w_i$ for all i, and thus $xw_i \leq u_i\backslash c_i$. Multiplying both sides by ${\sim}c_i$ we obtain $xw_i({\sim}c_i) \leq (u_i\backslash c_i)({\sim}c_i) \leq {\sim}u_i$. From this by cyclicity we get $xw_i({\sim}c_i) \leq -u_i$ and therefore $xw_i({\sim}c_i)u_i \leq (-u_i)u_i \leq 0$. By assumptions on $B^{\star}$ we have that ${\sim}c_i \in B^{\star}$ and so $w_i({\sim}c_i)u_i$ belongs to M. Since x was arbitrary, we get $Xw_i({\sim}c_i)u_i \leq 0$ and by the property above we conclude $zw_i({\sim}c_i)u_i \leq 0$. Now, using cyclicity several times we carry out the following calculation

$$\begin{aligned} zw_i({\sim}c_i)u_i \leq 0 \quad &\text{iff} \quad zw_i \leq 0/({\sim}c_i)u_i \\ &\text{iff} \quad zw_i \leq ({\sim}c_i)u_i\backslash 0 \\ &\text{iff} \quad zw_i \leq u_i\backslash({\sim}c_i\backslash 0) \end{aligned}$$

$$\begin{aligned} &\text{iff} \quad zw_i \leq u_i \backslash {\sim}{\sim} c_i \\ &\text{iff} \quad zw_i \leq u_i \backslash c_i \\ &\text{iff} \quad u_i z w_i \leq c_i \end{aligned}$$

Therefore $z \in (u_i, w_i, c_i)^{\lhd}$ for every i and thus $z \in X$. This proves left-involution. The argument for right-involution is symmetric. □

The identities $x \wedge {\sim}x \leq 0$ and $x \wedge -x \leq 0$ are also of certain interest. They are sometimes known as *pseudo-complementation*, especially in zero-bounded algebras. The next lemma also generalizes a result from [KO06].

LEMMA 6.50. *Subvarieties of* $\mathsf{FL_i}$ *defined, respectively, by* $x \wedge {\sim}x \leq 0$ *and* $x \wedge -x \leq 0$ *have FEP.*

PROOF. Suppose $z \in X \cap {\sim}X$. Thus $z \in X$ and $Xz \leq 0$. Therefore $z^2 \leq 0$, so $z \leq {\sim}z$. Hence $z = z \wedge z \leq z \wedge {\sim}z \leq 0$. It follows that $X \cap {\sim}X \subseteq (1, 1, 0)^{\lhd}$ as claimed. The proof of the other inequality is the same. □

Since the proofs of the lemmas above employ essentially the same construction of $\mathbf{D}$, they can be freely combined. All identities preserved by nuclei are also preserved in $\mathbf{D}$. We sum it up in the next theorem. Exercise 19 asks the reader to verify that the last condition is preserved as well.

THEOREM 6.51. *Let* $\mathcal{V}$ *be a subvariety of* $\mathsf{FL_i}$ *defined by any combination of the following identities:*

(1) *commutativity*
(2) *idempotency*
(3) *representability*
(4) *cyclicity*
(5) *cyclicity + involution*
(6) *pseudo-complementation*
(7) $0 = 1$
(8) $0 \leq x$

Then $\mathcal{V}$ *has FEP and thus its universal theory is decidable.*

COROLLARY 6.52. [MT44] *The variety* HA *has the FEP.*

In particular, the varieties MTL, IMTL and SMTL (subvarieties of $\mathsf{RFL_{ew}}$ well-known in fuzzy logic community, cf. [EGGC03]) are covered by the above theorem. Below are two examples of varieties with FEP not covered by Theorem 6.51.

THEOREM 6.53. *The variety generated by Sugihara algebras has FEP. Therefore, the consequence relation of the relevant logic* **RM** *is decidable.*

PROOF. By Lemma 6.47, it suffices to show that subdirectly irreducible Sugihara algebras have FEP. But subdirectly irreducible Sugihara algebras are chains with an additional property that both $a \cdot b$ and $a \to b$ belong to $\{a, b\}$. Therefore any finite subset B of a subdirectly irreducible Sugihara algebra can be expanded to a finite Sugihara algebra with the universe B. □

Another important result not covered by Theorem 6.51 was proved, using different techniques, by Agliano, Ferreirim and Montagna. Since their proof is quite involved, we will not reproduce it here.

THEOREM 6.54. [AFM99] *The variety* BL *has the FEP.*

We also do not reproduce the proof of the following result that was implicit in [Wój73]. It was shown explicitly in [BF00].

THEOREM 6.55. *The variety* MV *has the FEP.*

6.5.3. Counterexamples for FEP. For contrast, we now present two counterexamples for FEP. The first, coming from [BvA02], tracks lack of FEP down to the existence of certain infinite algebra. Let $\mathbb{Z}$ be the ℓ-group of the integers. Clearly, $\mathbb{Z}$ an FL_e-algebra and it falsifies the quasi-identity

$$x \geq 1 \text{ and } xy = 1 \Rightarrow x = 1$$

which says that the only positive invertible element is the unit. But the quasi-identity above is satisfied by every finite FL_e-algebra. To see it observe that if a positive element x is invertible, so is its square, and thus $x^2 > x$ for every such element. In a finite algebra this yields $x^n = \top$ for some n, and $\top$ is not invertible.

Exactly the same argument works with $\mathbb{Z}$ expanded with an additional constant 0, set to be any element of $\mathbb{Z}$. It also works with $\mathbf{T}_1[\mathbb{Z}]$, the generalized ordinal sum of $\mathbf{T}_1$ and $\mathbb{Z}$ on the set $\mathbb{Z} \cup \{\bot, \top\}$, (possibly expanded by 0) in place of $\mathbb{Z}$, so the presence or absence of bounds is irrelevant.

THEOREM 6.56. *Any subvariety of* FL, $\mathsf{FL_o}$, *or* RL *containing* $\mathbb{Z}$ *or* $\mathbf{T}_1[\mathbb{Z}]$ *lacks FEP. In particular, the varieties* $\mathsf{FL_e}$, $\mathsf{FL_o}$, CRL *as well as* IMALL *and* MALL *lack FEP.*

Thus, the varieties FL, $\mathsf{FL_e}$, $\mathsf{FL_o}$, as well as RL and CRL, are examples of varieties generated by finite members but lacking FEP.

Our second example is integral, but less general in scope than the previous one. It shows, however, that the result stated in Theorem 6.54 is near-optimal. Consider the variety of integral GBL-algebras. This is the subvariety of $\mathsf{FL_i}$ satisfying divisibility, i.e., the condition $x \wedge y = x(x\backslash y) = (y/x)x$. Any negative cone of a linearly ordered non-Abelian ℓ-group is an integral GBL-chain, so noncommutative examples exist. We also need the following lemma.

LEMMA 6.57. [JM06] *Finite integral GBL-algebras are commutative. The same holds for finite pseudo-BL-algebras.*

PROOF. Let **A** be a finite integral subdirectly irreducible GBL-algebra (or pseudo-BL-algebra). Notice that if $a \in A$ is an idempotent element, then since $ax \leq a \wedge x = a(a\backslash x) = a^2(a\backslash x) = a(a \wedge x) \leq ax$, we get that a is central (i.e., commutes with every element of A, cf. Section 3.6.3). As **A** is subdirectly irreducible, by Lemma 3.51, there is a unique largest central idempotent element $c \in A \setminus \{1\}$ such that $[1]_\mu = \{u \in A\colon u \geq c\}$, where μ stands for the monolith of **A**. By finiteness, $[1]_\mu$ must contain a coatom a. Let $S = \{a^k\colon k \in \mathbb{N}\}$. We claim that S is upward closed in A.

We first show that if $a^i > a^{i+1}$, then a^i covers a^{i+1}. Suppose the contrary; then $a^i > b > a^{i+1}$ and by divisibility $a^i(a^i\backslash b) = b$. Since $b > a^{i+1}$, we get $1 > a^i\backslash b \geq a^i\backslash a^{i+1} \geq a$. As a is a coatom, $a^i\backslash b = a$ and thus $a^{i+1} = b$, a contradiction.

Then, suppose $c \notin S$ but $c > a^j$, for some j. We can assume that for all $k < j$ the element a^k is not idempotent. By integrality, there is a smallest m (possibly $m = 0$) with the property that $a^m \geq c \not\geq a^{m+1}$. Notice that $m < j$, so a^m is not idempotent. Since a^m covers a^{m+1} we obtain $a^{m+1} \vee c = a^m$. Thus, $a^m = a^{m+1} \vee c = aa^m \vee c = a(a^{m+1} \vee c) \vee c = a^{m+2} \vee ac \vee c = a^{m+2} \vee c$. Iterating the argument, we obtain $a^m = a^j \vee c = a^j$, so a^m is idempotent contradicting the assumption.

Exercise 13 asks the reader to complete the argument. □

THEOREM 6.58. [JM06] *The varieties of integral GBL-algebras and pseudo-BL-algebras lack the FEP.*

PROOF. Let **A** be a noncommutative algebra in one of the varieties above, and **B** a finite partial subalgebra of **A** containing the elements a and b such that ab and ba are defined in B but $ab \neq ba$. By Lemma 6.57 any finite algebra **C** in one of our varieties is commutative, so **B** cannot be embedded into **C**. □

For much more trivial reasons, following by the decomposition of every GBL-algebra as the direct product of an ℓ-group and an integral GBL-algebra, given in [GT05], we obtain the following.

THEOREM 6.59. *The variety of GBL-algebras lacks the FEP.*

Exercises

(1) Verify that the maps e and d of Lemma 6.1 are embeddings.
(2) Verify the details of Theorem 6.2(1).
(3) Give an example of a lattice **L** whose canonical extension is not a compact lattice. This will prove the assertion on page 290 that compactness of $(e, \mathbf{C})$ is not the property of **C** alone. Hint: next exercise.

(4) Give an example of a *compact* lattice $\mathbf{L}$ whose canonical extension is not compact. Hint: try an $\mathbf{L}$ in which every chain is finite but there is no bound for the length of chains.
(5) Let $\mathbf{N}$ be the Boolean algebra of finite and cofinite subsets of $\mathbb{N}$. Define $\gamma\colon \mathcal{P}(\mathbf{N}) \to \mathcal{P}(\mathbf{N})$ by $\gamma(X) = \mathrm{Ig}(X)$. Show that $\mathcal{P}(\mathbf{N})_\gamma$ is isomorphic to $\mathcal{P}(\mathbb{N})$ but it is not the Dedekind-MacNeille completion of $\mathbf{N}$ under the natural embedding $x \mapsto {\downarrow}x$.
(6) Prove Lemmas 6.7 and 6.8.
(7) Prove that canonical extensions commute with subalgebras and homomorphic images.
(8) Expand the sketch of proof of Lemma 6.13.
(9) Prove that if f^σ preserves arbitrary non-empty joins in each coordinate, then it preserves upward directed joins.
(10) Show that negation is smooth in canonical extensions of Boolean algebras.
(11) Let $\mathbf{A}$ be a residuated lattice. Prove that the set ${\uparrow}(fg)$ is a filter, for all filters f, g of the lattice reduct of $\mathbf{A}$.
(12) Prove that canonical extensions do not commute with infinite products. Hint: consider $\mathbf{2}^{\mathbb{N}}$, where $\mathbf{2}$ is the two-element Boolean algebra.
(13) Complete the proof of Lemma 6.57. Hint: the subdirectly irreducible algebra $\mathbf{A}$ of that lemma has a subalgebra with the universe $(A \setminus S) \cup \{1\}$ with strictly fewer elements than A. Noncommutativity could only happen in this subalgebra. Notice that the presence of the constant 0 satisfying $0 \leq x$ does not interfere with this subalgebra, so the argument holds for pseudo-BL-algebras as well.
(14) Prove that an ℓ-group is bounded if and only if it is trivial.
(15) Prove that the construction of a finite algebra out of a partial subalgebra of a Heyting algebra, described on page 311 preserves existing residuals.
(16) Prove that the structure $(F(k), \cdot, \sqsubseteq)$, defined just above Lemma 6.42 is an integral pomonoid.
(17) Show that the variety $\mathsf{E_n}$ of n-potent FL_{ew}-algebras has FEP. Base your argument on the one for Heyting algebras on page 311. Instead of $\mathbf{D}$, use the subreduct in the signature $\{\vee, \cdot, 0, 1\}$. First show that it is locally finite, by distributivity of multiplication over join. Then expand the resulting finite algebra to the full signature.
(18) Continuing the previous exercise, prove that the universal theory of $\mathsf{E_n}$ is decidable. Use FEP and work out in detail the argument outlined on page 311.
(19) Prove that the condition $0 \leq x$ (in Theorem 6.51) is preserved by the finite embedding construction from Section 6.5.2.
(20) *Open problem:* is there a canonical variety of FL-algebras without canonical axiomatization?

Notes

(1) The abstract notion of canonical extension used in the first part of the chapter is due to Tarski and Jónsson [JT51, JT52], later extended to distributive lattices by Jónsson and Gehrke [GJ04], to arbitrary lattices by Gehrke and Harding [GH01] and to perfect posets (with certain connection to substructural logics) by Dunn, Gehrke and Palmigiano [DGP05]. For Heyting algebras and Boolean algebras with operators (as far as these fit in our present framework, e.g., for symmetric relation algebras), this notion of canonicity coincides with the notion of canonical extension that has its origins in modal logic.

(2) It is somehow misleading that the relation N was defined between M and $M^2 \times B$ in Section 6.5.1. It would be better to replace M^2 by the set S_M of all unary (one variable) linear (only one occurrence of each variable) polynomials over the monoid $\mathbf{M}$. A typical element f of S_M is a polynomial on one variable (say x) of the form $f(x) = uxw$, for $u, w \in M$. Now the relation N between M and $S_M \times B$ is defined by $c\ N\ (f, b)$ iff $f(c) \leq b$. This definition extends naturally to the non-associative case, where $\mathbf{M}$ is a groupoid (with unit); we simply take S_M to be again all unary linear polynomials over $\mathbf{M}$ (which are now more complicated due to the need for parentheses). The proof of finiteness, though, for the non-associative case is a bit more involved. For details, the reader can see the original proof in [BvA02] and its simplifications in terms of the relation N in [GO] and [GJ].

(3) FEP was implicitly known as early as 40s, but its real importance was recognized by Evans, who proved in [Eva69] that it implied solvability of word problem. In the context of residuated structures, FEP was studied in by Blok in [Blo99], Blok and Ferreirim in [BF00], by Blok and van Alten in [BvA02] and in [BvA05], where an important general construction was devised, and also by Agliano, Ferreirim and Montagna in [AFM99].

CHAPTER 7

Algebraic aspects of cut elimination

In this chapter we give a purely algebraic proof of the cut elimination theorem for various sequent systems. The basic idea of an algebraic proof is to introduce mathematical structures, which we call *Gentzen matrices*, for a given sequent system *without cut*, and then to show the completeness of the sequent system without cut with respect to the class of algebras for the logic determined by the sequent system with cut. In this completeness proof, we will use a *quasi-completion* of Gentzen matrices, which is a generalization of the Dedekind-MacNeille completion.

Our approach and several of the results given in this chapter originally appeared in a paper by Belardinelli, Jipsen and Ono [BJO04], for the commutative case. Our presentation here is more general and is also influenced by [GO], where the the results are obtained without the assumptions of commutativity or associativity, although we do assume associativity in this chapter and in most of the book. Moreover, the cut-free Gentzen system for (not necessarily commutative) involutive FL-algebras and the resulting decidability of **InFL** are taken from [GJ]; there the results are presented using residuated frames rather than Gentzen matrices.

The method presented here is closely related to those discussed by Maehara [Mae91] and independently by Okada [Oka96], [Oka99], in which semi-algebraic proofs of cut elimination are given, and is also inspired by the paper [JT02]. Moreover, the finite model property is obtained for many cases, by modifying our completeness proof. This is an algebraic presentation of the proof of the finite model property discussed by Lafont [Laf97] and Okada-Terui [OT99]. In addition, our method can be modified so as to prove the completeness theorem of tableau systems with respect to algebraic semantics.

Our motivation of giving an algebraic proof of the cut elimination theorem is to clarify the meaning of cut elimination from an algebraic point of view, and to give a proof of cut elimination attractive to algebraists, avoiding heavy syntactic arguments which are used in the standard cut elimination procedure. This approach is of general interest since cut elimination offers a useful tool for proving decidability. Our goal is to clarify the algebraic aspects of cut elimination and its consequences.

We note that there already exist several ways of proving cut elimination by semantical methods. For instance, in 1960 Schütte [Sch60] introduced the notion of *semi valuations* (or, Schütte's valuations in [Gir87b]) to prove the cut elimination theorem for higher order classical sequent system in a semantical way. Also, to show completeness of tableau systems for some modal logics and intuitionistic logic, which is essentially equivalent to cut elimination for them, Fitting introduced *consistency properties* in [Fit73]. But, proofs in these papers and also [Avi01], except [Mae91] and [Oka96], are not of an algebraic character in our sense.

Our method works well for a wide variety of sequent systems of nonclassical logics, both in propositional and predicate cases, including Gentzen's systems **LK** and **LJ** in [Gen35] for classical and intuitionistic logic, respectively. To explain our basic idea, we take first the sequent system **FL** as an example, and give a proof of cut elimination for it. In Section 7.3, we briefly show how our method can be applied to some other sequent systems of nonclassical logics, including modal logics.

By a slight modification of our completeness proof, we show in Section 7.4 the finite model property of some nonclassical logics. This is an algebraic presentation of the proof of the finite model property, discussed by Lafont [Laf97] and Okada-Terui [OT99]. The proof will show how the finiteness of proof-search procedures, which is of purely proof-theoretic character, is related to an algebraic property such as the finite model property.

7.1. Gentzen matrices for the sequent calculus FL

As shown in Chapter 2, FL-algebras are the appropriate algebraic structures for the sequent calculus **FL**. In the present section we will introduce structures for **FL** *without cut*, which we call *Gentzen matrices* for **FL**.

We begin with a simple observation.

PROPOSITION 7.1. *The following three conditions are mutually equivalent:*

(1) $\alpha_1, \ldots, \alpha_m \Rightarrow \beta$ *is provable in* **FL**,
(2) $\Rightarrow \alpha_m\backslash(\alpha_{m-1}\backslash(\ldots(\alpha_1\backslash\beta)\ldots))$ *is provable in* **FL**,
(3) $\alpha_1 \cdot \ldots \cdot \alpha_m \Rightarrow \beta$ *is provable in* **FL**.

This proposition says that commas in sequents can be interpreted as fusions in **FL**. But, to derive the the sequent $\alpha_1, \ldots, \alpha_m \Rightarrow \beta$ from the sequent $\alpha_1 \cdot \ldots \cdot \alpha_m \Rightarrow \beta$, we need the cut rule, in general. Thus, in **FL** without cut it is necessary to interpret commas as they stand.

Now, for a given nonempty set B, let B^* be the set of all (finite, possibly empty) sequences whose elements are in B. The empty sequence is denoted by ε in the following. For members x and y of B^*, xy denotes the sequence concatenation of x and y. As usual, the sequence consisting of elements $a_1, \ldots, a_m \in B$ is denoted by $(a_1, \ldots, a_m)$. Sometimes, we identify an element $c \in B$ with the singleton sequence (c) when no confusions will occur.

Thus, for example, when y is a singleton sequence (c), xy is written as xc. Obviously, B^* forms a monoid with respect to concatenation, whose unit is ε. In the following, x, y, z, u, v denote members of B^*, and a, b, c, d denote elements of B (and are also allowed to be ε if they occur by themselves in the right-hand side).

A *Gentzen matrix* for **FL** is a structure $\boldsymbol{\mathcal{B}} = (\mathbf{B}, \preceq)$ where $\mathbf{B} = (B, \wedge, \vee, \cdot, \backslash, /, 0, 1)$ is any algebra with binary operations $\wedge, \vee, \cdot, \backslash, /$, constants $0, 1 \in B$, and $\preceq$ is a subset of $B^* \times (B \cup \{\varepsilon\})$, which satisfies the following conditions:

$a \preceq a$	(refl)
$xy \preceq c$ implies $x1y \preceq c$	(1 left)
$\varepsilon \preceq 1$	(1 right)
$0 \preceq \varepsilon$	(0 left)
$x \preceq \varepsilon$ implies $x \preceq 0$	(0 right)
$xaby \preceq c$ implies $x(a \cdot b)y \preceq c$	($\cdot$ left)
$x \preceq a$ and $y \preceq b$ imply $xy \preceq a \cdot b$	($\cdot$ right)
$x \preceq a$ and $ybz \preceq c$ imply $yx(a\backslash b)z \preceq c$	($\backslash$ left)
$ax \preceq b$ implies $x \preceq a\backslash b$	($\backslash$ right)
$x \preceq a$ and $ybz \preceq c$ imply $y(b/a)xz \preceq c$	($/$ left)
$xa \preceq b$ implies $x \preceq b/a$	($/$ right)
$xay \preceq c$ and $xby \preceq c$ imply $x(a \vee b)y \preceq c$	($\vee$ left)
$x \preceq a$ implies $x \preceq a \vee b$	($\vee$ right)
$x \preceq b$ implies $x \preceq a \vee b$	($\vee$ right)
$xay \preceq c$ implies $x(a \wedge b)y \preceq c$	($\wedge$ left)
$xby \preceq c$ implies $x(a \wedge b)y \preceq c$	($\wedge$ left)
$x \preceq a$ and $x \preceq b$ imply $x \preceq a \wedge b$	($\wedge$ right)

Each of these conditions corresponds to either an instance of an initial sequent or an instance of a rule of inference of the Gentzen system for **FL**, if we replace $\preceq$ by $\Rightarrow$ and elements of B by formulas. Conversely, each rule of inference, except the cut rule, is represented by one of these conditions. It is easy to see that a Gentzen matrix for **FL** is just a matrix model of $\vdash_{\mathbf{FL}}$, as defined in Section 1.6.4.

Just as for FL-algebras, an assignment into a Gentzen matrix $\boldsymbol{\mathcal{B}}$ is defined as a homomorphism from the algebra of terms to the algebra reduct $\mathbf{B} = (B, \wedge, \vee, \cdot, \backslash, /, 0, 1)$. A sequent $s_1, \ldots, s_m \Rightarrow t$ is said to hold in $\boldsymbol{\mathcal{B}}$, ($\boldsymbol{\mathcal{B}} \models s_1, \ldots, s_m \Rightarrow t$, in symbols) if $(h(s_1), \ldots, h(s_m)) \preceq h(t)$ holds for every assignment h on $\mathbf{B}$. (Here we assume that $h(t) = \varepsilon$ when t is empty.) It is obvious that if a sequent $\alpha_1, \ldots, \alpha_m \Rightarrow \beta$ is provable in **FL** without using the cut rule then $\alpha_1, \ldots, \alpha_m \Rightarrow \beta$ holds in every Gentzen matrix for

FL. The converse of this implication is also true, and uses the *absolutely free* Gentzen matrix for **FL**. As shown below, the proof is essentially the same as, but much simpler than, the standard proof of completeness of **FL** with respect to the class of FL-algebras using the Lindenbaum algebra. Recall that **Fm** is the absolutely free term algebra (in the language of FL-algebras). The absolutely free Gentzen matrix is the structure $\mathcal{F}m = (\mathbf{Fm}, \preceq^{\mathcal{F}m})$ where the relation $\preceq$ is defined as follows:

> $(\alpha_1, \ldots, \alpha_m) \preceq^{\mathcal{F}m} \beta$ holds if and only if the corresponding sequent $\alpha_1, \ldots, \alpha_m \Rightarrow \beta$ is provable in **FL** without using the cut rule, for $m > 0$. Also, $\varepsilon \preceq^{\mathcal{F}m} \beta$ holds if and only if the sequent $\Rightarrow \beta$ is provable in **FL** without using the cut rule.

Recall here that the correspondence between formulas and terms is bijective. The structure $\mathcal{F}m$ thus obtained is a Gentzen matrix for **FL** with the property that if a sequent is not provable in **FL** without cut, then the corresponding sequent does not hold under the trivial assignment, i.e. the assignment h satisfying $h(t) = t$ for any term t. Thus we have the following.

LEMMA 7.2. *A sequent $\alpha_1, \ldots, \alpha_m \Rightarrow \beta$ is provable in* **FL** *without using the cut rule if and only if $\alpha_1, \ldots, \alpha_m \Rightarrow \beta$ holds in every Gentzen matrix for* **FL**.

Next we show that every FL-algebra can be regarded as a particular Gentzen matrix for **FL**. First, suppose that an FL-algebra **A** is given. Let $\leq$ be the lattice order of **A**. Define a subset $\preceq$ of $A^* \times (A \cup \{\varepsilon\})$ by the condition that $(a_1, \ldots, a_m) \preceq c$ holds if and only if $a_1 \cdot \ldots \cdot a_m \leq c$ holds in **A**, when $c \in A$. (Let $a_1 \cdot \ldots \cdot a_m = 1$ when $m = 0$. Also, define $(a_1, \ldots, a_m) \preceq \varepsilon$ if and only if $a_1 \cdot \ldots \cdot a_m \leq 0$ holds in **A**.) Then $\boldsymbol{\mathcal{A}} = (\mathbf{A}, \preceq)$ is a Gentzen matrix for **FL**. Moreover, the following *strong transitivity*, a property corresponding to the cut rule, holds in $\boldsymbol{\mathcal{A}}$:

$$z \preceq a \text{ and } xay \preceq c \text{ imply } xzy \preceq c.$$

Conversely, suppose that a Gentzen matrix $\boldsymbol{\mathcal{B}}$ for **FL** with a strongly transitive relation $\preceq$ is given. Let $\preceq_0$ be the restriction of $\preceq$ to $B \times B$. We note here that $\preceq$ is strongly transitive if and only if the following two conditions hold:

> the relation $\preceq_0$ is transitive,
> $(a_1, \ldots, a_m) \preceq c$ if and only if $a_1 \cdot \ldots \cdot a_m \preceq_0 c$.

Moreover, in **B** we have that $ax \preceq b$ if and only if $x \preceq a\backslash b$. To see this, it is enough to show the if-part since the only-if part follows from ($\backslash$ right) in the definition of Gentzen matrix. From $a \preceq a$ and $b \preceq b$ we deduce $a \cdot (a\backslash b) \preceq b$ by ($\backslash$ left). So $x \preceq a\backslash b$ implies $ax \preceq a \cdot (a\backslash b)$ by ($\cdot$ right). Then, by the strong transitivity of $\preceq$, we have $ax \preceq b$. This gives the following result.

LEMMA 7.3. *Let $\boldsymbol{\mathcal{B}} = (\mathbf{B}, \preceq)$ be any Gentzen matrix for* **FL** *with a strongly transitive $\preceq$. If the restriction $\preceq_0$ of $\preceq$ to $B \times B$ is moreover antisymmetric* *and therefore a partial order then* **B** *is an FL-algebra with the lattice order* $\preceq_0$.

The assumption that $\preceq_0$ is antisymmetric is not essential. For, if $\preceq_0$ is both reflexive and transitive, by using the congruence relation $\sim$ determined by $\preceq_0$ we can introduce a quotient algebra, in which the relation $\preceq_0$ congruent modulo $\sim$ becomes a partial order. On the other hand, we cannot take a quotient structure of a Gentzen matrix in general, since it lacks transitivity.

In conclusion, we can roughly say that any Gentzen matrix with a strongly transitive relation can be identified with an FL-algebra, and vice versa.

7.2. Quasi-completions and cut elimination

Results in the last part of the previous section tell us that each FL-algebra can be regarded as a particular Gentzen matrix for **FL**. Therefore, for all terms $s_1, \ldots, s_m$ and t, if a sequent $s_1, \ldots, s_m \Rightarrow t$ holds in every Gentzen matrix for **FL** then the corresponding inequality $s_1 \cdot \ldots \cdot s_m \leq t$ holds in every FL-algebra. Our Theorem 7.4, proved in the present section, says that the converse is also true, and turns out to be equivalent to cut elimination (see Lemma 7.7). This is the algebraic content of cut elimination for **FL**, and leads directly to the main result in Theorem 7.8.

THEOREM 7.4. *For all terms* $s_1, \ldots, s_m$ *and* t, *if* $\mathbf{A} \models s_1 \cdot \ldots \cdot s_m \leq t$ *for every FL-algebra* $\mathbf{A}$, *then* $\boldsymbol{\mathcal{B}} \models s_1, \ldots, s_m \Rightarrow t$ *for every Gentzen matrix* $\boldsymbol{\mathcal{B}}$ *for* **FL**.

We devote most of this section to proving the above result. Taking the contraposition, suppose that $s_1, \ldots, s_m \Rightarrow t$ fails to hold in a Gentzen matrix $\boldsymbol{\mathcal{B}}$ under an assignment h, i.e. $(h(s_1), \ldots, h(s_m)) \preceq h(t)$ does not hold in $\boldsymbol{\mathcal{B}}$. Our goal is to construct a FL-algebra $\mathbf{A}$ in which $s_1 \cdot \ldots \cdot s_m \leq t$ does not hold.

Since B^* is a monoid, we have that $\mathcal{P}(B^*)_\gamma$ is a complete bounded FL-algebra for any nucleus γ on $\mathcal{P}(B^*)$, by Lemma 3.34. We now define a particular nucleus γ_N on $\mathcal{P}(B^*)$ by constructing a nuclear relation $N \subseteq B^* \times (B^{*2} \times B)$ from the Gentzen matrix. The relation N is defined by

$$w \; N \; (x, y, a) \quad \text{iff} \quad xwy \preceq a$$

where $w, x, y \in B^*$ and $a \in B$. This gives a nuclear relation since concatenation is associative:

$$uw \; N \; (x, y, a) \;\; \text{iff} \;\; xuwy \preceq a \;\; \text{iff} \;\; u \; N \; (x, wy, a) \;\; \text{iff} \;\; w \; N \; (xu, y, a).$$

Note the strong similarity of the relation N with the one used in Section 6.5.1. By Lemma 3.36 it follows that the closure operator $\gamma = \gamma_N$, induced by N, is a nucleus and thus, by Corollary 6.28 (see also Section 3.4.11), $\mathcal{P}(B^*)_\gamma$ is a complete FL-algebra, called the *quasi-completion* of $\boldsymbol{\mathcal{B}}$ and denoted by $\mathbf{R}(\boldsymbol{\mathcal{B}})$.

It would be nice to get an embedding from $\mathbf{B}$ to $\mathbf{R}(\boldsymbol{\mathcal{B}})$. But we cannot expect that much, since $\mathbf{B}$ has only a weak mathematical structure. Still we can prove the following theorem that confirms the existence of a *quasi-embedding* from $\mathbf{B}$ to $\mathbf{R}(\boldsymbol{\mathcal{B}})$, which will be shown to be sufficient for our purpose.

Define a map $k : B \to \gamma(\mathcal{P}(B^*))$ by $k(a) = \{x \in B^* : x \preceq a\} = \{(\varepsilon, \varepsilon, a)\}^{\triangleleft}$ (recall from Lemma 3.8 that $\{w\}^{\triangleleft} = \{x : xNw\}$ is a basic closed set of the nucleus induced by the relation N). Note again the similarities of k with the embedding in Section 6.5.1, used to obtain the finite embeddability property. The next result follows the proof of Lemma 7.3 in [JT02] (see also Maehara [Mae91] and Okada [Oka96]).

THEOREM 7.5. *Suppose that $a, b \in B$ and that U and V are arbitrary γ-closed subsets of B^* such that $a \in U \subseteq k(a)$ and $b \in V \subseteq k(b)$. Then for each $\star \in \{\wedge, \vee, \cdot, \backslash, /\}$, $a \star b \in U \star_\gamma V \subseteq k(a \star b)$, where $\star_\gamma$ denotes $\cap, \vee_\gamma, \cdot_\gamma$ and $\backslash, /$, respectively. Thus, in particular $a \star b \in k(a) \star_\gamma k(b) \subseteq k(a \star b)$.*

PROOF. First note that, since the sets of the form $(y, z, c)^{\triangleleft}$ form a basis for the nucleus γ, the following is a necessary and sufficient condition for a given subset X of B^* to be γ-closed: for any $x \in B^*$,

$x \in X$ whenever $XN(y, z, c)$ implies $xN(y, z, c)$ for all $y, z \in B^*$ and $c \in B \cup \{\varepsilon\}$.

We give here a proof of the result for the two cases when $\star$ is $\vee$ and $\backslash$. The remaining cases are similar, with each one using the ($\star$ left) and ($\star$ right) rules of the Gentzen matrix definition.

First let $\star$ be $\vee$. We need to show that $a \vee b \in U \vee_\gamma V = \gamma(U \vee V)$, i.e. $(U \cup V)N(y, z, c)$ implies $(a \vee b)N(y, z, c)$ for all $y, z \in B^*$ and $c \in B \cup \{\varepsilon\}$. So assume $(U \cup V)N(y, z, c)$. Since $a \in U$ and $b \in V$, we have that $yaz \preceq c$ and $ybz \preceq c$, hence $y(a \vee b)z \preceq c$ by ($\vee$ left). Therefore $(a \vee b)N(y, z, c)$ as required. Now we show $U \vee_\gamma V \subseteq k(a \vee b)$, which reduces to $U \cup V \subseteq k(a \vee b)$ since $k(a \vee b)$ is γ-closed. By definition $x \in k(a)$ implies $x \preceq a$, so by ($\vee$ right) $x \preceq a \vee b$, which gives $x \in k(a \vee b)$. By assumption, $U \subseteq k(a)$ and $V \subseteq k(b)$, so we conclude $U \cup V \subseteq k(a \vee b)$.

Next suppose that $\star$ is $\backslash$. We show that $a \backslash b \in U \backslash V$, i.e. $U \cdot_\gamma \{a \backslash b\} \subseteq V$. For this, it is enough to show that $k(a) \cdot \{a \backslash b\} \subseteq V$, since V is γ-closed and $U \subseteq k(a)$ by our assumption. Take any element $w \in k(a)$ and any (y, z, c) such that $VN(y, z, c)$. Then $w \preceq a$ and $bN(y, z, c)$, i.e. $ybz \preceq c$ holds since b is an element of V. Therefore, $yw(a \backslash b)z \preceq c$ by ($\backslash$ left) and hence $(w(a \backslash b))N(y, z, c)$. Since V is γ-closed, this implies $w(a \backslash b) \in V$. Thus we have $k(a) \cdot \{a \backslash b\} \subseteq V$. We show next that $U \backslash V \subseteq k(a \backslash b)$. Take any

$w \in U \backslash V$. Then $U \cdot \{w\} \subseteq V \subseteq k(b)$. Since $a \in U$, this implies $aw \preceq b$ and hence $w \preceq a \backslash b$. Thus $w \in k(a \backslash b)$. □

A map such as k, which has the properties described in the above theorem, is called a *quasi-embedding*. As shown later, the notion of quasi-embedding can be regarded as a generalization of complete embedding.

Recall that we assumed $(h(s_1), \ldots, h(s_m)) \preceq h(t)$ does not hold in our Gentzen matrix $\boldsymbol{\mathcal{B}}$ where h is an assignment on B. Now we show that the corresponding inequality $(s_1 \cdot \ldots \cdot s_m) \leq t$ does not hold in the quasi-completion $\mathbf{R}(\boldsymbol{\mathcal{B}})$ of $\mathbf{B}$. We define an assignment g on $\mathbf{R}(\boldsymbol{\mathcal{B}})$ by $g(q) = k(h(q))$ for each propositional variable q, where k is the quasi-embedding. Note that for any $a \in B$, $a \in \gamma(\{a\}) \subseteq k(a)$, and that $h(0) = 0, h(1) = 1, g(0) = \gamma(\{0\})$ and $g(1) = \gamma(\{1\})$. So the lemma below holds for propositional variables and logical constants. For arbitrary terms t we proceed by induction on the length of the term, using Theorem 7.5.

LEMMA 7.6. *For any term t, $h(t) \in g(t) \subseteq k(h(t))$.*

Now suppose to the contrary that $s_1 \cdot \ldots \cdot s_m \leq t$ holds in $\mathbf{R}(\boldsymbol{\mathcal{B}})$. Then, $g(s_1) \cdot_\gamma \ldots \cdot_\gamma g(s_m) \subseteq g(t)$ must hold in particular. Since $h(s_i) \in g(s_i)$ for each i by Lemma 7.6, $(h(s_1), \ldots, h(s_m)) \in (g(s_1) \cdot_\gamma \ldots \cdot_\gamma g(s_m))$ and hence $(h(s_1), \ldots, h(s_m)) \in g(t) \subseteq k(h(t))$ hold. (Recall that the monoid operation on B^* is concatenation.) But this implies that $(h(s_1), \ldots, h(s_m)) \preceq h(t)$, which is a contradiction. Thus, $s_1 \cdot \ldots \cdot s_m \leq t$ does not hold in $\mathbf{R}(\boldsymbol{\mathcal{B}})$. This concludes the proof of Theorem 7.4.

The following lemma is an immediate consequence of Corollary 6.28 and Lemma 7.2.

LEMMA 7.7. *The statement of Theorem 7.4 is equivalent to the statement that cut elimination holds for* **FL**.

Hence we have shown our main result.

THEOREM 7.8. *Cut elimination holds for* **FL**. *In other words, the sequent system* **FL** *without the cut rule is complete with respect to the class of all FL-algebras.*

In algebraic terms, our theorem says that for all terms $s_1, \ldots, s_m$ and t, the inequality $s_1 \cdot \ldots \cdot s_m \leq t$ holds in all FL-algebras if and only if the relation $s_1, \ldots, s_m \Rightarrow t$ can be derived by using only conditions described in the definition of Gentzen matrices for **FL** (with all $\preceq$ replaced by $\Rightarrow$).

The cut elimination theorem says that the cut rule is admissible in the system obtained from **FL** by deleting the cut rule. In other words, if both $s_1, \ldots, s_m \Rightarrow t_0$ and $t_1, \ldots, t_0, \ldots, t_n \Rightarrow r$ hold in any Gentzen matrix for **FL**, then $t_1, \ldots, s_1, \ldots, s_m, \ldots, t_n \Rightarrow r$ holds also in any Gentzen matrix for **FL**. This should be distinguished from the fact that the strong transitivity condition, i.e. $z \preceq a$ and $xay \preceq c$ imply $xzy \preceq c$, does not always hold in a Gentzen matrix for **FL**, which is equivalent to the non-derivability of the cut rule in **FL**.

An actual example witnessing the non-derivability of cut can be constructed as follows. Take a sequent system **FL**$'$ obtained from cut-free **FL** by adding two axioms:

$$p \Rightarrow q \quad \text{and} \quad q \Rightarrow r \qquad \text{for distinct variables } p, q, r.$$

Obviously, $p \Rightarrow r$ is not provable in **FL**$'$ (because every formula in the upper sequent of each rule of cut-free **FL** appears in the lower sequent as a subformula of some formula). Note that the deducibility relation of **FL** determines a Gentzen matrix for **FL**.

It should be remarked also that properties of the monoid operation of B^* and of the relation $\preceq$ alone determine the structure of the FL-algebra $\mathcal{P}(B^*)_\gamma$. In other words, the structure is not affected by properties of any algebraic operation or constant, related to *logical connectives* and *logical constants*. This can be regarded as an intrinsic algebraic feature of substructural logics. Also, as can be seen in the proof of Theorem 7.5, we use only conditions concerning a given operation $\star$ when proving our theorem for $\star$. From these observation, we can derive an algebraic proof of the next theorem.

Let $\mathcal{K}$ be a nonempty sublanguage of the language of FL-algebras. Recall that a term t is a $\mathcal{K}$*-term* if it consists only of symbols in $\mathcal{K}$ and variables. Also, a $\mathcal{K}$*-Gentzen matrix* for **FL** is a structure defined similarly as a usual Gentzen matrix for **FL**, but by restricting the structure and conditions to those related only to members of $\mathcal{K}$. In logical terms, $\mathcal{K}$-Gentzen matrices for **FL** are precisely the Gentzen matrices for the $\mathcal{K}$-fragment of **FL**. Now, we have the following theorem on the conservativity of each fragment of **FL**. The theorem is usually proved syntactically as a consequence of the *subformula property* of **FL**, which in turn is one of the most important consequences of cut elimination of **FL**.

THEOREM 7.9. *Let $\mathcal{K}$ be any nonempty sublanguage of the language of FL-algebras. For $\mathcal{K}$-terms s and t, the following three conditions are mutually equivalent:*

(1) $\mathbf{A} \models s \leq t$ *for any FL-algebra* $\mathbf{A}$,

(2) $\boldsymbol{\mathcal{B}} \models' s \Rightarrow t$ *for any $\mathcal{K}$-Gentzen matrix $\boldsymbol{\mathcal{B}}$ for* **FL**,

(3) $\mathbf{A}' \models s \leq t$ *for any $\mathcal{K}$-reduct $\mathbf{A}'$ of FL-algebras.*

Actually, a stronger statement involving proofs *with assumptions* and quasiequations is also true; see [vAR04] and [GO] for a related discussion with respect to Hilbert systems.

In the rest of this section, we show how quasi-completions are related to the Dedekind-MacNeille completions. As mentioned in the previous section, each FL-algebra $\mathbf{A} = (A, \wedge, \vee, \cdot, \backslash, /, 0, 1)$ with lattice order $\leq$ can be identified with a Gentzen matrix $\boldsymbol{\mathcal{A}}$ if we define the relation $\preceq$ for $\boldsymbol{\mathcal{A}}$ by

$$(a_1, \ldots, a_m) \preceq c \text{ holds if and only if } (a_1 \cdot \ldots \cdot a_m) \leq c \text{ holds.}$$

For each $z = (a_1, \ldots, a_m)$ in A^*, let $\tilde{z}$ denote an element $a_1 \cdot \ldots \cdot a_m$ in A. (Define $\tilde{z} = 1$ when $z = \varepsilon$.) The above relation enables us to identify each z of A^* with an element $\tilde{z}$ of A, and $\preceq$ with $\leq$. Under this identification, each set of the form $\{(x, y, a)\}^{\lhd}$ is regarded as a subset $\{d \in A : \tilde{x} \cdot d \cdot \tilde{y} \leq a\}$, or equivalently $\{d \in A : d \leq (\tilde{x}\backslash a)/\tilde{y}\}$. Since any element in A can be expressed by an element of the form $(\tilde{x}\backslash a)/\tilde{y}$ for some $x, y \in A^*$ and some $a \in A$, we can assume that our base of closed sets consists of sets of the form $\downarrow c$ for $c \in A$, where $\downarrow c = \{d \in A : d \leq c\}$. Now let us define an operation γ on $\mathcal{P}(A)$, instead of $\mathcal{P}(A^*)$, by

$$\gamma(X) = \bigcap\{\downarrow c : X \subseteq \downarrow c \text{ for an element } c \in A\} \quad \text{for each subset } X \text{ of } A.$$

Note that $\gamma = \gamma_{\leq}$ and that $\leq$ is a nuclear relation, by the residuation property of $\mathbf{A}$. Since the sets of the form $\downarrow c$ form a basis for γ, we have

$$z \in \gamma(X) \text{ if and only if } c \in X^u \text{ implies } z \leq c \text{ for any c,}$$

where X^u denotes the set of all upper bounds of X, and hence $\gamma(X) = X^{ul}$, where Y^l denotes the set of all lower bounds of Y. This means that $\gamma(X)$ is the closure operator discussed on page 177 and associated with the Dedekind-MacNeille completion of $\mathbf{A}$. Exercise 1 asks you to verify that the quasi-completion $\mathbf{R}(\boldsymbol{\mathcal{A}})$ of $\boldsymbol{\mathcal{A}}$ is isomorphic to the Dedekind-MacNeille completion of $\mathbf{A}$.

Recall that the quasi-embedding $k : \mathbf{A} \to \mathbf{R}(\boldsymbol{\mathcal{A}})$ is defined by $k(a) = \{x \in A^* : x \preceq a\}$, and $a \star b \in k(a) \star_\gamma k(b) \subseteq k(a \star b)$ by Theorem 7.5. If $\preceq$ is strongly transitive, $a \star b \in k(a) \star_\gamma k(b)$ implies that $k(a \star b) \subseteq k(a) \star_\gamma k(b)$. Hence $k(a \star b) = k(a) \star_\gamma k(b)$ holds. Since k is injective, this means that k is an embedding, and in fact this k is identified with the complete embedding of an FL-algebra $\mathbf{L}$ into its Dedekind-MacNeille completion described in Lemma 3.37. Thus we have the following.

COROLLARY 7.10. *The quasi-completion of any FL-algebra* $\mathbf{B}$, *considered as a Gentzen matrix with a strongly transitive relation, is isomorphic to its Dedekind-MacNeille completion, and the quasi-embedding of* $\mathbf{B}$ *into its quasi-completion is a complete embedding.*

We end this section with some brief remarks about the possibility of replacing Gentzen matrices for **FL** with bonafide first-order structures, so that $\preceq$ is a binary relation on B, rather than a relation from B^* to B. Such an approach is indeed possible, and permits Gentzen matrices to be defined by a short list of universal Horn sentences, where the sequence constructor (comma) is replaced by fusion. A notion of *algebraic Gentzen proof* can now be formulated as a restriction of the standard notion of quasi-equational proof, and derivations in this system are somewhat shorter since fusion-elimination steps are omitted. From an algebraic standpoint this provides an even tighter connection between proof theory and universal algebra. But our present approach also has some advantages. For example the semantics of Gentzen matrices capture exactly the provability relation for sequents of standard Gentzen systems (with or without cut). This has certain benefits when establishing the finite model property (Section 7.4), and the presence of comma-separated sequences also allows some distinctions to be made that cannot be expressed by the first-order language.

7.3. Cut elimination for other systems

Our algebraic proof of cut elimination works for various sequent systems. For example, an outline of an algebraic proof of cut elimination of Gentzen's sequent system **LJ** for intuitionistic logic is given in [Ono03c]. Moreover, it is not hard to modify our method to apply it to intuitionistic substructural logics like $\mathbf{FL_e}$ and $\mathbf{FL_{ec}}$. In this section, we explain briefly how to extend our method to other sequent systems and tableau systems.

7.3.1. Involutive substructural logics. We now consider sequent systems with sequents of the form $\Gamma \Rightarrow \Delta$ where Γ and Δ are sequences of formulas. A typical example is Gentzen's sequent system **LK** for classical logic. For **LK**, a semi-algebraic proof of cut elimination is given in the paper [Mae91] by Maehara. Also [Bel02] explored an algebraic proof of cut elimination of sequent systems for involutive commutative substructural logics, based on the methods elaborated here. In the following, by taking the sequent system **InFL** as an example, we will explain how to modify our proof of cut elimination for these sequent systems. Note that a syntactic proof of the cut elimination theorem for $\mathbf{InFL_{ew}}$ is given by Grishin in [Gri82], and an algebraic cut elimination proof for **CyInFL** is presented by Wille in [Wil05]. The sequent system **InFL** given here is a context free version of **LK** without structural rules, but with external connectives $^{\sim}$, $^{-}$ for the two linear negations $\sim$ and $-$, in the same spirit as comma is considered an external connective for fusion. As usual, capital Greek letters denote (possibly empty) sequences of formulas, and if $\Delta = (\delta_1, \ldots, \delta_n)$ then $\Delta^{\sim} = (\delta_n^{\sim}, \ldots, \delta_1^{\sim})$ and similarly for Δ^{-}. The structural rules (with the *superscripts* $^{\sim}$ and $^{-}$) in the last two lines are *bidirectional* (i.e. invertible) and serve to move formulas back and forth between the two sides of a sequent thus enabling the application of the remaining rules.

$$\text{Axioms:} \qquad \alpha \Rightarrow \alpha \qquad \text{and} \qquad \Rightarrow 1$$

$$\frac{\alpha, \beta \Rightarrow \Delta}{\alpha \cdot \beta \Rightarrow \Delta}\ (\cdot \Rightarrow) \qquad \frac{\Gamma \Rightarrow \alpha \quad \Delta \Rightarrow \beta}{\Gamma, \Delta \Rightarrow \alpha \cdot \beta}\ (\Rightarrow \cdot) \qquad \frac{\Rightarrow \Delta}{1 \Rightarrow \Delta}\ (1 \Rightarrow)$$

$$\frac{\alpha \Rightarrow \Delta \quad \beta \Rightarrow \Delta}{\alpha \vee \beta \Rightarrow \Delta}\ (\vee \Rightarrow) \qquad \frac{\Gamma \Rightarrow \alpha}{\Gamma \Rightarrow \alpha \vee \beta}\ (\Rightarrow \vee) \qquad \frac{\Gamma \Rightarrow \beta}{\Gamma \Rightarrow \alpha \vee \beta}\ (\Rightarrow \vee)$$

$$\frac{\alpha \Rightarrow \Delta}{\alpha \wedge \beta \Rightarrow \Delta}\ (\wedge \Rightarrow) \qquad \frac{\beta \Rightarrow \Delta}{\alpha \wedge \beta \Rightarrow \Delta}\ (\wedge \Rightarrow) \qquad \frac{\Gamma \Rightarrow \alpha \quad \Gamma \Rightarrow \beta}{\Gamma \Rightarrow \alpha \wedge \beta}\ (\Rightarrow \wedge)$$

$$\frac{\Gamma \Rightarrow \alpha}{\Gamma, {\sim}\alpha \Rightarrow}\ ({\sim} \Rightarrow) \qquad \frac{\alpha, \Gamma \Rightarrow}{\Gamma \Rightarrow {\sim}\alpha}\ (\Rightarrow {\sim})$$

$$\frac{\Gamma \Rightarrow \alpha}{-\alpha, \Gamma \Rightarrow}\ (- \Rightarrow) \qquad \frac{\Gamma, \alpha \Rightarrow}{\Gamma \Rightarrow -\alpha}\ (\Rightarrow -)$$

$$\frac{\Gamma \Rightarrow \Delta, \Lambda}{\Gamma, \Lambda^{\sim} \Rightarrow \Delta}\ ({}^{\sim}\Rightarrow) \qquad \frac{\Gamma, \Delta \Rightarrow \Lambda}{\Delta \Rightarrow \Gamma^{\sim}, \Lambda}\ (\Rightarrow^{\sim})$$

$$\frac{\Gamma \Rightarrow \Delta, \Lambda}{\Delta^{-}, \Gamma \Rightarrow \Lambda}\ ({}^{-}\Rightarrow) \qquad \frac{\Gamma, \Delta \Rightarrow \Lambda}{\Gamma \Rightarrow \Lambda, \Delta^{-}}\ (\Rightarrow^{-})$$

An important feature of **InFL** is that the sequents ${\sim}{-}\alpha \Rightarrow \alpha$ and $-{\sim}\alpha \Rightarrow \alpha$ are provable in it:

$$\cfrac{\cfrac{\cfrac{\cfrac{\alpha \Rightarrow \alpha}{\alpha^{-}, \alpha \Rightarrow}\ ({}^{-}\Rightarrow)}{\alpha^{-} \Rightarrow -\alpha}\ (\Rightarrow -)}{\alpha^{-}, {\sim}{-}\alpha \Rightarrow}\ ({\sim} \Rightarrow)}{{\sim}{-}\alpha \Rightarrow \alpha}\ ({}^{-}\Rightarrow)$$

Note that the sequents $\alpha \Rightarrow {\sim}{-}\alpha$ and $\alpha \Rightarrow -{\sim}\alpha$ are provable already in **FL** since ${\sim}\alpha$ is defined as $\alpha\backslash 0$ and $-\alpha$ is defined as $0/\alpha$ in **FL**. Our presentation of **InFL** here does not contain residuals explicitly because α/β and $\beta\backslash\alpha$ are definable in terms of ${\sim}, -$ by the formulas $-[\beta \cdot ({\sim}\alpha)]$ and ${\sim}[(-\alpha) \cdot \beta]$. An algebraic version of this is proved in Lemma 3.16. Algebraic structures for **InFL** are *involutive* FL-algebras, i.e. they satisfy ${\sim}{-}x = x = -{\sim}x$.

A modification is necessary in the definition of $\preceq$ when we introduce *Gentzen matrices for* **InFL**. That is, $\preceq$ must be defined as a binary relation on sequences of elements of B. In fact we need the free bi-involutive monoid generated by B as carrier for the $\preceq$ relation. This object is defined as follows.

For an element $b \in B$, let $b^{(\sim^n)}$ be the sequence starting with b and followed by n copies of $^\sim$. Similarly $b^{(-^n)}$ is b followed by n copies of $^-$; for example $b^{(-^2)} = b^{--}$. We define $B^{\simeq} = B \cup \{b^{(\sim^n)} : b \in B,\ n > 0\} \cup \{b^{(-^n)} : b \in B,\ n > 0\}$ and we let $B^{\simeq *} = (B^{\simeq})^*$ be the set of all finite sequences of elements from $B^{\simeq}$. This set is of course a monoid under concatenation (denoted by juxtaposition) with the empty sequence as unit, but in addition it supports two unary operations $^\sim, ^-$ defined by

$$b^{(\sim^n)\sim} = b^{(\sim^{n+1})} \qquad b^{(\sim^n)-} = b^{(\sim^{n-1})}$$

$$b^{(-^n)-} = b^{(-^{n+1})} \qquad b^{(-^n)\sim} = b^{(-^{n-1})}$$

and for sequences of elements from $B^{\simeq}$, $(b_1 b_2 \ldots b_m)^\sim = b_m^\sim \ldots b_2^\sim b_1^\sim$ and $(b_1 b_2 \ldots b_m)^- = b_m^- \ldots b_2^- b_1^-$. It is easy to see that these operations satisfy the identities $x^{\sim -} = x = x^{-\sim}$, $(xy)^\sim = y^\sim x^\sim$, $(xy)^- = y^- x^-$ and $\varepsilon^\sim = \varepsilon = \varepsilon^-$.

Now we define a *Gentzen matrix for* **InFL** to be a structure $\boldsymbol{\mathcal{B}} = (\mathbf{B}, \preceq)$ where $\mathbf{B} = (B, \wedge, \vee, \cdot, \sim, -, 0, 1)$ is any algebra with binary operations $\wedge, \vee, \cdot$, unary operations $\sim, -$, constants $0, 1 \in B$, and $\preceq$ is a subset of $B^{\simeq *} \times B^{\simeq *}$, which satisfies the following conditions:

$a \preceq a$	(refl)
$\varepsilon \preceq z$ implies $1 \preceq z$	(1 left)
$\varepsilon \preceq 1$	(1 right)
$0 \preceq \varepsilon$	(0 left)
$x \preceq \varepsilon$ implies $x \preceq 0$	(0 right)
$ab \preceq z$ implies $a \cdot b \preceq z$	($\cdot$ left)
$x \preceq a$ and $y \preceq b$ imply $xy \preceq a \cdot b$	($\cdot$ right)
$a \preceq z$ and $b \preceq z$ imply $a \vee b \preceq z$	($\vee$ left)
$x \preceq a$ implies $x \preceq a \vee b$	($\vee$ right)
$x \preceq b$ implies $x \preceq a \vee b$	($\vee$ right)
$a \preceq z$ implies $a \wedge b \preceq z$	($\wedge$ left)
$b \preceq z$ implies $a \wedge b \preceq z$	($\wedge$ left)
$x \preceq a$ and $x \preceq b$ imply $x \preceq a \wedge b$	($\wedge$ right)
$x \preceq a$ implies $x(\sim a) \preceq \varepsilon$	($\sim$ left)
$ax \preceq \varepsilon$ implies $x \preceq \sim a$	($\sim$ right)
$xy \preceq z$ if and only if $y \preceq (x^\sim)z$	($^\sim$)
$x \preceq a$ implies $(-a)x \preceq \varepsilon$	($-$ left)
$xa \preceq \varepsilon$ implies $x \preceq -a$	($-$ right)
$xy \preceq z$ if and only if $x \preceq z(y^-)$	($^-$)

Note that each of these conditions corresponds to either an instance of an initial sequent or an instance of a rule of inference of the Gentzen system for **InFL**, if we replace $\preceq$ by $\Rightarrow$ and elements of B by formulas.

We now turn to the construction of $\mathbf{R}(\boldsymbol{\mathcal{B}})$. As in the case of **FL**, this is a nucleus image of a powerset algebra, but in this case the Galois relation N is identical with the matrix relation $\preceq$, and this relation is nuclear because of the matrix conditions $({}^{\sim})$ and $({}^{-})$ that correspond to the bidirectional Gentzen rules $(\Rightarrow^{\sim})$ and $(\Rightarrow^{-})$. Thus the closed base is given by the sets $\{y\}^{\lhd} = \{x \in B^{\simeq *} : x \preceq y\}$, and the associated closure operator $\gamma(X) = X^{\rhd\lhd}$ is a nucleus.

As usual, the unit element E of $\mathbf{R}(\boldsymbol{\mathcal{B}})$ is defined by $E = \gamma(\{\varepsilon\})$, and the fusion of two subsets X, Y of $B^{\simeq *}$ is $\gamma(XY)$ where $XY = \{xy : x \in X, y \in Y\}$. The two negations are defined by ${\sim}X = X^{\rhd\sim}$ and $-X = X^{\rhd-}$. The next lemma shows that the collection of closed subsets form an involutive FL-algebra.

Lemma 7.11. [GJ] *Let $\boldsymbol{\mathcal{B}}$ be a Gentzen matrix and let X, Y, Z be Galois closed subsets of $B^{\simeq *}$. Then*

(1) $-X = X^{\rhd-} = X^{-\lhd}$ *and* ${\sim}X = X^{\rhd\sim} = X^{\sim\lhd}$

(2) $XY \subseteq Z$ *iff* $Y \subseteq {\sim}((-Z)X)$ *iff* $X \subseteq -(Y({\sim}Z))$.

Proof. (1) Applying the matrix conditions $({}^{-})$ and $({}^{\sim})$ with $z = \varepsilon$ shows that $x \preceq y^{-}$ is equivalent to $y \preceq x^{\sim}$. Hence, for all $x \in B^{\simeq *}$, we have $x \in X^{-\lhd}$ iff $x \mathrel{N} X^{-}$ iff $X \mathrel{N} x^{\sim}$ iff $x^{\sim} \in X^{\rhd}$ iff $x \in X^{\rhd -}$.

(2) We have $XY \subseteq Z = Z^{\rhd\lhd}$ iff $XY \mathrel{N} Z^{\rhd}$ iff $Y \mathrel{N} X^{\sim}Z^{\rhd}$ iff $Y \subseteq (X^{\sim}Z^{\rhd})^{\lhd}$. Also, $(X^{\sim}Z^{\rhd})^{\lhd} = (X^{\sim}Z^{\rhd})^{-\sim\lhd} = (Z^{\rhd-}X^{\sim-})^{\sim\lhd} = {\sim}((-Z)X)$, by (1). The second equivalence is proved similarly. □

Taking $Y = E$ in (2), we have $X \subseteq Z$ iff $X \subseteq -{\sim}Z$ and taking $X = E$, we have $Y \subseteq Z$ iff $Y \subseteq -{\sim}Z$ hence ${\sim}{-}Z = Z = -{\sim}Z$

Now let $0 = {\sim}E$ ($= -E$ since $\varepsilon^{\sim} = \varepsilon = \varepsilon^{-}$). Then $X \backslash Z = {\sim}((-Z)X)$, so $X \backslash 0 = {\sim}((-0)X) = {\sim}X$. Similarly $0/X = -X$. Hence we conclude that $\mathbf{R}(\boldsymbol{\mathcal{B}})$ is an involutive FL-algebra.

We show next how to extend Theorem 7.5 to cover the negations $\sim, -$. As before we define a map $k : B \to \gamma(\mathcal{P}(B^{\simeq *}))$ by $k(a) = \{x \in B^{\simeq *} : x \preceq a\}$.

Theorem 7.12. [GJ] *Suppose that $a \in B$ and that U is a γ-closed subset of $B^{\simeq *}$ such that $a \in U \subseteq k(a)$. If ${\sim}a$ and $-a$ are defined then ${\sim}a \in {\sim}U \subseteq k({\sim}a)$ and $-a \in -U \subseteq k(-a)$. So in particular ${\sim}a \in {\sim}k(a) \subseteq k({\sim}a)$ and $-a \in -k(a) \subseteq k(-a)$*

Proof. The proof is similar to the argument for Theorem 7.5, but using the Gentzen matrix left and right rules for $\sim$ and $-$. Assume ${\sim}a$ is defined in **B** and $a \in U \subseteq k(a)$. To see that ${\sim}a \in {\sim}U$, observe that $U \subseteq k(a)$ iff $u \preceq a$ for all $u \in U$. By $(\Rightarrow \sim)$ it follows that $u({\sim}a) \preceq \varepsilon$, and thus ${\sim}a \preceq u^{\sim}$ for all $u \in U$ by $(\Rightarrow^{-})$ with $z = \varepsilon$. Hence ${\sim}a \in U^{\sim\lhd} = {\sim}U$.

To see that $\sim U \subseteq k(\sim a)$, let $x \in \sim U = U^{\sim\lhd}$. Then $x \preceq u^{\sim}$ for all $u \in U$. Since $a \in U$, we have $x \preceq a^{\sim}$, hence $ax \preceq \varepsilon$ by ($^{\sim}$). Using ($\Rightarrow \sim$) we conclude that $x \preceq \sim a$, whence $x \in k(\sim a)$. □

Lemma 7.6 also extends to cover the negations. Recall that for an assignment h into a Gentzen matrix $\boldsymbol{\mathcal{B}}$, one defines an assignment g into the (involutive) FL-algebra $\mathbf{R}(\boldsymbol{\mathcal{B}})$ by $g(q) = k(h(q))$ where q is any variable and $g(1) = E$.

LEMMA 7.13. [GJ] *For any term* r, $h(r) \in g(r) \subseteq k(h(r))$.

We are now ready for the central result of this section about **InFL**.

THEOREM 7.14. [GJ] *Cut elimination holds for the sequent system* **InFL**.

PROOF. A sequent of the form $s_1, \dots, s_m \Rightarrow t_1, \dots, t_n$ is said to be *valid* in an involutive FL-algebra $\mathbf{A}$ if for any assignment h, $(h(s_1) \cdot \ldots \cdot h(s_m)) \le (h(t_1) + \cdots + h(t_n))$ holds in $\mathbf{A}$. Here $a+b$ is defined by $a + b = \sim(-b \cdot -a)$ $(= -(\sim b \cdot \sim a)$ by Lemma 3.17(3)) for any $a, b \in A$.

We assume that the sequent $s_1, \dots, s_m \Rightarrow t_1, \dots, t_n$ is not provable in the Gentzen system for **InFL** without cut. Hence it is not valid in a Gentzen matrix $\boldsymbol{\mathcal{B}}$ for **InFL** under an assignment h, which means that $(h(s_1), \dots, h(s_m)) \preceq (h(t_1), \dots, h(t_n))$ does not hold. Moreover, we suppose to the contrary that it is valid in $\mathbf{R}(\boldsymbol{\mathcal{B}})$. Then, in particular

$$g(s_1) \cdot_\gamma \cdots \cdot_\gamma g(s_m) \subseteq \sim(-g(t_n) \cdot_\gamma \cdots \cdot_\gamma -g(t_1)).$$

By Lemma 7.13, $h(s_i) \in g(s_i)$ holds for each i and $g(t_j) \subseteq k(h(t_j))$ holds for each j. Therefore,

$$(h(s_1), \dots, h(s_m)) \in \sim(-k(h(t_n)) \cdot_\gamma \cdots \cdot_\gamma -k(h(t_1)))$$

holds. That is, for any x,

(1) if $x^- \in -k(h(t_n)) \cdot_\gamma \cdots \cdot_\gamma -k(h(t_1))$ then $(h(s_1), \dots, h(s_m)) \preceq x^{-\sim} = x$.

On the other hand, if $u \in k(h(t_j)) = \{h(t_j)\}^{\lhd}$ then $u \preceq h(t_j)$, and hence $h(t_j) \in \{u\}^{\rhd}$. It follows that $h(t_j)^- \in -k(h(t_j))$. Hence,

(2) $\quad (h(t_1), \dots, h(t_n))^- \in -k(h(t_n)) \cdot_\gamma \cdots \cdot_\gamma -k(h(t_1))$.

From (1) and (2) it follows that $(h(s_1), \dots, h(s_m)) \preceq (h(t_1), \dots, h(t_n))$. But this is a contradiction. □

The Gentzen system for **InFL** can be modified to cover some of the extensions of **InFL**. For example we can add *cyclicity* by replacing $\sim, -$ with a single negation $\neg$, thus obtaining a Gentzen system for cyclic involutive **FL** (denoted by **CyInFL**). In this case a Gentzen matrix $\boldsymbol{\mathcal{B}}$ gives rise to an involutive monoid $B^{\neg *}$. For $\mathbf{ClFL_e}$ we need to add exchange to the Gentzen system, and replace $B^{\neg *}$ by the commutative monoid generated by $B^{\neg}$. For *weakening* we add axioms $\alpha \Rightarrow 1$ and $x \preceq 1$ to the definition of the

Gentzen system and Gentzen matrix. In each case the resulting Gentzen system satisfies cut elimination.

COROLLARY 7.15. [GJ] *For the logics* **InFL**, **CyInFL**, $\mathbf{InFL_e}$, $\mathbf{CyInFL_w}$ *and* $\mathbf{InFL_{ew}}$ *it is decidable whether a given sequent is provable. Hence the equational theories of involutive FL-algebras (with or without cyclicity, exchange and/or weakening) are decidable.*

PROOF. Note that all the Gentzen rules satisfy the subformula property and, with the exception of the bidirectional rules, they all have premises with lower complexity than the conclusion. Hence, if we can show that any proof search needs to only consider a bounded number of applications of the bidirectional rules, decidability will follow. Suppose $\alpha_1, \ldots, \alpha_m \Rightarrow \beta_1, \ldots, \beta_n$ is a sequent without any external negations (i.e. without superscript $^\sim, ^-$). Applications of the $({}^-\Rightarrow)$ and $(\Rightarrow^\sim)$ rules produces sequents of the form

$$\beta_i^-, \ldots, \beta_1^-, \alpha_1, \ldots, \alpha_m \Rightarrow \beta_{i+1}, \ldots, \beta_n$$

$$\alpha_1, \ldots, \alpha_j \Rightarrow \beta_1, \ldots, \beta_n, \alpha_m^-, \ldots, \alpha_{j+1}^-$$

$$\beta_i^-, \ldots, \beta_1^-, \alpha_1, \ldots, \alpha_j \Rightarrow \beta_{i+1}, \ldots, \beta_n, \alpha_m^-, \ldots, \alpha_{j+1}^-$$

and similarly for the other two bidirectional rules. Hence there are only $2mn$ many sequents derived by these rules that contain at least one of the α_k or β_k without an external negation. Similar bounds apply to sequents that already contain some external negations. Since the remaining Gentzen rules of **InFL** can apply only to formulas without an external negation, it follows that a proof search only has to examine at each step only a bounded number of possible forms of the given sequence. Therefore the original proof search reduces to a finite one. □

7.3.2. Cyclic substructural modal logics. It is also possible to discuss cut elimination of sequent systems for cyclic substructural modal logics, in which case we need to introduce $\Diamond X$ for a subset X of $B^{\neg *}$ by

$$\Diamond X = \bigcap\{\{\Diamond y\}^{\lhd} : X \subseteq \{y\}^{\lhd}\},$$

where $\Diamond x$ is the element $(\Diamond a_1, \ldots, \Diamond a_m)$, of $B^{\neg *}$ when x is $(a_1, \ldots, a_m)$. By using this, we can get an algebraic proof of cut elimination of sequent systems of some of the basic modal logics, including **K**, **KT** and **S4**.

7.3.3. Completeness of tableau systems. It is easy to see that we can apply our method to one-sided sequent systems and tableau systems. The idea of introducing one-sided sequent systems is based on the fact that in involutive *cyclic* systems (hence also in $\mathbf{InFL_e}$), the provability of a sequent $\Gamma \Rightarrow \Delta$ is equivalent to that of $\Rightarrow \neg\Gamma, \Delta$, where $\neg\Gamma$ denotes $\neg\alpha_m, \ldots, \neg\alpha_1$ when Γ is $\alpha_1, \ldots, \alpha_m$. Furthermore, we write simply $\neg\Gamma, \Delta$ instead of $\Rightarrow \neg\Gamma, \Delta$. By this translation, the initial sequent $\alpha \Rightarrow \alpha$ becomes $\neg\alpha, \alpha$. In such a formal system, it is convenient to take the negation $\neg$ as a primitive symbol, for which we take the rule $(\neg)$ shown below. Also, the rule $(\Rightarrow \wedge)$ and the cut rule, for instance, will be expressed as follows.

$$\frac{\alpha, \Gamma}{\neg\neg\alpha, \Gamma}\ (\neg) \qquad \frac{\alpha, \Gamma \quad \beta, \Gamma}{\alpha \wedge \beta, \Gamma}\ (\wedge) \qquad \frac{\neg\alpha, \Gamma \quad \alpha, \Sigma}{\Gamma, \Sigma}\ (cut)$$

Gentzen matrices for these systems can be defined in the same way as before. In these cases, we may take a subset Pr of B^* instead of using a relation $\preceq$, and write $x \in \mathsf{Pr}$ whenever $\varepsilon \preceq x$. Then, we can show cut elimination for these one-sided sequent systems, as before.

Tableau systems can be defined as *duals* of one-sided sequent systems (without the cut rule, by definition). In this case, a sequent $\Gamma \Rightarrow \Delta$ is represented as $\Gamma, \neg\Delta$ in tableau systems. In other words, instead of searching for a proof of a sequent $\Rightarrow \alpha$, we try to show that $\neg\alpha$ is *refutable* in a tableau system. By a standard convention, rules in tableau systems are written upside down, using the separator $|$. For example, $(\neg)$ and $(\wedge)$ in tableau systems are expressed as follows:

$$\frac{\neg\neg\alpha, \Gamma}{\alpha, \Gamma}\ (\neg) \qquad \frac{\neg(\alpha \wedge \beta), \Gamma}{\neg\alpha, \Gamma \mid \neg\beta, \Gamma}\ (\wedge)$$

Gentzen matrices for tableau systems can be defined in the same way as those for one-sided sequent systems. This time, we take a subset Ref of B^* and define $x \in \mathsf{Ref}$ when $x \preceq \varepsilon$. Thus, conditions on Ref corresponding to initial sequents and the above two rules become as follows.

- $(\neg a, a) \in \mathsf{Ref}$,
- $ax \in \mathsf{Ref}$ implies $(\neg\neg a)x \in \mathsf{Ref}$,
- $(\neg a)x \in \mathsf{Ref}$ and $(\neg b)x \in \mathsf{Ref}$ imply $\neg(a \wedge b)x \in \mathsf{Ref}$.

Using quasi-completions, we can show the completeness of these tableau systems. Now let us define Con to be the complement of Ref with respect to B^*. Then, the above conditions can be obviously rewritten as follows.

- $(\neg a, a) \notin \mathsf{Con}$,
- $(\neg\neg a)x \in \mathsf{Con}$ implies $ax \in \mathsf{Con}$,
- $\neg(a \wedge b)x \in \mathsf{Con}$ implies $(\neg a)x \in \mathsf{Con}$ or $(\neg b)x \in \mathsf{Con}$.

When B is the set of formulas, such a set Con that satisfies these conditions is called a *consistency property* in Fitting [Fit73]. In the paper, it is shown that any member s of a consistency property is satisfiable, by constructing a Kripke model in which s is true. In this way, the completeness of these tableau systems with respect to *Kripke semantics* is obtained. It would be interesting to see whether there exists a relation between Fitting's construction of Kripke frames from consistency properties and our construction of algebras given here, in particular, in the case of modal logics.

7.4. Finite model property

In this section, we will give a proof of the finite model property of the logic **FL**. By the finite model property of **FL**, we mean that if a sequent $\Gamma \Rightarrow \delta$ is not provable in **FL**, then there exists a *finite* FL-algebra in which this sequent does not hold.

Our proof of the finite model property of **FL** given below is of algebraic character, and it is given by modifying our algebraic proof of the cut elimination theorem. We owe the idea of the present proof to ones by Lafont [Laf97] and Okada-Terui [OT99], though the presentation is different from them.

Suppose that $\boldsymbol{\mathcal{B}} = (B, \wedge, \vee, \cdot, \backslash, /, 0, 1, \preceq)$ is a Gentzen matrix for **FL** and that the basis $\{\{a\}^{\lhd} : a \in B\}$ for the nucleus associated with $\preceq$ is finite. Then, the set of all closed subsets is also finite, since each closed subset is obtained as an intersection of some of members of the basis. Thus we have the following lemma, by observing how $\mathbf{R}(\boldsymbol{\mathcal{B}})$ is constructed in the previous section.

LEMMA 7.16. *Let $\boldsymbol{\mathcal{B}} = (\mathbf{B}, \preceq)$ be a Gentzen matrix for* **FL** *such that the basis for the nucleus associated to $\preceq$ is finite. Then the quasi-completion $\mathbf{R}(\boldsymbol{\mathcal{B}})$ of the matrix $\boldsymbol{\mathcal{B}}$ is also finite.*

Now consider a Gentzen matrix $\boldsymbol{\mathcal{B}} = (\mathbf{B}, \preceq)$ for **FL**, and let (x, a) be a fixed member of the set $B^* \times (B \cup \{\varepsilon\})$. We define a subset $\mathcal{P}_{(x,a)}$ of $B^* \times (B \cup \{\varepsilon\})$ as follows. Each member of $\mathcal{P}_{(x,a)}$ is called a *predecessor* of (x, a).

1. $(x, a) \in \mathcal{P}_{(x,a)}$.
2. Suppose that $(w, b) \in \mathcal{P}_{(x,a)}$. If "$u \preceq c$ implies $w \preceq b$" is an instance of one of conditions for $\preceq$ in a Gentzen matrix for **FL** for some $u \in B^*$ and $c \in B \cup \{\varepsilon\}$, then (u, c) is a member of $\mathcal{P}_{(x,a)}$. Similarly, if "$u \preceq c$ and $v \preceq d$ imply $w \preceq b$" is an instance of one of conditions for $\preceq$ for some $u, v \in B^*$ and $c, d \in B \cup \{\varepsilon\}$, then both (u, c) and (v, d) are members of $\mathcal{P}_{(x,a)}$.
3. Every member of $\mathcal{P}_{(x,a)}$ is obtained in this way.

An intuitive proof-theoretic meaning of the set $\mathcal{P}_{(x,a)}$ is the set of all "sequents" which may appear in a cut-free proof of the "sequent" $x \preceq a$. For a finite subset S of $B^* \times (B \cup \{\varepsilon\})$, let $\mathcal{P}_S$ be the union of $\mathcal{P}_{(x,a)}$ such that $(x, a) \in S$. We say that the set S is *finitely based*, when $\mathcal{P}_S$ is finite. The following lemma shows that any finitely based subset of $B^* \times (B \cup \{\varepsilon\})$ can be *embedded* into a Gentzen matrix for **FL** with the same underlying set B such that the basis determined by it is finite.

LEMMA 7.17. *Suppose that $\boldsymbol{\mathcal{B}} = (\mathbf{B}, \preceq)$ is a Gentzen matrix for* **FL** *and that S is a finitely based subset of $B^* \times (B \cup \{\varepsilon\})$. Then, there exists a subset $\preceq^\star$ of $B^* \times (B \cup \{\varepsilon\})$ which satisfies the following conditions.*

(1) *if* $(w, b) \in S$ *then* $w \preceq^\star b$ *iff* $w \preceq b$,
(2) *the structure* $\boldsymbol{\mathcal{B}}^\star = (\mathbf{B}, \preceq^\star)$ *forms a Gentzen matrix for* **FL**,
(3) $\mathbf{R}(\boldsymbol{\mathcal{B}}^\star)$ *is finite.*

PROOF. Note that the set $\mathcal{P}_S$ is finite by our assumption. We define a subset $\preceq^\star$ of $B^* \times (B \cup \{\varepsilon\})$ as follows. For $w \in B^*$ and $b \in B \cup \{\varepsilon\}$, if $(w, b) \in \mathcal{P}_S$ then $w \preceq^\star b$ iff $w \preceq b$, and otherwise $w \preceq^\star b$ holds always; so, $\preceq^\star = \preceq \cup (\mathcal{P}_S)^c$. Clearly, the relation $\preceq^\star$ satisfies the first condition of the lemma.

To show that $\boldsymbol{\mathcal{B}}^\star$ is a Gentzen matrix, it is enough to check that $\preceq^\star$ satisfies all the conditions in the definition of Gentzen matrices for **FL**. Let us assume that one of conditions (for $\preceq^\star$), say ($\sharp$), is of the form "$u \preceq^\star c$ and $v \preceq^\star d$ implies $w \preceq^\star b$". To show that this holds for $\preceq^\star$, we suppose that $w \preceq^\star b$ does not hold. By the definition of $\preceq^\star$, this happens only when $(w, b) \in \mathcal{P}_S$ but $w \preceq b$ does not hold. In this case, both (u, c) and (v, d) must belong to $\mathcal{P}_S$. Since "$u \preceq c$ and $v \preceq d$ implies $w \preceq b$" is the condition ($\sharp$) for $\preceq$ which must be true, at least one of $u \preceq c$ and $v \preceq d$ does not hold. Therefore, at least one of $u \preceq^\star c$ and $v \preceq^\star d$ does not hold either. This means that the condition ($\sharp$) holds for $\preceq^\star$. In this way, we can show that all the conditions holds for $\preceq^\star$. Thus $\boldsymbol{\mathcal{B}}^\star$ is a Gentzen matrix for **FL**.

Let $B_1 = B^{*2} \times B$. We define a nucleus relation $N^\star$ on $B^* \times B_1$ by $wN^\star(x, y, a)$ iff $xwy \preceq^\star a$. Note that since we assume S is finitely based, the relation $N^\star$ is cofinite (i.e. has finite complement) in $B^* \times B_1$. This ensures that $B_1^{\triangleleft}$ is cofinite in B^* and, since any $\gamma_{N^\star}$-closed element must contain $B_1^{\triangleleft}$, it follows that there are at most finitely many $\gamma_{N^\star}$-closed elements. Hence $\mathbf{R}(\boldsymbol{\mathcal{B}}^\star)$ is finite. □

Now we are ready to provide a proof of the following theorem, originally due to Okada and Terui.

THEOREM 7.18. [OT99] *The logic* **FL** *has the finite model property.*

PROOF. Suppose that a sequent $\alpha_1, \ldots, \alpha_m \Rightarrow \beta$ is not provable in **FL**. Obviously, this is not provable in **FL** without using cut. Then, by the proof of Lemma 7.2, $(\alpha_1, \ldots, \alpha_m) \preceq \beta$ does not hold in the *absolutely free* Gentzen matrix $\boldsymbol{\mathcal{B}}^+$ for **FL** whose universe B^+ is the set of all terms, under the assignment i which is the identity mapping on the set of all term variables. (When $m = 0$, $(\alpha_1, \ldots, \alpha_m)$ denotes the empty sequence ε. Also, the term β denotes ε when the right hand side of the sequent is empty.) We show that the singleton set $\{((\alpha_1, \ldots, \alpha_m), \beta)\}$ is finitely based. To see this, define the "length" of any element of $(B^+)^* \times (B^+ \cup \{\varepsilon\})$ as follows. Let $\ell(\delta)$ denote the length of a given term δ. For an element $((\gamma_1, \ldots, \gamma_m), \delta)$ of $(B^+)^* \times (B^+ \cup \{\varepsilon\})$, its length is defined to be the sum $\ell(\gamma_1) + \cdots + \ell(\gamma_m) + \ell(\delta)$. Then we can show that if $((\gamma_1, \ldots, \gamma_m), \gamma)$ is a predecessor of $((\alpha_1, \ldots, \alpha_m), \beta)$, then the length is smaller than or equal to the length of $((\alpha_1, \ldots, \alpha_m), \beta)$ and moreover, any of $\gamma_1, \ldots, \gamma_m$ and δ is a subterm of any one of $\alpha_1, \ldots, \alpha_m, \beta$. Thus, the number of predecessors of $((\alpha_1, \ldots, \alpha_m), \beta)$ must be finite.

Then by Lemma 7.17, $\{((\alpha_1, \ldots, \alpha_m), \beta)\}$ is embedded into a Gentzen matrix $(\mathcal{B}^+)^\star$ for **FL** with relation $\preceq^\star$ such that the basis for the nucleus determined by $\preceq^\star$ is finite. Moreover, $(\alpha_1, \ldots, \alpha_m) \preceq^\star \beta$ does not hold in $(\mathcal{B}^+)^\star$. Now, using Lemma 7.16, the quasi-completion $\mathbf{R}(\mathcal{B}^+)^\star$ is finite. Since $(\alpha_1, \ldots, \alpha_m) \preceq^\star \beta$ does not hold in $(\mathcal{B}^+)^\star$, $(\alpha_1 \cdot \ldots \cdot \alpha_m) \leq \beta$ does not hold either in $\mathbf{R}(\mathcal{B}^+)^\star$, which is a finite FL-algebra, as shown just above Theorem 7.8. This completes the proof of the finite model property. □

By modifying the definition of a Gentzen matrix, so that it becomes a matrix model for $\vdash^{\mathbf{FL_e}}$, and implementing minor modifications to the proofs, we can obtain the FMP for $\mathbf{FL_e}$. In [OT99] the FMP is shown for more logics, by modifying the above arguments. For decidability and the FMP of **FL** (and its non-associative version) extended by structural rules, see aslo [GO].

THEOREM 7.19. [OT99] *All basic substructural logics except for* $\mathbf{FL_c}$ *have the finite model property.*

A key of our proof of the finite model property given here is the fact that the set $\mathcal{P}_{(x,a)}$ of all predecessors of any given (x, a) is finite. In syntactic terms, this means that the *proof search tree* of any sequent in the cut-free sequent system **FL** is always finite. Here, by a proof search tree of a given sequent $\Gamma \Rightarrow \delta$ we mean a proof search procedure represented in a tree-like form which searches for a (cut-free) proof of $\Gamma \Rightarrow \delta$ and can always find it as long as it is provable. Thus the finiteness of the proof search procedure means that after finitely many steps of the proof search we can always see whether a given sequent is provable or not. We have constructed a finite algebra in which a given unprovable sequent does not hold, by using the finite proof search tree of the sequent. Thus, our proof of the finite model property also works for other logics with cut-free sequent systems, as long as the proof search tree of any sequent in them is always finite.

Usually, the finite model property is proved in order to derive the decidability. On the other hand, as mentioned above our method uses the existence of a decision procedure to prove the finite model property. This may sound strange, but it is not unusual in the study of substructural logics, where decidability results for most of the basic substructural logics are obtained as simple consequences of cut elimination and therefore can be proved much earlier than the finite model property (see [Ono98a, MO94, Laf97, OT99]).

Our algebraic proof of cut elimination and its application to the finite model property seems to work well for various sequent systems. But this does not mean that we can prove most of the consequences of cut elimination algebraically. For instance, though induction on the length of formulas is a basic tool in syntactic arguments, it sometimes happens that we cannot find any substitute for it in algebraic arguments, because in algebra mathematical objects are not always distinguished from their representations. Thus sometimes it becomes necessary to introduce algebraic substitutes for syntactic objects, like free Gentzen matrices, to fill this gap.

Exercises

(1) Verify that the quasi-completion $\mathbf{R}(\boldsymbol{\mathcal{A}})$ of the Genzen matrix $\boldsymbol{\mathcal{A}}$ associated with an FL-algebra $\mathbf{A}$, defined on page 7.2, is isomorphic to the Dedekind-MacNeille completion of $\mathbf{A}$.

(2) Show that the following two rules, and their analogues for the other negation, are derivable in **InFL**.

$$\frac{\Gamma, \alpha^{\sim}, \Delta \Rightarrow \Lambda}{\Gamma, {\sim}\alpha, \Delta \Rightarrow \Lambda} \qquad \frac{\Lambda \Rightarrow \Gamma, \alpha^{\sim}, \Delta}{\Lambda \Rightarrow \Gamma, {\sim}\alpha, \Delta}$$

(3) Give an algebraic proof of cut elimination for $\mathbf{FL_e}^{n \rightsquigarrow 1}$ for $n > 1$ (see Exercises of Chapter 4).

(4) Give an algebraic proof of cut elimination for $\mathbf{FL_e}$ with the k-mingle rule (see also Exercises of Chapter 4).

(5) Show the finite model property and the decidability of $\mathbf{FL_e}$ with the k-mingle rule.

(6) Give an algebraic proof of cut elimination for $\mathbf{FL_{gc}}$, i.e. **FL** with the global contraction rule. See Section 4.1.

Notes

(1) Our algebraic proof of cut elimination works well for some standard sequent calculi of modal logics, as we mentioned before. This works also for substructural predicate logics (see [BJO04]), by generalizing the idea by Maehara [Mae91] who gave an algebraic proof of cut elimination for both predicate **LK** and **LJ**. Interestingly enough, the finite model property and the decidability of both predicate $\mathbf{FL_e}$ and $\mathbf{FL_{ew}}$ (without function symbols) are also shown in [BJO04] by using the method developed in this chapter. Such an algebraic proof can also be applied successfully to other kinds of sequent calculi, like a calculus for action logic [Bus].

(2) As we mentioned at the beginning of Section 7.3, our algebraic proof of cut elimination for **FL** can be easily extended to $\mathbf{FL_e}$, $\mathbf{FL_{ec}}$ and so on. In fact, this is shown by checking that the quasi-completion $\mathbf{R}(\boldsymbol{\mathcal{B}})$ of a Gentzen matrix $\boldsymbol{\mathcal{B}}$ is commutative (commutative and square-increasing) if $\boldsymbol{\mathcal{B}}$ is a Gentzen matrix for $\mathbf{FL_e}$ ($\mathbf{FL_{ec}}$, respectively). So, it is an interesting problem when such a property is "preserved"

under the quasi-completion. An important result in this direction is obtained in [CT06], in which necessary and sufficient conditions for a sequent calculus (with some restrictions) to admit cut elimination are discussed. In the same direction is the recent work by Ciabattoni, Galatos and Terui on the expressive power of structural rules of **FL** and [GO] for the non-associative case.

(3) The sequent calculus **InFL** introduced in Section 7.3 has four bidirectional rules, using extra operations $\Delta^{\sim}$ and Δ^{-} on a sequence Δ of formulas. This kind of technique is fully developed in *display calculi*. For more information on display calculi, see [Gore98].

(4) It has been hinted that the nuclear relations, as well as the (quasi) embeddings, mentioned in the proofs of the cut elimination and of the finite embeddability are related, not just superficially. In fact it is explained in [GO] that the the embedding in the proof of FEP can be viewed as an embedding of a Gentzen matrix. There, other applications of the same idea, including the strong separation of a Hilbert system, are obtained by defining an appropriate Gentzen matrix and considering the associated (quasi)embedding. The method is extended even more in [GJ] in the setting of (residuated) frames, structures that form relational semantics for substructural logics. There it is shown that all of these applications are instances of a single construction on a frame.

CHAPTER 8

Glivenko theorems

It is well known that classical propositional logic can be interpreted in intuitionistic propositional logic. In particular Glivenko's theorem states that a formula is provable in the former iff its double negation is provable in the latter. In this chapter that is taken from [GO06b], we extend Glivenko's theorem and show that for every involutive substructural logic there exists a minimum substructural logic that contains the first via a double negation interpretation. Our presentation is algebraic and is formulated in the context of residuated lattices. In the last part of the paper, we also discuss some extended forms of the Kolmogorov translation and we compare it to the Glivenko translation.

8.1. Overview

The following theorem, due to Glivenko [Gli29], shows that classical propositional logic can be interpreted in intuitionistic propositional logic.

THEOREM 8.1 (Glivenko). *A formula φ is provable in classical propositional logic iff the formula $\neg\neg\varphi$ is provable in intuitionistic propositional logic.*

Nevertheless, as it is shown in the following theorem, intuitionistic logic is not the only substructural logic into which classical logic can be interpreted by a double negation translation. Moreover, there are other pairs of substructural logics that are related in the same way. We denote by **SBL** the extension of Hájek basic logic **BL** by the axiom $(\chi \cdot (\chi \to \neg\chi)) \to \psi$. Note that **SBL** is incomparable with intuitionistic logic.

THEOREM 8.2. [CT03] [CT04]

(1) *A formula φ is provable in classical propositional logic iff the formula $\neg\neg\varphi$ is provable in* **SBL**.
(2) *A formula φ is provable in Łukasiewicz infinite-valued logic iff the formula $\neg\neg\varphi$ is provable in Hájek basic logic.*
(3) *Let* **L** *be an extension of* $\mathbf{FL_{ew}}$ *that contains the axiom* $\neg\neg(\neg\neg\psi \to \psi)$ *and let* **In**(**L**) *be the extension of* **L** *by the axiom* $\neg\neg\psi \to \psi$. *Then a formula φ is provable in* **In**(**L**) *iff the formula $\neg\neg\varphi$ is provable in* **L**.

Note that the above theorems establish the interpretability of an *involutive* substructural logic in another logic. Also, the previous results are limited to the case, where the rules of exchange and weakening are present.

If **K** and **L** are substructural logics, we say that the *Glivenko property* holds for **K** relative to **L** iff, for all formulas φ over the language of **FL**,

$$\vdash_{\mathbf{L}} \varphi \text{ iff } \vdash_{\mathbf{K}} -{\sim}\varphi \text{ iff } \vdash_{\mathbf{K}} {\sim}{-}\varphi,$$

where $\vdash_{\mathbf{M}}$ denotes the consequence relation associated with the logic **M**.

In view of the fact that the subvarieties of FL of serve as *equivalent algebraic semantics* for substructural logics over **FL** (see Theorem 2.29) the Glivenko property can also be reformulated in algebraic terms. If $\mathcal{W}$ and $\mathcal{V}$ are subvarieties of FL, we say that the *Glivenko property* holds for $\mathcal{W}$ relative to $\mathcal{V}$ iff, for every term t over the language of FL,

$$\models_{\mathcal{V}} 1 \leq t \text{ iff } \models_{\mathcal{W}} 1 \leq -{\sim}t \text{ iff } \models_{\mathcal{W}} 1 \leq {\sim}{-}t.$$

It follows from the algebraization of **FL** that the Glivenko property holds for **K** relative to **L** iff it holds for $\mathsf{V}(\mathbf{K})$ relative to $\mathsf{V}(\mathbf{L})$.

It is natural to consider the following strengthening of the Glivenko property. If $\mathcal{W}$ and $\mathcal{V}$ are subvarieties of FL, we say that the *equational Glivenko property* holds for $\mathcal{W}$ relative to $\mathcal{V}$ iff, for all terms s, t over the language of FL,

$$\models_{\mathcal{V}} s \leq t \text{ iff } \models_{\mathcal{W}} -{\sim}s \leq -{\sim}t \text{ iff } \models_{\mathcal{W}} {\sim}{-}s \leq {\sim}{-}t.$$

On the other hand, staying within the setting of logic, we can strengthen the Glivenko property in a different direction. We say that the *deductive Glivenko property* holds for **K** relative to **L** iff, for every set of formulas $\Sigma \cup \{\varphi\}$ over the language of **FL**,

$$\Sigma \vdash_{\mathbf{L}} \varphi \text{ iff } -{\sim}\Sigma \vdash_{\mathbf{K}} -{\sim}\varphi \text{ iff } {\sim}{-}\Sigma \vdash_{\mathbf{K}} {\sim}{-}\varphi,$$

where $-{\sim}\Sigma = \{-{\sim}\sigma \mid \sigma \in \Sigma\}$. In algebraic terms, the *deductive Glivenko property* holds for $\mathcal{W}$ relative to $\mathcal{V}$ iff, for all sets of terms $\{t_i \mid \in I\} \cup \{t\}$ over the language of FL,

$$D \models_{\mathcal{V}} 1 \leq t \text{ iff } -{\sim}D \models_{\mathcal{W}} 1 \leq -{\sim}t \text{ iff } {\sim}{-}D \models_{\mathcal{W}} 1 \leq {\sim}{-}t,$$

where $D = \{1 \leq t_i \mid \in I\}$ and $-{\sim}D = \{1 \leq -{\sim}t_i \mid i \in I\}$.

A common strengthening of all these three types of Glivenko property is the following. If $\mathcal{W}$, $\mathcal{V}$ are subvarieties of FL, we say that the *deductive equational Glivenko property* holds for $\mathcal{W}$ relative to $\mathcal{V}$ iff, for all sets of equations $E \cup \{s = t\}$ over the language of FL,

$$E \models_{\mathcal{V}} s \leq t \text{ iff } -{\sim}E \models_{\mathcal{W}} -{\sim}s \leq -{\sim}t \text{ iff } {\sim}{-}E \models_{\mathcal{W}} {\sim}{-}s \leq {\sim}{-}t,$$

where $-{\sim}E = \{-{\sim}u = -{\sim}v \mid (u = v) \in E\}$.

We will prove that the equational Glivenko property and the deductive equational Glivenko property are equivalent and they imply that **L** (or $\mathcal{V}$) is involutive; recall that a variety $\mathcal{V}$ is called *involutive*, if it satisfies ${\sim}{-}x = x = -{\sim}x$. The other properties are not equivalent in general – see

Proposition 8.16 and Proposition 8.17 – but under the assumption that $\mathbf{L}$ (or $\mathcal{V}$) is involutive all of the properties mentioned above are equivalent.

We show that for every involutive substructural logic $\mathbf{L}$, there exists a substructural logic $\mathbf{G}(\mathbf{L})$, called the *Glivenko logic* of $\mathbf{L}$, such that some/any Glivenko property holds for a substructural logic $\mathbf{K}$ relative to $\mathbf{L}$ iff $\mathbf{G}(\mathbf{L}) \subseteq \mathbf{K} \subseteq \mathbf{L}$, see Corollary 8.14. Thus, $\mathbf{G}(\mathbf{L})$ is the smallest substructural logic for which Glivenko property holds relative to $\mathbf{L}$. Given an axiomatization of a logic $\mathbf{L}$, we provide an axiomatization for the logic $\mathbf{G}(\mathbf{L})$, see Corollary 8.19, which is finite if $\mathbf{L}$ is finitely axiomatized, see Corollary 8.20, and show that $\mathbf{G}$ is an interior operator on the lattice of substructural logics, see Lemma 8.5(1). This answers the question: *Given an involutive substructural logic* $\mathbf{L}$*, for which logics does some Glivenko property hold relative to* $\mathbf{L}$*?*

We continue by addressing a question in the other direction: *Given a substructural logic* $\mathbf{K}$*, when and relative to which logics does a Glivenko property hold for* $\mathbf{K}$*?* We call two substructural logics (or subvarieties of FL) *Glivenko equivalent* if they contain the same negated formulas (equations, respectively), see Lemma 8.3. The equivalence classes determined by Glivenko equivalence are intervals of the lattice of substructural logics (the lattice of subvarieties of FL) of the form $[\mathbf{G}(\mathbf{K}), \mathbf{M}(\mathbf{K})]$ (or $[\mathbf{M}(\mathcal{W}), \mathbf{G}(\mathcal{W})]$). This implies that the maximum logic (or, the minimum variety) exists in each equivalence class. It turns out that two logics $\mathbf{K}_1$ and $\mathbf{K}_2$ are Glivenko equivalent iff $\mathbf{G}(\mathbf{K}_1) = \mathbf{G}(\mathbf{K}_2)$; see Lemma 8.5(3). In this case, any of the Glivenko properties holds for $\mathbf{K}_2$ relative to some logic iff the property holds for $\mathbf{K}_1$ relative to the same logic. Additionally, any of the Glivenko properties above holds for $\mathbf{K}$ relative to some substructural logic iff it holds relative to $\mathbf{M}(\mathbf{K})$; see Propositions 8.8, 8.11 and 8.13.

For each of the three Glivenko properties, we describe different degrees of involutiveness that $\mathbf{M}(\mathbf{K})$ has to possess. The existence of an *involutive* substructural logic (or subvariety of FL) relative to which some/any of the Glivenko properties holds for $\mathbf{K}$ (or $\mathcal{W}$) is equivalent to the condition that $\mathbf{K}$ contains the *Glivenko logic* $\mathbf{G}$ (or $\mathcal{W}$ being contained in the *Glivenko variety* $\mathcal{G}$); see Theorem 8.25 and the comments following it.

In each of the three Glivenko properties discussed above, there are three statements involved, which are stipulated to be equivalent. For example, we have

$$\models_{\mathcal{V}} s \leq t \text{ iff } \models_{\mathcal{W}} -{\sim}s \leq -{\sim}t \text{ iff } \models_{\mathcal{W}} {\sim}-s \leq {\sim}-t.$$

in the equational Glivenko property. One can consider finer versions of the properties by stipulating that the first two statements are equivalent (*left* version), or the first and the last statement are equivalent (*right* version of the property). So, for example, the *left equational Glivenko property* holds for $\mathcal{W}$ relative to $\mathcal{V}$ iff for all terms s, t over the language of FL,

$$\models_{\mathcal{V}} s \leq t \text{ iff } \models_{\mathcal{W}} -{\sim}s \leq -{\sim}t.$$

Our analysis respects these more detailed considerations and provides left and right versions of each of the results that we show. For simplicity we state most of the results in their left versions, but the *opposite* statements, where "left" is replaced by "right" and the terms and equations in the statement are replaced by their opposite, hold, as well.

In Section 8.5, we consider Glivenko properties for some special cases, from which Theorems 8.1 and 8.2 follow. Moreover, we obtain simplified axiomatizations for the largest integral subvariety of FL for which Glivenko property holds relative to a given integral involutive variety. Also, we study the case where a Glivenko property holds relative to the variety of Boolean algebras and the case of the subvarieties of GBL-algebras.

Finally, in Section 8.6, we discuss briefly a translation that generalizes the Kolmogorov translation to substructural logics and compare it to the Glivenko translation.

8.2. Glivenko equivalence

We introduce the notion of *Glivenko equivalence* between subvarieties of FL. Via the algebraization theorem, Theorem 2.29, there is an associated relation between substructural logics, for which we will use the same name. Glivenko equivalence will serve as a unifying concept that will connect the different Glivenko properties we will be consider.

LEMMA 8.3. *Let $\mathcal{W}$, $\mathcal{V}$ be subvarieties of* FL *and let r, s, t, s_i, t_i terms over* FL, *for $i \in I$. We consider the sets $E = \{s_i = t_i | i \in I\}$, $D = \{1 \leq t_i | i \in I\}$, ${\sim}E = \{{\sim}s_i = {\sim}t_i \mid i \in I\}$ and ${\sim}D = \{1 \leq {\sim}t_i \mid i \in I\}$. The following statements are equivalent.*

(1) $\mathcal{V} \models {\sim}s = {\sim}t$ *iff* $\mathcal{W} \models {\sim}s = {\sim}t$.
(2) $\mathcal{V} \models -s = -t$ *iff* $\mathcal{W} \models -s = -t$.
(3) $\mathcal{V} \models {\sim}s \leq {\sim}t$ *iff* $\mathcal{W} \models {\sim}s \leq {\sim}t$.
(4) $\mathcal{V} \models r \leq {\sim}t$ *iff* $\mathcal{W} \models r \leq {\sim}t$.
(5) $\mathcal{V} \models 1 \leq {\sim}t$ *iff* $\mathcal{W} \models 1 \leq {\sim}t$.
(6) $D \models_{\mathcal{V}} 1 \leq {\sim}t$ *iff* $D \models_{\mathcal{W}} 1 \leq {\sim}t$.
(7) ${\sim}D \models_{\mathcal{V}} 1 \leq {\sim}t$ *iff* ${\sim}D \models_{\mathcal{W}} 1 \leq {\sim}t$.
(8) $E \models_{\mathcal{V}} {\sim}s = {\sim}t$ *iff* $E \models_{\mathcal{W}} {\sim}s = {\sim}t$.
(9) ${\sim}E \models_{\mathcal{V}} {\sim}s = {\sim}t$ *iff* ${\sim}E \models_{\mathcal{W}} {\sim}s = {\sim}t$.

PROOF. Assume that (1) holds. Then, $\mathcal{V}$ satisfies $-s = -t$ iff it satisfies ${\sim}{-s} = {\sim}{-t}$, by Lemma 2.8(6). By (1), this is true iff $\mathcal{W}$ satisfies ${\sim}{-s} = {\sim}{-t}$; i.e., iff $\mathcal{W}$ satisfies $-s = -t$. Therefore, (1) implies (2); the converse is obtained by interchanging the two negation operations.

Note that ${\sim}s \leq {\sim}t$ iff ${\sim}s = {\sim}(s \vee t)$, by Lemma 2.8(1); also ${\sim}s = {\sim}t$ is equivalent to the conjunction of ${\sim}s \leq {\sim}t$ and ${\sim}t \leq {\sim}s$. The equivalence between (1) and (3) follows from these two facts.

Obviously, (4) implies both (3) and (5). (5) implies (4), since $r \leq {\sim}t$ iff $1 \leq r\backslash{\sim}t$ iff $1 \leq {\sim}(tr)$, by Lemma 2.8(8), and (3) implies (4), since $r \leq {\sim}t$ iff ${\sim}{-}r \leq {\sim}t$, by Lemma 2.8(2,3,4). Consequently, (1)-(5) are all equivalent. Moreover, the same argument shows that (9) implies (7). It is clear that (8) implies (9) and that (7) implies (5). We will show that (5) implies (6) and that (6) implies (8).

Assume that (5) holds. We have $\{1 \leq t_i \mid i \in I\} \models_{\mathcal{V}} 1 \leq {\sim}t$ iff there exists a natural number n and iterated conjugates γ_k over a sequence of terms, such that $\models_{\mathcal{V}} 1 \leq \prod_{k=1}^{n} \gamma_k(t_{i_k})\backslash{\sim}t = {\sim}[t \cdot \prod_{k=1}^{n} \gamma_k(t_{i_k})]$, by using Theorem 2.29 and Theorem 2.14. By (5) the same equation holds in $\mathcal{W}$ for the same n and the same iterated conjugates, hence $\{1 \leq t_i \mid i \in I\} \models_{\mathcal{W}} 1 \leq {\sim}t$; consequently, (6) holds.

Assume, now, that (6) holds. We have ${\sim}s = {\sim}t$ iff ${\sim}s \leq {\sim}t$ and ${\sim}t \leq {\sim}s$, iff $1 \leq {\sim}s\backslash{\sim}t = {\sim}(t \cdot {\sim}s)$ and $1 \leq {\sim}(s \cdot {\sim}t)$, by Lemma 2.8(8), iff $1 \leq {\sim}(t \cdot {\sim}s) \wedge {\sim}(s \cdot {\sim}t)$, iff $1 \leq {\sim}p$, where $p = (t \cdot {\sim}s) \vee (s \cdot {\sim}t)$, by Lemma 2.8(1). Moreover, $s = t$ iff $1 \leq s\backslash t \wedge t\backslash s$. So, (8) can be written in a form that is a special case of (6). □

If any of the equivalent statements of the previous lemma holds, we say that the variety $\mathcal{V}$ is *Glivenko equivalent* to the variety $\mathcal{W}$. Glivenko equivalence coincides with the notion of *negative equivalence* of S. Odintsov, for the special cases considered in [Odi04]. Obviously, Glivenko equivalence is an equivalence relation on $\mathbf{\Lambda}(\mathsf{FL})$. It is clear that if $\mathcal{V}$, $\mathcal{W}$ are Glivenko equivalent and $\mathcal{V} \subseteq \mathcal{U} \subseteq \mathcal{W}$, then $\mathcal{U}$ is Glivenko equivalent to $\mathcal{V}$ and $\mathcal{W}$. So, the equivalence classes of the Glivenko equivalence relation are convex.

We say that the substructural logics $\mathbf{K}$ and $\mathbf{L}$ are *Glivenko equivalent*, if for all formulas φ,

$$\vdash_{\mathbf{K}} {\sim}\varphi \text{ iff } \vdash_{\mathbf{L}} {\sim}\varphi$$

By Lemma 8.3, $\mathbf{K}$ and $\mathbf{L}$ are Glivenko equivalent iff $\mathsf{V}(\mathbf{K})$ and $\mathsf{V}(\mathbf{L})$ are Glivenko equivalent. It follows from Lemma 8.3 that ${\sim}$ can be replaced by $-$ in the above definition.

For every variety $\mathcal{V}$ of FL-algebras, let $\mathbf{G}(\mathcal{V})$ be the subvariety of FL axiomatized by the equations ${\sim}s = {\sim}t$, where s, t range over all pairs of terms such that the equation $s = t$ holds in $\mathcal{V}$. The variety $\mathbf{G}(\mathcal{V})$ is called the *Glivenko variety of* $\mathcal{V}$.

LEMMA 8.4. *For every subvariety $\mathcal{V}$ of FL, the variety $\mathbf{G}(\mathcal{V})$ is also axiomatized by the equations $-s = -t$, where $s = t$ holds in $\mathcal{V}$.*

PROOF. Consider the variety $\mathbf{G}'(\mathcal{V})$ axiomatized by the equations $-s = -t$, where $s = t$ holds in $\mathcal{V}$; we will show that $\mathbf{G}'(\mathcal{V}) = \mathbf{G}(\mathcal{V})$. For every equation $s = t$ valid in $\mathcal{V}$, the equation $-s = -t$ is valid in $\mathcal{V}$, as well. So, $\mathbf{G}(\mathcal{V})$ satisfies the equation ${\sim}{-}s = {\sim}{-}t$, hence it also satisfies the equation $-{\sim}{-}s = -{\sim}{-}t$. In view of Lemma 2.8(4), we have that $-s = -t$

holds in $\mathbf{G}(\mathcal{V})$. Thus, $\mathbf{G}(\mathcal{V}) \subseteq \mathbf{G}'(\mathcal{V})$. Likewise, we obtain the converse inclusion. □

LEMMA 8.5. *Let $\mathcal{U}$, $\mathcal{V}$, $\mathcal{W}$ be subvarieties of* FL.

(1) $\mathbf{G}$ *is a closure operator on* $\mathbf{\Lambda}(\mathsf{FL})$*; i.e.*
   (a) $\mathcal{V} \subseteq \mathbf{G}(\mathcal{V})$,
   (b) *if* $\mathcal{V} \subseteq \mathcal{W}$, *then* $\mathbf{G}(\mathcal{V}) \subseteq \mathbf{G}(\mathcal{W})$, *and*
   (c) $\mathbf{G}(\mathbf{G}(\mathcal{V})) = \mathbf{G}(\mathcal{V})$.
(2) *The varieties* $\mathcal{V}$ *and* $\mathbf{G}(\mathcal{V})$ *are Glivenko equivalent.*
(3) *The varieties* $\mathcal{V}$ *and* $\mathcal{W}$ *are Glivenko equivalent iff* $\mathbf{G}(\mathcal{V}) = \mathbf{G}(\mathcal{W})$.
(4) *The variety* $\mathbf{G}(\mathcal{V})$ *is the largest subvariety of* FL *that is Glivenko equivalent to* $\mathcal{V}$.

PROOF. (1) For (a), note that if ${\sim}s = {\sim}t$ is an axiom of $\mathbf{G}(\mathcal{V})$, namely $s = t$ is valid in $\mathcal{V}$, then ${\sim}s = {\sim}t$ is valid in $\mathcal{V}$. Thus, $\mathcal{V}$ is a subvariety of $\mathbf{G}(\mathcal{V})$. The fact that $\mathbf{G}$ is increasing is clear from its definition. To show that $\mathbf{G}(\mathbf{G}(\mathcal{V})) \subseteq \mathbf{G}(\mathcal{V})$, let ${\sim}s = {\sim}t$ be an axiom of $\mathbf{G}(\mathcal{V})$. Then, $-{\sim}s = -{\sim}t$ holds in $\mathbf{G}(\mathcal{V})$, hence ${\sim}-{\sim}s = {\sim}-{\sim}t$ holds in $\mathbf{G}(\mathbf{G}(\mathcal{V}))$. Thus, $\mathbf{G}(\mathbf{G}(\mathcal{V}))$ satisfies ${\sim}s = {\sim}t$, by Lemma 2.8(4).

(2) If $\mathcal{V}$ satisfies $-s = -t$, then $\mathbf{G}(\mathcal{V})$ satisfies ${\sim}-s = {\sim}-t$, hence it satisfies $-s = -t$, by Lemma 2.8(6). Conversely, if $-s = -t$ holds in $\mathbf{G}(\mathcal{V})$, then it also holds in $\mathcal{V}$, by (1a). Thus, $\mathcal{V}$ and $\mathbf{G}(\mathcal{V})$ are Glivenko equivalent.

(3) If $\mathcal{V}$ and $\mathcal{W}$ are Glivenko equivalent, then $\mathbf{G}(\mathcal{V})$ and $\mathbf{G}(\mathcal{W})$ are Glivenko equivalent, by (2). Thus, if ${\sim}s = {\sim}t$ is an axiom of $\mathbf{G}(\mathcal{V})$, then it is valid in $\mathbf{G}(\mathcal{W})$. So, $\mathbf{G}(\mathcal{W}) \subseteq \mathbf{G}(\mathcal{V})$. The other inclusion is obtained in a similar way, hence $\mathbf{G}(\mathcal{V}) = \mathbf{G}(\mathcal{W})$. Conversely, if $\mathbf{G}(\mathcal{V}) = \mathbf{G}(\mathcal{W})$, then $\mathcal{V}$, $\mathcal{W}$ are Glivenko equivalent, by (2).

(4) If $\mathcal{V}$, $\mathcal{W}$ are Glivenko equivalent, then $\mathbf{G}(\mathcal{V}) = \mathbf{G}(\mathcal{W})$, by (3). Since $\mathcal{W} \subseteq \mathbf{G}(\mathcal{W})$, by (1), we have $\mathcal{W} \subseteq \mathbf{G}(\mathcal{V})$. So, in view of (2), $\mathbf{G}(\mathcal{V})$ is the largest subvariety of FL that is Glivenko equivalent to $\mathcal{V}$. □

For a substructural logic $\mathbf{L}$ we define the *Glivenko logic* of $\mathbf{L}$ to be $\mathbf{G}(\mathbf{L}) = \mathbf{L}(\mathbf{G}(\mathsf{V}(\mathbf{L})))$. It follows from the preceding theorem and from Theorem 2.28 that $\mathbf{G}(\mathbf{L})$ is the smallest substructural logic that is Glivenko equivalent to $\mathbf{L}$.

By definition $\mathbf{G}(\mathsf{V}(\mathbf{L}))$ is axiomatized by the equations ${\sim}s = {\sim}t$, where $s = t$ ranges over all equations valid in $\mathsf{V}(\mathbf{L})$. Recalling that $s = t$ is valid in $\mathsf{V}(\mathbf{L})$ iff the formula $s\backslash t \wedge t\backslash s$ is in $\mathbf{L}$ iff both of $s\backslash t$ and $t\backslash s$ are in $\mathbf{L}$ we have that $\mathbf{G}(\mathbf{L})$ is axiomatized by the formulas ${\sim}s\backslash{\sim}t \wedge {\sim}t\backslash{\sim}s$, where $s\backslash t$ and $t\backslash s$ are in $\mathbf{L}$. Therefore, $\mathbf{G}(\mathbf{L}) = \mathbf{FL} + \{{\sim}\psi\backslash{\sim}\varphi \mid \varphi\backslash\psi \in \mathbf{L}\}$. The following proposition gives an alternative axiomatization for $\mathbf{G}(\mathbf{L})$.

PROPOSITION 8.6. *If* $\mathbf{L}$ *is a substructural logic, then* $\mathbf{G}(\mathbf{L})$ *is axiomatized relative to* $\mathbf{FL}$ *by either one of the sets* $\{-{\sim}\varphi \mid \varphi \in \mathbf{L}\}$ *and* $\{{\sim}-\varphi \mid \varphi \in \mathbf{L}\}$*; i.e.,* $\mathbf{G}(\mathbf{L}) = \mathbf{FL} + \{-{\sim}\varphi \mid \varphi \in \mathbf{L}\} = \mathbf{FL} + \{{\sim}-\varphi \mid \varphi \in \mathbf{L}\}$.

PROOF. We will show that $\mathbf{G}(\mathbf{L}) = \mathbf{M}$, where $\mathbf{M} = \mathbf{FL} + \{-{\sim}\varphi \mid \varphi \in \mathbf{L}\}$.

If $-{\sim}\varphi$ is an axiom of $\mathbf{M}$ for $\varphi \in \mathbf{L}$, then $1\backslash\varphi \in \mathbf{L}$; hence

$$-{\sim}\varphi = {\sim}\varphi\backslash 0 = {\sim}\varphi\backslash{\sim}1 \in \mathbf{G}(\mathbf{L}).$$

Consequently, $\mathbf{M} \subseteq \mathbf{G}(\mathbf{L})$.

Conversely, suppose that ${\sim}\psi\backslash{\sim}\varphi$ is an axiom of $\mathbf{G}(\mathbf{L})$ for $\varphi\backslash\psi \in \mathbf{L}$. So, $-{\sim}(\varphi\backslash\psi) \in \mathbf{M}$ and $\rho_\varphi(-{\sim}(\varphi\backslash\psi)) \in \mathbf{M}$.

By Lemma 2.6(4), we have $(0/\psi)\varphi(\varphi\backslash\psi) \leq 0 = {\sim}1$, so

$$(0/\psi)\varphi[-{\sim}(\varphi\backslash\psi)] \leq 0,$$

by Lemma 2.8(13). Hence, $\varphi[-{\sim}(\varphi\backslash\psi)] \leq (0/\psi)\backslash 0 = ({\sim}\psi)\backslash 0$. Since

$$\rho_\varphi(-{\sim}(\varphi\backslash\psi))\varphi = [\varphi(-{\sim}(\varphi\backslash\psi))/\varphi \wedge 1]\varphi \leq \varphi(-{\sim}(\varphi\backslash\psi)),$$

we have $\rho_\varphi(-{\sim}(\varphi\backslash\psi))\varphi \leq ({\sim}\psi)\backslash 0$. So,

$$\rho_\varphi(-{\sim}(\varphi\backslash\psi)) \leq (({\sim}\psi)\backslash 0)/\varphi = ({\sim}\psi)\backslash({\sim}\varphi),$$

by Lemma 2.6(7). By (mp_ℓ) of Theorem 2.17, we have $({\sim}\psi)\backslash({\sim}\varphi) \in \mathbf{M}$. Consequently, $\mathbf{G}(\mathbf{L}) \subseteq \mathbf{M}$. □

An axiomatization of $\mathbf{G}(\mathbf{L})$ is also given in [Odi04], for the special case of extensions of Johansson's logic.

We know form Theorem 8.5(4) that given a logic $\mathbf{L}$ there exists a smallest logic $\mathbf{G}(\mathbf{L})$ that is Glivenko equivalent to $\mathbf{L}$. For every substructural logic $\mathbf{L}$, we define the logic

$$\mathbf{M}(\mathbf{L}) = \mathbf{FL} + \{\varphi \mid -{\sim}\gamma(\varphi) \in \mathbf{L}, \text{ for every } \gamma \in \Gamma\},$$

where Γ denotes the set of all iterated conjugates.

THEOREM 8.7. *For every substructural logic* $\mathbf{L}$, *the logic* $\mathbf{M}(\mathbf{L})$ *is the greatest element of the Glivenko equivalence class of* $\mathbf{L}$.

PROOF. If $\psi \in \mathbf{L}$, then $\gamma(\psi) \in \mathbf{L}$, by (pn) and (adju) of Theorem 2.17, and $-{\sim}\gamma(\psi) \in \mathbf{L}$, by Lemma 2.8(3) and (mp_ℓ) of Theorem 2.17; hence $\psi \in \mathbf{M}(\mathbf{L})$. Therefore, $\mathbf{L} \subseteq \mathbf{M}(\mathbf{L})$.

If $-\psi \in \mathbf{M}(\mathbf{L})$, then $\mathbf{FL} + \Phi \vdash_{\mathbf{FL}} -\psi$, where

$$\Phi = \{\varphi \mid -{\sim}\gamma(\varphi) \in \mathbf{L}, \text{ for every } \gamma \in \Gamma\}.$$

By Corollary 2.4, we have $\Sigma_{\mathcal{L}}(\Phi) \vdash_{\mathbf{FL}} -\psi$. We will show that $\Sigma_{\mathcal{L}}(\Phi) = \Phi$.

Assume that $\varphi \in \Phi$, $\sigma \in \Sigma_{\mathcal{L}}$ and $\gamma \in \Gamma$. Let γ' be the iterated conjugate obtained from γ by replacing all common variables x_i of γ and φ in γ by new variables y_i not appearing in γ or φ. Also, let σ' be the substitution that maps the variables y_i to the variables x_i and otherwise behaves like σ. It is easy to see that $\sigma'(-{\sim}\gamma'(\varphi)) = -{\sim}\gamma(\sigma(\varphi))$. Since $\varphi \in \Phi$, we have $-{\sim}\gamma'(\varphi) \in L$ and hence $-{\sim}\gamma(\sigma(\varphi)) = \sigma'(-{\sim}\gamma'(\varphi)) \in L$. Thus, $\sigma(\varphi) \in \Phi$.

Consequently, $\Phi \vdash_{\mathbf{FL}} -\psi$, so there are $\varphi_i \in \Phi$ and iterated conjugates γ_i, for $i \in \{1, 2, \ldots, n\}$, for some non-negative integer n, such that

$$\prod_{i=1}^{n} \gamma_i(\varphi_i) \leq -\psi$$

By Lemma 3.50, we have

$$\prod_{i=n}^{1} -{\sim}\gamma_i'(\varphi_i) \leq -\psi$$

Since $\varphi_i \in \Phi$, for all i, we have that $-{\sim}\gamma_i'(\varphi_i) \in \mathbf{L}$, for all i; hence $\prod_{i=n}^{1} -{\sim}\gamma_i'(\varphi_i) \in \mathbf{L}$, by (pn) of Theorem 2.17 and $-\psi \in \mathbf{L}$, by (mp_ℓ) of the same theorem. Consequently, $\mathbf{L}$ and $\mathbf{M}(\mathbf{L})$ are Glivenko equivalent.

Now, assume that $\mathbf{K}$ is a substructural logic that is Glivenko equivalent to $\mathbf{L}$. If $\psi \in \mathbf{K}$, then, by Theorem 2.17, we have that $\sigma(\psi) \in \mathbf{K}$, for every substitution σ, $\gamma(\sigma(\psi)) \in \mathbf{K}$, for every iterated conjugate γ, and $-{\sim}\gamma(\sigma(\psi)) \in \mathbf{K}$. By the Glivenko equivalence, $-{\sim}\gamma(\sigma(\psi)) \in \mathbf{L}$, for all $\sigma \in \Sigma_{\mathcal{L}}$ and for all $\gamma \in \Gamma$; thus $\psi \in \mathbf{M}(\mathbf{L})$. Consequently, $\mathbf{K}$ is contained in $\mathbf{M}(\mathbf{L})$. □

A similar result is shown in [Odi04] for the special case considered in the paper.

The definitions and results in this as well as in the following sections can be transferred from subvarieties of FL to substructural logics over $\mathbf{FL}$ and vice versa. For example, for a subvariety $\mathcal{V}$ of FL we define $\mathbf{M}(\mathcal{V}) = \mathsf{V}(\mathbf{M}(\mathbf{L}(\mathcal{V})))$.

It follows, by Lemma 8.5(4) and Theorem 8.7, that the Glivenko equivalence classes are intervals in $\mathbf{\Lambda}(\mathsf{FL})$ of the form $[\mathbf{M}(\mathcal{V}), \mathbf{G}(\mathcal{V})]$. Also, the classes of the Glivenko equivalence between logics are intervals of the form $[\mathbf{G}(\mathbf{L}), \mathbf{M}(\mathbf{L})]$.

8.3. Glivenko properties

In this section we discuss when a Glivenko property holds for a substructural logic $\mathbf{K}$ relative to a substructural logic $\mathbf{L}$. As mentioned in the introduction, we consider three types of Glivenko properties. We provide a characterization for each of them in terms of different types of involutiveness that we introduce below.

A substructural logic $\mathbf{L}$ is called *left involutive*, if $\vdash_{\mathbf{L}} {\sim}{-}\varphi \backslash \varphi$, for every φ. We say that $\mathbf{L}$ is *left weakly involutive*, if ${\sim}{-}\varphi \vdash_{\mathbf{L}} \varphi$, for every φ, and that it is *left Glivenko involutive*, if $\vdash_{\mathbf{L}} {\sim}{-}\varphi$ implies $\vdash_{\mathbf{L}} \varphi$, for every φ. Clearly, left involutiveness is the strongest and left Glivenko involutiveness is the weakest among the three properties. Propositions 8.16 and 8.17 in the present section show that the associated implications are strict.

8.3.1. The Glivenko property. We say that the *left Glivenko property* holds for $\mathbf{K}$ relative to $\mathbf{L}$, or that $\mathbf{K}$ has the left Glivenko property relative to $\mathbf{L}$, if $\vdash_{\mathbf{L}} \varphi$ iff $\vdash_{\mathbf{K}} \sim-\varphi$, for all φ. The opposite of the left Glivenko property (obtained by interchanging $\sim$ and $-$) is the *right Glivenko property* and the conjunction of the two is the *Glivenko property.* We define the Glivenko property for subvarieties of FL by referring to their corresponding substructural logics. The following result then can be reformulated for subvarieties of FL in the obvious way.

PROPOSITION 8.8. *If* $\mathbf{L}$ *and* $\mathbf{K}$ *are substructural logics, then the following are equivalent.*

(1) *The left Glivenko property holds for* $\mathbf{K}$ *relative to* $\mathbf{L}$.
(2) $\mathbf{K}$ *and* $\mathbf{L}$ *are Glivenko equivalent and* $\mathbf{L}$ *is left Glivenko involutive.*
(3) $\mathbf{L} = \mathbf{M}(\mathbf{K})$ *and* $\mathbf{M}(\mathbf{K})$ *is left Glivenko involutive.*

PROOF. We first establish the equivalence of (1) and (2). By setting $\sim\varphi$ for φ in (1), it follows by Lemma 2.8(4) that $\mathbf{K}$ and $\mathbf{L}$ are Glivenko equivalent. In particular, $\vdash_{\mathbf{K}} \sim-\varphi$ iff $\vdash_{\mathbf{L}} \sim-\varphi$; hence $\vdash_{\mathbf{L}} \sim-\varphi$ iff $\vdash_{\mathbf{L}} \varphi$. for every φ, by (1). Conversely, if (2) holds, then $\vdash_{\mathbf{K}} \sim-\varphi$ iff $\vdash_{\mathbf{L}} \sim-\varphi$ iff $\vdash_{\mathbf{L}} \varphi$, by the assumption that $\mathbf{L}$ is left Glivenko involutive.

Obviously, (3) implies (2). For the converse implication, note that $\mathbf{L} \subseteq \mathbf{M}(\mathbf{K})$, since $\mathbf{K}$ and $\mathbf{L}$ are Glivenko equivalent. Moreover, if $\vdash_{\mathbf{M}(\mathbf{K})} \varphi$, then $\vdash_{\mathbf{M}(\mathbf{K})} \sim-\varphi$. By Glivenko equivalence, we have $\vdash_{\mathbf{L}} \sim-\varphi$, so $\vdash_{\mathbf{L}} \varphi$, since $\mathbf{L}$ is left Glivenko involutive. Thus, $\mathbf{M}(\mathbf{K}) \subseteq \mathbf{L}$. □

It follows from Proposition 8.8 that in every Glivenko equivalence class $[\mathbf{G}(\mathbf{K}), \mathbf{M}(\mathbf{K})]$ there is at most one left Glivenko involutive logic and it is equal to $\mathbf{M}(\mathbf{K})$, when it exists. The corresponding statement and the analogue of Proposition 8.8 hold for subvarieties of FL. Conditions on the existence of the left Glivenko involutive logic in a Glivenko equivalence class will be discussed in Section 8.4.2 after Theorem 8.25.

We show that there exists a substructural logic $\mathbf{K}$ for which the Glivenko property does not hold (relative to any logic $\mathbf{L}$). We will state and prove this result in the terminology of algebra; i.e. we will show that there is a subvariety $\mathcal{V}$ of FL, for which the Glivenko property does not hold. By Proposition 8.8, it is enough to show that $\mathbf{M}(\mathcal{V})$ is not left Glivenko involutive.

We define an order relation on the set $A = \{\bot, u, 1, \top\}$, by $\bot < u < 1 < \top$. Moreover, we define an idempotent multiplication, for which $\bot$ is an absorbing and 1 a unit element, by $\top u = \top$ and $u\top = u$. It is easy to check that multiplication preserves order, hence it preserves arbitrary joins, as well, since $\mathbf{A}$ is totally ordered. Therefore, multiplication is residuated with respect to the order and it can be easily checked that it is associative. We denote by $\mathbf{A}$ the associated FL-algebra, where $0 = \bot$.

PROPOSITION 8.9. *The subvariety of* FL *that is generated by* **A** *does not enjoy the left Glivenko property. Similarly, the variety generated by* $\mathbf{A}^{op}$ *does not enjoy the right Glivenko property.*

PROOF. It is easy to see that **A** does not have any subalgebras or homomorphic images other than the trivial and the universal. As we will see in Theorem 9.3, this implies that the variety $\mathcal{W}$ generated by **A** is an atom, and consequently $\mathbf{M}(\mathcal{W}) = \mathcal{W}$. To show that $\mathcal{W}$ is not left Glivenko involutive, it suffices to show that there is a term t such that $\mathbf{A} \models 1 \leq {\sim}{-}t$, but not $\mathbf{A} \models 1 \leq t$. Such a term is $t(x) = 1/[x \vee (x\backslash 1)]$. Indeed, it is not hard to verify that $t^{\mathbf{A}}(x) = u$ and ${\sim}{-}t^{\mathbf{A}}(x) = \top$, if $x \neq 1$, $t^{\mathbf{A}}(1) = 1$ and ${\sim}{-}t^{\mathbf{A}}(1) = \top$. □

On the other hand, we have the following result.

PROPOSITION 8.10. $\mathbf{M}(\mathbf{K})$ *is Glivenko involutive whenever* **K** *is a substructural logic that contains* $\mathbf{FL_{ew}}$*. Thus, the Glivenko property holds for every substructural logic over* $\mathbf{FL_{ew}}$*.*

PROOF. It follows from Theorem 8.7, and the fact that conjugates do not contribute anything in the commutative integral case, that $\mathbf{M}(\mathbf{K}) = \mathbf{FL_{ew}} + \{\varphi \mid {\sim}{\sim}\varphi \in \mathbf{K}\}$. We will show that $\mathbf{M}(\mathbf{K})$ is Glivenko involutive. If $\vdash_{\mathbf{M}(\mathbf{K})} {\sim}{\sim}\varphi$, then, by Theorem 2.14 and integrality, there exist φ_i $(i \in I)$ such that $\vdash_{\mathbf{K}} {\sim}{\sim}\varphi_i$ and $\vdash_{\mathbf{FL_{ew}}} (\prod \varphi_i) \to {\sim}{\sim}\varphi$. It follows from Lemma 3.50, commutativity and integrality that $\vdash_{\mathbf{FL_{ew}}} (\prod {\sim}{\sim}\varphi_i) \to {\sim}{\sim}\varphi$; alternatively, it follows from the fact that λ is a *nucleus* in the commutative case according to Lemma 3.35. Therefore, $\vdash_{\mathbf{K}} {\sim}{\sim}\varphi$, by (mp_ℓ) of Theorem 2.17; hence $\vdash_{\mathbf{M}(\mathbf{K})} \varphi$. □

8.3.2. The deductive Glivenko property. We say that the *left deductive Glivenko property* holds for **K** relative to **L**, if $\Sigma \vdash_{\mathbf{L}} \varphi$ iff $\Sigma \vdash_{\mathbf{K}} {\sim}{-}\varphi$, for all $\Sigma \cup \{\varphi\}$. The *right deductive Glivenko property* is defined as the opposite statement to the left deductive Glivenko property; the *deductive Glivenko property* is the conjunction of the two properties.

PROPOSITION 8.11. *If* **L** *and* **K** *are substructural logics and* $\Phi \cup \{\psi\}$ *are formulas, then the following are equivalent.*

(1) *The* left deductive Glivenko property *holds for* **K** *relative to* **L***.*
(2) $\Phi \vdash_{\mathbf{L}} \psi$ *iff* ${\sim}{-}\Phi \vdash_{\mathbf{K}} {\sim}{-}\psi$*, for all* $\Phi \cup \{\psi\}$*.*
(3) **K** *and* **L** *are Glivenko equivalent and* **L** *is left weakly involutive.*
(4) $\mathbf{L} = \mathbf{M}(\mathbf{K})$ *and* $\mathbf{M}(\mathbf{K})$ *is left weakly involutive.*

PROOF. (1) $\Rightarrow$ (2). Assume that (1) holds and let ${\sim}{-}\Phi \vdash_{\mathbf{K}} {\sim}{-}\psi$, for some $\Phi \cup \{\psi\}$. Since $\varphi \vdash_{\mathbf{K}} {\sim}{-}\varphi$ for all $\varphi \in \Phi$, we get $\Phi \vdash_{\mathbf{K}} {\sim}{-}\psi$, by the transitivity of $\vdash_{\mathbf{K}}$. By (1), we obtain $\Phi \vdash_{\mathbf{L}} \psi$. Conversely, let $\Phi \vdash_{\mathbf{L}} \psi$, for some $\Phi \cup \{\psi\}$. Taking $\{{\sim}{-}\chi\}$ for Σ and χ for φ in (1), we obtain ${\sim}{-}\chi \vdash_{\mathbf{L}} \chi$, for every χ. So, $\sim-\Phi \vdash_{\mathbf{L}} \varphi$, for all $\varphi \in \Phi$; hence $\sim-\Phi \vdash_{\mathbf{L}} \psi$, by transitivity. By (1), we get $\sim-\Phi \vdash_{\mathbf{K}} \sim-\psi$.

(2) $\Rightarrow$ (3). Recall that $\sim-\sim-\psi = \sim-\psi$; so, for $\Phi = \{\sim-\psi\}$, (2) yields $\sim-\psi \vdash_{\mathbf{L}} \psi$. Moreover, by substituting the empty set for Φ and $\sim\psi$ for ψ in (2), we obtain $\vdash_{\mathbf{L}} \sim\psi$ iff $\vdash_{\mathbf{K}} \sim\psi$, for all ψ, by Lemma 2.8(4). Consequently, **K** and **L** are Glivenko equivalent.

(3) $\Rightarrow$ (1). Since **K** and **L** are Glivenko equivalent, we have $\Phi \vdash_{\mathbf{K}} \sim-\psi$ iff $\Phi \vdash_{\mathbf{L}} \sim-\psi$. Moreover, since ψ and $\sim-\psi$ are mutually deducible in **L**, i.e. $\psi \vdash_{\mathbf{L}} \sim-\psi$ and $\sim-\psi \vdash_{\mathbf{L}} \psi$, we have that $\Phi \vdash_{\mathbf{L}} \sim-\psi$ is equivalent to $\Phi \vdash_{\mathbf{L}} \psi$.

Obviously, (4) implies (3). For the converse, if **L** is weakly involutive, then it is Glivenko involutive, so $\mathbf{L} = \mathbf{M}(\mathbf{K})$, by Proposition 8.8. □

It is easy to see that the opposite of Proposition 8.11 is valid, so we obtain a characterization for the deductive Glivenko property, as well.

8.3.3. The equational Glivenko property. In view of Theorem 2.29, it is clear that **K** has the left Glivenko property relative to **L** iff, for every term t,

$$\mathsf{V}(\mathbf{L}) \models 1 \leq t \text{ iff } \mathsf{V}(\mathbf{K}) \models 1 \leq -\sim t.$$

It is natural to consider the stronger property given by condition 5 of the following proposition.

LEMMA 8.12. *Let $\mathcal{W}$, $\mathcal{V}$ be subvarieties of* FL *and let s, t be terms over* FL. *Then, the following statements are equivalent.*

(1) $\mathcal{V} \models s = t$ *iff* $\mathcal{W} \models -\sim s = -\sim t$, *for all* s, t.
(2) $\mathcal{V} \models s = t$ *iff* $\mathcal{W} \models \sim s = \sim t$, *for all* s, t.
(3) $\mathcal{V} \models s \leq t$ *iff* $\mathcal{W} \models -\sim s \leq -\sim t$, *for all* s, t.
(4) $\mathcal{V} \models s \leq t$ *iff* $\mathcal{W} \models \sim s \geq \sim t$, *for all* s, t.
(5) $\mathcal{V} \models s \leq t$ *iff* $\mathcal{W} \models s \leq -\sim t$, *for all* s, t.

The opposite statements are pairwise equivalent, as well.

PROOF. The equivalences of (1) to (2) and of (3) to (4) follow from the fact that their right hand sides are equivalent, by Lemma 2.8. The same holds for the equivalence of (3) and (5), since by Lemma 2.8, $-\sim s \leq -\sim t$ iff $s \leq -\sim t$. Moreover, it is clear that (3) implies (1). To show the converse it is enough to show that (2) implies (3). We assume that (2) holds. The inequality $s \leq t$ is valid in $\mathcal{V}$ iff the equation $s \vee t = t$ is valid in $\mathcal{V}$. By (2) this is the case exactly when $\mathcal{W}$ satisfies the equation $\sim(s \vee t) = \sim t$; i.e. by Lemma 2.8(1), when $\mathcal{W}$ satisfies $\sim s \wedge \sim t = \sim t$. The last equation is in turn equivalent to $\sim t \leq \sim s$, which, by Lemma 2.8(2) and (4), is equivalent to $-\sim s \leq -\sim t$.

To show that (1) implies (3), assume that (1) holds. The inequality $s \leq t$ is valid in $\mathcal{V}$ iff the equation $s \vee t = t$ is valid in $\mathcal{V}$. By (1) this is the case exactly when $\mathcal{W}$ satisfies $-\sim(s \vee t) = -\sim t$. By Lemma 2.8(5), the last equation is equivalent to $\sim(s \vee t) = \sim t$ and, by Lemma 2.8(1), it is equivalent to $\sim s \wedge \sim t = \sim t$. The last equation is in turn equivalent to $\sim t \leq \sim s$, which, by Lemma 2.8(2) and (4), is equivalent to $-\sim s \leq -\sim t$. □

Let $\mathcal{W}$ and $\mathcal{V}$ be subvarieties of FL. We say that the *left (right) equational Glivenko property holds for $\mathcal{W}$ relative to $\mathcal{V}$*, if any of the statements (1)-(5) (the opposite statements of (1)-(5), respectively) of the previous lemma holds. If both the left and the right equational Glivenko property hold for $\mathcal{W}$ relative to $\mathcal{V}$, we say that the *equational Glivenko property holds for $\mathcal{W}$ relative to $\mathcal{V}$*. The definition of the equational Glivenko property for substructural logics refers to the corresponding definition for varieties. The following results have obvious analogues for substructural logics.

PROPOSITION 8.13. *Let $\mathcal{W}$ and $\mathcal{V}$ be subvarieties of FL. Then, the following statements are equivalent.*

(1) *The left equational Glivenko property holds for $\mathcal{W}$ relative to $\mathcal{V}$.*
(2) *$\mathcal{W}$ and $\mathcal{V}$ are Glivenko equivalent and $\mathcal{V}$ is left involutive.*
(3) *$\mathcal{V} = \mathbf{M}(\mathcal{W})$ and $\mathbf{M}(\mathcal{W})$ is left involutive.*

PROOF. Assume that the left equational Glivenko property holds for $\mathcal{W}$ relative to $\mathcal{V}$. By Lemma 2.8, $\mathcal{W}$ satisfies $-\sim-\sim x = -\sim x$, so $\mathcal{V}$ satisfies $-\sim x = x$; i.e., $\mathcal{V}$ is left involutive. Moreover, the variety $\mathcal{V}$ satisfies $-s = -t$ iff $\mathcal{W}$ satisfies $-\sim-s = -\sim-t$, by (1) of Lemma 8.12, iff $\mathcal{W}$ satisfies $-s = -t$, by Lemma 2.8(4). Thus, (2) of Lemma 8.3 holds; i.e., $\mathcal{V}$ and $\mathcal{W}$ are Glivenko equivalent.

Conversely, assume that $\mathcal{V}$ is left involutive and that statement (2) of Lemma 8.3 of holds; i.e. assume that

$$\mathcal{V} \models -s = -t \text{ iff } \mathcal{W} \models -s = -t.$$

We will show that (1) of Lemma 8.12 holds, as well. If $\mathcal{V}$ satisfies $s = t$, then it also satisfies $-\sim s = -\sim t$. Thus, $\mathcal{W}$ satisfies $-\sim s = -\sim t$, by (2) of Lemma 8.3. Conversely, if $\mathcal{W}$ satisfies $-\sim s = -\sim t$, then $\mathcal{V}$ satisfies $-\sim s = -\sim t$, by (2) Lemma 8.3. So, $\mathcal{V}$ satisfies $s = t$, since $\mathcal{V}$ is left involutive.

Obviously, (3) implies (2). Conversely, if $\mathcal{V}$ is left involutive, then it is left Glivenko involutive, so by the algebraic analogue of Proposition 8.8, we have $\mathcal{V} = \mathbf{M}(\mathcal{W})$. □

We summarize the previous results in the following corollary.

COROLLARY 8.14. *Let $\mathcal{W}$, $\mathcal{V}$ and $\mathcal{U}$ be subvarieties of FL and consider the three properties for $\mathcal{W}$ relative to $\mathcal{V}$ – the (left) Glivenko property, deductive Glivenko property and equational Glivenko property – and the corresponding notions of involutiveness for $\mathcal{V}$ – (left) Glivenko involutive, weakly involutive and involutive.*

(1) *A Glivenko property holds for $\mathcal{W}$ relative to $\mathcal{V}$ iff $\mathcal{V}$ possesses the corresponding type of involutiveness and $\mathcal{V} \subseteq \mathcal{W} \subseteq \mathbf{G}(\mathcal{V})$.*
(2) *If $\mathcal{V}$ and $\mathcal{W}$ are Glivenko equivalent, then the left version of a Glivenko property holds for $\mathcal{V}$ relative to $\mathcal{U}$ iff it holds for $\mathcal{W}$ relative to $\mathcal{U}$.*
(3) *If $\mathcal{V}$ possesses the left version of a type of involutiveness, then the corresponding right Glivenko property for $\mathcal{W}$ relative to $\mathcal{V}$ implies the corresponding left Glivenko property for $\mathcal{W}$ relative to $\mathcal{V}$.*
(4) *In particular, if $\mathcal{V}$ possesses both the left and right versions of a type of involutiveness, then the left and right versions of the corresponding Glivenko property for $\mathcal{W}$ relative to $\mathcal{V}$ are mutually equivalent.*

PROOF. All statements are clear, if one recalls that, by Proposition 8.8 and the algebraization result, if $\mathcal{V}$ is even (left) Glivenko involutive, then $\mathbf{M}(\mathcal{V}) = \mathcal{V}$. □

COROLLARY 8.15. *If $\mathcal{V}$ is a left involutive or right involutive subvariety of FL and there exists a variety $\mathcal{W}$ with a decidable equational theory, such that $\mathcal{V} \subseteq \mathcal{W} \subseteq \mathbf{G}(\mathcal{V})$, then $\mathcal{V}$ has a decidable equational theory, as well.*

We have shown that the equational Glivenko property implies the deductive Glivenko property; also, the later implies the Glivenko property. We will provide examples that show that the converse of these implications do not hold. We say that a property holds for a subvariety of FL or a substructural logic, if the property holds for the variety or the logic with respect to some variety or logic.

It follows from Proposition 8.8, Proposition 8.11 and Proposition 8.13 that a variety or logic has a certain type of involutiveness iff it satisfies the corresponding Glivenko property with respect to itself. We will make use of this remark in the proofs of the following two propositions.

PROPOSITION 8.16. *The variety $\mathsf{FL_e}$ (the logic $\mathbf{FL}_e$) is Glivenko involutive, but not weakly involutive. In other words, the Glivenko property holds for $\mathsf{FL_e}$ ($\mathbf{FL}_e$), but the deductive Glivenko property fails.*

PROOF. Note that $\sim\sim\varphi$ is provable in $\mathbf{FL}_e$ iff φ is provable in $\mathbf{FL}_e$. This follows from the cut elimination theorem for $\mathbf{FL}_e$. In detail, if $\Rightarrow (\varphi \to 0) \to 0$ is the last sequent in a proof in $\mathbf{FL}_e$, then the only possibility for the upper sequent of the last rule is $\varphi \to 0 \Rightarrow 0$. In turn, the only possibilities for the upper sequents of the next to the last rule are $\Rightarrow \varphi$ and $0 \Rightarrow 0$. Consequently, $\mathsf{FL_e}$ is Glivenko involutive. Therefore, the Glivenko property holds for $\mathsf{FL_e}$ ($\mathbf{FL}_e$), relative to itself, by Proposition 8.8.

On the other hand, if $\mathbf{FL}_e$ satisfies the deductive Glivenko property, then $\sim\sim p \vdash_{\mathbf{FL}_e} p$, where p is a propositional variable. By the local deduction theorem, we have that for some n, $\vdash_{\mathbf{FL}_e} (\sim\sim p \wedge 1)^n \to p$. Nevertheless, there is a commutative FL-algebra that does not satisfy the identity $(\sim\sim x \wedge 1)^n \leq x$, for any n. Indeed, consider the two-element residuated lattice on the set

$\{\bot, 1\}$, where $\bot < 1$, multiplication is idempotent and commutative, and 1 is the unit element. If we chose $0 = 1$, then we have $({\sim}{\sim}\bot \wedge 1)^n = (1 \wedge 1)^n = 1$, for all n. □

PROPOSITION 8.17. *The variety* $\mathsf{FL_e} \cap \mathrm{Mod}(({\sim}{\sim}x)^2 \leq x)$ *(the logic* $\mathbf{FL_e} + ({\sim}{\sim}p)^2 \to p$*) is weakly involutive, but not involutive. In other words, The deductive Glivenko property holds for it, but the equational Glivenko property fails.*

PROOF. It follows from the local deduction theorem that the variety $\mathcal{V} = \mathsf{FL_e} \cap \mathrm{Mod}(({\sim}{\sim}x)^2 \leq x)$ is weekly involutive. Consequently, $\mathcal{V} = \mathbf{M}(\mathcal{V})$ and the deductive Glivenko property holds for $\mathcal{V}$ relative to itself, by Proposition 8.11.

The equational Glivenko property holds for $\mathcal{V} = \mathbf{M}(\mathcal{V})$ iff it is involutive, by Proposition 8.13. The variety $\mathcal{V}$ is not involutive, since there exists a commutative FL-algebra that satisfies the identity $({\sim}{\sim}x)^2 \leq x$ but is not involutive. Indeed, consider the residuated lattice on the set $\{\bot, a, 1\}$, where $\bot < a < 1$, 1 is the unit, $\bot$ is an absorbing element and $a^2 = \bot$. If we chose $0 = a$, then ${\sim}{\sim}\bot = a$, so the algebra is not involutive. Nevertheless, $({\sim}{\sim}x)^2 = x$, for every $x \in \{\bot, a, 1\}$. □

8.3.4. An axiomatization for the Glivenko variety of an involutive variety. Next, given an equational basis of an involutive variety $\mathcal{V}$, we show how to obtain an explicit axiomatization of the Glivenko variety $\mathbf{G}(\mathcal{V})$ of $\mathcal{V}$. Recall that $\mathbf{G}(\mathcal{V})$ was defined on page 349. Equivalently, given an axiomatization of an involutive substructural logic $\mathbf{L}$, we give an explicit axiomatization of the Glivenko logic $\mathbf{G}(\mathbf{L})$ of $\mathbf{L}$.

The subvariety Gl of FL axiomatized by the equations

$$\text{(Gl)} \qquad {\sim}(x \star y) = {\sim}(-{\sim}x \star -{\sim}y)$$

where $\star \in \{\wedge, \cdot, \backslash, /\}$, is called the *left Glivenko variety*. Also, the subvariety Gr of FL axiomatized by the equations

$$\text{(Gr)} \qquad -(x \star y) = -({\sim}-x \star {\sim}-y),$$

where $\star \in \{\wedge, \cdot, \backslash, /\}$, is called the *right Glivenko variety*. The variety $\mathsf{G} = \mathsf{Gl} \cap \mathsf{Gr}$ is called the *Glivenko variety*. We will show, see Proposition 8.18(3), that the (left-, right-) Glivenko variety is the (left-, right-) Glivenko variety of the largest (left-, right-) involutive subvariety of FL; see the beginning of Section 8.4.2 for the definition. Note that, by Lemma 2.6(3) and Lemma 2.8(4), the equations (Gl) and (Gr) for $\star = \vee$ hold in all subvarieties of FL, thus we do not include them in the axiomatization of the left Glivenko and right Glivenko variety.

For every subvariety $\mathcal{V}$ of FL, and for every equational basis $B = \{s_i = t_i \mid i \in I\}$ of $\mathcal{V}$ relative to FL, let $\mathcal{V}^B$ ($\mathcal{V}_B$, respectively) be the subvariety of Gl (Gr, respectively) axiomatized by the equations ${\sim}s_i = {\sim}t_i$ ($-s_i = -t_i$, respectively) for $i \in I$. We will show that if $\mathcal{V}$ is a subvariety of $\mathcal{G}$, then $\mathcal{V}_B$ and $\mathcal{V}^B$ are equal to $\mathbf{G}(\mathcal{V})$. Thus, we obtain an explicit axiomatization of $\mathbf{G}(\mathcal{V})$ relative to FL.

For every FL-algebra $\mathbf{A}$, define the binary relations λ, ρ and θ, by $x \mathrel{\lambda} y$ iff ${\sim}x = {\sim}y$, $x \mathrel{\rho} y$ iff $-x = -y$, and $x \mathrel{\theta} y$ iff both $x \mathrel{\lambda} y$ and $x \mathrel{\rho} y$, for all $x, y \in A$. Obviously, λ, ρ and θ are equivalence relations on A.

Proposition 8.18. *Assume that $\mathcal{V}$ is a subvariety of* FL, *B an equational basis of $\mathcal{V}$ and $\mathbf{A}$ a FL-algebra.*

1. *The implications* (a) $\Rightarrow$ (b) $\Rightarrow$ (c) *hold for the following statements.*
   - (a) *$\mathbf{A}$ is in $\mathcal{V}^B$ ($\mathcal{V}_B$, $\mathcal{V}^B \cap \mathcal{V}_B$, respectively).*
   - (b) *λ (ρ, θ, respectively) is a congruence relation on $\mathbf{A}$ and $\mathbf{A}/\lambda$ ($\mathbf{A}/\rho$, $\mathbf{A}/\theta$, respectively) is in $\mathcal{V}$,*
   - (c) *$\mathbf{A}$ is in $\mathbf{G}(\mathcal{V})$.*

   Consequently, $\mathcal{V}_B \subseteq \mathbf{G}(\mathcal{V})$ and $\mathcal{V}^B \subseteq \mathbf{G}(\mathcal{V})$.
2. *If $\mathbf{A}$ is in $\mathbf{G}(\mathcal{V})$ and λ (ρ, θ, respectively) is a congruence relation on $\mathbf{A}$, then $\mathbf{A}/\lambda$ ($\mathbf{A}/\rho$, $\mathbf{A}/\theta$, respectively) is in $\mathcal{V}$.*
3. *If $\mathcal{V}$ is a subvariety of* Gl (Gr, G), *then the corresponding statements in* (a)-(c) *of (1) are equivalent. In particular, $\mathcal{V}^B = \mathbf{G}(\mathcal{V})$ ($\mathcal{V}_B = \mathbf{G}(\mathcal{V})$, $\mathcal{V}_B = \mathcal{V}^B = \mathbf{G}(\mathcal{V})$, respectively).*
4. *If $\mathcal{V}$ is a finitely axiomatized subvariety of* Gl *or of* Gr, *then so is $\mathbf{G}(\mathcal{V})$.*

Proof. For the first implication in (1), assume that $\mathbf{A}$ is in $\mathcal{V}^B$. By the definitions of the relation λ and the variety $\mathcal{V}^B$ and by (1) and (4) of Lemma 2.8, it is clear that λ is a congruence on $\mathbf{A}$. Consequently, $\mathbf{A}/\lambda$ is a residuated lattice. Note that $\mathbf{A}/\lambda$ satisfies all the equations in B, by the definition of the variety $\mathcal{V}^B$; hence $\mathbf{A}/\lambda \in \mathcal{V}$.

For the second implication, assume that λ is a congruence relation on $\mathbf{A}$ and $\mathbf{A}/\lambda$ is in $\mathcal{V}$. If the equation $s = t$ holds in $\mathcal{V}$, then it also holds in $\mathbf{A}/\lambda$; hence, the equation ${\sim}s = {\sim}t$ is valid in $\mathbf{A}$. Consequently, $\mathbf{A}$ is in $\mathbf{G}(\mathcal{V})$.

For (2), let $\mathbf{A}$ be in $\mathbf{G}(\mathcal{V})$ and let λ be a congruence on $\mathbf{A}$. If $s = t$ holds in $\mathcal{V}$, then ${\sim}s = {\sim}t$ holds in $\mathbf{G}(\mathcal{V})$, hence also in $\mathbf{A}$. So, $s = t$ holds in $\mathbf{A}/\lambda$. Thus, $\mathbf{A}/\lambda \in \mathcal{V}$.

In view of (1), it suffices to show the implication (c) $\Rightarrow$ (a) in order to establish (3). If $\mathcal{V}$ is a subvariety of Gl, then it satisfies the equations (Gl). Since, $\mathbf{G}(\mathcal{V})$ is Glivenko equivalent to $\mathcal{V}$, it satisfies the equations (Gl), as well. Consequently, $\mathbf{G}(\mathcal{V})$ satisfies all the equations in the axiomatization of $\mathcal{V}^B$; thus, $\mathbf{G}(\mathcal{V}) \subseteq \mathcal{V}^B$. Finally, the statement (4) follows from (3). ☐

We define the *left Glivenko logic* $\mathbf{Gl} = \mathbf{L}(\mathsf{Gl})$, the *right Glivenko logic* $\mathbf{Gr} = \mathbf{L}(\mathsf{Gr})$ and the *Glivenko logic* $\mathbf{G} = \mathbf{L}(\mathsf{G})$. We restate the main result in Proposition 8.18 in the terminology of logic.

COROLLARY 8.19. *If a logic* $\mathbf{L}$ *is an extension of* $\mathbf{Gl}$ *axiomatized by a set of formulas* Φ, *then* $\mathbf{G}(\mathbf{L})$ *is axiomatized by*

$$\{\sim-\varphi \mid \varphi \in \Phi\} \cup \{(\sim(\varphi \star \psi))/(\sim(-\sim\varphi \star -\sim\psi)) \mid \star \in \{\wedge, \cdot, \backslash, /\}\},$$

or by the opposite formulas.

PROOF. Note that $\mathsf{V}(\mathbf{L})$ is a subvariety of Gl axiomatized by $\{\varphi = \varphi \vee 1 \mid \varphi \in \Phi\}$. By Proposition 8.18, $\mathbf{G}(\mathsf{V}(\mathbf{L}))$ is axiomatized by $\{\sim\varphi = \sim(\varphi \vee 1) \mid \varphi \in \Phi\} \cup (\mathrm{Gl})$, or equivalently by $\{1 \leq -\sim\varphi \mid \varphi \in \Phi\} \cup (\mathrm{Gl})$. □

COROLLARY 8.20. *If a logic* $\mathbf{L}$ *is a finitely axiomatized extension of* $\mathbf{Gl}$, *then* $\mathbf{G}(\mathbf{L})$ *is also finitely axiomatized.*

For example, if $\mathbf{L}$ is left involutive and finitely axiomatized, we can give an explicit axiomatization of the smallest logic for which the Glivenko property holds relative to $\mathbf{L}$. We will give some interesting such examples in Section 8.5.

8.4. More on the equational Glivenko property

8.4.1. The deductive equational Glivenko property. In this section, we show that the deductive form of the equational Glivenko property is equivalent to the equational Glivenko property.

LEMMA 8.21. *Let* $\mathcal{W}$ *and* $\mathcal{V}$ *be subvarieties of* FL *and let* $E \cup \{s = t\}$ *be a set of equations in the language of* FL. *Then, the following statements are equivalent.*

(1) $E \models_{\mathcal{V}} s = t$ *iff* $E \models_{\mathcal{W}} -\sim s = -\sim t$.
(2) $E \models_{\mathcal{V}} s = t$ *iff* $E \models_{\mathcal{W}} \sim s = \sim t$.
(3) $E \models_{\mathcal{V}} s \leq t$ *iff* $E \models_{\mathcal{W}} -\sim s \leq -\sim t$.
(4) $E \models_{\mathcal{V}} s \leq t$ *iff* $E \models_{\mathcal{W}} \sim s \geq \sim t$.
(5) $E \models_{\mathcal{V}} s \leq t$ *iff* $E \models_{\mathcal{W}} s \leq -\sim t$.

The opposite statements are pairwise equivalent, as well.

PROOF. The proof is similar to the proof of Lemma 8.12, which is a special case for $E = \emptyset$, and is left to the reader. □

Let $\mathcal{W}$ and $\mathcal{V}$ be subvarieties of FL. We say that the *left (right) deductive equational Glivenko property holds for* $\mathcal{W}$ *relative to* $\mathcal{V}$, if, for every set $E \cup \{s = t\}$ of equations in the language of FL, the statements (the opposite of the statements, respectively) in the first set of Lemma 8.21 hold. We say that the *deductive equational Glivenko property holds for* $\mathcal{W}$ *relative to* $\mathcal{V}$, if both left and right deductive equational Glivenko properties hold.

Recall the FL-algebra terms (or term operations) $\lambda(x) = -\sim x$ and $\rho(x) = \sim-x$. Also, recall the binary relation λ defined in the previous section and note that, if $\mathbf{A}$ is a FL-algebra and $x, y \in A$, then $x \mathrel{\lambda} y$ iff $\lambda(x) = \lambda(y)$. In other words, we use the same symbol for the map and its kernel.

THEOREM 8.22. *Let $\mathcal{V}$ and $\mathcal{W}$ be subvarieties of* FL. *Then, the following statements are equivalent.*

(1) *The left equational Glivenko property holds for $\mathcal{W}$ relative to $\mathcal{V}$.*
(2) *The left deductive equational Glivenko property holds for $\mathcal{W}$ relative to $\mathcal{V}$.*
(3) $E \models_{\mathcal{V}} s = t$ *iff* $-{\sim}E \models_{\mathcal{W}} -{\sim}s = -{\sim}t$, *for all* E, s, t.

PROOF. By taking E to be the empty set in (3), we obtain the left equational Glivenko property; so (3) implies (1).

We will show that (2) implies (3). Suppose that (2) holds and assume that $E \models_{\mathcal{V}} s = t$. Then, $-{\sim}E \models_{\mathcal{V}} s = t$, since $\mathcal{V}$ is left involutive by Proposition 8.13. By (2), we have $-{\sim}E \models_{\mathcal{W}} -{\sim}s = -{\sim}t$. Conversely, assume that $-{\sim}E \models_{\mathcal{W}} -{\sim}s = -{\sim}t$. Since $E \models_{\mathcal{W}} -{\sim}u = -{\sim}v$, for all $(u = v) \in E$, we have $E \models_{\mathcal{W}} -{\sim}s = -{\sim}t$. By (2), we get $E \models_{\mathcal{W}} s = t$. Thus, (2) implies (3). We will prove that (1) implies (2).

We will first show that, for all $\mathbf{A} \in \mathcal{W}$, λ is a homomorphism from $\mathbf{A}$ onto $\mathbf{A}_\lambda$. By Proposition 8.13, Proposition 8.18(3) and Lemma 2.8, $\mathbf{G}(\mathcal{V})$ satisfies the equations ${\sim}(x \star y) = {\sim}(-{\sim}x \star -{\sim}y)$, hence also the equations $-{\sim}(x \star y) = -{\sim}(-{\sim}x \star -{\sim}y)$ where $\star \in \{\wedge, \cdot, \backslash, /\}$. Consequently, the latter set of equations holds in $\mathbf{A}$, since $\mathcal{W} \subseteq \mathbf{G}(\mathcal{V})$; the last inclusion follows from Lemma 8.5(4) and the fact that, by Proposition 8.13, the varieties $\mathcal{V}$ and $\mathcal{W}$ are Glivenko equivalent. For $\star = \cdot$, we have $\lambda(x) \cdot \lambda(y) \le \lambda(\lambda(x) \cdot \lambda(y)) = \lambda(xy)$, for all $x, y \in A$; so, in view of Lemma 3.35, λ is a nucleus from $\mathbf{A}$ to $\mathbf{A}_\lambda$. Thus,

$$\lambda(xy) = \lambda(x) \cdot_{\mathbf{A}_\lambda} \lambda(y),\ \lambda(1) = 1_{\mathbf{A}_\lambda},\ \lambda(0) = 0_{\mathbf{A}_\lambda} \text{ and}$$
$$\lambda(x \vee y) = \lambda(x) \vee_{\mathbf{A}_\lambda} \lambda(y),$$

for all $x, y \in A$, by Lemma 3.35. By the same lemma, $\mathbf{A}_\lambda$ is closed under the meet and division operations of $\mathbf{L}$. So, for $\star \in \{\wedge, \backslash, /\}$ we have

$$\lambda(x \star y) = \lambda(\lambda(x) \star \lambda(y)) = \lambda(\lambda(x) \star_{\mathbf{A}_\lambda} \lambda(y)) = \lambda(x) \star_{\mathbf{A}_\lambda} \lambda(y).$$

Now, assume that $E \models_{\mathcal{V}} s = t$, where $E = \{s_i(\bar{x}) = t_i(\bar{x}) \mid i \in I\}$. To show that $E \models_{\mathcal{W}} -{\sim}s = -{\sim}t$, let $\mathbf{A} \in \mathcal{W}$ and assume that, for all $i \in I$, $s_i^{\mathbf{A}}(\bar{a}) = t_i^{\mathbf{A}}(\bar{a})$, where $\bar{a}$ is an element of the appropriate power of $\mathbf{A}$. We will show that $\lambda(s^{\mathbf{A}}(\bar{a})) = \lambda(t^{\mathbf{A}}(\bar{a}))$. Since $\lambda(s_i^{\mathbf{A}}(\bar{a})) = \lambda(t_i^{\mathbf{A}}(\bar{a}))$, for all $i \in I$, and since λ is a homomorphism, we obtain $s_i^{\mathbf{A}_\lambda}(\lambda(\bar{a})) = t_i^{\mathbf{A}_\lambda}(\lambda(\bar{a}))$, where $\lambda(\bar{a})$ denotes the sequence consisting of the λ images of the terms in $\bar{a}$. Note that $\mathbf{A}_\lambda$ is in $\mathcal{W}$, since it is a homomorphic image of an algebra in $\mathcal{W}$; hence $\mathbf{A}_\lambda \in \mathbf{G}(\mathcal{V})$, since $\mathcal{W} \subseteq \mathbf{G}(\mathcal{V})$. If $u = v$ is an equation valid in $\mathcal{V}$, then the equation ${\sim}u = {\sim}v$ holds in $\mathbf{G}(\mathcal{V})$ and hence it holds in $\mathbf{A}_\lambda$. By Lemma 3.35, $\mathbf{A}_\lambda$ is left involutive, so the equation $s = t$ holds in $\mathbf{A}_\lambda$. Consequently, $\mathbf{A}_\lambda$ is in $\mathcal{V}$, hence it satisfies $s^{\mathbf{A}_\lambda}(\lambda(\bar{a})) = t^{\mathbf{A}_\lambda}(\lambda(\bar{a}))$. Thus, $\lambda(s^{\mathbf{A}}(\bar{a})) = \lambda(t^{\mathbf{A}}(\bar{a}))$, since λ is a homomorphism. Consequently, $E \models_{\mathcal{W}} -{\sim}s = -{\sim}t$. Conversely, suppose that $E \models_{\mathcal{W}} -{\sim}s = -{\sim}t$. By

Corollary 8.14, $\mathcal{V} \subseteq \mathcal{W}$, so $E \models_{\mathcal{V}} -{\sim}s = -{\sim}t$. Since $\mathcal{V}$ is left involutive, we have $E \models_{\mathcal{V}} s = t$. □

COROLLARY 8.23. *If $\mathcal{V}$ is an involutive subvariety of* FL *and the quasi-equational theory of a variety $\mathcal{W}$, where $\mathcal{V} \subseteq \mathcal{W} \subseteq \mathbf{G}(\mathcal{V})$, is decidable, then the quasi-equational theory of $\mathcal{V}$ is decidable, as well.*

8.4.2. An alternative characterization for the equational Glivenko property. We have obtained a characterization of the Glivenko properties in terms of the type of involutiveness that the minimal variety of the Glivenko equivalence class has to possess. Here we describe varieties whose subvarieties are exactly varieties for which the equational Glivenko property holds.

Let lIFL (respectively, IrFL) be the variety of left- (right-) involutive FL-algebras, i.e. the subvariety of FL axiomatized by the equation $-{\sim}x = x$ (respectively, ${\sim}{-}x = x$). Also, let $\mathsf{InFL} = \mathsf{lIFL} \cap \mathsf{IrFL}$. By Proposition 8.18(3), it follows that $\mathsf{Gl} = \mathbf{G}(\mathsf{lIFL})$, $\mathsf{Gr} = \mathbf{G}(\mathsf{IrFL})$ and $\mathsf{G} = \mathbf{G}(\mathsf{InFL})$.

Note that, by Proposition 8.18, $\mathbf{A} \in \mathsf{Gl}$ ($\mathbf{A} \in \mathsf{Gr}$, $\mathbf{A} \in \mathsf{G}$) iff λ (ρ, θ, respectively) is a congruence on $\mathbf{A}$ and $\mathbf{A}/\lambda$ is left involutive ($\mathbf{A}/\rho$ is right involutive, $\mathbf{A}/\theta$ is involutive, respectively).

For every subvariety $\mathcal{V}$ of FL, set $\mathbf{Il}(\mathcal{V}) = \mathsf{lIFL} \cap \mathcal{V}$, $\mathbf{Ir}(\mathcal{V}) = \mathsf{IrFL} \cap \mathcal{V}$ and $\mathbf{In}(\mathcal{V}) = \mathsf{InFL} \cap \mathcal{V}$ – the largest left involutive, right involutive and involutive subvariety of $\mathcal{V}$, respectively. Note that $\mathbf{Il}$, $\mathbf{Ir}$ and $\mathbf{In}$ are interior operators on $\mathbf{\Lambda}(\mathsf{FL})$. A notion related to $\mathbf{In}(\mathcal{V})$ is also discussed in [CT03].

LEMMA 8.24. *Let $\mathcal{V}$ and $\mathcal{W}$ be subvarieties of* FL.

(1) *If $\mathcal{V}$ and $\mathcal{W}$ are Glivenko equivalent, then $\mathbf{Il}(\mathcal{V}) = \mathbf{Il}(\mathcal{W})$, $\mathbf{Ir}(\mathcal{V}) = \mathbf{Ir}(\mathcal{W})$, and $\mathbf{In}(\mathcal{V}) = \mathbf{In}(\mathcal{W})$.*
(2) $\mathbf{Il}(\mathcal{V}) = \mathbf{Il}(\mathbf{G}(\mathcal{V}))$, $\mathbf{Ir}(\mathcal{V}) = \mathbf{Ir}(\mathbf{G}(\mathcal{V}))$, *and* $\mathbf{In}(\mathcal{V}) = \mathbf{In}(\mathbf{G}(\mathcal{V}))$.
(3) *If $\mathcal{V}$ is left involutive (right involutive, involutive), then $\mathbf{Il}(\mathbf{G}(\mathcal{V})) = \mathcal{V}$ ($\mathbf{Ir}(\mathbf{G}(\mathcal{V})) = \mathcal{V}$, $\mathbf{In}(\mathbf{G}(\mathcal{V})) = \mathcal{V}$, respectively).*
(4) *The varieties $\mathbf{Il}(\mathcal{V})$, $\mathbf{Ir}(\mathcal{V})$ and $\mathbf{In}(\mathcal{V})$ are subvarieties of $\mathbf{M}(\mathcal{V})$.*
(5) $\mathbf{G}(\mathbf{Il}(\mathcal{V})) = \mathsf{Gl} \cap \mathbf{G}(\mathcal{V})$, $\mathbf{G}(\mathbf{Ir}(\mathcal{V})) = \mathsf{Gr} \cap \mathbf{G}(\mathcal{V})$, *and* $\mathbf{G}(\mathbf{In}(\mathcal{V})) = \mathsf{G} \cap \mathbf{G}(\mathcal{V})$.

PROOF. For (1), assume that $\mathcal{V}$ and $\mathcal{W}$ are Glivenko equivalent. If $\mathcal{V}$ satisfies $s = t$, then it also satisfies ${\sim}s = {\sim}t$. So, $\mathcal{W}$ satisfies ${\sim}s = {\sim}t$, hence it satisfies $-{\sim}s = -{\sim}t$. Consequently, $\mathbf{Il}(\mathcal{W})$ satisfies $s = t$. Thus, $\mathbf{Il}(\mathcal{W}) \subseteq \mathcal{V}$, so $\mathbf{Il}(\mathcal{W}) \subseteq \mathbf{Il}(\mathcal{V})$. Likewise, we show the other inclusion, so $\mathbf{Il}(\mathcal{W}) = \mathbf{Il}(\mathcal{I})$.

Statement (2) follows from (1) and Lemma 8.5(2). Statement (3) is a direct consequence of (2). For statement (4), note that since $\mathcal{V}$ and $\mathbf{M}(\mathcal{V})$ are Glivenko equivalent, using (1) we obtain $\mathbf{Il}(\mathcal{V}) = \mathbf{Il}(\mathbf{M}(\mathcal{V})) \subseteq \mathbf{M}(\mathcal{V})$.

To show (5), note that if a FL-algebra $\mathbf{A}$ is in $\mathsf{Gl} \cap \mathbf{G}(\mathcal{V}) = \mathbf{G}(\mathsf{lIFL}) \cap \mathbf{G}(\mathcal{V})$, then, by taking lIFL for $\mathcal{V}$ in Proposition 8.18(3), we have that

λ is a congruence relation on $\mathbf{A}$ and $\mathbf{A}/\lambda$ is in IlFL. So by applying Proposition 8.18(2) to $\mathcal{V}$, we get that $\mathbf{A}/\lambda$ is in $\mathcal{V}$. Therefore, $\mathbf{A}/\lambda$ is in $\mathsf{IlFL} \cap \mathcal{V} = \mathbf{Il}(\mathcal{V})$. By Proposition 8.18(1) we get $\mathbf{A}$ is in $\mathbf{G}(\mathbf{Il}(\mathcal{V}))$.

Conversely, if $\mathbf{A}$ is in $\mathbf{G}(\mathbf{Il}(\mathcal{V}))$, then λ is a congruence relation on $\mathbf{A}$ and $\mathbf{A}/\lambda$ is in $\mathbf{Il}(\mathcal{V}) = \mathsf{IlFL} \cap \mathcal{V}$, by Proposition 8.18(3). By applying Proposition 8.18(1) to $\mathcal{V}$ and to IlFL, we have $\mathbf{A} \in \mathbf{G}(\mathcal{V})$ and $\mathbf{A} \in \mathbf{G}(\mathsf{IlFL})$. So, $\mathbf{A} \in \mathbf{G}(\mathcal{V}) \cap \mathbf{G}(\mathsf{IlFL}) = \mathbf{G}(\mathbf{Il}(\mathcal{V}))$. □

The following theorem shows that the equational Glivenko property holds for a subvariety $\mathcal{V}$ of FL iff $\mathcal{V}$ is contained in the Glivenko variety.

THEOREM 8.25. *The following are equivalent.*

1. (1) $\mathcal{V}$ *is a subvariety of* Gl *(*Gr, G*).*
2. (2) $\mathbf{G}(\mathcal{V})$ *is a subvariety of* Gl *(*Gr, G, *respectively).*
3. (3) $\mathbf{M}(\mathcal{V})$ *is a subvariety of* Gl *(*Gr, G, *respectively).*
4. (4) $\mathbf{M}(\mathcal{V})$ *is equal to* $\mathbf{Il}(\mathcal{V})$ *(*$\mathbf{Ir}(\mathcal{V})$, $\mathbf{In}(\mathcal{V})$, *respectively).*
5. (5) *The left equational Glivenko property holds for* $\mathcal{V}$ *relative to* $\mathbf{Il}(\mathcal{V})$ *(*$\mathbf{Ir}(\mathcal{V})$, $\mathbf{In}(\mathcal{V})$, *respectively).*
6. (6) *The left equational Glivenko property holds for* $\mathcal{V}$ *relative to some variety.*
7. (7) $\mathbf{M}(\mathcal{V})$ *is a subvariety of* IlFL *(*IrFL, InFL, *respectively).*

PROOF. We will establish the implications (1) ⇒ (2) ⇒ (3) ⇒ (4) ⇒ (5) ⇒ (6) ⇒ (1), and the equivalence (7) ⇔ (4). The implications (2) ⇒ (3) and (5) ⇒ (6) are clear and the equivalence (7) ⇔ (4) follows from Lemma 8.24(4) and the fact that $\mathbf{M}(\mathcal{V}) \subseteq \mathcal{V}$.

(1) ⇒ (2): If $\mathcal{V} \subseteq \mathsf{Gl} = \mathbf{G}(\mathsf{IlFL})$, then $\mathbf{G}(\mathcal{V}) \subseteq \mathbf{G}(\mathbf{G}(\mathsf{IlFL})) = \mathsf{Gl}$.

(3) ⇒ (4): Since $\mathbf{M}(\mathcal{V}) \subseteq \mathbf{G}(\mathcal{V})$, using Lemma 8.24(5) we obtain $\mathbf{M}(\mathcal{V}) \subseteq \mathbf{G}(\mathbf{Il}(\mathcal{V}))$. Moreover, $\mathbf{Il}(\mathcal{V}) \subseteq \mathbf{M}(\mathcal{V})$ by Lemma 8.24(4) so the left equational Glivenko property holds for $\mathbf{M}(\mathcal{V})$ relative to $\mathbf{Il}(\mathcal{V})$ by Corollary 8.14(1). Consequently, by Proposition 8.13, $\mathbf{M}(\mathcal{V})$ and $\mathbf{Il}(\mathcal{V})$ are Glivenko equivalent. Since $\mathbf{M}(\mathcal{V})$ is Glivenko equivalent to $\mathcal{V}$, and $\mathcal{V}$ and $\mathbf{Il}(\mathcal{V})$ are Glivenko equivalent, we obtain $\mathbf{M}(\mathcal{V}) = \mathbf{Il}(\mathcal{V})$ by Proposition 8.13.

(4) ⇒ (5): Since $\mathbf{Il}(\mathcal{V}) = \mathbf{M}(\mathcal{V})$ and $\mathbf{Il}(\mathcal{V})$ is involutive, (5) follows from Proposition 8.13.

(6) ⇒ (1): If the left equational Glivenko property holds for $\mathcal{V}$ relative to some variety $\mathcal{U}$, then $\mathcal{U} \subseteq \mathcal{V} \subseteq \mathbf{G}(\mathcal{U})$ and $\mathcal{U}$ is left involutive, by Lemma 8.14(1). So, $\mathcal{U} = \mathbf{Il}(\mathcal{U})$ and $\mathcal{V} \subseteq \mathbf{G}(\mathbf{Il}(\mathcal{U}))$. Since $\mathbf{Il}(\mathcal{U}) \subseteq \mathsf{IlFL}$, we have $\mathbf{G}(\mathbf{Il}(\mathcal{U})) \subseteq \mathbf{G}(\mathsf{IlFL}) = \mathsf{Gl}$, by Lemma 8.5(1b). Thus, $\mathcal{V} \subseteq \mathsf{Gl}$. □

The equivalence of statements (2) and (3) of the preceding theorem implies that the Glivenko equivalence class $[\mathbf{M}(\mathcal{V}), \mathbf{G}(\mathcal{V})]$ of $\mathcal{V}$ is either contained in the principal order ideal of $\mathbf{\Lambda}(\mathsf{FL})$ generated by Gl (Gr, G, respectively), or it is completely disjoint from it. In the first case and only then the least variety $\mathbf{M}(\mathcal{V})$ (equivalently some variety) of the interval is contained in IlFL (IrFL, InFL, respectively), or equivalently it is equal to $\mathbf{Il}(\mathcal{V})$ ($\mathbf{Ir}(\mathcal{V})$,

$\mathbf{In}(\mathcal{V})$ respectively). In the second case, the equational Glivenko property fails to hold for every variety in the interval.

COROLLARY 8.26. $\mathbf{G}(\mathbf{Il}(\mathcal{V}))$ *is the largest subvariety of* $\mathbf{G}(\mathcal{V})$ *for which the left equational Glivenko property holds relative to some variety.*

PROOF. By Theorem 8.25, the left equational Glivenko property holds for a subvariety $\mathcal{W}$ of $\mathbf{G}(\mathcal{V})$ relative to some variety iff $\mathcal{W} \subseteq \mathbf{G}(\mathcal{V})$ and $\mathcal{W} \subseteq \mathsf{Gl}$. By Lemma 8.24(5), this is equivalent to $\mathcal{W} \subseteq \mathbf{G}(\mathbf{Il}(\mathcal{V}))$. □

8.5. Special cases

In this section we discuss some special cases for which the Glivenko properties holds and describe how Theorems 8.1 and 8.2 follow from our results.

As we have seen the left Glivenko (right Glivenko, Glivenko) variety Gl (Gr, G, respectively) is axiomatized by the equations $\sim(x \star y) = \sim(-\sim x \star -\sim y)$ ($-(x \star y) = -(\sim -x \star \sim -y)$, the combination of both sets of equations, respectively), where $\star \in \{\wedge, \cdot, \backslash, /\}$. Given an axiomatization of a left involutive (right involutive, involutive) variety $\mathcal{V}$, an axiomatization of the Glivenko variety $\mathbf{G}(\mathcal{V})$ of $\mathcal{V}$ – the largest subvariety of FL for which the left Glivenko (right Glivenko, Glivenko, respectively) property holds relative to $\mathcal{V}$ – consists of the axiomatization of Gl (Gr, G, respectively) mentioned above plus the left (right, left and right, respectively) negations of the equations in the axiomatization of $\mathcal{V}$.

Moreover, for every subvariety $\mathcal{W}$ of FL, an axiomatization of the variety $\mathbf{G}_{\mathcal{W}}(\mathcal{V}) = \mathbf{G}(\mathcal{V}) \cap \mathcal{W}$ – the largest subvariety of $\mathcal{W}$ for which the left Glivenko (right Glivenko, Glivenko) property holds relative to the left involutive (right involutive, involutive, respectively) variety $\mathcal{V}$ – is obtained by combining an axiomatization of $\mathbf{G}(\mathcal{V})$ with one of $\mathcal{W}$. Below we give a number of varieties $\mathcal{W}$ for which the axiomatization of $\mathbf{G}_{\mathcal{W}}(\mathcal{V})$ relative to $\mathcal{W}$ can be simplified. In particular, we obtain simpler axiomatizations for $\mathsf{Gl}_{\mathcal{W}} = \mathsf{Gl} \cap \mathcal{W}$, $\mathsf{Gr}_{\mathcal{W}} = \mathsf{Gr} \cap \mathcal{W}$ and $\mathsf{G}_{\mathcal{W}} = \mathsf{G} \cap \mathcal{W}$ relative to $\mathcal{W}$. Moreover, we study the cases where $\mathcal{V}$ is the variety of Boolean algebras. In this way we show how known results on the Glivenko theorem can be derived from our result.

8.5.1. The cyclic case. Let $\mathsf{CyFL} = \mathrm{Mod}(\sim x = -x)$ be the *cyclic* subvariety of FL. Before we proceed, we point out that the equation for one of the division operations in (Gl) can be simplified.

LEMMA 8.27. *In every cyclic FL-algebra the equations*

$$1 \leq \sim\sim(\sim\sim y \backslash y) \text{ and } \sim(x \backslash y) = \sim(\sim\sim x \backslash \sim\sim y)$$

are equivalent. The same holds for the equations

$$1 \leq \sim\sim(y / \sim\sim y) \text{ and } \sim(y/x) = \sim(\sim\sim y / \sim\sim x).$$

PROOF. Assume that $\sim(\sim\sim x\backslash\sim\sim y) = \sim(x\backslash y)$ holds. Then,

$$\sim\sim(\sim\sim x\backslash\sim\sim y) = \sim\sim(x\backslash y)$$

holds as well. By setting $x = \sim\sim y$ and using (3), (2) and (4) of Lemma 2.8, we obtain

$$1 \leq \sim\sim 1 \leq \sim\sim(\sim\sim y\backslash\sim\sim y) = \sim\sim(\sim\sim\sim\sim y\backslash\sim\sim y) = \sim\sim(\sim\sim y\backslash y).$$

For the converse, note that $(x\backslash\sim\sim y)(\sim\sim y\backslash y) \leq x\backslash y \leq \sim\sim(x\backslash y)$, by Lemma 2.6(4,5) and by Lemma 2.8(4); so $\sim\sim y\backslash y \leq (x\backslash\sim\sim y)\backslash\sim\sim(x\backslash y)$, hence

$$\sim\sim(\sim\sim y\backslash y) \leq \sim\sim[(x\backslash\sim\sim y)\backslash\sim\sim(x\backslash y)].$$

Moreover, by Lemma 2.8(12),

$$\sim\sim[(x\backslash\sim\sim y)\backslash\sim\sim(x\backslash y)] = (x\backslash\sim\sim y)\backslash\sim\sim(x\backslash y)$$

and, by hypothesis, $1 \leq \sim\sim(\sim\sim y\backslash y)$, so we have $1 \leq (x\backslash\sim\sim y)\backslash\sim\sim(x\backslash y)$, that is $x\backslash\sim\sim y \leq \sim\sim(x\backslash y)$. Now, since $\sim\sim x\backslash\sim\sim y \leq x\backslash\sim\sim y$, we have $\sim\sim x\backslash\sim\sim y \leq \sim\sim(x\backslash y)$, hence $\sim(\sim\sim x\backslash\sim\sim y) \geq \sim(x\backslash y)$. The converse inequality follows from (11) of Lemma 2.8. □

Note that $\mathsf{G}_{\mathsf{CyFL}} = \mathsf{Gl}_{\mathsf{CyFL}} = \mathsf{Gr}_{\mathsf{CyFL}}$

COROLLARY 8.28. *The variety* $\mathsf{G}_{\mathsf{CyFL}}$ *is axiomatized relative to* CyFL *by the conjunction of the following equations*

(1) $1 \leq \sim\sim(y/\sim\sim y)$,
(2) $1 \leq \sim\sim(\sim\sim y\backslash y)$,
(3) $\sim(x \wedge y) = \sim(\sim\sim x \wedge \sim\sim y)$.

PROOF. Recall that $\mathsf{G}_{\mathsf{CyFL}}$ is axiomatized by the equations (Gl) in Section 8.3.4; see page 358. We will show that the identity for multiplication is redundant. For every x, y in an algebra in the variety $\mathsf{G}_{\mathsf{CyFL}}$, we have

$$\begin{aligned} \sim(xy) &= y\backslash\sim x && \text{(Lemma 2.8(8))} \\ &= y\backslash -x && (\sim x = -x) \\ &= -\sim y\backslash -\sim -x && \text{(Lemma 2.8(4,10))} \\ &= \sim\sim y\backslash\sim\sim\sim x && (\sim x = -x) \\ &= \sim(\sim\sim x \cdot \sim\sim y) && \text{(Lemma 2.8(8))} \end{aligned}$$

By Lemma 8.27, the identities for the division operations follow from (1) and (2). □

Recall that by $\mathsf{FL_e}$ we denote the variety of commutative FL-algebras.

COROLLARY 8.29. $\mathsf{G}_{\mathsf{FL_e}}$ *is axiomatized relative to* $\mathsf{FL_e}$ *by the equations* $1 \leq \sim\sim(\sim\sim y \to y)$ *and* $\sim(x \wedge y) = \sim(\sim\sim x \wedge \sim\sim y)$.

Let IFL denote the variety of integral FL-algebras and set ICyFL $=$ CyFL $\cap$ IFL.

COROLLARY 8.30. *The variety* $\mathsf{G}_{\mathsf{ICyFL}}$ *is axiomatized relative to* ICyFL *by the equations* $1 \leq \sim\sim(y/\sim\sim y)$ *and* $1 \leq \sim\sim(\sim\sim y\backslash y)$.

PROOF. We will show that the equation for the meet operation in Corollary 8.28 is redundant. Using integrality we have,

$$(x/\sim\sim x)(\sim\sim x \wedge \sim\sim y)(\sim\sim y\backslash y) \leq (x/\sim\sim x)\sim\sim x \leq x.$$

Similarly,

$$(x/\sim\sim x)(\sim\sim x \wedge \sim\sim y)(\sim\sim y\backslash y) \leq y,$$

so

$$(x/\sim\sim x)(\sim\sim x \wedge \sim\sim y)(\sim\sim y\backslash y) \leq x \wedge y \leq \sim\sim(x \wedge y).$$

By applying Lemma 2.8(13) twice, we obtain

$$\sim\sim(x/\sim\sim x)(\sim\sim x \wedge \sim\sim y)\sim\sim(\sim\sim y\backslash y) \leq \sim\sim(x \wedge y).$$

Since $1 \leq \sim\sim(x/\sim\sim x)$ and $1 \leq \sim\sim(\sim\sim y\backslash y)$, we have

$$\sim\sim x \wedge \sim\sim y \leq \sim\sim(x \wedge y).$$

By (2) and (4) of Lemma 2.8, we obtain

$$\sim(x \wedge y) \leq \sim(\sim\sim x \wedge \sim\sim y).$$

The converse inequality follows from (3) and (2) of Lemma 2.8. □

The following corollary can also be obtained from Theorem 5.1 of [CT04].

COROLLARY 8.31. $\mathsf{G}_{\mathsf{FL_{ei}}}$ *is axiomatized relative to* $\mathsf{FL_{ei}}$ *by the equation* $1 \leq \sim\sim(\sim\sim y \to y)$.

As a consequence we obtain Theorem 8.2(3).

In [CT04] Glivenko's Theorem 8.1 is generalized to logics containing BCK-logic. In algebraic terminology and in our notation it is shown that if $\mathcal{W}$ is a subquasivariety of a *natural expansion* of the quasivariety of bounded BCK-algebras that satisfies the equation $1 \leq \sim\sim(\sim\sim y \to y)$, then the Glivenko property holds for $\mathcal{W}$ relative to $\mathbf{In}(\mathcal{W})$. This result extends the original theorem to expansions of quasivarieties, but is limited to the integral, commutative case, where the negation constant 0 is the least element. The result presented here has exactly the opposite attributes and it extends to the stronger equational Glivenko property. Both results include the extension of Glivenko's theorem for commutative, integral, bounded FL-algebras where 0 is the least element, given by Corollary 8.31.

8.5.2. The classical case. Recall that a *Brouwerian algebra* is (term equivalent to) a residuated lattice that satisfies the law $xy = x \wedge y$ and a *Heyting algebra* is (term equivalent to) a FL-algebra, whose residuated lattice reduct is a Brouwerian algebra and 0 is its least element.

LEMMA 8.32. *An FL-algebra is (term equivalent to) a Boolean algebra iff it satisfies the equations $xy = x \wedge y$ and $x\backslash y = {\sim}x \vee y$.*

PROOF. Setting $x = 1$ into the second equation, we have $0 \leq y$. So, in view of the first equation the FL-algebra is a Heyting algebra; hence it has a distributive lattice reduct. To show that it is a Boolean algebra, it suffices to show that every element has a complement. For every x, we have $x \wedge {\sim}x = x(x\backslash 0) \leq 0$, so $x \wedge {\sim}x = 0$. Also, by the second equation ${\sim}x \vee x = x\backslash x = 1$, since every Heyting algebra is integral. □

In the case when the variety relative to which a Glivenko property holds is the variety BA of Boolean algebras, we obtain a simpler axiomatization.

COROLLARY 8.33. $\mathbf{G}(\mathsf{BA})$ *is axiomatized by the following equations.*

(1) ${\sim}(x \wedge y) = {\sim}(xy)$
(2) ${\sim}(x\backslash y) = {\sim}(-x \vee y)$
(3) $-(x\backslash y) = -({\sim}x \vee y)$
(4) ${\sim}(x\backslash y) = {\sim}(-{\sim}x\backslash -{\sim}y)$
(5) ${\sim}(x/y) = {\sim}(-{\sim}x/-{\sim}y)$

PROOF. Given the axiomatization of BA in Lemma 8.32, an axiomatization of $\mathbf{G}(\mathsf{BA})$ consists of the equations (1), (2), (4) and (5) plus the equations ${\sim}(x \cdot y) = {\sim}(-{\sim}x \cdot -{\sim}y)$ and ${\sim}(x \wedge y) = {\sim}(-{\sim}x \wedge -{\sim}y)$. We will verify that these two equations follow from the proposed list. We have

$$\begin{aligned} {\sim}(-{\sim}x \wedge -{\sim}y) &= {\sim}-({\sim}x \vee {\sim}y) && \text{(Lemma 2.8(1))} \\ &= {\sim}-(x\backslash {\sim}y) && (3) \\ &= {\sim}-{\sim}(yx) && \text{(Lemma 2.8(8))} \\ &= {\sim}(yx) && \text{(Lemma 2.8(4))} \\ &= {\sim}(x \wedge y) && (1) \end{aligned}$$

Consequently, we have ${\sim}(xy) = {\sim}(x \wedge y) = {\sim}(-{\sim}x \wedge -{\sim}y) = {\sim}(-{\sim}x \cdot -{\sim}y)$, as well. □

COROLLARY 8.34. $\mathbf{G}_{\mathsf{CyFL}}(\mathsf{BA})$ *is axiomatized relative to* CyFL *by the following equations.*

(1) ${\sim}(x \wedge y) = {\sim}(xy)$
(2) ${\sim}(y/x) = {\sim}({\sim}x \vee y)$
(3) ${\sim}(x\backslash y) = {\sim}({\sim}x \vee y)$

Alternatively, (2) *and* (3) *can be replaced respectively by*

(4) $1 \leq {\sim}{\sim}({\sim}{\sim}x\backslash x)$
(5) $1 \leq {\sim}{\sim}(x/{\sim}{\sim}x)$.

PROOF. For the first axiomatization, in view of Corollary 8.33, it suffices to show that the equations for the division operations are redundant. We have

$$\begin{aligned}
\sim(y/x) &= \sim(\sim x \vee y) && (2)\\
&= \sim\sim x \wedge \sim y && \text{(Lemma 2.8(1))}\\
&= \sim\sim\sim\sim x \wedge \sim\sim\sim y && \text{(Lemma 2.8(4))}\\
&= \sim(\sim\sim\sim x \vee \sim\sim y) && \text{(Lemma 2.8(1))}\\
&= \sim(\sim\sim y/\sim\sim x) && (2)
\end{aligned}$$

Likewise, $\sim(x\backslash y) = \sim(\sim\sim x\backslash\sim\sim y)$. Therefore, (1) and (2) form an axiomatization for $\mathbf{G}_{\mathsf{CyFL}}(\mathsf{BA})$ relative to CyFL.

Finally, we will show that (2) and (4) are equivalent; the equivalence of (3) and (5) follows in a similar way. If (2) holds, we have

$$\begin{aligned}
\sim\sim(\sim\sim x\backslash x) &= \sim\sim(\sim\sim\sim x \vee x) && (2)\\
&= \sim\sim(\sim x \vee x) && \text{(Lemma 2.8(4))}\\
&= \sim(\sim\sim x \wedge \sim x) && \text{(Lemma 2.8(1))}\\
&= \sim(\sim\sim x \cdot \sim x) && (1)\\
&\geq \sim 0 = 1
\end{aligned}$$

Conversely, assume that (4) holds. We have, $(x\backslash\sim\sim y)(\sim\sim y\backslash y) \leq x\backslash y$, by Lemma 2.6(13) and $x\backslash y \leq \sim\sim(x\backslash y)$, by Lemma 2.8(3), so

$$(x\backslash\sim\sim y)\cdot(\sim\sim y\backslash y)\cdot\sim(x\backslash y) \leq 0.$$

By Lemma 2.9(3), we obtain $\sim\sim(x\backslash\sim\sim y)\cdot\sim\sim(\sim\sim y\backslash y)\cdot\sim\sim\sim(x\backslash y) \leq 0$, i.e. $\sim\sim(x\backslash\sim\sim y)\cdot\sim\sim(\sim\sim y\backslash y) \leq \sim\sim(x\backslash y)$. Using (4) and the fact that $\sim\sim x\backslash\sim\sim y \leq x\backslash\sim\sim y$, we obtain

$$\sim\sim(\sim\sim x\backslash\sim\sim y) \leq \sim\sim(x\backslash y),$$

thus $\sim(x\backslash y) \leq \sim(\sim\sim x\backslash\sim\sim y)$.

On the other hand, $x(x\backslash y) \leq y \leq \sim\sim y$, so $\sim\sim x\cdot\sim\sim(x\backslash y) \leq \sim\sim y$, by Lemma 2.9(4). Consequently, we have $\sim\sim(x\backslash y) \leq \sim\sim x\backslash\sim\sim y$ and $\sim(\sim\sim x\backslash\sim\sim y) \leq \sim(x\backslash y)$. Therefore, $\sim(\sim\sim x\backslash\sim\sim y) = \sim(x\backslash y)$

Additionally, we have

$$\begin{aligned}
\sim(\sim x \vee y) &= \sim\sim\sim(\sim x \vee y) && \text{(Lemma 2.8(4))}\\
&= \sim\sim(\sim\sim x \wedge \sim y) && \text{(Lemma 2.8(1))}\\
&= \sim\sim(\sim y \wedge \sim\sim x)\\
&= \sim\sim(\sim y \cdot \sim\sim x) && (1)\\
&= \sim(\sim\sim x\backslash\sim\sim y) && \text{(Lemma 2.8(8)).}
\end{aligned}$$

Consequently, $\sim(x\backslash y) = \sim(\sim x \vee y)$. □

COROLLARY 8.35. *The variety* $\mathbf{G}_{\mathsf{ICyFL}}(\mathsf{BA})$ *is axiomatized relative to* ICyFL *by the equations:*

(1) $\sim(x \wedge y) = \sim(xy)$
(2) $1 \leq \sim\sim(\sim\sim x\backslash x)$
(3) $1 \leq \sim\sim(x/\sim\sim x)$.

Alternatively, (1) *can be replaced by either one of the equations*

(4) $x \wedge {\sim}x \leq 0$

(5) ${\sim}(x^2) = {\sim}x$.

PROOF. We will show that (1), (4) and (5) are all equivalent. First assume that (1) holds. We have $x \cdot {\sim}x \leq 0$, so $1 \leq (x \cdot {\sim}x)\backslash 0 = {\sim}(x \cdot {\sim}x) = {\sim}(x \wedge {\sim}x)$; hence $x \wedge {\sim}x \leq 0$.

Assume, now, that (4) holds. We have $x \cdot (x \backslash {\sim}x) \leq x$, by integrality and $x \cdot (x \backslash {\sim}x) \leq {\sim}x$. So, $x \cdot (x \backslash {\sim}x) \leq x \wedge {\sim}x \leq 0$. Therefore, $x \backslash {\sim}x \leq {\sim}x$, so ${\sim}(x^2) = x \backslash (x \backslash 0) \leq {\sim}x$. On the other hand, $x^2 \leq x$, by integrality, so ${\sim}x \leq {\sim}(x^2)$.

Finally, if (5) holds, then we have ${\sim}(x \wedge y) \leq {\sim}(xy)$, by integrality, and ${\sim}(xy) \leq {\sim}[(x \wedge y)(x \wedge y)] = {\sim}(x \wedge y)^2 = {\sim}(x \wedge y)$, by (4). □

COROLLARY 8.36. $\mathbf{G}_{\mathsf{FL_e}}(\mathsf{BA})$ *is axiomatized relative to* $\mathsf{FL_e}$ *by the equations:*

(1) ${\sim}(x \wedge y) = {\sim}(xy)$

(2) ${\sim}(x \to y) = {\sim}({\sim}x \vee y)$.

Alternatively, (2) *can be replaced by*

(3) $1 \leq {\sim}{\sim}({\sim}{\sim}x \to x)$

COROLLARY 8.37. *The variety* $\mathbf{G}_{\mathsf{FL_{ei}}}(\mathsf{BA})$ *is axiomatized relative to* $\mathsf{FL_{ei}}$ *by the equations:*

(1) ${\sim}(x \wedge y) = {\sim}(xy)$

(2) $1 \leq {\sim}{\sim}({\sim}{\sim}x \to x)$.

Alternatively, (1) *can be replaced by either one of the equations*

(3) $x \wedge {\sim}x \leq 0$

(4) ${\sim}(x^2) = {\sim}x$.

The algebraic version of Theorem 8.1 follows from the following corollary.

COROLLARY 8.38. *The (equational) Glivenko property holds for* HA *relative to* BA.

PROOF. Note that HA satisfies ${\sim}{\sim}x \cdot x \leq x$ by integrality, so $x \leq {\sim}{\sim}x \to x$ and ${\sim}({\sim}{\sim}x \to x) \leq {\sim}x = {\sim}{\sim}{\sim}x$. Consequently, ${\sim}{\sim}x \cdot {\sim}({\sim}{\sim}x \to x) \leq 0 \leq x$, so ${\sim}({\sim}{\sim}x \to x) \leq {\sim}{\sim}x \to x$, hence ${\sim}({\sim}{\sim}x \to x) \wedge {\sim}({\sim}{\sim}x \to x) \leq 0$. Therefore, ${\sim}({\sim}{\sim}x \to x) \leq 0$; thus $1 \leq {\sim}{\sim}({\sim}{\sim}x \to x)$.

By Corollary 8.31, $\mathsf{HA} \subseteq \mathsf{G}$, so the equational Glivenko property holds for HA relative to $\mathbf{In}(\mathsf{HA}) = \mathsf{BA}$, by Theorem 8.25. □

For every bounded residuated lattice $\mathbf{A}$, consider the FL-algebra $\mathbf{A}'$, that is obtained by appending to $\mathbf{A}$ a new bottom element $\bot$ and setting $0 = \bot$. That $\mathbf{A}'$ is an FL-algebra can be easily verified; actually $\mathbf{A}'_r = \mathbf{2}[\mathbf{A}]$. Note that ${\sim}a = -a = 0$, for all $a \in A$, and ${\sim}0 = -0 = 1$. Using Corollary 8.33, it is easy to see that $\mathbf{A}' \in \mathbf{G}(\mathsf{BA})$. As an example we verify

Corollary 8.33(2), for $x = 0$ and $y \in A$; we have $\sim(0\backslash y) = \sim\top = 0$ and $\sim(-0 \vee y) = \sim(\top \vee y) = \sim\top = 0$. Therefore, the variety $\mathbf{G}(\mathsf{BA})$ is neither integral, nor commutative, nor contractive.

8.5.3. The basic logic case. Recall that a BL-algebra is an integral, commutative FL-algebra that satisfies the equations

$$0 \le x,\ x(x \to y) = x \wedge y \text{ and } (x \to y) \vee (y \to x) = 1.$$

Also, an MV-algebra is an integral, commutative FL-algebra that satisfies the equations $0 \le x$ and $(x \to y) \to y = (y \to x) \to x$. Recall from Section 3.4.6 that a *(pointed) generalized BL-algebra* or *(pointed) GBL-algebra* is a (pointed) residuated lattice that satisfies the equation

$$\text{(GBL)} \qquad y(y\backslash x \wedge 1) = x \wedge y = (1 \wedge x/y)y$$

and from Section 3.4.5 that a *(pointed) generalized MV-algebra* or *(pointed) GMV-algebra* is a (pointed) residuated lattice that satisfies

$$\text{(GMV)} \qquad x/(y\backslash x \wedge 1) = x \vee y = (1 \wedge x/y)\backslash x.$$

The above equations are equivalent to their homonymous ones given in Sections 3.4.5 and 3.4.6, so we will use the same name for them; the proof of the equivalence follows the ideas of by Exercises 33 and 34 of Chapter 3. In the same chapter it was mentioned that commutative, representable (as a subdirect product of totally ordered algebras), bounded, pointed GBL-algebras in which 0 is the least element are term equivalent to BL-algebras. Similarly, MV-algebras are just commutative, bounded, pointed GMV-algebras in which 0 is the least element. In both cases, integrality follows from the fact that the algebras are bounded. We denote the varieties of BL-algebras, pointed GBL-algebras, MV-algebras and pointed GMV-algebras by BL, GBL^0, MV, and GMV^0, respectively. Recall that (pointed) GMV-algebras are (pointed) GBL-algebras. Under the assumption of involutiveness the converse is true as well.

LEMMA 8.39. *Involutive pointed GBL-algebras are pointed GMV-algebras.*

PROOF. For every x, y in an involutive pointed GBL-algebra, we have

$$\begin{aligned}
x \vee y &= \sim-(x \vee y) && (x = \sim-x)\\
&= \sim((-x) \wedge (-y)) && \text{(Lemma 2.8(1))}\\
&= \sim[(-x)((-x)\backslash(-y) \wedge 1)] && \text{(GBL)}\\
&= \sim[(-x)((\sim-x)/y \wedge 1)] && \text{(Lemma 2.8(9))}\\
&= \sim[-x(x/y \wedge 1)] && (x = \sim-x)\\
&= (x/y \wedge 1)\backslash\sim-x && \text{(Lemma 2.8(8))}\\
&= (x/y \wedge 1)\backslash x && (x = \sim-x).
\end{aligned}$$

Likewise we obtain the opposite equation. □

It is observed in [CT03] that BL satisfies the equation $1 \le \sim\sim(\sim\sim y \to y)$, and that MV coincides with the variety $\mathbf{In}(\mathsf{BL})$, so, by Theorem 8.25 and

Corollary 8.31, Glivenko's theorem holds for BL relative to MV. We will obtain a generalization by dropping the assumption of representability and by replacing the commutativity assumption by cyclicity. We first establish the following non-commutative generalization of a property observed in [CT03]; the proof is essentially the same.

LEMMA 8.40. *Every cyclic pointed GBL-algebra in which 0 is the least element satisfies the equations* $1 = {\sim}{\sim}(x/{\sim}{\sim}x)$ *and* $1 = {\sim}{\sim}({\sim}{\sim}x\backslash x)$.

PROOF. First note that every pointed bounded GBL-algebra in which 0 is the least element is integral; see [BCG+03] or [GT05] for details. We have

$${\sim}x \cdot {\sim}{\sim}x = {\sim}x({\sim}x\backslash 0) \leq 0 \leq x,$$

so ${\sim}x \leq x/{\sim}{\sim}x$, hence ${\sim}(x/{\sim}{\sim}x) \leq {\sim}{\sim}x$. Consequently, we have

$$\begin{aligned}
{\sim}(x/{\sim}{\sim}x) &= {\sim}{\sim}x \wedge {\sim}(x/{\sim}{\sim}x) \\
&= ({\sim}{\sim}x)(({\sim}{\sim}x)\backslash{\sim}(x/{\sim}{\sim}x)) && \text{(GBL)} \\
&= ({\sim}{\sim}x) \cdot {\sim}((x/{\sim}{\sim}x)({\sim}{\sim}x)) && \text{(Lemma 2.8(8))} \\
&= ({\sim}{\sim}x) \cdot {\sim}({\sim}{\sim}x \wedge x) && \text{(GBL)} \\
&= ({\sim}{\sim}x) \cdot {\sim}x \leq 0 && \text{(Lemma 2.8(3))}
\end{aligned}$$

Thus, ${\sim}{\sim}(x/{\sim}{\sim}x) = 1$. Similarly we prove the other equation. □

It follows from Corollary 8.30 and Lemma 8.40 that the variety $\mathsf{CyGBL}^{\perp}$ of cyclic pointed bounded GBL-algebras where 0 is the least element is contained in the Glivenko variety G. Moreover, $\mathbf{In}(\mathsf{CyGBL}^{\perp})$ is equal to the variety $\mathsf{CyGMV}^{\perp}$ of cyclic pointed bounded GMV-algebras where 0 is the least element, by Lemma 8.39 and the fact that pointed GMV-algebras are pointed GBL-algebras. Thus, in view of Theorem 8.25, we have the following corollary, which is implies Theorem 8.2(2).

COROLLARY 8.41. *The (equational) Glivenko property holds for* $\mathsf{CyGBL}^{\perp}$ *relative to* $\mathsf{CyGMV}^{\perp}$. *Consequently, the (equational) Glivenko property holds for* BL *relative to* MV, *as well.*

Following [CT03], a *SBL-algebra* is a BL-algebra that satisfies $x \wedge {\sim}x = 0$. The last equation can be replaced by either of the equations $x(x^2 \to {\sim}y) \leq {\sim}y$ and ${\sim}(x^2) = {\sim}x$. We denote the variety of all SBL-algebras by SBL. In [CT03] it is shown that the Glivenko property holds for SBL relative to BA, a fact that also follows from Lemma 8.40 and Corollary 8.37. We generalize this result by dropping representability and replacing commutativity with cyclicity. The following corollary is a consequence of Corollary 8.35.

COROLLARY 8.42. *The (equational) Glivenko property holds for* $\mathsf{CyGBL}^{\perp} \cap \mathrm{Mod}({\sim}(x^2) = {\sim}x)$ *relative to* BA. *Consequently, the (equational) Glivenko property holds for* SBL *relative to* BA.

Note that Theorem 8.2(1) follows from the preceding corollary.

8.6. Generalized Kolmogorov translation

Propositional intuitionistic logic can be interpreted in propositional classical logic via the Glivenko double negation translation as well as via the Kolmogorov negative translation. Having studied generalizations of the former property, we now discuss the latter.

Let γ be a unary residuated lattice or FL-algebra term. For a residuated lattice (or FL-algebra) term t, its γ-Kolmogorov translation $K_\gamma(t)$ is defined inductively on the complexity of t as follows: $K_\gamma(1) = \gamma(1)$, $K_\gamma(0) = \gamma(0)$, $K_\gamma(x) = \gamma(x)$ for every variable x, and $K_\gamma(s \star r) = \gamma(K_\gamma(s) \star K_\gamma(r))$, where $\star \in \{\wedge, \vee, \cdot, \backslash, /\}$. Note that the standard Kolmogorov negative translation discussed by A. N. Kolmogorov in 1925 is obtained for $\gamma(x) = {\sim}{\sim}x$ (see [Kol61]).

For every variety $\mathcal{V}$ of residuated lattices or FL-algebras, let $\mathcal{V}_\gamma$ be the subvariety of $\mathcal{V}$ axiomatized relative to $\mathcal{V}$ by the equation $\gamma(x) = x$. We say that the *γ-Kolmogorov translation holds for $\mathcal{W}$ relative to $\mathcal{V}$*, if for every set of equations $E \cup \{s = t\}$ in the language of residuated lattices or FL-algebras, $E \models_{\mathcal{V}} s = t$ iff $K_\gamma[E] \models_{\mathcal{W}} K_\gamma(s) = K_\gamma(t)$, where $K_\gamma[E] = \{K_\gamma(u) = K_\gamma(v) \mid (u = v) \in E\}$. Also, if $\mathbf{K}$ and $\mathbf{L}$ are substructural logics, we say that the *γ-Kolmogorov translation holds for* $\mathbf{K}$ *relative to* $\mathbf{L}$, if for every set of formulas $\Phi \cup \{\psi\}$ in the language of (pointed) residuated lattices, $\Phi \models_{\mathbf{K}} \psi$ iff $K_\gamma[\Phi] \models_{\mathbf{L}} K_\gamma(\psi)$, where $K_\gamma[\Phi] = \{K_\gamma(\varphi) \mid \varphi \in \Phi\}$. Corollary 8.48 shows the connection between these two definitions.

Note that if $\mathcal{W}$ is a subvariety of FL and γ is a unary (pointed) residuated lattice term, then $\gamma^{\mathbf{A}}$ is a nucleus on $\mathbf{A}$ for all $\mathbf{A} \in \mathcal{W}$ iff $\mathcal{W}$ satisfies the equations

$$\text{(nuc)} \qquad x \leq \gamma(x), \gamma(x) \leq \gamma(x \vee y), \gamma(\gamma(x)) = \gamma(x), \gamma(x)\gamma(y) \leq \gamma(xy)$$

THEOREM 8.43. *Let $\mathcal{V}$ be a variety of residuated lattices or FL-algebras, and γ a term that contains only the connectives $\wedge, \backslash, /$, and also the constant 0 only if $\gamma(0) = 0$ holds in $\mathcal{V}$. Moreover assume that for every algebra $\mathbf{A}$ in $\mathcal{V}$, $\gamma^{\mathbf{A}}$ is a nucleus on $\mathbf{A}$; equivalently, assume that $\mathcal{V}$ satisfies the equations (nuc). Then, the γ-Kolmogorov translation holds for $\mathcal{V}$ relative to $\mathcal{V}_\gamma$.*

PROOF. Using induction on the length of t, we can show that if t is a (pointed) residuated lattice term, $\mathbf{A} \in \mathcal{V}$ and $\bar{a}$ is an element of an appropriate power of A, then

$$(*) \qquad K_\gamma(t)^{\mathbf{A}}(\bar{a}) = t^{\mathbf{A}_\gamma}(\gamma^{\mathbf{A}}(\bar{a})),$$

where we have abbreviated $\mathbf{A}_{\gamma^{\mathbf{A}}}$ to $\mathbf{A}_\gamma$; see Lemma 3.35 for the definition of $\mathbf{A}_{\gamma^{\mathbf{A}}}$. To see this note that $K_\gamma(t)^{\mathbf{A}}(\bar{a})$ is just the application on $\gamma^{\mathbf{A}}(\bar{a})$ of the term function that corresponds to the term t, where every application of an operation is followed by γ; on the other hand $t^{\mathbf{A}_\gamma}(\gamma^{\mathbf{A}}(\bar{a}))$ is the the application on $\gamma^{\mathbf{A}}(\bar{a})$ of the term function that corresponds to the term t, where every operation is computed in $\mathbf{A}_\gamma$. The operations $\cdot, \vee$ and 1, when computed in $\mathbf{A}_\gamma$ are, by definition, equal to the corresponding operations on $\mathbf{A}$ followed by γ. The same holds for the other operations trivially, because the result of those operations on elements of $\mathbf{A}_\gamma$ is already an element of $\mathbf{A}_\gamma$, so the application or not of γ does not make any difference. For example, for the term $t = x \cdot y$, we have $K_\gamma(t)^{\mathbf{A}}(a, b) = \gamma(\gamma(a) \cdot^{\mathbf{A}} \gamma(b))$ and $t^{\mathbf{A}_\gamma}(\gamma(a), \gamma(b)) = \gamma(a) \cdot_\gamma \gamma(b) = \gamma(\gamma(a) \cdot^{\mathbf{A}} \gamma(b))$.

Recall that γ contains only the connectives $\wedge, \backslash, /$, and also the constant 0 only if $\gamma(0) = 0$ holds in $\mathcal{V}$, so $\gamma^{\mathbf{A}_\gamma}(a) = \gamma^{\mathbf{A}}(a)$, for every element a of $\mathbf{A}_\gamma$. Moreover, $\gamma^{\mathbf{A}}(a) = a$, since γ is a nucleus. Hence $\gamma^{\mathbf{A}_\gamma}(a) = a$, for all $a \in \mathbf{A}_\gamma$, and $\mathbf{A}_\gamma \in \mathcal{V}_\gamma$.

We will show that the γ-Kolmogorov translation holds for $\mathcal{V}$ relative to $\mathcal{V}_\gamma$. First suppose that $E \models_{\mathcal{V}_\gamma} s = t$; we will show that $K_\gamma(E) \models_{\mathcal{V}} K_\gamma(s) = K_\gamma(t)$. Let $\mathbf{A}$ be in $\mathcal{V}$ and $\bar{a}$ be an element of an appropriate power of A, such that $K_\gamma(u)^{\mathbf{A}}(\bar{a}) = K_\gamma(v)^{\mathbf{A}}(\bar{a})$, for all $(u = v) \in E$. Then, $u^{\mathbf{A}_\gamma}(\gamma^{\mathbf{A}}(\bar{a})) = v^{\mathbf{A}_\gamma}(\gamma^{\mathbf{A}}(\bar{a}))$ by (*), and $\mathbf{A}_\gamma \in \mathcal{V}_\gamma$. So, by assumption, $s^{\mathbf{A}_\gamma}(\gamma^{\mathbf{A}}(\bar{a})) = t^{\mathbf{A}_\gamma}(\gamma^{\mathbf{A}}(\bar{a}))$, hence $K_\gamma(s)^{\mathbf{A}}(\bar{a}) = K_\gamma(t)^{\mathbf{A}}(\bar{a})$.

Conversely, if $E \not\models_{\mathcal{V}_\gamma} s = t$, then there exists an algebra $\mathbf{B} \in \mathcal{V}_\gamma \subseteq \mathcal{V}$ and a sequence $\bar{b}$ of elements of B such that $u^{\mathbf{B}}(\bar{b}) = v^{\mathbf{B}}(\bar{b})$ for all $(u = v) \in E$, but $s^{\mathbf{B}}(\bar{b}) \neq t^{\mathbf{B}}(\bar{b})$. Since $\mathbf{B}$ satisfies $\gamma(x) = x$, we have $\mathbf{B} = \mathbf{B}_\gamma$. For every $(u = v) \in E$, we have $K_\gamma(u)^{\mathbf{B}}(\bar{b}) = u^{\mathbf{B}_\gamma}(\gamma^{\mathbf{B}}(\bar{b})) = u^{\mathbf{B}}(\bar{b}) = v^{\mathbf{B}}(\bar{b}) = K_\gamma(v)^{\mathbf{B}}(\bar{b})$ and $K_\gamma(s)^{\mathbf{B}}(\bar{b}) = s^{\mathbf{B}}(\bar{b}) \neq t^{\mathbf{B}}(\bar{b}) = K_\gamma(t)^{\mathbf{B}}(\bar{b})$. Consequently, $K_\gamma(E) \not\models_{\mathcal{V}} K_\gamma(s) = K_\gamma(t)$. □

Recall the definition of λ from Section 8.4. As shown in Lemma 3.35, if we assume cyclicity, then λ is a nucleus.

COROLLARY 8.44. *If $\mathcal{V}$ is cyclic, then the λ-Kolmogorov translation holds for $\mathcal{V}$ relative to $\mathcal{V}_\lambda$.*

The λ-Kolmogorov translation is simply called the *Kolmogorov translation* in the literature.

The following observation shows one of the differences between the Kolmogorov translation and the Glivenko property.

THEOREM 8.45. *Assume that $\mathcal{W}$ is a subvariety of* FL, *that γ is a unary residuated lattice (or Kolmogorov translation) term that contains only the connectives $\wedge, \backslash, /$, and also the constant 0 only if $\gamma(0) = 0$ holds in $\mathcal{W}$, and that $\mathcal{W}$ satisfies the equations (nuc). Then the γ-Kolmogorov translation holds for $\mathcal{W}$ relative to $\mathcal{V}$ iff $\mathcal{W}_\gamma = \mathcal{V}$.*

PROOF. One direction follows from Theorem 8.43. For the forward direction, suppose that the γ-Kolmogorov translation holds for $\mathcal{W}$ relative to $\mathcal{V}$. Then, for all sets of equations $E \cup \{s = t\}$, $E \models_{\mathcal{V}} s = t$ iff $K_\gamma(E) \models_{\mathcal{W}} K_\gamma(s) = K_\gamma(t)$. On the other hand, by Theorem 8.43, we have $E \models_{\mathcal{W}_\gamma} s = t$ iff $K_\gamma(E) \models_{\mathcal{W}} K_\gamma(s) = K_\gamma(t)$. Thus, $E \models_{\mathcal{V}} s = t$ iff $E \models_{\mathcal{W}_\gamma} s = t$; hence $\mathcal{V} = \mathcal{W}_\gamma$. □

COROLLARY 8.46. *The variety* $\mathsf{InFL_{ew}}$ *is the only subvariety of* FL *relative to which the* λ*-Kolmogorov translation holds for* $\mathsf{FL_{ew}}$.

THEOREM 8.47. *Assume that* $\mathcal{V}$ *and* $\mathcal{W}$ *are two subvarieties of* FL *and* γ *is a unary residuated lattice (or FL-algebra) term such that* $\mathcal{W}$ *satisfies the equations (nuc). The following are equivalent.*

(1) *For every set of equations* $E \cup \{s = t\}$ *in the language of residuated lattices (or FL-algebras),*
$$E \models_{\mathcal{V}} s = t \text{ iff } K_\gamma[E] \models_{\mathcal{W}} K_\gamma(s) = K_\gamma(t).$$
(2) *For every set of formulas* $\Phi \cup \{\psi\}$ *in the language of residuated lattices (or FL-algebras),*
$$\{1 \leq \varphi \mid \varphi \in \Phi\} \models_{\mathcal{V}} 1 \leq \psi \text{ iff } \{1 \leq K_\gamma(\varphi) \mid \varphi \in \Phi\} \models_{\mathcal{W}} 1 \leq K_\gamma(\varphi).$$

PROOF. We first show that for every FL-algebra term s, the variety $\mathcal{W}$ satisfies

$$(*) \qquad 1 \leq K_\gamma(s) \Leftrightarrow K_\gamma(1) = K_\gamma(s \wedge 1)$$

and

$$(**) \qquad 1 \leq K_\gamma(s\backslash t) \Leftrightarrow K_\gamma(s) \leq K_\gamma(t).$$

For (*), if $1 \leq K_\gamma(s)$ then $K_\gamma(1) = \gamma(1) \leq \gamma(K_\gamma(s)) = K_\gamma(s)$, so $K_\gamma(1) = K_\gamma(s) \wedge K_\gamma(1)$, hence $K_\gamma(1) = \gamma(K_\gamma(1)) = \gamma(K_\gamma(s) \wedge K_\gamma(1)) = K_\gamma(s \wedge 1)$. Conversely, if $K_\gamma(1) = K_\gamma(s \wedge 1)$, then $1 \leq \gamma(1) = K_\gamma(1) = K_\gamma(s \wedge 1) = \gamma(K_\gamma(s) \wedge K_\gamma(1)) \leq \gamma(K_\gamma(s)) = K_\gamma(s)$.

For (**), we have $K_\gamma(s\backslash t) = \gamma(K_\gamma(s)\backslash K_\gamma(t)) \leq \gamma(K_\gamma(s))\backslash\gamma(K_\gamma(t)) = K_\gamma(s)\backslash K_\gamma(t)$, since $\mathcal{W}$ satisfies the equations (nuc), so if $1 \leq K_\gamma(s\backslash t)$ then $1 \leq K_\gamma(s)\backslash K_\gamma(t)$; hence $K_\gamma(s) \leq K_\gamma(t)$. Conversely, if $K_\gamma(s) \leq K_\gamma(t)$, then $1 \leq K_\gamma(s)\backslash K_\gamma(t)$; hence $1 \leq \gamma(1) \leq \gamma(K_\gamma(s)\backslash K_\gamma(t)) = K_\gamma(s\backslash t)$, by the definition of K_γ.

Assume that (1) holds. Note that $\{1 \leq \varphi \mid \varphi \in \Phi\} \models_{\mathcal{V}} 1 \leq \psi$ is equivalent to $\{1 = 1 \wedge \varphi \mid \varphi \in \Phi\} \models_{\mathcal{V}} 1 = 1 \wedge \psi$, and, by (1), to $\{K_\gamma(1) = K_\gamma(1 \wedge \varphi) \mid \varphi \in \Phi\} \models_{\mathcal{W}} K_\gamma(1) = K_\gamma(1 \wedge \psi)$. By (*), this is equivalent to $\{1 \leq K_\gamma(\varphi) \mid \varphi \in \Phi\} \models_{\mathcal{W}} 1 \leq K_\gamma(\varphi)$.

Now, assume that (2) holds. We have $E \models_{\mathcal{V}} s = t$ iff $\{1 \leq u\backslash v, 1 \leq v\backslash u | (u = v) \in E\} \models_{\mathcal{V}} \{1 \leq s\backslash t, 1 \leq t\backslash s\}$ iff $\{1 \leq K_\gamma(u\backslash v), 1 \leq K_\gamma(v\backslash u) | (u = v) \in E\} \models_{\mathcal{W}} \{1 \leq K_\gamma(s\backslash t), 1 \leq K_\gamma(t\backslash s)\}$, by (2), iff $\{K_\gamma(u) \leq K_\gamma(v), 1 \leq K_\gamma(v) \leq K_\gamma(u) \mid (u = v) \in E\} \models_{\mathcal{W}} \{K_\gamma(s) \leq K_\gamma(t), K_\gamma(t) \leq K_\gamma(s)\}$, by (**), iff $\{K_\gamma(u) = K_\gamma(v) \mid (u = v) \in E\} \models_{\mathcal{W}} K_\gamma(s) = K_\gamma(t)$. □

The following corollary is a direct consequence of Theorem 2.29 and of Theorem 8.47.

COROLLARY 8.48. *Let* $\mathbf{K}$ *and* $\mathbf{L}$ *be substructural logics. The* γ*-Kolmogorov translation holds for* $\mathbf{K}$ *relative to* $\mathbf{L}$ *iff it holds for* $\mathsf{V}(\mathbf{K})$ *relative to* $\mathsf{V}(\mathbf{L})$.

Exercises

(1) Show that for every sequent $\Gamma \Rightarrow \Delta$ of **LK**, $\Gamma \Rightarrow \Delta$ is provable in **LK** iff $\Gamma, \neg\Delta \Rightarrow$ is provable in **LJ**. Here, $\neg\Delta$ denotes $\neg\beta_1, \ldots, \neg\beta_m$ when Δ is $\beta_1, \ldots, \beta_m$. (Hint. Use the induction on the length of a given proof of $\Gamma \Rightarrow \Delta$ in **LK**.)

(2) Derive Glivenko's theorem (Theorem 8.1) from the above.

(3) Show that an FL_{ew}-algebra satisfies the equation $x \wedge \neg x = 0$ iff it satisfies $\neg x^2 \leq \neg x$.

(4) Let $\mathbf{L}_1$ be the sequent calculus obtained from $\mathbf{FL_{ew}}$ by adding all the initial sequents of the form $\neg\alpha^2 \Rightarrow \neg\alpha$, and $\mathbf{L}_2$ be the sequent calculus obtained from $\mathbf{FL_{ew}}$ by adding the following restricted form of contraction rule:

$$\frac{\alpha, \alpha, \Gamma \Rightarrow}{\alpha, \Gamma \Rightarrow}$$

Show that a sequent is provable in $\mathbf{L}_1$ iff it is provable in $\mathbf{L}_2$.

(5) Show that for every sequent $\Gamma \Rightarrow \Delta$ of **LK**, $\Gamma \Rightarrow \Delta$ is provable in **LK** iff $\Gamma, \neg\Delta \Rightarrow$ is provable in $\mathbf{L}_2$.

(6) By using the above two, show that the Glivenko property holds for $\mathbf{L}_1$ relative to classical logic.

(7) Let K be the standard Kolmogorov translation, i.e. for each formula α of classical logic, $K(\alpha) = K_\gamma(\alpha)$ with $\gamma(x) = \neg\neg x$. Show that for every sequent $\Gamma \Rightarrow \Delta$ of **LK**, $\Gamma \Rightarrow \Delta$ is provable in **LK** iff $K(\Gamma), \neg K(\Delta) \Rightarrow$ is provable in **LJ**. Here, $K(\Gamma)$ denotes $K(\beta_1), \ldots, K(\beta_m)$ when Γ is $\beta_1, \ldots, \beta_m$.

(8) Show that for each formula α, α is provable in **LK** iff $K(\alpha)$ is provable in **LJ**.

Notes

(1) In this Chapter we discussed two translations between logics: the Glivenko and the Kolmogorov translations. In Section 9.6 will will also mention briefly another simple translation, dealing with the addition or removal of the bottom element of an FL_o-algebra, and its implications on the structure of the subvariety lattice of FL. Moreover, various other translations, including ones involving modal substructural logics (not defined in this book), are currently under investigation. We hope that this chapter will serve as a starting point and as inspiration for the study of other translations between logics (or varieties) and will shed even more light into the structure of the lattices of subvarieties and of substructural logics.

CHAPTER 9

Lattices of logics and varieties

In this chapter we discuss substructural logics and varieties of FL-algebras not in themselves, but as elements of the lattices they form. As it has already been mentioned, see Theorem 2.28(2), the lattice $\mathbf{\Lambda}(\mathbf{FL})$ of substructural logics of **FL** is dually isomorphic to the lattice $\mathbf{\Lambda}(\mathsf{FL})$ of subvarieties of FL, therefore we will study only the latter. Recall that RL is a subvariety of FL satisfying $0 = 1$ (this is our somewhat artificial way of making 0 disappear from the type). Since the structure of the whole $\mathbf{\Lambda}(\mathsf{FL})$ is likely to be rather complex, it is reasonable to look at very small subvarieties of RL first. This is known in universal algebra as "description of the bottom of the lattice of subvarieties". Clearly, the rule of thumb here is: the fewer small subvarieties the better chances for a good description. Such a description naturally begins at the level of atoms, and then—if possible—proceeds to covers of atoms, and further on.

The lattice $\mathbf{\Lambda}(\mathsf{FL})$ is complete, its greatest element being FL and least element the trivial variety consisting of all one-element FL-algebras. Note that the trivial variety is axiomatized by the identity $x = y$. We begin by looking at the atoms—also called *minimal varieties*—of $\mathbf{\Lambda}(\mathsf{FL})$, in fact, even of $\mathbf{\Lambda}(\mathsf{RL})$. It is worth mentioning in this context that logics corresponding to minimal varieties are characterized by the property that their proper extensions are all inconsistent. Such logics are known as *maximally consistent* or *Post complete*.

Every variety of FL-algebras is axiomatized by a countable set of equations (only countably many equations can be written using a finite number of fundamental operations and a countable number of variables). Therefore, there are at most continuum many elements in $\mathbf{\Lambda}(\mathsf{FL})$. On the other hand, Theorem 1.59 states that its sublattice $\mathbf{\Lambda}(\mathsf{HA})$ has continuum many elements. Therefore, $\mathbf{\Lambda}(\mathsf{FL})$ has exactly continuum many elements. We will also show that $\mathbf{\Lambda}(\mathsf{RL})$ has exactly continuum many elements, and actually both $\mathbf{\Lambda}(\mathsf{FL})$ and $\mathbf{\Lambda}(\mathsf{RL})$ have continuum many atoms. So even the lowest nontrivial levels of $\mathbf{\Lambda}(\mathsf{FL})$ and $\mathbf{\Lambda}(\mathsf{RL})$ are as complicated as they can be. Having shown that, we will focus on lattices of subvarieties of certain subvarieties of FL and determine the cardinality of their set of atoms. We summarize these results in Tables 9.1 and 9.2.

We proceed to investigate parts of the level above atoms and focus on *almost minimal* varieties (varieties of height 2). In particular we look at covers of the variety of Boolean algebras in the subvariety lattice of $\mathsf{FL_{ew}}$. This is of special interest since BA is the only atom in $\mathbf{\Lambda}(\mathsf{FL_{ew}})$. We will also look at the covers of the atoms $\mathsf{V}(\mathbf{To}_n)$. Table 9.3 summarizes the results about almost minimal varieties.

We then move to more global studies of the subvariety lattice. Meets of subvarieties of FL are simply intersections. Given axiomatizations for two subvarieties or of two substructural logics they are each axiomatized by the union of the two axiomatizations. Nevertheless, joins of varieties and meets of logics are in general difficult to describe. We give an axiomatization of the join of two subvarieties of FL, equivalently of the meet (intersection) of two substructural logics, and show that under certain conditions finite axiomatizability is preserved. In particular, the join of two finitely axiomatized varieties of FL_e-algebras, or of two commutative substructural logics, is also finitely axiomatized.

Finally, we investigate certain areas of the subvariety lattice that look similar. For example, we show that the subvariety lattices of LG^- and LG are isomorphic. Moreover, there is a copy of the subvariety lattice of IRL sitting above the atomic variety BA of Boolean algebras and copies of the subvariety lattice of RL sitting above each of the atoms $\mathsf{V}(\mathbf{To}_n)$.

9.1. General facts about atoms

As we mentioned before, a (non-trivial) variety $\mathcal{V}$ is called *minimal* iff $\mathcal{V}$ has only one, trivial, proper subvariety; i.e. it is an atom in the lattice of subvarieties (of a bigger variety). Clearly, a minimal variety must be generated by a single subdirectly irreducible algebra; equally clearly, the converse is not true. No general characterization of these subdirectly irreducible algebras which generate minimal varieties exists at present, although quite a lot is known in this direction (cf. e.g., Szendrei [Sze92]).

A non-trivial algebra $\mathbf{A}$ is called *strictly simple*, if it lacks non-trivial proper subalgebras and congruences. Whereas for finite algebras the notion of proper subalgebra is straightforward, for infinite ones we say that a subalgebra $\mathbf{B}$ of $\mathbf{A}$ is proper if $\mathbf{B}$ is not isomorphic to $\mathbf{A}$. Thus, it may (and does) happen that $\mathbf{B}$ is not a proper subalgebra of $\mathbf{A}$ even though B is a proper subset of A. Some examples of such algebras occur later on, but we invite the reader to think of an example now as an exercise.

Recall that by Theorem 3.47, congruences on residuated lattices correspond to convex normal subalgebras. So, the absence of non-trivial proper subalgebras in a residuated lattice is enough to establish strict simplicity. This, of course, is not the case for FL-algebras, where, due to the presence of the constant 0, the appropriate notion is that of a congruence class of 1. So, on the one hand 0 complicates matters, but on the other it may make things easier, if it can be used to define bounds. To make that observation precise and general, we will now define the class of FL-algebras with *nearly term definable lower bound.* Let us deconstruct that ugly phrase. An element $b \in A$ is a nearly term definable lower bound, if b is the bottom element of A and there is an n-ary term-operation $t(\overline{x})$ on A such that $t(\overline{a}) = b$ holds unless $\overline{a} = \underbrace{(1,\dots,1)}_{n\text{-times}}$. Obviously, if b is term definable then it is nearly term definable. In particular, every FL_o-algebra has nearly term definable lower bound. The next result shows that varieties generated by strictly simple FL-algebras with nearly term definable lower bound are atoms. In the following lemma we denote by $\bot$ the bottom element of an algebra that happens to be bounded, but we do not assume that $\bot$ is in the type.

LEMMA 9.1. *Let* $\mathbf{A}$ *be a strictly simple FL-algebra or residuated lattice with bottom element* $\bot$ *nearly term definable by an* n*-ary term* t *that does not involve the constant* 0*. Then,* $\mathsf{V}(\mathbf{A})$ *is a minimal variety. Moreover, if* $\mathbf{A}'$ *is a strictly simple FL-algebra or residuated lattice with bottom element nearly term definable by the same term* t*, then* $\mathsf{V}(\mathbf{A}) \subseteq \mathsf{V}(\mathbf{A}')$ *if and only if* $\mathbf{A}$ *and* $\mathbf{A}'$ *are isomorphic.*

PROOF. Let $\mathcal{V}$ be the variety generated by $\mathbf{A}$. By Jónsson's Lemma, for congruence distributive varieties, the subdirectly irreducible algebras of $\mathcal{V}$ are contained in $\mathsf{HSP_U}(\mathbf{A})$. So, if $\mathbf{D} \in \mathcal{V}_{\mathsf{SI}}$, there exists an ultrapower $\mathbf{B} = \mathbf{A}^I/U$ and a non-trivial subalgebra $\mathbf{C}$ of $\mathbf{B}$ such that $\mathbf{D} = f(\mathbf{C})$, for some homomorphism f. In the following we will identify A with its isomorphic copy $\{(a) : a \in A\}$ in $\mathbf{B}$, where (a) denotes the constant function with value a. Since $\mathbf{A}$ is strictly simple, thus generated by any of its non-identity elements, $\mathbf{A}$ is generated by $\bot$. Note that $\mathbf{A}$ satisfies the first order formula[1]

$$(\forall x_1,\dots,x_n)\big((x_1 \neq 1 \text{ or } \dots \text{ or } x_n \neq 1) \Rightarrow (t(x_1,\dots,x_n) = \bot)\big).$$

Therefore, by properties of ultraproducts, $\mathbf{B}$ satisfies it, too, with $\bot$ being the element $(\bot : i \in I)/U$, i.e., the equivalence class of the sequence with all terms equal to $\bot$. Abusing notation slightly, we will call that element $\bot$ as well; obviously $\bot$ is the least element of $\mathbf{B}$. Since the formula above is universal, it is also satisfied by each subalgebra of $\mathbf{B}$. It follows that each nontrivial subalgebra $\mathbf{C}$ of $\mathbf{B}$ contains $\bot$ and thus is bounded. Since $\mathbf{A}$ is a subalgebra of $\mathbf{B}$ generated by $\bot$, it follows that every nontrivial subalgebra of $\mathbf{B}$, in particular $\mathbf{C}$, in turn contains $\mathbf{A}$ as a subalgebra. Consider the homomorphism $f\colon C \to D$. Suppose $f(u) = f(v)$ for some distinct elements u and v of $A \subseteq C$. Then, since $\mathbf{A}$ is simple and $\bot \in A$ we obtain $f(\bot) = 1$. But $\bot$ is also the bottom element of $\mathbf{C}$ and therefore $\mathrm{Cg}^{\mathbf{C}}(u,v) = \mathrm{Cg}^{\mathbf{C}}(\bot,1)$ is the full congruence. It follows that $f(C) = D$ is a singleton, which

[1]To save space we use the shorthand $x_i \neq 1$ instead of our official not $x_i = 1$.

contradicts $\mathbf{D}$ being subdirectly irreducible. Therefore, f is injective on A and thus $\mathbf{A}$ is a subalgebra of $\mathbf{D}$. It follows that $\mathbf{A}$ is isomorphic to a subalgebra of every subdirectly irreducible member of $\mathcal{V}$; hence $\mathcal{V}$ is an atom.

For the nontrivial direction of the 'moreover' part, observe that if the varieties $\mathsf{V}(\mathbf{A})$ and $\mathsf{V}(\mathbf{A}')$ are equal, then, by the previous argument, $\mathbf{A}$ is isomorphic to a subalgebra of every subdirectly irreducible member of $\mathsf{V}(\mathbf{A}')$. Hence, $\mathbf{A}$ is embedded in $\mathbf{A}'$. But $\mathbf{A}$ is generated by $\bot$ and so is $\mathbf{A}'$, therefore the embedding is onto. □

One use of the lemma above will be for certain varieties of FL-algebras. If such a variety $\mathcal{V}$ is generated by a strictly simple FL-algebra $\mathbf{A}$ which happens to be bounded and which does not identify 0 with 1, then $\mathbf{A}$ is generated by 0 and thus $\bot = t(0, \ldots, 0)$, for some term t. So, $\mathcal{V}$ must be minimal.

COROLLARY 9.2. *Let $\mathbf{A}$ and $\mathbf{B}$ be strictly simple bounded FL-algebras, such that $0 \neq 1$ holds in both. Then $\mathsf{V}(\mathbf{A})$ and $\mathsf{V}(\mathbf{B})$ are minimal varieties. Moreover, $\mathsf{V}(\mathbf{A}) = \mathsf{V}(\mathbf{B})$ if and only if $\mathbf{A}$ and $\mathbf{B}$ are isomorphic.*

Even for varieties of residuated lattices, Lemma 9.1 can be used to show that finitely generated varieties are atoms.

COROLLARY 9.3. *Let $\mathcal{V}$ be a finitely generated variety. Then $\mathcal{V}$ is an atom in $\mathbf{\Lambda}(\mathsf{RL})$ iff $\mathcal{V} = \mathsf{V}(\mathbf{L})$, for some finite strictly simple residuated lattice $\mathbf{L}$. Moreover, if $\mathbf{L}$ and $\mathbf{M}$ are non-isomorphic finite strictly simple residuated lattices, then the minimal varieties $\mathsf{V}(\mathbf{L})$ and $\mathsf{V}(\mathbf{M})$ are distinct.*

PROOF. Let $\mathcal{V}$ be a minimal variety generated by a finite algebra $\mathbf{K}$. If $\mathbf{K}$ is not strictly simple, then there is a minimal non-trivial subalgebra $\mathbf{L}$ of $\mathbf{K}$. Since $\mathcal{V}$ is an atom, it is generated by $\mathbf{L}$. The converse is a direct consequence of Lemma 9.1; the term nearly defining the bottom element exists because $\mathbf{L}$ is strictly simple and finite; see Exercise 1. □

9.2. Minimal subvarieties of RL

We begin at the atomic level of $\mathbf{\Lambda}(\mathsf{RL})$. Recall that RL is the subvariety of FL defined by the identity $0 = 1$. Since this identity, as we may put it, makes one constant disappear, constructions from this section may be a little harder than in the more general case of FL. In a number of cases we will recycle them for use in FL, too. First, we construct infinitely many commutative atoms and continuum many non-commutative, representable, bounded atoms that satisfy the identity $x^4 = x^3$. Later, by means of quite a different construction, we obtain a continuum of non-commutative, representable, idempotent atoms; the generating algebras of these atoms are not bounded. We will also show that there are only two commutative idempotent atoms and only two cancellative atoms.

9.2.1. Commutative, representable atoms. Recall the definitions and results about generalized Boolean algebras, Brouwerian algebras and generalized MV-algebras from Section 3.4 (page 156). Using Proposition 3.23(5), it is not difficult to show that $\mathsf{GBA} = \mathsf{Br} \cap \mathsf{GMV}$. Moreover, by Corollary 9.3 and Proposition 3.23 the variety GBA is an atom. It is easy to see that it is the only atom below Br.

Being in search of finite strictly simple residuated lattices, in view of Corollary 9.3, we note that every such residuated lattice different from $\mathbf{2}_r$ has to have a top element different from 1; otherwise $\{\bot, 1\}$, where $\bot$ is the least element of the lattice, defines a subalgebra isomorphic to $\mathbf{2}_r$. In Figure 9.1 we give several examples of finite strictly simple residuated lattices. In each case we give only the multiplication operation, since the residuals are determined by it. Also, in all cases we have $x0 = 0 = 0x$, $x1 = x = 1x$ and $x\top = x = \top x$ for $x \neq 1$. The first two examples on the left describe in-

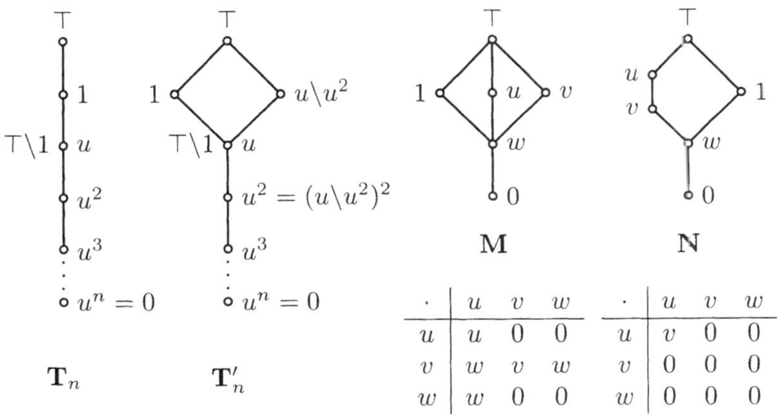

$\cdot$	u	v	w
u	u	0	0
v	w	v	w
w	w	0	0

$\cdot$	u	v	w
u	v	0	0
v	0	0	0
w	0	0	0

FIGURE 9.1. Some strictly simple residuated lattices.

finite lists of finite, commutative residuated lattices that are strictly simple and, consequently, generate distinct atoms in $\mathbf{\Lambda}(\mathsf{RL})$. We will now analyse the first example on the left in more detail.

For every natural number n, set $T_n = \{\top, 1\} \cup \{u_k : k \in \mathbb{N}_n^+\}$, where $\mathbb{N}_n^+ = \{1, 2, \ldots, n\}$. Define an order relation on T_n, by $u_k \leq u_l$ iff $k \geq l$, and $u_k < 1 < \top$, for all positive integers $k, l \leq n$; see Figure 9.1. Also, define multiplication by $x\top = \top x = x$, for all $x \neq 1$; $u_k u_l = u_{min\{n,k+l\}}$, for all $k, l \in \mathbb{N}_n^+$; and two division operations by $x/y = \bigvee\{z \in T_n : zy \leq x\}$ and $y\backslash x = \bigvee\{z \in T_n : yz \leq x\}$. It is easy to verify that $\mathbf{T}_n = (T_n, \wedge, \vee, \cdot, \backslash, /, 1)$ is a residuated lattice. We set $u = u_1$. Note that $u_k = u^k$ for all $k \leq n$.

LEMMA 9.4. [Gal05] *The variety* $\mathsf{V}(\mathbf{T}_n)$ *is an atom in the subvariety lattice of* RL, *for every natural number* n.

PROOF. Note that the residuated lattice $\mathbf{T}_n$ is generated by any of its non-identity elements. If $x < 1$, then $1/x = \top$; moreover, $1/\top = u_1$ and $u_k = u_1^k$, for all $k \leq n$. So, $\mathbf{T}_n$ is strictly simple; hence it generates an atom by Corollary 9.3. □

COROLLARY 9.5. [Gal05] *There are infinitely many commutative, representable atoms in the subvariety lattice of* RL.

Since the varieties $\mathsf{V}(\mathbf{T}_n)$ are finitely generated and congruence distributive, they are finitely based by Baker's Finite Basis Theorem (see e.g. [BS81]). But we can provide a particularly neat finite basis for each of them. Define the terms $\overline{\top}(x) = x \vee 1/x$ and $\overline{u}(x) = 1/\overline{\top}(x)$. Note that if $a \in T_n \setminus \{1\}$, then $\overline{\top}(a) = \top$ and $\overline{u}(a) = u$.

For every non-zero natural number n, we denote by $\mathbf{C}_n$ the integral residuated lattice defined by the monoid on the set $\{1, a, a^2, \ldots, a^{n-1}\}$, under the linear order $a^{n-1} < \cdots < a < 1$. It is easy to see that $\mathbf{C}_n$ is an MV-algebra.

PROPOSITION 9.6. [Gal05] *For every positive natural number n, the following list of equations is an equational basis for* $\mathsf{V}(\mathbf{T}_n)$, *relative to* RL.

(1) $\lambda_z(x/(x \vee y)) \vee \rho_w(y/(x \vee y)) = 1$
(2) $x^{n+1} = x^n$
(3) $(x \vee 1)^2 = (x \vee 1)$
(4) $1/((x \vee 1)\backslash 1) = x \vee 1$
(5) $(\overline{u}(x))^n \cdot x = (\overline{u}(x))^n$
(6) $x \wedge y \wedge 1 = ((y \wedge 1)/(x \wedge 1) \wedge 1)(x \wedge 1)$
(7) $(x \wedge 1)^n = (x \wedge 1)^n/((x \wedge 1)^n \backslash (y \wedge 1)^n \wedge 1) \wedge 1$
(8) $[(x \wedge 1)^n/(y \wedge 1)^n \wedge 1]^2 = (x \wedge 1)^n/(y \wedge 1)^n \wedge 1$
(9) $xy = yx$
(10) $(\overline{u}(x))^n/(\overline{u}(x))^{n-1} \wedge 1 = \overline{u}(x)$

PROOF. It is easy to check that $\mathsf{V}(\mathbf{T}_n)$ satisfies all the equations. Conversely, let $\mathbf{L}$ be a subdirectly irreducible residuated lattice that satisfies the equations. $\mathbf{L}$ has to be a chain, because of the first equation and Theorem 9.75. Consider first the negative cone $\mathbf{L}^-$ of $\mathbf{L}$. We will show that it is isomorphic to $\mathbf{C}_m$, for some $m \leq n$.

By (2), $\mathbf{L}^-$ is n-potent. Equations (6) and (9) imply that $\mathbf{L}^-$ is an integral commutative GBL-algebra. Note that the negative idempotent elements are of the form x^n and that they form a subalgebra of $\mathbf{L}^-$. Indeed, they are closed under division by (8), obviously closed under the lattice operations, and the product of any two such elements is their meet – if $a \leq b$, then $a = a^2 \leq ab \leq a$. By the identity (7) and Proposition 3.23(6), this subalgebra is a generalized Boolean algebra. Since it is also totally ordered it is isomorphic to $\mathbf{2}$.

We will show now that $\mathbf{L}^-$ is generated by a single element as a monoid. Assume that there are non-identity elements $a \leq b$ that are not powers of a common element. Define $a_1 = a$, $b_1 = b$, $a_{k+1} = b_k \wedge b_k\backslash^{\mathbf{L}^-} a_k$ and $b_{k+1} = b_k \vee b_k\backslash^{\mathbf{L}^-} a_k$. Note that for every positive integer k, $a_k \leq b_k$; $a_k = a_{k+1}b_{k+1}$, because of (9) and (6); and b_k is an increasing sequence. So, $a = a_1 = a_2b_2 = a_3b_3b_2 = \cdots = a_{n+1}b_{n+1}b_nb_{n-1}\cdots b_3b_2 \leq 1(b_{n+1})^n = (b_{n+1})^n$. Since there are only two idempotent elements in $\mathbf{L}$, either $(b_{n+1})^n = 0$, or $(b_{n+1})^n = 1$. In the first case, $a = 0 = b^n$, so both a and b are powers of b. In the second case $b_{n+1} = 1$. Since $b_{n+1} = b_n \vee b_n\backslash^{\mathbf{L}^-} a_n = b_{n-1} \vee b_{n-1}\backslash^{\mathbf{L}^-} a_{n-1} \vee b_n\backslash^{\mathbf{L}^-} a_n = \cdots = b \vee b_1\backslash^{\mathbf{L}^-} a_1 \vee \ldots \vee b_n\backslash^{\mathbf{L}^-} a_n$, we have $b_l\backslash^{\mathbf{L}^-} a_l = 1$, for some l. Thus, $b_l \leq a_l$, so $a_l = b_l$. Using the fact that $b_k \in \{a_{k+1}, b_{k+1}\}$ and $a_k = a_{k+1}b_{k+1}$, for all k, and induction, it is not hard to see that both b and a are powers of b^l. So, $L^- = \{1, u, u^2, \ldots, u^m = \bot\}$, for some $m \leq n$.

Observe that $\mathbf{L}$ has a strictly positive element a. Otherwise, $\mathbf{L}$ would be integral, so $1/x = 1$, for all $x \in L$, hence $\overline{\top}(x) = 1$, i.e. $\overline{u}(x) = 1$. In that case, (5) would imply $1 \cdot x = 1$, for all $x \in A$, a contradiction. By (3), we get $a^2 = a$. For every strictly positive element b of L, we have $u = eu \leq bu$. If $1 \leq bu$, we have $1 \leq b1u \leq bbuu \leq \cdots \leq b^nu^m = b^m\bot = \bot$, a contradiction. So $bu = u$, hence $b\backslash 1 = u$. Using equation (4), we have $b = 1/(b\backslash 1) = 1/u$, so there is a unique strictly positive element $\top = 1/u$ in $\mathbf{L}$. Finally, note that if $m < n$, then $u^n = u^{n-1}$, thus $u^n/u^{n-1} \wedge 1 = 1$, a fact that contradicts (10). Thus, $\mathbf{L}$ is isomorphic to $\mathbf{T}_n$. □

The cardinality of the class of all representable, commutative atoms in $\mathbf{\Lambda}(\mathsf{RL})$ remains unknown to us. We conjecture, though, that there are only countably many such atoms (see open problem 17). Some justification for this conjecture is provided by the next proposition, which states that finite, commutative, strictly simple, residuated chains have properties resembling those of the algebras $\mathbf{T}_n$.

PROPOSITION 9.7. [Gal05] *Let* $\mathbf{L}$ *be a finite, commutative, representable, strictly simple residuated lattice and let* $\top$ *be its top element. Then we have* $x\top = x$, *for all* $x \neq 1$. *Moreover,* $\top$ *covers* 1 *and* 1 *covers* $1/\top$, *if* $\top \neq 1$.

PROOF. Obviously, $\mathbf{L}$ is a subdirectly irreducible element of RRL, so $\mathbf{L}$ is chain. If $\mathbf{L} \cong \mathbf{2}$, then the conclusion is obvious. Otherwise, $\mathbf{L}$ has a top element $\top \neq 1$. If $1 = 1/\top$, then $1 \leq 1/\top$, i.e., $\top \leq 1$, a contradiction. So, $1 \neq 1/\top$. Note that $1/\top = 1/\top^2 = (1/\top)/\top$, by Lemma 2.6(6), so $1/\top \leq (1/\top)/\top$; hence $(1/\top)\top \leq 1/\top$. On the other hand, $1/\top \leq (1/\top)\top$, since $1 \leq \top$; so $(1/\top)\top = (1/\top)$.

It is easy to show that if $x\top = x$ and $y\top = y$, then $xy\top = xy$, $(x/y)\top = x/y$ and $(1/x)\top = 1/x$. By the assumption of strict simplicity, for every element of $a \neq 1$ of $\mathbf{L}$, there exists a term t_a, such that $a = t_a(\top)$. By induction on the complexity of t_a, it can be shown that $x\top = x$, for all $x \neq 1$.

To show that 1 is covered by $\top$, note that if $x > 1$, then $\top \leq \top x = x$. It is obvious that $1/\top \leq 1$. If $x < 1$, then $\top x \leq 1$, so $x \leq 1/\top$, hence 1 is a cover of $1/\top$. □

9.2.2. Cancellative atoms. It is well known and easy to observe that the variety $\mathsf{V}(\mathbb{Z})$ generated by the ℓ-group of the integers under addition is the only atom in the subvariety lattice of the variety LG of ℓ-groups. It is also easy to prove, and it will follow from Theorem 9.86, that the variety $\mathsf{V}(\mathbb{Z}^-)$, generated by the negative cone of $\mathbb{Z}$, is the only atom below the variety LG^- of negative cones of ℓ-groups. Both of these atoms are cancellative. Below we note that they actually are the only atoms below the variety of cancellative residuated lattices.

PROPOSITION 9.8. [Gal05] *For every cancellative residuated lattice, either it has $\mathbb{Z}^-$ as a subalgebra or it is an ℓ-group.*

PROOF. Let $\mathbf{L}$ be a cancellative residuated lattice. Since division is order reversing in the denominator, for every negative element a, $1 \leq 1/a$. Hence, either there exists a strictly negative element a of L such that $1/a = 1$, or for every strictly negative element x of L, $1 < 1/x$. It is easy to see that in the first case the subalgebra generated by a is isomorphic to $\mathbb{Z}^-$. Since $a < 1$, we get $a^{n+1} \leq a^n$, for every natural number n. Actually, $a^{n+1} < a^n$, because otherwise, we would get $a = 1$, by cancellativity. Moreover, $a^{k+m}/a^m = a^k$ and $a^m/a^{m+k} = 1$, for all natural numbers m, k. Thus, the set of all powers of a defines a subalgebra of $\mathbf{L}$ isomorphic to $\mathbb{Z}^-$.

In the second case for every element a of L, consider the element $x = (1/a)a$; we have $x \leq 1$, by Lemma 2.6(4). It cannot be strictly negative because $1/x = 1/(1/a)a = (1/a)/(1/a) = 1$, by Lemma 2.6(6) and cancellativity; so $x = 1$. Thus, $\mathbf{L}$ is an ℓ-group. □

COROLLARY 9.9. [Gal05] *The varieties $\mathsf{V}(\mathbb{Z})$ and $\mathsf{V}(\mathbb{Z}^-)$ are the only cancellative atoms in the subvariety lattice of* RL.

9.2.3. Bounded, 3-potent, representable atoms. In view of Corollary 9.9, it makes sense to investigate the other end of the spectrum of atoms, i.e., varieties that are n-potent, for some natural number n. We produce a continuum of 3-potent residuated chains that generate distinct atoms of $\mathbf{\Lambda}(\mathsf{RL})$. Let S be any subset of $\mathbb{N}$. The algebra $\mathbf{J}_S$ is based on the set $\{\bot, a, b, 1, \top\} \cup \{c_i : i \in \mathbb{N}\} \cup \{d_i : i \in \mathbb{N}\}$, with the following linear order:

$$\bot < a < b < c_0 < c_1 < c_2 < \cdots < \cdots < d_2 < d_1 < d_0 < 1 < \top$$

The operation $\cdot$ is defined by (1) $1x = x = x1$, (2) if $x \neq 1$ then $\top x = x = x\top$, and (3) if $x \notin \{1, \top\}$ then $\bot x = \bot = x\bot$, $ax = \bot = xa$, and

$\cdot$	$\top$	1	d_0	d_1	d_2	d_3	$\dots$	$\dots$	c_3	c_2	c_1	c_0	b	a	$\bot$
$\top$	$\top$	$\top$	d_0	d_1	d_2	d_3	$\dots$	$\dots$	c_3	c_2	c_1	c_0	b	a	$\bot$
1	$\top$	1	d_0	d_1	d_2	d_3	$\dots$	$\dots$	c_3	c_2	c_1	c_0	b	a	$\bot$
d_0	d_0	d_0	b	b	b	b	$\dots$	$\dots$	b	b	b	$\bot$	$\bot$	$\bot$	$\bot$
d_1	d_1	d_1	b	b	b	b	$\dots$	$\dots$	b	b	$\bot$	$\bot$	$\bot$	$\bot$	$\bot$
d_2	d_2	d_2	b	b	b	b	$\dots$	$\dots$	b	$\bot$	$\bot$	$\bot$	$\bot$	$\bot$	$\bot$
d_3	d_3	d_3	b	b	b	b	$\dots$	$\dots$	$\bot$	$\bot$	$\bot$	$\bot$	$\bot$	$\bot$	$\bot$
$\vdots$	$\vdots$	$\vdots$	$\vdots$	$\vdots$	$\vdots$	$\vdots$			$\vdots$	$\vdots$	$\vdots$	$\vdots$	$\vdots$	$\vdots$	$\vdots$
$\vdots$	$\vdots$	$\vdots$	$\vdots$	$\vdots$	$\vdots$	$\vdots$			$\vdots$	$\vdots$	$\vdots$	$\vdots$	$\vdots$	$\vdots$	$\vdots$
c_3	c_3	c_3	b	b	$\mathbf{s}_2$	a	$\dots$	$\dots$	$\bot$	$\bot$	$\bot$	$\bot$	$\bot$	$\bot$	$\bot$
c_2	c_2	c_2	b	$\mathbf{s}_1$	a	$\bot$	$\dots$	$\dots$	$\bot$	$\bot$	$\bot$	$\bot$	$\bot$	$\bot$	$\bot$
c_1	c_1	c_1	$\mathbf{s}_0$	a	$\bot$	$\bot$	$\dots$	$\dots$	$\bot$	$\bot$	$\bot$	$\bot$	$\bot$	$\bot$	$\bot$
c_0	c_0	c_0	a	$\bot$	$\bot$	$\bot$	$\dots$	$\dots$	$\bot$	$\bot$	$\bot$	$\bot$	$\bot$	$\bot$	$\bot$
b	b	b	$\bot$	$\bot$	$\bot$	$\bot$	$\dots$	$\dots$	$\bot$	$\bot$	$\bot$	$\bot$	$\bot$	$\bot$	$\bot$
a	a	a	$\bot$	$\bot$	$\bot$	$\bot$	$\dots$	$\dots$	$\bot$	$\bot$	$\bot$	$\bot$	$\bot$	$\bot$	$\bot$
$\bot$	$\bot$	$\bot$	$\bot$	$\bot$	$\bot$	$\bot$	$\dots$	$\dots$	$\bot$	$\bot$	$\bot$	$\bot$	$\bot$	$\bot$	$\bot$

FIGURE 9.2. Multiplication table for $\mathbf{J}_S$.

$bx = \bot = xb$. Furthermore, (4) for all $i, j \in \mathbb{N}$, $c_i c_j = \bot$, $d_i d_j = b$,

$$c_i d_j = \begin{cases} \bot & \text{if } i < j \\ a & \text{if } i = j \text{ or } (i = j+1 \text{ and } j \in S) \\ b & \text{otherwise} \end{cases} \qquad d_i c_j = \begin{cases} \bot & \text{if } i \geq j \\ b & \text{otherwise.} \end{cases}$$

This information is given in the form of an operation table in Figure 9.2. Depending on the chosen subset S, the elements $\mathbf{s}_i$ in the table are either a (if $i \in S$) or b (if $i \notin S$). It is easy to check that this gives an associative operation since $xyz = \bot$ whenever $1, \top \notin \{x, y, z\}$. Now $\top = \bot\backslash\bot$, $d_0 = \top\backslash 1$, $c_i = d_i\backslash\bot$ and $d_{i+1} = c_i\backslash\bot$, so the algebra is generated by $\bot$. Further, $\bot$ is nearly term definable by $t(x) = (x\backslash 1 \wedge x)^3$ and it is also easy to verify that distinct subsets S of $\mathbb{N}$ produce (a continuum of) non-isomorphic algebras. Then, the result below follows by Lemma 9.1.

THEOREM 9.10. [JT02] *There are continuum many atoms in* $\mathbf{\Lambda}(\mathsf{RL})$ *that satisfy the identity* $x^3 = x^4$.

9.2.4. Idempotent, commutative atoms. We will now focus on idempotent atoms. The first thing we show about them is that if to idempotency we add commutativity, or even the weaker condition $1/x = x\backslash 1$, we get only two atoms. Both turn out to be representable (and commutative, for that matter), although in general idempotency plus commutativity do not imply representability.

THEOREM 9.11. [Gal05] *The varieties* $\mathsf{V}(\mathbf{2})$ *and* $\mathsf{V}(\mathbf{T}_1)$ *are the only atoms below the variety* $\mathrm{Mod}\,\{x^2 = x, 1/x = x\backslash 1\}$.

PROOF. Assume $\mathbf{A}$ is a non trivial member of $\mathrm{Mod}\,\{x^2 = x, 1/x = x\backslash 1\}$ and let a be a negative element of $\mathbf{A}$. Since $a \leq 1$, we have $1 \leq 1/a$. If $1/a = 1$, then $\{1, a\}$ is a subuniverse of $\mathbf{A}$. If $1 < 1/a$, set $T = 1/a$ and $b = 1/T$. We will show that $\{b, e, T\}$ is a subuniverse of $\mathbf{A}$. Note that $aT = a(1/a) = a(a\backslash 1) \leq 1$. So, $a \leq 1/T = b$. We have $b \leq bT = (1/T)T \leq 1$ and $bT = bbT \leq b$, so $bT = b$. Since $1/T = T\backslash 1$, we also get $Tb = b$. Additionally, $T \leq 1/b$. Also, $(1/b)a \leq (1/b)b \leq 1$, so $1/b \leq 1/a = T$; thus, $T = 1/b$. Moreover, $b \leq b/T \leq (b/T)T \leq b$, so $b/T = b$. Also, $a \leq aa \leq ba \leq a$, so $T/b = (1/a)/b = 1/ba = 1/a = T$. □

COROLLARY 9.12. [Gal05] *The varieties* $\mathsf{V}(\mathbf{2})$ *and* $\mathsf{V}(\mathbf{T}_1)$ *are the only idempotent commutative atoms.*

9.2.5. Idempotent, representable atoms. Since the two atoms of Corollary 9.12 are representable, it is natural to relax the commutativity condition and replace it by representability. Under these requirements the cardinality of the set of atoms jumps all the way up to continuum. These atoms cannot be commutative, of course.

For every set of integers S, set $N_S = \{a_i : i \in \mathbb{Z}\} \cup \{b_i : i \in \mathbb{Z}\} \cup \{1\}$. We define an order on N_S, by $b_i < b_j < 1 < a_k < a_l$, for all $i, j, k, l \in \mathbb{Z}$, such that $i < j$ and $k > l$; see Figure 9.3. Obviously, this is a total order on N_S. We also define a multiplication operation by

$$a_i a_j = a_{min\{i,j\}}, \qquad b_i b_j = b_{min\{i,j\}}$$

and

$$b_j a_i = \begin{cases} b_j & \text{if } j < i \text{ or } i = j \in S \\ a_i & \text{if } j > i \text{ or } i = j \notin S \end{cases}, \qquad a_i b_j = \begin{cases} a_i & \text{if } i < j \text{ or } i = j \in S \\ b_j & \text{if } i > j \text{ or } i = j \notin S \end{cases}$$

In other words $x_i y_i$ is the leftmost if $i \in S$ and the rightmost if $i \notin S$. Finally, we define two division operations on N_S, by $x/y = \bigvee\{z : zy \leq x\}$ and $y\backslash x = \bigvee\{z : yz \leq x\}$; note that the joins exist.

It is easy to see that multiplication is associative and residuated by the division operations. So, we can define a residuated lattice $\mathbf{N}_S$ with underlying set N_S and operations the ones described above. We will investigate for which sets S the variety generated by $\mathbf{N}_S$ is an atom in the subvariety lattice of residuated lattices.

We define the following residuated lattice terms:

$$\ell(x) = x\backslash 1,\ r(x) = 1/x,$$

$$t(x) = 1/x \vee x\backslash 1,$$

$$m(x) = \ell\ell(x) \wedge \ell r(x) \wedge r\ell(x) \wedge rr(x),$$

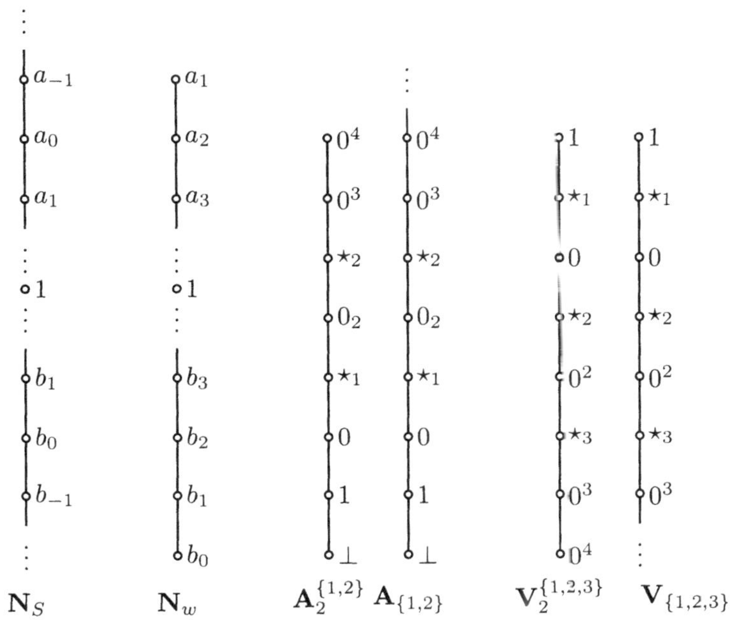

FIGURE 9.3. Chains generating minimal varieties.

$$p(x) = \ell\ell(x) \vee \ell r(x) \vee r\ell(x) \vee rr(x).$$

Moreover, we consider three binary relations defined by,

$$x \xrightarrow{r} y \Leftrightarrow r(x) = y,$$

$$x \underset{\ell}{\rightarrow} y \Leftrightarrow \ell(x) = y,$$

$$x \rightarrow y \Leftrightarrow r(x) = y \text{ or } \ell(x) = y.$$

In the following we write $x \prec y$ for the fact that x is covered by y: i.e., $x < y$ and for every z, if $x \leq z \leq y$, then $z = x$ or $z = y$.

LEMMA 9.13. [Gal05] *The following properties hold for* $\mathbf{N}_S$, *for every* $S \subseteq \mathbb{Z}$.

(1) *For all* $i \in \mathbb{Z}$, $m(b_i) = b_{i-1}$, $p(b_i) = b_{i+1}$, $m(a_i) = a_{i+1}$, $p(a_i) = a_{i-1}$. *Moreover,* $t(a_i) = b_i$ *and* $t(b_i) = a_i$.

(2) *For all* x, y, $x \leq y$ *or* $y \leq x$.

(3) *For every* x, $\{xt(x), t(x)x\} = \{x, t(x)\}$.

(4) *If* $x < 1 < y$, *then* $m(x) \prec x \prec p(x) < 1 < m(y) \prec y \prec p(y)$ *and* $t(y) < 1 < t(x)$.

(5) *For every* x, $m(t(x)) = t(p(x))$, $p(t(x)) = t(m(x))$, $m(p(x)) = p(m(x)) = x$ *and* $t(t(x)) = x$.

(6) *If* x *is negative, then* $xy = yx = \begin{cases} x & \text{for } x \leq y < t(x) \\ y & \text{for } y \leq x \text{ or } t(x) < y. \end{cases}$

If x *is positive, then* $xy = yx = \begin{cases} x & \text{for } t(x) < y \leq x \\ y & \text{for } y < t(x) \text{ or } x \leq y. \end{cases}$

(7) *For all* x, y; $x \wedge y, x \vee y, xy \in \{x, y\}$.

(8) *For all* x, y; $x/y, y\backslash x \in \{x, m(x), p(x), t(x), m(t(x)), t(y), m(t(y)), p(t(y))\}$.

(9) *For every finite word* v *there exists a universal first order formula* φ_v, *such that* v *is not a subword of* w_S *iff* φ_v *is satisfied in* $\mathbf{N}_S$.

PROOF. It is easy to see that

$$b_{i-1} \underset{\ell}{\leftarrow} a_i \underset{\ell}{\overset{r}{\rightleftarrows}} b_i \overset{r}{\rightarrow} a_{i+1} \qquad (i \in S)$$

$$b_{i-1} \overset{r}{\leftarrow} a_i \underset{\ell}{\overset{r}{\leftrightarrows}} b_i \underset{\ell}{\rightarrow} a_{i+1} \qquad (i \notin S)$$

It follows directly that $t(b_i) = a_i \vee a_{i+1} = a_i$ and $t(a_i) = b_{i-1} \vee b_i = b_i$. Moreover,

$$\{r(r(b_i)), r(\ell(b_i)), \ell(r(b_i)), \ell(\ell(b_i))\} = \{b_{i-1}, b_i, b_{i+1}\},$$

so $m(b_i) = b_{i-1}$ and $p(b_i) = b_{i+1}$. Similarly, $m(a_i) = a_{i+1}$ and $p(a_i) = a_{i-1}$; so, (1) holds. Moreover, (2) is obvious from the definition; (3)-(7) follow from (1); and (8) is routine to check. Finally for (9), the first order formula associated to a finite word v is φ_v defined as

$$\begin{aligned} \forall x_1, \ldots, x_n, y_1, \ldots, y_n \big(x_1 \prec x_2 \prec \cdots \prec x_n < 1 < y_n \prec \cdots \prec y_1 \ \mathsf{and} \\ t(x_1) = y_1 \ \mathsf{and} \ \ldots \ \mathsf{and} \ t(x_n) = y_n \big) \Rightarrow \\ \mathsf{not} \ (x_1 y_1 = s_1 \ \mathsf{and} \ \ldots \ \mathsf{and} \ x_n y_n = s_n), \end{aligned}$$

where n is the length of v; $s_i = x_i$, if $v(i) = 1$; and $s_i = y_i$, if $v(i) = 0$. Note that φ_v is equivalent to a universally quantified first-order formula in the language of residuated lattices. Indeed, we can replace $x_i \prec x_{i+1}$ by $x_i = m(x_{i+1})$ and $y_{i+1} \prec y_i$ by $y_i = p(y_{i+1})$. □

COROLLARY 9.14. [Gal05] *The residuated lattice* $\mathbf{N}_S$ *is strictly simple, for every set of integers* S.

PROOF. For all $a, b \in N_S \setminus \{1\}$, (a, b) is in the transitive closure of the relation $\rightarrow$ defined above. Thus, $\mathbf{N}_S$ is strictly simple. □

LEMMA 9.15. [Gal05] *Every non-trivial, one-generated subalgebra of an ultrapower of* $\mathbf{N}_S$ *is isomorphic to* $\mathbf{N}_{S'}$, *for some set of integers* S'.

PROOF. Every first order formula true in $\mathbf{N}_S$ is also true in an ultrapower of it. Since properties (2)–(8) of Lemma 9.13 can be expressed as first order formulas, they hold in every ultrapower of $\mathbf{N}_S$.

By property (2), any ultrapower $\mathbf{B}$ of $\mathbf{N}_S$ is totally ordered, so the same holds for every subalgebra of $\mathbf{B}$. Let $\mathbf{A}$ be a non-trivial one-generated subalgebra of $\mathbf{B}$ and let b be a generator for $\mathbf{A}$. The element b can be taken to be negative, since if b is positive, $t(b)$ is negative, by property (4), and it generates $\mathbf{A}$, because, by property (5), we have $t(t(b)) = b$.

By properties (7) and (8), A is the set of evaluations of the terms composed by the terms m, p, t and the constant term 1. By property (5), these compositions reduce to one of the forms $m^n(x), p^n(x), p^n(t(x))$ and $m^n(t(x))$, for n a natural number.

Set $b_{-n} = m^n(b), b_n = p^n(b), a_{-n} = p^n(t(b))$ and $a_n = m^n(t(b))$, for all natural numbers n. By the remark above, A consists of exactly these elements together with 1. Define a subset S' of $\mathbb{Z}$, by $m \in S'$ iff $b_m a_m = b_m$ and consider the map $f : A \to N_{S'} = \{b'_i : i \in \mathbb{Z}\} \cup \{a'_i : i \in \mathbb{Z}\} \cup \{1'\}$, defined by $f(b_i) = b'_i$, $f(a_i) = a'_i, f(1) = 1'$. By property (4), f is an order isomorphism and, consequently, a lattice isomorphism. Moreover, it is easy to check that f is a monoid homomorphism, using properties (3) and (6). Every lattice isomorphism preserves existing joins, so f preserves the two division operations. Thus, $\mathbf{A}$ is isomorphic to $\mathbf{N}_{S'}$. □

A *word* over $\{0, 1\}$ is a function $w : A \to \{0, 1\}$, where A is a subinterval of $\mathbb{Z}$; A is called the *support*, supp(w), of w. We call w *finite* (*infinite*, *bi-infinite*) if $|A| < \mathbb{N}$ ($A = \mathbb{N}^+$, $A = \mathbb{Z}$, respectively); actually we can restrict finite words to the case where $A = \{1, 2, \ldots, n\}$, for some n. If w is a word and v a finite word, we say v is a *subword* of w, if there exists an integer k, such that $v(i) = w(i + k)$ for all $i \in \text{supp}(v)$. Note that the characteristic function w_S of a subset S of $\mathbb{Z}$ is a bi-infinite word; $w_S(i) = 1$ iff $i \in S$. For two words w_1, w_2, define $w_1 \leq w_2$ iff every finite subword of w_1 is a subword of w_2. Obviously, $\leq$ is a preorder. Define $w_1 \cong w_2$ iff $w_1 \leq w_2 \leq w_1$. We call a bi-infinite word w *minimal* with respect to the preorder $\leq$, if, for every bi-infinite word w', $w' \leq w$ implies $w \cong w'$.

THEOREM 9.16. [Gal05] *Let* $\mathbf{A}$ *be a one-generated residuated lattice and* S *a subset of* $\mathbb{Z}$. *Then,* $\mathbf{A} \in \mathsf{HSP_U}(\mathbf{N}_S)$ *iff* $\mathbf{A} \cong \mathbf{N}_{S'}$, *for some* S' *such that* $w_{S'} \leq w_S$.

PROOF. Let S' be a set of integers, such that $w_{S'} \leq w_S$. Also, let $\mathbf{B} = (\mathbf{N}_S)^{\mathbb{N}}/U$, where U is a nonprincipal ultrafilter over $\mathbb{N}$, and $N_S = \{b_i : i \in \mathbb{Z}\} \cup \{a_i : i \in \mathbb{Z}\} \cup \{1\}$. We will show that $\mathbf{N}_{S'} \in \mathsf{ISP_U}(\mathbf{N}_S)$.

For every natural number n, define the finite approximations v_n of the bi-infinite word $w_{S'}$, by $v_n(i) = w_{S'}(i)$, for all $i \in \text{supp}(v_n) = [-n, n]_{\mathbb{Z}}$. Since, $w_{S'} \leq w_S$, the words v_n are subwords of w_S, so for every natural number n there exists an integer K_n, such that $v_n(i) = w_S(K_n + i)$, for all $i \in [-n, n]_{\mathbb{Z}}$.

Let $\bar{b} = (b_{K_n})_{n\in\mathbb{N}}$, where $b_{K_n} \in N_S$. By Lemma 9.15, the subalgebra of $\mathbf{B}$ generated by $\tilde{b} = [\bar{b}]$, the equivalence class of $\bar{b}$ under U, is isomorphic to $\mathbf{N}_{\tilde{S}}$, with $N_{\tilde{S}} = \{\tilde{b}_i : i \in \mathbb{Z}\} \cup \{\tilde{a}_i : i \in \mathbb{Z}\} \cup \{\tilde{1}\}$, for some subset $\tilde{S}$ of $\mathbb{Z}$. We identify the subalgebra generated by $\tilde{b}$ with $\mathbf{N}_{\tilde{S}}$ and, without loss of generality, we choose $\tilde{S}$ such that $\tilde{b}_0 = \tilde{b}$. We will show that $\tilde{S} = S'$.

We pick representatives $\bar{b}_m$ and $\bar{a}_m$, for $\tilde{b}_m$ and $\tilde{a}_m$, respectively, for all $m \in \mathbb{Z}$, and we adopt a double subscript notation for their coordinates. So, there exist $\bar{b}_{mn}$ and $\bar{a}_{mn}$ in $\mathbf{N}_S$, such that $\tilde{b}_m = [\bar{b}_m] = [(\bar{b}_{mn})_{n\in\mathbb{N}}]$ and $\tilde{a}_m = [\bar{a}_m] = [(\bar{a}_{mn})_{n\in\mathbb{N}}]$.

It is easy to prove that $\tilde{b}_m = [(b_{K_n+m})_{n\in\mathbb{N}}]$ and $\tilde{a}_m = [(a_{K_n+m})_{n\in\mathbb{N}}]$, using the definition of $\tilde{b}$, Lemma 9.13(1), basic induction and the following facts:

$$\tilde{a}_m = t(\tilde{b}_m) = t([(\bar{b}_{mn})_{n\in\mathbb{N}}]) = [(t(\bar{b}_{mn}))_{n\in\mathbb{N}}]$$
$$\tilde{b}_{m+1} = p(\tilde{b}_m) = p([(\bar{b}_{mn})_{n\in\mathbb{N}}]) = [(p(\bar{b}_{mn}))_{n\in\mathbb{N}}]$$
$$\tilde{b}_{m-1} = m(\tilde{b}_m) = m([(\bar{b}_{mn})_{n\in\mathbb{N}}]) = [(m(\bar{b}_{mn}))_{n\in\mathbb{N}}]$$

Now, for $|m| < n$, i.e., $m \in \mathrm{supp}(v_n)$, we have

$$\begin{aligned} K_n + m \in S &\Leftrightarrow w_S(K_n + m) = 1 \\ &\Leftrightarrow v_n(m) = 1 \\ &\Leftrightarrow w_{S'}(m) = 1 \\ &\Leftrightarrow m \in S'. \end{aligned}$$

Since, $b_{K_n+m}a_{K_n+m} = b_{K_n+m}$ exactly when $K_n + m \in S$, we get that if $|m| < n$, then $b_{K_n+m}a_{K_n+m} = b_{K_n+m}$ is equivalent to $m \in S'$.

In other words,

$$\{n : |m| < n\} \subseteq \{n : b_{K_n+m}a_{K_n+m} = b_{K_n+m} \Leftrightarrow m \in S'\}.$$

Since the first set is in U, the second one is in U, as well. It is not hard to check that this means that: $\{n : b_{K_n+m}a_{K_n+m} = b_{K_n+m}\} \in U$ is equivalent to $m \in S'$. So, $\tilde{b}_m\tilde{a}_m = \tilde{b}_m$ is equivalent to $m \in S'$; hence $m \in \tilde{S}$ iff $m \in S'$. Thus, $\tilde{S} = S'$.

For the converse, we will prove the implication for $\mathbf{A} \in \mathsf{SP_U}(\mathbf{N}_S)$. This is sufficient since under a homomorphism every one generated subalgebra will either map isomorphically or to the identity element, because of the strictly simple nature of the algebras $\mathbf{N}_{S'}$. Let $\mathbf{A}$ be a subalgebra of an ultrapower of $\mathbf{N}_S$. By Lemma 9.15, $\mathbf{A}$ is isomorphic to $\mathbf{N}_{S'}$, for some subset S' of $\mathbb{Z}$.

To show that $w_{S'} \leq w_S$ it suffices to show that, for every finite word v, if v is not a subword of w_S, then it is not a subword of $w_{S'}$ either. If v is not a subword of w_S, then $\mathbf{N}_S$ satisfies φ_v of Lemma 9.13(9); hence so does every ultrapower of $\mathbf{N}_S$. Since φ_v is a universal formula it is also satisfied by any subalgebra of an ultrapower of $\mathbf{N}_S$ and in particular by $\mathbf{N}_{S'}$. Thus, v is not a subword of $w_{S'}$. □

COROLLARY 9.17. [Gal05] *Let S, S' be sets of integers, then*

(1) $\mathsf{V}(\mathbf{N}_{S'}) \subseteq \mathsf{V}(\mathbf{N}_S)$ *if and only if* $w_{S'} \leq w_S$, *and*
(2) *if* w_S *is minimal with respect to* $\leq$, *then* $\mathcal{V} = \mathsf{V}(\mathbf{N}_S)$ *is an atom in the subvariety lattice of* RL.

PROOF. If $w_{S'} \leq w_S$ then, by Theorem 9.16, $\mathbf{N}_{S'} \in \mathsf{HSP_U}(\mathbf{N}_S) \subseteq \mathsf{V}(\mathbf{N}_S)$, so $\mathsf{V}(\mathbf{N}_{S'}) \subseteq \mathsf{V}(\mathbf{N}_S)$. Conversely, if $\mathsf{V}(\mathbf{N}_{S'}) \subseteq \mathsf{V}(\mathbf{N}_S)$, then $\mathbf{N}_{S'} \in \mathsf{V}(\mathbf{N}_S)$. Since $\mathbf{N}_{S'}$ is subdirectly irreducible, by Lemma 9.14, so, by Jónsson's Lemma, $\mathbf{N}_{S'} \in \mathsf{HSP_U}(\mathbf{N}_S)$. By Theorem 9.16 then, $w_{S'} \leq w_S$. This proves (1).

For (2) we reason as follows. If $\mathbf{L}$ is a subdirectly irreducible algebra from $\mathcal{V}$, then $\mathbf{L} \in \mathsf{HSP_U}(\mathbf{N}_S)$, by Jónsson's Lemma. Every one-generated subalgebra $\mathbf{A}$ of $\mathbf{L}$ is a member of $\mathsf{SHSP_U}(\mathbf{N}_S) \subseteq \mathsf{HSP_U}(\mathbf{N}_S)$, because $\mathsf{SH} \leq \mathsf{HS}$; so, by Theorem 9.16, $\mathbf{A}$ is isomorphic to some $\mathbf{N}_{S'}$, where $w_{S'} \leq w_S$. Since w_S is minimal with respect to the preorder $\leq$, we have $w_{S'} \cong w_S$; hence $\mathsf{V}(\mathbf{N}_{S'}) = \mathsf{V}(\mathbf{N}_S)$, by (1). Thus, $\mathcal{V} = \mathsf{V}(\mathbf{N}_{S'}) = \mathsf{V}(\mathbf{A}) \subseteq \mathsf{V}(\mathbf{L}) \subseteq \mathcal{V}$. Since $\mathcal{V} = \mathsf{V}(\mathbf{L})$, for every subdirectly irreducible $\mathbf{L}$ in $\mathcal{V}$, we get that $\mathcal{V}$ is an atom. □

COROLLARY 9.18. [Gal05] *There are continuum many atoms in the subvariety lattice of* $\mathsf{RRL} \cap \mathrm{Mod}\,\{x^2 = x\}$.

PROOF. In [Lot02] one can find a study on infinite and bi-infinite words. Among other things, words that are minimal with respect to the $\leq$ preorder are constructed. Such words have been re-discovered in different areas of mathematics and have numerous applications.

One way to construct such a word is to consider a line ℓ on the plane. The *lower mechanical word* w_ℓ corresponding to the line ℓ is obtained by approximating the line from below by a broken line, see Figure 9.4. The admissible line segments of the approximating broken line have to have endpoints (x, y) and $(x+1, y+k_x)$, such that x, y are integers and $k_x = 0$ or $k_x = 1$. The word w_ℓ is defined by $w_\ell(x) = k_x$, for all $x \in \mathbb{Z}$. It is shown in [Lot02] that if ℓ has irrational slope then w_ℓ is minimal. Moreover, if we consider only lines that contain the origin and have irrational slope, we obtain a class of cardinality continuum such that all words are minimal and pairwise incomparable.

For a more precise definition of the lower mechanical word associated with a line and for the proofs of the facts mentioned above, the reader is referred to [Lot02]. □

9.3. Minimal subvarieties of FL

We begin by recycling some of the previous constructions, with smaller or larger modifications. These will produce continua of atoms in $\mathbf{\Lambda}(\mathsf{FL_o})$ and

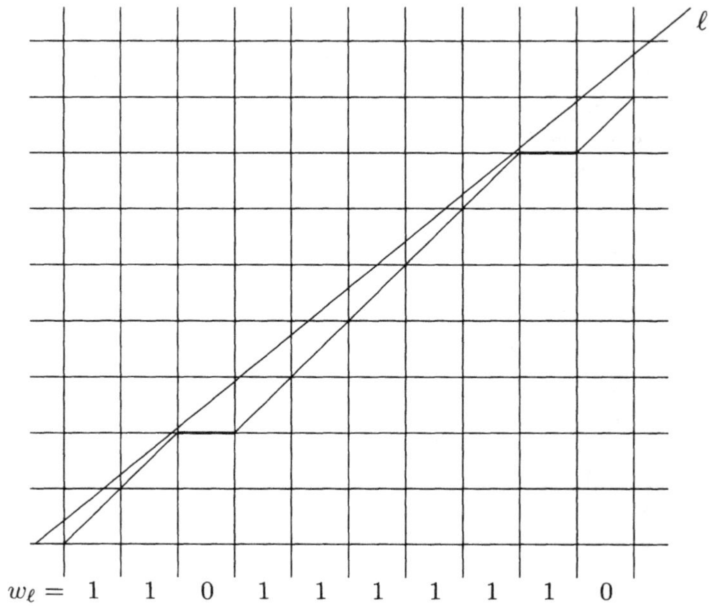

FIGURE 9.4. The word w_ℓ corresponding to a line ℓ.

$\mathbf{\Lambda}(\mathsf{FL_i})$. Continuum atoms in $\mathbf{\Lambda}(\mathsf{FL_e})$ will require a new construction. By recycling again, we will get countably many atoms in $\mathbf{\Lambda}(\mathsf{FL_{ei}})$ and $\mathbf{\Lambda}(\mathsf{FL_{eo}})$. We will also show that there are only two atoms in $\mathbf{\Lambda}(\mathsf{FL_{eco}})$.

9.3.1. Minimal subvarieties of $\mathsf{FL_o}$ and $\mathsf{FL_i}$. Consider the algebras $\mathbf{J}_S$ defined prior to Theorem 9.10 (see Figure 9.2). They admit two rather straightforward modifications. Firstly, nothing in the construction of $\mathbf{J}_S$ compels us to keeping the identity $0 = 1$ intact. We can just as well set $0 = \bot$, thus making $\bot$ a constant in the type. With this modification, each $\mathbf{J}_S$ becomes an FL_o-algebra. Let us call these algebras $\mathbf{Jo}_S$. Clearly, each $\mathsf{V}(\mathbf{Jo}_S)$ is minimal, and $\mathsf{V}(\mathbf{Jo}_S) = \mathsf{V}(\mathbf{Jo}_T)$ iff $S = T$. The next result is then immediate.

COROLLARY 9.19. *There are continuum many atoms in* $\mathbf{\Lambda}(\mathsf{FL_o})$.

Secondly, inspection of the operations of $\mathbf{J}_S$ makes it apparent that the role played by the element $\top$ amounts to preventing $\{\bot, 1\}$ from being a subuniverse. This is necessary, because $\mathbf{J}_S \models 0 = 1$. When we give up on that identity, we can set (say) $0 = a$, dispose $\top$ and restrict the original multiplication and residuation accordingly. Let us call the resulting algebras $\mathbf{Ji}_S$.

LEMMA 9.20. *For each* $S \subseteq \mathbb{N}$, *the algebra* $\mathbf{Ji}_S$ *is an* FL_i*-algebra and generates a minimal variety. Moreover,* $\mathsf{V}(\mathbf{Ji}_S) = \mathsf{V}(\mathbf{Ji}_T)$ *iff* $S = T$.

PROOF. We only give a sketch and leave the rest of the proof as an exercise (see Exercise 5). Note that $0^2 = \bot$, $d_0 = 0\backslash\bot$, $c_i = d_i\backslash\bot$, $d_{i+1} = c_i\backslash\bot$ and $d_i d_j = b$. Also, $\bot$ is nearly term definable by the term $t(x) = x^3$. □

COROLLARY 9.21. *There are continuum many atoms in* $\mathbf{\Lambda}(\mathsf{FL_i})$.

Recall that $\mathsf{FL_i} \cap \mathsf{FL_o} = \mathsf{FL_w}$. At the atomic level $\mathsf{FL_i}$ and $\mathsf{FL_o}$ are however almost disjoint. Note that $\mathbf{2}$ is a subalgebra of every non-trivial algebra in $\mathsf{FL_w}$; hence we have the following result.

PROPOSITION 9.22. *There is only one atom in* $\mathbf{\Lambda}(\mathsf{FL_w})$, *namely* $\mathsf{V}(\mathbf{2})$.

The atoms from Corollary 9.19 satisfy $x^4 = x^3$, but are not idempotent. Below we construct a continuum of idempotent, representable atoms in $\mathbf{\Lambda}(\mathsf{FL_o})$. By the same token, they are atoms of $\mathsf{FL_{co}}$. Our atoms will be varieties generated by algebras similar to the ones of the form $\mathbf{N}_S$.

For every infinite word w, set $N_w = \{a_i : i \in \mathbb{N}^+\} \cup \{b_i : i \in \mathbb{N}\} \cup \{1\}$, where $\mathbb{N}^+$ denotes the set of positive integers. We define an order on N_w, by $b_i < b_j < 1 < a_k < a_l$, for all $i, j \in \mathbb{N}$ and $k, l \in \mathbb{N}^+$, such that $i < j$ and $k > l$; see Figure 9.3. Obviously, this is a total order on N_w. Let $S_w = \{n \in \mathbb{N}^+ : w(n) = 1\}$ and define multiplication on N_w, by

$$a_i a_j = a_{\min\{i,j\}}, \qquad b_i b_j = b_{\min\{i,j\}}$$

and

$$b_j a_i = \begin{cases} b_j & \text{if } j < i \text{ or } i = j \in S_w \\ a_i & \text{if } j > i \text{ or } i = j \notin S_w \end{cases} \qquad a_i b_j = \begin{cases} a_i & \text{if } i < j \text{ or } i = j \in S_w \\ b_j & \text{if } i > j \text{ or } i = j \notin S_w \end{cases}$$

Finally, we define two division operations on N_S, by $x/y = \bigvee\{z : zy \leq x\}$ and $y\backslash x = \bigvee\{z : yz \leq x\}$; note that the joins exist.

It is easy to see that multiplication is associative and residuated by the division operations. So, we can define a residuated lattice $\mathbf{N}_w$ with underlying set N_w and operations the ones described above.

For uniformity, if w is a bi-infinite word and $S_w = \{n \in \mathbb{Z} : w(n) = 1\}$, we set $\mathbf{N}_w = \mathbf{N}_{S_w}$. The following result complements Corollary 9.17(1).

COROLLARY 9.23. [Gal05]

(1) *Assume that* w' *is a bi-infinite word and* w *an infinite word. Then,*
   (a) $\mathsf{V}(\mathbf{N}_{w'}) \subseteq \mathsf{V}(\mathbf{N}_w)$ *if and only if* $w' \leq w$; *and*
   (b) *if* $w \cong w'$, *then* $\mathsf{V}(\mathbf{N}_{w'})$ *is covered by* $\mathsf{V}(\mathbf{N}_w)$.
(2) *If* w' *and* w *are distinct infinite words then* $\mathsf{V}(\mathbf{N}_{w'})$ *and* $\mathsf{V}(\mathbf{N}_w)$ *are distinct incomparable varieties.*

PROOF. (Sketch: see Exercise 7) For (1a) note first that $\mathbf{N}_w$ is strictly simple and, consequently, subdirectly irreducible. Following the lines of the proof of Lemma 9.15, we can show that the one-generated subalgebras of an ultrapower of $\mathbf{N}_w$, where w is an infinite word, are either isomorphic to $\mathbf{N}_w$ or of the form $\mathbf{N}_S$, for $S \subseteq \mathbb{Z}$. Moreover, we can show that $w_S \leq w$, using the formula of Lemma 9.13(9). Homomorphic images preserve these facts, because $\mathbf{N}_w$ and $\mathbf{N}_S$ are strictly simple. Conversely, if $w_S \leq w$, then we can show that $\mathbf{N}_S$ is isomorphic to a subalgebra of the ultrapower $(\mathbf{N}_w)^{\mathbb{N}}$, mimicking the proof of Theorem 9.16.

For (1b) it is clear from the above that the only subdirectly irreducible algebras in $\mathsf{V}(\mathbf{N}_w)$ are $\mathbf{N}_w$ and $\mathbf{N}_S$, for $w_S \leq w$. Moreover, we know that the only subdirectly irreducible algebras in $\mathsf{V}(\mathbf{N}_{w'})$ are isomorphic to $\mathbf{N}_S$, for $w_S \leq w'$, i.e. $w_S \leq w$. Thus, $\mathsf{V}(\mathbf{N}_{w'})$ is covered by $\mathsf{V}(\mathbf{N}_w)$.

Finally (2) follows from the analysis of subdirectly irreducibles from the previous paragraph. □

For every infinite word w, let $\mathbf{No}_w$ be the FL_o-algebra whose 0-free reduct is $\mathbf{N}_w$ and for which $0 = b_0$.

THEOREM 9.24. [Gal05] *The varieties* $\mathsf{V}(\mathbf{No}_w)$, *where* w *is an infinite word, form a class of continuum many atoms in* $\mathbf{\Lambda}(\mathsf{FL_{co}})$.

PROOF. (Sketch: see Exercise 7) We work as in Corollary 9.17. We focus on the zero-generated subalgebra of an ultrapower of $\mathbf{No}_w$ and show that it is isomorphic to $\mathbf{No}_w$. Moreover, it has to be contained in every subalgebra of the ultrapower. If a homomorphic image of a subalgebra of an ultrapower collapses any two elements of the isomorphic copy of $\mathbf{No}_w$, it has to collapse the whole subalgebra. To see that, first recall that $\mathbf{No}_w$ is strictly simple, so we may assume that $h(b_0) = h(b_1)$ under the homomorphism h. Then, $h(b_0) = h(b_0/b_1 \wedge 1) = h(b_0)/h(b_1) \wedge 1 = h(b_1)/h(b_1) \wedge 1 = 1$. Since the multiplicative identity is the least element only in a one-element residuated lattice, we have that the homomorphic image is trivial. The rest of the proof proceeds as in Corollary 9.17. □

In [TW06], Tsinakis and Wille discuss cyclic, involutive FL-algebras. By applying a product-like construction they embed the algebras $\mathbf{J}_S$, defined in Section 9.2.3, to cyclic involutive FL-algebras. We mention the follwoing result without giving a proof.

THEOREM 9.25. [TW06] *There are continuum-many atoms in the subvariety lattice of* CyInFL.

9.3.2. Minimal subvarieties of representable $\mathsf{FL_{ec}}$ and $\mathsf{FL_{ei}}$. Recall that we have constructed countably many minimal varieties of commutative representable residuated lattices, including the varieties $\mathsf{V}(\mathbf{T}_n)$. Although our list does not exhaust minimal varieties of commutative representable residuated lattices, we conjectured that there are only countably many such varieties. Here we will demonstrate how the presence of 0 may change things. We will construct continuum minimal varieties of commutative, square-increasing and representable FL-algebras, i.e., representable FL_{ce}-algebras. These give directly maximal consistent (positive) relevance logics; i.e. extensions of the positive fragment $\mathbf{R}^+$ of the relevance logic $\mathbf{R}$.

For any subset S of $\mathbb{N} \setminus \{0\}$ and $k \in \mathbb{N}$, let $\mathbf{A}_k^S = \langle A_k^S, \wedge, \vee, \cdot, \to, 0, 1\rangle$, be an algebra defined as follows. We put $A_k^S = \{\bot\} \cup \{0^i : 0 \le i \le k+1\} \cup \{\star_i : 0 < i < k,\ i \in S\}$, with 0^0 and 0^1 being respectively the constants 1 and 0. On A_k^S we define a linear ordering, setting $\bot < 1 < 0 < 0^2 \cdots < 0^{k+1} = \top$ and $0^i < \star_i < 0^{i+1}$, for $i \in S$; see Figure 9.3. Then, we define multiplication as follows

$$\bot \cdot x = x \cdot \bot = \bot, \text{ for all } x$$

$$0^m \cdot 0^n = 0^{\min\{m+n,k+1\}} = \star_n \cdot \star_m$$

where the right-hand side only makes sense for $n, m \in S$, and

$$\star_n \cdot 0^m = 0^m \cdot \star_n = \begin{cases} \star_n & \text{if } m = 0 \\ 0^{\min\{n+m,k+1\}} & \text{if } m > 0 \end{cases}$$

It is not difficult to verify that multiplication is associative. It is also easily checked that, for any $x, y \in A_k^S$, the set $\{z \in A_k^S : zx \le y\}$ has the greatest element, thus, multiplication is residuated. Below are the explicit values.

$$\bot \to x = 0^{k+1} (= \top)$$

$$x \to \bot = \begin{cases} 0^{k+1} (= \top) & \text{if } x = \bot \\ \bot & \text{otherwise} \end{cases}$$

$$0^i \to 0^j = 0^i \to \star_j = \star_i \to \star_j = \begin{cases} 0^{j-i} & \text{if } i \le j \le k \text{ and } j - i \notin S \\ \star_{j-i} & \text{if } i \le j \le k \text{ and } j - i \in S \\ 0^{k+1} & \text{if } i \le j = k+1 \\ \bot & \text{if } j < i \end{cases}$$

$$\star_i \to 0^j = \begin{cases} 0^{j-i} & \text{if } i < j \le k \text{ and } j - i \notin S \\ \star_{j-i} & \text{if } i < j \le k \text{ and } j - i \in S \\ 0^{k+1} & \text{if } i < j = k+1 \\ \bot & \text{if } j \le i \end{cases}$$

Note that since $0 \notin S$, we always get $0^i \to 0^i = \star_i \to \star_i = 1$ for $i < k+1$. On the other hand, $0^{k+1} \to 0^{k+1} = 0^{k+1}$, so $\top \to \top = \top$ as it should.

Let S be a subset of $\mathbb{N}$ and U a nonprincipal ultrafilter over $\mathbb{N}$. Consider the ultraproduct $\prod_{k \in \mathbb{N}} \mathbf{A}_k^S / U$ of the family $\{\mathbf{A}_k^S : k \in \mathbb{N}\}$. Then, let $\mathbf{A}_S$ be the zero-generated subalgebra of $\prod_{k \in \mathbb{N}} \mathbf{A}_k^S / U$, i.e., the subalgebra generated by the constants 1 and 0. See Figure 9.3 again.

LEMMA 9.26. *The algebra $\mathbf{A}_S$ is a strictly simple FL_{ec}-algebra.*

PROOF. It is routine to check that each $\mathbf{A}_k^S$ is an FL-algebra. Commutativity and being square-increasing are immediate. Further, the negative cone of $\mathbf{A}_k^S$ is exactly $\{\bot, 1\}$. These properties are all expressible by equations, except for the last one, which is expressible by a universal first order sentence (see Exercise 8). Thus, they all carry over to ultraproducts and subalgebras. In particular, the negative cone of $\mathbf{A}_S$ is also $\{\bot, 1\}$ and so any nontrivial congruence on $\mathbf{A}_S$ collapses the negative cone to a single point. Thus, it collapses the whole algebra to a single point and therefore $\mathbf{A}_S$ is a simple FL_e-algebra. Since it is zero-generated, it is also strictly simple. □

LEMMA 9.27. *For distinct subsets S, T of $\mathbb{N}$, the algebras $\mathbf{A}_T$ and $\mathbf{A}_S$ are non-isomorphic.*

PROOF. (Sketch: see Exercise 9) Firstly, observe that the lattice structure of $\mathbf{A}_S$, for any $S \subseteq \mathbb{N}$, is a chain of order type $\mathbb{N} + 1$. Further, unlike in the finite algebras, $\top$ is multiplication-irreducible (i.e., $xy = \top$ implies $x = \top$ or $y = \top$). Moreover, $0^i \in A_S$ for any $i \in \mathbb{N}$. Therefore, for any $l \in S$, the element $\star_l$ gets generated by $0 \to 0^{l+1}$. Now, for distinct S and T suppose f is an isomorphism between $\mathbf{A}_S$ and $\mathbf{A}_T$. Then f must have $f(0^i) = 0^i$ for every $i \in \mathbb{N}$. Take the smallest $l \in S \setminus T$ (by symmetry of the situation we can assume it exists). We get $f(\star_l) = f(0 \to 0^{l+1}) = f(0) \to f(0^{l+1}) = 0 \to 0^{l+1} = 0^l = f(0^l)$. So f is not injective, hence cannot be an isomorphism. □

Now the result below follows by an application of Corollary 9.2.

THEOREM 9.28. *There are continuum many minimal varieties of representable FL_{ce}-algebras satisfying $1 \leq 0$.*

As the reader will recall from Chapter 2, the interpretation of the constants 1 and 0 in logic is that the former stands for the weakest true proposition and the latter for the strongest false one. In this context, the condition $1 < 0$ may look somewhat suspiciously, for it amounts to stating that *truth implies falsity* but not *vice versa.* Yet, such logics are not as pathological as it may seem. Namely, a logic $\mathbf{L}$ is called *paraconsistent* if $\varphi \wedge \neg\varphi$ is a theorem of $\mathbf{L}$ at least for some φ, but $\mathbf{L}$ itself is consistent, i.e., not equal to the set of all formulas. This means, that the principle *ex contradictione quodlibet* does not apply to $\mathbf{L}$. In a somewhat loose way, we may say that in a paraconsistent logic we may safely assume contradictions, because they do not overfill the system. The logics corresponding to the algebras $\mathbf{A}_S$ are clearly paraconsistent.

Returning to the less pathological realm, we will now exhibit a construction in a certain sense dual to the one just presented. However, contrary to what one might expect, it produces only countably many minimal subvarieties of $\mathsf{RFL}_{\mathsf{ei}}$. By itself this result does not add much to our knowledge, there are other ways to produce countable families of such minimal varieties (see Exercise 13). We present it here mainly to demonstrate certain pitfalls that may befall the reckless. For any subset S of $\mathbb{N}\setminus\{0\}$ and $k\in\mathbb{N}$, let $\mathbf{V}_k^S = \langle V_k^S, \wedge, \vee, \cdot, \to, 0, 1\rangle$, be an algebra defined as follows. We put $V_k^S = \{0^i\colon 0\leq i\leq k+1\}\cup\{\star_i\colon 0<i<k,\ i\in S\}$, with 0^0 and 0^1 being respectively the constants 1 and 0. On V_k^S we define a linear ordering, setting $\bot = 0^{k+1} < \cdots 0^2 < 0 < 1 = \top$ and $0^i < \star_i < 0^{i-1}$, for $i\in S$; see Figure 9.3 yet again. Then, we define multiplication putting

$$0^m\cdot 0^n = 0^{\min\{m+n,k+1\}} = \star_n\cdot\star_m$$

where the right-hand side only makes sense for $n,m\in S$, and

$$\star_n\cdot 0^m = 0^m\cdot\star_n = \begin{cases}\star_n & \text{if } m=0\\ 0^{\min\{n+m,k+1\}} & \text{if } m>0\end{cases}$$

Notice that $\mathbf{V}_k^S$ may be viewed as obtained from $\mathbf{A}_k^S$ by first stripping off $\bot$, then chopping A_k^S into blocks $\{0^i,\star_i\}$ for $i\in S$ and $\{0^i\}$ for $i\notin S$, and finally reversing the ordering between blocks, but not within them (hence the name: V is A upside-down with some adjustments). Clearly, each $\mathbf{V}_k^\emptyset$ is strictly simple (see Exercise 10 for an extension of this observation) so the next theorem is immediate.

THEOREM 9.29. *There are at least $\aleph_0$ minimal varieties of representable, integral FL_e-algebras.*

Now one can mimic the ultraproduct argument used before to produce an uncountable family of pairwise non-isomorphic algebras $\mathbf{V}_S$. Namely, define $\mathbf{V}_S$ to be the zero-generated subalgebra of an ultraproduct $\prod_{k\in\mathbb{N}}\mathbf{V}_k^S/U$ for a nonprincipal U. See Figure 9.3 for what $\mathbf{V}_S$ looks like. However, an analogue of Lemma 9.26 fails to hold and that leads to the following somewhat unwelcome result, which we ask the reader to prove in Exercise 11.

PROPOSITION 9.30. *If the variety $\mathsf{V}(\mathbf{V}_S)$ is distinct from $\mathsf{V}(\mathbf{V}_\emptyset)$ then $\mathsf{V}(\mathbf{V}_S)$ is not minimal.*

9.3.3. Minimal subvarieties of $\mathsf{FL_e}$ with term-definable bounds. In this section we will construct another continuum of minimal varieties of commutative FL-algebras, so, in purely algebraic terms, we are not strengthening the result (or at least not much). However, the algebras we construct here differ from the ones in previous section in that they are all bounded, with term-definable bounds.

The construction itself will follow the lines of the two previous ones, so we will allow ourselves to be very sketchy. For any $S\subseteq 2\mathbb{N}+1$ and $k\in\mathbb{N}$, we define an algebra $\mathbf{B}_k^S = \langle B_k^S, \wedge, \vee, \cdot, \to, 0, 1\rangle$, as follows. For the universe, we put $B_k^S = \{\top,\bot\}\cup\{0^i\colon 0\leq i\leq 2k+1\}\cup\{\star_i\colon 0<i<k, i\in S\}$, with 0^0 and 0^1 being respectively the constants 1 and 0. Then we define

an ordering on B_k^S, setting $\bot < 1 < 0^2 < 0^4 < \cdots < 0^{2k} < \top$ and $\bot < 0 < \star_1 < 0^3 < \star_2 < \cdots < 0^{2i+1} < \star_{2i+1} < 0^{2i+3} < \cdots < 0^{2k+1} < \top$. Notice that this is in fact a lattice ordering with $x \wedge y = \bot$ and $x \vee y = \top$ for all pairs (x, y) of incomparable elements. By the same token, the constants $\bot$ and $\top$, although not in the type, are definable by $\bot = 0 \wedge 1$ and $\top = 0 \vee 1$. Further, multiplication gets defined by

$$1 \cdot x = x \cdot 1 = x, \text{ for all } x$$

$$\bot \cdot x = x \cdot \bot = \bot, \text{ for all } x$$

$$\top \cdot x = x \cdot \top = \top, \text{ for all } x \neq \bot$$

$$0^m \cdot 0^n = \begin{cases} 0^{m+n} & \text{if } m + n \leq 2k + 1 \\ 0^{2k} & \text{if } m + n > 2k + 1 \text{ and } m + n \text{ is even} \\ 0^{2k+1} & \text{if } m + n > 2k + 1 \text{ and } m + n \text{ is odd} \end{cases}$$

$$\star_n \cdot \star_m = \star_n \cdot 0^m = 0^m \cdot \star_n = 0^m \cdot 0^n \text{ for } n, m > 0$$

where the relevant equations in the last line are defined only for the appropriate values of n and m. This definition is exhaustive ($m = 0$ implies $0^0 = 1$, and $\star_0$ does not exist), and verifying that so defined multiplication is associative is tedious but straightforward. It is also not difficult to see that, for any $x, y \in B_k^S$, the set $\{z \in B_k^S : zx \leq y\}$ has the greatest element, thus, multiplication is residuated. We will not give all the values explicitly, but observe that

$$0^{2i} \to 0^{2l+1} = \begin{cases} 0^{2k+1} & \text{if } l = k \\ 0^{2(l-1)+1} & \text{if } i \leq l < k \text{ and } l - i + 1 \notin S \\ \star_{l-i+1} & \text{if } i \leq l < k \text{ and } l - i + 1 \in S \\ \bot & \text{if } l < i \end{cases}$$

and

$$0^{2i+1} \to 0^{2l} = \begin{cases} 0^{2k+1} & \text{if } l = k \\ 0^{2(l-i)-1} & \text{if } i < l < k \text{ and } l - i \notin S \\ \star_{l-i} & \text{if } i < l < k \text{ and } l - i \in S \\ \bot & \text{if } l \leq i \end{cases}$$

hold. These suffice to make sure that all the existing $\star_j$ elements are generated by the constants.

Again as before, for any $S \subseteq \mathbb{N}$ and U a nonprincipal ultrafilter over $\mathbb{N}$, we take the ultraproduct $\prod_{k \in \mathbb{N}} \mathbf{B}_k^S / U$ and define $\mathbf{B}_S$ to be the zero-generated subalgebra of $\prod_{k \in \mathbb{N}} \mathbf{B}_k^S / U$.

LEMMA 9.31. *The algebra $\mathbf{B}_S$ is a strictly simple FL_e-algebra.*

PROOF. A tedious but straightforward case-by-case verification ensures that each $\mathbf{B}_k^S$ is an FL_e-algebra. Its negative part is $\{\bot, 1\}$. Now we finish the proof by an argument paralleling the proof of Lemma 9.26. □

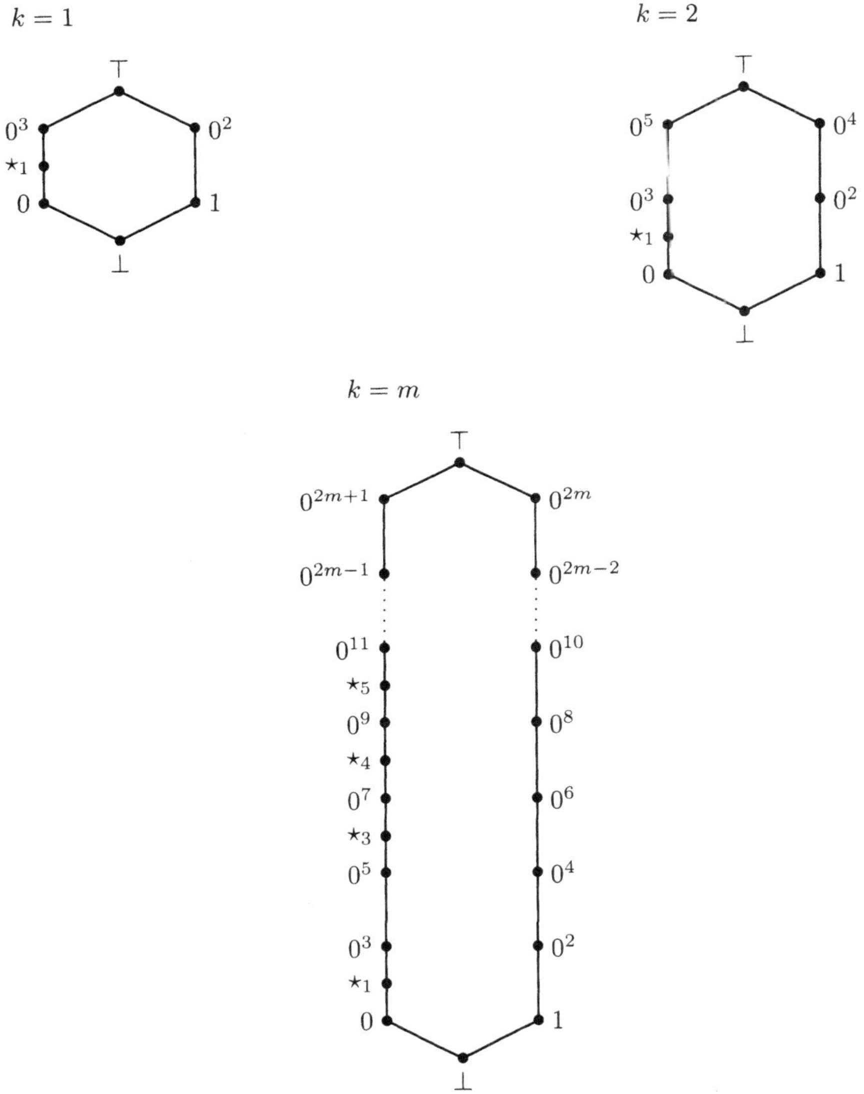

FIGURE 9.5. The algebras $\mathbf{B}_k^S$ for $S = \{1, 3, 4, 5\}$.

We leave it for the reader to verify (see Exercise 12) that distinct subsets S, T of $\mathbb{N}$ yield non-isomorphic algebras $\mathbf{B}_T$ and $\mathbf{B}_S$. Then the result below follows by an application of Corollary 9.2.

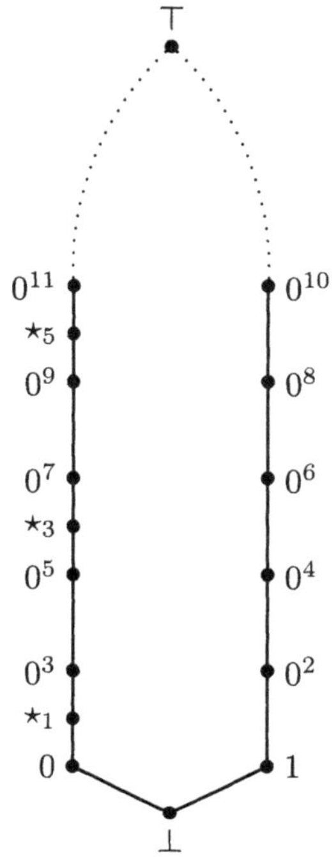

FIGURE 9.6. $\mathbf{B}_S$ for $S = \{n\colon n \text{ is odd}\}$.

THEOREM 9.32. [Kih06] *There are continuum many minimal varieties of FL_e-algebras satisfying* $0 \wedge 1 \leq x \leq 0 \vee 1$.

9.3.4. Minimal subvarieties of $\mathsf{FL}_{\mathsf{eco}}$. We have seen that $\mathbf{\Lambda}(\mathsf{FL}_{\mathsf{co}})$ and $\mathbf{\Lambda}(\mathsf{FL}_{\mathsf{e}})$ each have continuum atoms. In this section we will show that they have only two atoms in common. Let $\mathbf{A}$ be a FL_{eco}-algebra and recall that $\mathbf{A}^-$ denotes the negative cone of $\mathbf{A}$.

LEMMA 9.33. *For any FL_{eco}-algebra $\mathbf{A}$, its negative cone $\mathbf{A}^-$ is a Heyting algebra. Moreover* $\mathbf{Con\,A}$ *and* $\mathbf{Con\,A}^-$ *are isomorphic via the natural restriction map.*

PROOF. Since $\mathbf{A}$ is square-increasing, its negative cone consists entirely of idempotent elements and therefore is a Heyting algebra and hence it has a distributive lattice reduct. The rest follows by Lemma 3.49. □

LEMMA 9.34. *Let $\mathbf{A}$ be a strictly simple FL_{eco}-algebra. Then, $\mathbf{A}$ is isomorphic to $\mathbf{2}$ or $\mathbf{To}_1$.*

PROOF. By Lemma 9.33, $\mathbf{A}^-$ is a simple Heyting algebra, so it must be isomorphic to $\mathbf{2}$. Thus, the negative cone of A is $\{1, 0\}$. Since $\mathbf{A}$ is strictly

$\to$	$\top$	1	0
$\top$	$\top$	?	0
1	$\top$	1	0
0	$\top$	$\top$	$\top$

FIGURE 9.7. Partial table for residuation.

Variety	# of atoms	namely	reference
$\mathsf{P_3RRL}$	$2^{\aleph_0}$	$\mathsf{V}(\mathbf{J}_S)$	Thm. 9.10
$\mathsf{P_1RRL}$	$2^{\aleph_0}$	$\mathsf{V}(\mathbf{N}_S)$ (bi-infinite)	Cor. 9.18
CRRL	at least $\aleph_0$	$\mathsf{V}(\mathbf{T}_n)$	Cor. 9.5
$\mathsf{P_1CRRL}$	2	$\mathsf{V}(\mathbf{2}_r)$, $\mathsf{V}(\mathbf{T}_1)$	Thm. 9.11
CanRL	2	$\mathsf{V}(\mathbb{Z})$, $\mathsf{V}(\mathbb{Z}^-)$	Cor. 9.9
GBL	at least 3	$\mathsf{V}(\mathbf{2}_r)$, $\mathsf{V}(\mathbb{Z}^-)$, $\mathsf{V}(\mathbb{Z})$	above
ICRRL	at least 2	$\mathsf{V}(\mathbf{2}_r)$, $\mathsf{V}(\mathbb{Z}^-)$	above

TABLE 9.1. Minimal subvarieties (atoms) of RL.

simple, it is generated by 0. Note that $0 \to 1 = 0 \to 0 = \top$ is the top element. If $\top = 1$, then $\{1, 0\}$ is a subuniverse of $\mathbf{A}$ and thus equal to A. In that case $\mathbf{A} = \mathbf{2}$. If $\top > 1$, then the chain $\{0, 1, \top\}$ is closed under multiplication and almost closed under residuation, namely, we have the table in Figure 9.7, with only the entry for $\top \to 1$ missing. Note that $a = \top \to 1$ is a negative element, since $a \leq \top a \leq 1$; hence $a \in \{0, 1\}$. On the other hand $a \neq 1$, since otherwise we would have $\top = \top a \leq 1$, a contradiction. Thus, $a = 0$. This completes the table and shows that $\mathbf{A}$ is then isomorphic to $\mathbf{To}_1$. □

COROLLARY 9.35. *There are exactly two atoms in* $\mathbf{\Lambda}(\mathsf{FL_{eco}})$.

9.4. Almost minimal subvarieties of $\mathsf{FL_{ew}}$

Now we move one level up in the subvariety lattice, namely to covers of atoms, which we call *almost minimal*. In particular, we will look at almost minimal subvarieties of varieties whose subvariety lattices have only one atom. Two classical examples of such varieties are: the variety HA of Heyting algebras, where the unique atom is the variety BA of Boolean algebras, and the variety of ℓ-groups, where the unique atom is the variety of Abelian ℓ-groups. Now, in $\mathbf{\Lambda}(\mathsf{HA})$ there is only one almost minimal variety, generated by the three element Heyting algebra $\mathbf{H}_3$, as shown in Lemma 1.58, whereas in $\mathbf{\Lambda}(\mathsf{LG})$ there are continuum many almost minimal varieties (see e.g., [GH89]). One natural generalization of $\mathbf{\Lambda}(\mathsf{HA})$ is the lattice $\mathbf{\Lambda}(\mathsf{FL_{ew}})$, whose unique atom is also BA. We will begin by looking at this lattice.

Variety	# of atoms	namely	reference
$\mathsf{P_3RFL_o}$	$2^{\aleph_0}$	$\mathsf{V}(\mathbf{Jo}_S)$	Cor. 9.19
$\mathsf{P_3RFL_i}$	$2^{\aleph_0}$	$\mathsf{V}(\mathbf{Ji}_S)$	Cor. 9.21
$\mathsf{P_1FL_c}$	$2^{\aleph_0}$	$\mathsf{V}(\mathbf{N}_S)$	Cor. 9.18
$\mathsf{P_1RFL_o}$	$2^{\aleph_0}$	$\mathsf{V}(\mathbf{No}_w)$	Thm. 9.24
$\mathsf{FL_e}$	$2^{\aleph_0}$	$\mathsf{V}(\mathbf{B}_S)$	Thm. 9.32
$\mathsf{RFL_{ce}}$	$2^{\aleph_0}$	$\mathsf{V}(\mathbf{A}_S)$	Thm. 9.28
$\mathsf{RFL_{ei}}$	at least $\aleph_0$	$\mathsf{V}(\mathbf{V}_n^{\emptyset})$	Thm. 9.29
$\mathsf{RFL_{eo}}$	at least $\aleph_0$	$\mathsf{V}(\mathbf{To}_n)$	Ex. 6
$\mathsf{FL_{eco}}$	2	$\mathsf{V}(\mathbf{2})$, $\mathsf{V}(\mathbf{To}_1)$	Cor. 9.35
$\mathsf{FL_{ci}}$	2	$\mathsf{V}(\mathbf{2})$, $\mathsf{V}(\mathbf{2}_r)$	Ex. 4
$\mathsf{FL_w}$	1	$\mathsf{V}(\mathbf{2})$	Prop. 9.22
CyInFL	$2^{\aleph_0}$		Thm. 9.25

TABLE 9.2. Minimal subvarieties (atoms) of FL.

9.4.1. General facts about almost minimal varieties. Let $\mathbf{A}$ be a subdirectly irreducible FL_{ew}-algebra. We say that $\mathbf{A}$ is *stiff* if (1) $\mathbf{A}$ has exactly one proper nontrivial quotient isomorphic to the two element Boolean algebra, (2) $\mathbf{A}$ has no proper subalgebras other than the two element Boolean algebra.

Recall from Chapter 3 that $\mathsf{E_n}$ is the variety of integral n-potent residuated lattices. In particular, $\mathsf{E_n} \cap \mathsf{FL_{ew}} = \mathsf{P_nFL_{ew}}$.

LEMMA 9.36. [KKU06] *Let $\mathbf{A}$ be a stiff FL_{ew}-algebra, with monolith μ. If $\mathbf{A} \in \mathsf{E_n}$ for some positive integer n, the following conditions hold and can be expressed by first-order formulas:*

(1) *there is precisely one $b \in A \setminus \{0,1\}$, with $b^2 = b$;*
(2) *for any $a \in A$, $a \in [1]_\mu$ iff $a \geq b$;*
(3) *for any $a \in A$, $a \in [1]_\mu$ or $\neg a \in [1]_\mu$;*
(4) ***Con A** is a three-element chain.*

PROOF. For (1) observe first that any non-unit element $u \in [1]_\mu$, we have $u^{n+1} = u^n$. Putting $b = u^n$, we get that b is the unique idempotent element in $A \setminus \{1,0\}$. Since (1) is already a first-order formula, the claim follows. For (2) note that by (1) b is a definable constant; moreover, by EDPC for $\mathsf{E_n}$, we can express '$a \in [1]_\mu$' by a first-order formula. If (3) did not hold, then $\mathbf{A}/\mu$ would fail to be a Boolean algebra. Finally, (4) holds in $\mathbf{A}$ by definition, and is equivalent to the following: for any $a \in A$, if $a \notin [1]_\mu$, then the congruence generated by $(a,1)$ is the full congruence. This, again by EDPC, can be expressed as a first-order formula for $\mathbf{A}$. □

LEMMA 9.37. [KKU06] *Let $\mathbf{A} \in \mathsf{E_n}$ be stiff. For any equation $\sigma = \tau$, we have: $\sigma = \tau$ is true in the two-element Boolean algebra iff $\sigma \to \tau \in [1]_\mu$ and*

$\tau \to \sigma \in [1]_\mu$ iff $\mathbf{A} \models \varphi(\sigma, \tau)$, where $\varphi(\sigma, \tau)$ is a certain first-order formula definable from σ and τ. In particular, the fact that $\mathbf{A}/\mu$ is isomorphic to the two-element Boolean algebra is expressible by a first-order formula.

PROOF. The first equivalence follows from the definition of stiff FL_{ew}-algebra and the fact that $a \to b \in [1]_\mu$ iff $a/\mu \leq b/\mu \in \mathbf{A}/\mu$. For the second it suffices to recall that '$a \in [1]_\mu$' is first-order definable. For the last statement note that '$\mathbf{A}/\mu$ is isomorphic to the two-element Boolean algebra' can be restated as the conjunction of '$\mathbf{A}/\mu$ has precisely two elements' and 'all equations that hold in Boolean algebras hold in $\mathbf{A}/\mu$'. The former is expressible by a first-order sentence, regardless of the type (as long as identity is present). The latter is further equivalent to 'all axioms of Boolean algebras hold in $\mathbf{A}/\mu$' and that, by finite axiomatizability of Boolean algebras, is a finite conjunction of first-order formulas. □

LEMMA 9.38. [KKU06] *Let $\mathbf{A} \in \mathsf{E_n}$ be stiff. All the claims from Lemmas 9.36 and 9.37 hold for any ultrapower $\mathbf{A}^I/U$ of $\mathbf{A}$.*

PROOF. As $\mathbf{A}$ and $\mathbf{A}^I/U$ are elementarily equivalent, $\mathbf{A}^I/U$ satisfies all the properties of $\mathbf{A}$ expressible by first-order formulas. Since $\mathbf{A} \in \mathsf{E_n}$, all the claims from Lemma 9.36 are such. Thus, the congruence lattice of $\mathbf{A}^I/U$ is a three-element chain, and therefore $\mathbf{A}^I/U$ is subdirectly irreducible. Let ν stand for its monolith. Then, '$a \in [1]_\nu$' is first-order definable on $\mathbf{A}^I/U$, which suffices to conclude that the claims from Lemma 9.37 also hold. □

LEMMA 9.39. [KKU06] *Let $\mathbf{A} \in \mathsf{E_n}$ be stiff, and let $\mathbf{C}$ be a subdirectly irreducible algebra in $\mathsf{V}(\mathbf{A})$ non-isomorphic to the two-element Boolean algebra. Then, $\mathbf{A}$ is a subalgebra of $\mathbf{C}$.*

PROOF. By Jónsson's Lemma, $\mathbf{C}$ belongs to $\mathsf{HSP_U}(\mathbf{A})$. By congruence extension property, $\mathbf{C} \in \mathsf{SHP_U}(\mathbf{A})$. Thus, there is an ultrapower $\mathbf{A}^I/U$, with I any set of indices and U an ultrafilter on I, and a congruence ψ on $\mathbf{A}^I/U$, such that $\mathbf{C} \subseteq (\mathbf{A}^I/U)/\psi$. By Lemma 9.38, $\mathbf{A}^I/U$ is subdirectly irreducible, and the congruence lattice of $\mathbf{A}^I/U$ is isomorphic to the three-element chain. Thus, our congruence ψ can only be trivial. It follows that $\mathbf{C} \subseteq \mathbf{A}^I/U$. By Lemma 9.36 again, there is precisely one element $b \in (A^I/U) \setminus \{0, 1\}$, with $b^2 = b$, and, moreover, for any $a \in (A^I/U) \setminus \{0, 1\}$ we have either $a^n = b$ or $(\neg a)^n = b$, by Lemma 9.36(3). Since the canonical embedding of $\mathbf{A}$ into $\mathbf{A}^I/U$ sends idempotent elements to idempotent elements and $\mathbf{A}$ is stiff, we have that b generates $\mathbf{A}$ as a subalgebra of $\mathbf{A}^I/U$. Thus, $\mathbf{A} \subseteq \mathbf{C}$. □

LEMMA 9.40. [KKU06] *Each stiff member of $\mathsf{E_n}$ generates an almost minimal variety. If $\mathbf{A} \in \mathsf{E_n}$ and $\mathbf{B} \in \mathsf{E_n}$ are non-isomorphic stiff algebras, then $V(\mathbf{A}) \neq V(\mathbf{B})$.*

PROOF. Let $\mathbf{A}$ be a stiff algebra in $\mathsf{E_n}$, and $\mathbf{C}$ be any non-Boolean subdirectly irreducible algebra in $V(\mathbf{A})$. Clearly, $V(\mathbf{A}) \supseteq V(\mathbf{C})$. By Lemma 9.39, $\mathbf{A} \subseteq \mathbf{C}$. Thus, $V(\mathbf{A}) = V(\mathbf{C})$, which proves the first statement.

For the second, by way of contradiction assume that $V(\mathbf{A}) = V(\mathbf{B})$. Then, $\mathbf{A} \in V(\mathbf{B})$ and $\mathbf{B} \subseteq \mathbf{A}$, by Lemma 9.39. Since $\mathbf{A}$ is stiff, it has only two subalgebras: the two-element Boolean algebra and itself. Since the two-element Boolean algebra is not stiff, we get that $\mathbf{B} \cong \mathbf{A}$, a contradiction. □

As there are only countably many finite stiff FL_{ew}-algebras, our purpose requires constructing infinite ones. This, however, is much less difficult than it may seem, at least for varieties with EDPC, since being stiff is a first-order property and so carries over to ultraproducts. We will assume that we work within a subvariety $\mathcal{V}$ of $\mathsf{E_k}$, for some positive integer k. Suppose that for each $n \in \mathbb{N}$, we have a finite stiff FL_{ew}-algebra $\mathbf{A}_n$, with $|A_n| > n$. Recall from Lemma 9.36 that each of these has a unique idempotent element different from 0 and 1. We will reserve the symbol $\star$ for it. Let now $\mathbf{A} = \prod_{n\in\mathbb{N}} \mathbf{A}_n/U$, be an ultraproduct of all the $\mathbf{A}_n$ by a free ultrafilter U on $\mathbb{N}$. Then, let $\mathbf{A}^\star$ be the subalgebra of $\mathbf{A}$ generated by $\star = \langle \star_n : n \in \mathbb{N}\rangle/U$.

LEMMA 9.41. [KKU06] *The algebra $\mathbf{A}^\star$ is an infinite stiff FL_{ew}-algebra that belongs to $\mathsf{E_k}$.*

PROOF. That $\mathbf{A} \in \mathsf{E_k}$ is clear as it is a subalgebra of an ultraproduct of algebras $\mathbf{A}_n$ that belong to $\mathsf{E_k}$. Thus, the ultraproduct $\mathbf{A}$ is subdirectly irreducible, and its congruence lattice is a three-element chain. The same goes for $\mathbf{A}^\star$, and to show that it is stiff we only need to prove that every $a \in A^\star \setminus \{1,0\}$ generates $\mathbf{A}^\star$. Clearly, it suffices to show that a generates $\star$. If $a \in [1]_\mu$, where μ is the monolith of $\mathbf{A}^\star$, then $a^n = \star$. Otherwise, $\neg a \in [1]_\mu$, and thus $(\neg a)^n = \star$.

Now to show that $\mathbf{A}^\star$ is infinite, note first that the elements generated by $\star$ in $\mathbf{A}$ are congruence classes of elements generated in $\prod_{n\in\mathbb{N}} \mathbf{A}_n$ by $\langle \star_n : n \in \mathbb{N}\rangle$. As usual we view the generation process as follows: for any $n \in \mathbb{N}$, let $G_n^0 = \{\star_n\}$, and $G_n^{k+1} = \{F(g) : g \in G_n^k, F - \text{a basic operation}\}$. Suppose $\mathbf{A}^\star$ is finite. Then, by properties of ultraproducts, any element x generated by $\langle \star_n : n \in \mathbb{N}\rangle$ at every $k+1$-th stage of generation "later" than some fixed m, belongs to G_i^k at almost all coordinates i. This implies that $G_i^{k+1} = G_i^k$, for almost all i. Since the type is finite, $|G_i^k|$ is bounded by a finite number that does not depend on i (its very rough estimate is $\underbrace{4(4(\cdots(4}_{k-1}\cdot\overbrace{3^{2})^2\ldots)^2}^{k-1}$, for $k \geq 2$). It follows that infinitely many $\mathbf{A}_i$ have sizes bounded by the same finite number, which contradicts the assumption that $|A_n| > n$ for each n. □

It follows from Lemma 9.41 that one way of constructing continuum many almost minimal varieties of FL_{ew}-algebras can go as follows. Working within $\mathsf{E_k}$, for some positive integer k, construct continuum many families of finite stiff algebras, each family containing algebras of unbounded size. Say, for each $S \subseteq \mathbb{N}$, construct a family $(\mathbf{A}_{S,n})_{n\in\mathbb{N}}$, with $|\mathbf{A}_{S,n}| \geq n$, for each $n \in \mathbb{N}$. Then, for each S, the algebra $\mathbf{A}_S^\star$, is stiff. The task will be completed if we manage to show that $\mathbf{A}_S^\star$ is non-isomorphic to $\mathbf{A}_T^\star$, whenever $S \neq T$.

We will present three constructions employing the general strategy outlined above, the first in some detail, the other two rather briefly.

9.4.2. Almost minimal subvarieties of $\mathsf{InFL_{ew}}$. Let S be any subset of $\mathbb{N}$. For any positive integer K, define C_K^S be the disjoint union of the sets: $A = \{a_0, \dots, a_K\}$, $B = \{b_{-1}, b_0, \dots, b_K\}$, $M = \{m_{-1}, m_0, \dots, m_K\}$, $N = \{n_0, \dots, n_K\}$, and $E = \{0, 1\}$. C_K^S is partially ordered (see Figure 9.9); the order relation is specified by the conditions:

- $1 > a_0$, $n_0 > 0$;
- $a_i > a_j$ iff $i < j$, $b_i > b_j$ iff $i < j$;
- $m_i > m_j$ iff $i > j$, $n_i > n_j$ iff $i > j$;
- $a_K > b_{-1}$, $m_{-1} > n_K$;
- $b_0 > m_K$, $b_K > m_0$.

Multiplication on C_K^S is defined by putting:

- $1 \cdot x = x$, $0 \cdot x = 0$, $x \cdot y = y \cdot x$, for all $x, y \in C_K^S$;
- $a_i \cdot a_j = \begin{cases} b_{\max\{i,j\}-1}, & \text{if } i \neq j,\ \max\{i,j\} \in S, \text{ or} \\ & \quad \min\{i,j\} = 0, \\ b_{\max\{i,j\}}, & \text{if } \min\{i,j\} > 0,\ \max\{i,j\} \notin S, \text{ or} \\ & \quad i = j > 0; \end{cases}$
- $a_i \cdot b_j = b_K$;
- $a_i \cdot n_j = \begin{cases} 0, & \text{if } i \geq j, \\ n_0, & \text{if } i < j; \end{cases}$
- $a_i \cdot m_j = \begin{cases} n_0, & \text{if } i > j, \\ n_1, & \text{if } i = j \in \{0,1\}, \text{ or} \\ & \quad i = j > 1,\ j \notin S, \\ n_j, & \text{if } i = j > 1,\ j \in S, \text{ or} \\ & \quad 0 < i < j,\ j \notin S, \\ n_{j+1}, & \text{if } 0 = i < j < K, \text{ or} \\ & \quad 0 < i < j < K,\ j \in S, \\ m_{-1}, & \text{if } 0 = i < j = K, \text{ or} \\ & \quad 0 < i < j = K,\ K \notin S; \end{cases}$
- $b_i \cdot b_j = b_K$;
- $b_i \cdot m_j = \begin{cases} 0, & \text{if } i \geq j, \\ n_0, & \text{if } i < j; \end{cases}$
- $b_i \cdot n_j = 0$;

- $m_i \cdot m_j = \begin{cases} 0, & \text{if } \min\{i,j\} \leq 0, \\ n_0, & \text{otherwise;} \end{cases}$
- $m_i \cdot n_j = 0$;
- $n_i \cdot n_j = 0$.

Residuation on C_K^S is defined by putting:

- $x \to y = \bigvee\{z \in C_K^S : z \cdot x \leq y\}$.

Finally, we define $\mathbf{C}_K^S$ to be the algebra $\langle C_K^S, \vee, \wedge, \cdot, \to, 0, 1\rangle$.

LEMMA 9.42. [KKU06] *The algebra $\mathbf{C}_K^S$ is an FL_{ew}-algebra.*

PROOF. Because of finiteness, it suffices to verify two things: that multiplication above, commutative by definition, is also associative, and that multiplication distributes over join, i.e., $x \cdot (y \vee z) = (x \cdot y) \vee (x \cdot z)$ holds. For associativity we have to consider several cases, of which only one is not straightforward and is dealt with below:

$$(a_i \cdot a_j) \cdot m_k = \begin{cases} b_{j-1} \cdot m_k, & \text{if } i = 0, \\ b_{i-1} \cdot m_k, & \text{if } j = 0, \\ b_{j-1} \cdot m_k, & \text{if } 0 < i < j,\ j \in S, \\ b_i \cdot m_k, & \text{if } 0 < i = j, \\ b_j \cdot m_k, & \text{if } 0 < i < j,\ j \notin S, \\ b_i \cdot m_k, & \text{if } 0 < i > j,\ i \notin S; \end{cases}$$

which yields further:

$$(a_i \cdot a_j) \cdot m_k = \begin{cases} 0, & \text{if } j-1 \geq k,\ i = 0, \text{ or} \\ & \quad i-1 \geq k,\ j = 0, \text{ or} \\ & \quad j-1 \geq k,\ 0 < i < j,\ j \in S, \text{ or} \\ & \quad i-1 \geq k,\ 0 < j < i,\ i \in S, \text{ or} \\ & \quad i \geq k,\ 0 < i = j, \text{ or} \\ & \quad j \geq k,\ 0 < i < j,\ j \notin S, \text{ or} \\ & \quad i \geq k,\ 0 < j < i,\ i \notin S; \\ n_0, & \text{if } j-1 < k,\ i = 0, \text{ or} \\ & \quad i-1 < k,\ j = 0, \text{ or} \\ & \quad j-1 < k,\ 0 < i < j,\ j \in S, \text{ or} \\ & \quad i-1 < k,\ 0 < j > i,\ i \in S, \text{ or} \\ & \quad i < k,\ 0 < i = j, \text{ or} \\ & \quad j < k,\ 0 < i < j,\ j \notin S, \text{ or} \\ & \quad i < k,\ 0 < j < i,\ i \notin S. \end{cases}$$

Then, changing the bracketing, we get:

$$a_i \cdot (a_j \cdot m_k) = \begin{cases} a_i \cdot n_0, & \text{if } j > k, \\ a_i \cdot n_1 & \text{if } j = k \in \{0,1\}, \text{ or} \\ & \quad j = k > 1,\ k \notin S, \\ a_i \cdot n_k, & \text{if } j = k > 1,\ k \in S, \text{ or} \\ & \quad 0 < j < k,\ k \notin S, \\ a_i \cdot n_{k+1}, & \text{if } 0 = j < k < K, \text{ or} \\ & \quad 0 < j < k < K,\ k \in S, \text{ or} \\ a_i \cdot m_{-1}, & \text{if } 0 = j < k = K, \text{ or} \\ & \quad 0 < j < k = K,\ K \in S. \end{cases}$$

which finally yields:

$$a_i \cdot (a_j \cdot m_k) = \begin{cases} 0, & \text{if } j > k, \text{ or} \\ & \quad i \geq 1,\ j = k \in \{0,1\}, \text{ or} \\ & \quad i \geq 1,\ j = k > 1,\ k \notin S, \text{ or} \\ & \quad i \geq k,\ j = k > 1,\ k \in S, \text{ or} \\ & \quad i \geq k,\ 0 < j < k,\ k \notin S, \text{ or} \\ & \quad i \geq k+1,\ 0 = j < k < K, \text{ or} \\ & \quad i \geq k+1,\ 0 < j < k < K,\ k \in S; \\ n_0, & \text{if } i = 0,\ j = k \in \{0,1\}, \text{ or} \\ & \quad i = 0,\ j = k > 1,\ k \notin S, \text{ or} \\ & \quad i < k,\ j = k > 1,\ k \in S, \text{ or} \\ & \quad i < k,\ 0 < j < k,\ k \notin S, \text{ or} \\ & \quad i < k+1,\ 0 = j < k < K, \text{ or} \\ & \quad i < k+1,\ 0 < j < k < K,\ k \in S, \text{ or} \\ & \quad 0 = j < k = K, \text{ or} \\ & \quad 0 < j < k = K,\ K \in S. \end{cases}$$

The proof now reduces to a series of tedious case-by-case calculations confirming that $a_i \cdot (a_j \cdot m_k) = 0$ if and only if $(a_i \cdot a_j) \cdot m_k = 0$.

For distributivity of multiplication over join, we proceed in two steps. First, we show that multiplication is monotone, i.e., $x \leq y$ implies $z \cdot x \leq z \cdot y$. Out of a number of cases only two deserve attention:

The first is $z = a_i$, $x = a_j$, $y = a_k$. As $a_j \leq a_k$, we have $j \geq k$. We may assume $j > k$. Then, if $i \geq j > k$, both $a_i \cdot a_j$ and $a_i \cdot a_k$ are equal to either of b_i, b_{i-1}. The only dubious case arises when $a_i \cdot a_j = b_{i-1}$. This, however, can happen only if $i \in S$, and then $a_i \cdot a_k = b_{i-1}$ as well. If $j > i \geq k$, we get $a_i \cdot a_j \in \{b_{j-1}, b_j\}$ and $a_i \cdot a_k \in \{b_{i-1}, b_i\}$. Since $j > i$, we have $b_{j-1} \leq b_i$, which establishes the claim for all subcases. Finally, if $j > k > i$, we have

$a_i \cdot a_j \in \{b_{j-1}, b_j\}$ and $a_i \cdot a_k \in \{b_{k-1}, b_k\}$. As $j > k$, the previous reasoning applies.

The second is $z = a_i$, $x = m_j$, $y = m_k$. Here we have $j \leq k$, and we may also assume $j < k$. Then, if $i \leq j$, we have $a_i \cdot m_j \in \{n_1, n_j, n_{j+1}\}$, and only the cases (1) $a_i \cdot m_j = n_j$ and (2) $a_i \cdot m_j = n_{j+1}$ are not straightforward. Case (1) splits into two: (1a) $i = j > 1$ and $j \in S$, in which case $k > i > 1$, and thus $a_i \cdot m_k \in \{n_j, n_{j+1}, m_{-1}\}$, proving the claim; (1b) $0 < i < j$ and $j \notin S$, in which case $a_i \cdot m_k \in \{n_k, n_{k+1}, m_{-1}\}$, proving the claim as well, for $j < k$, by the assumption. Case (2) also splits into two: (2a) if $0 = i < j < K$, then $a_i \cdot m_k \in \{n_{k+1}, m_{-1}\}$, and since $j < k$, the claim is proved; (2b) if $0 < i < j < K$ and $j \in S$, then $a_i \cdot m_k \in \{n_k, n_{k+1}, m_{-1}\}$; now the assumption guarantees that $j + 1 \leq k$, therefore the claim holds here as well.

Having established monotonicity, we can approach the proof of distributivity. Consider $x(y \vee z)$. If y and z are compatible, say $y \geq z$, then $x(y \vee z) = xy = xy \vee xz$, by monotonicity. Suppose y and z are incompatible. We may take $y = b_i$ ($i \in \{1, \dots, K\}$) and $z = m_j$ ($j \in \{1, \dots, K\}$). Then, $x(y \vee z) = x(b_i \vee m_j) = xb_0$, and $xy \vee xz = xb_i \vee xm_j$. If $x = 1$, or $x = 0$, then the desired equality obviously holds. Assume $x \notin \{1, 0\}$. Then, if $x \geq b_K$, we have: $x(y \vee z) = xb_0 = b_K$, and $xy \vee xz = xb_i \vee xm_j = b_K$, since $xm_j \leq m_0$. Thus, the equality holds here, too. If $x \leq m_K$, then two cases should be distinguished: (1) $x \leq m_0$, in which case $x(y \vee z) = xb_0 = 0$, and $xy \vee xz = xb_i \vee xm_j = 0 \vee 0 = 0$; (2) $x = m_k$, with $k \in \{1, \dots, K\}$, in which case $x(y \vee z) = m_k b_0 = n_0$, and $xy \vee xz = m_k b_i \vee m_k m_j = n_0$, as well. This finishes the whole proof. □

LEMMA 9.43. [KKU06] *The FL_{ew}-algebra $\mathbf{C}_K^S$ is stiff. Moreover, it belongs to E_3 and satisfies $\neg\neg x = x$.*

PROOF. It is clear from the construction that $\mathbf{C}_K^S$ satisfies $x^4 = x^3$. It is also clear, that the only nontrivial and non-full congruence on $\mathbf{C}_K^S$ is the one associated with the filter $F = \{x \in C_K^S : x \geq b_K\}$. The quotient of this congruence is the two-element Boolean algebra. We also have: $\neg a_i = n_i$, $\neg b_i = m_i$, $\neg n_i = a_i$, and $\neg m_i = b_i$; we leave out the detailed calculations. This shows that $\neg\neg x = x$ holds. Moreover, for any $x \in C_K^S \setminus \{1, 0\}$, we have either $x^3 = b_K$, or $(\neg x)^3 = b_K$. Thus, to prove that $\mathbf{C}_K^S$ is stiff and finish the whole proof, it suffices to show that $\mathbf{C}_K^S$ is generated by b_K. Take $b_K \vee \neg b_K = b_K \vee m_K = b_0$. This generates a_0, since $b_0 \rightarrow b_K = a_0$. Further, $a_0 a_0 = b_{-1}$. Then, $a_0 \rightarrow b_0 = a_1$, and $a_1^2 = b_1$. Suppose $a_0, \dots, a_n$, b_{-1}, $b_0, \dots, b_n$ have been generated. Then, $a_0 \rightarrow b_n = a_{n+1}$, and $a_{n+1}^2 = b_{n+1}$. This shows that all the elements in A and B get generated. Then the sets N and M are generated by negation, and this finishes the generation process. Observe that this process is independent from the set S, in the sense that, had S been chosen differently, the operations involved in the generation would still yield precisely the same results. □

Consider the family $\{\mathbf{C}^S_K\}_{K\in\mathbb{N}^+}$, where $\mathbb{N}^+ = \mathbb{N}\setminus\{0\}$. Take any nonprincipal ultrafilter U on $\mathbb{N}$ and let $\mathbf{C}_S$ stand for the ultraproduct $\prod_{K\in\mathbb{N}^+}\mathbf{C}^S_K/U$. Further, let $\mathbf{C}^\star_S$ be the subalgebra of $\mathbf{C}_S$ generated by $\star$. Clearly, the assumptions of Lemma 9.41 apply to our construction. Thus, $\mathbf{C}^\star_S$ is an infinite stiff FL_{ew}-algebra, and as such it generates an almost minimal variety.

LEMMA 9.44. [KKU06] *For any S, T subsets of $\mathbb{N}\setminus\{0,1\}$, we have: if $S\neq T$, then $\mathbf{C}^\star_S$ is not isomorphic to $\mathbf{C}^\star_T$.*

PROOF. Before we embark on the proof, let us dwell for a while on what $\mathbf{C}_S$ and $\mathbf{C}_T$ look like. We will refer to certain elements of these ultraproducts by the names of the elements of factor algebras, for instance, a_i in the appropriate context, will stand for $\langle e(n)\colon n\in\mathbb{N}^+\rangle/U$, where $e(n) = a_i(n)$, if a_i exists in C^S_n, or is arbitrary otherwise. In particular, it is helpful to think of K used in $b_K = \langle b_K(n)\colon n\in\mathbb{N}^+\rangle/U$ as an infinitely large natural number, so that $K > n$, for any $n\in\mathbb{N}$ (just like in nonstandard models for arithmetic). Notice that $\star$ is always unambiguous, being a definable constant.

Now, suppose S, T are subsets of $\mathbb{N}\setminus\{0,1\}$ and $S\neq T$, yet $\mathbf{C}^\star_S$ and $\mathbf{C}^\star_T$ are isomorphic. We can therefore identify the lattices underlying $\mathbf{C}^\star_S$ and $\mathbf{C}^\star_T$. Now, to obtain the desired contradiction, we look at multiplication and residuation induced on these lattices.

Let i be the smallest number such that $i\in S$ but $i\notin T$; we can always assume such a number exists, if not we just swap S and T. Consider the elements a_1, a_i; note that we use these names unambiguously, because, as the generation process from the previous proof does not depend on the choice of S and T, we may use a_1 as shorthand for $((\star\vee\neg\star)\to\star)\to(\star\vee\neg\star)$, and a_i as a shorthand for something similar, only much longer. Now, consider $a_1\cdot^S a_i$. Since $0<1<i\in S$, the first clause in the definition of '$\cdot$' applies at almost all coordinates, yielding in the ultraproduct: $a_1\cdot^S a_i = b_{i-1}$. Then, for $a_1\cdot^T a_i$, we have $0<1<i\notin T$ and the second clause applies, yielding: $a_1\cdot^T a_i = b_i$. However, b_i and b_{i-1} are different in the ultraproduct, hence, $a_1\cdot^S a_i$ and $a_1\cdot^T a_i$ produce different results, contradicting the assumption that $\mathbf{C}^\star_S$ and $\mathbf{C}^\star_T$ are isomorphic. □

THEOREM 9.45. [KKU06] *There are continuum many almost minimal subvarieties of $\mathsf{P_3InFL_{ew}}$.*

For contrast we have the next theorem.

THEOREM 9.46. [KKU06] *There are two almost minimal subvarieties of $\mathsf{P_2InFL_{ew}}$.*

PROOF. We will show that if $\mathbf{U}$ is a subdirectly irreducible algebra generating an almost minimal subvariety of $\mathsf{P_2InFL_{ew}}$, then $\mathbf{U}$ is isomorphic to

$\cdot$	1	a	0
1	1	a	0
a	a	0	0
0	0	0	0

$\cdot$	1	a	b	0
1	1	a	b	0
a	a	a	0	0
b	b	0	0	0
0	0	0	0	0

FIGURE 9.8. Algebras $\mathbf{C}_3$ and $\mathbf{U}_4$ generating the two almost minimal subvarieties of $\mathsf{P_2InFL_{ew}}$.

either $\mathbf{C}_3$ or $\mathbf{U}_4$; see Figure 9.8. Let $\mathbf{U}$ be such an algebra. By Lemma 3.60, $\mathbf{U}$ must have a unique coatom c.

Suppose first that $\mathbf{U}$ is simple. Then, we must have $c^2 = 0$, as otherwise $c^3 = c^2$ and $\{u \in U : u \geq c^2\}$ is a congruence class, contradicting simplicity. It follows that $\neg c = c$ and therefore $\{1, c, 0\}$ is a subuniverse, hence $U = \{1, c, 0\}$ as otherwise $\mathbf{U}$ does not generate an almost minimal variety. Thus, $\mathbf{U}$ is isomorphic to $\mathbf{C}_3$ and the variety it generates is MV_2.

If $\mathbf{U}$ is subdirectly irreducible but not simple, we must have $c^2 > 0$ and thus $\neg c < c$. Moreover, $(\neg c^2)^2 \leq c^2$ as otherwise the set $\{u \in U : u \geq \neg c^2)^2\}$ defines a congruence incomparable with the monolith: a contradiction. Further, we get $c^2(\neg c^2)^2 = (\neg c^2)^2$ because $\neg c^2$ is idempotent; but, $c^2(\neg c^2)^2 \leq c^2 \cdot \neg c^2 = 0$, so $(\neg c^2)^2 = 0$. Suppose $c^2 \vee \neg c^2 > c^2$. Then, $\neg c^2(c^2 \vee \neg c^2) = \neg c^2 \cdot c^2 \vee (\neg c^2)^2 = 0$ and therefore $\neg\neg c^2 \geq c^2 \vee \neg c^2 > c^2$ contradicting involutiveness. It follows that $\neg c^2 \leq c^2$ and since c^2 is idempotent they cannot be equal, so $c^2 > \neg c^2$. Altogether, we obtain $1 > c \geq c^2 > \neg c^2 \geq \neg c > 0$. Since $c^2 \to \neg c^2 = \neg c^2$, the set $\{1, c^2, \neg c^2, 0\}$ is a subuniverse of U. The algebra with this universe is isomorphic to $\mathbf{U}_4$ from Figure 9.8. This ends the argument. □

Multiplication tables for fusions of the two generating algebras are shown in Figure 9.8. Lattice operations in these algebras are determined by the linear ordering in which the elements occur on the left-hand side of the table. The algebra on the left is of course $\mathbf{C}_3$.

9.4.3. Almost minimal subvarieties of representable $\mathsf{FL_{ew}}$. For any S with $0 \in S \subseteq \mathbb{N}$. For any positive integer K define L_K^S to be the disjoint union of the sets: $B = \{b_0, \dots, b_{K+1}\}$, $A = \{a_0, a_1, a_2\}$, $N = \{n_0, \dots, n_{K+1}\}$, $C = \{0, 1\}$, and $D = \{d_s : s < K,\ s \in S\} \cup \{e\}$. L_K^S is totally ordered (see Figure 9.9); the order relation is specified by the conditions

- $1 > b_0 > b_{K+1} > a_0 > a_1 > a_2 > n_{K+1} > d_0 > e > 0$;
- $b_i > b_{i+1}$, $n_{i+1} > n_i$, for all $i \in \{0, \dots, K\}$;
- $d_r > d_s$ iff $r < s$ in the natural ordering of S.

For any $i \in \mathbb{N}$, let $\ell(i)$ stand for the largest $s \in S$ with $s \leq i$. Such an s always exists, for $0 \in S$. With this notation at hand, we define multiplication on C_K^S, putting:

- $1 \cdot x = x$, $0 \cdot x = 0$, $x \cdot y = y \cdot x$, for all $x, y \in L_K^S$;
- $b_i \cdot b_j = \begin{cases} a_0, & \text{if } \min\{i,j\} = 0, \\ a_1, & \text{if } 0 < i,\ 0 < j < K, \text{ or} \\ & \quad 0 < j,\ 0 < i < K, \text{ or} \\ & \quad i = j = K, \\ a_2 & \text{if } i \in \{K, K+1\},\ j = K+1, \text{ or} \\ & \quad j \in \{K, K+1\},\ i = K+1; \end{cases}$
- $b_i \cdot a_j = a_2 = a_i \cdot a_j$
- $b_i \cdot n_j = \begin{cases} d_{\ell(i)}, & \text{if } j > i+1, \\ e, & \text{if } j = i+1, \\ 0, & \text{if } j \leq i; \end{cases}$
- $x \cdot y = 0$, in all other cases.

Residuation is defined as previously:

- $x \to y = \bigvee\{z \in L_K^S : z \cdot x \leq y\}$.

Then, let $\mathbf{L}_K^S$ be the algebra $\langle L_K^S, \vee, \wedge, \cdot, \to, 0, 1\rangle$.

LEMMA 9.47. [KKU06] *The algebra $\mathbf{L}_K^S$ is an FL_{ew}-algebra.*

We leave the proof for the reader.

LEMMA 9.48. [KKU06] *The FL_{ew}-algebra $\mathbf{L}_K^S$ is stiff. Moreover, it belongs to E_3 and satisfies $x \to y \vee y \to x = 1$.*

PROOF. The only thing that is not immediately seen from the construction is that $\mathbf{L}_K^S$ has no subalgebras apart from the two-element Boolean algebra and itself. To show this it suffices to prove that the element a_2 generates $\mathbf{L}_K^S$ and that every element generates a_2. For the first claim we have: $\neg a_2 = n_{K+1}$ and $\neg n_{K+1} = b_{K+1}$. Then, $b_{K+1} \to a_2 = b_K$. Further, $b_K \cdot n_{K+1} = e$, and $\neg b_K = n_K$, $n_K \to e = b_{K-1}$. Then, by backward induction on i, $\neg b_i = n_i$, $n_i \to e = b_{i-1}$ and thus we have generated all the elements, except a_0, a_1 and all the elements d_s, for $s \in S \cap \{0, \dots, K-1\}$. To get these, we may, for instance, employ: $b_0^2 = a_0$, $b_1^2 = a_1$, and finally $b_{K+1} \cdot n_s = d_s$, for all suitable s. To show that every element generates a_2, we use the facts: $\neg n_i = b_i$, $\neg d_s = \neg e = b_0$, $b_i^3 = a_2$ and $a_i^2 = a_2$. □

As in the previous case, let $\mathbf{L}_S$ be the ultraproduct $\prod_{K \in \mathbb{N}^+} \mathbf{L}_K^S / U$ by a nonprincipal ultrafilter U on $\mathbb{N}$ and $\mathbf{L}_S^\star$ its subalgebra generated by $\star$. Again, Lemma 9.41 applies, thus $\mathbf{L}_S^\star$ is an infinite stiff algebra.

LEMMA 9.49. [KKU06] *Let S, T be subsets of $\mathbb{N}$, each containing 0. If $S \neq T$, then $\mathbf{L}_S^\star$ is not isomorphic to $\mathbf{L}_T^\star$.*

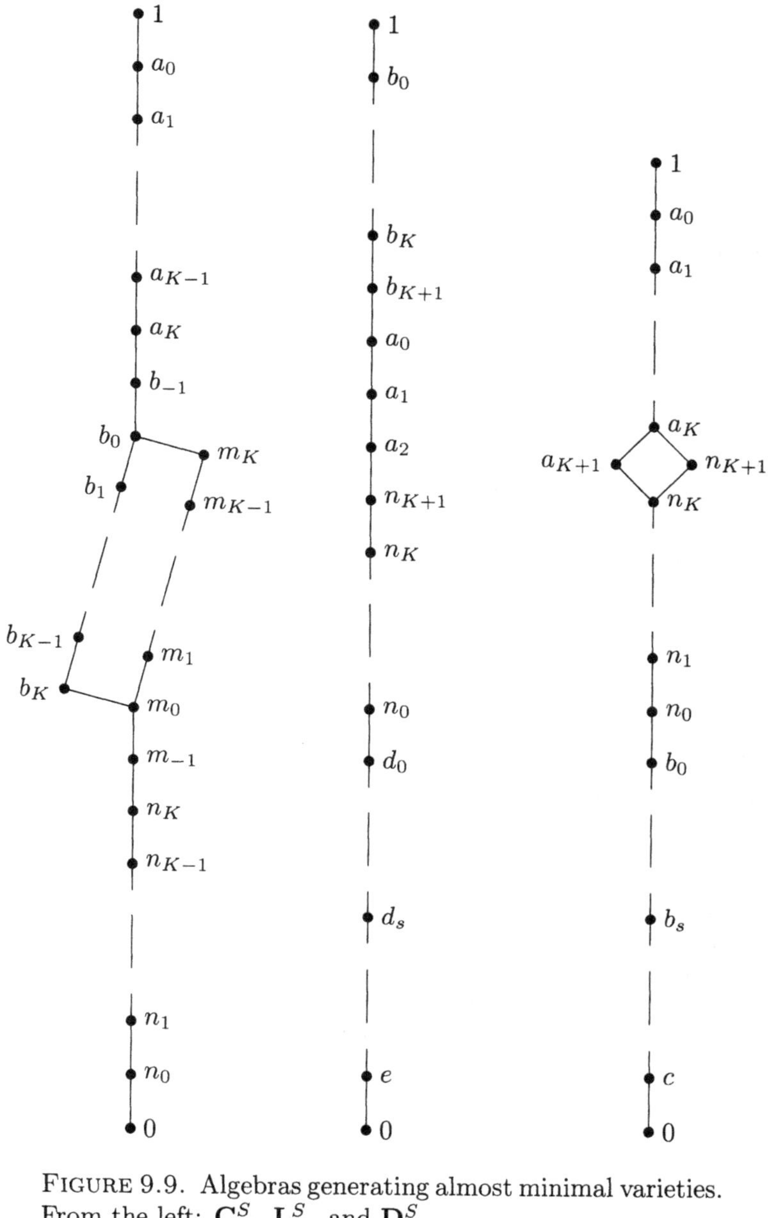

FIGURE 9.9. Algebras generating almost minimal varieties. From the left: $\mathbf{C}_K^S$, $\mathbf{L}_K^S$, and $\mathbf{D}_K^S$.

PROOF. We adopt all the conventions from Lemmas 9.43 and 9.44. Suppose i is the smallest number with $i \in S \setminus T$. Observe that in both $\mathbf{L}_S^\star$ and $\mathbf{L}_T^\star$ we have a finite decreasing sequence $d_0, d_{\ell(1)}, \ldots, d_{\ell(i-1)}$, with k distinct terms for some $k \leq i$. Now since K is infinite (recall the convention adopted in the proof of Lemmas 9.44), we have $K+1 > i+1$ for any finite i, and therefore $b_i \cdot^S n_{K+1} = d_{\ell(i)} = d_i \neq d_{\ell(i-1)} = b_{i-1} \cdot^S n_{K+1}$, since $i \in S$; but $b_i \cdot^T n_{K+1} = d_{\ell(i)} = d_{\ell(i-1)} = b_{i-1} \cdot^T n_{K+1}$, as $i \notin T$. □

THEOREM 9.50. [KKU06] *There are continuum many almost minimal subvarieties of* $\mathsf{P_3RFL_{ew}}$.

Again, we contrast it with the following.

THEOREM 9.51. [KKU06] *There are six almost minimal subvarieties of* $\mathsf{P_2RFL_{ew}}$.

PROOF. Suppose $\mathbf{U} \in \mathsf{P_2RFL_{ew}}$ is subdirectly irreducible and generates an almost minimal variety. Let c be the coatom of $\mathbf{U}$. If $\mathbf{U}$ is simple, $c^2 = 0$ and then $\neg c = c$ as c is a coatom, so $\mathbf{U}$ is isomorphic to $\mathbf{C}_3$.

Suppose that $\mathbf{U}$ is subdirectly irreducible but not simple. Let a be the smallest element of $[1]_\mu$, with μ the monolith. Then $a^2 = a$. We verify that the chain $1 > \neg\neg a \to a > \neg\neg a > a > \neg a > \neg(\neg\neg a \to a) > (\neg\neg a \to a)\cdot\neg a > 0$ is "freely generated" by a subject to the conditions $a^2 = a$, 2-potency, and linearity. To begin with, $\neg\neg a \geq a$ always holds, so by freeness we get $\neg\neg a > a$. Then, observe that $\neg a \geq a$ forces $a^2 = 0$, because $\neg a \cdot a = 0$, so we must have $a > \neg a$. By 2-potency we must have $u^2 = a$ for each u with $1 > u \geq a(\neg\neg a)^2 = a$ for otherwise $\{u \in U : u \geq (\neg\neg a)^2\}$ is a congruence class strictly contained in $[1]_\mu$. Further, since $a \to \neg a = a^2 \to 0 = a \to 0 = \neg a$, we obtain that all elements $u \to w$ with $u \geq a$ and $w \leq \neg a$ must themselves be below $\neg a$, so the universe U splits into the top half $T \geq a$ and the bottom half $B \leq \neg a$. Moreover, every u below $\neg a$ must square to 0, since u is then below both a and $\neg a$. It follows that $tt' = a$ for $t, t' \in T$ and $bb' = 0$ for $b, b' \in B$. On the one hand $\neg\neg a \to \neg a \geq \neg a$ by integrality, on the other $\neg\neg a \to \neg a \leq \neg a$ since $\neg\neg a \in T$; hence equality. As $(\neg\neg a)^2 = a$, we obtain $\neg\neg a \to a \geq \neg\neg a$ because $\neg\neg a \to a$ is the greatest c such that $\neg\neg a \cdot c \leq a$. Freeness then yields $\neg\neg a \to a > \neg\neg a$ and this element must be the coatom of $\mathbf{U}$ since if we have $1 > c > \neg\neg a \to a$ then $c \cdot \neg\neg a \leq a$ contradicting $\neg\neg a \to a$ being the greatest with this property. Again by freeness, we obtain $\neg a > \neg(\neg\neg a \to a)$. Then, the element $(\neg\neg a \to a)\cdot\neg a$ is nonzero by freeness yet again: all the other products of elements generated so far must be either 0 or a. Finally, $(\neg\neg a \to a)\cdot\neg a \leq (\neg\neg a \to a) \to 0 = \neg(\neg\neg a \to a)$ if and only if $(\neg\neg a \to a)^2 \cdot \neg a \leq 0$, and the latter holds because $(\neg\neg a \to a)^2 = a$, we get by freeness $\neg(\neg\neg a \to a) > (\neg\neg a \to a)\cdot\neg a > 0$.

Now we have to verify that the generation process ends at this stage. First, it is easy to see now that all products of non-unit elements generated so far are either equal to a or 0. Second, we show that $t \to t' = \neg\neg a \to a$ for $t > t' \in T$. For $t = \neg\neg a$ and $t' = a$ it is clear. To that end it suffices to prove that $\neg\neg a \to a \leq (\neg\neg a \to a) \to a$. But this is equivalent to $(\neg\neg a \to a)^2 \leq a$ and that holds, as we have established. Third, using order preserving/reversing

$\cdot$	1	a	0
1	1	a	0
a	a	a	0
0	0	0	0

$\cdot$	1	a	b	c	0
1	1	a	b	c	0
a	a	b	b	0	0
b	b	b	b	0	0
c	c	0	0	0	0
0	0	0	0	0	0

$\cdot$	1	a	b	c	d	e	0
1	1	a	b	c	d	e	0
a	a	c	c	c	d	0	0
b	b	c	c	c	0	0	0
c	c	c	c	c	0	0	0
d	d	e	0	0	0	0	0
e	e	0	0	0	0	0	0
0	0	0	0	0	0	0	0

$\cdot$	1	a	b	c	d	e	f	0
1	1	a	b	c	d	e	f	0
a	a	c	c	c	f	0	0	0
b	b	c	c	c	0	0	0	0
c	c	c	c	c	0	0	0	0
d	d	f	0	0	0	0	0	0
e	e	0	0	0	0	0	0	0
f	f	0	0	0	0	0	0	0
0	0	0	0	0	0	0	0	0

FIGURE 9.10. Four more algebras, $\mathbf{H}_3$, $\mathbf{U}_5$, $\mathbf{U}_7$, and $\mathbf{U}_8$, generating almost minimal subvarieties of $\mathsf{E}_2 \cap \mathsf{FL}_{ew}$.

properties of $\to$ together with what we already established it is not difficult to show that $t \to b = \neg a$ for each $t \in T$ and each $b \in B \setminus \{0\}$. Fourth, by similar arguments we verify that for $b > b' \in B$ we have $b \to b' = \neg\neg a \to a$.

Our eight element chain is therefore closed under the operations. Let $\mathbf{U}_8$ be the algebra with this universe. Notice that this is precisely the eight element algebra from Figure 9.10. As a by-product, we have obtained that $\mathbf{U}_8$ is stiff and thus generates an almost minimal variety.

To complete the proof, we ask the reader (see Exercise 15) to walk through the generation process again, altering strict inequalities to equalities in all possible combinations. This will produce the remaining three algebras of Figure 9.10 and two from Figure 9.8. □

The algebras generating these six varieties are the two from Figure 9.8 and the four from Figure 9.10. The algebra in the upper left-hand corner of the latter picture is the three-element Heyting algebra.

9.4.4. Almost minimal subvarieties of 2-potent DFL_{ew}. We finish off with yet one more construction of continuum many almost minimal varieties, this time descending as low as E_2 in the subvarieties of FL_{ew}. However, as all the proofs proceed exactly as in previous cases, none will be presented.

Let S be a subset of $\mathbb{N}$ with with $0 \in S$. For any positive integer K, let D^S_K be the disjoint union of the sets: $A = \{a_0, \ldots, a_{K+1}\}$, $N = \{n_0, \ldots, n_{K+1}\}$, $C = \{0, 1\}$, $B = \{b_s : s < K,\ s \in S\} \cup \{c\}$. D^S_K is partially ordered (see Figure 9.9); the order relation is specified by the by the conditions:

- $1 > a_0$, $n_0 > b_0$, $c > 0$;
- $x > c$, for all $x \neq 0$;
- $a_i > a_j$, $n_j > n_i$ iff $i < j$ $(i, j \in \{0, \dots, K+1\})$;
- $a_K > n_{K+1}$, $a_{K+1} > n_K$;
- $b_r > b_s$ iff $r < s$ in the natural ordering of S.

For any $i \in \mathbb{N}$, let $\ell(i)$ stand, as previously, for the largest $s \in S$ with $s \leq i$. Define multiplication on D_K^S, putting:

- $1 \cdot x = x$, $0 \cdot x = 0$, $x \cdot y = y \cdot x$, for all $x, y \in D_K^S$;
- $a_i \cdot a_j = a_{K+1}$;
- $a_i \cdot n_j = \begin{cases} b_{\ell(i)}, & \text{if } j > i+1, \\ c, & \text{if } j = i+1, \\ 0, & \text{if } j \leq i; \end{cases}$
- $x \cdot y = 0$, in all other cases.

LEMMA 9.52. [KKU06] *The algebra $\mathbf{D}_K^S$ is a stiff FL_{ew}-algebra. Moreover, it belongs to E_2 and is distributive as a lattice.*

PROOF. To prove that the algebra is stiff we first show that a_{K+1} generates the whole algebra by using the facts: $\neg a_{K+1} = n_{K+1}$, $a_{K+1} \vee n_{K+1} = a_K$, $a_K n_{K+1} = c$, $\neg a_i = n_i$, $n_i \rightarrow c = a_{i-1}$ and $b_s = a_s n_{K+1}$. Then we show that every element generates a_{K+1} using the facts: $\neg c = \neg b_s = a_0$, $\neg n_i = a_i$, $a_i^2 = a_{K+1}$. Clearly, the lattice reduct of the algebra is distributive, since it is clearly a subdirect product of chains, and obviously for every element x, we have $x^3 = x^2$. □

The next lemma can be proved using an idea paralleling the one used in the proof of Lemma 9.49 (see Exercise 14).

LEMMA 9.53. [KKU06] *Let S, T be subsets of $\mathbb{N}$, each containing 0. If $S \neq T$, then $\mathbf{D}_S^\star$ is not isomorphic to $\mathbf{D}_T^\star$.*

THEOREM 9.54. [KKU06] *There are continuum many almost minimal subvarieties of $\mathsf{E}_2 \cap \mathsf{FL_{ew}}$ satisfying distributivity.*

9.5. Almost minimal varieties of BL-algebras

As we have seen, even the representability restriction on FL_{ew}-algebras is not enough to force the number of almost minimal subvarieties below continuum many. On the other hand, we know from [Kom81] that there are only countably many almost minimal varieties of MV-algebras. We can do slightly better than that, namely, we can show that the same holds true about varieties of BL-algebras. We begin with an easy observation, whose proof we leave to the reader.

LEMMA 9.55. [KKU06] *Let* **A** *be a subdirectly irreducible BL-algebra. If there is an idempotent element* $a \in A\backslash\{0,1\}$, *then the three-element Heyting algebra* $\mathbf{H}_3$ *is a subalgebra of* **A**.

PROOF. It is enough to show that $\neg a = 0$. We have $\neg a = a(a \to \neg a) = a(a \to (a \to 0)) = a(a^2 \to 0) = a(a \to 0) \leq 0$. □

Thus, if **A** is to generate an almost minimal variety different from $V(\mathbf{H}_3)$, it cannot contain idempotents different from 0 and 1.

LEMMA 9.56. [BF00] *Suppose* **A** *generates almost minimal variety different from* $V(\mathbf{H}_3)$ *and that* **A** *is finite. Then,* **A** *is isomorphic to the MV-algebra* $\mathbf{C}_{p+1}$, *for a prime* p.

For the next four lemmas, we assume that **A** is infinite and has no idempotents beside 0 and 1. Let U be the filter on A that corresponds to the monolith congruence. By the assumption, U has no smallest element, and since the $\{\to,\cdot,1\}$-reduct of **A** is a hoop (in fact, a *basic* hoop, we can conclude, again by [BF00], that $(U,\cdot,\to,1)$ is a simple Wajsberg hoop, therefore isomorphic to $\mathbb{Z}^-$ (cf. Figure 2.5), which we will here present as $(\{a^i : i \in \mathbb{N}\}, \wedge, \vee, \cdot, \to, 1)$, where $1 = a^0$, $a^j \cdot a^k = a^{j+k}$, and $a^j \to a^k = a^{k-j}$ if $k < j$ and $a^j \to a^k = 1$ otherwise. Thus, by linearity, we obtain that A is the disjoint union of U and a certain set L such that for each $u \in U$ and $l \in L$ we have $u > l$.

LEMMA 9.57. [KKU06] *For all* $u \in U$, *we have* $\neg u \in L$ *and* $\neg\neg u \in U$.

PROOF. Suppose $\neg u \in U$ for some $u \in U$. Then $u = a^j$ and $\neg u = a^k$, for some $j, k \in \mathbb{N}$. Therefore $0 = u \cdot \neg u = a^j \cdot a^k = a^{j+k}$, a contradiction. □

LEMMA 9.58. [KKU06] *Let* $M = \{l \in L : l = \neg a^k$ *for some* $k \in \mathbb{N}\}$. *Then,* $U \cup M$ *is a subuniverse of* **A**.

PROOF. By case analysis, of which we will present the two non-obvious cases. First, we establish closure under multiplication. Suppose $d \in U$ and $b \in M$. Then, $d = a^j$ and $b = \neg a^k$, for some $j, k \in \mathbb{N}$, therefore $bd = a^j \cdot \neg a^k$. If $j \geq k$, then $a^j \leq a^k$ and thus $a^j \cdot \neg a^k = 0 = \neg a^0 \in M$. If $j \leq k$, then $a^j \cdot \neg a^k = a^j \cdot \neg a^{k-j+j} = a^j(a^j \to \neg a^{k-j})$. This, by the hoop axiom, equals $a^j \wedge \neg a^{k-j}$ and that, by linearity, equals further $\neg a^{k-j}$, which belongs to M.

Then, we verify closure under residuation. Suppose $d, b \in M$, i.e., $d = \neg a^j$ and $b = \neg a^k$, for some $j, k \in \mathbb{N}$. If $k \geq j$, then $\neg a^j \leq \neg a^k$ and thus $\neg a^j \to \neg a^k = 1 = a^0 \in U$. If $k < j$, then a^{j-k} is well defined and strictly smaller than 1. We get $a^{j-k} \cdot \neg a^j = a^{j-k} \cdot \neg a^{j-k+k} = a^{j-k}(a^{j-k} \to \neg a^k) = a^{j-k} \wedge \neg a^k = \neg a^k$. So, $a^{j-k} \cdot \neg a^j = \neg a^k$ and this by residuation yields $a^{j-k} \leq \neg a^j \to \neg a^k$. Therefore, $\neg a^j \to \neg a^k \in U$. □

LEMMA 9.59. [KKU06] *If* $\neg a^j = \neg a^{j+1}$, *for some* $j \in \mathbb{N}$, *then* $\neg a^k = 0$, *for all* $k \in \mathbb{N}$.

Variety	#	namely	reference
$\mathsf{P_3InFL_{ew}}$	$2^{\aleph_0}$	$\mathsf{V}(\mathbf{C}_S^*)$	Thm. 9.45
$\mathsf{P_2InFL_{ew}}$	2	MV_2, $\mathsf{V}(\mathbf{U}_4)$	Thm. 9.46
$\mathsf{P_3RFL_{ew}}$	$2^{\aleph_0}$	$\mathsf{V}(\mathbf{L}_S^*)$	Thm. 9.50
$\mathsf{P_2RFL_{ew}}$	6	MV_2, GA_3, $\mathsf{V}(\mathbf{U}_4)$, $\mathsf{V}(\mathbf{U}_5)$, $\mathsf{V}(\mathbf{U}_7)$, $\mathsf{V}(\mathbf{U}_8)$	Thm. 9.51
$\mathsf{P_2DFL_{ew}}$	$2^{\aleph_0}$	$\mathsf{V}(\mathbf{D}_S^*)$	Thm. 9.54
MV	$\aleph_0$	MV_p (p prime), $\mathsf{MV}_1^{\varepsilon}$	Fig. 3.3
BL	$\aleph_0$	MV_p (p prime), $\mathsf{MV}_1^{\varepsilon}$, GA_3, $\mathsf{V}(\mathbb{Z}_0^-)$	Fig. 3.3
SRA	≥ 13		Thm. 3.31

TABLE 9.3. Almost minimal subvarieties of $\mathsf{FL_w}$.

PROOF. Induction on k. For $k = 0$ the claim holds trivially, for $k = 1$ we have $\neg a = \neg a \wedge a^j = a^j(a^j \to \neg a) = a^j(a^j \to (a \to 0)) = a^j \cdot \neg a^{j+1} = a^j \cdot \neg a^j = 0$. Then, in the inductive step we obtain $\neg a^{k+1} = a \to \neg a^k$, which by inductive assumption equals further $a \to 0 = \neg a = 0$. ☐

LEMMA 9.60. [KKU06] *If $\neg a^j \neq \neg a^k$, for all $j \neq k$, then $\neg\neg a^k = a^k$, for all $k \in \mathbb{N}$.*

PROOF. Suppose the contrary, and let k be the smallest such that $\neg\neg a^k > a^k$. Since $\neg\neg a^k \in U$, by Lemma 9.57, we obtain that $\neg\neg a^k = a^j$ for some $j < k$. Thus, $\neg\neg a^j = a^j$, as k is the smallest number for which it fails. Therefore, $\neg\neg a^j = \neg\neg a^k$, which yields $\neg a^j = \neg\neg\neg a^j = \neg\neg\neg a^k = \neg a^k$, contradicting the assumption. ☐

THEOREM 9.61. [KKU06] [KO06] *There are exactly countably many almost minimal varieties of BL-algebras. They are: all minimal varieties of MV-algebras, the only almost minimal variety of Heyting algebras, and the variety of product logic algebras.*

PROOF. If $\mathbf{A}$ is finite, then by Lemmas 9.55 and 9.56 it is either $\mathbf{H}_3$ or $\mathbf{C}_{p+1}$ for some prime p. If $\mathbf{A}$ is infinite, then by Lemmas 9.59 and 9.60 the set M (defined just before Lemma 9.58; it contains the negations of the elements of the deductive filter U) can be either a singleton, or bijective with U with negation being the bijection. In the former case $\mathbf{A}$ is $\mathbb{Z}_0^-$ and generates the variety of product logic algebras; in the latter $\mathbf{A}$ is Chang's MV-algebra $\mathbf{C}_\infty$ and generates $\mathsf{MV}_1^{\varepsilon}$. ☐

9.6. Translations of subvariety lattices

In this section, we comment on height 2 varieties in $\mathbf{\Lambda}(\mathsf{FL_o})$, which we still call *almost minimal*. In particular, we explain why minimal varieties of $\mathbf{\Lambda}(\mathsf{RL})$ give rise to almost minimal varieties of $\mathbf{\Lambda}(\mathsf{FL_o})$ that are covers of atoms like $\mathsf{V}(\mathbf{2})$ or $\mathsf{V}(\mathbf{T}_1)$.

9.6.1. Generalized ordinal sums. In Section 3.6.7, we reviewed two constructions from [GR04], where given a residuated lattice $\mathbf{L}$ and $\bot, \top \notin L$ we defined a residuated lattice $\mathbf{b}(\mathbf{L})$ on the set $\mathbf{b}(L) = L \cup \{\bot, \top\}$ by extending the operations of $\mathbf{L}$ to $\mathbf{b}(L)$ as follows. For all $a \in L$, $b \in L \cup \{\top\}$ and $c \in \mathbf{b}(L)$, the operations $\wedge, \vee$ are given by $\bot < a < \top$; multiplication is defined by $b\top = \top b = \top$, $\bot c = c\bot = \bot$; and the division operations by $\top/b = b\backslash\top = c/\bot = \bot\backslash c = \top$, $a/\top = \top\backslash a = \bot/b = b\backslash\bot = \bot$. The algebra $\mathbf{b}(\mathbf{L}) = (\mathbf{b}(L), \wedge, \vee, \cdot, \backslash, /, e, \bot)$ was shown to be an FL_o-algebra.

Moreover, if $\mathbf{L}$ is an integral residuated lattice, we can modify the above construction and obtain an FL_w-algebra $\mathbf{d}(\mathbf{L})$, where $\mathbf{d}(L) = L \cup \{\bot\}$ and the extension of the operations is defined in a similar way.

The constructions obtained using the operators $\mathbf{b}$ and $\mathbf{d}$ produce FL_o-algebras or (bounded) residuated lattices (depending on whether we include $\bot$ in the type), and are special cases of a more general construction that we describe below.

We call an element a in an algebra $\mathbf{A}$ *irreducible* with respect to an n-ary operation f of $\mathbf{A}$, if, for all $a_1, a_2, \ldots, a_n \in A$, $f(a_1, a_2, \ldots, a_n) = a$ implies $a_i = a$, for some i. We call a *totally irreducible*, if it is irreducible with respect to all the operations of $\mathbf{A}$.

Let $\mathbf{K}$ and $\mathbf{L}$ be residuated lattices and assume that the identity element of $\mathbf{K}$ is totally irreducible; also, assume that either $\mathbf{L}$ is integral, or there is no element $k \in K$ such that $1/k = 1$ or $k\backslash 1 = 1$. In this case we say that $\mathbf{K}$ is *admissible* by $\mathbf{L}$. A class $\mathcal{K}$ is *admissible* by a class $\mathcal{L}$, if every algebra of $\mathcal{K}$ is admissible by every algebra of $\mathcal{L}$.

Let[2] $K[L] = (K \setminus \{1_\mathbf{K}\}) \cup L$, and extend the operations of $\mathbf{K}$ and $\mathbf{L}$ to $K[L]$ by $l \star k = 1_\mathbf{K} \star_\mathbf{K} k$ and $k \star l = k \star_\mathbf{K} 1_\mathbf{K}$, for $k \in (K \setminus \{1_\mathbf{K}\})$, $l \in L$ and $\star \in \{\wedge, \vee, \cdot, \backslash, /\}$. We will show that $\mathbf{K}[\mathbf{L}] = (K[L], \wedge, \vee, \cdot, \backslash, /, 1_\mathbf{L})$ is a residuated lattice. If $\mathbf{K}$ is an FL-algebra we extend the operator $\mathbf{K}$ on classes of algebras that admit $\mathbf{K}$ and we use the same notation; the results we will prove about $\mathbf{K}[\mathbf{L}]$ transfer directly to this case, as well. Note that for every residuated lattice $\mathbf{L}$, $\mathbf{b}(\mathbf{L}) = \mathbf{T}_1[\mathbf{L}]$; if $\mathbf{L}$ is integral, then $\mathbf{d}(\mathbf{L}) = \mathbf{2}[\mathbf{L}]$.

The residuated lattices $\mathbf{T}_n$, for n a positive natural number, and $\mathbf{N}_w$, for w an infinite or bi-infinite word are examples of algebras that are admissible by all residuated lattices.

Note that if both $\mathbf{K}$ and $\mathbf{L}$ are integral, then $\mathbf{K}$ is admissible by $\mathbf{L}$. In this case, the algebra $\mathbf{K}[\mathbf{L}]$ is called the *ordinal sum* of $\mathbf{L}$ and $\mathbf{K}$ and is sometimes denoted by $\mathbf{L}\oplus\mathbf{K}$. This is the reason that without the assumption of integrality the algebra $\mathbf{K}[\mathbf{L}]$ can be considered as a *generalized* ordinal sum of the two algebras. Without going into details, we mention that the ordinal sum construction has been used for BL-algebras and hoops.

---

[2]Note that $\backslash$ is used for both the left division operation and the set-theoretic difference.

This notion is different from the usual ordinal sum of two posets $\mathbf{P}$ and $\mathbf{Q}$, which is defined to be the poset with underlying set $P \cup Q$ and order relation containing the orders of $\mathbf{P}$ and $\mathbf{Q}$, and setting every element of P less than every element of Q; so the new order is $\leq_{\mathbf{P}} \cup \leq_{\mathbf{Q}} \cup (P \times Q)$. Also, it is different than the *coalesced ordinal sum* of two posets $\mathbf{P}$ and $\mathbf{Q}$ ($\mathbf{P}$ is assumed to have a top element and $\mathbf{Q}$ a bottom element), which is defined by identifying in the usual ordinal sum of $\mathbf{P}$ and $\mathbf{Q}$ the top element of $\mathbf{P}$ with the bottom element of $\mathbf{Q}$.

LEMMA 9.62. [Gal05] *Assume that* $\mathbf{K}$, $\mathbf{L}$ *are residuated lattices such that* $\mathbf{K}$ *is admissible by* $\mathbf{L}$. *Then,*

(1) *the algebra* $\mathbf{K}[\mathbf{L}]$ *is a residuated lattice,*
(2) *the congruence lattice of* $\mathbf{K}[\mathbf{L}]$ *is isomorphic to the coalesced ordinal sum of the congruence lattice of* $\mathbf{L}$ *and the congruence lattice of* $\mathbf{K}$. *Thus,* $\mathbf{K}[\mathbf{L}]$ *is subdirectly irreducible iff* $\mathbf{L}$ *is subdirectly irreducible, or* $\mathbf{L} \cong \mathbf{1}$ *and* $\mathbf{K}$ *is subdirectly irreducible.*

PROOF. 1) The proof for $\mathbf{K} = \mathbf{T}_1$ and $\mathbf{K} = \mathbf{2}$ can be found in [GR04]. The proof for an arbitrary admissible $\mathbf{K}$ is straight-forward.

2) By Theorem 3.47, the congruence lattice of $\mathbf{K}[\mathbf{L}]$ is isomorphic to the lattice of convex subalgebras of its $\perp$-free reduct. Note that L is normal in $\mathbf{K}[\mathbf{L}]$, because for all $a \in L$ and $k \in K \setminus \{1\}$, $\rho_k(a) = ka/k \wedge 1 = k/k \wedge 1 = 1$ and similarly $\lambda_k(a) = 1$. Actually, $\mathbf{L}$ is a convex normal subalgebra of the $\perp$-free reduct of $\mathbf{K}[\mathbf{L}]$. Assume that $\mathbf{M}$ is a convex (normal) subalgebra of the $\perp$-free reduct of $\mathbf{K}[\mathbf{L}]$, such that $M \not\subseteq L$. Without loss of generality, we can chose $k \in M \setminus L$, such that $k < 1$. Thus, $1 \leq e/k$. If $1 < 1/k$, then $k, e/k \in K$, and $k < e < e/k$. So, $k < l < e/k$, for all $l \in L$. Consequently, $L \subseteq M$, since $k, e/k \in M$. If $1/k = 1$, then $\mathbf{L}$ is integral, since it is admissible by $\mathbf{K}$, and $k \in K$. Thus, $k < l \leq e$ for all $l \in L$. So, $L \subseteq M$, since $k \in M$. Consequently, for every convex (normal) subalgebra $\mathbf{M}$ of $\mathbf{K}[\mathbf{L}]$, either $L \subseteq M$ or $M \subseteq L$. Additionally, it can be verified that if $\mathbf{M}$ is a convex normal subalgebra of $\mathbf{K}[\mathbf{L}]$ such that $L \subseteq M$, then $(M \setminus L) \cup \{1\}$, defines a convex normal subalgebra of $\mathbf{K}$. Conversely, for every convex normal subalgebra $\mathbf{N}$ of $\mathbf{K}$, $(N \setminus \{1\}) \cup L$ defines a convex normal subalgebra of $\mathbf{K}[\mathbf{L}]$. Moreover, these correspondences are mutually inverse lattice isomorphisms. Consequently, the lattice of convex normal subalgebras of $\mathbf{K}[\mathbf{L}]$ is isomorphic to the coalesced ordinal sum of the lattices of convex normal subalgebras of $\mathbf{L}$ and $\mathbf{K}$. Statement (2), then, follows directly from Theorem 3.47. □

LEMMA 9.63. [Gal05] *Assume that* $\mathcal{L}$ *is a class of residuated lattices and that* $\mathbf{K}$ *is a finite strictly simple residuated lattice admissible by* $\mathcal{L}$. *Then,*

(1) $\mathsf{O}(\mathbf{K}[\mathcal{L}]) = \mathbf{K}[\mathsf{O}(\mathcal{L})]$, *where the operator* O *is any of the operators* $\mathsf{IP_u}$, S *or* H. *(We abuse notation by using the same symbol* O *for the operators on subclasses of* RL *and* $\mathsf{FL_o}$*).*

(2) $(\mathsf{V}(\mathbf{K}[\mathcal{L}]))_{SI} = \mathbf{K}[(\mathsf{V}(\mathcal{L}))_{SI}] \cup \mathsf{I}(\mathbf{K})$.

PROOF. 1) Assume first that $\mathsf{O} = \mathsf{IP_u}$. If $\mathbf{L}_i \in \mathcal{L}$, $i \in I$ and U is an ultrafilter over I, then $(\prod_{i\in I} \mathbf{K}[\mathbf{L}_i])/U \cong \mathbf{K}[(\prod_{i\in I} \mathbf{L}_i)/U]$. This is true, because every element of the the left hand side has a representative (I-sequence) that has all or none of its terms in $K \setminus \{1\}$. Thus we can separate the elements in ones of type $\mathbf{K}$ and in ones of type $\mathcal{L}$. The elements of type $\mathcal{L}$ form a subalgebra of the left hand side isomorphic to $(\prod_{i\in I} \mathbf{L}_i)/U$. On the other hand, every element of type $\mathbf{K}$ is of the form $[(k)_{i\in I}]_U$, for $k \in K \setminus \{1\}$, since $\mathbf{K}$ is finite. Thus, the left hand side is isomorphic to $\mathbf{K}[(\prod_{i\in I} \mathbf{L}_i)/U]$.

For the case $\mathsf{O} = \mathsf{S}$, note that every subalgebra (computed in $\mathsf{FL_o}$) of $\mathbf{K}[\mathbf{L}]$, for $\mathbf{L} \in \mathcal{L}$, is equal to $\mathbf{K}$ of a subalgebra (computed in RL) of $\mathbf{L}$, by the very construction of $\mathbf{K}[\mathbf{L}]$ and the fact that $\mathbf{K}$ is strictly simple. Conversely, for every subalgebra $\mathbf{L}'$ (computed in RL) of $\mathbf{L}$, $\mathbf{K}[\mathbf{L}']$ is a subalgebra (computed in $\mathsf{FL_o}$) of $\mathbf{K}[\mathbf{L}]$.

Finally, if $\mathsf{O} = \mathsf{H}$, we will show that every homomorphism h on $\mathbf{K}(\mathbf{L})$, where $\mathbf{L}$ is an algebra of $\mathcal{L}$, maps only elements of $\mathbf{K}$ to the elements of $\mathbf{K}$ of its image. To see this assume that $h(a) = h(k)$, for some $a \in L$ and $k \in K \setminus \{1\}$. Then, $h(a \vee k) = h(a) = h(a \wedge k)$ and, by the construction of $\mathbf{K}[\mathbf{L}]$, $a \wedge k \in K \setminus \{1\}$ or $a \vee k \in K \setminus \{1\}$. Thus, we may assume that $h(a) = h(k)$, for some $a \in L$ and $k \in K \setminus \{1\}$, and $k < a$ or $a < k$. In the first case $k < e$ and

$$h(k) = h(k/a \wedge 1) = h(k)/h(a) \wedge 1 = h(a)/h(a) \wedge 1 = 1.$$

In the second case $1 < k$, $1/k < 1$ and

$$h(1/k) = h(a/k \wedge 1) = h(a)/h(k) \wedge 1 = h(a)/h(a) \wedge 1 = 1.$$

Since $\mathbf{K}$ lacks non-trivial subalgebras, for every element $k' \in K \setminus \{1\}$ there is a term $\overline{k'}(x)$ such that $\overline{k'}(x) = k'$, for $x \neq 1$. Thus, for every $k' \in K \setminus \{1\}$, $h(k') = 1$, i.e. the image of h is trivial. Consequently, the restriction of h to the subalgebra (computed in RL) $\mathbf{L}$ of the $\{\bot\}$-free reduct of $\mathbf{K}[\mathbf{L}]$ maps to a subalgebra $\mathbf{M}$ of its image $\mathbf{K}[\mathbf{M}]$. Thus, the homomorphic image of $\mathbf{K}[\mathbf{L}]$ is in $\mathbf{K}[\mathsf{H}(\mathcal{L})]$. Conversely, if $\mathbf{M}$ is a homomorphic image (computed in RL) of an algebra $\mathbf{L}$ of $\mathcal{L}$, then $\mathbf{K}[\mathbf{M}]$ is a homomorphic image (computed in $\mathsf{FL_o}$) of $\mathbf{K}[\mathbf{L}]$.

2) By Jónsson's Lemma for congruence distributive varieties and (1), we have $(\mathsf{V}(\mathbf{K}[\mathcal{L}]))_{SI} \subseteq \mathbf{HSP}_u(\mathbf{K}[\mathcal{L}]) = \mathbf{K}[\mathbf{HSP}_u(\mathcal{L})]$. Assume that $\mathbf{L} \in (\mathsf{V}(\mathbf{K}[\mathcal{L}]))_{SI}$. Then $\mathbf{L} = \mathbf{K}[\mathbf{M}]$, for some $\mathbf{M} \in \mathbf{HSP}_u(\mathcal{L}) \subseteq \mathsf{V}(\mathcal{L})$. Moreover, since $\mathbf{K}[\mathbf{M}]$ is subdirectly irreducible, we have that $\mathbf{M}$ is subdirectly irreducible or $\mathbf{M} \cong \mathbf{1}$, by Lemma 9.62(3). Consequently, $\mathbf{L} \in \mathbf{K}[(\mathsf{V}(\mathcal{L}))_{SI}] \cup \mathsf{I}(\mathbf{K}[\mathbf{1}])) = \mathbf{K}[(\mathsf{V}(\mathcal{L}))_{SI}] \cup \mathsf{I}(\mathbf{K})$. Conversely, if $\mathbf{L} \in \mathbf{K}[(\mathsf{V}(\mathcal{L}))_{SI}] \cup \mathsf{I}((\mathbf{K})$, then $\mathbf{L}$ is subdirectly irreducible by Lemma 9.62(3). Note that $\mathbf{K}[(\mathsf{V}(\mathcal{L}))_{SI}] \cup \mathsf{I}(\mathbf{K}) \subseteq \mathbf{K}[\mathbf{HSP}_u(\mathcal{L})] = \mathbf{HSP}_u(\mathbf{K}[\mathcal{L}]) \subseteq \mathsf{V}(\mathbf{K}[\mathcal{L}])$. Thus, $\mathbf{L} \in (\mathsf{V}(\mathbf{K}[\mathcal{L}]))_{SI}$. □

Note that $\mathbf{K}$ does not commute with the operator P (see Exercise 16).

Let $\mathcal{S}$ be a class of residuated lattices and $\mathbf{K}$ a finite, strictly simple FL_o-algebra admissible by $\mathcal{S}$. We define $\mathcal{S}_{\mathbf{K}} = \mathsf{V}(\mathbf{K}[\mathcal{S}])$.

THEOREM 9.64. *Let $\mathbf{K}$ be a finite, strictly simple FL_o-algebra admissible by RL. The subvariety lattice $\mathbf{\Lambda}(\mathsf{RL})$ of RL is isomorphic to the interval $[\mathsf{V}(\mathbf{K}), \mathsf{RL}_{\mathbf{K}}]$ of $\mathbf{\Lambda}(\mathsf{FL_o})$ via the map $\mathcal{V} \mapsto \mathcal{V}_{\mathbf{K}}$.*

PROOF. Let $\mathcal{V}$ be a subvariety of RL. Employing Lemma 9.63(2), we have

$$(\mathcal{V}_{\mathbf{K}})_{\mathsf{SI}} = (\mathsf{V}(\mathbf{K}[\mathcal{V}]))_{\mathsf{SI}} = \mathbf{K}[(\mathsf{V}(\mathcal{V}))_{\mathsf{SI}}] \cup \mathbf{I}(\mathbf{K}) = \mathbf{K}[\mathcal{V}_{\mathsf{SI}}] \cup \mathsf{I}(\mathbf{K}).$$

Thus, $\mathsf{V}(\mathbf{K}) \subseteq \mathcal{V}_o \subseteq \mathsf{RL}_o$, and the map is order preserving. Moreover, if $\mathcal{V}_{\mathbf{K}} \subseteq \mathcal{U}_{\mathbf{K}}$, then $\mathbf{K}[\mathcal{V}_{\mathsf{SI}}] \cup \mathsf{I}(\mathbf{K}) \subseteq \mathbf{K}[\mathcal{U}_{\mathsf{SI}}] \cup \mathsf{I}(\mathbf{K})$ and $\mathcal{V}_{\mathsf{SI}} \subseteq \mathcal{U}_{\mathsf{SI}}$; hence $\mathcal{V} \subseteq \mathcal{U}$ and the map reflects the order.

If $\mathcal{W}$ be a subvariety of $\mathsf{RL_o}$, then $\mathcal{W}_{\mathsf{SI}} \subseteq (\mathsf{RL}_{\mathbf{K}})_{\mathsf{SI}} = \mathbf{K}[\mathsf{RL}_{\mathsf{SI}}] \cup \mathsf{I}(\mathbf{K})$, so $\mathcal{W}_{\mathsf{SI}} = \mathbf{K}[\mathcal{S}] \cup \mathsf{I}(\mathbf{K})$, for some $\mathcal{S} \subseteq \mathsf{RL}_{\mathsf{SI}}$. Clearly $\mathcal{W} = \mathsf{V}(\mathbf{K}[\mathcal{S}] \cup \{\mathbf{K}\})$ and $\mathcal{W}_{\mathsf{SI}} = (\mathsf{V}(\mathbf{K}[\mathcal{S}] \cup \{\mathbf{K}\}))_{\mathsf{SI}} = (\mathsf{V}(\mathbf{K}[\mathcal{S}]))_{\mathsf{SI}} \cup \{\mathbf{K}\} = \mathbf{K}[(\mathsf{V}(\mathcal{S}))_{\mathsf{S}}] \cup \mathbf{I}(\mathbf{K}) = \mathbf{K}[(\mathsf{V}(\mathsf{V}(\mathcal{S})))_{\mathsf{SI}}] \cup \mathbf{I}(\mathbf{K}) = (\mathsf{V}(\mathbf{K}[\mathsf{V}(\mathcal{S})]))_{\mathsf{SI}} = ((\mathsf{V}(\mathcal{S}))_{\mathbf{K}})_{\mathsf{SI}}$; hence $\mathcal{W} = (\mathsf{V}(\mathcal{S}))_{\mathbf{K}}$ and the map is onto. □

COROLLARY 9.65. *The subvariety lattice $\mathbf{\Lambda}(\mathsf{RL})$ of RL is isomorphic to the interval $[\mathsf{V}(\mathbf{To}_n), \mathsf{RL}_{\mathbf{To}_n}]$ of $\mathbf{\Lambda}(\mathsf{FL_o})$ via the map $\mathcal{V} \mapsto \mathcal{V}_{\mathbf{To}_n}$, for every n.*

The same argument works for the algebra $\mathbf{2}$, so we have the following result.

THEOREM 9.66. *The subvariety lattice $\mathbf{\Lambda}(\mathsf{IRL})$ of IRL is isomorphic to the interval $[\mathsf{V}(\mathbf{K}), \mathsf{IRL}_{\mathbf{2}}]$ of $\mathbf{\Lambda}(\mathsf{FL_w})$ via the map $\mathcal{V} \mapsto \mathcal{V}_{\mathbf{2}}$.*

LEMMA 9.67. *The variety $\mathsf{IRL}_{\mathbf{2}}$ is axiomatized relative to $\mathsf{FL_w}$ by the set of identities $\gamma_1(\sim x) \vee \gamma_2(\sim\sim x) = 1$, where γ_1 and γ_2 range over iterated conjugates. Also, $\mathsf{ICRL}_{\mathbf{2}}$ is axiomatized relative to $\mathsf{FL_{ew}}$ by the identity $\neg x \vee \neg\neg x$.*

PROOF. First note that an algebra $\mathbf{A}$ is of the form $\mathbf{2}[\mathbf{B}]$ iff $A^* = A \setminus \{0\}$ is a 0-free subalgebra of $\mathbf{A}$. Since the join and residual of two elements of A^* is always in A^* and since closure under multiplication implies closure under meet, this is equivalent to A^* being closed under product.

We claim that this is in turn equivalent to the stipulation that $\mathbf{A}$ satisfies the first order formula $\forall x(x = 0 \text{ or } \sim x = 0)$. Indeed, let $x, y \in A^*$ be such that $xy = 0$ and suppose that $\mathbf{A}$ satisfies the first order formula. Then $y \to x\backslash 0 = \sim x$ and $\sim x = 0$, a contradiction. Conversely, if A^* is closed under product then, since $x(\sim x) = 0$, we have $x = 0$ or $\sim x = 0$.

By Lemma 9.63(2), the subdirectly irreducible algebras in $\mathsf{IRL}_{\mathbf{2}}$ are exactly $\mathbf{2}[(\mathsf{IRL})_{\mathsf{SI}}] \cup \mathsf{I}(\mathbf{2})$. In view of Lemma 9.62, these are exactly the subdirectly irreducible algebras in $\mathsf{FL_w}$ that satisfy the first order formula $\forall x(x = 0 \text{ or } \sim x = 0)$.

By Corollary 9.74, to be proved in the next section, the subvariety of $\mathsf{FL_w}$ whose subdirectly irreducible algebras satisfy the positive universal formula $\forall x(x = 0 \text{ or } {\sim}x = 0)$, or equivalently to the formula $\forall x(1 \leq x\backslash 0$ or $1 \leq (x\backslash 0)\backslash 0)$, is axiomatized by the set of identities $\gamma_1({\sim}x) \vee \gamma_2({\sim}{\sim}x) = 1$, where γ_1 and γ_2 range over iterated conjugates. In the commutative case, the conjugates are not needed. □

Recall that the logic **KC** of weak excluded middle is the extension of intuitionistic logic axiomatized by the formula $\neg p \vee \neg\neg p$. We denote the corresponding subvariety of HA by KC.

The following result holds because $\mathsf{Br}_{\mathbf{2}} = \mathsf{KC}$ and is essentially due to Jankov [Jan68], who proved that **KC** is the biggest superintuitionistic logic whose positive fragment coincides with the positive fragment of intuitionistic logic.

COROLLARY 9.68. *The subvariety lattice $\mathbf{\Lambda}(\mathsf{Br})$ of the variety Br of Brouwerian algebras is isomorphic to the interval $[\mathsf{BA}, \mathsf{KC}]$ of $\mathbf{\Lambda}(\mathsf{HA})$ via the map $\mathcal{V} \mapsto \mathcal{V}_{\mathbf{2}}$.*

COROLLARY 9.69. [Gal05] *If $\mathcal{V}$ is an atom in the subvariety lattice of RL and $\mathbf{K}$ a finite, strictly simple FL_o-algebra admissible by $\mathcal{V}$, then $\mathsf{V}(\mathbf{K}[\mathcal{V}])$ is a cover of $\mathsf{V}(\mathbf{K})$ in $\mathbf{\Lambda}(\mathsf{FL_o})$. Moreover, this correspondence is injective.*

COROLLARY 9.70. [Gal05] *The varieties $\mathsf{V}(\mathbf{T}_1[\mathbf{N}_S])$, where w_S is minimal, form a class of continuum many idempotent, representable, almost minimal varieties of $\mathbf{\Lambda}(\mathsf{FL_o})$.*

9.7. Axiomatizations for joins of varieties and meets of logics

Given a positive universal formula in the language of residuated lattices, we will construct a recursive basis of equations for a variety, such that a subdirectly irreducible residuated lattice is in the variety exactly when it satisfies the positive universal formula. We use this correspondence to prove, among other things, that the join of two finitely based varieties of commutative residuated lattices is also finitely based. This implies that the intersection of two finitely axiomatized substructural logics over $\mathbf{FL_e}$ is also finitely axiomatized. Finally, we give examples of cases where the join of two varieties is their Cartesian product.

9.7.1. Varieties of residuated lattices generated by positive universal classes. An *open positive universal formula* in a given language is an open first order formula that can be written as a disjunction of conjunctions of equations in the language. A *(closed) positive universal formula* is the universal closure of an open one. A *positive universal class* is the collection of all models of a set of positive universal formulas.

LEMMA 9.71. [Gal04] *Every open (closed) positive universal formula, φ, in the language of residuated lattices is equivalent to (the universal closure of)* *a disjunction φ' of equations of the form $1 = r$, where the evaluation of the term r is negative in all residuated lattices.*

PROOF. Every equation $t = s$ in the language of residuated lattices, where t, s are terms, is equivalent to the conjunction of the two inequalities $t \leq s$ and $s \leq t$, which in turn is equivalent to the conjunction of the inequalities $1 \leq s/t$ and $1 \leq t/s$. Moreover, a conjunction of a finite number of inequalities of the form $1 \leq t_i$, for $1 \leq i \leq n$ is equivalent to the inequality $1 \leq t_1 \wedge \ldots \wedge t_n$. So, a conjunction of a finite number of equations is equivalent to a single inequality of the form $1 \leq p$, which in turn is equivalent to the equation $1 = r$, where $r = p \wedge 1$. □

We now recall some facts on conjugates.Let $\mathbf{L}$ be a residuated lattice and Y a set of variables. For $y \in Y$ and $x \in L \cup Y \cup \{1\}$ language of residuated lattices, the polynomials

$$\rho_x(y) = xy/x \wedge 1 \text{ and } \lambda_x(y) = x\backslash yx \wedge 1,$$

are, respectively, the right and left conjugate of y with respect to x. An iterated conjugate is a composition of a number of left and right conjugates. For any X, A subsets of $L \cup Y \cup \{1\}$, and for $m \in \mathbb{N}$, we define the sets

$$\Gamma_X^0 = \{\lambda_1\},$$

$$\Gamma_X^m = \{\gamma_{x_1} \circ \gamma_{x_2} \circ \ldots \gamma_{x_m} : \gamma_{x_i} \in \{\lambda_{x_i}, \rho_{x_i}\},\ x_i \in X \cup \{1\},\ i \in \mathbb{N}\},$$

$$\Gamma_X^m(A) = \{\gamma(a) : \gamma \in \Gamma_X^m,\ a \in A\},$$

$$\Gamma_X = \bigcup\{\Gamma_X^n : n \in \mathbb{N}\},$$

$$\Gamma_X(A) = \bigcup\{\Gamma_X^n(A) : n \in \mathbb{N}\}.$$

Now, for a positive universal formula $\varphi(\bar{x})$ and a countable set of variables Y disjoint from $\bar{x}$, we define the sets of residuated-lattice equations

$$B_Y^m(\varphi'(\bar{x})) = \{1 = \gamma_1(r_1(\bar{x})) \vee \ldots \vee \gamma_n(r_n(\bar{x})) \mid \gamma_i \in \Gamma_Y^m\}$$

and

$$B_Y(\varphi'(\bar{x})) = \bigcup\{B_Y^m(\varphi'(\bar{x})) : m \in \mathbb{N}\},$$

where $m \in \mathbb{N}$ and

$$\varphi'(\bar{x}) = (r_1(\bar{x}) = 1) \vee \ldots \vee (r_n(\bar{x}) = 1)$$

is the formula equivalent to $\varphi(\bar{x})$, given in Lemma 9.71.

It is clear that $B_Y^m(\varphi'(\bar{x}))$ is an infinite set for $m \geq 1$. Nevertheless, if we enumerate $Y = \{y_i : i \in I\}$ and insist that the indices of the conjugating elements of Y in $\gamma_1, \gamma_2, \ldots, \gamma_n$ appear in the natural order and they form an initial segment of the natural numbers, then we obtain a finite subset of $B_Y^m(\varphi'(\bar{x}))$, which is equivalent to the latter. In that respect $B_Y^m(\varphi'(\bar{x}))$ is essentially finite.

LEMMA 9.72. [Gal04] *Let* $\mathbf{L}$ *be a residuated lattice and* $A_1, \ldots, A_n$ *finite subsets of* L. *If* $a_1 \vee \ldots \vee a_n = 1$, *for all* $a_i \in A_i$, $i \in \{1, \ldots, n\}$, *then for all* $i \in \{1, \ldots, n\}$, $n_i \in \mathbb{N}$, *and for all* $a_{i1}, a_{i2}, \ldots, a_{in_i} \in A_i$, *we have* $p_1 \vee \ldots \vee p_n = 1$, *where* $p_i = a_{i1}a_{i2} \cdots a_{in_i}$.

PROOF. The proof is a simple induction argument. For the basis of induction and for the induction step we use Lemma 2.6(1). If $a \vee b = a \vee c = 1$, then $1 = (a \vee b)(a \vee c) = a^2 \vee ac \vee ba \vee bc \leq a \vee bc \leq a \vee b = 1$. So, $a \vee bc = 1$. □

THEOREM 9.73. [Gal04] *Let* φ *be an open positive universal formula in the language of residuated lattices and* $\mathbf{L}$ *a residuated lattice.*

(1) *If* $\mathbf{L}$ *satisfies* $(\forall \bar{x})(\varphi(\bar{x}))$, *then* $\mathbf{L}$ *satisfies* $(\forall \bar{x}, \bar{y})(\varepsilon(\bar{x}, \bar{y}))$, *for all* $\varepsilon(\bar{x}, \bar{y})$ *in* $B_Y(\varphi'(\bar{x}))$ *and* $\bar{y} \in Y^l$, *for some appropriate* $l \in \mathbb{N}$.
(2) *If* $\mathbf{L}$ *is subdirectly irreducible, then* $\mathbf{L}$ *satisfies* $(\forall \bar{x})(\varphi(\bar{x}))$ *iff* $\mathbf{L}$ *satisfies the equation* $(\forall \bar{x}, \bar{y})(\varepsilon(\bar{x}, \bar{y}))$, *for all* $\varepsilon(\bar{x}, \bar{y})$ *in* $B_Y(\varphi'(\bar{x}))$ *and* $\bar{y} \in Y^l$.

PROOF. For (1), let $\mathbf{L}$ be a residuated lattice that satisfies $(\forall \bar{x})(\varphi(\bar{x}))$. Moreover, let $\varepsilon(\bar{x}, \bar{y})$ be an equation in $B_Y(\varphi'(\bar{x}))$, $\bar{c} \in L^k$ and $\bar{d} \in L^l$, for some appropriate $k, l \in \mathbb{N}$. We will show that $\varepsilon(\bar{c}, \bar{d})$ holds in $\mathbf{L}$. Since $\mathbf{L}$ satisfies $(\forall \bar{x})(\varphi(\bar{x}))$, $\varphi'(\bar{c})$ holds in $\mathbf{L}$. So, $r_i(\bar{c}) = 1$, for some $i \in \{1, 2, \ldots, n\}$; hence $\gamma(r_i(\bar{c})) = 1$, for all $\gamma \in \Gamma_Y$. Consequently, $\varepsilon(\bar{c}, \bar{d})$ holds.

For (2), let $\mathbf{L}$ be a subdirectly irreducible residuated lattice that satisfies $B_Y(\varphi'(\bar{x}))$, and let $\bar{c} \in L^k$ and $a_i = r_i(\bar{c})$. We will show that $a_i = 1$ for some i.

Let $b \in M(a_1) \cap \cdots \cap M(a_n)$, where $M(x)$ symbolizes the convex normal submonoid of the negative cone generated by x. Using Theorem 3.47(3), we have that for all $i \in \{1, 2, \ldots, n\}$, $\prod_{j=1}^{s_i} g_{ij} \leq b \leq 1$, for some $s_1, s_2, \ldots, s_n \in \mathbb{N}$ and $g_{i1}, g_{i2}, \ldots, g_{is_i} \in \Gamma_L(a_i)$. So,

$$\prod_{j=1}^{s_1} g_{1j} \vee \prod_{j=1}^{s_2} g_{2j} \vee \ldots \vee \prod_{j=1}^{s_n} g_{2j} \leq b \leq 1.$$

On the other hand,

$$\gamma_1(a_1) \vee \gamma_2(a_2) \vee \ldots \vee \gamma_n(a_n) = 1,$$

for all $\gamma_i \in \Gamma_L$, since every equation of $B_Y(\varphi'(\bar{x}))$ holds in $\mathbf{L}$. So, for all $i \in \{1, 2, \ldots, n\}$ and $g_i \in \Gamma_L(a_i)$, we have $g_1 \vee g_2 \vee \ldots \vee g_n = 1$ and, by Lemma 9.72,

$$\prod_{j=1}^{s_1} g_{1j} \vee \prod_{j=1}^{s_2} g_{2j} \vee \ldots \vee \prod_{j=1}^{s_n} g_{2j} = 1.$$

Thus, $b = 1$ and $M(a_1) \cap \cdots \cap M(a_n) = \{1\}$.

Using the lattice isomorphisms of Theorem 3.47, we obtain

$$\Theta(a_1, 1) \cap \Theta(a_2, 1) \cap \cdots \cap \Theta(a_n, 1) = \Delta,$$

where $\Theta(a, 1)$ denotes the principal congruence generated by $(a, 1)$ and Δ denotes the diagonal congruence. Since $\mathbf{L}$ is subdirectly irreducible, this implies that $\Theta(a_i, 1) = \Delta$, i.e., $a_i = 1$, for some i. Thus, $(\forall \bar{x})(\varphi'(\bar{x}))$ holds in $\mathbf{L}$. □

COROLLARY 9.74. [Gal04] *Let $\{\varphi_i : i \in I\}$ be a collection of positive universal formulas. Then, $\bigcup\{B(\varphi_i') : i \in I\}$ is an equational basis for the variety generated by the (subdirectly irreducible) residuated lattices that satisfy φ_i, for every $i \in I$.*

PROOF. By the previous theorem a subdirectly irreducible residuated lattice satisfies φ_i iff it satisfies all the equations in $B(\varphi_i')$, so we have

$$\begin{aligned}(\mathrm{Mod}(\bigcup\{\varphi_i : i \in I\}))_{SI} &= \bigcap\{(\mathrm{Mod}(\varphi_i))_{SI} : i \in I\}\\ &= \bigcap\{(\mathrm{Mod}(B(\varphi_i')))_{SI} : i \in I\}\\ &= (\mathrm{Mod}(\bigcup\{B(\varphi_i') \mid i \in I\}))_{SI},\end{aligned}$$

Consequently, we obtain

$$\begin{aligned}\mathsf{V}((\mathrm{Mod}(\bigcup\{\varphi_i : i \in I\}))_{SI}) &= \mathsf{V}((\mathrm{Mod}(\bigcup\{B(\varphi_i') : i \in I\}))_{SI})\\ &= \mathrm{Mod}(\bigcup\{B(\varphi_i') : i \in I\})\end{aligned}$$

□

Note that the equational basis for the variety generated by the models of a recursive positive universal class is recursive. In particular, the equational basis is recursive if the positive universal class is defined by a single formula.

The basis given in Theorem 9.73 is by no means of minimal cardinality. It is always infinite, while, as it can be easily seen, for commutative subvarieties it simplifies to the conjunction of commutativity and the equation of $B^0(\varphi')$. So, for example, the variety generated by the commutative residuated lattices, whose underlying set is the union of its positive and negative cone, is axiomatized by $xy = yx$ and $1 = (x \wedge 1) \vee (1/x \wedge 1)$.

Recall that the RRL of representable residuated lattices is generated by the class of all totally ordered residuated lattices.

COROLLARY 9.75. [BT03] [JT02] *The variety RRL is axiomatized by the 4-variable identity $\lambda_z((x \vee y)\backslash x) \vee \rho_w((x \vee y)\backslash y) = 1$.*

PROOF. The variety RRL is clearly generated by the class of all subdirectly irreducible totally ordered residuated lattices. A subdirectly irreducible residuated lattice is totally ordered if it satisfies the universal first-order formula

$$\forall x, y(x \leq y \text{ or } y \leq x).$$

The first order formula can also be written as

$$\forall x, y(1 = [(x \vee y)\backslash x] \wedge 1 \text{ or } 1 = [(x \vee y)\backslash y)] \wedge 1).$$

By Corollary 9.74, RRL is axiomatized by the identities

$$1 = \gamma_1([(x \vee y)\backslash x] \wedge 1) \vee \gamma_2([(x \vee y)\backslash y] \wedge 1),$$

where γ_1 and γ_2 range over arbitrary iterated conjugates. Actually, since $\gamma(t \wedge 1) \leq \gamma(t)$, for every iterated conjugate γ, if $1 = \gamma_1([(x \vee y)\backslash x] \wedge 1) \vee \gamma_2([(x \vee y)\backslash y] \wedge 1)$ holds, then $1 = \gamma_1((x \vee y)\backslash x) \vee \gamma_2((x \vee y)\backslash y)$ holds, as well. The converse is also true if γ_1 and γ_2 range over arbitrary iterated conjugates, since for example $\lambda_1(t) = t \wedge 1$. Therefore, RRL is axiomatized by the identities

$$1 = \gamma_1((x \vee y)\backslash x) \vee \gamma_2((x \vee y)\backslash y),$$

where γ_1 and γ_2 range over arbitrary iterated conjugates.

Therefore, RRL satisfies the identity

$$\lambda_z((x \vee y)\backslash x) \vee \rho_w((x \vee y)\backslash y) = 1.$$

Conversely, the variety axiomatized by this identity clearly satisfies the implications

$$x \vee y = 1 \Rightarrow \lambda_z(x) \vee y = 1 \qquad x \vee y = 1 \Rightarrow x \vee \rho_w(y) = 1.$$

By repeated applications of this implications on the identity

$$\lambda_z((x \vee y)\backslash x) \vee \rho_w((x \vee y)\backslash y) = 1$$

we can obtain

$$1 = \gamma_1((x \vee y)\backslash x) \vee \gamma_2((x \vee y)\backslash y),$$

for any pair of iterated conjugates γ_1 and γ_2. □

9.7.2. Equational basis for joins of varieties. In what follows, we apply the correspondence established above to obtain an equational basis for the join of a finite number of residuated lattice varieties. Moreover, we provide sufficient conditions for a variety of residuated lattices, in order for the join of any two of its finitely based subvarieties to be finitely based, as well.

COROLLARY 9.76. [Gal04] *If the varieties $\mathcal{V}_1, \mathcal{V}_2, \ldots, \mathcal{V}_n$ are axiomatized by the sets of equations $B_1, B_2, \ldots B_n$, where the sets of all variables in each B_i are pairwise disjoint, then $\bigcup\{B(\varphi_i') : i \in I\}$ is an equational basis for the join $\mathcal{V}_1 \vee \mathcal{V}_2 \vee \ldots \vee \mathcal{V}_n$, where φ_i ranges over all possible disjunctions of n equations, one from each of $B_1, B_2, \ldots, B_n$.*

PROOF. By Jónsson's Lemma, a subdirectly irreducible residuated lattice in the join of finitely many varieties is in one of the varieties. Moreover, by the definition of φ_i, it is clear that a subdirectly irreducible residuated lattice satisfies φ_i, for all $i \in I$, if and only if it is in one of the varieties

$\mathcal{V}_1, \mathcal{V}_2, \ldots, \mathcal{V}_n$. So,

$$\begin{aligned}\mathcal{V}_1 \vee \mathcal{V}_2 \vee \ldots \vee \mathcal{V}_n &= \mathcal{V}((\mathcal{V}_1 \vee \mathcal{V}_2 \vee \ldots \vee \mathcal{V}_n)_{SI}) \\ &= \mathcal{V}((\mathcal{V}_1 \cup \mathcal{V}_2 \cup \cdots \cup \mathcal{V}_n)_{SI}) \\ &= \mathcal{V}(\mathrm{Mod}(\bigcup\{\varphi_i : i \in I\})_{SI}) \\ &= \mathrm{Mod}(\bigcup\{B(\varphi_i') : i \in I\}).\end{aligned}$$

□

COROLLARY 9.77. [Gal04] *The join of finitely many recursively based varieties of residuated lattices is recursively based.*

In the case of the join of finitely based varieties the situation is simpler.

COROLLARY 9.78. *If $B_1, B_2, \ldots B_n$ are finite equational bases for the varieties $\mathcal{V}_1, \mathcal{V}_2, \ldots, \mathcal{V}_n$, then $B(\varphi')$ is an equational basis for the join $\mathcal{V}_1 \vee \mathcal{V}_2 \vee \ldots \vee \mathcal{V}_n$ of the varieties, where $\varphi = (\bigwedge B_1 \vee \bigwedge B_2 \vee \cdots \vee \bigwedge B_n)$ and for every $i \in \{1, 2, \cdots, n\}$, $\bigwedge B_i$ denotes the conjunction of the equations in B_i.*

PROOF. Retaining the notation of Corollary 9.76, we see that $\bigcup\{\varphi_i : i \in I\}$ is equivalent to φ and $\bigcup\{B(\varphi_i') : i \in I\}$ is equivalent to $B(\varphi')$. □

For each $k \in \mathbb{N}$ we define the variety

$$\mathsf{C}_\mathsf{k}^- \mathsf{RL} = \mathrm{Mod}((x \wedge 1)^k y = y(x \wedge 1)^k).$$

We also define the variety

$$\mathsf{CanC}_1^- \mathsf{RL} = \mathsf{CanRL} \cap \mathsf{C}_1^- \mathsf{RL}.$$

For all natural numbers m, n, we set $B_n^m = B^m(\varphi_n')$, where φ_n' is the term $(x_1 = 1) \vee (x_2 = 1) \vee \ldots \vee (x_n = 1)$. We say that a residuated lattice $\mathbf{L}$ satisfies the implication $B_n^m \to B_n^k$ iff, for all $\bar{a} \in L^n$, if $1 = r(\bar{a})$ for all $(1 = r) \in B_n^m$, then $1 = r(\bar{a})$ for all $(1 = r) \in B_n^k$.

THEOREM 9.79. [Gal04]

(1) *If $\mathcal{V}_1$, $\mathcal{V}_2$ are finitely based varieties of residuated lattices that satisfy the implication $B_2^m \to B_2^{m+1}$, for some natural number m, then their join is also finitely based.*
(2) *The join of any pair of finitely based subvarieties of the variety $\mathsf{LG} \vee \mathsf{CanC}_1^- \mathsf{RL} \vee \mathsf{RRL} \vee \mathsf{C}_\mathsf{k}^- \mathsf{RL}$ is also finitely based, for every $k \geq 1$.*

PROOF. For (1) note first that, since B_2^m is equivalent to a finite set of equations and since the conjunction of a finite set of equations is equivalent to a single equation, B_2^m is equivalent to a single equation. All subdirectly irreducible algebras in the join $\mathcal{V}_1 \vee \mathcal{V}_2$ coincide with the subdirectly irreducible algebras in the union $\mathcal{V}_1 \cup \mathcal{V}_2$, so they satisfy the implication $B_2^m \to B_2^{m+1}$. Since all residuated lattices in the join $\mathcal{V}_1 \vee \mathcal{V}_2$ are subdirect products of subdirectly irreducible algebras, and quasi-equations are preserved under products and subalgebras, the join satisfies the implication. Moreover, if B_1, B_2 are finite equational basis for $\mathcal{V}_1$, $\mathcal{V}_2$, respectively, then $B(\varphi')$ is an equational basis for $\mathcal{V}_1 \vee \mathcal{V}_2$, where $\varphi = \bigwedge B_1 \vee \bigwedge B_2$. So, the implication is a consequence of a finite subset B of $B(\varphi')$, by compactness. It is clear then, that $B \cup B^m(\varphi')$ is a finite equational basis for $\mathcal{V}_1 \vee \mathcal{V}_2$.

For (2) observe that LG satisfies $\lambda_z(\lambda_w(x)) = \lambda_{wz}(x)$ and $\rho_z(x) = \lambda_{z^{-1}}(x)$, where $z^{-1} = z\backslash 1$ since

$$\begin{aligned}\lambda_z(\lambda_w(x)) &= z\backslash(w\backslash xw \wedge 1)z \wedge 1 \\ &= z^{-1}(w^{-1}xw \wedge 1)z \wedge 1 \\ &= z^{-1}w^{-1}xwz \wedge z^{-1}z \wedge 1 \\ &= (wz)^{-1}xwz \wedge 1 \\ &= wz\backslash xwz \wedge 1 \\ &= \lambda_{wz}(x)\end{aligned}$$

and

$$\rho_z(x) = zx/z \wedge 1 = zxz^{-1} \wedge 1 = z^{-1}\backslash xz^{-1} \wedge 1 = \lambda_{z^{-1}}(x).$$

So, $\lambda_z(\lambda_w(x \wedge 1)) = \lambda_{wz}(x \wedge 1)$ and $\rho_z(x \wedge 1) = \lambda_{z^{-1}}(x \wedge 1)$ hold in LG. The same two equations hold in $\mathsf{CanC_1^-RL}$, since for any negative element a and any element b,

$$\lambda_b(a) = b\backslash ab \wedge 1 = b\backslash ba \wedge 1 = a \wedge 1 = a$$

and $\rho_b(a) = a \wedge 1 = a$. Thus, LG and $\mathsf{CanC_1^-RL}$ satisfy $B_2^1 \to B_2^2$.

On the other hand, the variety RRL satisfies the implication

$$x \vee y = 1 \Rightarrow \lambda_z(x) \vee \rho_w(y) = 1,$$

by Theorem 9.75. We will show that the same implication holds in $\mathsf{C_k^-RL}$. If $x \vee y = 1$, then, by Lemma 9.72, $x^k \vee y^k = 1$. Since, $x \leq 1$, we have $x^k \leq x \leq 1$; so, for all z, $x^kz = zx^k$, hence $x^k \leq z\backslash x^kz$ and $x^k \leq zx^k/z$. Since $x^k \leq 1$, this implies

$$x^k \leq z\backslash x^kz \wedge 1 \text{ and } x^k \leq zx^k/z \wedge 1,$$

i.e., $x^k \leq \lambda_z(x^k)$ and $x^k \leq \rho_z(x^k)$, for all z. Thus, $\lambda_z(x^k) \vee \rho_w(y^k) = 1$. Moreover, left and right conjugates are increasing in their arguments, so $\lambda_z(x) \vee \rho_w(y) = 1$. So, RRL and $\mathsf{C_k^-RL}$ satisfy the implication $B_2^0 \to B_2^1$, hence also the implication $B_2^1 \to B_2^2$.

Using the same argumentation as in the proof of (1), it is easy to see that the join $\mathsf{LG} \vee \mathsf{CanC_1^-RL} \vee \mathsf{RRL} \vee \mathsf{C_k^-RL}$ of the four varieties satisfies the implication $B_2^1 \to B_2^2$. Consequently, every subvariety of the join satisfies the implication, as well. Statement (2) then follows from (1). □

COROLLARY 9.80. [Gal04] *The join of two finitely based commutative varieties of residuated lattices is finitely based.*

It is an open problem whether the join of two finitely based varieties of residuated lattices is finitely based.

COROLLARY 9.81. [Gal04] *The intersection of two finitely axiomatized substructural logics over* $\mathbf{FL_e}$ *is finitely axiomatized, as well. In particular, if two* φ *and* ψ *that do not share any common variables axiomatize two commutative logics then* $(\varphi \wedge 1) \vee (\psi \wedge 1)$ *is an axiomatization of their intersection. For logics over* $\mathbf{FL_{ei}}$ *the axiomatization simplifies to* $\varphi \vee \psi$.

9.7.3. Direct product decompositions. Certain pairs of subvarieties of RL are so different that their join decomposes into their Cartesian product—the class of all Cartesian products of algebras of the two varieties up to isomorphism. Such a pair is the variety of ℓ-groups and the variety of their negative cones. The following proposition is in the folklore of the subject and allows us to obtain such decompositions given two projection terms.

PROPOSITION 9.82. [Gal04] *Let* $\mathcal{V}_1, \mathcal{V}_2$ *be subvarieties of* RL *with equational bases* B_1 *and* B_2*, respectively, and let* $\pi_1(x), \pi_2(x)$ *be unary terms, such that* $\mathcal{V}_1$ *satisfies* $\pi_1(x) = x$ *and* $\pi_2(x) = 1$ *and* $\mathcal{V}_2$ *satisfies* $\pi_1(x) = e$ *and* $\pi_2(x) = x$*. Then* $\mathcal{V}_1 \vee \mathcal{V}_2 = \mathcal{V}_1 \times \mathcal{V}_2$ *and the following list,* $B_1 * B_2$*, of equations is an equational basis for the variety* $\mathcal{V}_1 \vee \mathcal{V}_2$.

(1) $\pi_1(x) \cdot \pi_2(x) = x$
(2) $\pi_i(\pi_i(x)) = \pi_i(x)$ *and* $\pi_i(\pi_j(x)) = 1$*, for* $i \neq j$*;* $i, j \in \{1, 2\}$
(3) $\pi_i(x \star y) = \pi_i(x) \star \pi_i(y)$*, where* $\star \in \{\wedge, \vee, \cdot, /, \backslash\}$ *and* $i \in \{1, 2\}$
(4) $\varepsilon(\pi_1(x_1), \ldots, \pi_1(x_n))$*, for all equations* $\varepsilon(x_1, \ldots, x_n)$ *of* B_1
(5) $\varepsilon(\pi_2(x_1), \ldots, \pi_2(x_n))$*, for all equations* $\varepsilon(x_1, \ldots, x_n)$ *of* B_2

The same decomposition holds for the join of any pair of subvarieties of $\mathcal{V}_1, \mathcal{V}_2$*. Note that if* B_1, B_2 *are finite, then so is* $B_1 * B_2$.

PROOF. It is easy to see that the equations in $B_1 * B_2$ hold both in $\mathcal{V}_1$ and $\mathcal{V}_2$, hence they hold in $\mathcal{V}_1 \vee \mathcal{V}_2$, also. Moreover, $\mathcal{V}_1 \times \mathcal{V}_2 \subseteq \mathcal{V}_1 \vee \mathcal{V}_2$. Finally, suppose that the residuated lattice A satisfies the equations $B_1 * B_2$; we will show that A is in $\mathcal{V}_1 \times \mathcal{V}_2$.

Define $A_1 = \{x \in A \mid \pi_2(x) = 1\}$ and $A_2 = \{x \in A \mid \pi_1(x) = 1\}$. Using (3) and (1), it is easy to see that A_1 and A_2 are subalgebras of A. Define the map $f : A \to A_1 \times A_2$, by $f(x) = (\pi_1(x), \pi_2(x))$. It is easy to check that f is well defined, using (2); that it is a homomorphism, using (3); one-to-one, using (1); and onto, using (3) and (1). Thus, A is isomorphic to $A_1 \times A_2 \in \mathcal{V}_1 \times \mathcal{V}_2$. □

The first example of a pair of two varieties whose join is their Cartesian product is the variety of ℓ-groups and the variety of integral residuated lattices.

COROLLARY 9.83. [Gal04] *The join of* LG *and* IRL *is equal to their product. Moreover, if* $B_1 = \{(1/x)x = 1\}$ *and* $B_2 = \{1 \wedge x = x\}$*, then* $B_1 * B_2$ *is an equational basis for* $\mathsf{LG} \vee \mathsf{IRL}$.

PROOF. Let $\pi_1(x) = 1/(1/x)$ and $\pi_2(x) = (1/x)x$. It is easy to see that LG satisfies $1/(1/x) = 1(1x^{-1})^{-1} = x$ and $(1/x)x = x^{-1}x = 1$. Moreover, IRL satisfies $1/x = 1$, so it also satisfies $(1/x)x = 1x = x$ and $1/(1/x) = 1$. □

It is shown in [BCG+03] that the class LG^-, consisting of the negative cones of ℓ-groups, is a variety. As an application of the previous example we have $\mathsf{LG} \vee \mathsf{LG}^- = \mathsf{LG} \times \mathsf{LG}^-$.

Moreover, the class $\mathsf{IGMV} = \mathrm{Mod}(x/(y\backslash x) = x \vee y = (x/y)\backslash x)$ of integral generalized MV-algebras is easily shown to be a subvariety of IRL. So, Corollary 9.83 provides an equational basis for $\mathsf{LG} \vee \mathsf{IGMV} = \mathsf{LG} \times \mathsf{IGMV}$. In [GT05], generalized MV-algebras are studied and an alternative, simpler equational basis is given for $\mathsf{LG} \vee \mathsf{IGMV}$.

The second example involves the minimal varieties $\mathsf{V}(\mathbf{T}_n)$ that we have met before (cf. Figure 9.1) and cancellative integral residuated lattices.

COROLLARY 9.84. [Gal04] *Let $\mathcal{V}$ be a variety of residuated lattices that satisfies the identities $(1/(1/x))^n \leq x$ and $(x \wedge 1)^n = (x \wedge 1)^{n+1}$, for some $n \in \mathbb{N}$. Then, $\mathsf{CanIRL} \vee \mathcal{V} = \mathsf{CanIRL} \times \mathcal{V}$.*

PROOF. Let $\pi_1(x) = ((1 \wedge x)^{n+1}/(1 \wedge x)^n) \wedge 1$ and $\pi_2(x) = (1/(1/x))^n \vee x$. Note that CanIRL satisfies

$$\pi_1(x) = ((1 \wedge x)^{n+1}/(1 \wedge x)^n) \wedge 1 = (1 \wedge x) \wedge 1 = x$$

and

$$\pi_2(x) = (1/(1/x))^n \vee x = 1^n \vee x = 1.$$

On the other hand, $\mathcal{V}$ satisfies

$$\pi_2(x) = (1/(1/x))^n \vee x = x$$

and

$$\pi_1(x) = ((1 \wedge x)^{n+1}/(1 \wedge x)^n) \wedge 1 = ((1 \wedge x)^n/(1 \wedge x)^n) \wedge 1 = 1.$$

□

COROLLARY 9.85. [Gal04] *The join of the varieties $\mathcal{V} = \mathsf{V}(\mathbf{T}_{i_1}, \mathbf{T}_{i_2}, \ldots, \mathbf{T}_{i_k})$ and CanIRL is equal to their Cartesian product, for all $k, i_1, \ldots, i_k \in \mathbb{N}$.*

PROOF. In view of the last corollary, we need only verify that $\mathcal{V}$ satisfies the identities $(x \wedge 1)^n = (x \wedge 1)^{n+1}$ and $(1/(1/x))^n \leq x$, for some $n \in \mathbb{N}$.

If $n \geq m$, then $(1 \wedge x)^{n+1} = (1 \wedge x)^m = (1 \wedge x)^n$, for $x \in \mathbf{T}_m$. Moreover, $(1/(1/T))^n \leq T$, $(1/(1/1))^n = 1$ and $(1/(1/x))^n = (1/T)^n = u^m \leq x$, for $x < 1$. If $n \geq \max\{i_1, \ldots, i_k\}$, then $\mathcal{V} = \mathsf{V}(\mathbf{T}_{i_1}, \mathbf{T}_{i_2}, \ldots, \mathbf{T}_{i_k})$ satisfies both identities. □

Note that $\mathsf{IRL} \vee \mathsf{V}(\mathbf{T}_1) \neq \mathsf{IRL} \times \mathsf{V}(\mathbf{T}_1)$, since $\mathbf{A}$ is in $\mathbf{S}(\mathbf{2} \times \mathbf{T}_1)$ but not in $\mathsf{IRL} \times \mathsf{V}(\mathbf{T}_1)$, where $A = \{(1, T), (1, 1), (1, u_1), (0, u_1)\}$ and $2 = \{0, 1\}$.

9.8. The subvariety lattices of LG and LG$^-$

We now extend the map $^-: \mathsf{LG} \to \mathsf{LG}^-$ to subclasses of LG, and in particular to the lattice of subvarieties $\mathbf{\Lambda}(\mathsf{LG})$. We show that the image of a variety is always a variety, that every subvariety of LG$^-$ is obtained in this way and that the map is an order isomorphism, hence a lattice isomorphism (see Figure 9.11).

In the second part of this section we show how equational bases can be translated back and forth between corresponding varieties of LG and LG$^-$. We conclude the section by indicating how these results are related to R. McKenzie's general characterization of categorical equivalence [McK96].

Recall that for a class $\mathcal{K}$ of residuated lattices, $\mathsf{H}(\mathcal{K})$, $\mathsf{S}(\mathcal{K})$, $\mathsf{P}(\mathcal{K})$ and $\mathcal{K}^-$ denote, respectively, the class of homomorphic images, subalgebras, products and negative cones of members of $\mathcal{K}$.

THEOREM 9.86. [BCG+03] *The map $\mathcal{K} \mapsto \mathcal{K}^-$, defined on classes of ℓ-groups, commutes with the operators* H, S *and* P, *and restricts to a lattice isomorphism between the subvariety lattices of* LG *and* LG$^-$.

PROOF. Consider any subclass $\mathcal{K}$ of LG. To see that $\mathsf{S}(\mathcal{K}^-) = (\mathsf{S}\mathcal{K})^-$, let $\mathbf{G}^-$ be a member of $\mathcal{K}^-$ and let $\mathbf{H}'$ be a subalgebra of $\mathbf{G}^-$. By the definition of $\mathcal{K}^-$, there exists $\mathbf{G} \in \mathcal{K}$ whose negative cone is $\mathbf{G}^-$. We show that $\mathbf{H}'$ is the negative cone of a subalgebra $\mathbf{H}$ of $\mathbf{G}$.

Consider the set $H = \{ab^{-1} : a, b \in H'\}$. We establish that H is the subgroup of $\mathbf{G}$ generated by H' by proving that H is closed under products in $\mathbf{G}$. Let ab^{-1} and cd^{-1} be in H. Using Theorem 3.30, we can show that

$$ab^{-1}cd^{-1} = ab^{-1}cbb^{-1}d^{-1} = a(b\backslash cb)(db)^{-1} = ac_b(db)^{-1},$$

which is again an element of H since $\mathbf{H}'$ is a subalgebra of $\mathbf{G}^-$. What we know so far is that H is a partially ordered group with respect to the order of $\mathbf{G}$. To show that H is a subalgebra (ℓ-subgroup) of $\mathbf{G}$, it will suffice to show that for $a, b \in H'$ the join $ab^{-1} \vee 1$ (in $\mathbf{G}$) is an element of H. Let $a, b \in H'$, and note that by using Theorem 3.30 we obtain $ab^{-1} \vee^{\mathbf{G}} 1 = (a \vee^{\mathbf{G}^-} b)b^{-1}$. Since $\mathbf{H}'$ is a subalgebra of $\mathbf{G}^-$, we have $a \vee^{\mathbf{G}^-} b = a \vee^{\mathbf{H}'} b$. It follows that $ab^{-1} \vee^{\mathbf{G}} 1 \in H$, as was to be shown.

Since $\mathbf{G}$ is an algebra in $\mathcal{K}$, with subalgebra $\mathbf{H}$, this establishes that $\mathsf{S}(\mathcal{K}^-) \subseteq (\mathsf{S}\mathcal{K})^-$. The reverse inclusion is immediate.

The fact that $\mathsf{P}(\mathcal{K}^-) = (\mathsf{P}\mathcal{K})^-$ follows from the observation that for a collection of residuated lattices $\{\mathbf{G}_i\}_{i\in I}$ in $\mathcal{K}$,

$$\prod_{i\in I} \mathbf{G}_i^- = (\prod_{i\in I} \mathbf{G}_i)^-.$$

Finally we must show that $\mathsf{H}(\mathcal{K}^-) = (\mathsf{H}\mathcal{K})^-$. So let $\mathbf{G}^- \in \mathcal{K}^-$ and consider a surjective residuated lattice homomorphism $\varphi : \mathbf{G}^- \to \mathbf{H}'$. As suggested by the notation, $\mathbf{G}^-$ is the negative cone of an ℓ-group $\mathbf{G} \in \mathcal{K}$, and since LG$^-$ is a variety, we may assume that $\mathbf{H}'$ is the negative cone of an

ℓ-group $\mathbf{H}$. We extend φ to $\overline{\varphi}:\mathbf{G}\to\mathbf{H}$ by defining $\overline{\varphi}(ab^{-1})=\varphi(a)\varphi(b)^{-1}$. We need to show that this map is a well-defined ℓ-group homomorphism onto $\mathbf{H}$.

To show that $\overline{\varphi}$ is well-defined, consider two different representations for the same element of $\mathbf{G}$: $ab^{-1}=cd^{-1}$. This is the case if and only if $ad_b=cb$, implying that $\varphi(a)\varphi(d)_{\varphi(b)}=\varphi(c)\varphi(b)$ since φ is a homomorphism. This in turn implies that $\varphi(a)\varphi(b)^{-1}=\varphi(c)\varphi(d)^{-1}$.

The map $\overline{\varphi}$ is clearly onto, since φ is onto and every element $x\in H$ can be written as ab^{-1} for some a and b in H'.

To show that $\overline{\varphi}$ is a group homomorphism, let ab^{-1} and cd^{-1} be elements of G. Recall that $ab^{-1}cd^{-1}=ab^{-1}cbb^{-1}d^{-1}=ac_b(db)^{-1}$ so that $\overline{\varphi}(ab^{-1}cd^{-1})=\overline{\varphi}(ac_b(db)^{-1})=\varphi(ac_b)\varphi(db)^{-1}=\varphi(a)\varphi(c)_{\varphi(b)}\varphi(b)^{-1}\varphi(d)^{-1}$, since φ is a homomorphism. But this gives

$$\varphi(a)\varphi(b)^{-1}\varphi(c)\varphi(b)\varphi(b)^{-1}\varphi(d)^{-1}=\overline{\varphi}(ab^{-1})\overline{\varphi}(cd^{-1}).$$

Finally, we show that $\overline{\varphi}$ preserves joins. As before, it suffices to consider joins of the form $ab^{-1}\vee e$. Recalling that $ab^{-1}\vee 1=(a\vee b)b^{-1}$, we have

$$\overline{\varphi}(ab^{-1}\vee 1)=\overline{\varphi}((a\vee b)b^{-1})=\varphi(a\vee b)\varphi(b)^{-1}=(\varphi(a)\vee\varphi(b))\varphi(b)^{-1},$$

since φ preserves joins. But this last product is $\varphi(a)\varphi(b)^{-1}\vee e=\overline{\varphi}(ab^{-1})\vee 1$. Thus $\overline{\varphi}$ is a lattice homomorphism as well.

This shows that $\mathbf{H}\in\mathsf{H}\mathcal{K}$, hence $\mathbf{H}'=\mathbf{H}^-\in(\mathsf{H}\mathcal{K})^-$, as desired. Again the reverse inclusion is straightforward.

The preceding considerations immediately imply that if $\mathcal{V}$ is a variety of ℓ-groups, then $\mathcal{V}^-$ is a subvariety of LG^-. Moreover, the map $\mathcal{V}\mapsto\mathcal{V}^-$ is clearly order preserving. To show it is onto, let $\mathcal{W}\in\mathbf{\Lambda}(\mathsf{LG}^-)$ and consider the class $\mathcal{K}=\{\mathbf{L}\in\mathsf{LG}:\mathbf{L}^-\in\mathcal{W}\}$. Then $\mathcal{K}^-=\mathcal{W}$ and $\mathcal{K}$ is a variety since $\mathbf{L}\in\mathsf{HSP}\mathcal{K}$ implies $\mathbf{L}^-\in(\mathsf{HSP}\mathcal{K})^-=\mathsf{HSP}(\mathcal{K}^-)=\mathcal{W}$, so $\mathbf{L}\in\mathcal{K}$.

Finally, we show that the map is one-to-one and reflects the order as well. Indeed, an ℓ-group is determined up to isomorphism by its negative cone (the preceding map $\overline{\varphi}$ is an isomorphism when φ is an isomorphism), so if $\mathcal{V}\not\leq\mathcal{V}'$ and $G\in\mathcal{V}\setminus\mathcal{V}'$, then $G^-\in\mathcal{V}^-\setminus\mathcal{V}'^-$. It follows that $\mathcal{V}\mapsto\mathcal{V}^-$ is a lattice isomorphism between the two subvariety lattices. □

COROLLARY 9.87. [BCG+03] *The variety* $\mathsf{V}(\mathbb{Z}^-)$ *consists of all negative cones of Abelian* ℓ*-groups.*

It was proved above that there is a one-to-one correspondence between subvarieties of LG and LG^-. Since the proof made use of the HSP characterization of varieties, it gave no insight into how one might find an equational basis for $\mathcal{V}^-$ given a basis for $\mathcal{V}$, and vice versa. We proceed to do that in the remainder of this section.

9.8.1. From subvarieties of LG^- to subvarieties of LG. In this direction, the translation is derived essentially from the definition of the negative

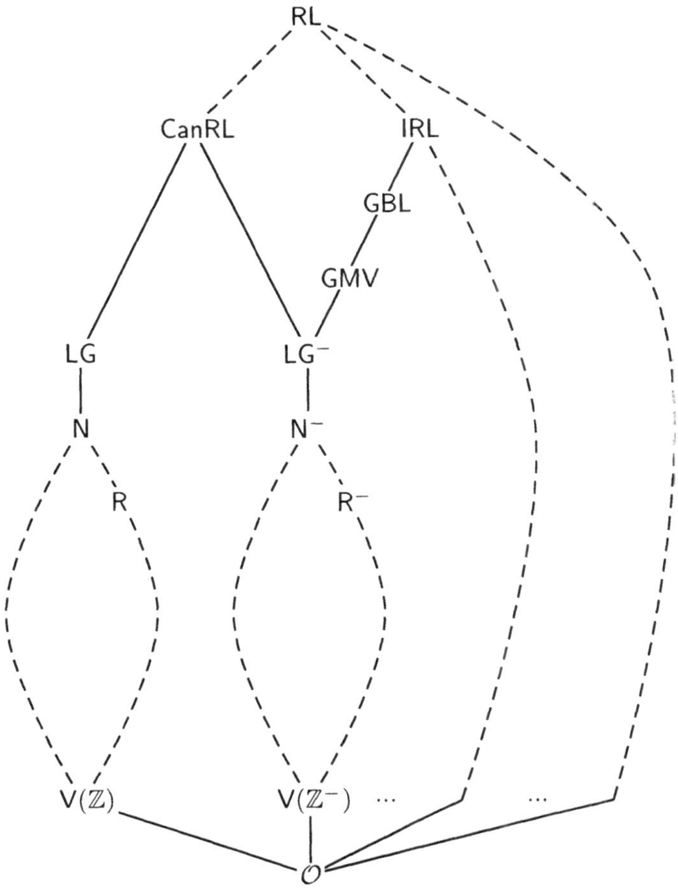

FIGURE 9.11. Inclusions between some subvarieties of RL.

cone. For a residuated lattice term t, we define a translated term t^- by

$$
\begin{array}{lll}
x^- = x \wedge 1 & 1^- = 1 & \\
(s/t)^- = s^-/t^- \wedge 1 & (s\backslash t)^- = s^-\backslash t^- \wedge 1 & \\
(st)^- = s^-t^- & (s \vee t)^- = s^- \vee t^- & (s \wedge t)^- = s^- \wedge t^-
\end{array}
$$

LEMMA 9.88. [BCG+03] *Let* $\mathbf{L} \in \mathsf{RL}$ *and consider any* RL *term* t. *For any* $a_1, \ldots, a_n \in L$,

$$t^{-\mathbf{L}}(a_1, \ldots, a_n) = t^{\mathbf{L}^-}(a_1 \wedge e, \ldots, a_n \wedge 1).$$

PROOF. By definition this is true for variables and the constant term 1. Assume the statement holds for terms s and t. Then

$$\begin{aligned}(s/t)^{-\mathbf{L}}(a_1,\dots,a_n) &= (s^{-\mathbf{L}}(a_1,\dots,a_n)/^{\mathbf{L}}t^{-\mathbf{L}}(a_1,\dots,a_n)) \wedge 1\\ &= (s^{\mathbf{L}^-}(a_1\wedge 1,\dots,a_n\wedge 1)/^{\mathbf{L}}t^{\mathbf{L}^-}(a_1\wedge 1,\dots,a_n\wedge 1))\wedge 1\\ &= (s/t)^{\mathbf{L}^-}(a_1\wedge 1,\dots,a_n\wedge 1)\end{aligned}$$

and similar inductive steps for $\backslash$, $\cdot$, $\vee$, $\wedge$ complete the proof. □

LEMMA 9.89. [BCG+03] *For any $\mathbf{L} \in \mathsf{RL}$, $\mathbf{L}^- \models s \approx t$ iff $\mathbf{L} \models s^- \approx t^-$.*

PROOF. Suppose $\mathbf{L}^- \models s \approx t$, and let $a_1,\dots,a_n \in L$. By the preceding lemma, $s^{-\mathbf{L}}(a_1,\dots,a_n) = s^{\mathbf{L}^-}(a_1\wedge 1,\dots,a_n\wedge 1) = t^{\mathbf{L}^-}(a_1\wedge 1,\dots,a_n\wedge 1) = t^{-\mathbf{L}}(a_1,\dots,a_n)$, hence $\mathbf{L} \models s^- \approx t^-$. The reverse implication is similar and uses the observation that for $a_i \in L^-$, $a_i = a_i \wedge 1$. □

THEOREM 9.90. [BCG+03] *Let $\mathcal{V}$ be a subvariety of LG^-, defined by a set E of identities and let $\mathcal{W} = \mathrm{Mod}\,(E^-) \cap \mathsf{LG}$, where $E^- = \{s^- \approx t^- : (s \approx t) \in E\}$. Then $\mathcal{W}^- = \mathcal{V}$.*

PROOF. Consider $\mathbf{M} \in \mathcal{W}^-$, which means there exists an $\mathbf{L} \in \mathcal{W}$ such that $\mathbf{M}$ is isomorphic to $\mathbf{L}^-$. Then $\mathbf{L} \models E^-$, and by the previous lemma this is equivalent to $\mathbf{L}^- \models E$, which in turn is equivalent to $\mathbf{L}^- \in \mathcal{V}$. Hence $\mathbf{M} \in \mathcal{V}$.

Conversely, let $\mathbf{M} \in \mathcal{V}$. Then there exists an ℓ-group $\mathbf{L}$ such that $\mathbf{M}$ is isomorphic to $\mathbf{L}^-$. Using the previous lemma again, we get that $\mathbf{L} \models E^-$, hence $\mathbf{M} \in \mathcal{W}^-$. □

As an example, consider the variety $\mathcal{N}^-$ that is defined by the identity $x^2y^2 \leq yx$ relative to LG^-. The corresponding identity for the variety $\mathcal{N}$ of normal valued ℓ-groups is $(x \wedge 1)^2(y \wedge 1)^2 \leq (y \wedge 1)(x \wedge 1)$.

9.8.2. From subvarieties of LG to subvarieties of LG$^-$. Note that since $\cdot$ and $^{-1}$ distribute over $\vee$ and $\wedge$, any LG identity is equivalent to a conjunction of two identities of the form $1 \leq p(g_1,\dots,g_n)$, where p is a lattice term and $g_1,\dots,g_n$ are group terms. Since ℓ-groups are distributive, this can be further reduced to a finite conjunction of inequalities of the form $1 \leq g_1 \vee \cdots \vee g_n$.

For a term $t(x_1,\dots,x_m)$ and a variable z distinct from $x_1,\dots,x_m$, let

$$\bar{t}(z,x_1,\dots,x_m) = t(z^{-1}x_1,\dots,z^{-1}x_m).$$

LEMMA 9.91. [BCG+03] *Let $\mathbf{L}$ be an ℓ-group, and t an ℓ-group term. Then*

$$\begin{gathered}\mathbf{L} \models e \leq t(x_1,\dots,x_m) \ \textit{iff}\\ \mathbf{L} \models x_1 \vee \cdots \vee x_m \vee z \leq 1 \Rightarrow 1 \leq \bar{t}(z,x_1,\dots,x_m).\end{gathered}$$

PROOF. In the forward direction this is obvious. To prove the reverse implication, assume the right hand side holds and let $a_1,\dots,a_m \in L$. Define $c = a_1^{-1} \wedge \cdots \wedge a_m^{-1} \wedge 1$ and $b_i = ca_i$ for $i = 1,\dots,m$. Then $c \leq e$ and

$c \leq a_i^{-1}$, hence $b_i \leq 1$. Now by assumption, $1 \leq t(c^{-1}b_1, \ldots, c^{-1}b_m) = t(a_1, \ldots, a_m)$. □

LEMMA 9.92. [BCG+03] *Let* $\mathbf{L} \in \mathsf{LG}$. *For any group term g, there exist an* RL *term $\hat{g}$ such that* $(g \wedge 1)^{\mathbf{L}}|_{L^-} = \hat{g}^{\mathbf{L}^-}$.

PROOF. Essentially we have to rewrite group terms so that all the variables with inverses appear at the beginning of the term. This is done using conjugation: $xy^{-1} = y^{-1}(yxy^{-1}) = y^{-1}(yx/y)$. Note that L^- is closed under conjugation by arbitrary elements, since $x \leq 1$ implies $yxy^{-1} \leq 1$. If we also have $y \leq 1$, then $yx \in L^-$ and $yx \leq y$, hence $yx/^{\mathbf{L}^-}y = yx/^{\mathbf{L}}y$.

To describe the translation of an arbitrary group term, we may assume that it is of the form $p_1q_1^{-1}p_2q_2^{-1}\cdots p_nq_n^{-1}$ where the p_i and q_i are products of variables (without inverses). By using conjugation, we write this term in the form

$$q_1^{-1}q_2^{-1}\cdots q_n^{-1}(q_n(\cdots(q_2(q_1p_1/q_1)p_2/q_2)\cdots)p_n/q_n).$$

So we can take $\hat{g} = s\backslash t$ where

$$s = q_n\cdots q_2q_1 \text{ and } t = q_n(\cdots(q_2(q_1p_1/q_1)p_2/q_2)\cdots)p_n/q_n.$$

□

COROLLARY 9.93. [BCG+03] *Let $g_1, \ldots, g_n$ be group terms with variables among $x_1, \ldots, x_m$. For any ℓ-group* $\mathbf{L}$,

$$\mathbf{L}^- \models \hat{g}_1 \vee \ldots \vee \hat{g}_n = 1 \textit{ iff } \mathbf{L} \models x_1 \vee \ldots \vee x_m \leq 1 \Rightarrow 1 \leq g_1 \vee \ldots \vee g_n.$$

For the next result, recall the discussion about identities in ℓ-groups, and the definition of $\bar{t}$ at the beginning of this subsection.

THEOREM 9.94. [BCG+03] *Let $\mathcal{V}$ be a subvariety of* LG, *defined by a set E of identities, which we may assume are of the form $1 \leq g_1 \vee \ldots \vee g_n$. Let*

$$\bar{E} = \{e = \widehat{g_1} \vee \ldots \vee \widehat{g_n} : e \leq g_1 \vee \ldots \vee g_n \textit{ is in } E\}.$$

Then $\bar{E}$ is an equational basis for $\mathcal{V}^-$ relative to LG$^-$.

PROOF. By construction, any member of $\mathcal{V}^-$ satisfies all the identities in $\bar{E}$. On the other hand, if $\mathbf{M} \in \mathsf{LG}^-$ is a model of the identities in $\bar{E}$, then $\mathbf{M}$ is the negative cone of some $\mathbf{L} \in \mathsf{LG}$. From the reverse directions of Corollary 9.93 and Lemma 9.91 we infer that $\mathbf{L}$ satisfies the equations in E, hence $\mathbf{M} \in \mathcal{V}^-$. □

For example consider the variety R of *representable ℓ-groups* which (by definition) is generated by the class of totally ordered groups (see [AF88] for more details). An equational basis for this variety is given by $1 \leq x^{-1}yx \vee y^{-1}$ (relative to LG). Applying the translation above, we obtain $1 = zx\backslash(zy/z)x \vee y\backslash z$ as as equational basis for $\mathcal{R}^-$.

COROLLARY 9.95. [BCG+03] *The map $\mathcal{V} \mapsto \mathcal{V}^-$ from $\mathbf{\Lambda}(\mathsf{LG})$ to $\mathbf{\Lambda}(\mathsf{LG}^-)$ sends finitely based subvarieties of* LG *to finitely based subvarieties of* LG$^-$.

9.8.3. Categorical equivalence and the functor $\mathbf{L} \mapsto \mathbf{L}^-$. The connection between LG and LG^- is actually a special case of a categorical equivalence. In the algebraic setting such equivalences were characterized by R. McKenzie in [McK96] by combinations of the following two constructions.

Let $\mathbf{A}$ be an algebra, and let T be the set of all terms in the language of $\mathbf{A}$. Given a unary term σ we define a new algebra called the *σ-image* of $\mathbf{A}$ by $\mathbf{A}(\sigma) = \langle \sigma(A), \{t_\sigma : t \in T\}\rangle$, where $t_\sigma^{\mathbf{A}(\sigma)}(x_1, \ldots, x_n) = \sigma^{\mathbf{A}}(t^{\mathbf{A}}(x_1, \ldots, x_n))$.

The second construction is the *matrix power* of $\mathbf{A}$. Let T_k be the set of k-ary terms. For a positive integer n we define

$$\mathbf{A}^{[n]} = \langle A^n, \{m_t : t \in (T_{kn})^n \text{ for some } k > 0\}\rangle,$$

where $m_t : (A^n)^k \to A^n$ is given by $m_t(\bar{x}_1, \ldots, \bar{x}_k)_i = t_i^{\mathbf{A}}(x_{11}, \ldots, x_{kn})$.

For a class $\mathcal{K}$ of algebras we let $\mathcal{K}(\sigma)$ and $\mathcal{K}^{[n]}$ be the classes of σ-images and n-th matrix powers respectively. A term σ is *idempotent in $\mathcal{K}$* if $\mathcal{K} \models \sigma(\sigma(x)) = \sigma(x)$, and it is *invertible in $\mathcal{K}$* if there exist unary terms $t_1, \ldots, t_n$ and an n-ary term t (for some $n > 0$) such that

$$\mathcal{K} \models x = t(\sigma(t_1(x)), \ldots, \sigma(t_n(x))).$$

THEOREM 9.96. [McK96] *Two varieties $\mathcal{V}$ and $\mathcal{W}$ are categorically equivalent if and only if there is an $n > 0$ and an invertible idempotent term σ for $\mathcal{V}^{[n]}$ such that $\mathcal{W}$ is term-equivalent to $\mathcal{V}^{[n]}(\sigma)$.*

In the setting of ℓ-groups and their negative cones, we can see an instance of this result. The term $\sigma(x) = x \wedge 1$ is certainly idempotent, and it is invertible (with $n = 2$) since $x = (x \wedge 1)(x^{-1} \wedge 1)^{-1}$ holds in all ℓ-groups. Of course $\mathbf{L}(\sigma)$ is not of the same type as $\mathbf{L}^-$, but with the help of Lemma 9.92 it is easy to see that they are term equivalent. In the other direction, every member of LG^- can be mapped to a τ-image of a matrix square that is term-equivalent to an ℓ-group. In general, the term τ is given by $\tau(\bar{x}) = \langle \sigma t_1 t(\bar{x}), \ldots, \sigma t_n t(\bar{x})\rangle$, which reduces to $\tau(\langle x, y\rangle) = \langle x/y, y/x\rangle$ for negative cones.

Exercises

(1) Complete the proof of Corollary 9.3.
(2) Show that GBA is the only atom below Br.
(3) Show that $\mathsf{V}(\mathbb{Z})$ is the only atom in $\mathbf{\Lambda}(\mathsf{LG})$ and $\mathsf{V}(\mathbb{Z}^-)$ is the only atom in $\mathbf{\Lambda}(\mathsf{LG}^-)$.
(4) Show that $\mathbf{2}$ and $\mathbf{2}_r$ (recall that $\mathbf{2}_r$ is the two-element residuated lattice, equivalently, the zero-free reduct of $\mathbf{2}$) generate the only two atoms in $\mathbf{\Lambda}(\mathsf{FL_{ci}})$.
(5) Prove Lemma 9.20.
(6) Show that there are at least $\aleph_0$ minimal subvarieties of $\mathsf{FL_{eo}}$. Hint: let $\mathbf{To}_n$ be an algebra defined from $\mathbf{T}_n$ (see Figure 9.1) by augmenting the type by the constant $0 = \bot$. Then, $\mathsf{V}(\mathbf{To}_n)$ are the varieties we want.

(7) Supply the details in the proofs of Corollary 9.23 and Theorem 9.24.

(8) For some $S \subseteq \mathbb{N}$ and $k \in \mathbb{N} \setminus 0$ consider the algebra $\mathbf{A}_k^S$ defined just above Lemma 9.26. Write explicitly a first-order sentence expressing the fact that the negative cone of $\mathbf{A}_k^S$ is precisely the set $\{\perp, 1\}$.

(9) Prove Lemma 9.27 in a detailed way.

(10) Find some nonempty $S \subseteq \mathbb{N} \setminus \{0\}$ and $k \in \mathbb{N}$ for which the algebra $\mathbf{V}_k^S$ is strictly simple. A toy open question: find a necessary and sufficient condition for $\mathbf{V}_k^S$ to be strictly simple.

(11) Prove Proposition 9.30. Hint: the algebra $\mathbf{V}_S$ is not simple for any S. Look closer at its homomorphic images.

(12) Prove that the algebras algebras $\mathbf{B}_T$ and $\mathbf{B}_S$ are non-isomorphic for distinct $S, T \subseteq 2\mathbb{N} + 1$.

(13) Prove that relaxing the condition $0 \leq x$ for simple MV-algebras $\mathbf{C}_n$ and "moving the zero up" produces infinitely many minimal subvarieties of $\mathsf{RFL_{ei}}$. Doing that with $\mathbf{C}_\infty$ does not produce a minimal variety. Why?

(14) Prove Lemma 9.53. Base your argument on the proof of Lemma 9.49. Remember that K is infinite.

(15) Complete the proof of Theorem 9.51.

(16) Consider the construction $\mathbf{K}[\mathcal{L}])$ from Lemma 9.63. Find an example of a residuated lattice $\mathbf{K}$ and a class $\mathcal{L}$ such that $\mathsf{P}(\mathbf{K}[\mathcal{L}]) \neq \mathbf{K}[\mathsf{P}(\mathcal{L})]$.

(17) *Open problem*: What is the cardinality of the "set" of atoms in CRRL? We conjecture that it is $\aleph_0$. See Lemma 9.7 for some justification of that conjecture.

(18) *Open problem*: How many atoms are there in $\mathbf{\Lambda}(\mathsf{RFL_{ei}})$? By Theorem 9.29 at least $\aleph_0$, but here we are reluctant to conjecture anything.

(19) *Open problem*: Same question for $\mathbf{\Lambda}(\mathsf{RFL_{eo}})$. Do Exercise 6 first.

Notes

(1) Lattice structures have been studied extensively for the lattice of all superintuitionistic logics and the lattice of all extensions of Łukasiewicz's infinitely many valued logic. As we mentioned in Chapter 2, Hosoi introduced the notion of slices in [Hos67] and classified the set of superintuitionistic logics except inconsistent one into countably many subsets, called n-th slices, for $1 \leq n \leq \omega$. The first slice consists only of classical logic, and the second slice forms a countable chain of type ω whose greatest element is the logic characterized by the three element Heyting algebra (see [HO70]). On the other hand, Kuznetsov [Kuz75] showed that the third slice has continuum many elements. The classification of superintuitionistic logics by slices is quite useful for understanding the lattice structure, since it is closely related to splittings discussed in Chapter 10. For splittings in lattices of modal logics, see Kracht [Kra99].

(2) Lattices of modal logics have also been extensively studied. Blok [Blo78] studied lattices of varieties modal algebras systematically, obtaining important results on so-called degree of incompleteness. As for minimal varieties of modal algebras, Makinson [Mak71] shows that there are two. Among varieties of tense algebras there are continuum many, as shown in [Kow98].
(3) Komori [Kom81] proved many important results on extensions of Łukasiewicz's infinitely many valued logic, including finite axiomatizability and decidability of them. Moreover, he gave a complete description of the lattice structure.
(4) It is well-known that if superintuitionistic logics $\mathbf{L}_1$ and $\mathbf{L}_2$ are axiomatized over intuitionistic logic by axioms α and β, respectively, then the intersection $\mathbf{L}_1 \cap \mathbf{L}_2$ is axiomatized by the axiom $\alpha \vee \beta$. It is pointed out in [Ono01] that the same result holds also for extensions of $\mathbf{FL_{ew}}$.

CHAPTER 10

Splittings

The idea of splitting up a lattice into a disjoint pair of an ideal and a filter was investigated by Whitman in [Whi43] and later taken up by McKenzie in [McK72] where he applied the notion to the lattice of varieties of lattices, generalizing the well-known division of that lattice into a modular and a nonmodular part. About a decade earlier, Jankov [Jan63a, Jan63b] used the same notion (although under a different name) to analyze the lattice of superintuitionistic logics. Jankov's ideas were developed further for modal logics and brought about a variety of results in that area. The usefulness of such *splitting methods*, as they came to be known, depends however on the existence of many splittings, and the more splittings there are, the better they work. In this chapter we will prove that $\mathbf{\Lambda}(\mathsf{FL_{ew}})$ fares pretty badly in that respect there being only one, somewhat trivial, splitting.

10.1. Splittings in general

A pair (a, b) of elements of a lattice $\mathbf{L}$ is said to *split* $\mathbf{L}$ if $a \not\leq b$ and for any $c \in L$, either $a \leq c$ or $c \leq b$. In other words, b is the largest element not above a. Pairs that split $\mathbf{L}$ are called *splitting pairs*. Splitting pairs enable us to present $\mathbf{L}$ as a disjoint union of the principal filter $\uparrow a$ and the principal ideal $\downarrow b$. Since for an $x \in \uparrow a$ and $y \in \downarrow b$ we have $x \vee y = x \vee a \vee y$ and $x \wedge y = x \wedge b \wedge y$, we can recover the lattice structure of $\mathbf{L}$ from $\uparrow a$ and $\downarrow b$ as long as we know what the joins $a \vee y$ and meets $b \wedge x$ are for all $x \in \downarrow a$ and $y \in \uparrow b$. Thus, splittings provide a divide and conquer strategy for analyzing the structure of $\mathbf{L}$. In this chapter, based on [KO00b], we investigate the splittings of the lattice of subvarieties of $\mathsf{FL_{ew}}$. Actually, we show that the variety of Boolean algebras is its only splitting.

The following proposition, whose proof we leave to the reader as an exercise (Exercise 1), characterizes splitting pairs in complete lattices.

LEMMA 10.1. *In a complete lattice the following are equivalent:*

(1) *(a, b) is a splitting pair,*
(2) *b is completely meet prime and $a = \bigwedge\{z \in L : z \not\leq b\}$,*
(3) *a is completely join prime and $b = \bigvee\{z \in L : z \not\geq a\}$.*

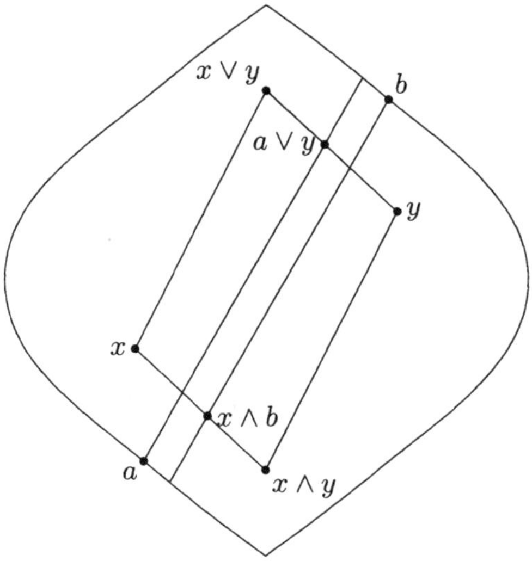

FIGURE 10.1. How splittings work.

10.2. Splittings in varieties of algebras

Let $\mathcal{V}$ be a variety of algebras and $\mathbf{\Lambda}(\mathcal{V})$ the lattice of subvarieties of $\mathcal{V}$. If $(\mathcal{V}_1, \mathcal{V}_2)$ is a splitting pair in $\mathbf{\Lambda}(\mathcal{V})$, then since

$$\mathcal{V}_2 = \operatorname{Mod}\operatorname{Th}_e(\mathcal{V}_2) = \bigcap\{\operatorname{Mod}(\varepsilon) : \varepsilon \in \operatorname{Th}_e(\mathcal{V}_2)\}$$

Since $\mathcal{V}_2$ is completely meet prime by Lemma 10.1, $\mathcal{V}_2 = \operatorname{Mod}(\varepsilon_0)$ for some $\varepsilon_0 \in \operatorname{Th}_e(\mathcal{V}_2)$ and so $\mathcal{V}_2$ is axiomatized by a single identity. The identity ε_0 axiomatizing $\mathcal{V}_2$ is known as the *splitting identity*. Because every variety is generated by its finitely generated subdirectly irreducible members (cf. [BS81]), we have

$$\mathcal{V}_1 = \bigvee\{\mathsf{V}(\mathbf{A}) : \mathbf{A} \in \mathcal{V}_{\mathsf{SI}}, \mathbf{A} \text{ finitely generated}\}$$

and since $\mathcal{V}_1$ is completely join prime again by Lemma 10.1, we obtain that $\mathcal{V}_1$ is generated by a single finitely generated subdirectly irreducible algebra $\mathbf{A}$. Such an $\mathbf{A}$ is called a *splitting algebra* in $\mathcal{V}$.

We saw one example of splitting in Chapter 9. Namely, it follows from Proposition 9.8 that $\mathbb{Z}^-$ is a splitting algebra in CanRL and that $\mathsf{V}(\mathbb{Z}^-)$ and LG form a splitting pair in $\mathbf{\Lambda}(\mathsf{CanRL})$.

As another example we mention the variety LG itself. There, we have exactly two splitting pairs, namely $(\mathsf{V}(\mathbb{Z}), \mathsf{Triv})$ and $(\mathsf{LG}, \mathsf{N})$, where Triv is the trivial variety and N is the variety of *normal valued* ℓ-groups. This follows from the fact that $\mathbf{\Lambda}(\mathsf{LG})$ has the unique atom $\mathsf{V}(\mathbb{Z})$ and the unique coatom N.

LEMMA 10.2. [McK72] *If $\mathcal{V}$ is congruence distributive and generated by its finite members, then every splitting algebra in $\mathcal{V}$ is finite and uniquely determined by the splitting pair.*

PROOF. Let $\mathbf{A}$ be a splitting algebra in $\mathcal{V}$. Thus, $\mathbf{A}$ is subdirectly irreducible and $\mathsf{V}(\mathbf{A})$ is completely join prime. Since $\mathcal{V}$ is generated by its finite members, $\mathcal{V}$ is the join of all of its subvarieties finitely generated by subdirectly irreducibles. (For joins of varieties, see Chapter 9). As $\mathsf{V}(\mathbf{A})$ is a completely join prime subvariety of $\mathcal{V}$, it must be included in one of these and thus be generated by a finite subdirectly irreducible algebra. Let $\mathbf{B}$ be a finite subdirectly irreducible algebra generating $\mathsf{V}(\mathbf{A})$. Then by Jónsson's Lemma (cf. Theorem 1.31 and Corollary 1.32) $\mathbf{A} \in \mathsf{HS}(\mathbf{B})$, so $\mathbf{A}$ is finite. Similarly, $\mathbf{B} \in \mathsf{HS}(\mathbf{A})$ and then a simple cardinality argument shows that $\mathbf{A}$ and $\mathbf{B}$ are isomorphic. □

Blok, Köhler and Pigozzi proved in [BKP84] that if $\mathcal{V}$ is of finite type and has EDPC, then every finite subdirectly irreducible member of $\mathcal{V}$ is splitting and thus splitting algebras in such varieties are all and only the finite subdirectly irreducible ones. The following proposition is a corollary. Recall that $\mathsf{P_nFL_{ew}}$ is the variety of n-potent FL_{ew}-algebras.

PROPOSITION 10.3. *For any $n \in \mathbb{N}$, all finite subdirectly irreducible algebras in $\mathsf{P_nFL_{ew}}$ split $\mathbf{\Lambda}(\mathsf{P_nFL_{ew}})$.*

Finally, let us state the following easy observation:

PROPOSITION 10.4. *The two-element Boolean algebra $\mathbf{2}$ splits $\mathbf{\Lambda}(\mathsf{FL_{ew}})$.*

PROOF. Since $\mathbf{2}$ is a subalgebra of any nontrivial FL_{ew}-algebra, the largest subvariety of $\mathsf{FL_{ew}}$ not containing $\mathsf{V}(\mathbf{2})$ is the trivial subvariety Triv. Thus, $(\mathsf{V}(\mathbf{2}), \mathsf{Triv})$ is the relevant splitting pair. □

The rest of the chapter is devoted to proving that this is the only splitting in $\mathbf{\Lambda}(\mathsf{FL_{ew}})$.

10.3. Algebras describing themselves

Our argument will employ a generalized version of the technique introduced by Jankov in [Jan63a]. In fact, Jankov used it to prove for Heyting algebras the opposite of what we are going to show for FL_{ew}-algebras, namely that every finite subdirectly irreducible Heyting algebra splits the variety HA of Heyting algebras. The technique was built upon an observation that the language of Heyting algebras is rich enough to describe finite subdirectly irreducible Heyting algebras, or, as we may also put it, finite subdirectly irreducible Heyting algebras can speak about themselves.

10.3.1. Jankov terms. Let $\mathbf{A}$ be a subdirectly irreducible Heyting algebras and let X be a set of $|A|$-many distinct variables indexed by the elements of A, so that x_a, x_b are distinct iff $a \neq b$. We will use $\star$ to denote the smallest element of the congruence class $1/\mu$ where μ it the monolith congruence on $\mathbf{A}$. By properties of Heyting algebras, $\star$ is the unique coatom of $\mathbf{A}$ (cf. Lemma 3.60). Notice that the term $x_a \leftrightarrow x_b$ will evaluate to 1 if and only if x_a and x_b evaluate to one and the same element. Thus, under the *intended* assignment v sending x_a to a for every $a \in \mathbf{A}$ we have $v(x_a \leftrightarrow x_b) = 1$ iff $a = b$. We then build a term $D_\mathbf{A}$ called the *diagram* of $\mathbf{A}$ by setting $D_\mathbf{A} = \bigwedge\{x_{f(a,b)} \leftrightarrow f(x_a, x_b) : a, b \in A, f \in \{\wedge, \vee, \cdot, \rightarrow, 0, 1\}\}$. In particular for $f \in \{0, 1\}$ we get conjuncts $x_0 \leftrightarrow 0$ and $x_1 \leftrightarrow 1$.

LEMMA 10.5. *Let $\mathbf{B}$ be a Heyting algebra and $\mathbf{A}$ a finite subdirectly irreducible Heyting algebra. Then, $\mathbf{A} \in \mathsf{IS}(\mathbf{B})$ iff there is an assignment v in $\mathbf{B}$, such that $\mathbf{B}, v \models D_\mathbf{A} = 1$ and $x_\star \neq 1$.*

PROOF. For the 'only if' part, observe that in $\mathbf{A}$ we have $v(D_\mathbf{A}) = 1$ and $v(x_\star) \neq 1$ under the intended assignment $v(x_a) = a$. Since $\mathbf{A}$ can be embedded in $\mathbf{B}$, the same properties hold for $\mathbf{B}$ (under the composition of v with the embedding).

For the 'if' part, suppose we have an assignment v with the required properties. Then, define a map $h : \mathbf{A} \to \mathbf{B}$, putting $h(a) = v(x_a)$. This map is a homomorphism into $\mathbf{B}$. Let us verify it for implication and the constant 1, leaving other cases as an exercise. We have $h(a \to b) = v(x_{a\to b})$. Since $x_{a\to b} \leftrightarrow (x_a \to x_b)$ is a conjunct of $D_\mathbf{A}$, we get $1 = v(x_{a\to b} \leftrightarrow (x_a \to x_b)) = v(x_{a\to b}) \leftrightarrow (v(x_a) \to v(x_b))$. Therefore, $v(x_{a\to b}) = v(x_a) \to v(x_b) = h(a) \to h(b)$ as required. For 1, we have $h(1) = v(x_1)$ and since $v(x_1 \leftrightarrow 1) = 1$ we get $h(1) = v(x_1) = 1$. Having established that h is a homomorphism, we then observe that the assumption $1 \neq v(x_\star)$ yields $h(1) = 1 \neq v(x_\star) = h(\star)$. Therefore, as $\mathbf{A}$ is subdirectly irreducible and every nontrivial congruence on $\mathbf{A}$ contains $(1, \star)$, the kernel of h must be the trivial congruence on $\mathbf{A}$. Hence, h is an embedding. □

Let now $J_\mathbf{A} = D_\mathbf{A} \to x_\star$. This has been known among logicians as *Jankov formula* or *characteristic formula* for $\mathbf{A}$. At variance with this usage, but in accord with our conventions, we will call it *Jankov term*, since we use it in an algebraic setting. In the next section we will generalize Jankov terms somewhat, to suit our purpose. Now we state an essential property of $J_\mathbf{A}$, sometimes referred to as Jankov's Lemma.

LEMMA 10.6. *Let $\mathbf{B}$ be a Heyting algebra and $\mathbf{A}$ a finite subdirectly irreducible Heyting algebra. Then $\mathbf{A} \notin \mathsf{SH}(\mathbf{B})$ iff $\mathbf{B} \models J_\mathbf{A} = 1$.*

PROOF. Under the intended assignment $v(x_a) = a$ the algebra $\mathbf{A}$ falsifies $J_\mathbf{A} = 1$, so if $\mathbf{A} \in \mathsf{SH}(\mathbf{B})$, then $\mathbf{B} \not\models J_\mathbf{A} = 1$.

Conversely, if $\mathbf{B} \not\models J_{\mathbf{A}} = 1$, then there is an assignment v such that $v(D_{\mathbf{A}}) \not\leq v(x_\star)$, in particular $v(x_\star) \neq 1$. Take the filter $\uparrow v(D_{\mathbf{A}})$; by Theorem 1.22 it is a deductive filter. Let θ be the congruence corresponding to $\uparrow v(D_{\mathbf{A}})$ and let $\mathbf{C} = \mathbf{B}/\theta$. Define $e : A \to C$ by putting $e(a) = v(x_a)/\theta$. As in the previous lemma we can show that e is a homomorphism. Now suppose e is not one-one. Then since $\mathbf{A}$ is subdirectly irreducible we have $e(\star) = v(x_\star)/\theta = 1$. It follows that $v(x_\star) \in \uparrow v(D_{\mathbf{A}})$ which amounts to $v(D_{\mathbf{A}}) \leq v(x_\star)$. This contradicts the assumption about the assignment v. □

THEOREM 10.7. *Every finite subdirectly irreducible Heyting algebra* $\mathbf{A}$ *splits* $\mathbf{\Lambda}(\mathsf{HA})$. *The largest variety of Heyting algebras not containing* $\mathsf{V}(\mathbf{A})$ *is axiomatized by* $J_{\mathbf{A}} = 1$.

PROOF. The first claim follows from the second, so we only need to prove that for every variety $\mathcal{W}$, we have $\mathcal{W} \models J_{\mathbf{A}} = 1$ iff $\mathbf{A} \notin \mathcal{W}$. The forward direction is clear since $\mathbf{A}$ does not satisfy $J_{\mathbf{A}} = 1$. Conversely, if there is an algebra $\mathbf{D} \in \mathcal{W}$ falsifying $J_{\mathbf{A}} = 1$, then by Lemma 10.6 we get $\mathbf{A} \in \mathsf{SH}(\mathbf{D})$, so $\mathbf{A} \in \mathcal{W}$. □

10.3.2. Example of Jankov term and diagram. To illustrate the inner workings of Jankov term machinery let us consider an example. Take the three-element Heyting algebra $\mathbf{H}_3$ with the universe $\{0, \star, 1\}$. It is linearly ordered, hence subdirectly irreducible. Its diagram $D_{\mathbf{H}_3}$ written in full is

$$
\begin{aligned}
&(x_0 \leftrightarrow 0) \land \\
&(x_0 \leftrightarrow (x_0 \land x_0)) \land (x_0 \leftrightarrow (x_0 \land x_\star)) \land (x_0 \leftrightarrow (x_0 \land x_1)) \land \\
&(x_\star \leftrightarrow (x_\star \land x_\star)) \land (x_\star \leftrightarrow (x_\star \land x_1)) \land (x_1 \leftrightarrow (x_1 \land x_1)) \land \\
&(x_0 \leftrightarrow (x_0 \lor x_0)) \land (x_\star \leftrightarrow (x_0 \lor x_\star)) \land (x_1 \leftrightarrow (x_0 \lor x_1)) \land \\
&(x_\star \leftrightarrow (x_\star \lor x_\star)) \land (x_1 \leftrightarrow (x_\star \lor x_1)) \land (x_1 \leftrightarrow (x_1 \lor x_1)) \land \\
&(x_1 \leftrightarrow (x_0 \to x_0)) \land (x_1 \leftrightarrow (x_0 \to x_\star)) \land (x_1 \leftrightarrow (x_0 \to x_1)) \land \\
&(x_0 \leftrightarrow (x_\star \to x_0)) \land (x_1 \leftrightarrow (x_\star \to x_\star)) \land (x_1 \leftrightarrow (x_\star \to x_1)) \land \\
&(x_0 \leftrightarrow (x_1 \to x_0)) \land (x_\star \leftrightarrow (x_1 \to x_\star)) \land (x_1 \leftrightarrow (x_1 \to x_1)) \land \\
&(x_1 \leftrightarrow 1)
\end{aligned}
$$

and it is immediately clear that it contains many redundancies. Getting rid of some most obvious ones produces

$$
\begin{aligned}
&(x_0 \leftrightarrow 0) \land \\
&(x_0 \leftrightarrow (x_0 \land x)) \land (x_0 \leftrightarrow (x_0 \land x_1)) \land \\
&(x \leftrightarrow (x \land x_1)) \land \\
&(x \leftrightarrow (x_0 \lor x)) \land (x_1 \leftrightarrow (x_0 \lor x_1)) \land \\
&(x_1 \leftrightarrow (x \lor x_1)) \land
\end{aligned}
$$

$$(x_1 \leftrightarrow (x_0 \to x)) \wedge (x_1 \leftrightarrow (x_0 \to x_1)) \wedge$$
$$(x_0 \leftrightarrow (x \to x_0)) \wedge (x_1 \leftrightarrow (x \to x_1)) \wedge$$
$$(x_0 \leftrightarrow (x_1 \to x_0)) \wedge (x \leftrightarrow (x_1 \to x)) \wedge$$
$$(x_1 \leftrightarrow 1)$$

where we also substituted plain x for $x_\star$. This term can be further reduced, with the help of identities of Heyting algebras such as $x \leftrightarrow (x \wedge y) = x \to y$, to something like

$$(x_0 \leftrightarrow 0) \wedge$$
$$(x_0 \to x) \wedge (x_0 \to x_1) \wedge (x \to x_1) \wedge$$
$$(x_1 \leftrightarrow (x_0 \to x)) \wedge (x_1 \leftrightarrow (x_0 \to x_1)) \wedge$$
$$(x_0 \leftrightarrow (x \to x_0)) \wedge (x_1 \leftrightarrow (x \to x_1)) \wedge$$
$$(x_0 \leftrightarrow (x_1 \to x_0)) \wedge (x \leftrightarrow (x_1 \to x)) \wedge$$
$$(x_1 \leftrightarrow 1)$$

or even somewhat shorter. More to the point is the following observation: the presence of $(x_0 \leftrightarrow 0)$ and $(x_1 \leftrightarrow 1)$ in the diagram make it tempting to say that x_0 is to be 0 and x_1 is to be 1 in some yet unspecified sense of "is to be". It turns out that this temptation is justified and we can simply replace x_0 by 0 and x_1 by 1 everywhere. Doing that, and making use of some more identities of HA, we finally reduce the diagram $D_{\mathbf{H}_3}$ to

$$(x \to 0) \to 0$$

and, thus, $J_{\mathbf{H}_3}$ turns out to be equivalent to

$$((x \to 0) \to 0) \to x.$$

To see that it indeed works this way, consider $J_{\mathbf{H}_3} = D_{\mathbf{H}_3} \to x$. Suppose a Heyting algebra $\mathbf{H}$ falsifies $J_{\mathbf{H}_3} = 1$. Then, for some assignment v we have $v(D_{\mathbf{H}_3}) \not\leq v(x)$ in $\mathbf{H}$. Taking the deductive filter $\uparrow(v(D_{\mathbf{H}_3}))$ and the congruence θ corresponding to it we get that in the (nontrivial) quotient algebra $\mathbf{H}' = \mathbf{H}/\theta$, under the quotient assignment $v' = v/\theta$, we have $\mathbf{H}', v \models D_{\mathbf{H}_3} = 1$ and $x \neq 1$.

Under that assignment we then must have $v'(x_0) = 0$ and $v'(x_1) = 1$ so $v'(D_{\mathbf{H}_3}) = 0 \leftrightarrow (v'(x) \to 0) = (v'(x) \to 0) \to 0 = 1$. In particular then, $v'((x \to 0) \to 0) \neq v'(x)$. $\mathbf{H}' \not\models (x \to 0) \to 0 = x$, which would be typically written using negation, as $\mathbf{H}' \not\models \neg\neg x = x$. Therefore, $\mathbf{H} \not\models \neg\neg x = x$.

Conversely, suppose $\mathbf{H}$ verifies $J_{\mathbf{H}_3} = 1$. Then, it verifies also its substitution instance with x_0 substituted by 0 and x_1 by 1. This substitution instance reduces to $((x \to 0) \to 0) \to x = 1$, which in turn is equivalent to $\neg\neg x \to x = 1$ and further to $\neg\neg x = x$. It follows that $J_{\mathbf{H}_3} = 1$ is equivalent to double negation, and the largest variety of Heyting algebras not containing $\mathbf{H}_3$ is the variety BA of Boolean algebras. Since $\mathsf{V}(\mathbf{H}_3)$ contains

BA, it also shows that $\mathsf{V}(\mathbf{H}_3)$ covers BA in the lattice of varieties of Heyting algebras.

10.3.3. Generalized Jankov terms. Let $\mathbf{A}$ be a finite subdirectly irreducible FL_{ew}-algebra. As before, we fix a set X of $|A|$-many distinct variables and index them by the elements of A, so that x_a, x_b are distinct iff $a \neq b$. The *diagram* of $\mathbf{A}$ is defined exactly as before, by $D_{\mathbf{A}} = \bigwedge\{x_{f(a,b)} \leftrightarrow f(x_a, x_b) : a, b \in A, f \in \{\wedge, \vee, \cdot, \rightarrow, 0, 1\}\}$. Then we define *Jankov term of order* n for $\mathbf{A}$ as $J^{(n)}_{\mathbf{A}} = (D_{\mathbf{A}})^n \rightarrow x_\star$, where $\star \in A$ is the smallest member of the class $1/\mu$, with μ being the monolith of $\mathbf{A}$. The proof of the next lemma goes exactly as that of Lemma 10.5 and is left to the reader as an exercise.

LEMMA 10.8. [KO00b] *Let $\mathbf{B} \in \mathsf{FL}_{\mathsf{ew}}$. Then, for a finite subdirectly irreducible $\mathbf{A}$, $\mathbf{A} \in \mathsf{IS}(\mathbf{B})$ iff there is an assignment v, such that $\mathbf{B}, v \models D_{\mathbf{A}} = 1$ and $x_\star \neq 1$.*

Unfortunately, the full analogy with Heyting algebras ends here. The best analogue of Lemma 10.6 we can get is the following.

LEMMA 10.9. [KO00b] *Let $\mathbf{B}$ be a FL_{ew}-algebra and $\mathbf{A}$ a finite subdirectly irreducible FL_{ew}-algebra. Then $\mathbf{A} \notin \mathsf{SHP}(\mathbf{B})$ if and only if there exists a $k \in \mathbb{N}$ such that $\mathbf{B} \models J^{(k)}_{\mathbf{A}} = 1$.*

PROOF. Since $\mathbf{A}$ falsifies $J^{(k)}_{\mathbf{A}} = 1$ for any $k \in \mathbb{N}$, if $\mathbf{A} \in \mathsf{SHP}(\mathbf{B})$ then also $\mathbf{B} \not\models J^{(k)}_{\mathbf{A}} = 1$ for any $k \in \mathbb{N}$.

For the converse suppose $\mathbf{B}$ falsifies $J^{(k)}_{\mathbf{A}} = 1$ for any $k \in \mathbb{N}$. Then for each $k \in \mathbb{N}$ there is an assignment v_k such that $\mathbf{B} \models v_k(D_{\mathbf{A}})^k \not\leq v_k(x_\star)$. It is easy to see that in the direct power $\mathbf{B}^{\mathbb{N}}$ we have that $\big(v_i(D_{\mathbf{A}})\colon i \in \mathbb{N}\big)^k \not\leq \big(v_i(x_\star)\colon i \in \mathbb{N}\big)$, for any $k \in \mathbb{N}$. Let θ be the congruence corresponding to the deductive filter generated by $\big(v_i(D_{\mathbf{A}})\colon i \in \mathbb{N}\big)$. In the quotient algebra $\mathbf{B}^{\mathbb{N}}/\theta$ we obtain then $v(D_{\mathbf{A}}) = 1$ and $v(x_\star) \neq 1$, with v being the quotient assignment $(v_i\colon i \in \mathbb{N})/\theta$. Then by Lemma 10.8 $\mathbf{A} \in \mathsf{IS}(\mathbf{B}^{\mathbb{N}}/\theta)$ and thus $\mathbf{A} \in \mathsf{ISHP}(\mathbf{B}) = \mathsf{SHP}(\mathbf{B})$ as claimed. □

Notice that by CEP for $\mathsf{FL}_{\mathsf{ew}}$ the lemma above shows also that $\mathbf{A} \notin \mathsf{V}(\mathbf{B})$ if and only if $\mathbf{B} \models J^{(k)}_{\mathbf{A}} = 1$ for some $k \in \mathbb{N}$. We are now ready to characterize non-splitting algebras in $\mathsf{FL}_{\mathsf{ew}}$.

LEMMA 10.10. [KO00b] *For a finite subdirectly irreducible $\mathbf{A}$ in $\mathsf{FL}_{\mathsf{ew}}$, the following are equivalent:*

(1) *$\mathbf{A}$ is not a splitting algebra in $\mathsf{FL}_{\mathsf{ew}}$,*
(2) *for all $i \in \mathbb{N}$ there exists an algebra $\mathbf{B} \in \mathsf{FL}_{\mathsf{ew}}$ such that $\mathbf{A} \notin \mathsf{V}(\mathbf{B})$ and $\mathbf{B} \not\models J^{(i)}_{\mathbf{A}} = 1$.*

(3) *for all* $i \in \mathbb{N}$ *there exists an algebra* $\mathbf{B} \in \mathsf{FL_{ew}}$ *and a natural number* $j > i$ *such that* $\mathbf{B} \models J^{(j)}_{\mathbf{A}} = 1$ *and* $\mathbf{B} \not\models J^{(i)}_{\mathbf{A}} = 1$.

PROOF. To prove the implication from (2) to (1) take a sequence of algebras $\mathbf{B}_i$, $(i \in \mathbb{N})$ such that $\mathbf{B}_i \not\models J^{(i)}_{\mathbf{A}} = 1$ and $\mathbf{A} \notin \mathsf{V}(\mathbf{B}_i)$. Let $k = |A|$. Choose a sequence $\overline{b(i)} = (b(i)_0, \ldots, b(i)_{k-2}, s(i))$ of elements of $\mathbf{B}_i$ such that $J^{(i)}_{\mathbf{A}}(b(i)_0, \ldots, b(i)_{k-2}, s(i)) < 1$ in $\mathbf{B}_i$. The choice ensures that $D^i_{\mathbf{A}}(b(i)_0, \ldots, b(i)_{k-2}) \not\leq s(i)$. Take $\mathbf{B} = \prod_{i\in\mathbb{N}} \mathbf{B}_i$, and consider the element $\bar{b} = \big((b(i)_0 \colon i \in \mathbb{N}), \ldots, (b(i)_{k-2} \colon i \in \mathbb{N}), (s(i) \colon i \in \mathbb{N})\big) \in B^k$. Let s stand for $(s(i) \colon i \in \mathbb{N})$. It is readily seen that for all $i \in \mathbb{N}$ we have $D^i_{\mathbf{A}}(\bar{b}) \not\leq s$ in $\mathbf{B}$. Thus, the deductive filter F generated by $D_{\mathbf{A}}(\bar{b})$ does not contain s. Hence, taking the congruence θ determined by F, we obtain that $D_{\mathbf{A}}(\bar{b}/\theta) = 1$ and $s/\theta \neq 1$ in the quotient $\mathbf{B}/\theta$. By Lemma 10.8, $\mathbf{A}$ embeds in $\mathbf{B}/\theta$; hence, $\mathbf{A} \in \mathsf{V}(\mathbf{B})$. Therefore, $\mathsf{V}(\mathbf{A})$ is contained in the join of the varieties $\mathsf{V}(\mathbf{B}_i)$, for $i \in I$, but it is not contained in any one of them. Consequently, $\mathsf{V}(\mathbf{A})$ is not join prime and $\mathbf{A}$ is not a splitting algebra.

To show that (1) implies (2) we will argue the contrapositive. Assume there is an $i \in \mathbb{N}$ such that for any $\mathbf{B} \in \mathsf{FL_{ew}}$ we have that $\mathbf{A} \notin \mathsf{V}(\mathbf{B})$ implies $\mathbf{B} \models J^{(i)}_{\mathbf{A}} = 1$. Let m be the smallest number with this property. We will show that the subvariety $\mathcal{W}$ of $\mathsf{FL_{ew}}$ defined by the identity $J^{(m)}_{\mathbf{A}} = 1$ is the largest such that it does not contain $\mathsf{V}(\mathbf{A})$. Obviously, $\mathsf{V}(\mathbf{A}) \not\subseteq \mathcal{W}$, as otherwise we would have $\mathbf{A} \models J^{(m)}_{\mathbf{A}} = 1$ which cannot be the case. Take any subvariety $\mathcal{V}$ of $\mathsf{FL_{ew}}$, with $\mathbf{A} \notin \mathcal{V}$. Let $\mathbf{F}$ be the free countably generated algebra in $\mathcal{V}$. Thus, $\mathcal{V} = \mathsf{V}(\mathbf{F})$. This, by our assumption, implies that $\mathbf{F} \models J^{(m)}_{\mathbf{A}} = 1$. Thus, $\mathcal{V} \models J^{(m)}_{\mathbf{A}} = 1$ and thus $\mathcal{W} \supseteq \mathcal{V}$ as desired.

Finally, (3) is equivalent to (2) by Lemma 10.9 and the observation that if $\mathbf{B}$ falsifies $J^{(i)}_{\mathbf{A}} = 1$, then it also falsifies $J^{(j)}_{\mathbf{A}} = 1$, for all $j \leq i$ (exercise). □

10.4. Construction that excludes splittings

To employ Lemma 10.10 in showing that a certain FL_{ew}-algebra $\mathbf{A}$ does not split $\mathbf{\Lambda}(\mathsf{FL_{ew}})$, we have to produce, for any given $i \in \mathbb{N}$, another FL_{ew}-algebra $\mathbf{B}$ with the properties required by (2) or (3). Our proof will in fact make use of (2), so strictly speaking (3) is unnecessary. We feel however that (3) demonstrates better than (2) that the task we face is rather subtle. An outline of the whole argument goes as follows. Taking $\mathbf{A}$ as a starting point, we will first construct, for each $i \in \mathbb{N}$, a certain large FL_{ew}-algebra $\mathbf{G}$, which falsifies $x^i = x^{i+1}$. The algebra $\mathbf{G}$, however, will contain $\mathbf{A}$ as a subalgebra, so we will distort it by means of another construction and arrive at an algebra we want. If one had a taste for paradox, they might say that this "expand and twist" technique provides a constructive *nonexistence* proof.

10.4.1. An introductory example. In this section we present a relatively simple concrete example, which we believe will help the reader to go smoothly over the intricacies of the constructions in subsequent sections. Suppose we want to show that the three-element MV-algebra $\mathbf{C}_3$ (see Figure 2.5) does not split $\mathbf{\Lambda}(\mathsf{FL_{ew}})$. We will produce an algebra with the desired properties for $i = 2$. We start with expanding C_3, by adding two new elements: d and e, and stipulating $0 < e < c < d < 1$. Further, we set $d^2 = c$, $d^3 = e$, $d^4 = 0$. Then, we set all the other multiplications so that the resulting structure becomes the five-element MV-algebra $\mathbf{C}_5$. Clearly, $\mathbf{C}_3 \subseteq \mathbf{C}_5$; moreover, $\mathbf{C}_5$ falsifies $x^2 = x^3$, verified by $\mathbf{C}_3$. Then, we proceed to distort $\mathbf{C}_5$ in a suitable way. To do this, we take the simple MV-algebra $\mathbf{C}_{p+1}$, for the smallest prime p greater or equal to the number of elements in our original algebra. As the original algebra was $\mathbf{C}_3$, we take $\mathbf{C}_4$. Our target algebra will be called $\mathbf{C}_5 \odot \mathbf{C}_4$. As its universe, we take a subset of the direct product $C_5 \times C_4$, namely $C_5 \odot C_4 = (C_5 \times C_4) \setminus (\{(1, x) : x \neq 1\} \cup \{(y, 1) : y \neq 1\})$. All the operations, except residuation, are defined as they would have been in the direct product, coordinatewise. Residuation is then implicitly defined by the finiteness of the universe and the distributivity of multiplication over join. The algebras $\mathbf{C}_3$, $\mathbf{C}_5$, and $\mathbf{C}_5 \odot \mathbf{C}_4$ are pictured in Figure 10.2, where the white dots in C_5 mark the new elements inserted between the elements of C_3. The grid on the right-hand side is the Hasse diagram of the lattice structure of $\mathbf{C}_5 \times \mathbf{C}_4$, the thick lines and black dots depict $\mathbf{C}_5 \odot \mathbf{C}_4$. To compute an implication $x \to y$ in $\mathbf{C}_5 \odot \mathbf{C}_4$ you compute it in the direct product and then if it comes out of the bold area you move down one step to the closest element of $C_5 \odot C_4$ (there always is such). Explicit definitions for the general case will come in Section 10.4.4. We follow the notational convention from there in viewing the elements of $C_5 \odot C_4$ as elements from the first coordinate indexed by the elements from the second. Thus, for instance, d_q is the unique coatom of $\mathbf{C}_5 \odot \mathbf{C}_4$, $(d_q)^2 = d^2_{q^2} = c_{q^2}$, etc.

As $\mathbf{C}_5 \odot \mathbf{C}_4$ has a unique coatom it is subdirectly irreducible. In fact, $\mathbf{C}_5 \odot \mathbf{C}_4$ is even simple, for $(d_q)^4 = d^4_{q^4} = 0$. Now, if $\mathbf{C}_3$ belonged to $\mathsf{V}(\mathbf{C}_5 \odot \mathbf{C}_4)$, we would have $\mathbf{C}_3 \in \mathsf{HS}(\mathbf{C}_5 \odot \mathbf{C}_4)$, by Jónsson's Lemma, and $\mathbf{C}_3 \in \mathsf{SH}(\mathbf{C}_5 \odot \mathbf{C}_4)$, by CEP for $\mathsf{FL_{ew}}$ (see Section 3.6.5). Since $\mathbf{C}_5 \odot \mathbf{C}_4$ is simple, $\mathbf{C}_3$ cannot be a subalgebra of any proper homomorphic image of $\mathbf{C}_5 \odot \mathbf{C}_4$. Thus, the only possibility is $\mathbf{C}_3 \subseteq \mathbf{C}_5 \odot \mathbf{C}_4$. However, since $\mathbf{C}_{p+1}$ is one-generated for any prime p, any purported embedding will produce at least 4 distinct elements in $\mathbf{C}_5 \odot \mathbf{C}_4$. Hence, $\mathbf{C}_3 \notin \mathsf{V}(\mathbf{C}_5 \odot \mathbf{C}_4)$.

It remains to show that $\mathbf{C}_5 \odot \mathbf{C}_4$ falsifies $J^{(2)}_{\mathbf{C}_3}$. Now, even though $\mathbf{C}_3$ has only three elements, its full diagram can be unwieldy, as we saw in Section 10.3.1. We will write down only those conjuncts which will not be equal to 1 under the intended assignment. These are $x_0 \leftrightarrow 0$ and $x_0 \leftrightarrow x_c \cdot x_c$ (we ask the reader to verify this as an exercise). The intended assignment v is: $v(x_1) = 1_1$, $v(x_c) = c_q$, $v(x_0) = 0_q$. Under this assignment we have:

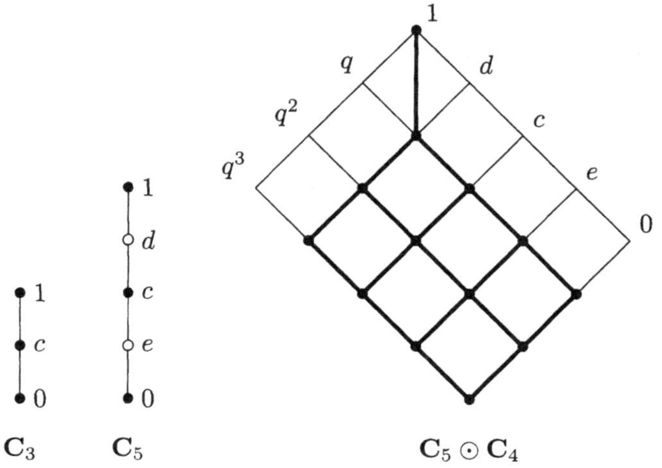

FIGURE 10.2. Simple example of "expand and twist".

$v(D_{\mathbf{C}_3}) = v((x_0 \leftrightarrow 0) \wedge (x_0 \leftrightarrow x_c \cdot x_c)) = (0_{q^2} \leftrightarrow 0_{q^3}) \wedge (0_q \leftrightarrow 0_{q^3}) = d_q \wedge d_{q^2} = d_{q^2}$. Thus, $v(D^2_{\mathbf{C}_3}) = (d_{q^2})^2 = d^2_{q^3} = c_{q^3}$. But $\star = 0$ in $\mathbf{C}_3$ and thus $v(x_\star) = v(x_0) = 0_q$. Therefore, $v(J^{(2)}_{\mathbf{C}_3}) = c_{q^3} \rightarrow 0_q = c_q \neq 1$.

In the example above we were very lucky at the first stage of the construction. It turned out that inserting an extra element between x and cx, for all $x \in \{1, c, 0\}$ with $cx < x$, produced a suitable FL_{ew}-algebra. Unfortunately, this cannot work in general. There are simple examples of FL_{ew}-algebras for which this will destroy the lattice structure, or residuation (or both). We will need something more refined. The next section deals with developing an appropriate construction.

10.4.2. Expansions. We begin the construction by fixing a finite subdirectly irreducible FL_{ew}-algebra $\mathbf{A}$. Such an algebra has a unique coatom c, by Lemma 3.59. Take the set $A_0 = \{a \in A : ca < a\}$, and let D be any set disjoint from A, with $|D| = |A_0|$. Thus, by means of any bijection, we can index the elements of D by the elements of A_0, getting $D = \{d_a : a \in A_0\}$. Let $P = A \cup D$. We will proceed to define an order relation and a multiplication on P. For $x, y \in P$, we put $x \leq y$ if either of the following holds:

$$(10.1)\qquad \begin{aligned} & x, y \in A \text{ and } x \leq^{\mathbf{A}} y, \\ & x = d_a \in D, y \in A \text{ and } a \leq^{\mathbf{A}} y, \\ & x \in A, y = d_a \in D \text{ and } x \leq^{\mathbf{A}} ca, \\ & x = d_a, y = d_b \in D \text{ and } a \leq^{\mathbf{A}} b. \end{aligned}$$

Notice that we have $ca < d_a < a$ whenever $ca < a$. Then, we define a binary operation $\cdot$ on P. To avoid overloaded notation, we will abbreviate $x \cdot^{\mathbf{A}} y$ everywhere by xy. Thereby, we commit ourselves to never abbreviating the new operation $x \cdot y$, within the present section. For any $x, y \in P$, let:

$$x \cdot y = y \cdot x = \begin{cases} xy & \text{if } x, y \in A, \\ d_{ay} & \text{if } x = d_a \in D,\ y \in A,\ cay < ay, \\ ay & \text{if } x = d_a \in D,\ y \in A,\ cay = ay, \\ cab & \text{if } x = d_a, y = d_b \in D. \end{cases} \tag{10.2}$$

LEMMA 10.11. [KO00b] *The structure* $\mathbf{P} = (P, \cdot, 1, 0, \leq)$ *is a partially ordered, bounded, commutative, integral monoid. Moreover, multiplication is order-preserving.*

PROOF. Reflexivity of the relation $\leq$ is obvious. For antisymmetry, observe that if $x \in A$, $y \in D$, and $x \leq y$, or if $x \in D$, $y \in A$, and $x \leq y$, then $y \not\leq x$; in other cases it follows from the antisymmetry of $\leq^{\mathbf{A}}$. Transitivity for triples $x, y, z \in A$ or $x, y, z \in D$ is obvious as well. There remain six cases to consider, of which we take one as an example. Suppose $x = d_a \leq y \in A$, $y \leq z = d_b$. Then, $a \leq^{\mathbf{A}} y$, $y \leq^{\mathbf{A}} cb \leq^{\mathbf{A}} b$. By transitivity of $\leq^{\mathbf{A}}$, $a \leq^{\mathbf{A}} b$. By definition of $\leq$, $d_a \leq d_b$ as needed. The top element 1 and the bottom element 0 of $\mathbf{A}$ remain the top and the bottom of $\mathbf{P}$ under the ordering $\leq$.

Commutativity of multiplication is clear from its definition. Its associativity is obvious for triples $x, y, z \in A$. With commutativity granted, the number of remaining cases reduces to three, of which we again take one as an example. Let $x = d_a, y \in A, z = d_b$. We have four cases:

(1) $cyb < yb$, $cay < ay$. Then, $d_a \cdot (y \cdot d_b) = d_a \cdot d_{yb} = cayb$. Starting from the other end, we get $(d_a \cdot y) \cdot d_b = d_{ay} \cdot d_b = cayb$, as needed.
(2) $cyb < yb$, $cay = ay$. Observe that the latter forces $cayb = ayb$. Then, $d_a \cdot (y \cdot d_b) = d_a \cdot d_{yb} = cayb$. From the other end, $(d_a \cdot y) \cdot d_b = ay \cdot d_b = ayb = cayb$, as needed.
(3) $cyb = yb$, $cay < ay$. The former forces $cayb = ayb$. Then, $d_a \cdot (y \cdot d_b) = d_a \cdot yb = ayb$; and $(d_a \cdot y) \cdot d_b = d_{ay} \cdot d_b = cayb = ayb$, as well.
(4) $cyb = yb$, $cay = ay$. Then $cayb = ayb$. Further, $d_a \cdot (y \cdot d_b) = d_a \cdot yb = ayb$; also $(d_a \cdot y) \cdot d_b = ay \cdot d_b = ayb$.

Since it readily seen that $1 \cdot x = x$ for all $x \in P$, the unit element of the monoid coincides with the top of the partial order, which is another way of saying that the monoid in question is integral. Thus, we have shown the first part of Lemma 10.11.

For the 'moreover' part, we again take only one of eight possible cases. Suppose $x, y \in A$ with $x \leq y$ and $z = d_a \in D$. Then, out of another four cases, we pick the trickiest, namely, when $x \cdot d_a = ax$ and $y \cdot d_a = d_{ay}$. We must thus have, $cax = ax$ and $cay < ay$. As $x \leq y$, and multiplication in $\mathbf{A}$

is order-preserving, we get $ax = cax \leq cay < ay$. By definition, $cay < d_{ay}$, and thus, $x \cdot d_a = ax \leq cay < d_{ay} = y \cdot d_a$, as required. □

So far, we have been dealing with two 'sorts' of elements: members of A, and members of D. Now we want to merge them together smoothly under a uniform multiplication. The intuitive idea is that we need only one member of D, namely d_1, all the others are obtained by multiplying d_1 by some member of A. Let us write d for d_1 (notice that d_1 always exists, i.e., is in D, for $1 \in A_0$ since $c1 = c < 1$) and state:

LEMMA 10.12. [KO00b] *For any $x, y \in A$, the following hold:*

1. *if $cx < x$, then $d \cdot x = d_x$, otherwise $d \cdot x = x$,*
2. *$d \cdot x \geq y$ iff $cx \geq y$,*
3. *$d \cdot x \cdot d \cdot y = cxy$,*
4. *$d \cdot x \leq d \cdot y$ iff $d \cdot x \leq y$ iff $x \leq y$.*

PROOF. The only point which is not an immediate consequence of the definitions is (4). To prove it notice first that if $x \leq y$, then $d \cdot x \leq y$ and $d \cdot x \leq d \cdot y$, since multiplication is order-preserving. Now, if $d \cdot x \leq d \cdot y$, then, as $d \cdot y \leq y$, we get, $d \cdot x \leq y$. It remains to show that this forces $x \leq y$. Suppose $d \cdot x \leq y$. If $d \cdot x = d_x$, then $x \leq y$ follows by definition; otherwise, i.e., when $d \cdot x = x$, it follows trivially. □

To state the next observation, it will be convenient to view $\mathbf{P}$ as a partial algebra $(P, \wedge, \vee, \rightarrow, \cdot, 0, 1)$, with the operations $\wedge$, $\vee$, and $\rightarrow$ only partially defined. Namely, $\wedge$, $\vee$ and $\rightarrow$ are respectively equal to meet, join and residuation whenever they exist and are undefined otherwise.

LEMMA 10.13. [KO00b] $\mathbf{A}$ *is a subalgebra of* $\mathbf{P}$.

PROOF. Obviously, the $\{\cdot, 0, 1\}$-reduct of $\mathbf{A}$ is a subalgebra of the $\{\cdot, 0, 1\}$-reduct of $\mathbf{P}$. It remains to verify the same for the three partial operations.

Take any $a, b \in A$. We claim that $a \wedge^{\mathbf{A}} b$, and $a \vee^{\mathbf{A}} b$ are, respectively, the infimum and supremum of $\{a, b\}$ in $\mathbf{P}$. If a z from A has $z \leq a$, $z \leq b$, then, clearly, $z \leq a \wedge^{\mathbf{A}} b$. The dual argument works for join. Take now an $s \in D$ with $s \leq a$, $s \leq b$. We have, $s = d \cdot e$, for some $e \in A$, thus, by Lemma 10.12(4), $e \leq a$, $e \leq b$. As $e \in A$, we get $e \leq a \wedge^{\mathbf{A}} b$. Thus, $s = d \cdot e \leq a \wedge^{\mathbf{A}} b$, as required. Then, suppose $s \in D$ and $s \geq a$, $s \geq b$. As previously, $s = d \cdot e$ with $e \in A$. By Lemma 10.12(2) then, $ce \geq a$ and $ce \geq b$, from which it follows that $ce \geq a \vee^{\mathbf{A}} b$ and, thus, by Lemma 10.12(2) again, $d \cdot e = s \geq a \vee^{\mathbf{A}} b$ as needed. Therefore, $a \wedge^{\mathbf{A}} b = a \wedge^{\mathbf{P}} b$, and $a \vee^{\mathbf{A}} b = a \vee^{\mathbf{P}} b$.

Next, we claim that $a \rightarrow^{\mathbf{A}} b$ is the residual of (a, b). We have to show that $s \cdot a \leq b$ iff $s \leq a \rightarrow^{\mathbf{A}} b$. This is obvious, when $s \in A$. Suppose $s \notin A$, i.e., $s \in D$, and thus, $s = d \cdot e$, for some $e \in A$. As an easy consequence of 10.2, we have $d \cdot e \cdot a = d \cdot ea$. Thus, we get: $s \cdot a = d \cdot ea \leq b$ iff $ea \leq b$ iff

$e \leq a \to^{\mathbf{A}} b$ iff $d \cdot e = s \leq a \to^{\mathbf{A}} b$, where the first and the last equivalence follow by Lemma 10.12(4). Therefore, $a \to^{\mathbf{A}} b = a \to^{\mathbf{P}} b$, as required. $\square$

Despite Lemma 10.13 above, we cannot expect $\mathbf{P}$ to be a total FL_{ew}-algebra. Indeed, simple examples show that $\mathbf{P}$ might be neither residuated nor a lattice (see Exercises 10 and 11). To remedy this unwelcome situation, we will resort to what will turn out to be our familiar nuclear completion of $\mathbf{P}$ (see Chapters 3 and 6). To use it, we have to produce a nucleus on $\mathcal{P}(P)$ tailored to our needs. That, however, turns out to be a little complicated, so we will proceed step by step. First, we define a closure operator. Namely, we call a subset X of P closed if it satisfies the following four conditions:

$$\begin{aligned} &0^{\mathbf{A}} \in X,\\ &\text{if } x \in X \text{ and } y \leq x, \text{ then } y \in X,\\ &\text{if } x, y \in A, x \in X \text{ and } y \in X, \text{ then } x \vee y \in X,\\ &\text{if } x, y \in A, d \cdot x \in X \text{ and } d \cdot y \in X, \text{ then } d \cdot (x \vee y) \in X. \end{aligned} \tag{10.3}$$

Since, as it is easy to verify, the intersection of any family of closed sets in the above sense is itself closed, we can define $\gamma : \mathcal{P}(P) \to \mathcal{P}(P)$ to be the map sending each $X \subseteq P$ to the smallest closed subset of P containing X. We will we denote it by $\gamma(X)$ and call the closure of X. Of course the reason for for the symbol γ is that it will turn out to be a nucleus. To justify this terminology, we will go through a series of lemmas. The first of these has an entirely straightforward proof (see Exercise 12).

LEMMA 10.14. [KO00b] *The map γ defined above is a closure operator on P.*

For a closed $X \subseteq P$, define $\hat{x}$ to be $\bigvee^{\mathbf{A}} X$, and $\dot{x}$ to be $\bigvee^{\mathbf{A}}\{z \in A\colon d \cdot z \in X\}$. The joins here are taken in P, but, by Lemma 10.13 and finiteness, joins of elements of A exist, and are again in A. Thus, the definitions are legitimate.

LEMMA 10.15. [KO00b] *If $X \subseteq P$ is closed, then $X = {\downarrow}\hat{x} \cup {\downarrow}(d \cdot \dot{x})$. Moreover, if $a \in A$, then both ${\downarrow}a$ and ${\downarrow}(d \cdot a)$ are closed.*

PROOF. For the first statement, observe that we have $\hat{x} \in A$ and $\dot{x} \in A$. By closedness of X, $\hat{x} \in X$. Now, $d \cdot \dot{x} = d \cdot \bigvee\{x \in A : d \cdot z \in X\} = d \cdot (x_0 \vee \cdots \vee x_{k-1})$, for some $k \in \mathbb{N}$. Further, $d \cdot x_0, \ldots, d \cdot x_{k-1} \in X$. From this, again, by closedness of X, it follows that $d \cdot \dot{x} \in X$. This proves $X \supseteq {\downarrow}\hat{x} \cup {\downarrow}(d \cdot \dot{x})$.

To show the other inclusion, notice firstly that if $z \in X \cap A$, then $z \leq \hat{x}$. Secondly, if $z \in X \setminus A$, then $z = d \cdot y$, for some $y \in A$, and $y \leq \dot{x}$. Thus, $z = d \cdot y \leq d \cdot \dot{x}$. Altogether, $z \in {\downarrow}\hat{x} \cup {\downarrow}(d \cdot \dot{x})$, as needed.

Then, to see that ${\downarrow}a$ is closed, we only need to check that if $d \cdot x$, $d \cdot y \leq a$, then $d \cdot (x \vee y) \leq a$, the three other conditions being trivially satisfied by

$\downarrow a \cap A$ being an ideal of **A**. Since $a \in A$, we have $x, y \leq a$, hence, $x \vee y \leq a$. Thus, $d \cdot (x \vee y) \leq a$.

Finally, to show that $\downarrow(d \cdot a)$ is closed, we notice that $\downarrow(d \cdot a)$ restricted to A is an ideal of **A**, so we only have to verify the fourth condition. If $d \cdot x$ and $d \cdot y$ are in $\downarrow(d \cdot a)$, for some $x, y \in A$, then $x \vee y \leq a$; hence, $d \cdot (x \vee y) \leq d \cdot a$, i.e., $d \cdot (x \vee y) \in \downarrow(d \cdot a)$ as required. □

For $X, Y \subseteq P$ we lift multiplication to the powerset, putting $X \circ Y = \{x \cdot y : x \in X, y \in Y\}$ and define residuation as usual (cf. Chapter 3), by $X \Rightarrow Y = \{z \in P : z \cdot x \in Y, \text{ for all } x \in X\}$.

LEMMA 10.16. [KO00b] *Let X, Y be closed subsets of P, and let $Q = \gamma(X \circ Y)$. The following hold:*

(1) $\hat{q} = c\dot{x}\dot{y} \vee \hat{x}\hat{y}$, $\dot{q} = \dot{x}\hat{y} \vee \hat{x}\dot{y}$,
(2) $\gamma(X \circ Y) = \downarrow(c\dot{x}\dot{y} \vee \hat{x}\hat{y}) \cup \downarrow(d \cdot (\dot{x}\hat{y} \vee \hat{x}\dot{y}))$,
(3) $X \Rightarrow Y$ *is closed.*

PROOF. For (1), we will sketch the proof of the second equality. Take $\dot{q}$. This, by definition, equals $\bigvee\{a \in A : d \cdot a \in Q\}$. Since $d \cdot \dot{x}\hat{y}$ and $d \cdot \dot{y}\hat{x} = d \cdot \hat{x}\dot{y}$ are both in $X \circ Y$, we get $d \cdot (\dot{x}\hat{y} \vee \dot{y}\hat{x}) \in C(X \circ Y) = Q$. Thus, $\dot{x}\hat{y} \vee \dot{y}\hat{x} \leq \dot{q}$.

To establish the converse inequality, take any $a \in A$ with $d \cdot a \in Q$, and such that $d \cdot a$ is a maximal element of Q. The second closure clause from 10.3 is taken care of by the maximality requirement, the first and the third cannot produce a genuine (i.e., not majorized in Q by an element of A) maximal $d \cdot a$, so only the fourth clause remains. Let $a = x \vee y$, with $d \cdot x, d \cdot y \in X \circ Y$. Here again there are cases to consider, but, as usual by now, we will take only one as an example. Let $d \cdot x = d \cdot u \cdot w$, with $d \cdot u \in X$, $w \in Y$; and $d \cdot y = d \cdot s \cdot t$, with $d \cdot s \in Y$, $t \in X$. Then, $d \cdot u \leq d \cdot \dot{x}$, $w \leq \hat{y}$, $d \cdot s \leq d \cdot \dot{y}$, $t \leq \hat{x}$. As multiplication is order-preserving, it follows that: $d \cdot x \leq d \cdot \dot{x} \cdot \hat{y}$, and $d \cdot y \leq d \cdot \dot{y} \cdot \hat{x}$. By Lemma 10.12, we obtain: $x \leq \dot{x}\hat{y}$, and $y \leq \dot{y}\hat{x}$. It follows that: $a = x \vee y \leq \dot{x}\hat{y} \vee \dot{y}\hat{x}$, precisely as required.

Clause (2) follows from (1) by Lemma 10.15.

For (3), let us firstly show that $X \Rightarrow Y$ is downward closed. Let $b \leq a$ and $a \in X \Rightarrow Y$, i.e., $a \cdot x \in Y$ for all $x \in X$. As multiplication is order-preserving and Y is downward closed, we get: $b \cdot x \in Y$ for all $x \in X$, i.e., $b \in X \Rightarrow Y$. Notice also, that $X \Rightarrow Y \supseteq Y$, and thus, $X \Rightarrow Y$ contains $0^{\mathbf{A}}$, for Y does.

For the other two closure clauses, we have—yet again—to proceed case by case. For an instance, let us take: $d \cdot a \in X \Rightarrow Y$ and $d \cdot b \in X \Rightarrow Y$. We have to show that for any $x \in X$, $(d \cdot (a \vee b)) \cdot x \in Y$. We have: $(d \cdot (a \vee b)) \cdot x = d \cdot ((a \vee b) \cdot x)$, by associativity, and here the reasoning splits into two further cases:

(1) if $x \in X \cap A$, then as $a, b \in A$ by assumption, we have distributivity of multiplication over join, since we remain inside A. Thus, $d \cdot ((a \vee b) \cdot x) =$

$d\cdot(ax\vee bx)$. Now, since $d\cdot a$ and $d\cdot b$ are in $X \Rightarrow Y$, we get: $d\cdot a\cdot x = d\cdot ax \in Y$ and $d\cdot b\cdot x = d\cdot bx \in Y$. By closedness of Y, $d\cdot(ax \vee bx) \in Y$.

(2) if $x \in X\backslash A$, then $x = d_y = d\cdot y$, for some $y \in A$. Thus, $(d\cdot(a\vee b))\cdot x = (d\cdot(a\vee b))\cdot(d\cdot y)$. By associativity, this equals $(d\cdot d)\cdot((a\vee b)\cdot y)$, and we may again employ the distributivity of multiplication over join within A; furthermore, $d\cdot d = c$ by definition. Altogether, we obtain: $(d\cdot(a\vee b))\cdot x = cay\vee cby$. Now, since $d\cdot a \in X \Rightarrow Y$, we have: $d\cdot a\cdot d\cdot y = cay \in Y$. Similarly, for $d\cdot b\cdot d\cdot y = cby$. By closedness of Y, we finally get: $(d\cdot(a\vee b))\cdot x = cay \vee cby \in Y$. □

Let $\mathbf{G}$ be the algebra $(\gamma(\mathcal{P}(P)), \vee, \wedge, \cdot, \rightarrow, 1, 0)$, with operations defined by the usual clauses (cf. Chapter 3), which we recall below.

$$\begin{aligned} X \wedge Y &= X \cap Y, \\ X \vee Y &= \gamma(X \cup Y), \\ X \cdot Y &= \gamma(X \circ Y), \\ X \rightarrow Y &= X \Rightarrow Y, \\ 1^{\mathbf{G}} &= P, \\ 0^{\mathbf{G}} &= \{0^{\mathbf{A}}\}. \end{aligned} \tag{10.4}$$

The algebra $\mathbf{G}$ will be called an *expansion* of $\mathbf{A}$.

LEMMA 10.17. [KO00b] *The expansion $\mathbf{G}$ of $\mathbf{A}$ is a FL_{ew}-algebra, and $\mathbf{A}$ is a subalgebra of $\mathbf{G}$.*

PROOF. That $\mathbf{G}$, with the lattice operations defined as above is indeed a (complete, hence bounded) lattice, follows from the fact that γ is a closure operator. Of the other things needed to conclude that $\mathbf{G}$ is a FL_{ew}-algebra, we only show two: that $\cdot$ is associative, and that $\rightarrow$ is indeed its residual.

For the former, we let Q stand for $Y\cdot Z$, and, employing Lemma 10.16, calculate: $X\cdot(Y\cdot Z) = X\cdot Q = {\downarrow}(c\dot{x}\dot{q}\vee\hat{x}\hat{q}) \cup {\downarrow}(d\cdot(\dot{x}\hat{q}\vee\hat{x}\dot{q})) = {\downarrow}\big(c\dot{x}(\dot{y}\hat{z}\vee\hat{y}\dot{z})\vee\hat{x}(c\dot{y}\dot{z}\vee\hat{y}\hat{z})\big)\cup{\downarrow}\big(d\cdot(\dot{x}(c\dot{y}\dot{z}\vee\hat{y}\hat{z})\vee\hat{x}(\dot{y}\hat{z}\vee\hat{y}\dot{z}))\big) = {\downarrow}\big(c\dot{x}\dot{y}\hat{z}\vee c\dot{x}\hat{y}\dot{z}\vee c\hat{x}\dot{y}\dot{z}\vee\hat{x}\hat{y}\hat{z}\big)\cup{\downarrow}\big(d\cdot(c\dot{x}\dot{y}\dot{z}\vee\dot{x}\hat{y}\hat{z}\vee\hat{x}\dot{y}\hat{z}\vee\hat{x}\hat{y}\dot{z})\big)$. This, by the obvious symmetries of the situation (swap z and x and go backwards), equals $(X\cdot Y)\ Z$.

For the latter, we need to show that $X\cdot Y \subseteq Z$ iff $X \subseteq Y \Rightarrow Z$. For the left-to-right direction, suppose $x \in X$. Then, take any $y \in Y$. We have $x\cdot y \in X\circ Y \subseteq X\cdot Y \subseteq Z$. Thus, $x\cdot y \in Z$, for any $y \in Y$, which definitionally means $x \in Y \Rightarrow Z$ as needed. For the other direction, observe, that, by the assumption, we have $X \subseteq \{u \in P\colon u\cdot y \in Z \text{ for all } y \in Y\}$, which amounts to: $X\circ Y \subseteq Z$. Since Z is closed, we further get: $X\cdot Y = \gamma(X\circ Y) \subseteq \gamma(Z) = Z$, as needed.

To show that $\mathbf{A} \subseteq \mathbf{G}$, we put $e : A \rightarrow G$ to be the map defined by $e(a) = {\downarrow}a$. This is clearly one-one. To prove that it is a homomorphism, we need to show that the following equalities hold:

(1) ${\downarrow}a \vee {\downarrow}b = {\downarrow}(a \vee^{\mathbf{A}} b)$,

(2) $\downarrow a \wedge \downarrow b = \downarrow(a \wedge^{\mathbf{A}} b)$,
(3) $\downarrow a \to \downarrow b = \downarrow(a \to^{\mathbf{A}} b)$,
(4) $\downarrow a \cdot \downarrow b = \downarrow(a \cdot^{\mathbf{A}} b)$,
(5) $\downarrow 1 = P$, and $\downarrow 0^{\mathbf{A}} = \{0^{\mathbf{A}}\}$.

Of these, let us only verify (3). We have: $\downarrow a \to \downarrow b = \{z \in P \colon z \cdot x \leq b$, for all $x \leq a\}$. Since $(a{\to}^{\mathbf{A}} b)\cdot^{\mathbf{A}} a$, equal by our convention to $(a{\to}^{\mathbf{A}} b)a$, is smaller than b we get: $(a{\to}^{\mathbf{A}} b)\cdot x \leq b$ for all $x \leq a$. Hence, $a{\to}^{\mathbf{A}} b \in \downarrow a{\to}\downarrow b$, i.e., $\downarrow a \to \downarrow b \supseteq \downarrow(a \to^{\mathbf{A}} b)$. For the reverse inclusion, take any $z \in \downarrow a \to \downarrow b$. Then, $z \cdot x \leq b$ for all $x \leq a$. In particular, $z \cdot a \leq b$. If $z \in A$, then $z \leq a \to^{\mathbf{A}} b$, since $\mathbf{A}$ is a subalgebra of (the partial algebra) $\mathbf{P}$. Suppose $z = d_y = d\cdot y$, for some $y \in A$. Then, $b \geq z\cdot a = d\cdot y\cdot a$ and, by Lemma 10.12, $b \geq y \cdot a = ya$. Thus, $y \leq a \to^{\mathbf{A}} b$, hence, $z = d_y \leq a \to^{\mathbf{A}} b$, as well. □

Since $\mathbf{G}$ is finite (hence, complete) it follows (see Exercise 12 in Chapter 3) that γ is a nucleus, as we have announced. Thus, an alternative proof of Lemma 10.17 can go as follows: (1) prove directly that γ is a nucleus on P, (2) prove that A (as a subset of P) is a basis for γ and conclude by Theorem 6.29 that $\mathbf{A}$ is a subalgebra of $\mathbf{G}$. We ask the reader in Exercise 15 to develop a proof along these lines.

10.4.3. Iterated expansions. Let $\mathbf{A}$ be a finite subdirectly irreducible FL_{ew}-algebra with monolith μ. To avoid the mouthful of *deductive filter corresponding to the monolith*, we call $\mathrm{F_c}(\mu)$ the *monolithic filter* of $\mathbf{A}$. We will say that the filter $\mathrm{F_c}(\mu)$ is *of depth n* iff n is the smallest natural number for which $c^n = c^{n+1} = \star$, where c is the unique coatom of $\mathbf{A}$.

LEMMA 10.18. [KO00b] *Let $\mathbf{A}$ be a finite subdirectly irreducible FL_{ew}-algebra with monolithic filter $F(\mu)$ of depth n, and $\mathbf{G}$ be its expansion. Then, the following hold:*

(1) $\mathbf{A} \subseteq \mathbf{G}$,
(2) $\mathbf{G}$ *is subdirectly irreducible,*
(3) $F(\nu)$, *the monolithic filter of* $\mathbf{G}$, *is of depth* $2n$,
(4) $F(\mu) = F(\nu)|_{\mathbf{A}}$,
(5) $\mathbf{G}/\nu$ *is isomorphic to* $\mathbf{A}/\mu$.

PROOF. Point (1) has already been proved. To prove (2), it suffices to prove that $\downarrow d$ is the unique coatom of $\mathbf{G}$. It follows from the construction, that d is the unique coatom of $\mathbf{P}$. Thus $\downarrow d$ is the largest proper subset of $\mathbf{P}$. As such, it contains any closed proper subset of P, which proves the claim.

To prove the remaining two claims, observe first, that in the intermediary structure $\mathbf{P}$, we have: $d \cdot d = c$, $d \cdot d \cdot c = c \cdot c = c^2$, and, inductively, $d \cdot d \cdot c^k = c^{k+1} = d^{2(k+1)}$. Let n be the depth of $F(\mu)$. Then $c^n = cc^{n-1} = d \cdot d \cdot c^{n-1} < d \cdot c^{n-1} < c^{n-1}$ and $d \cdot d \cdot c^{n-1} = d^{2n}$. Further, we will prove that for any $0 < k \leq n$ the equality: $(\downarrow d)^{2k} = \downarrow(d^{2k}) = \downarrow(c^k)$, holds. For $k = 1$, we proceed as follows: Take $\downarrow d$ and name it X, for convenience. By Lemma 10.15, X is closed. We will calculate X^2. Observe, that we have: $\dot{x} = 1$, and $\hat{x} = c$. By Lemma 10.16, $X^2 = \gamma(X \circ X)$ equals $\downarrow(c11 \vee c^2) \cup \downarrow(d \cdot c) = \downarrow c \cup \downarrow(d \cdot c)$. As $d \cdot c \leq c$ by definition, we finally get $X^2 = (\downarrow d)^2 = \downarrow(d^2) = \downarrow c$. For $k+1$, we get: $(\downarrow d)^{2k+2} = (\downarrow d^{2k}) \cdot (\downarrow d)^2$, which, by inductive hypothesis, equals $(\downarrow c^k) \cdot (\downarrow c)$. Now, as previously, let Y, Z, stand, respectively, for $\downarrow(c^k)$ and $\downarrow c$. These are also closed, by Lemma 10.15. By definition, we get $\dot{y} = \bigvee\{y \in A : d \cdot y \leq c^k\}$, and, since by Lemma 10.12(4) $d \cdot y \leq c^k$ iff $y \leq c^k$, we obtain: $\dot{y} = c^k$. In similar fashion, we get: $\hat{y} = c^k$, $\dot{z} = c$, and $\hat{z} = c$. Then, $(\downarrow c^k) \cdot (\downarrow c) = \gamma(Y \circ Z)$, and this, by Lemma 10.16 and the line above, equals $\downarrow(cc^k c \vee c^k c) \cup \downarrow(d \cdot (c^k c \vee c^k c)) = \downarrow(c^{k+2} \vee c^{k+1}) \cup \downarrow(d \cdot c^{k+1}) = \downarrow(c^{k+1})$, as we claimed. In particular, for $k = n$, we get: $(\downarrow d)^{2n} = \downarrow(d^{2n}) = \downarrow(c^n) = \downarrow\star$, which proves (3).

Point (4) follows from (3) and (1). It remains to show (5). Notice first that $\nu = \mathrm{Cg}^{\mathbf{G}}(\downarrow d, P)$, and take any closed $X \subseteq P$. By Lemma 10.15, $X = \downarrow\hat{x} \cup \downarrow(d \cdot \dot{x})$. Now, we claim that $(X, \downarrow\hat{x}) \in \nu$. This will prove (5), for, since $\hat{x} \in A$ (i.e., $\downarrow\hat{x} \in e(A)$), it will follow that every class of ν contains an element of A. To prove our claim, it suffices to show that $X \Rightarrow \downarrow\hat{x} \supseteq \downarrow\star$, which in turn, by downward closedness, amounts to $\star \in X \Rightarrow \downarrow\hat{x}$. By definition, $X \Rightarrow \downarrow\hat{x} = \{z \in P\colon z \cdot x \in \downarrow\hat{x} \text{ for all } x \in X\}$. Since multiplication is order-preserving, it is then enough to show that $\star \cdot \hat{x} \leq \hat{x}$ and $\star \cdot (d \cdot \dot{x}) \leq \hat{x}$. The first is trivially true. For the second, we have: $\star \cdot (d \cdot \dot{x}) = (d \cdot \star) \cdot \dot{x} = \star \cdot \dot{x} \leq c \cdot \dot{x}$. Writing $\dot{x}$ in full, we obtain: $c \cdot \dot{x} = c \cdot \bigvee\{z \in A \mid d \cdot z \in X\} = \bigvee\{c \cdot z\colon z \in A \text{ and } d \cdot z \in X\}$. Then, since $c \cdot z \leq d \cdot z$ and $c \cdot z \in A$, we get that $c \cdot z \leq \hat{x}$, for all $z \in A$ such that $d \cdot z \in X$. Hence, $\bigvee\{c \cdot z\colon z \in A \text{ and } d \cdot z \in X\} \leq \hat{x}$, which proves the claim. □

LEMMA 10.19. [KO00b] *Let* $\mathbf{A}$ *be a finite subdirectly irreducible* FL_{ew}*-algebra with the monolithic filter* $F(\mu)$ *of depth* n*, and* k *be any natural number. Then, there is a finite subdirectly irreducible* FL_{ew}*-algebra* $\mathbf{E}$ *with monolith* ν*, such that:*

(1) $\mathbf{A} \subseteq \mathbf{E}$,
(2) $F(\nu)$ *is of depth greater than* k,
(3) $F(\mu) = F(\nu)|_{\mathbf{A}}$,
(4) $\mathbf{E}/\nu$ *is isomorphic to* $\mathbf{A}/\mu$.

PROOF. Let $m \in \mathbb{N}$ be the smallest such that $2^m n \geq k$. Now, observe that the construction of $\mathbf{G}$ may be iterated. We may 'expand' $\mathbf{G}$ in the same way, as we did it with $\mathbf{A}$, and get an algebra $\mathbf{D}$, such that $\mathbf{A}$ is a subalgebra of $\mathbf{D}$, the restriction to $\mathbf{A}$ of the monolithic filter F of $\mathbf{D}$ is precisely $F(\mu)$, the depth of F equals $4n$, and $\mathbf{D}/\theta(F) \cong \mathbf{G}/\nu \cong \mathbf{A}/\mu$ (ν is the monolith of $\mathbf{G}$, as in Lemma 10.18). The m-th iteration of this construction yields the algebra $\mathbf{E}$ satisfying all the claims of the lemma. □

10.4.4. Twisted products. Having finished the "expand" stage of the construction, we are now ready for the "twist" stage. Since in our somewhat limited case of $\mathsf{FL_{ew}}$ it always works the way we saw in Section 10.4.1, the exposition here will be rather brief.

Let **A**, **B** be finite subdirectly irreducible FL_{ew}-algebras with coatoms c and q, respectively. The *twisted product* $\mathbf{A} \odot \mathbf{B}$ of **A** and **B** is the algebra with the universe $A \odot B = \big((A \setminus \{1\}) \times (B \setminus \{1\})\big) \cup \{(1,1)\}$ and operations defined a few lines below. To make the definition more readable, we will write a_i, instead of (a,i), and the hyper-correct 1_1, 0_0 we will further abbreviate to 1, 0, whenever it will not cause confusion. In other words, we view the elements of $A \odot B$ as elements of A indexed by elements of B. Now, the operations on $A \odot B$ are defined by:

$$\begin{aligned}
1 &= 1_1,\\
0 &= 0_0,\\
a_i \wedge b_j &= (a \wedge b)_{i \wedge j},\\
a_i \vee b_j &= (a \vee b)_{i \vee j},\\
a_i \cdot b_j &= (a \cdot b)_{i \cdot j},\\
a_i \to b_j &= \begin{cases} (a \to b)_{i \to j} & \text{if } a \not\leq b,\ i \not\leq j,\\ (a \to b)_q & \text{if } a \not\leq b,\ i \leq j,\\ 1_1 & \text{if } a \leq b,\ i \leq j,\\ c_{i \to j} & \text{if } a \leq b,\ i \not\leq j. \end{cases}
\end{aligned}$$

LEMMA 10.20. [KO00b] $\mathbf{A} \odot \mathbf{B}$ *is a subdirectly irreducible* FL_{ew}*-algebra.*

PROOF. First, notice that $A \odot B$, viewed as a subset of $A \times B$ is closed under meet, join, and multiplication, and in fact, these have been defined coordinatewise in the clauses above. Thus, $\mathbf{A} \odot \mathbf{B}$ is a lattice ordered commutative monoid, with multiplication distributive over join. Moreover, the first and the third clauses in the definition of $\to$ are precisely what they would have been in the full product, thus, they indeed define residuation. All that remains is to verify that the second and fourth clauses do so as well.

Let $a \not\leq b$, and $i \leq j$. Then, $(a_i \to b_j) \cdot a_i = (a \to b)_q \cdot a_i = ((a \to b) \cdot a)_{q \cdot i}$. Since $((a \to b) \cdot a) \leq b$ and $q \cdot i \leq i \leq j$, we obtain $(a_i \to b_j) \cdot a_i \leq b_j$, as needed. Now, take an element s_k with $s_k \cdot a_i \leq b_j$. Thus, $(s \cdot a)_{k \cdot i} \leq b_j$, and, in particular, $s \cdot a \leq b$. Hence, $s \leq a \to b$, and since $k \leq q$ is always the case, we obtain $s_k \leq (a \to b)_q$, as needed.

Let then $a \leq b$ and $i \not\leq j$. We get: $(a_i \to b_j) \cdot a_i = c_{i \to j} \cdot a_i = (c \cdot a)_{(i \to j) \cdot i}$. As $(i \to j) \cdot i \leq j$ and $c \cdot a \leq a \leq b$, we have $(a_i \to b_j) \cdot a_i \leq b_j$, as desired. Take a s_k with $s_k \cdot a_i \leq b_j$, i.e., with $(s \cdot a)_{k \cdot i} \leq b_j$. Then, in particular, $k \cdot i \leq j$, and thus, $k \leq i \to j$. As $s \leq c$ is always the case, we have $s_k \leq c_{i \to j} = a_i \to b_j$, as desired.

Being subdirectly irreducible is obvious, the monolith is the congruence generated by $(c_q, 1)$. □

Now, let $\mathbf{A}$, μ, $\mathbf{E}$, $n \leq k \in \mathbb{N}$ be as in Lemma 10.19, and let $m \geq k$ be the depth of ν. Consider $\mathbf{E} \odot \mathbf{C}_{p+1}$, where $\mathbf{C}_{p+1}$ is the simple MV-algebra with $p+1$ elements, for the first prime number p greater or equal to $|E|$. Let also c, d, and q stand for the unique coatoms of $\mathbf{A}$, $\mathbf{E}$, and $\mathbf{C}_{p+1}$, respectively.

LEMMA 10.21. [KO00b] *For any nontrivial congruence θ on $\mathbf{E} \odot \mathbf{C}_{p+1}$, $(\mathbf{E} \odot \mathbf{C}_{p+1})/\theta$ is isomorphic to a proper homomorphic image of $\mathbf{A}$.*

PROOF. As $p \geq m$, we get $(d_q)^p = d^p_{q^p} = \star_0$. Thus, every nontrivial θ has $1\theta\star_0$. Since $a_q \to a_0 = d_{q\to 0} \geq \star_{q\to 0} \geq \star_0$, we also have $a_i\theta a_j$, for any $i, j \in C_{p+1} \setminus \{1\}$. Consider the relation $\sim$ on $(E \odot C_{p+1}) \times (E \odot C_{p+1})$, defined as $\{(x, y) : x = a_i, y = a_j, \text{for some } a \in E \text{ and some } i, j \in C_{p+1}\}$. So defined, $\sim$ is a congruence on the lattice reduct of $\mathbf{E} \odot \mathbf{C}_{p+1}$. Clearly, $\theta \geq \sim$. From the construction, it is readily seen that $((E \odot C_{p-1})/\sim, \wedge, \vee)$ and $(E, \wedge, \vee)$ are isomorphic as lattices (cf. Figure 10.3). Now the conclusion will follow by Lemma 10.19(4), if we show that θ collapses the ('projection' of the) monolith ν of $\mathbf{E}$. This happens, if $\theta \ni (\star_0, 1)$, and precisely this has been established at the beginning of the proof. □

Figure 10.3 shows what $\mathbf{E} \odot \mathbf{C}_{p+1}$ looks like. Notice that the number of iterations required for an element d_i, with any index i, to descend below $\star_q$ depends only on the 'distance' between the line of d's and the line of $\star$'s; in other words, on the depth of ν.

LEMMA 10.22. [KO00b] *If $\mathbf{A}$ is not isomorphic to $\mathbf{2}$, then $\mathbf{A} \notin \mathsf{SH}(\mathbf{E} \odot \mathbf{C}_{p+1})$; therefore, $\mathbf{A} \notin \mathsf{V}(\mathbf{E} \odot \mathbf{C}_{p+1})$.*

PROOF. By the previous fact it follows that $|E \odot C_{p+1}/\theta| < |A|$, and thus $\mathbf{A}$ cannot be a subalgebra of any proper homomorphic image of $\mathbf{E} \odot \mathbf{C}_{p+1}$.

Suppose that there is a map $e : A \to (E \odot C_{p+1})$ which embeds $\mathbf{A}$ into $\mathbf{E} \odot \mathbf{C}_{p+1}$. Assume that for some $a \neq b$, $e(a) = v_i$ and $e(b) = u_i$. Without loss of generality, let $a \to b \neq 1$; then $v_i \to u_i \neq 1$ and $v \not\leq u$, so $e(a \to b) = v_i \to u_i = (v \to u)_q$. Let w stand for $v \to u$. Consider the chain $w_q, (w_q)^2, \ldots, (w_q)^p$. By definition, its elements are respectively equal to $w_q, w^2_{q^2}, \ldots, w^p_{q^p}$. As $q, q^2, \ldots, q^p$ are all distinct by the construction, the chain above has p distinct elements. This contradicts the assumptions about $\mathbf{A}$. Thus, each (image of an) element of A must belong to a different 'layer' in $E \odot C_{p+1}$. Since $\mathbf{A}$ is not isomorphic to $\mathbf{2}$, A has at least three elements. Thus, for some $a \in A$ we must have $e(a) = a_i$ with $i \neq 0, 1$. However, since in $\mathbf{C}_{p+1}$ each non-zero and non-unit element generates the whole algebra, we obtain that a_i generates p distinct elements in $E \odot W$. This leads to the same contradiction again.

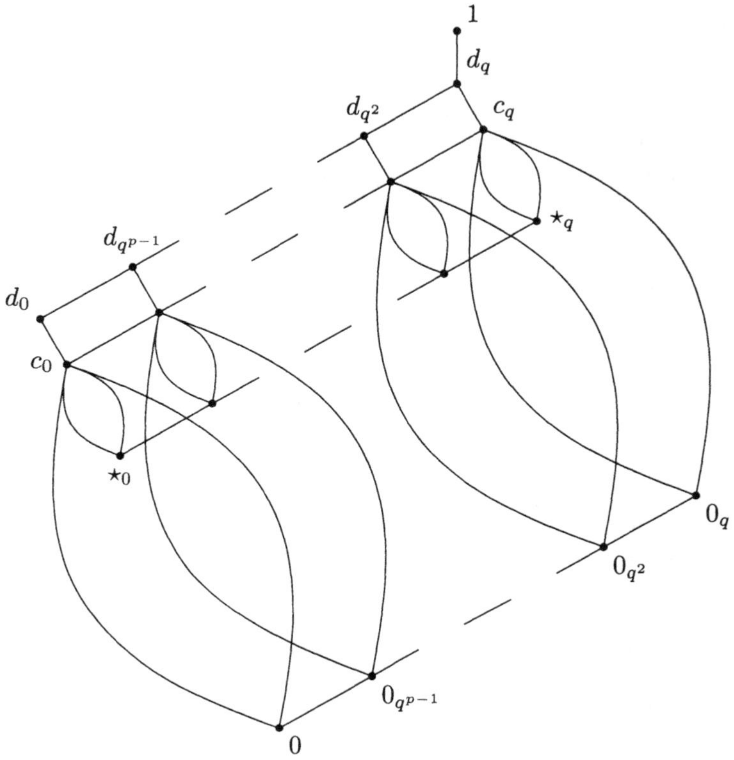

FIGURE 10.3. The algebra $\mathbf{E} \odot \mathbf{C}_{p+1}$.

The rest follows by Jónsson's Lemma from the fact that $\mathsf{FL_{ew}}$ has CEP, $\mathbf{A}$ is subdirectly irreducible, Lemma 10.21 and $\mathbf{E} \odot \mathbf{C}_{p+1}$ is finite. □

LEMMA 10.23. [KO00b] *There is an assignment v such that $\mathbf{E} \odot \mathbf{C}_{p+1}, v \not\models J_{\mathbf{A}}^{(k)} = 1$, for any $k < p$.*

PROOF. Take the assignment v with $v(x_1) = 1$ and $v(x_a) = a_q$, if $a \neq 1$. For $\diamond \in \{\vee, \wedge, \rightarrow\}$, we have $v(x_{a\diamond b} \leftrightarrow x_a \diamond x_b) = (a \diamond b)_q \leftrightarrow (a_q \diamond b_q) = (a \diamond b)_q \leftrightarrow (a \diamond b)_q = 1$. In the case of multiplication, we have: $v(x_{a \cdot b} \leftrightarrow x_a \cdot x_b) = (a{\cdot}b)_q \leftrightarrow (a_q{\cdot}b_q) = (a{\cdot}b)_q \leftrightarrow (a{\cdot}b)_{q\cdot q} = (a{\cdot}b)_q {\rightarrow} (a{\cdot}b)_{q \cdot q} = d_{q \rightarrow (q \cdot q)} = d_q$. For 1 we get $v(x_1 \leftrightarrow 1) = v(x_1) \leftrightarrow v(1) = 1 \leftrightarrow 1 = 1$. Finally for 0 we obtain $v(x_0 \leftrightarrow 0) = v(x_0) \leftrightarrow v(0) = 0_q \leftrightarrow 0_{q^p} = 0_{q^{p-1}}$. Altogether, $v(D_{\mathbf{A}}) = d_{q^{p-1}}$. Now, since the depth of the monolith of $\mathbf{E}$ is m, the properties of $\mathbf{E} \odot \mathbf{C}_{p+1}$ guarantee that $(d_{q^{p-1}})^k \not\leq \star_q$ for any $k < m$. To see that, recall that the lattice operations and the multiplication in $\mathbf{E} \odot \mathbf{C}_{p+1}$ behave exactly as they would have in the direct product $\mathbf{E} \times \mathbf{C}_{p+1}$. Thus, for any k with $1 < k < m$, we have $(d_{q^{p-1}})^k = d^k_{(q^{p-1})^k} = d^k_0$. Moreover, $d^k_0 = d^k_q \wedge d_0$, and thus, as multiplication and meet are computed coordinatewise, we obtain: $d_0^k = d_q^k \wedge d_0 \leq \star_q$ iff $d^k \wedge d \leq \star$ in $\mathbf{E}$ and $q \wedge 0 \leq q$ in $\mathbf{C}_{p+1}$. By construction of $\mathbf{E}$, however, we have $d^k \wedge d = d^k > d^m = \star$ in $\mathbf{E}$; hence $d_0^k \not\leq \star_q$, which is the desired conclusion. For $k = 1$ the argument is analogous, with the index q^{p-1} in place of 0.

Thus, we have $v(J_{\mathbf{A}}^{(k)}) = v(D_{\mathbf{A}}^k \to x_\star) = (d_{q^{p-1}})^k \to \star_q = d_0^k \to \star_q \neq 1$. Hence, $\mathbf{E} \odot \mathbf{C}_{p+1}, v \not\models J_{\mathbf{A}}^{(k)} = 1$. □

10.5. Only one splitting

We are now ready to prove the main result of the chapter.

THEOREM 10.24. [KO00b] *The only algebra that splits $\mathbf{\Lambda}(\mathsf{FL_{ew}})$ is the two-element Boolean algebra* $\mathbf{2}$.

PROOF. That $\mathbf{2}$ splits $\mathbf{\Lambda}(\mathsf{FL_{ew}})$ follows by Proposition 10.4. Let $\mathbf{A}$ be a finite subdirectly irreducible FL_{ew}-algebra different from $\mathbf{2}$. Take any $k \in \mathbb{N}$. Let $\mathbf{E}$ be the algebra whose existence is guaranteed by Lemma 10.19. Then, by Lemma 10.23, $\mathbf{E} \odot \mathbf{C}_{p+1}$ falsifies $J_{\mathbf{A}}^{(k)} = 1$. By Lemma 10.22, $\mathbf{A} \notin \mathsf{V}(\mathbf{E} \odot \mathbf{C}_{p+1})$. Together, these constitute precisely the condition (2) of Lemma 10.10 (with $\mathbf{E} \odot \mathbf{C}_{p+1}$ as $\mathbf{B}$) by which the theorem follows. □

Naturally, one would like to know how far the above result would go. Certainly, for any variety $\mathcal{V}$ *containing* $\mathsf{FL_{ew}}$ it limits chances for splittings in $\mathbf{\Lambda}(\mathcal{V})$ to finite subdirectly irreducible algebras in $\mathcal{V} \setminus \mathsf{FL_{ew}}$. This is of particular interest for natural varieties containing $\mathbf{FL_{ew}}$. Problem 17 asks whether there are any splitting algebras of natural super-varieties of $\mathsf{FL_{ew}}$. A preliminary step in this direction was made in [Kow04], where it was shown that no finite subdirectly irreducible FL_e-algebra $\mathbf{A}$ with $|A| > 2$ and such that 1 is comparable with every element splits $\mathbf{\Lambda}(\mathsf{FL_e})$.

On the other hand, splittings in subvarieties of $\mathsf{FL_{ew}}$ are also of interest. This question is only settled for subvarieties with EDPC, where the splitting algebras are exactly all finite subdirectly irreducible algebras. For other subvarieties of $\mathsf{FL_{ew}}$ it remains open and the methods used in the proof do not seem to carry over.

Exercises

(1) Prove Lemma 10.1.
(2) Using Proposition 9.8 prove that $\mathbb{Z}^-$ is a splitting algebra in CanRL and that $\mathsf{V}(\mathbb{Z}^-)$ and LG form a splitting pair in $\mathbf{\Lambda}(\mathsf{CanRL})$.
(3) Prove Lemma 10.2 in a detailed way.
(4) Supply the missing details in the proof of Lemma 10.5.
(5) Prove that Lemmas 10.5 and 10.6 as well as Theorem 10.7 still hold if we redefine diagram of $\mathbf{A}$ to be $D'_{\mathbf{A}} = \bigwedge\{x_{\neg a} \leftrightarrow \neg x_a : a \in A\} \wedge \bigwedge\{x_{a \circ b} \leftrightarrow$

$x_a \diamond x_b : a, b \in A, \diamond \in \{\vee, \wedge, \cdot, \to\}\}$. What happens if we leave out negation?

(6) Prove Lemma 10.8.

(7) Prove that if $\mathbf{B} \not\models J^{(i)}_{\mathbf{A}} = 1$, then $\mathbf{B} \not\models J^{(j)}_{\mathbf{A}} = 1$ for all $j \leq i$.

(8) Write the diagram of $\mathbf{C}_3$ in full and work out the details of the example from Section 10.4.1. Precise definition of residuation in the grid algebra is given in Section 10.4.4.

(9) Work through the missing cases in the proof of Lemma 10.11.

(10) Let $\mathbf{P}$ be the structure of defined prior to Lemma 10.11, for some FL_{ew}-algebra $\mathbf{A}$. Find an example of an $\mathbf{A}$ for which $\mathbf{P}$ is not a lattice. [Hint: let a and b be incomparable, $ca < a$, $cb < b$ and $c(a \wedge b) < ca \wedge cb$.]

(11) Let $\mathbf{P}$ be as above. Find an example of an $\mathbf{A}$ for which Dedekind-MacNeille completion of $\mathbf{P}$ does not preserve residuation from $\mathbf{A}$. [Hint: something must go wrong with the fourth condition in 10.3.]

(12) Prove Lemma 10.14.

(13) Provide the missing details in proofs of Lemmas 10.15 and 10.16.

(14) Provide the missing details in proof of Lemma 10.17.

(15) Carry out the alternative proof of Lemma 10.17 suggested on page 454. [Hint: do not expect significant gains in simplicity of the argument.]

(16) Consider subdirectly irreducible FL_e-algebras such that 1 is comparable with every element. Define diagrams and Jankov terms as before, but with $x \leftrightarrow y = (x \to y) \wedge (y \to x) \wedge 1$. Produce suitable counterparts of expansions and twisted products for these. Hint: in expansions leave the positive part intact.

(17) *Open problem*: Are there any splitting algebras in $\mathsf{FL} \setminus \mathsf{FL_{ew}}$ or in other natural super-varieties of $\mathsf{FL_{ew}}$?

(18) *Open problem*: Investigate splittings in subvarieties of $\mathsf{FL_{ew}}$.

(19) *Open problem*: Which subvarieties of $\mathsf{FL_{ew}}$ without EDPC have non-Boolean splitting algebras?

Notes

(1) Apart from universal algebra, splittings and related notions have been used rather extensively in modal logic. In particular, Jankov's idea of *characteristic formulas*, was taken up by Fine (cf. [Fin74]) and applied to modal logics. This led to successive refinements of the notion, first by Fine in 1970s and later by Zakharyaschev in a series of papers in 1980s. We refer the interested reader to Chagrov and Zakharyaschev's [CZ97] (especially Chapter 9) for an excellent survey of these. See also [Kra99]. In fact, splitting methods and results from modal logic had some impact on universal algebra, too. For instance, Blok proved in [Blo78] that each modal logic axiomatized by a characteristic formula induces a splitting of the lattice of modal logics.

This certainly influenced later results of his, Köhler's and Pigozzi's (cf. [BP82], [BKP84]) on splittings in varieties with EDPC.

(2) As to open problem 17, on the one hand, even the two element Boolean algebra is not guaranteed to be a splitting in FL or $\mathsf{FL_e}$, since it no longer generates a unique atom. On the other hand, for the case of $\mathsf{FL_e}$ it does not seem unreasonable to hope that the constructions used in the proof, or some modifications of them (cf. [Kow04] or Exercise 16) will work, providing a proof that there are very few splitting algebras in $\mathsf{FL_e}$.

(3) As to open problems 18 and 19, there seems to be a certain tension between the "expand" and "twist" parts of the construction. Consider $\mathsf{DFL_{ew}}$. There, the "expand" part does not have much chance for survival, as completions are notorious for destroying distributivity. But consider $\mathsf{RFL_{ew}}$. Expansions seem to work all right, but twisting is doomed, as it never produces anything linear, except in trivial cases.

CHAPTER 11

Semisimplicity

As in the previous chapter, we deal exclusively with FL_{ew}-algebras. The main results are two: that free FL_{ew}-algebras are semisimple, and that semisimple varieties of FL_{ew}-algebras are discriminator varieties. Although both results deal with semisimple algebras and both have more universal algebraic interest rather than logical, they are quite independent from one another. We precede each part with an outline, as the proofs are rather long and may be quite unreadable otherwise.

11.1. Semisimplicity, discriminator, EDPC

Recall that an algebra $\mathbf{A}$ is called *semisimple* if it has a subdirect representation with simple factors. A variety $\mathcal{V}$ is semisimple if every algebra $\mathbf{A} \in \mathcal{V}$ is semisimple. Equivalently, $\mathcal{V}$ is semisimple if all its subdirectly irreducible members are simple. Notice that $V(\mathbf{A})$ does not have to be semisimple even if $\mathbf{A}$ is. This is exactly what is the case for $\mathsf{FL_{ew}}$ — the free algebra generating $\mathsf{FL_{ew}}$ is semisimple, as we will show in this chapter, yet $\mathsf{FL_{ew}}$ is not.

It will also turn out that a subvariety of $\mathsf{FL_{ew}}$ is semisimple if and only if it satisfies the identity $x \vee \neg x^n = 1$, for some positive integer n. These varieties we will call $\mathsf{EM_n}$. Another result we prove about semisimple subvarieties of $\mathsf{FL_{ew}}$ is that they coincide with discriminator varieties. Recall that these are varieties with ternary discriminator as a term operation on all subdirectly irreducibles (cf. Chapter 1). In the context of FL_{ew}-algebras, another function, known as *unary discriminator*, is useful. It is defined by:

$$u(x) = \begin{cases} 0, & \text{for } x \neq y, \\ 1, & \text{for } x = 1. \end{cases}$$

Its usefulness follows from the lemma below, whose easy proof we leave as an exercise to the reader (Exercise 2). Recall that $x \leftrightarrow y$ is simply equal to $(x \to y) \wedge (y \to x)$, in the context of commutativity and integrality.

LEMMA 11.1. *Let u be a unary discriminator term on an FL_{ew}-algebra $\mathbf{A}$. Then the term $(u(x \leftrightarrow y) \wedge z) \vee (\neg u(x \leftrightarrow y) \wedge x)$, is the ternary discriminator on $\mathbf{A}$.*

It follows from Theorem 3.55 that inside $\mathsf{FL_{ew}}$ both DPC and EDPC coincide with the property of being n-potent for some $n \in \mathbb{N}$. We state it explicitly below.

THEOREM 11.2. *For a variety $\mathcal{V}$ of FL_{ew}-algebras the following conditions are equivalent:*

(1) *$\mathcal{V}$ satisfies $x^{n+1} = x^n$, for some $n \in \mathbb{N}$;*
(2) *$\mathcal{V}$ has EDPC;*
(3) *$\mathcal{V}$ has DPC;*
(4) *$\mathcal{V} \subseteq \mathsf{E_n}$, for some $n \in \mathbb{N}$.*

11.1.1. Some connections to logics. Logics corresponding to varieties $\mathsf{EM_n}$ are these that satisfy a weak form of excluded middle, to be explicit, a logic $\mathbf{L}$ satisfies *iterated excluded middle* if $\varphi \vee \neg(\varphi^n) \in \mathbf{L}$ for some n. If $n = 1$, then $\mathbf{L}$ is the classical propositional logic (and if $n = 0$, $\mathbf{L}$ is inconsistent).

Logics corresponding to varieties $\mathsf{E_n}$ are these that satisfy a weak form of contraction. They also enjoy a rather strong version of deduction theorem. To give it a more precise formulation, let us say that a logic $\mathbf{L}$ over $\mathbf{FL_{ew}}$ is *n-contractive*, if $\varphi^n \to \varphi^{n+1} \in \mathbf{L}$. Notice that the converse is a theorem of $\mathbf{FL_{ew}}$ for any n. If $\mathbf{L}$ is n-contractive for some positive integer n, it will be called *weakly contractive*. The following theorem holds.

THEOREM 11.3. *Suppose that $\Sigma \cup \{\delta_1, \ldots, \delta_k, \varphi\}$ is a set of formulas and $\mathbf{L}$ is an n-contractive logic over $\mathbf{FL_{ew}}$ for some positive integer n. Then*

$$\Sigma, \delta_1, \ldots, \delta_k \vdash_{\mathbf{L}} \varphi \quad \textit{iff} \quad \Sigma \vdash_{\mathbf{L}} (\prod_{i=1}^{k} \delta_i)^n \to \varphi.$$

In particular, for $k = 1$, writing δ for δ_1 we obtain

$$\Sigma, \delta \vdash_{\mathbf{L}} \varphi \quad \textit{iff} \quad \Sigma \vdash_{\mathbf{L}} \delta^n \to \varphi.$$

To see what is involved the reader should compare the above theorem to Theorem 2.14 and the series of corollaries following it. The most important difference is that the outer exponent n is independent of the set $\{\delta_1, \ldots, \delta_k, \varphi\}$ which is not the case in general. Another useful comparison is to deductive filter generation (cf. Theorem 3.47, recall that by integrality convex normal subalgebras coincide with deductive filters). Let Σ be empty, and think of $\{\delta_1, \ldots, \delta_k, \varphi\}$ as a subset of the free FL_{ew}-algebra. Then, $\varphi \in \mathrm{Fg}\{\delta_1, \ldots, \delta_k\}$ if and only if $(\delta_1 \cdot \ldots \cdot \delta_k)^n \leq \varphi$.

It is not difficult to show that if a logic satisfies iterative excluded middle then it is n-contractive, for some positive integer n (see Exercise 1). Also, if a logic $\mathbf{L}$ has either of the two properties for n, then $\mathbf{L}$ has it for any $m \geq n$. These observations lead to a natural classification of logics over $\mathbf{FL_{ew}}$ (cf. [Ued00]). We will show that a parallel classification of subvarieties of $\mathsf{FL_{ew}}$ is also natural for purely algebraic reasons.

11.2. Free FL_{ew}-algebras are semisimple: outline

Our proof that free FL_{ew}-algebras are semisimple follows in footsteps of Grišin, who in [Gri82] proved similar result for free involutive FL_{ew}-algebras. Two crucial insights here are these: (1) that semisimplicity can be characterized by a second-order equational condition, and (2) that this condition can be shown to hold in free algebras by viewing them as Lindenbaum algebras of formulas and applying a proof-theoretical method, namely, cut-elimination theorem. The second part proceeds by a series of subtle modifications of the standard sequent system for $\mathbf{FL_{ew}}$, which twist it into a more manageable form while not destroying admissibility of cut. This proof comes from [KO00a].

11.3. A characterization of semisimple FL_{ew}-algebras

To reduce clutter we will call deductive filters simply filters, throughout. Let Φ be the set of all maximal filters of a FL_{ew}-algebra $\mathbf{A}$. Recall that a filter F is called maximal iff F is maximal among proper filters of $\mathbf{A}$. It is easy to see that a filter F is maximal iff for each $u \in A$ either $u \in F$ or there is a $k \geq 1$ such that $\neg u^k \in F$. Define the *radical* Rad_A of $\mathbf{A}$ by $\mathrm{Rad}_A = \bigcap_{F\in\Phi} F$. Then, since maximal filters correspond to maximal congruences, it follows immediately that a FL_{ew}-algebra $\mathbf{A}$ is semisimple if and only if $\mathrm{Rad}_A = \{1\}$.

For elements $a, b \in A$, set $a + b$ to be $\neg(\neg a \cdot \neg b)$. Note that this operation, while commutative, is usually not associative. Moreover, while $a+a = \neg(\neg a \cdot \neg a) = \neg(\neg a)^2$ holds, neither $a+(a+a)$, nor $(a+a)+a$ yields the desired $\neg(\neg a)^3$. To remedy that, for any positive integer m we define $\widetilde{m}a$ to be $\neg(\neg a)^m$. Note that $\widetilde{1}a = \neg\neg a$.

Lemma 11.4. [KO00a] *For any x in a given FL_{ew}-algebra $\mathbf{A}$, $x \in \mathrm{Rad}_A$ if and only if for any $n \geq 1$ there exists $m \geq 1$ such that $\widetilde{m}(x^n) = 1$.*

Proof. Assume that for any $n \geq 1$ there exists $m \geq 1$ such that $\widetilde{m}(x^n) = 1$. Suppose $x \notin \mathrm{Rad}_A$. Thus, there is a maximal filter F with $x \notin F$. Since F is maximal, there is a $k \geq 1$ such that $\neg x^k \in F$. Further, by the assumption, there is an $l \geq 1$ for which $\widetilde{l}(x^k) = 1$, i.e., $(\neg x^k)^l = 0$. This implies $0 \in F$, which contradicts the fact that F is proper.

Conversely, suppose that there exists $n \geq 1$ such that $\widetilde{m}(x^n) \neq 1$ for any m. If $(\neg(x^n))^m = 0$ then $\widetilde{m}(x^n) = \neg(\neg(x^n))^m = 1$, which is a contradiction. Thus, $(\neg(x^n))^m > 0$ for any m. Let $z = \neg(x^n)$ and H be the filter generated by z. Clearly, H is proper as $z^m > 0$ for any m. By Zorn lemma, there exists a maximal filter G such that $H \subseteq G$. Now, suppose that $x \in G$. Then x^n must be also in G. But this is a contradiction, since $z = \neg(x^n) \in G$. Hence, $x \notin G$ and therefore it does not belong to Rad_A. □

Corollary 11.5. [KO00a] *An FL_{ew}-algebra $\mathbf{A}$ is semisimple iff for every $a \in A\backslash\{1\}$ there is an $N \geq 1$ such that for any $m \geq 1$, we have: $\widetilde{m}(a^N) < 1$.*

11.4. Sequent calculi for $\mathbf{FL_{ew}}$

Before we put Corollary 11.5 to better use in the context of free FL_{ew}-algebras, we prepare some tools for reasoning about them. Our tools come in the form of two sequent calculi for this logic. First, we will introduce a convenient modification of the usual sequent calculus for $\mathbf{FL_{ew}}$. In the modified calculus, called $\mathbf{SFL_{ew}}$, we will slightly change the form of initial sequents and generalize the form of rules for fusion. This will allow us to drop the weakening rule. Further, we will view sequents as multisets of formulas, rather than lists of them, which will let us drop the exchange rule as well, so that no structural rules get in the way. In our new rules for fusion, we will allow formulas of the form $\alpha_1 \cdot \ldots \cdot \alpha_m$ for any $m \geq 2$, with no conventions about implicit bracketing. This is harmless, since the associativity of fusion is provable in $\mathbf{FL_{ew}}$.

To be more formal, a sequent of $\mathbf{SFL_{ew}}$ is of the form $\Gamma \Rightarrow \beta$, where Γ is a possibly empty multiset of formulas. Note that the right-hand side of $\Rightarrow$ must always exist. As usual, when Γ is a multiset $\{\alpha_1, \ldots, \alpha_m\}$, the sequent $\Gamma \Rightarrow \beta$ is also expressed as $\alpha_1, \ldots, \alpha_m \Rightarrow \beta$. Also, the multiset union of multisets Γ and Δ (and of $\{\alpha\}$ and Δ) is denoted by Γ, Δ (and α, Δ, respectively). In what follows, both Γ and Δ denote arbitrary multisets of formulas. As before, we assume that the negation $\neg\alpha$ of a formula α is defined by $\alpha \to 0$. The system $\mathbf{SFL_{ew}}$ consists of the following initial sequents:

$$0, \Gamma \Rightarrow \gamma \qquad\qquad x, \Gamma \Rightarrow x \quad \text{for any variable } x$$

and the following rules of inference:
Cut rule:

$$\frac{\Gamma \Rightarrow \alpha \quad \alpha, \Delta \Rightarrow \gamma}{\Gamma, \Delta \Rightarrow \gamma}$$

Rules for logical connectives:

$$\frac{\alpha, \Gamma \Rightarrow \beta}{\Gamma \Rightarrow \alpha \to \beta}\ (\Rightarrow \to) \qquad\qquad \frac{\Gamma \Rightarrow \alpha \quad \beta, \Delta \Rightarrow \gamma}{\alpha \to \beta, \Gamma, \Delta \Rightarrow \gamma}\ (\to \Rightarrow)$$

$$\frac{\Gamma \Rightarrow \alpha}{\Gamma \Rightarrow \alpha \vee \beta}\ (\Rightarrow \vee 1) \qquad\qquad \frac{\Gamma \Rightarrow \beta}{\Gamma \Rightarrow \alpha \vee \beta}\ (\Rightarrow \vee 2)$$

$$\frac{\alpha, \Gamma \Rightarrow \gamma \quad \beta, \Gamma \Rightarrow \gamma}{\alpha \vee \beta, \Gamma \Rightarrow \gamma}\ (\vee \Rightarrow)$$

$$\frac{\Gamma \Rightarrow \alpha \quad \Gamma \Rightarrow \beta}{\Gamma \Rightarrow \alpha \wedge \beta}\ (\Rightarrow \wedge)$$

$$\frac{\alpha, \Gamma \Rightarrow \gamma}{\alpha \wedge \beta, \Gamma \Rightarrow \gamma} (\wedge 1 \Rightarrow) \qquad \frac{\beta, \Gamma \Rightarrow \gamma}{\alpha \wedge \beta, \Gamma \Rightarrow \gamma} (\wedge 2 \Rightarrow)$$

$$\frac{\Gamma_1 \Rightarrow \alpha_1 \quad \dots \quad \Gamma_m \Rightarrow \alpha_m}{\Gamma_1, \dots, \Gamma_m \Rightarrow \alpha_1 \cdot \dots \cdot \alpha_m} (\Rightarrow \cdot) \qquad \frac{\alpha_1, \dots, \alpha_m, \Gamma \Rightarrow \gamma}{\alpha_1 \cdot \dots \cdot \alpha_m, \Gamma \Rightarrow \gamma} (\cdot \Rightarrow)$$

As usual, we say that a formula α is provable in $\mathbf{SFL_{ew}}$ if the sequent $\Rightarrow \alpha$ is provable in it. Since the cut elimination theorem holds for the calculus $\mathbf{SFL_{ew}}$, the cut rule is admissible. We can prove cut elimination for $\mathbf{SFL_{ew}}$ by modifying the proof of cut elimination for $\mathbf{FL_{ew}}$. (For the latter, see e.g. [OK85].)

Next, we will introduce yet another sequent calculus $\mathbf{SFL^{+}_{ew}}$, which will then be shown to be equivalent to $\mathbf{SFL_{ew}}$. We will say that α is a *fusion-formula* if the outermost connective of α is fusion. Now, $\mathbf{SFL^{+}_{ew}}$ is the system obtained from $\mathbf{SFL_{ew}}$ by deleting the cut rule and by adding the following condition (+) in any application of the rules $(\Rightarrow \cdot)$ and $(\cdot \Rightarrow)$:

$$\textit{none of } \alpha_i s \textit{ are fusion-formulas.} \qquad (+)$$

For double precision we note here that when we set out to show the equivalence of $\mathbf{SFL^{+}_{ew}}$ to $\mathbf{SFL_{ew}}$, we will identify each formula of the form $\alpha_1 \cdot \dots \cdot \alpha_m$ of $\mathbf{SFL^{+}_{ew}}$ with an arbitrary formula of $\mathbf{SFL_{ew}}$ obtained from $\alpha_1 \cdot \dots \cdot \alpha_m$ by introducing any proper bracketing in it. As we have already said, this does not cause problems since all such formulas are equivalent to one another over $\mathbf{FL_{ew}}$.

LEMMA 11.6. [KO00a] *A sequent $\Gamma \Rightarrow \delta$ is provable in $\mathbf{SFL_{ew}}$ if and only if it is provable in $\mathbf{SFL^{+}_{ew}}$.*

PROOF. It is enough to show that if a sequent $\Gamma \Rightarrow \delta$ is provable in $\mathbf{SFL_{ew}}$ then it is provable in $\mathbf{SFL^{+}_{ew}}$. In other words, it suffices to show that any application of $(\Rightarrow \cdot)$ and $(\cdot \Rightarrow)$ not satisfying the condition (+) in a given cut-free proof P of $\Gamma \Rightarrow \delta$ can be replaced by one with (+). Let β be any formula of the form $\beta_1 \cdot \dots \cdot \beta_k$ $(k > 0)$, where none of β_js are fusion-formulas. Then we define the *degree* $d(\beta)$ of β by $d(\beta) = k - 1$. i.e. the number of outermost occurrences of fusion in β. Next, when I is either an application of $(\cdot \Rightarrow)$ of the form

$$\frac{\alpha_1, \dots, \alpha_m, \Gamma \Rightarrow \gamma}{\alpha_1 \cdot \dots \cdot \alpha_m, \Gamma \Rightarrow \gamma}$$

or an application of $(\Rightarrow \cdot)$ of the form

$$\frac{\Gamma_1 \Rightarrow \alpha_1 \quad \dots \quad \Gamma_m \Rightarrow \alpha_m}{\Gamma_1, \dots, \Gamma_m \Rightarrow \alpha_1 \cdot \dots \cdot \alpha_m}$$

we define the *degree* $d(I)$ of I by $d(I) = d(\alpha_1) + \ldots + d(\alpha_m)$. It is obvious that an application I of either $(\cdot \Rightarrow)$ or $(\Rightarrow \cdot)$ satisfies the condition (+) if and only if $d(I) = 0$.

Assume first that there exists an application of $(\cdot \Rightarrow)$ not satisfying the condition (+) in P. Let us take one of the uppermost applications among them, which we call J. Obviously, $d(J) > 0$. We suppose that the lower sequent of J is $\alpha_1 \cdot \ldots \cdot \alpha_m, \Gamma \Rightarrow \gamma$ with the principal formula $\alpha_1 \cdot \ldots \cdot \alpha_m$. We will show by induction on the degree $d(J)$ that $\alpha_1 \cdot \ldots \cdot \alpha_m, \Gamma \Rightarrow \gamma$ has a cut-free proof in which every application of $(\cdot \Rightarrow)$ satisfies the condition (+).

By our assumption, the degree of one of α_is must be nonzero. Without loss of generality, we can assume that α_1 is of the form $\delta_1 \cdot \ldots \cdot \delta_s$ for $s > 1$. Let Q be the proof of $\alpha_1 \cdot \ldots \cdot \alpha_m, \Gamma \Rightarrow \gamma$ which is a subproof of P. We will trace back ancestors of the auxiliary formula α_1 of J in all branches of Q. Then, we can see that in each of these branches, α_1 is introduced somewhere either as the principal formula of an application of $(\cdot \Rightarrow)$ (with (+)), or as a side formula of an initial sequent. (Note that x must be a variable in initial sequents of the second type.) Now, we replace any ancestor α_1 by the multiset $\delta_1, \ldots, \delta_s$. If one of such α_1 is introduced by an application of $(\cdot \Rightarrow)$ then its lower sequent becomes identical with the upper sequent by this replacement. In such a case, we eliminate this $(\cdot \Rightarrow)$. On the other hand, if it is introduced as a side formula of an initial sequent, the sequent obtained by this replacement remains still an initial sequent. Hence, the proof Q′ obtained from Q by this replacement remains a correct proof of $\alpha_1 \cdot \ldots \cdot \alpha_m, \Gamma \Rightarrow \gamma$ whose last inference is an application J' of $(\cdot \Rightarrow)$, with the upper sequent $\delta_1, \ldots, \delta_s, \alpha_2, \ldots, \alpha_n, \Gamma \Rightarrow \gamma$. Since $d(J') = d(J) - d(\alpha_1) < d(J)$, by the hypothesis of induction, $\alpha_1 \cdot \ldots \cdot \alpha_n, \Gamma \Rightarrow \gamma$ has a cut-free proof in which every application of $(\cdot \Rightarrow)$ satisfies (+). In this way, we have a cut-free proof P′ of $\Gamma \Rightarrow \delta$ where every application of $(\cdot \Rightarrow)$ satisfies the condition (+).

For example, consider the following (sub)proof whose last inference is an application of $(\cdot \Rightarrow)$ without the condition (+).

$$\dfrac{\dfrac{x \cdot y, z, z \to (x \to t) \Rightarrow z \qquad \dfrac{x, y, z, z \to (x \to t) \Rightarrow t}{x \cdot y, z, z \to (x \to t) \Rightarrow t}}{x \cdot y, z, z \to (x \to t) \Rightarrow z \wedge t}}{x \cdot y \cdot z, z \to (x \to t) \Rightarrow z \wedge t}$$

The replacement mentioned above will change the above proof into the following one.

$$\dfrac{\dfrac{x, y, z, z \to (x \to t) \Rightarrow z \qquad x, y, z, z \to (x \to t) \Rightarrow t}{x, y, z, z \to (x \to t) \Rightarrow z \wedge t}}{x \cdot y \cdot z, z \to (x \to t) \Rightarrow z \wedge t}$$

Next, we will remove any application of $(\Rightarrow \cdot)$ not satisfying the condition (+) in P′. Suppose that there exists such an application. Similarly to the above, take one of the uppermost applications of $(\Rightarrow \cdot)$ not satisfying (+), called J, which is of the form given above. We will show by induction on $d(J)$ that the lower sequent $\Gamma_1, \ldots, \Gamma_m \Rightarrow \alpha_1 \cdot \ldots \cdot \alpha_m$ has a cut-free proof in which every application of $(\Rightarrow \cdot)$ satisfies (+). Without loss of generality, we can assume that $d(\alpha_1) = k > 0$ and that α_1 is of the form $\delta_1 \cdot \ldots \cdot \delta_k$, where none of δ_js are fusion-formulas. Let R be the proof of $\Gamma_1 \Rightarrow \alpha_1$, which is a subproof of P′. We will trace back the branches in R which consist of sequents having α_1 in the conclusion (such an α_1 must clearly be an ancestor of the α_1 from $\Gamma_1 \Rightarrow \alpha_1$) to the places where this α_1 is introduced. There are two possibilities. It is introduced either by means of an initial sequent of the form: $0, \Delta \Rightarrow \alpha_1$, or as the principal formula of an application of $(\Rightarrow \cdot)$ of the form:

$$\frac{\Delta_1 \Rightarrow \delta_1 \quad \ldots \quad \Delta_k \Rightarrow \delta_k}{\Delta_1, \ldots, \Delta_k \Rightarrow \alpha_1}$$

We will modify the proof R as follows. If α_1 is introduced as the succedent of an initial sequent, we replace this sequent by $0, \Delta, \Gamma_2, \ldots, \Gamma_m \Rightarrow \alpha_1 \cdot \ldots \cdot \alpha_m$, which is still an initial sequent. In the other case, we replace it by:

$$\frac{\Delta_1 \Rightarrow \delta_1 \quad \ldots \quad \Delta_k \Rightarrow \delta_k \quad \Gamma_2 \Rightarrow \alpha_2 \quad \ldots \quad \Gamma_m \Rightarrow \alpha_m}{\Delta_1, \ldots, \Delta_k, \Gamma_2, \ldots, \Gamma_m \Rightarrow \alpha_1 \cdot \alpha_2 \cdot \ldots \cdot \alpha_m}$$

and put the subproof of each $\Gamma_i \Rightarrow \alpha_i$ in R over it for each $i = 2, \ldots, m$. Note that the degree of the above application of $(\Rightarrow \cdot)$ is $d(J) - d(\alpha_1)$, which is smaller than $d(J)$. Therefore, by the hypothesis of induction the lower sequent of this inference has a cut-free proof in which every application of $(\Rightarrow \cdot)$ satisfies (+). Finally, we replace every sequent $\Sigma \Rightarrow \alpha_1$ in a branch which we have traced, by the sequent $\Sigma, \Gamma_2, \ldots, \Gamma_m \Rightarrow \alpha_1 \cdot \ldots \cdot \alpha_m$. Then, after this replacement, we get a proof R′, which is in fact a proof of $\Gamma_1, \Gamma_2, \ldots, \Gamma_m \Rightarrow \alpha_1 \cdot \ldots \cdot \alpha_m$, in which every application of $(\Rightarrow \cdot)$ satisfies (+). (Note that when $(\rightarrow \Rightarrow)$ is used somewhere in these branches, this replacement will be done only for the right upper sequent. Thus, this replacement never cause unnecessary duplications of $\Gamma_2, \ldots, \Gamma_m$.) By repeating this, we have a proof of $\Gamma \Rightarrow \delta$ in $\mathbf{SFL^+_{ew}}$. □

For example, let us consider the following proof R.

$$\cfrac{\cfrac{\cfrac{x \Rightarrow x \quad y \Rightarrow y}{x, y \Rightarrow x \cdot y}\,(\Rightarrow \cdot)}{x \wedge w, y \Rightarrow x \cdot y}\,(\wedge \Rightarrow) \qquad z, u \Rightarrow z}{x \wedge w, y, z, u \Rightarrow x \cdot y \cdot z}\,(J)$$

The above transformation gives us the following proof in $\mathbf{SFL^+_{ew}}$.

$$\cfrac{\cfrac{x \Rightarrow x \quad y \Rightarrow y \quad z, u \Rightarrow z}{x, y, z, u \Rightarrow x \cdot y \cdot z}\,(\Rightarrow \cdot)}{x \wedge w, y, z, u \Rightarrow x \cdot y \cdot z}\,(\wedge \Rightarrow)$$

11.5. Semisimplicity of free $\mathbf{FL}_{ew}$-algebras

As we already announced, we will employ the characterization of semisimple FL_{ew}-algebras from Corollary 11.5. For each formula α, let $l(\alpha)$ denote the length of α (as a sequence of symbols). For a sequence Γ of formulas $\alpha_1, \ldots, \alpha_m$, the length $l(\Gamma)$ of Γ is defined by $l(\Gamma) = l(\alpha_1) + \ldots + l(\alpha_m)$. Also, $\{\neg\alpha^N\}^m$ will stand for the multiset $\neg\alpha^N, \ldots, \neg\alpha^N$ with m times $\neg\alpha^N$.

LEMMA 11.7. [KO00a] *Suppose that a formula α is not provable in $\mathbf{SFL}^+_{\mathbf{ew}}$ and that N is any positive integer greater than $l(\alpha)$. Then, for any sequent $\Gamma \Rightarrow \gamma$ such that $l(\Gamma, \gamma) \leq l(\alpha)$ and any positive integer m, if $\{\neg\alpha^N\}^m, \Gamma \Rightarrow \gamma$ is provable in $\mathbf{SFL}^+_{\mathbf{ew}}$ then $\Gamma \Rightarrow \gamma$ is provable in $\mathbf{SFL}^+_{\mathbf{ew}}$.*

PROOF. The proof will proceed by double induction on $(m, l(\Gamma, \gamma))$. So, we assume that the lemma holds for $m' < m$ and it also holds for $(m, l(\Delta, \delta))$, whenever $l(\Delta, \delta) < l(\Gamma, \gamma)$. We suppose that α is not provable but the sequent $\{\neg\alpha^N\}^m, \Gamma \Rightarrow \gamma$ is provable in $\mathbf{SFL}^+_{\mathbf{ew}}$. First, let $\{\neg\alpha^N\}^m, \Gamma \Rightarrow \gamma$ be an initial sequent. Then, either γ is a propositional variable which occurs also in Γ, or 0 occurs in Γ. It is clear that $\Gamma \Rightarrow \gamma$ is provable in either case.

Next, suppose that the sequent $\{\neg\alpha^N\}^m, \Gamma \Rightarrow \gamma$ is the lower sequent of an inference rule I. We first assume that the principal formula of I is either in Γ or in γ. Then, (each of) upper sequent(s) of I is of the form $\{\neg\alpha^N\}^{m_i}, \Delta_i \Rightarrow \delta_i$ with $m_i \leq m$ and $l(\Delta_i, \delta_i) < l(\Gamma, \gamma)$, by the subformula property and the fact that $\mathbf{SFL}^+_{\mathbf{ew}}$ has no contraction rule. Thus, by the hypothesis of induction, (each) $\Delta_i \Rightarrow \delta_i$ is provable. Then $\Gamma \Rightarrow \gamma$ is also provable by applying the same inference rule I.

Finally suppose that the principal formula of I is one of $\neg\alpha^N$. Then, I must be an application of $(\to \Rightarrow)$. Recall here that a formula $\neg\beta$ is the abbreviation of $\beta \to 0$. The upper sequents must be of the form $\{\neg\alpha^N\}^{m_1}, \Gamma_1 \Rightarrow \alpha^N$ and $0, \{\neg\alpha^N\}^{m_2}, \Gamma_2 \Rightarrow \gamma$ such that $m_1 + m_2 = m - 1$ and Γ_1, Γ_2 is equal to Γ. Now consider the proof R of the left upper sequent $\{\neg\alpha^N\}^{m_1}, \Gamma_1 \Rightarrow \alpha^N$. As we did in the proof of Lemma 11.6, we trace back branches in R, which consists of sequents having α^N in the conclusion, to the places where these α^N are introduced. It is easy to see that each α^N is introduced either as an initial sequent of the form $0, \Delta \Rightarrow \alpha^N$, or by an application of $(\Rightarrow \cdot)$. Suppose that in at least one place, α^N is introduced by an application J of $(\Rightarrow \cdot)$, whose lower sequent is of the form $\{\neg\alpha^N\}^k, \Sigma \Rightarrow \alpha^N$. Clearly, $k \leq m_1$. We assume here that α is of the form $\delta_1 \cdot \ldots \cdot \delta_w$ such that none of δ_j are fusion-formulas. Without loss of generality, we can also assume that $\delta_1, \ldots, \delta_w$ are all distinct. Then, I must have $N \cdot w$ upper sequents, each of which is of the form $\{\neg\alpha^N\}^{t_i}, \Xi_i \Rightarrow \delta_{n_i}$, where $1 \leq n_i \leq w$, $k = t_1 + \ldots + t_{N \cdot w}$ and the multiset $\Xi_1, \ldots, \Xi_{N \cdot w}$ is equal to Σ. For each j such that $1 \leq j \leq w$, there exist exactly N sequents with the conclusion δ_j among these sequents. We enumerate them as ${S^j}_1, \ldots, {S^j}_N$. Next, for each

h such that $1 \leq h \leq N$, take $S^1{}_h, \ldots, S^w{}_h$ for upper sequents and apply $(\Rightarrow \cdot)$ to them. Then, we can get a sequent of the form $\{\neg\alpha^N\}^{u_h}, \Pi_h \Rightarrow \alpha$ for $1 \leq h \leq N$ such that $k = u_1 + \ldots + u_N$ and the multiset $\Pi_1, \ldots, \Pi_N$ is equal to Σ. Now, $l(\Sigma) \leq l(\Gamma_1) \leq l(\Gamma, \gamma) \leq l(\alpha) < N$. If $l(\Pi_h) > 0$ for any h such that $1 \leq h \leq N$ then $l(\Sigma) \geq N$, which is a contradiction. Thus, Π_h must be empty for some h. Let it be f. Then, $\{\neg\alpha^N\}^{u_f} \Rightarrow \alpha$ is provable. By our assumption that α is not provable, u_f must be positive. Taking $\emptyset$ and α for Γ and γ, respectively, $\Rightarrow \alpha$ is provable by the hypothesis of induction, since $u_f \leq m_1 \leq m - 1 < m$ and $l(\alpha) \leq l(\alpha)$. This is a contradiction.

Thus, we have shown that in any place α^N is introduced as an initial sequent of the form $0, \Delta \Rightarrow \alpha^N$. We will modify the proof R of $\{\neg\alpha^N\}^{m_1}, \Gamma_1 \Rightarrow \alpha^N$ as follows. We replace every sequent $\Lambda \Rightarrow \alpha^N$ in a branch which we have traced in R, including initial sequents of the form $0, \Delta \Rightarrow \alpha^N$ mentioned above, by the sequent $\Lambda, \Gamma_2 \Rightarrow \gamma$. Then we will get a proof R^* whose end sequent is $\{\neg\alpha^N\}^{m_1}, \Gamma \Rightarrow \gamma$. Note that $m_1 \leq m-1 < m$. Hence, by the hypothesis of induction, $\Gamma \Rightarrow \gamma$ is provable. This completes the proof. □

LEMMA 11.8. [KO00a] *If a formula α is not provable in* $\mathbf{FL_{ew}}$, *there exists a number $N(\geq 1)$ such that $\widetilde{m}(\alpha^N)$ is not provable in* $\mathbf{FL_{ew}}$ *for any $m \geq 1$.*

PROOF. Follows immediately from Lemma 11.7 by taking the sequent $\Rightarrow 0$ for $\Gamma \Rightarrow \gamma$, since $\Rightarrow 0$ is not provable in $\mathbf{SFL^+_{ew}}$. □

THEOREM 11.9. [KO00a] *Free FL_{ew}-algebras are semisimple.*

PROOF. By Lemma 11.8 and Corollary 11.5, using algebraizability of $\mathbf{FL_{ew}}$. □

11.6. Inside $\mathsf{FL_{ew}}$ semisimplicity implies discriminator: outline

Two sequences of subvarieties of $\mathsf{FL_{ew}}$ will be of importance in the remaining part of this chapter. The varieties $\mathsf{E_n}$, defined for any $n \in \mathbb{N}$, by the identity:

$$x^n = x^{n+1} \tag{E_n}$$

and the varieties $\mathsf{EM_n}$, defined for any $n \in \mathbb{N}$, by the identity:

$$x \vee \neg x^n = 1. \tag{EM_n}$$

We have seen (cf. Theorem 11.2) that will show that the subvarieties of E_n are precisely those varieties of FL_{ew}-algebras that have EDPC. We will now show that the subvarieties of $\mathsf{EM_n}$ are precisely those varieties of FL_{ew}-algebras that are semisimple. This will in fact fall out of the proof of that a variety of FL_{ew}-algebras is semisimple if and only if it is a discriminator variety. Since the argument is a little convoluted we outline the strategy of proof here. A theorem announced in [BP82] and proved, as Corollary 3.4, in [BKP84] states that if $\mathcal{V}$ is a congruence permutable variety, then $\mathcal{V}$

is a discriminator variety if and only if $\mathcal{V}$ is semisimple and has EDPC. Therefore, in our setting, if we manage to show that semisimplicity forces EDPC, then we will have proved that semisimplicity is equivalent to being a discriminator variety, which is exactly what we are after. Since we know, by the first correspondence, that satisfying (E_n) for some $n \in \mathbb{N}$ is equivalent to EDPC, it will suffice to show that semisimplicity implies satisfying (E_n) for some $n \in \mathbb{N}$. This we will indeed do. The proof comes from [Kow04], but is a modification of an argument developed earlier (published later, in [KK06]) for Boolean algebras with operators.

11.7. A characterization of semisimple subvarieties of $\mathsf{FL_{ew}}$

We characterized semisimple FL_{ew}-*algebras* in Section 11.3. Characterizing semisimple *varieties* is a different proposition, for semisimple algebras can occur in non-semisimple varieties, as the example of free FL_{ew}-algebras shows.

From now on we will assume that $\mathcal{V}$ is a nontrivial semisimple subvariety of $\mathsf{FL_{ew}}$. Consider an algebra $\mathbf{A}$ from $\mathcal{V}$ and an element $a \in A \setminus \{1\}$ such that $a^n > 0$ for all $n \in \mathbb{N}$. Note that such an algebra and element can always be found. For instance, we can take $\mathbf{A}$ to be the free algebra in $\mathcal{V}$ freely generated by a.

11.7.1. Finding subdirectly irreducibles. Take an algebra $\mathbf{A}$ and an element $a \in A$ as above. Put $\alpha = \mathrm{Cg}^{\mathbf{A}}(a, 1)$. By assumptions on $\mathbf{A}$, the congruence α is non-zero and non-full. Since α is principal, there must be a congruence β with $\beta \prec \alpha$.

LEMMA 11.10. [Kow05] *There is a positive integer m such that:*

(1) $a^{m+1} \equiv_\beta a^m$,
(2) $\neg a^m \equiv_\beta (\neg a^m)^2$,
(3) $a^m \equiv_\beta \neg\neg a^m$.

PROOF. Consider the set $\Gamma = \{\theta \in \mathrm{Con}\mathbf{A} \colon \theta \geq \beta,\ \theta \not\geq \alpha\}$. Observe that if $\Gamma = \{\beta\}$, then $\mathbf{A}/\beta$ is subdirectly irreducible but not simple, and thus we must have a $\theta \neq \beta,\ \theta \in \Gamma$. By congruence distributivity, γ, defined as $\bigvee \Gamma$, is a member of Γ. Therefore, $\mathbf{A}/\gamma$ is subdirectly irreducible; hence simple. From this and congruence permutability it follows that $\alpha \circ \gamma = 1$. Thus, $(1, 0) \in \alpha \circ \gamma$, and there must be a $c \in A$ with $(1, c) \in \alpha$ and $(c, 0) \in \gamma$; hence also $(\neg c, 1) \in \gamma$. Now, $(1, c) \in \alpha$ if and only if for some $m \in \mathbb{N}$ we have $a^m \leq c$. Thus, $\neg a^m \geq \neg c$ and therefore $(\neg a^m, 1) \in \gamma$ as well. We can then assume $c = a^m$. By definition we have $\alpha \cap \gamma = \beta$, from which by 1-regularity we obtain $1/\alpha \cap 1/\gamma = 1/\beta$.

To prove (1) consider $a^{m+1} \vee \neg a^m$. By the previous paragraph we have $(\neg a^m, 1) \in \gamma$ and thus $\neg a^m \in 1/\gamma$. Also, $a^{m+1} \in 1/\alpha$. It follows that $a^{m+1} \vee \neg a^m$ belongs to $1/\alpha \cap 1/\gamma = 1/\beta$ and thus we obtain $a^{m+1} \vee \neg a^m \equiv_\beta 1$.

Therefore, $(a^{m+1} \vee \neg a^m)a^m \equiv_\beta a^m$ as well. Then, distributing the left-hand side yields $a^{2m+1} \vee (\neg a^m)a^m \equiv_\beta a^m$, and further $a^{2m+1} \equiv_\beta a^m$. Since $a^{2m+1} \leq a^{m+1} \leq a^m$, we obtain $a^{m+1} \equiv_\beta a^m$ as needed.

For (2) take $a^m \vee (\neg a^m)^2$. This belongs to $1/\beta$, too, and therefore $a^m \vee (\neg a^m)^2 \equiv_\beta 1$. As before, we multiply both sides by the same factor to get $(a^m \vee (\neg a^m)^2)(\neg a^m) \equiv_\beta \neg a^m$. After distributing the left-hand side, we obtain $(\neg a^m)^3 \equiv_\beta \neg a^m$, from which the required $(\neg a^m)^2 \equiv_\beta \neg a^m$ follows.

Finally, for (3) the argument is a little less straightforward. To begin with, we show that a^m/β and $\neg a^m/\beta$ are lattice complements in $\mathbf{A}/\beta$. We need to prove that $a^m \vee \neg a^m \equiv_\beta 1$ and $a^m \wedge \neg a^m \equiv_\beta 0$. The former is clear. For the latter take $\neg(a^m \vee \neg a^m) \equiv_\beta \neg 1 = 0$; from this we get $\neg a^m \wedge \neg\neg a^m \equiv_\beta 0$. Now, as $\neg\neg a^m \geq a^m$ we obtain $\neg a^m \wedge a^m \equiv_\beta 0$ as needed. Secondly, we prove that if c and d are complements, then $\neg c = d$. Since $dc \leq d \wedge c = 0$, by residuation we have $d \leq c \to 0 = \neg c$. For the other direction take $\neg c = (\neg c)1 = (\neg c)(c \vee d) = (\neg c)c \vee (\neg c)d = (\neg c)d$. Thus, we have $(\neg c)d = \neg c \geq \neg c \wedge d$, but on the other hand $(\neg c)d \leq \neg c \wedge d$. Hence, $\neg c = (\neg c)d = \neg c \wedge d$, which means that $d \geq \neg c$. Therefore $\neg c = d$. Now taking a^m/β for d and $\neg a^m/\beta$ for c we obtain $\neg\neg a^m/\beta = a^m/\beta$, which yields the desired $a^m \equiv_\beta \neg\neg a^m$. $\square$

11.7.2. A necessary condition for semisimplicity. We will exhibit an equational property that all semisimple *varieties* of FL_{ew}-algebras possess. Consider the following condition on $\mathcal{V}$:

($\star$) For every $k \in \mathbb{N}$ there are $r, l \in \mathbb{N}$ such that $\mathcal{V} \models x \geq (\neg(\neg x^r)^k)^l$.

Certainly, for an arbitrary subvariety of $\mathsf{FL_{ew}}$ the condition ($\star$) may or may not hold. It turns out however that for semisimple subvarieties ($\star$) always holds. Notice the similarity of ($\star$) to the characterization of semisimple *algebras* in Section 11.3. Despite the similarity, a direct proof of ($\star$) from that characterization has eluded us. We have a different proof, but before we present it we need a certain technicality.

Suppose that $\mathcal{V}$ falsifies ($\star$). Then, there is a $k \in \mathbb{N}$ such that for all $r, l \in \mathbb{N}$ our variety $\mathcal{V}$ falsifies $x \geq (\neg(\neg x^r)^k)^l$. Let K be the smallest such k; note that for all $k' \geq K$ the variety $\mathcal{V}$ also falsifies $x \geq (\neg(\neg x^r)^{k'})^l$.

Let $\mathbf{F}$ be the free algebra in $\mathcal{V}$ on one free generator x. Thus, in $\mathbf{F}$, we have $x \not\geq (\neg(\neg x^r)^K)^l$, for all $r, l \in \mathbb{N}$. For each $r \in \mathbb{N}$ define $\theta_r = \mathrm{Cg}^{\mathbf{F}}(\neg(\neg x^r)^K, 1)$. Since $\neg(\neg x^r)^K \geq \neg(\neg x^{r+1})^K$, the family of congruences $\{\theta_r \colon r \in \mathbb{N}\}$ forms an increasing chain. Put $\alpha = \mathrm{Cg}^{\mathbf{F}}(x, 1)$ and $\Theta = \bigvee_{r \in \mathbb{N}} \theta_r$.

LEMMA 11.11. [Kow05] *The congruence Θ lies strictly between 0 and α.*

PROOF. If $\Theta = 0$, then, since the reasoning takes place in the one-generated free algebra in $\mathcal{V}$, we get that $\mathcal{V} \models \neg(\neg x^r)^K = 1$, for all $r \in \mathbb{N}$. In particular, substituting 0 for x we obtain $\mathcal{V} \models \neg(\neg 0^r)^K = 1$. This means $\mathcal{V} \models 0 = 1$, contradicting the nontriviality assumption.

To see that $\Theta \leq \alpha$ notice that $(x, 1) \in \alpha$ by definition; therefore, for any $r \in \mathbb{N}$ we have $x^r \equiv_\alpha 1$ and then it is easy to see that $(\neg(\neg x^r)^K, 1) \in \alpha$. This means that the pair generating θ_r belongs to α and thus $\theta_r \leq \alpha$ for any $r \in \mathbb{N}$. Hence, $\Theta = \bigvee_{r\in\mathbb{N}} \theta_r \leq \alpha$. Since α is principal, $\Theta = \alpha$ if and only if there is a finite set $S \subseteq \{\theta_r : r \in \mathbb{N}\}$ with $\bigvee S = \alpha$. However, as $\{\theta_r : r \in \mathbb{N}\}$ is an increasing chain, the existence of such an S amounts to the existence of an $r \in \mathbb{N}$ such that $\theta_r = \alpha$. Now, both θ_r and α are principal, and thus $\theta_r = \alpha$ if and only if there is an $l \in \mathbb{N}$ with $x \geq (\neg(\neg x^r)^K)^l$. Since $\mathbf{F}$ is free, it follows that $\mathcal{V}$ satisfies $x \geq (\neg(\neg x^r)^K)^l$. This contradicts the assumption that $\mathcal{V}$ falsifies $(\star)$. □

LEMMA 11.12. [Kow05] *The variety $\mathcal{V}$ satisfies $(\star)$.*

PROOF. Again, we proceed by *reductio.* Suppose $\mathcal{V}$ falsifies $(\star)$ and let K be the smallest number witnessing that, precisely as in Lemma 11.11. In view of that lemma, we can choose a congruence β such that $\Theta \leq \beta \prec \alpha$. Then, by Lemma 11.10, we obtain $\neg a^m \equiv_\beta (\neg a^m)^2$, for some $m \in \mathbb{N}$. Thus, $\neg a^m \equiv_\beta (\neg a^m)^K$ as well, and therefore, using Lemma 11.10 again, we obtain $a^m \equiv_\beta \neg\neg a^m \equiv_\beta \neg(\neg a^m)^K \equiv_{\theta_m} 1$. Since $\theta_m \leq \Theta \leq \beta$, this yields $a^m \equiv_\beta 1$. Therefore, also $a \equiv_\beta 1$ and thus $\beta \geq \alpha$ contradicting the choice of β. □

As our variety $\mathcal{V}$ apart from being semisimple is completely arbitrary, Lemma 11.12 states that all semisimple subvarieties of $\mathsf{FL_{ew}}$ satisfy $(\star)$. We have not been able to determine whether the converse is true, although we know that no finitely generated variety can be a counterexample. More precisely, the following is the case.

PROPOSITION 11.13. [Kow05] *If* $\mathbf{A}$ *is a finite subdirectly irreducible FL_{ew}-algebra satisfying $(\star)$, then* $\mathbf{A}$ *is simple.*

PROOF. Suppose for contradiction that $\mathbf{A}$ is subdirectly irreducible but not simple. By finiteness, the monolithic deductive filter F of $\mathbf{A}$ is generated by a coatom c, which is the only coatom in A. Since $\mathbf{A}$ is not simple, we have $c^{n+1} = c^n > 0$, for some positive integer n. Put $b = c^n$. Now, since $\mathbf{A}$ satisfies $(\star)$ there are positive integers r and l such that $x \geq (\neg(\neg x^r)^{n+1})^l$ holds. Applying it to b we get $b \geq (\neg(\neg b)^{n+1})^l$ since b is idempotent. Now observe that $c^n \cdot \neg b = b \cdot \neg b = 0$ and as $b > 0$ and c is the unique coatom we have $\neg b \leq c$. Therefore $(\neg b)^n \cdot \neg b \leq c^n \cdot \neg b = 0$. Thus, $(\neg b)^{n+1} = 0$ and so $(\neg(\neg b)^{n+1}) = 1$, which yields $b \geq (\neg(\neg b)^{n+1})^l = 1$. This is a contradiction. □

11.8. Semisimplicity forces discriminator

We now begin working toward the proof that all semisimple varieties of FL_{ew}-algebras are discriminator varieties.

11.8.1. An ultraproduct construction. From now on we will assume that our $\mathcal{V} \subseteq \mathsf{FL_{ew}}$ satisfies $(\star)$. Define a function $r: \mathbb{N} \to \mathbb{N}$ by taking $r(i)$ to be the smallest number such that there exists an $l \in \mathbb{N}$ with $\mathcal{V} \models x \geq (\neg(\neg x^{r(i)})^i)^l$.

LEMMA 11.14. [Kow05] *The function r is nondecreasing.*

PROOF. Suppose the contrary. Then, for a certain $i \in \mathbb{N}$, we have $r(i) > r(i+1)$. We have $(\neg x^{r(i+1)})^{i+1} \leq (\neg x^{r(i+1)})^i$ and thus $\neg(\neg x^{r(i+1)})^{i+1} \geq \neg(\neg x^{r(i+1)})^i$. Hence, by definition of the function r there is an $l \in \mathbb{N}$ such that $x \geq (\neg(\neg x^{r(i+1)})^{i+1})^l$ holds. Taking the l-th "power" of both sides of the previous inequality, we obtain $(\neg(\neg x^{r(i+1)})^{i+1})^l \geq (\neg(\neg x^{r(i+1)})^i)^l$, and this in turn yields $x \geq (\neg(\neg x^{r(i+1)})^i)^l$. That however is a contradiction, since $r(i)$ is by definition the smallest number for which a suitable l exists, yet $r(i+1)$ is strictly smaller. □

Now, define another function $l: \mathbb{N} \to \mathbb{N}$, this time taking $l(i)$ to be the smallest number such that $\mathcal{V} \models x \geq (\neg(\neg x^{r(i)})^i)^{l(i)}$. So defined, l depends on i via r.

LEMMA 11.15. [Kow05] *If $\mathcal{V}$ falsifies $x^{n+1} = x^n$ for all $n \in \mathbb{N}$, then, for each $i > 0$ there is a simple algebra $\mathbf{A}_i \in \mathcal{V}$ and element $a_i \in A_i$ such that $a_i^{r(i)} > 0$ but $a_i^{2r(i)} = 0$.*

PROOF. Suppose otherwise. Then, there is a $j > 0$ such that for each simple algebra $\mathbf{A} \in \mathcal{V}$ and each element $a \in A$ we have that $a^{2r(j)} = 0$ implies $a^{r(j)} = 0$. Take any $b < 1$. As $\mathbf{A}$ is simple we must have a $k \in \mathbb{N}$ with $b^k = 0 < b^{k-1}$. We will show that in fact $b^{r(j)} = 0$. If $k \leq 2r(j)$ this is clear. Suppose $k > 2r(j)$; then let s be the smallest integer such that $2s \cdot r(j) \geq k > k-1 \geq s \cdot r(j)$. Thus, $b^{2s \cdot r(j)} \leq b^k = 0 < b^{k-1} \leq b^{s \cdot r(j)}$ and it follows that $(b^s)^{2r(j)} = 0 < (b^s)^{r(j)}$. However, as the implication above is universally quantified, $(b^s)^{2r(j)} = 0$ forces $(b^s)^{r(j)} = 0$ which is a contradiction. Thus, we obtain that all simple algebras in $\mathcal{V}$ satisfy $b^{r(j)+1} = b^{r(j)} = 0$ for all non-unit elements. Since $1^{r(j)+1} = 1^{r(j)}$ trivially holds, it follows that $\mathcal{V}$ satisfies $x^{r(j)+1} = x^{r(j)}$ contrary to the assumption. □

Suppose that $\mathcal{V}$ indeed falsifies $x^{n+1} = x^n$ for all $n \in \mathbb{N}$. Therefore, for each $i \in \mathbb{N}$, we can produce an algebra $\mathbf{A}_i \in \mathcal{V}$ and an element $a_i \in A_i$ as in the lemma above. Put $b_i = \neg a_i^{r(i)}$ and let $f(i)$ be the function $l(i) \cdot 4r(i)$.

LEMMA 11.16. [Kow05] *For every positive integer k, the following holds: $b_i^k > 0$ and $(\neg b_i^k)^{f(k)} = 0$, for every $i \geq k$.*

PROOF. To prove the first statement suppose for contradiction that $b_i^k = 0$. This forces $b_i^i = 0$ as $k \leq i$. By definition of b_i then, we get $b_i^i = (\neg a_i^{r(i)})^i = 0$ and thus $\neg(\neg a_i^{r(i)})^i = 1$. Therefore $(\neg(\neg a_i^{r(i)})^i)^l = 1$ for any $l \in \mathbb{N}$; hence $a_i \not\geq (\neg(\neg a_i^{r(i)})^i)^l$ for any $l \in \mathbb{N}$. This contradicts the definition of r.

For the second statement, we can reason as follows. We have $r(i) \geq \lfloor r(i)/r(k) \rfloor \cdot r(k)$; therefore,

$$a_i^{r(i)} \leq a_i^{\lfloor r(i)/r(k)\rfloor \cdot r(k)},$$
$$(\neg a_i^{r(i)})^k \geq (\neg a_i^{\lfloor r(i)/r(k)\rfloor \cdot r(k)})^k,$$
$$\left(\neg(\neg a_i^{r(i)})^k\right)^{f(k)} \leq \left(\neg(\neg a_i^{\lfloor r(i)/r(k)\rfloor \cdot r(k)})^k\right)^{f(k)}.$$

On the other hand, from $(\neg(\neg x^{r(k)})^k)^{l(k)} \leq x$ we derive:

$$\left(\neg(\neg a_i^{\lfloor r(i)/r(k)\rfloor \cdot r(k)})^k\right)^{l(k)} \leq a_i^{\lfloor r(i)/r(k)\rfloor},$$
$$\left(\neg(\neg a_i^{\lfloor r(i)/r(k)\rfloor \cdot r(k)})^k\right)^{l(k)\cdot 4r(k)} \leq a_i^{\lfloor r(i)/r(k)\rfloor \cdot 4r(k)},$$
$$\left(\neg(\neg a_i^{\lfloor r(i)/r(k)\rfloor \cdot r(k)})^k\right)^{f(k)} \leq a_i^{\lfloor r(i)/r(k)\rfloor \cdot 4r(k)}.$$

By Lemma 11.14 the function r is nondecreasing, so we have $\lfloor r(i)/r(k)\rfloor \geq 1$ and therefore:

$$\begin{aligned}
\lfloor r(i)/r(k)\rfloor \cdot 4r(k) &= (\lfloor r(i)/r(k)\rfloor + \lfloor r(i)/r(k)\rfloor) \cdot 2r(k)\\
&\geq (\lfloor r(i)/r(k)\rfloor + 1) \cdot 2r(k)\\
&\geq \lceil r(i)/r(k)\rceil \cdot 2r(k)\\
&\geq (r(i)/r(k)) \cdot 2r(k)\\
&= 2r(i).
\end{aligned}$$

We obtain $a_i^{\lfloor r(i)/r(k)\rfloor \cdot 4r(k)} \leq a_i^{2r(i)} = 0$, and then putting all the inequalities together we finally get:

$$(\neg b_i^k)^{f(k)} = \left(\neg(\neg a_i^{r(i)})^k\right)^{f(k)} \leq a_i^{\lfloor r(i)/r(k)\rfloor \cdot 4r(k)} \leq a_i^{2r(i)} = 0,$$

as required by the claim. □

Now, let $\mathbf{B}$ be the ultraproduct $\prod_{i\in\mathbb{N}} \mathbf{A}_i/U$, for a nonprincipal ultrafilter U over $\mathbb{N}$. Consider the element $b = \langle b_i : i \in \mathbb{N}\rangle/U$. The next lemma follows from Lemma 11.16 by properties of ultraproducts.

LEMMA 11.17. [Kow05] *For any* $k \in \mathbb{N}$, *we have* $\mathbf{B} \models b^k > 0$ *and* $\mathbf{B} \models (\neg b^k)^{f(k)} = 0$.

11.8.2. Semisimplicity forces n-potency. Let us recall our assumptions briefly. We are working with a nontrivial, semisimple subvariety $\mathcal{V}$ of $\mathsf{FL_{ew}}$; by Lemma 11.12, our $\mathcal{V}$ satisfies $(\star)$.

LEMMA 11.18. [Kow05] *The variety* $\mathcal{V}$ *satisfies* $x^{n+1} = x^n$ *for some* $n \in \mathbb{N}$.

PROOF. Suppose $\mathcal{V}$ falsifies $x^{n+1} = x^n$ for all $n \in \mathbb{N}$. Given that, the previous section shows how to produce an algebra $\mathbf{B} \in \mathcal{V}$ and an element $b \in B$ such that for all $k \in \mathbb{N}$ it satisfies $b^k > 0$ and $(\neg b^k)^{f(k)} = 0$. Then, putting $\alpha = \mathrm{Cg}^{\mathbf{B}}(b, 1)$ and taking β and γ as in our setup of congruences, by Lemma 11.10 we get that there is an $m \in \mathbb{N}$ for which $\neg b^m \equiv_\beta (\neg b^m)^2 \equiv_\beta$.

Therefore, $\neg b^m \equiv_\beta (\neg b^m)^{f(m)} = 0$, and thus $b^m \equiv_\beta 1$. As $b^m \leq b$, it follows that $b \equiv_\beta 1$, which forces $\beta \geq \alpha$. This contradicts the choice of β as a subcover of α. □

THEOREM 11.19. [Kow05] *For a variety $\mathcal{V}$ of FL_{ew}-algebras the following conditions are equivalent:*

(1) *$\mathcal{V}$ verifies $x \vee \neg x^n = 1$, for some $n \in \mathbb{N}$;*
(2) *$\mathcal{V}$ is semisimple;*
(3) *$\mathcal{V}$ is a discriminator variety.*

PROOF. It is well-known that (3) implies (2). That (1) implies (3) is easy to see as follows. From (1) we get that in any subdirectly irreducible algebra $\mathbf{A} \in \mathcal{V}$ and any $a \in A$, either $a = 1$ or else $a^n = 0$. Then, it is easily checked that the term $t(x, y, z) = ((x \leftrightarrow y)^n \wedge z) \vee (\neg(x \leftrightarrow y)^n \wedge x)$ is a discriminator term for $\mathcal{V}$.

It remains to show that (2) implies (1). Take any simple algebra $\mathbf{A}$ from $\mathcal{V}$ and an element $a \in A$ with $a < 1$. By Lemma 11 18, there is an $n \in \mathbb{N}$ with $a^{n+1} = a^n$. If $a^n > 0$, then $\{x \in A \colon x \geq a^n\}$, being a filter closed under multiplication, defines a nontrivial and non-full congruence on $\mathbf{A}$, which is impossible since $\mathbf{A}$ is simple. Thus, $a^n = 0$ and as $\mathbf{A}$ and a have been chosen arbitrarily, we obtain that $\mathcal{V}$ verifies $x \vee \neg x^n = 1$ as required. □

Theorem 11.19 answers (positively) Question 1.1 from [KO01]. Furthermore, logics over $\mathbf{FL_{ew}}$ can be naturally classified according to whether they are n-contractive, or satisfy iterated excluded middle, for some $n \in \mathbb{N}$. See [Ono01, OU03] for some attempts in this direction. Theorems 3.55 and 11.19 yield algebraic counterparts of these notions.

Exercises

(1) Prove that if $\mathbf{L}$ satisfies iterated excluded middle then it is n-contractive for some $n \in \mathbb{N}$. Use algebraic counterparts.
(2) Prove that the unary discriminator implies the ternary one (Lemma 11.1).
(3) Prove that discriminator varieties satisfy CEP. Hint: otherwise the variety would contain a simple algebra with a non-simple subalgebra.
(4) Prove that a deductive filter F in a FL_{ew}-algebra is maximal iff for each $u \in A$ either $u \in F$ or there is a $k \geq 1$ such that $\neg u^k \in F$.
(5) Let $a + b = \neg(\neg a \cdot \neg b)$ (cf. Section 11.3). Prove that $+$ is commutative. Prove that $+$ is associative in involutive FL_{ew}-algebras. Find an example of a non-involutive algebra where $+$ is associative.
(6) A proof theoretical refresher. Prove cut elimination for $\mathbf{SFL_{ew}}$ along the lines suggested in Section 11.4.

(7) A universal algebra refresher. Prove that a compact element in an algebraic lattice must have a subcover. Conclude that a compact (in particular, principal) congruence must have a subcover.

(8) Another. Let $\mathbf{A}$ be congruence distributive. Let $\beta \prec \alpha$ in $\mathbf{Con\,A}$. Take $\Gamma = \{\theta \in \mathrm{Con}\,\mathbf{A} \colon \theta \geq \beta,\ \theta \not\geq \alpha\}$. Put $\gamma = \bigvee \Gamma$. Prove that $\gamma \in \Gamma$. Conclude that $\mathbf{A}/\gamma$ is subdirectly irreducible. What is its monolith?

(9) Prove Lemma 11.12 using Corollary 11.5 instead of Lemma 11.10. Hint: use the congruences Θ, α and β as defined prior to Lemma 11.12 and find a quotient of $\mathbf{F}$ that should be simple but cannot.

(10) *Open problem*: Generalize the proof of the fact that semisimple subvarieties are precisely discriminator subvarieties to some supervarieties of $\mathsf{FL_{ew}}$. Warning: do not try noncommutative case.

Notes

(1) The proof of semisimplicity of free algebras presented here seems to have quite some potential for generalizations. For example, Takamura proved in [Tak03], using the same technique, that free FL_w-algebras are semisimple; he also extended the result in different direction, to free *modal* FL_{ew}-algebras (cf. [Tak06])

(2) Applicability of our technique to subvarieties of $\mathsf{FL_{ew}}$ may, however, be rather limited, as it relies on cut-elimination. For example, it is known that every free MV-algebra is semisimple (see Chapter 3 of [CDM00]), but the proof of this fact employs a different technique, relying on much more detailed description of free MV-algebras by means of McNaughton functions.

(3) The proof of the fact that semisimple subvarieties of $\mathsf{FL_{ew}}$ are discriminator varieties can also be generalized. The essential properties in the proof are n-potency (for EDPC) and zero-boundedness (for the setup of congruences in Lemma 11.10), but integrality is not used. However, any reasonable generalizations must stay within the commutative case. This is because it is easy to come up with finite simple noncommutative residuated lattices falsifying CEP (cf. Sections 3.6.5 and 3.6.6). The varieties generated by these algebras are semisimple, but cannot be discriminator varieties (see Exercise 3).

(4) It is shown in [KK06], by essentially the same argument, that all semisimple varieties of Boolean algebras with operators of finite type are discriminator varieties.

Bibliography

[Abr90] V. M. Abrusci, *Sequent calculus for intuitionistic linear propositional logic*, Mathematical Logic (P. P. Petkov, ed.), Plenum Press, New York, 1990, pp. 223–242.

[Agl96] P. Agliano, *An algebraic investigation of linear logic*, Universitá di Siena, 1996.

[AFM99] P. Agliano, I. M. A. Ferreirim, and F. Montagna, *Basic hoops: An algebraic study of continuous t-norms* (1999), preprint.

[AM03] P. Agliano and F. Montagna, *Varieties of BL-algebras I: general properties*, Journal of Pure and Applied Algebras **181** (2003), 105–129.

[AB75] A. R. Anderson and N. D. Belnap, *Entailment. The logic of relevance and necessity*, Princeton University Press, 1975.

[ABD92] A. R. Anderson, Jr. Belnap N. D, and J. M. Dunn, *Entailment: The Logic of Relevance and Necessity II*, Princeton University Press, 1992.

[AF88] M. Anderson and T. Feil, *Lattice-Ordered Groups: an introduction*, D. Reidel, 1988.

[AGN97] H. Andréka, S. Givant, and I. Németi, *Decision Problems for Equational Theories of Relation Algebras*, Memoirs of the AMS, vol. 126, American Mathematical Society, Providence, Rhode Island, 1997.

[ANS01] H. Andréka, I. Németi, and I. Sain, *Algebraic Logic*, Handbook of Philosophical Logic, 2nd edition (D. Gabbay and F. Guenther, eds.), Vol. 2, Kluwer, Dordrecht, 2001.

[Avi01] J. Avigad, *Algebraic proofs of cut elimination*, Journal of Logic and Algorithmic Programming **49** (2001), 15–30.

[Avr88] A. Avron, *The semantics and proof theory of Linear Logic*, Theoretical Computer Science **57** (1988), 161–184.

[Bac75] P. D. Bacsish, *Amalgamation properties and interpolation theorems for equational theories*, Algebra Universalis **5** (1975), 45–55.

[BCG$^+$03] P. Bahls, J. Cole, N. Galatos, P. Jipsen, and C. Tsinakis, *Cancellative residuated lattices*, Algebra Universalis **50** (2003), no. 1, 83–106.

[BD74] R. Balbes and P. Dwinger, *Distributive Lattices*, University of Missouri Press, 1974.

[BO96] Bayu Surarso and H. Ono, *Cut-elimination in noncommutative substructural logics*, Reports on Mathematical Logic **30** (1996), 13–29.

[Bel02] F. Belardinelli, *Aspetti semantici delle logiche sottostrutturali*, Graduate Thesis, University of Pisa, 2002.

[BJO04] F. Belardinelli, P. Jipsen, and H. Ono, *Algebraic aspects of cut elimination*, Studia Logica **77** (2004), 209-240.

[BJ60] N. D. Belnap Jr., *Entailment and relevance*, Journal of Symbolic Logic **25** (1960), 144-146.

[BJW65] N. D. Belnap Jr. and J. R. Wallace, *A decision procedure for the system $E_{\overline{I}}$ of entailment with negation*, Zeitschrift für mathematische Logik und Grundlagen der Mathematik **11** (1965), 277–289.

[BH04] G Bezhanishvili and J Harding, *MacNeille completions of Heyting algebras*, Houston Journal of Mathematics **30** (2004), 937–952.

[Bir35] G. Birkhoff, *On the structure of abstract algebras* **31** (1935), 433–454.

[Bir44] G. Birkhoff, *Subdirect unions in universal algebra*, Bull. Amer. Math. Soc. **50** (1944), 764–768.

[Bir79] G. Birkhoff, *Lattice theory*, 3rd ed., American Mathematical Society Colloquium Publications, vol. 25, American Mathematical Society, Providence, R.I., 1979.

[Blo78] W. J. Blok, *On the degree of incompleteness in modal logic and the covering relation in the lattice of modal logics*, Technical Report 78–07, Department of Mathematics, University of Amsterdam, 1978.

[Blo99] W. J. Blok, *Varieties vs. quasivarieties in algebraic logic*, Algebra and Substructural Logics (Japan Advanced Institute of Science and Technology), November 10, 1999.

[BF93] W. J. Blok and I. M. A. Ferreirim, *Hoops and their implicational reducts (abstract)*, Algebraic Methods in Logic and Computer Science **28** (1993), 219–230.

[BF00] W. J. Blok and I. M. A. Ferreirim, *On the structure of hoops*, Algebra Universalis **43** (2000), 233–257.

[BJ06] W. J. Blok and B. Jonsson, *Equivalence of consequence operators*, Studia Logica **83** (2006), 91–110.

[BKP84] W. J. Blok, P. Köhler, and D. Pigozzi, *On the structure of varieties with equationally definable principal congruences II*, Algebra Universalis **18** (1984), 334–379.

[BP82] W. J. Blok and D. Pigozzi, *On the structure of varieties with equationally definable principal congruences I*, Algebra Universalis **15** (1982), 195–227.

[BP89] W. J. Blok and D. Pigozzi, *Algebraizable Logics*, Memoirs of the AMS, American Mathematical Society, Providence, Rhode Island, 1989.

[BP] W. J. Blok and D. Pigozzi, *Abstract algebraic logic and the deduction theorem*, The Bulletin of Symbolic Logic, to appear.

[BP94a] W. J. Blok and D. Pigozzi, *On the structure of varieties with equationally definable principal congruences III*, Algebra Universalis **32** (1994), 545–608.

[BP94b] W. J. Blok and D. Pigozzi, *On the structure of varieties with equationally definable principal congruences IV*, Algebra Universalis **31** (1994), 1–35.

[BR93] W. J. Blok and J. G. Raftery, *Failure of the congruence extension property in BCK-algebras and related structures*, Mathematica Japonica **38** (1993), 633–638.

[BR95] W. J. Blok and J. G. Raftery, *On the quasivariety of BCK-algebra and its subvarieties*, Algebra Univerisalis **33** (1995), 68–90.

[BR97] W. J. Blok and J. G. Raftery, *Varieties of commutative residuated integral pomonoids and their residuation subreducts*, Journal of Algebra **190** (1997), 280–328.

[BR04] W. J. Blok and J. G. Raftery, *Fragments of R-Mingle*, Studia Logica **78** (2004), 59-106.

[BvA02] W. J. Blok and C. J. van Alten, *The finite embeddability property for residuated lattices, pocrims and BCK-algebras*, Algebra Universalis **48** (2002), 253–271.

[BvA05] W. J. Blok and C. J. van Alten, *On the finite embeddability property for residuated ordered groupoids*, Trans. Amer. Math. Soc. **357** (2005), no. 10, 4141–4157.

[Blou99] K. Blount, *On the structure of residuated lattices*, Ph.D. thesis, Vanderbilt University, Nashville, Tennessee, 1999.

[BT03] K. Blount and C. Tsinakis, *The structure of residuated lattices*, International Journal of Algebra and Computation **13** (2003), no. 4, 437–461.

[Bly05] T. S. Blyth, *Lattices and Ordered Algebraic Structures*, Universitext, Springer, 2005.

[BJ72] T. S. Blyth and M. F. Janowitz, *Residuation Theory*, International Series of Monographs in Pure and Applied Mathematics, vol. 102, Pergamon Press, 1972.

[Boo48] G. Boole, *The mathematical analysis of logic; being an essay towards a calculus of deductuve reasoning*, B. Blackwell, Ocford, 1848.

[Boo54] G. Boole, *An investigation of the laws of thought on which are founded the mathematical theories of logic and probabilities*, Walton and Maberly, London, 1854.

[Bra90] R. T. Brady, *The Gentzenization and decidability of RW*, Journal of Philosophical Logic **19** (1990), 35–73.

[Bra91] R. T. Brady, *The Gentzenization and decidability of some contractionless relevant logics*, Journal of Philosophical Logic **20** (1991), 97–117.

[BS81] S. Burris and H. P. Sankappanavar, *A Course in Universal Algebra*, Graduate Texts in Mathematics, Springer, 1981, available online.

[Bus82] W. Buszkowski, *Some decision problems in the theory of syntactic categories*, Zeitschrift für mathematische Logik und Grundlagen der Mathematik **28** (1982), 539–548.

[Bus96] W. Buszkowski, *The finite model property for BCI and related systems*, Studia Logica **57** (1996), 303–323.

[Bus02] W. Buszkowski, *Finite models of some substructural logics*, Mathematical Logic Quarterly **48** (2002), no. 1, 63–72.

[Bus] W. Buszkowski, *On action logic: equational theories of action algebras*, Journal of Logic and Computation, to appear.

[BM02] W. Buszkowski and M. Moortgat, *Editorial introduction*, Studia Logica **71** (2002), 261–275.

[Cha92] A. Chagrov, *Continuality of the set of maximal superintuitionistic logics with the disjunction property*, Mathematical Notes **51** (1992), 188–193.

[CZ91] A. Chagrov and M. Zakharyaschev, *The disjunction property of intermediate propositional logics*, Studia Logica **50** (1991), 189–216.

[CZ93] A. Chagrov and M. Zakharyaschev, *The undecidability of the disjunction property of propositional logics and other related problems*, Journal of Symbolic Logic **58** (1993), 967–1002.

[CZ97] A. Chagrov and M. Zakharyaschev, *Modal Logic*, Clarendon Press, Oxford, 1997.

[Cha58] C. C. Chang, *Algebraic analysis of many-valued logics*, Transactions of the AMS **88** (1958), 467–490.

[Cha59] C. C. Chang, *A new proof of the completeness of the Łukasiewicz axioms*, Transactions of the AMS **93** (1959), 74–90.

[CK90] C. C. Chang and H. J. Keisler, *Model theory*, 3rd ed., Studies in Logic and the Foundations of Mathematics, vol. 73, North-Holland Publishing Co., Amsterdam, 1990.

[CT51] L. H. Chin and A. Tarski, *Distributive and modular laws in the arithmetic of relation algebras*, Univ. California Publ. Math. (N.S.) **1** (1951), 341–384.

[CT06] A. Ciabattoni and K. Terui, *Towards a semantic characterization of cut-elimination*, Studia Logica **82** (2006), no. 1, 95–119.

[CDM00] R. Cignoli, I. D'Ottaviano, and D. Mundici, *Algebraic Foundations of Many-valued Reasoning*, Trends in Logic, Kluwer, Dordrecht, 2000.

[CEGT00] R. Cignoli, F. Esteva, L. Godo, and A. Torrens, *Basic Fuzzy Logic is the logic of continuous t-norms and their residua*, Soft Computing (2000), no. 4, 106–112.

[CT00] R. Cignoli and A. Torrens, *An algebraic analysis of product logic*, Journal of Multi-Valued Logic and Soft Computing **5** (2000), no. 1, 45–65.

[CT03] R. Cignoli and A. Torrens, *Hájek basic fuzzy logic and Łukasievicz infinite-valued logic*, Arch. Math. Logic **42** (2003), 361–370.

[CT04] R. Cignoli and A. Torrens, *Glivenko like theorems in natural expansions of BCK-logic*, Math. Log. Quart. **50** (2004), no. 2, 111–125.

[Cin01] P. Cintula, *About axiomatic systems of product fuzzy logic*, Soft Computing **5** (2001), no. 3, 243–244.

[Cra57a] W. Craig, *Linear reasoning. a new form of the Herbrand-Gentzen theorem*, Journal of Symbolic Logic **22** (1957), 250–268.

[Cra57b] W. Craig, *Three uses of Herbrand-Gentzen theorem in relating model theory and proof theory*, Journal of Symbolic Logic **22** (1957), 269–285.

[Cur63] H. B. Curry, *Foundations of mathematical logic*, McGraw-Hill Book Co., Inc., New York, 1963.

[CF58] H. B. Curry and R. Feys, *Combinatory Logic I*, North Holland, Amsterdam, 1958.

[CHS72] H. B. Curry, R. Hindley, and G. Seldon, *Combinatory Logic II*, North Holland, 1972.

[Cze85] J. Czelakowski, *Algebraic aspects of deduction theorems*, Studia Logica **44** (1985), 369–387.

[Cze86] J. Czelakowski, *Local deduction theorems*, Studia Logica **45** (1986), 377–391.

[Cze01] J. Czelakowski, *Protoalgebraic logics*, Trends in Logic—Studia Logica Library, vol. 10, Kluwer Academic Publishers, Dordrecht, 2001.

[CD96] J. Czelakowski and W. Dziobiak, *The parameterized local deduction theorem for quasivarieties of algebras and its application*, Algebra Universalis **35** (1996), no. 3, 373–419.

[CP99] J. Czelakowski and Pigozzi, *Amalgamation and interpolation in abstract algebraic logic*, Models, Algebras, and Proofs (Bogota 1995) (X. Caicedo and C. H. Montenegro, ed.), Lecture Notes in Pure and Applied Mathematics, vol. 203, Marcel Dekker, Inc., 1999, pp. 187–265.

[CP04] J. Czelakowski and D. Pigozzi, *Fregean logics*, Ann. Pure Appl. Logic **127** (2004), no. 1-3, 17–76.

[Dar77] G. K. Dardžaniá, *Intuitionistic system without contraction*, Bulletin of the Section of Logic, Polish Academy of Sciences **6** (1977), 2–8.

[DP02] B. A. Davey and H. A. Priestley, *Introduction to lattices and order*, 2nd ed., Cambridge University Press, New York, 2002.

[Day72] A. Day, *Varieties of Heyting algebras. I, II* (1972), unpublished.

[DM56] A. De Morgan, *On the symbols of logic, the theory of the syllogism, and in particular of the copula, and the application of the theory of probabilities to some questions of the theory of evidence*, Transactions of the Cambridge Philosophical Society **9** (1856), 79–127.

[DM64] A. De Morgan, *On the syllogism III, IV*, Transactions of the Cambridge Philosophical Society **10** (1864), 173–230, 331–358.

[DNGI02] A. Di Nola, G. Georgescu, and A. Iorgulescu, *Pseudo-BL algebras I*, Mult. Val. Log. **8** (2002), no. 5-6, 673–714.

[Dil39] R. P. Dilworth, *Non-commutative residuated lattices*, Transactions of the American Mathematical Society **46** (1939), 426–444.

[Doš88] K. Došen, *Sequent systems and groupoid models I*, Studia Logica **47** (1988), 353–385.

[Doš89] K. Došen, *Sequent systems and groupoid models II*, Studia Logica **48** (1989), 41–65.

[Doš93] K. Došen, *A historical introduction to substructural logics*, Substructural Logics (K. Došen and P. Schroeder-Heister, eds.), Oxford Univ. Press, 1993, pp. 1–30.

[Dum59] M. Dummett, *A propositional calculus with denumerable matrix*, Journal of Symbolic Logic **24** (1959), 97–106.

[Dun86] J. M. Dunn, *Relevance Logic and Entailment*, Handbook of Philosophical Logic, 2nd edition (D. Gabbay and F. Guenther, eds.), Vol. III, Reidel, Dordrecht, 1986.

[Dun93] J. M. Dunn, *Partial Gaggles Applied to Logics with Restricted Structural Rules*, Substructural Logics (D. Schroeder-Heister, ed.), Oxford University Press, 1993, pp. 63–108.

[DGP05] J. M. Dunn, M. Gehrke, and A. Palmigiano, *Canonical extensions and relational completeness of some substructural logics*, J. Symbolic Logic **70** (2005), no. 3, 713–740.

[DH01] J. M. Dunn and G Hardegree, *Algebraic Methods in Philosophical Logic*, Oxford Logic Guides, vol. 41, Oxford University Press, 2001.

[DR02] J. M. Dunn and G. Restall, *Relevance Logic*, Handbook of Philosophical Logic, 2nd edition (D. Gabbay and F. Guenther, eds.), Vol. 6, Kluwer, Dordrecht, 2002, pp. 1-128.

[Dvu02] A. Dvurečenskij, *Pseudo MV-algebras are intervals in l-groups*, J. Aust. Math. Soc. **72** (2002), no. 3, 427–445.

[Dyc92] R. Dyckhoff, *Contraction-free sequent calculi for intuitionistic logic*, Journal of Symbolic Logic **57** (1992), 795–807.

[Dzi83] W. Dziobiak, *There are $2^{\aleph_0}$ logics with the relevance principle between R and RM*, Studia Logica **42** (1983), no. 1, 49–61.

[EG01] F. Esteva and L. Godo, *Monoidal t-norm based logic: towards a logic for left-continuous t-norms*, Fuzzy Sets and Systems **124** (2001), no. 3, 271–288.

[EGGC03] F. Esteva, L. Godo, and À. García-Cerdaña, *On the hierachy of t-norm based residuated fuzzy logics*, Beyond two: theory and applications of multiple-valued logic, Physica Verlag, 2003, pp. 251-272.

[Eva69] T. Evans, *Some connections between residual finiteness, finite embeddability and the word problem*, Journal of London Mathematics Society **2** (1969), no. 1, 399–403.

[Fer92] I. M. A. Ferreirim, *On varieties and quasivarieties of hoops and their reducts*, Ph.D. thesis, Vanderbilt University, Nashville, Tennessee, 1992.

[Fin74] K. Fine, *An ascending chain of S4 logics*, Theoria **40** (1974), 110–116.

[Fit73] M. Fitting, *Model existence theorems for modal and intuitionistic logics*, Journal of Symbolic Logic **38** (1973), 613–627.

[Fit83] M. C. Fitting, *Proof Methods for Modal and Intuitionistic Logics*, D.Reidel Publishing Co., Dordrecht, 1983.

[FGI01] P. Flondor, G. Georgescu, and A. Iorgulescu, *Pseudo-t-norms and pseudo-BL-algebras*, Soft Computing **5** (2001), 355–371.

[FJP03] J. M. Font, R. Jansana, and D. Pigozzi, *A survey of abstract algebraic logic*, Studia Logica **74** (2003), no. 1/2, 13–97.

[FRT84] J. M. Font, A. J. Rodríguez, and A. Torrens, *Wajsberg algebras*, Stochastica **8** (1984), 5–31.

[Fre80] R. Freese, *Free modular lattices*, Transactions of the AMS **261** (1980), 81–91.

[Fre79] G. Frege, *Begriffschrift, eine der arithmetischen nachgebildete Formalsprache des reinen Denkens*, L. Nebert, Halle, 1979.

[Fuc63] L. Fuchs, *Partially Ordered Algebraic Systems*, Pergamon Press, Oxford, 1963.

[GM05] D. M. Gabbay and L. L. Maksimova, *Interpolation and Definability: Modal and Intuitionistic Logics*, Oxford Logic Guides, vol. 46, Oxford University Press, 2005.

[Gal88] G. I. Galanter, *Halldén-completeness for superintuitionistic logics (Russian)*, Proceedings of IV Soviet-Finland Symposium for Mathematical Logic (1988), 81–89.

[Gal02] N. Galatos, *The undecidability of the word problem for distributive residuated lattices*, Ordered algebraic structures (J. Martinez, ed.), Kluwer, Dordrecht, 2002, pp. 231–243.

[Gal03] N. Galatos, *Varieties of residuated lattices*, Ph.D. thesis, Vanderbilt University, Nashville, Tennessee, 2003.

[Gal04] N. Galatos, *Equational bases for joins of residuated-lattice varieties*, Studia Logica **76** (2004), no. 2, 227–240.

[Gal05] N. Galatos, *Minimal varieties of residuated lattices*, Algebra Universalis **52** (2005), no. 2, 215–239.

[GJ] N. Galatos and P. Jipsen, *Residuated frames with applications to decidability.* manuscript.

[GO] N. Galatos and H. Ono, *Cut elimination and strong separation for substructural logics: an algebraic approach.* manuscript.

[GO06a] N. Galatos and H. Ono, *Algebraization, parametrized local deduction theorem and interpolation for substructural logics over* **FL**, Studia Logica **83** (2006), 279–308.

[GO06b] N. Galatos and H. Ono, *Glivenko theorems for substructural logics over* **FL**, Journal of Symbolic Logic **71** (2006), 1353–1384.

[GR04] N. Galatos and J. G. Raftery, *Adding Involution to Residuated Structures*, Studia Logica **77** (2004), no. 2, 181–207.

[GT] N. Galatos and C. Tsinakis, *Equivalence of closure operators: an order-theoretic and categorical perspective.* manuscript.

[GT05] N. Galatos and C. Tsinakis, *Generalized MV-algebras*, Journal of Algebra **283** (2005), no. 1, 254–291.

[GH01] M. Gehrke and J. Harding, *Bounded lattice expansions*, Journal of Algebra **238** (2001), no. 1, 345–371.

[GHV06] M. Gehrke, J. Harding, and Y. Venema, *MacNeille completions and canonical extensions*, Transactions of the American Mathematical Society **358** (2006), 573–590.

[GJ04] M. Gehrke and B. Jónsson, *Bounded distributive lattice expansions*, Math. Scand. **94** (2004), no. 1, 13–45.

[Gen35] G. Gentzen, *Untersuchungen über das logische Schließen I, II*, Mathematische Zeitschrift **39** (1935), no. 1, 176–210, 405–431.

[GI01] G. Georgescu and A. Iorgulescu, *Pseudo-MV algebras*, Mult.-Valued Log. **6** (2001), no. 1-2, 95–135.

[Gia85] S. Giambrone, *TW_+ and RW_+ are undecidable*, Journal of Philosophical Logic **14** (1985), 235–254.

[Gir87a] J.-Y. Girard, *Linear logic*, Theoretical Computer Science **50** (1987), 1–102.

[Gir87b] J.-Y. Girard, *Proof Theory and Logical Complexity*, Studies in Proof Theory, vol. I, Bibliopolis, 1987.

[GG83] *The word problem for lattice-ordered groups*, Trans. Amer. Math. Soc. **280** (1983), no. 1, 127–138.

[GH89] A. M. W. Glass and W. C. Holland (eds.), *Lattice-Ordered Groups*, Kluwer Academic Publishers, 1989.

[Gli29] V. Glivenko, *Sur quelques points de la logique de M Brouwer*, Bulletin Academie des Sciences de Belgique **15** (1929), 183–188.

[Göd33] K. Gödel, *Eine Interpretation des intuitionistischen Aussagenkalküls*, Erg. Math. Kolloqu. **4** (1933), 39–40.

[Gol00] R. Goldblatt, *Algebraic polymodal logic: a survey*, Log. J. IGPL **8** (2000), no. 4, 393–450.

[Gore98] R. Goré, *Substructural logics on display*, Logic Journal of IGPL **6** (1998), no. 3, 451–504.

[Gor98] V. A. Gorbunov, *Algebraic theory of quasivarieties*, Siberian School of Algebra and Logic, Kluwer, Dordrecht, 1998.

[Grä98] G. Grätzer, *General lattice theory*, 2nd ed., Birkhäuser Verlag, Basel, 1998. New appendices by the author with B. A. Davey, R. Freese, B. Ganter, M. Greferath, P. Jipsen, H. A. Priestley, H. Rose, E. T. Schmidt, S. E. Schmidt, F. Wehrung and R. Wille.

[Gri74] V. N. Grišin, *On a nonstandard logic and its application to the set theory (Russian)*, Isslédovanija po formalizovannym jazykam i néklassičéskim logikam (1974), 135–171.

[Gri76] V. N. Grišin, *On algebraic semantics for logic without the contraction rule (Russian)*, Isslédovanija po formalizovannym jazykam i néklassičéskim logikam (1976), 247–264.

[Gri82] V. N. Grišin, *Predicate and set-theoretical calculi based on logic without the contraction rule*, Mathematical USSR Izvestiya **18** (1982), 41–59; English transl., Izvéstia Akademii Nauk SSSR **45** (1981), 47–68.

[Gur67] Y. Gurevich, *Hereditary undecidability of a class of lattice-ordered Abelian groups*, Algebra i Logika Sem. **6** (1967), no. 1, 45–62.

[GL84] Y. Gurevich and H. R. Lewis, *The word problem for cancellation semigroups with zero*, Journal of Symbolic Logic **49** (1984), 184–191.

[Háj98] P. Hájek, *Metamathematics of Fuzzy Logic*, Trends in Logic, vol. 4, Kluwer, Dordrecht, 1998.

[Háj05] P. Hájek, *Fleas and fuzzy logic*, J. Mult.-Valued Logic Soft Comput. **11** (2005), no. 1-2, 137–152.

[HGE96] P. Hájek, L. Godo, and F. Esteva, *A complete many-valued logic with product-conjunction*, Archive for Mathematical Logic **35** (1996), 191–208

[Hal51] S. Halldén, *On the semantic non-completeness of certain Lewis calculi*, Journal of Symbolic Logic **16** (1951), 127–129.

[Har56] R. Harrop, *On disjunctions and existential statements in intuitionistic systems of logic*, Math. Annalen **5** (1956), 347–361.

[Har60] R. Harrop, *Concerning formulas of the types $A \to B \vee C$. $A \to \exists x B(x)$ in intuitionistic formal system*, J. of Symbolic Logic **25** (1960), 27–32.

[HRT02] J. Hart, L. Rafter, and C. Tsinakis, *The Structure of Commutative Residuated Lattices*, International Journal of Algebra and Computation **12** (2002), no. 4, 509–524.

[HMT85a] L. Henkin, J. D. Monk, and A. Tarski, *Cylindric algebras. Part I*, Studies in Logic and the Foundations of Mathematics, vol. 64, North-Holland Publishing Co., Amsterdam, 1985. With an introductory chapter: General theory of algebras; Reprint of the 1971 original.

[HMT85b] L. Henkin, J. D. Monk, and A. Tarski, *Cylindric algebras. Part II*, Studies in Logic and the Foundations of Mathematics, vol. 115, North-Holland Publishing Co., Amsterdam, 1985. MR **781930** **(86m:**03095b)

[His66] N. G. Hisamiev, *Universal theory of lattice-ordered Abelian groups*, Algebra i Logika Sem. **5** (1966), no. 3, 71–76.

[Hod93] W. Hodges, *Model theory*, Encyclopedia of Mathematics and its Applications, vol. 42, Cambridge University Press, Cambridge, 1993.

[Hod97] W. Hodges, *A shorter model theory*, Cambridge University Press, Cambridge, 1997.

[HV05] I. Hodkinson and Y. Venema, *Canonical varieties with no canonical axiomatisation*, Trans. Amer. Math. Soc. **357** (2005), no. 11, 4579–4605 (electronic).

[Höh95] U. Höhle, *Commutative residuated ℓ monoids*, Non-classical Logics and Their Applications to Fuzzy Subsets (U. Höhle and P. Klement, eds.), Kluwer, 1995, pp. 53–106.

[Hol63] C. Holland, *The lattice-ordered groups of automorphisms of an ordered set*, Michigan Math. J. **10** (1963), 399–408.

[HM79] W. C. Holland and S. H. McCleary, *Solvability of the word problem in free lattice-ordered groups*, Houston Journal of Mathematics **5** (1979), no. 1, 99–105.

[HOS94] R. Hori, H. Ono, and H. Schellinx, *Extending intuitionistic linear logic with knotted structural rules*, Notre Dame Journal of Formal Logic **35** (1994), 219–242.

[Hos67] T. Hosoi, *On Intermediate logics I*, J. Fac. Sci., Univ. Tokyo **14** (1967), 293–312.

[Hos74] T Hosoi, *On Intermediate logics III*, Journal of Tsuda College **6** (1974), 23–38.

[HO70] T. Hosoi and H. Ono, *The intermediate logics on the second slice*, Journal of the Faculty of Science, University of Tokyo **17** (1970), 457–461.

[HO73] T. Hosoi and H. Ono, *Intermediate propositional logics (a survey)*, Journal of Tsuda College **5** (1973), 67–82.

[Hud93] J. Hudelmaier, *An O(n log n)-space decision procedure for intuitionistic propositional logic*, Journal of Logic and Computation **3** (1993), 63–75.

[Idz84a] P. M. Idziak, *Lattice operations in BCK-algebras*, Ph.D. thesis, Jagiellonian University, Kraków, Poland, 1984.

[Idz84b] P. M. Idziak, *Lattice operations in BCK-algebras*, Mathematica Japonica **29** (1984), 839–846.

[ISW] Paweł. M. Idziak, K. Słomczyńska, and A. Wroński, *Equivalential algebras: a study of Fregean varieties.*

[II66] Y. Imai and K. Iseki, *On axiom systems of propositional calculi XIV*, Proceedings of the Japan Academy **42** (1966), 19–22.

[Isé79] K. Iséki, *BCK-algebras with condition (S)*, Mathematica Japonica **24** (1979), 107–119.

[IT78] K. Iseki and S. Tanaka, *An introduction to the theory of BCK-algebras*, Mathematica Japonica **23** (1978), 1–26.

[Jan63a] V. A. Jankov, *On connection between provability in intuitionistic propositional calculus and finite implicative structures*, Doklady Akademii Nauk SSSR **151** (1963), 1293–1294 (Russian).

[Jan63b] V. A. Jankov, *The relationship between deducibility in the intuitionistic propositional calculus and finite implicational structures*, Soviet Mathematics Doklady **4** (1963), 1203–1204.

[Jan68] V. A. Jankov, *The construction of a sequence of strongly independent superintuitionistic propositional calculi*, Soviet Mathematics Doklady **9** (1968), 806–807.

[JM02] S. Jenei and F. Montagna, *A Proof of Standard Completeness for Esteva and Godo's Logic MTL*, Studia Logica **70** (2002), no. 2, 183–192.

[JM01] J. Ježek and R. McKenzie, *The variety generated by equivalence algebras*, Algebra Universalis **45** (2001), no. 2-3, 211–219.

[Jip00] P. Jipsen, *A Gentzen system and decidability for residuated lattices* (2000), unpublished note, available online, available at `http://www1.chapman.edu/~jipsen/reslat/gentzenrl.pdf`.

[Jip03] P. Jipsen, *An overview of generalized basic logic algebras*, Neural Network World **13** (2003), no. 5, 491–500.

[JL94] P. Jipsen and E. Lukács, *Minimal relation algebras*, Algebra Universalis **32** (1994), no. 2, 189–203.

[JM06] P. Jipsen and F. Montagna, *On the structure of generalized BL-algebras*, Algebra Universalis **55** (2006), 226–237.

[JT02] P. Jipsen and C. Tsinakis, *A survey of residuated lattices*, Ordered Algebraic Structures (J. Martinez, ed.), Kluwer, Dordrecht, 2002, pp. 19–56.

[Joh36] I. Johansson, *Der Minimalkalkül ein reduzierter Intuitionistischer Formalismus*, Compositio Mathematica **4** (1936), 119–136.

[Jón65] B. Jónsson, *Extensions of relational strucures*, The Theory of Models: Proceedings of the 1963 Symposium at Berkley, North-Holland Publ. Co., 1965, pp. 146–157.

[Jón67] B. Jónsson, *Algebras whose congruence lattices are distributive*, Mathematica Scandinavica **21** (1967), 110–121.

[JT51] B. Jónsson and A. Tarski, *Boolean algebras with operators. I*, Amer. J. Math. **73** (1951), 891–939.

[JT52] B. Jónsson and A. Tarski, *Boolean algebras with operators. II*, Amer. J. Math. **74** (1952), 127–162.

[KW75] J. Kabziński and A. Wroński, *On equivalential algebras*, Proceedings of the 1975 International Symposium on Multiple-valued Logic, Indiana University, Bloomington, 1975, pp. 231–243.

[Kan57] S. Kanger, *Provability in Logic*, Stockholm Studies in Philosophy, vol. 1, øAlmqvist & Wiksell, Stockholm, 1957.

[KKU06] Y. Katoh, T. Kowalski, and M. Ueda, *Almost minimal varieties related to fuzzy logics*, Reports on Mathematical Logic **41** (2006), 173–194.

[Kih06] H. Kihara, *Commutative substructural logics - an algebraic study*, Ph.D. thesis, Japan Advanced Institute of Science and Technology, Nomi, Japan, 2006.

[KO07] H. Kihara and H. Ono, *Algebraic characterizations of variable separation and interpolation (tentative title)* (2007), in preparation.

[KO91] E. Kiriyama and H. Ono, *The contraction rule and decision problems for logics without structural rules*, Studia Logica **50** (1991), 299–319.

[Kle52] S. C. Kleene, *Introduction to metamathematics*, D. Van Nostrand Co., Inc., New York, N. Y., 1952.

[Kol61] A. N. Kolmogorov, *On the principle on tertium non datur*, From Frege to Gödel (J. van Heijenoort, ed.), Cambridge University Press, 1961, pp. 414–437.

[Kom75] Y. Komori, *The finite model property of the intermediate propositional logics on finite slices*, Journal of the Faculty of Science, University of Tokyo, Sec. IA **22** (1975), no. 2, 117–120.

[Kom78] Y. Komori, *Logics without Craig's interpolation property*, Proc. Japan Acad. **54** (1978), 46–48.

[Kom81] Y. Komori, *Super-Łukasiewicz propositional logics*, Nagoya Mathematical Journal **84** (1981), 119–133.

[Kom84] Y. Komori, *The class of BCC–algebras is not a variety*, Mathematica Japonica **29** (1984), 391–394.

[Kom86] Y. Komori, *Predicate logics without the structural rules*, Studia Logica **45** (1986), 393–404.

[Kop89] S. Koppelberg, *Handbook of Boolean algebras. Vol. 1*, North-Holland Publishing Co., Amsterdam, 1989. Edited by J. Donald Monk and Robert Bonnet.

[Kow95] T. Kowalski, *The bottom of the lattice of BCK-varieties*, Reports on Mathematical Logic **29** (1995), 87–93.

[Kow98] T. Kowalski, *Varieties of tense algebras*, Rep. Math. Logic (1998), no. 32, 53–95.

[Kow04] T. Kowalski, *Splittings in certain varieties of residuated lattices*, The 48th Annual Meeting of the Australian Mathematical Society (Melbourne, September 28, 2004), 2004, abstract.

[Kow05] T. Kowalski, *Semisimplicity, EDPC and discriminator varieties of residuated lattices*, Studia Logica **77** (2005), 255–265.

[KK06] T. Kowalski and M. Kracht, *Semisimple varieties of modal algebras*, Studia Logica **83** (2006), 351–363.

[KL07] T. Kowalski and T. Litak, *Completions of GBL-algebras: negative results*, Algebra Universalis (2007), forthcoming.

[KO00a] T. Kowalski and H. Ono, *The variety of residuated lattices is generated by its finite simple members*, Reports on Mathematical Logic **34** (2000), 57–75.

[KO00b] T. Kowalski and H. Ono, *Splittings in the variety of residuated lattices*, Algebra Universalis **44** (2000), 283–298.

[KO01] T. Kowalski and H. Ono, *Residuated Lattices: An algebraic glimpse at logics without contraction*, Japan Advanced Institute of Science and Technology, March 2001.

[KO06] T. Kowalski and H. Ono, *Fuzzy logics from substructural perspective*, Proceedings of the 26th Linz Seminar on Fuzzy Set Theory, 2006.

[Kra99] M. Kracht, *Tools and Techniques in modal logi*, Studies in Logics, vol. 142, Elsevier, Amsterdam, 1999.

[Kri59] S. A. Kripke, *The problem of entailment (abstract)*, Journal of Symbolic Logic **24** (1959), 324.

[Kru24] W. Krull, *Axiomatische Begründung der allgemeinen Idealtheorie*, Sitzungsberichte der physikalisch medizinischen Societät der Erlangen **56** (1924), 47–63.

[KZ77] P. S. Krzystek and S. Zachorowski, *Łukasiewicz logics have not the interpolation property*, Reports on Mathematical Logic **9** (1977), 39–40.

[Kuz75] A. V. Kuznetsov, *On superintuitionistic logics*, Proceedings of the International Congress of Mathematicians (1975), 243–249.

[Laf97] Y. Lafont, *The finite model property for various fragments of linear logic*, Journal of Symbolic Logic **62** (1997), 1202–1208.

[Lam58] J. Lambek, *The mathematics of sentence structure*, American mathematical Monthly **65** (1958), 154–170.

[Lam61] J. Lambek, *On the calculus of syntactic types*, Structure of Language and its Mathematical Aspects (R. Jakobson, ed.), American Mathematical Society, 1961, pp. 166–2178.

[Lam95] J. Lambek, *Some lattice models of bilinear logic*, Algebra Universalis **34** (1995), 541–550.

[Lem66] E. J. Lemmon, *A note on Halldén-incompleteness*, Notre Dame Journal of Formal Logic **7** (1966), 296–300.

[LMSS92] P. Lincoln, J. Mitchell, A. Scedrov, and N. Shankar, *Decision problems for propositional linear logic*, Ann. Pure Appl. Logic **56** (1992), no. 1-3, 239–311.

[Lip74] L. Lipshitz, *The undecidability of the word problems for projective geometries and modular lattices*, Transactions of the AMS **193** (1974), 171–180.

[Lot02] M. Lothaire, *Algebraic Combinatorics on Words*, Encyclopedia of Mathematics and its Applications, vol. 90, Cambridge University Press, 2002.

[Łu20] J. Łukasiewicz, *O logice trójwartościowej*, Polish Logic 1920–1939 (S. McCall, ed.), Clarendon Press, 1967, 1920.

[Lyn50] R. C. Lyndon, *The representation of relational algebras*, Ann. of Math. (2) **51** (1950), 707–729.

[Lyn56] R. C. Lyndon, *The representation of relation algebras. II*, Ann. of Math. (2) **63** (1956), 294–307.

[Mac37] H. M. MacNeille, *Partially ordered sets*, Transaction of the American mathematical society **42** (1937), 416–460.

[Mad98] J. Madarász, *Interpolation and amalgamation: Pushing the limits. Part I*, Studia Logica **61** (1998), 311–345.

[Mae54] S. Maehara, *Eine Darstellung der intuitionistische Logik in der klassischen*, Nagoya Mathematical Journal **7** (1954), 45–64.

[Mae60] S. Maehara, *On the interpolation theorem of Craig (Japanese)*, Sugaku **12** (1960), 235–237.

[Mae91] S. Maehara, *Lattice-valued representation of the cut-elimination theorem*, Tsukuba Journal of Mathematics **15** (1991), 509–521.

[Mak71] D. Makinson, *Some embedding theorems for modal logic*, Notre Dame Journal of Formal Logic **XII** (1971), no. 2, 252–254.

[Mak76] L. L. Maksimova, *The principle of separation of variables in propositional logics*, Algebra i Logika **15** (1976), 168–184.

[Mak77] L. L. Maksimova, *Craig's theorem in superintuitionistic logics and amalgamable varieties of pseudo-Boolean algebras*, Algebra i Logika **16** (1977), 643–681.

[Mak79a] L. L. Maksimova, *Interpolation properties of superintuitionistic logics*, Studia Logica **38** (1979), 419–428.

[Mak79b] L. L. Maksimova, *Interpolation theorems in modal logics and amalgamable varieties of topological Boolean algebras*, Algebra i Logika **18** (1979), 556–586.

[Mak86] L. L. Maksimova, *On maximal intermediate logics with the disjunction property*, Studia Logica **45** (1986), 69–75.

[Mak95] L. L. Maksimova, *On variable separation in modal and superintuitionistic logics*, Studia Logica **55** (1995), 99–112.

[Mak99] L. L. Maksimova, *Interrelation of algebraic, semantical and logical properties for superintuitionistic and modal logics*, Logic, Algebra and Computer Science, Banach Center Publications, vol. 46, Polish Academy of Science, Warszawa, 1999, pp. 159–168.

[Mak05] L. L. Maksimova, *Interpolation and definability in extensions of minimal logic*, Algebra Logika **44** (2005), no. 6, 726–750, 764–765 (Russian, with Russian summary); English transl., Algebra Logic **44** (2005), no. 6, 407–421.

[MSŠ79] L. L. Maksimova, D. P. Skvorcov, and V. B. Šehtman, *Impossibility of finite axiomatization of Medvedev's logic of finite problems*, Dokl. Akad. Nauk SSSR **245** (1979), no. 5, 1051–1054 (Russian).

[Mal71] A. I. Mal′cev, *The metamathematics of algebraic systems. Collected papers: 1936–1967*, North-Holland Publishing Co., Amsterdam, 1971. Translated, edited, and provided with supplementary notes by Benjamin Franklin Wells, III; Studies in Logic and the Foundations of Mathematics, Vol. 66.

[McK72] R. N. McKenzie, *Equational bases and non-modular lattice varieties*, Transactions of the American Mathematical Society **156** (1972), 1–43.

[McK87] R. N. McKenzie, *Finite equational bases for congruence modular varieties*, Algebra Universalis **24** (1987), 224–250.

[McK96] R. N. McKenzie, *An algebraic version of categorical equivalence for varieties and more general algebraic categories*, Logic and Algebra (A. Ursini and P. Aglianò, eds.), Marcel Dekker, 1996, pp. 211–244.

[MMT87] R. N. McKenzie, G. F McNulty, and W. F Tailor, *Algebras, Lattices, Varieties*, Vol. 1, Wadsworth & Brooks/Cole, Monterey, California, 1987.

[McKi40] J. C. C. McKinsey, *Postulates for the calculus of binary relations*, J. Symbolic Logic **5** (1940), 85–97. MR 0002513 (2,66b)

[MT44] J. C. C. McKinsey and A. Tarski, *The algebra of topology*, Ann. of Math. (2) **45** (1944), 141–191. MR 0009842 (5,211f)

[MT46] J. C. C. McKinsey and A. Tarski, *On closed elements in closure algebras*, Ann. of Math. (2) **47** (1946), 122–162. MR 0015037 (7,359e)

[MT48] J. C. C. McKinsey and A. Tarski, *Some theorems about the sentential calculi of Lewis and Heyting*, J. Symbolic Logic **13** (1948), 1–15.

[Mey66] R. K. Meyer, *Topics in modal and many-valued logic*, Doctoral dissertation, University of Pittsburgh, 1966.

[Mey73] R. K. Meyer, *Improved decision procedures for pure relevant logic*, 1973, unpublished.

[Mey75] R. K. Meyer, *Sugihara is a characteristic matrix for RM*, Entailment. The logic of relevance and necessity, Princeton University Press, 1975, pp. 393–420.

[Mey76] R. K. Meyer, *Metacompleteness*, Notre Dame Journal of Formal Logic **17** (1976), 501–516.

[MO94] R. K. Meyer and H. Ono, *The finite model property for BCK and BCIW*, Studia Logica **53** (1994), 107–118.

[MR72] R. K. Meyer and R. Routley, *Algebraic analysis of entailment*, Logique at Analyse **15** (1972), 407–428.

[MR73a] R. K. Meyer and R. Routley, *Classical relevant logics I*, Studia Logica **32** (1973), 51–66.

[MR73b] R. K. Meyer and R. Routley, *Classical relevant logics II*, Studia Logica **33** (1973), 183–194.

[MR73c] R. K. Meyer and R. Routley, *Semantics of entailment*, Truth, Syntax and Modality (H. Leblanc, ed.), North Holland, 1973, pp. 194–243.

[MS02] R. K. Meyer and J. K. Slaney, *A, still adorable*, Paraconsistency (São Sebastião, 2000), Lecture Notes in Pure and Appl. Math., vol. 228, Dekker, New York, 2002, pp. 241–260.

[Mon70] J. D. Monk, *Completions of Boolean algebras with operators*, Mathematishe Nachrichten **46** (1970), 47–55.

[Mon06] F. Montagna, *Interpolation and Beth's property in propositional many-valued logics: a semantic investigation*, Annals of Pure and Applied Logic **141** (2006), 148–179.

[Moo88] M. Moortgat, *Categorial Investigations: Logical and Linguistic Aspects of the Lambek Calculus*, Foris Publications, 1988.

[Mul86] C. J. Mulvey, *&*, Rendiconti del Circolo Matematico di Palermo, Ser. 2, Supplement **12** (1986), 94–104.

[Mun86] D. Mundici, *Interpretation of AF C*-algebras in Łukasiewicz sentential calculus*, Journal of Functional Analysis **65** (1986), no. 1, 15–63.

[NBSO98] H. Naruse, Bayu Surarso, and H. Ono, *A syntactic approach to Maksimova's principle of variable separation for some substructural logics*, Notre Dame Journal of Formal Logic **39** (1998), no. 1, 94–113.

[Odi04] S. Odintsov, *Negative equivalence of extensions of minimal logic*, Studia Logica **78** (2004), 417–442.

[OM57] M. Ohnishi and K. Matsumoto, *Gentzen method in modal calculi*, Osaka Mathematical Journal **9** (1957), 113–130.

[Oka96] M. Okada, *Phase semantics for higher order completeness, cut-elimination and normalization proofs (extended abstract)*, Electronic Notes in Theoretical Computer Science **3** (1996), 22 pp. (electronic).

[Oka98] M. Okada, *An introduction to linear logic: expressiveness and phase semantics*, Theories of Types and Proofs (M. Takahashi, M. Okada, and M. Dezani, eds.), MSJ-Memoir 2, Mathematical Society of Japan, 1998.

[Oka99] M. Okada, *Phase semantic cut-elimination and normalization proofs of first- and higher-order linear logic*, Theoretical Computer Science **227** (1999), 333–396.

[OT99] M. Okada and K. Terui, *The finite model property for various fragments of intuitionistic linear logic*, Journal of Symbolic Logic **64** (1999), 790–802.

[Ono72] H. Ono, *Some results on the intermediate logics*, Publications of the RIMS, Kyoto University **8** (1972), 117–130.

[Ono86] H. Ono, *Interpolation and the Robinson property for logics not closed under the Boolean operations*, Algebra Universalis **23** (1986), 111–122.

[Ono90] H. Ono, *Structural rules and a logical hierarchy*, Mathematical Logic (P. P. Petkov, ed.), Plenum, New York, 1990, pp. 95–104.

[Ono92] H. Ono, *Algebraic aspect of logics without structural rules*, Proceedings of the International Conference on Algebra honoring A. Malcev, Novosibirsk, 1989, Contemporary Mathematics (L. A. Bokut, Yu. L. Ershov, O. H. Kegel, and A. I. Kostrikin, eds.), Vol. 131 (Part 3), American Mathematical Society, 1992, pp. 601–621.

[Ono93] H. Ono, *Semantics for substructural logics*, Substructural Logics (K. Došen and P. Schröder-Heister, eds.), Oxford University Press, 1993, pp. 259–291.

[Ono98a] H. Ono, *Decidability and the finite model property of substructural logics*, Tbilisi Symposium on Logic, Language and Computation: Selected Papers (J. Ginzburg, Z. Khasidashvili, C. Vogel, Jean-Jacques Lévy, and Enric Vallduví, eds.), Studies in Logic, Language and Information, CSLI, Stanford, 1998, pp. 263–274.

[Ono98b] H. Ono, *Proof-theoretic methods for nonclassical logic: an introduction*, Theories of Types and Proofs (M. Takahashi, M. Okada, and M. Dezani-Ciancaglini, eds.), MSJ Memoirs, vol. 2, Mathematical Society of Japan, 1998, pp. 207–254.

[Ono01] H. Ono, *Logics without contraction rule and residuated lattices I* (2001), unpublished.

[Ono03a] H. Ono, *Substructural Logics and Residuated Lattices: an introduction*, 50 Years of Studia Logica (V. F. Hendricks and J. Malinowski, eds.), Trends in Logic, vol. 20, Kluwer, Dordrecht, 2003, pp. 177–212.

[Ono03b] H. Ono, *Closure operators and complete embeddings of residuated lattices*, Studia Logica **74** (2003), no. 3, 427–440.

[Ono03c] H. Ono, *Completions of algebras and completeness of modal and substructural logics*, Advances in modal logic. Vol. 4, King's Coll. Publ., London, 2003, pp. 335–353.

[OK85] H. Ono and Y. Komori, *Logics without the contraction rule*, Journal of Symbolic Logic **50** (1985), 169–201.

[OU03] H. Ono and M. Ueda, *A classification of logics over* $\mathbf{FL_{ew}}$ *and almost maximal logics*, Philosophical Dimensions of Logic and Science (A. Rojszczak, J. Cachro, and G. Kurczewski, eds.), Kluwer, 2003, pp. 3–13.

[Pao02] F. Paoli, *Substructural Logics: A Primer*, Trends in Logic, vol. 13, Kluwer, Dordrecht, 2002.

[Pav79] J. Pavelka, *On fuzzy logic II: Enriched residuated lattices and semantics of propositional calculi*, Zeitschrift für mathematische Logik und Grundlagen der Mathematik **25** (1979), 119–134.

[Pei70] C. S. Peirce, *Description of a notation for the logic of relatives, resulting from an amplification of the conception of Boole's calculus of logic*, Memoirs of the American Academy of Sciences **9** (1870), 317–378.

[Pei80] C. S. Peirce, *On the algebra of logic*, American Journal of Mathematics **3** (1880), 15–57.

[Pie73] K. R. Pierce, *Amalgamations of lattice ordered groups*, Trans. Am. Math. Soc. **172** (1973), 249–260.

[Pig72] D. Pigozzi, *Amalgamations, congruence-extension, and interpolation properties in algebras*, Algebra Universalis **1** (1972), 269–349.

[Pit92] A. M. Pitts, *On an interpretation of second order quantification in first order intuitionistic propositional logic*, Journal of Symbolic Logic **57** (1992), 33–52.

[Pru76] T. Prucnal, *Structural completeness of Medvedev's propositional calculus*, Reports on Mathematical Logic (1976), no. 6, 103–105.

[PW74] T. Prucnal and A. Wroński, *An algebraic characterization of the notion of structural completeness*, Bulletin of the Section of Logic of the Polish Academy of Sciences (1974), no. 3, 30–33.

[Raf06] J. G. Raftery, *Correspondences between Gentzen and Hilbert systems*, J. Symbolic Logic **71** (2006), no. 3, 903-957.

[Ras74] H. Rasiowa, *An Algebraic Approach to Non-classical Logics*, North-Holland Pub. Co., Amsterdam, 1974.

[RS63] H. Rasiowa and R. Sikorski, *The Mathematics of Metamathematics*, Państwowe Wydawnictwo Naukowe, 1963.

[Rau79] W. Rautenberg, *Klassische und nicht-klassische Aussagenlogik*, Logik und Grundlagen der Mathematik, vol. 22, Vieweg & Sohn, Braunschweig, 1979.

[Res98] G. Restall, *Displaying and deciding substructural logics 1: Logics with contraposition*, J. Philosophical Logic **27** (1998), 179-216.

[Res00] G. Restall, *An Introduction to Substructural Logics*, Routledge, 2000.

[RV01] A. Robinson and A. Voronkov, *Handbook of Automated Reasoning*, Vol. 1, Elsevier and The MIT Press, 2001.

[Ros90a] K. I. Rosenthal, *A note on Girard quantales*, Cahiers Topologie Géom. Différentielle Catég. **31** (1990), no. 1, 3–11 (English, with French summary).

[Ros90b] K. I. Rosenthal, *Quantales and Their Applications*, Pitman Research Notes in Mathematics, vol. 234, Longman, 1990.

[RS94] G. Rozenberg and A. Salomaa, *Cornerstones of Undecidability*, International Series in Computer Science, Prentice Hall, 1994.

[Sam95] G. Sambin, *Pretopologies and completeness proof*, Journal of Symbolic Logic **60** (1995), 861–878.

[ST77] J. Schmidt and C. Tsinakis, *Relative pseudo-complements, join-extensions and meet-retractions*, Mathematische Zeitschrift **157** (1977), 271–284.

[Sch95a] F. W. K. E. Schröder, *Note über die Algebra der binären Relative*, Mathematische Annalen (1895), 144-158.

[Sch95b] F. W. K. E. Schröder, *Vorlesungen über die Algebra der Logik (exakte Logik), Vol 3*, B. G. Teubner, Leibzig, 1895.

[SHD93] P. Schröder-Heister and K. Došen (eds.), *Substructural Logics*, Oxford University Press, 1993.

[Sch60] K. Schütte, *Syntactical and semantical properties of simple type theory*, Journal of Symbolic Logic **25** (1960), 305–325.

[Sch77] K. Schütte, *Proof Theory*, Springer-Verlag, 1977.

[Sek01] T. Seki, *Some remarks on Maehara's method*, Bulletin of the Section of Logic **30** (2001), 147–154.

[Sla84] J. Slaney, *A metacompleteness theorem for contraction-free relevant logics*, Studia Logica **43** (1984), 159–168.

[Sou07] D. Souma, *An algebraic approach to the disjunction property of substructural logics*, Notre Dame Journal of Formal Logic (2007), to appear.

[Sto36] M. H. Stone, *The theory of representations for Boolean algebras*, Trans. Amer. Math. Soc. **40** (1936), no. 1, 37–111.

[Suz89] N.-Y. Suzuki, *Intermediate logics characterized by a class of algebraic frames with infinite individual domain*, Bulletin of Section of Logic **18** (1989), no. 2, 63–71.

[Swi99] K Swirydowicz, *There exist exactly two strictly relevant extensions of relevant logic* **R**, Journal of Symbolic Logic **64** (1999), 1125–1154.

[Sze92] Á. Szendrei, *A survey on strictly simple algebras and minimal varieties*, Universal algebra and quasigroup theory (Jadwisin, 1989), Res. Exp. Math., vol. 19, Heldermann, Berlin, 1992, pp. 209–239.

[Tak03] H. Takamura, *Every free biresiduated lattice is semisimple*, Rep. Math. Logic (2003), no. 37, 125–133.

[Tak06] H. Takamura, *The variety of modal FL_{ew}-algebras is generated by its finite simple members*, Advances in Modal Logic 6 (G. Governatori, I. Hodkinson, and Y. Venema, eds.), College Publications, 2006, 2006, pp. 469–479.

[Tak87] G. Takeuti, *Proof Theory, 2nd edition*, North-Holland, 1987.

[Tam74] S. Tamura, *On a decision procedure for free lo-algebraic systems*, Technical Report of Mathematics, Yamaguchi University **9** (1974).

[Tar41] A. Tarski, *On the calculus of relations*, J. Symbolic Logic **6** (1941), 73–89.

[Tar46] A. Tarski, *A remark on functionally free algebras*, Ann. of Math. (2) **47** (1946), 163–165.

[Tax73] R. E. Tax, *On the intuitionistic equivalential calculus*, Notre Dame J. Formal Logic **14** (1973), 448–456.

[TMM88] P. B. Thistlewaite, M. A. McRobbie, and R. K. Meyer, *Automated Theorem-Proving in Non-Classical Logics*, Research Notes in Theoretical Computer Science, Pitman, John Wiley & Sons, Inc., 1988.

[Tro92] A. S. Troelstra, *Lectures on Linear Logic*, Lecture Notes, vol. 29, CSLI, Stanford, 1992.

[TS00] A. S. Troelstra and H. Schwichtenberg, *Basic proof theory*, 2nd ed., Cambridge Tracts in Theoretical Computer Science, vol. 43, Cambridge University Press, Cambridge, 2000.

[TvD88] A. S. Troelstra and D. van Dalen, *Constructivism in mathematics. Vol. I*, Studies in Logic and the Foundations of Mathematics, vol. 121, North-Holland Publishing Co., Amsterdam, 1988.

[TW06] C Tsinakis and A Wille, *Minimal varieties of involutive residuated lattices*, Studia Logica **83** (2006), 407–423.

[Ued00] M. Ueda, *A study of classification of residuated lattices and logics without contraction rule*, Master thesis, Japan Advanced Institute of Science and Technology, 2000.

[Ume55] T. Umezawa, *Über Zwischensysteme der Aussagenlogik*, Nagoya Mathematical Journal **9** (1955), 181–189.

[Urq78] A. Urquhart, *A topological representation theory for lattices*, Algebra Universalis **8** (1978), no. 1, 45–58. MR 0450150 (56 #8447)

[Urq84] A. Urquhart, *The undecidability of entailment and relevant implication*, Journal of Symbolic Logic **49** (1984), no. 4, 1059–1073.

[Urq93] A. Urquhart, *Failure of interpolation in relevance logics*, Journal of Philosophical Logic **22** (1993), 449–479.

[Urq95] A. Urquhart, *Decision problems for distributive lattice-ordered semigroups*, Algebra Universalis **33** (1995), 399–418.

[Urq98] A. Urquhart, *Failure of interpolation in positive relevance logics* (1998), unpublished.

[Urq99a] A. Urquhart, *Complexity of decision procedures in relevance logic II*, Journal of Symbolic Logic **64** (1999), 1774-1802.

[Urq99b] A. Urquhart, *Beth's definability theorem in relevant logics*, Logic at Work; Essays dedicated to the memory of Helena Rasiowa (E. Orlowska, ed.), Studies in Fuzziness and Soft Computing, vol. 24, Physica-Verlag, 1999, pp. 229–234.

[vA05a] C. J. van Alten, *The finite model property for knotted extensions of propositional linear logic*, J. Symbolic Logic **70** (2005), no. 1, 84–98.

[vA05b] C. J. van Alten, *Congruence properties in congruence permutable and in ideal determined varieties, with applications*, Algebra Universalis **53** (2005), 433–449.

[vAR97a] C. J. van Alten and J. G. Raftery, *On quasivariety semantics of fragments of Intuitionistic propositional logic without exchange and contraction rules*, Reports on Mathematical Logic (1997), no. 31, 3–56.

[vAR97b] C. J. van Alten and J. G. Raftery, *On the algebra of noncommutative residuation: polrims and left residuation algebras*, Mathematica Japonica (1997), no. 46, 29–46.

[vAR99] C. J. van Alten and J. G. Raftery, *The finite model property for the implicational fragment of IPC without exchange and contraction*, Studia Logica (1999), no. 63, 213–222.

[vAR04] C. J. van Alten and J. G. Raftery, *Rule Separation and Embedding Theorems for Logics Without Weakening*, Studia Logica (2004), no. 76, 241–274.

[vD02] D. van Dalen, *Intuitionistic Logic*, Handbook of Philosophical Logic, 2nd edition (D. Gabbay and F. Guenther, eds.), Vol. 5, Kluwer, Dordrecht, 2002.

[vD04] D. van Dalen, *Logic and Structure, 4rd edition*, Springer-Verlag, 2004.

[vN60] J. von Neumann, *Continuous Geometry*, Princeton University Press, 1960.

[Wan63] H. Wang, *A Survey of Mathematical Logic*, Studies in Logic and the Foundations of Mathematics (1963).

[War40] M. Ward, *Residuated distributive lattices*, Duke Mathematical Journal **6** (1940), 641–651.

[WD39] M. Ward and R. P. Dilworth, *Residuated lattices*, Transactions of the AMS **45** (1939), 335–354.

[Wei63] E. C. Weinberg, *Free lattice-ordered abelian groups*, Math. Ann. **151** (1963), 187–199.

[Weis86] V. Weispfenning, *The complexity of the word problem for abelian l-group*, Theoret. Comput. Sci. **48** (1986), no. 1, 127–132.

[Wer78] H. Werner, *Discriminator Algebras*, Akademie Verlag, Berlin, 1978.

[Whi43] Ph. M. Whitman, *Splittings of a lattice*, American Journal of Mathematics **65** (1943), 179–196.

[Wil05] A. M. Wille, *A Gentzen system for involutive residuated lattices*, Algebra Universalis **54** (2005), no. 4, 449–463.

[Wil07] A. M. Wille, *The Word Problem for Involutive Residuated Lattices and Related Structures*, Archiv der Mathematik (2007), to appear.

[Wój73] Wójcicki, *On matrix representations of consequence operations of Łukasiewiczs sentential calculi*, Zeitschr. f. math. Logik und Grundlag. d. Math. **10** (1973), 239–247.

[Wro73] A. Wroński, *Intermediate logics and the disjunction property*, Reports on Mathematical Logic **1** (1973), 39–51.

[Wro76] A. Wroński, *Remarks on Halldén-completeness of modal and intermediate logics*, Bulletin of the Section of Logic **5** (1976), no. 4, 126–129.

[Wro83] A. Wroński, *BCK-algebras do not form a variety*, Mathematica Japonica **28** (1983), 211–213.

[Wro84a] A. Wroński, *Interpolation and amalgamation properties of BCK-algebras*, Math. Japonica **29** (1984), 115–121.

[Wro84b] A. Wroński, *On a form of equational interpolation property*, Foundations of Logic and Linguistics. Problems and Solution. Selected contributions to the 7th International Congress, Plenum Press, 1984, pp. 23–29.

[Wro85] A. Wroński, *An algebraic motivation for BCK-algebras*, Mathematica Japonica **30** (1985), 187–19.

[Yet90] D. N. Yetter, *Quantales and (noncommutative) linear logic*, J. Symbolic Logic **55** (1990), no. 1, 41–64.

[ZWC01] M. Zakharyaschev, F. Wolter, and A. Chagrov, *Advanced modal logic*, Handbook of Philosophical Logic, 2nd edition (D. Gabbay and F. Guenther, eds.), Vol. III, Kluwer, Dordrecht, 2001.

Index

www.ingramcontent.com/pod-product-compliance
Lightning Source LLC
LaVergne TN
LVHW011858200926
848484LV00002B/474
9780444521415